Rick Osborne

翁景民
75.4.4.

HANDBOOK OF ECONOMETRICS

VOLUME II

HANDBOOKS IN ECONOMICS

Editors

KENNETH J. ARROW
MICHAEL D. INTRILIGATOR

NORTH-HOLLAND
AMSTERDAM · NEW YORK · OXFORD

HANDBOOK OF ECONOMETRICS

VOLUME II

Edited by

ZVI GRILICHES
Harvard University

and

MICHAEL D. INTRILIGATOR
University of California, Los Angeles

1984
NORTH-HOLLAND
AMSTERDAM · NEW YORK · OXFORD

ISBN for this volume 0 444 86186 6
ISBN for this set 0 444 86188 2

Publishers:
ELSEVIER SCIENCE PUBLISHERS B.V.
P.O. Box 1991
1000BZ Amsterdam
The Netherlands

Sole distributors for the U.S.A. and Canada:
ELSEVIER SCIENCE PUBLISHING COMPANY, INC.
52 Vanderbilt Avenue
New York, N.Y. 10017
U.S.A

Library of Congress Cataloging in Publication Data
Main entry under title:

Handbook of econometrics.

(Handbooks in economics; bk. 2)
Includes bibliographies.
1. Econometrics--Addresses, essays, lectures.
I. Griliches, Zvi, 1930- . II. Intriligator, Michael D. III. Series.
HB139.H36 1983 330'.028 83-2396
Vol. 1: ISBN 0 444 86185 8
Vol. 2: ISBN 0 444 86186 6

PRINTED IN THE NETHERLANDS

INTRODUCTION TO THE SERIES

The aim of the *Handbooks in Economics* series is to produce Handbooks for various branches of economics, each of which is a definitive source, reference, and teaching supplement for use by professional researchers and advanced graduate students. Each Handbook provides self-contained surveys of the current state of a branch of economics in the form of chapters prepared by leading specialists on various aspects of this branch of economics. These surveys summarize not only received results but also newer developments, from recent journal articles and discussion papers. Some original material is also included, but the main goal is to provide comprehensive and accessible surveys. The Handbooks are intended to provide not only useful reference volumes for professional collections but also possible supplementary readings for advanced courses for graduate students in economics.

局版臺業字第〇八五二號
發行人：張　澤　雲
住址：台北市羅斯福路四段六號
發行所：雙　葉　書　店
總經銷：雙葉書廊有限公司
地址：台北市羅斯福路四段六號
電話：三四一四一九八號
郵政劃撥：〇〇〇七一一六—七號
中華民國七十四年　月　日

CONTENTS OF THE HANDBOOK

VOLUME I

PREFACE TO THE HANDBOOK

Purpose

The *Handbook of Econometrics* aims to serve as a source, reference, and teaching supplement for the field of econometrics, the branch of economics concerned with the empirical estimation of economic relationships. Econometrics is conceived broadly to include not only econometric models and estimation theory but also econometric data analysis, econometric applications in various substantive fields, and the uses of estimated econometric models. Our purpose has been to provide reasonably comprehensive and up-to-date surveys of recent developments and the state of various aspects of econometrics as of the early 1980s, written at a level intended for professional use by economists, econometricians, and statisticians and for use in advanced graduate econometrics courses.

Econometrics is the application of mathematics and statistical methods to the analysis of economic data. Mathematical models help us to structure our perceptions about the forces generating the data we want to analyze, while statistical methods help us to summarize the data, estimate the parameters of our models, and interpret the strength of the evidence for the various hypotheses that we wish to examine. The evidence provided by the data affects our ideas about the appropriateness of the original model and may result in significant revisions of such models. There is, thus, a continuous interplay in econometrics between mathematical-theoretical modeling of economic behavior, data collection, data summarizing, model fitting, and model evaluation. Theory suggests data to be sought and examined; data availability suggests new theoretical questions and stimulates the development of new statistical methods. The examination of theories in light of data leads to their revision. The examination of data in the light of theory leads often to new interpretations and sometimes to questions about its quality or relevance and to attempts to collect new and different data.

In this volume we review only a subset of what might be called "econometrics". The mathematical-theoretical tools required for model building are discussed primarily in the *Handbook of Mathematical Economics*. Issues of sampling theory, survey design, data collection and editing, and computer programming, all important aspects of the daily life of a practicing econometrician, had, by and large, to be left out of the scope of this *Handbook*. We concentrate, instead, on statistical problems and economic interpretation issues associated with the modeling and estimation of economic behavioral relationships from already assembled and often badly collected data. If economists had access to good experimental data, or were able to design and to perform the relevant economic experiments, the topics to be covered in such a *Handbook* would be quite different. The fact that the generation and collection of economic data is mostly outside the hands of the econometrician is the cause of many of the inferential problems which are discussed in this *Handbook*.

Organization

The organization of the *Handbook* follows in relatively systematic fashion the way an econometric study would proceed, starting from basic mathematical and statistical methods and econometric models, proceeding to estimation and computation, through testing, and ultimately to applications and uses. The *Handbook* also includes a fairly detailed development of time series topics and many other special topics. In particular:

Part 1 summarizes some basic tools used repeatedly in econometrics, including linear algebra, matrix methods, and statistical theory.

Part 2 deals with econometric models, their relationship to economic models, their identification, and the question of model choice and specification analysis.

Part 3 takes up more advanced topics in estimation and computation theory such as non-linear regression methods, biased estimation, and computational algorithms in econometrics. This part also includes a series of chapters on simultaneous equations models, their specification and estimation, distribution theory for such models, and their Bayesian analysis.

Part 4 considers testing of econometric estimators, including Wald, likelihood ratio, and Lagrange multiplier tests; multiple hypothesis testing; distribution theory for econometric estimators and associated test statistics; and Monte Carlo experimentation in econometrics.

Part 5 treats various topics in time series analysis, including time series and spectral methods in econometrics, dynamic specification, inference and casuality in economic time series models, continuous time stochastic models, random and changing coefficient models, and the analysis of panel data.

Parts 6 and 7 present discussions of various special topics in econometrics, including latent variable, limited dependent variable, and discrete choice models; functional forms in econometric model building; economic data issues including longitudinal data issues; and disequilibrium, self selection, and switching models.

Finally, *Part 8* covers selected applications and uses of econometrics. Because of the extremely wide range of applications of econometrics, we could select only a few of the more prominent applications. (Other applications will be treated in later volumes in the "Handbooks in Economics" Series.) Applications discussed here include demand analysis, production and cost analysis, and labor economics. This part includes also chapters on evaluating the predictive accuracy of models, econometric approaches to stabilization policy, the formulation and estimation of models with actors having rational expectations, and the use of econometric models for economic policy formation.

A brief history of econometrics

A brief review of the history of econometrics will put this *Handbook* in perspective. The historical evolution of econometrics was driven both by the increased availability of data and by the desire of generations of scientists to analyze such data in a rigorous and coherent fashion. There are many historical precursors to that which became "econometrics" in this century. Attempts to interpret economic data "scientifically" go back at least as far as Sir William Petty's "political arithmetic" in the seventeenth century and Engel's studies of household expenditures in the nineteenth. The results of the latter became known as Engel's Law, stating that the proportion of total expenditures devoted to food falls as income rises. This "Law" has been tested extensively for many countries over various time periods, as discussed in Houthakker's (1957) centenary article.

The development of statistical theory has played a critical role in the history of econometrics since econometric techniques are, to a large extent, based on multivariable statistics. Modern statistical theory starts with the work of Legendre and Gauss on least squares, motivated by the attempt to remove errors of observation in astronomy and geodesy. The next great impulse from biology, in particular from evolutionary theory with, among others, Galton's work on regression (a term he invented). Later developments in mathematical statistics included Yule's work on multiple regression, Karl Pearson's formulation of the notions of probable error and of testing hypotheses, the more rigourous small-sample theory of Student and R. A. Fisher, R. A. Fisher's work on the foundations of statistical inference, and the Neyman–Pearson theory of hypothesis testing. All of these developments in mathematical statistics had a significant influence on the development of econometrics.

In the first half of the twentieth century the increased availability of price and quantity data and the interest in price indexes aided by the development of family expenditure surveys generated interest both in theoretical modeling of demand structures and their empirical estimation. Particularly noteworthy were the demand studies of Moore (1914, 1917), Marschak (1931), and Schultz (1928, 1938) and studies of family expenditure by Allen and Bowley (1935). This period also witnessed the initial formulation of the identification problem in econometrics in E. Working (1927); studies of production functions by Cobb and Douglas (1928) [see also Douglas (1948)], and Marschak and Andrews (1944); studies of price determination in agricultural markets by H. Working (1922), Wright (1925), Hanau (1928), Bean (1928), and Waugh (1929), among others; and the statistical modeling of business cycles by Slutsky (1927) and Frisch (1933). Macroeconometric modeling also began in the 1930s by Tinbergen (1935, 1939) and was given additional impetus by the development of National Income Accounts in the United States and other countries and by Keynes' theoretical work.

The growth of data availability and the development of economic and statistical theory generated a demand for more extensive, more rigorous and higher quality data analysis efforts, stimulating significant research into the methodology of economic data analysis. Of great importance in this respect was the founding of the Econometric Society in 1930 and the publication, starting in 1933, of its journal *Econometrica*. Ragnar Frisch played a key role as the first editor of this Journal.

There was a great flourishing of econometric theory and applications in the period after World War II, particularly due to the work of the Cowles Commission at the University of Chicago. The development of the simultaneous equations model in Haavelmo (1943, 1944, 1947), Koopmans (1950), Hood and Koopmans (1953), Theil (1954) and Basmann (1957) provided econometricians with tools designed specifically for them, rather than for by biologists and psychologists. The estimation of simultaneous equations and macroeconometric models in Klein (1950) and Klein and Goldberger (1955) started economic forecasting on a new path. This period also witnessed the important demand studies by Stone (1954a, 1954b) for the United Kingdom and Wold and Jureen (1953) for Sweden and the influential studies by Friedman (1957) of the consumption function and by Theil (1958) of economic forecasts and policy. [For collections of historically important papers in econometrics see Zellner (1968), Hooper and Nerlove (1970), and Dowling and Glahe (1970).]

The more recent period of the 1960s and 1970s has witnessed many important developments in econometric theory and applications. Econometric theory has been refined and extended in many ways. Of particular note is the Bayesian approach to econometrics and the study of special features of econometric models, such as limited dependent variables, latent variables, and non-linear models. Great progress was also made in the statistical analysis of time series. In addition, the development of electronic computers, the great increase in computing power, and the development of sophisticated econometric software packages made it possible to pursue much more ambitious data analysis strategies. These developments expanded the range of applications of econometric methods greatly beyond the earlier applications to household expenditure, demand functions, production and cost functions, and macroeconometric models. Econometrics is now used in virtually every field of economics, including public finance, monetary economics, labor economics, international economics, economic history, health economics, studies of fertility, and studies of criminal behavior, just to mention a few. In all of these fields the greater use of econometric techniques, based in part on increased data availability and more powerful estimation techniques, has led to greater precision in the specification, estimation, and testing of economic data-based models.

Most of the important developments in econometric methods during the 1960s and 1970s are discussed in this *Handbook*. The significant topics under development in this period and the chapters treating them include:

(1) *Bayesian econometrics*, using Bayesian methods in the specification and estimation of econometric models. These topics are discussed in Chapter 2 by Zellner and Chapter 9 by Drèze and Richard.

(2) *Time series methods*, including specialized techniques and problems arising in the analysis of economic time series, such as spectral methods, dynamic specification, and causality. These techniques and problems are discussed in Part 5 on "Time Series Topics," including Chapter 17 by Granger and Watson; Chapter 18 by Hendry, Pagan and Sargan; Chapter 19 by Geweke; Chapter 20 by Bergstrom; and Chapter 21 by Chow. Related issues are discussed in Chapter 33 by Fair.

(3) *Discrete choice models*, in which there is a discrete choice of alternatives available, e.g. buy/don't buy decisions, yes/no responses, or alternative possibilities for urban transportation. Such models are discussed specifically in Chapter 27 by Dhrymes and Chapter 24 by McFadden and are also treated in Chapter 22 by Chamberlain, Chapter 28 by Maddala, and Chapter 29 by Heckman and Singer.

(4) *Latent variables models*, in which certain unmeasurable variables systematically influence measured phenomena, such as ability influencing earnings. This topic is treated in Chapter 23 by Aigner, Hsiao, Kapteyn, and Wansbeek and reappears in various guises in Chapter 22 by Chamberlain and Chapter 32 by Heckman and MaCurdy, among others.

(5) *Specification analysis*, involving problems of model choice and their specification and identification. These issues are treated in Chapter 3 by Intriligator, Chapter 4 by Hsiao, Chapter 5 by Leamer, Chapter 26 by Lau, and Chapter 28 by Maddala. This topic, of course, pervades many other chapters in this *Handbook* and overlaps with chapters which deal with testing and distribution theory.

(6) *Non-linear models and methods*, in which models that are intrinsically nonlinear are specified and estimated. Such models are discussed in Chapter 6 by Amemiya and Chapter 12 by Quandt and surface also in many of the other chapters of this *Handbook*.

(7) *Data analysis issues*, involving various problems with data and how they can be treated. These issues are treated in Chapter 10 by Judge and Bock, in Chapter 11 by Krasker, Kuh, and Welsch, Chapter 22 by Chamberlain, Chapter 25 by Griliches, and Chapter 29 by Heckman and Singer, among others.

(8) *Testing and small sample theory*, including various test procedures and Monte Carlo experimentation. These topics are treated in Part 4 on "Testing," including Chapter 13 by Engle, Chapter 14 by Savin, Chapter 15 by Rothenberg, and Chapter 16 by Hendry. Related issues are discussed in Chapter 8 by Phillips.

(9) *Rational expectations* models which treat economic agents as forming expectations in an optimal fashion, given the information available to them,

impose cross-equation constraints on parameters and lead to new problems of identification and estimation. This topic is discussed in Chapter 34 by Taylor.

ZVI GRILICHES
Harvard University

MICHAEL D. INTRILIGATOR
University of California, Los Angeles

References

Allen, R. G. D. and A. L. Bowley (1935) *Family Expenditure*. London: P. S. King

Basmann, R. L. (1957) "A Generalized Classical Method of Linear Estimation of Coefficients in a Structural Equation", *Econometrica*, 25, 77–83.

Bean, L. H. (1928) "Some Interrelationships between the Supply, Price, and Consumption of Cotton", *USDA*, mimeographed.

Cobb, C. W. and P. H. Douglas (1928) "A Theory of Production", *American Economic Review*, 18 (supplement), 139–165.

Douglas, P. H. (1948) "Are There Laws of Production?", *American Economic Review*, 38, 1–41.

Dowling, J. M. and F. R. Glahe (eds.) (1970) *Readings in Econometric Theory*. Boulder: Colorado Associated University Press.

Friedman, M. (1957) *A Theory of the Consumption Function*, National Bureau of Economic Research. Princeton: Princeton University Press.

Frisch, R. (1933) "Propagation Problems and Impulse Problems in Dynamic Economics", in: *Economic Essays in Honor of Gustav Cassel*. London: George Allen & Unwin, pp. 171–205.

Haavelmo, T. (1943) "The Statistical Implications of a System of Simultaneous Equations", *Econometrica*, 11, 1–12.

Haavelmo, T. (1944) "The Probability Approach in Econometrics", *Econometrica*, 12 (supplement), 1–115.

Haavelmo, T. (1947) "Methods of Measuring the Marginal Propensity to Consume", *Journal of the American Statistical Association*, 42, 105–122 (reprinted in Hood and Koopmans (eds) (1953)).

Hanau, A. (1928) "Die Prognose der Schweinepreise", *Vierteljahrshefte zur Konjukturforschung*, Sonderheft 7, Berlin.

Hood, W. C. and T. C. Koopmans (eds.) (1953) *Studies in Econometric Method*, Cowles Commission Monograph No. 14, New York: John Wiley & Sons.

Hooper, J. W. and M. Nerlove (eds.) (1970) *Selected Readings in Econometrics from Econometrica*. Cambridge: MIT Press.

Houthakker, H. S. (1957) "An International Comparison of Household Expenditure Patterns, Commemorating the Centenary of Engel's Law", *Econometrica*, 25, 532–551.

Klein, L. R. (1950) *Economic Fluctuations in the United States, 1921–1941*, Cowles Commission Monograph No. 11. New York: John Wiley & Sons.

Klein, L. R. and A. S. Goldberger (1955) *An Econometric Model of the United States, 1929–1952*. Amsterdam: North-Holland Publishing Co.

Koopmans, T. C. (ed.) (1950) *Statistical Inference in Dynamic Economic Models*, Cowles Commission Monograph No. 10. New York: John Wiley & Sons.

Marschak, J. (1931) *Elastizität der Nachfrage*. Tübingen: J. C. B. Mohr.

Marschak, J. and W. H. Andrews (1944) "Random Simultaneous Equations and the Theory of Production", *Econometrica* 12, 143–205.

Moore, H. L. (1914) *Economic Cycles: Their Law and Cause*. New York: The Macmillan Company.

Moore, H. L. (1917) *Forecasting the Yield and Price of Cotton*. New York: The Macmillan Company.

Schultz, H. (1928) *Statistical Laws of Demand and Supply*. Chicago: University of Chicago Press.

Schultz, H. (1938) *The Theory and Measurement of Demand*, Chicago: University of Chicago Press.

Slutsky, E. (1927) "The Summation of Random Causes as the Source of Cyclic Processes" (Russian with English summary), in: *Problems of Economic Conditions*, vol. 3. Moscow: Rev. English edn., 1937; *Econometrica*, 5, 105–146.
Stone, R. (1954a) "Linear Expenditure Systems and Demand Analysis: An Application to the Pattern of British Demand", *Economic Journal*, 64, 511–527.
Stone, R. (1954b) *The Measurement of Consumers' Expenditure and Behavior in the United Kingdom, 1920–1938*. New York: Cambridge University Press.
Theil, H. (1954) "Estimation of Parameters of Econometric Models", *Bulletin of the International Statistics Institute*, 34, 122–128.
Theil, H. (1958) *Economic Forecasts and Policy*. Amsterdam: North-Holland Publishing Co. (Second Edition, 1961).
Tinbergen, J. (1935) "Quantitative Fragen der Konjunkturpolitik", *Weltwirtschaftliches Archiv*, 42, 316–399.
Tinbergen, J. (1939) *Statistical Testing of Business Cycle Theories*. Vol. 1: *A Method and its Application to Investment Activity*; Vol. 2: *Business Cycles in the United States of America, 1919–1932*. Geneva: League of Nations.
Waugh, F. V. (1929) *Quality as a Determinant of Vegetable Prices*. New York: Columbia University Press.
Wold, H. and L. Jureen (1953) *Demand Analysis*. New York: John Wiley & Sons.
Working, E. J. (1927) "What do Statistical 'Demand Curves' Show?", *Quarterly Journal of Economics*, 41, 212–235.
Working, H. (1922) "Factors Determining the Price of Potatoes in St. Paul and Minneapolis", University of Minnesota Agricultural Experiment Station Technical Bulletin 10.
Wright, S. (1925) *Corn and Hog Correlations*, Washington, USDA, Bul. 1300.
Zellner, A., Ed. (1968) *Readings in Economic Statistics and Econometrics*. Boston: Little, Brown.

CONTENTS OF VOLUME II

Part 5 – TIME SERIES TOPICS

Part 6 – SPECIAL TOPICS IN ECONOMETRICS – 1

Chapter 23

Latent Variable Models in Econometrics

PART 4

TESTING

Chapter 13

WALD, LIKELIHOOD RATIO, AND LAGRANGE MULTIPLIER TESTS IN ECONOMETRICS

ROBERT F. ENGLE*

University of California

Contents

*Research supported by NSF SOC 78-09476 and 80-08580.

Handbook of Econometrics, Volume II, Edited by Z. Griliches and M.D. Intriligator

1. Introduction

If the confrontation of economic theories with observable phenomena is the objective of empirical research, then hypothesis testing is the primary tool of analysis. To receive empirical verification, all theories must eventually be reduced to a testable hypothesis. In the past several decades, least squares based tests have functioned admirably for this purpose. More recently, the use of increasingly complex statistical models has led to heavy reliance on maximum likelihood methods for both estimation and testing. In such a setting only asymptotic properties can be expected for estimators or tests. Often there are asymptotically equivalent procedures which differ substantially in computational difficulty and finite sample performance. Econometricians have responded enthusiastically to this research challenge by devising a wide variety of tests for these complex models.

Most of the tests used are based either on the Wald, Likelihood Ratio or Lagrange Multiplier principle. These three general principles have a certain symmetry which has revolutionized the teaching of hypothesis tests and the development of new procedures. Essentially, the Lagrange Multiplier approach starts at the null and asks whether movement toward the alternative would be an improvement, while the Wald approach starts at the alternative and considers movement toward the null. The Likelihood ratio method compares the two hypotheses directly on an equal basis. This chapter provides a unified development of the three principles beginning with the likelihood functions. The properties of the tests and the relations between them are developed and their forms in a variety of common testing situations are explained. Because the Wald and Likelihood Ratio tests are relatively well known in econometrics, major emphasis will be put upon the cases where Lagrange Multiplier tests are particularly attractive. At the conclusion of the chapter, three other principles will be compared: Neyman's (1959) $C(\alpha)$ test, Durbin's (1970) test procedure, and Hausman's (1978) specification test.

2. Definitions and intuitions

Hypothesis testing concerns the question of whether data appear to favor or disfavor a particular description of nature. Testing is inherently concerned with one particular hypothesis which will be called the *null* hypothesis. If the data fall into a particular region of the sample space called the *critical region* then the test is said to *reject* the null hypothesis, otherwise it *accepts*. As there are only two possible outcomes, an hypothesis testing problem is inherently much simpler than an estimation problem where there are a continuum of possible outcomes. It is important to notice that both of these outcomes refer only to the null hypothesis —we either reject or accept it. To be even more careful in terminology, we either reject or fail to reject the null hypothesis. This makes it clear that the data may not contain evidence against the null simply because they contain very little information at all concerning the question being asked.

As there are only two possible outcomes, there are only two ways to make incorrect inferences. *Type I* errors are committed when the null hypothesis is falsely rejected, and *Type II* errors occur when it is incorrectly accepted. For any test we call α the *size* of the test which is the probability of Type I errors and β is the probability of Type II errors. The *power* of a test is the probability of rejecting the null when it is false, which is therefore $1-\beta$.

In comparing tests, the standard notion of optimality is based upon the size and power. Within a class of tests, one is said to be *best* if it has the maximum power (minimum probability of Type II error) among all tests with size (probability of Type I error) less than or equal to some particular level.

To make such conditions operational, it is necessary to specify how the data are generated when the null hypothesis is false. This is the *alternative* hypothesis and it is through careful choice of this alternative that tests take on the behavior desired by the investigator. By specifying an alternative, the critical region can be tailored to look for deviations from the null in the direction of the alternative. It should be emphasized here that rejection of the null does not require accepting the alternative. In particular, suppose some third hypothesis is the true one. It may be that the test would still have some power to reject the null even though it was not the optimal test against the hypothesis actually operating. Another case in point might be where the data would reject the null hypothesis as being implausible, but the alternative could be even more unlikely.

As an example of the role of the alternative, consider the diagnostic problem which is discussed later in Section 7. The null hypothesis is that the model is correctly specified while the alternative is a particular type of problem such as serial correlation. In this case, rejection of the model does not mean that a serial correlation correction is the proper solution. There may be an omitted variable or incorrect functional form which is responsible for the rejection. Thus the serial correlation test has some power against omitted variables even though it is not the optimal test against that particular alternative.

To make these notions more precise and set the stage for large sample results, let y be a $T\times 1$ random vector drawn from the joint density $f(y,\theta)$ where θ is a $k\times 1$ vector of unknown parameters and $\theta\in\Theta$, the parameter space. Under the null $\theta\in\Theta_0\subset\Theta$ and under the alternative $\theta\in\Theta_1\in\Theta$ with $\Theta_0\cap\Theta_1=\emptyset$. Frequently $\Theta_1=\Theta-\Theta_0$. Then for a critical region C_T, the size α_T is given by:

$$\alpha_T=\Pr(y\in C_T|\theta\in\Theta_0). \tag{1}$$

The power of the test is:

$$\pi_T(\theta) = \Pr(y \in C_T|\theta), \quad \text{for } \theta \in \Theta_1. \tag{2}$$

Notice that although the power will generally depend upon the unknown parameter θ, the size usually does not. In most problems where the null hypothesis is *composite* (includes more than one possible value of θ) the class of tests is restricted to those where the size does not depend upon the particular value of $\theta \in \Theta_0$. Such tests are called *similar* tests.

Frequently, there are no tests whose size is calculable exactly or whose size is independent of the point chosen within the null parameter space. In these cases, the investigator may resort to asymptotic criteria of optimality for tests. Such an approach may produce tests which have good finite sample properties and in fact, if there exist exact tests, the asymptotic approach will generally produce them. Let C_T be a sequence of critical regions perhaps defined by a sequence of vectors of statistics $s_T(y) \geq c_T$, where c_T is a sequence of constant vectors. Then the limiting size and power of the test are simply

$$\alpha = \lim_{T\to\infty} \alpha_T; \qquad \pi(\theta) = \lim_{T\to\infty} \pi_T(\theta), \quad \text{for } \theta \in \Theta_1. \tag{3}$$

A test is called *consistent* if $\pi(\theta) = 1$ for all $\theta \in \Theta_1$. That is, a consistent test will always reject the null when it is false; Type II errors are eliminated for large samples if a test is consistent.

As most hypothesis tests are consistent, it remains important to choose among them. This is done by examining the rate at which the power function approaches its limiting value. The most common limiting argument is to consider the power of the test to distinguish alternatives which are very close to the null. As the sample grows, alternatives ever closer to the null can be detected by the test. The power against such *local* alternatives for tests of fixed asymptotic size provides the major criterion for the optimality of asymptotic tests.

The vast majority of all testing problems in econometrics can be formulated in terms of a partition of the parameter space into two sub-vectors $\theta = (\theta_1', \theta_2')'$ where the null hypothesis specifies values, θ_1^0 for θ_1, but leaves θ_2 unconstrained. In a normal testing problem, θ_1 might be the mean and θ_2 the variance, or in a regression context, θ_1 might be several of the parameters while θ_2 includes the rest, the variance and the serial correlation coefficient, if the model has been estimated by Cochrane–Orcutt. Thus θ_1 includes the parameters of interest in the test.

In this context, the null hypothesis is simply:

$$H_0: \theta_1 = \theta_1^0, \qquad \theta_2 \text{ unrestricted}. \tag{4}$$

A sequence of local alternatives can be formulated as:

$$H_1\colon \theta_1^T = \theta_1^0 + \delta/T^{1/2}, \qquad \theta_2 \text{ unrestricted}, \tag{5}$$

for some vector δ. Although this alternative is obviously rather peculiar, it serves to focus attention on the portion of the power curve which is most sensitive to the quality of the test. The choice of δ determines in what direction the test will seek departures from the null hypothesis. Frequently, the investigator will chose a test which is equally good in all directions δ, called an *invariant* test.

It is in this context that the optimality of the likelihood ratio test can be established as is done in Section 6. It is asymptotically locally most powerful among all invariant tests. Frequently in this chapter the term *asymptotically optimal* will be used to refer to this characterization. Any tests which have the property that asymptotically they always agree if the data are generated by the null or by a local alternative, will be termed *asymptotically* equivalent. Two tests ξ_1 and ξ_2 with the same critical values will be asymptotically equivalent if $\text{plim}|\xi_1 - \xi_2| = 0$ for the null and local alternatives.

Frequently in testing problems non-linear hypotheses such as $g(\theta) = 0$ are considered where g is a $p \times 1$ vector of functions defined on Θ. Letting the true value of θ under the null be θ^0, then $g(\theta^0) = 0$. Assuming g has continuous first derivatives, expand this in a Taylor series:

$$g(\theta) = g(\theta^0) + G(\bar{\theta})(\theta - \theta^0),$$

where $\bar{\theta}$ lies between θ and θ^0 and $G(\cdot)$ is the first derivative matrix of g. For the null and local alternatives, θ approaches θ^0 so $G(\bar{\theta}) \to G(\theta^0) \equiv G$ and the restriction is simply this linear hypothesis:

$$G\theta = G\theta^0.$$

For any linear hypothesis one can always reparameterize by a linear non-singular matrix $A^{-1}\theta = \phi$ such that this null is H_0: $\phi_1 = \phi_1^0$, ϕ_2 unrestricted. To do this let A_2 have $K - p$ columns in the orthogonal complement of G so that $GA_2 = 0$. The remaining p columns of A say A_1, span the row space of G so that GA is non-singular. Then the null becomes:

$$G\theta^0 = G\theta = GA\phi = GA_1\phi_1 + GA_2\phi_2 = GA_1\phi_1,$$

or $\phi_1 = \phi_1^0$ with $\phi_1^0 = (GA_1)^{-1}G\theta^0$.

Thus, for local alternatives there is no loss of generality in considering only linear hypotheses, and in particular, hypotheses which have preassigned values for a subset of the parameter vector.

3. A general formulation of Wald, Likelihood Ratio, and Lagrange Multiplier tests

In this section the basic forms of the three tests will be given and interpreted. Most of this material is familiar in the econometrics literature in Breusch and Pagan (1980) or Savin (1976) and Berndt and Savin (1977). Some new results and intuitions will be offered. Throughout it will be assumed that the likelihood function satisfies standard regularity conditions which allow two term Taylor series expansions and the interchange of integral and derivative. In addition, it will be assumed that the information matrix is non-singular, so that the parameters are (locally) identified.

The simplest testing problem assumes that the data y are generated by a joint density function $f(y,\theta^0)$ under the null hypothesis and by $f(y,\theta)$ with $\theta \in R^k$ under the alternative. This is a test of a simple null against a composite alternative. The log-likelihood is defined as:

$$L(\theta, y) = \log f(y, \theta), \tag{6}$$

which is maximized at a value $\hat{\theta}$ satisfying:

$$\frac{\partial L}{\partial \theta}(\hat{\theta}, y) = 0.$$

Defining $s(\theta, y) = \partial L(\theta, y)/\partial\theta$ as the score, the MLE sets the score to zero. The variance of $\hat{\theta}$ is easily calculated as the inverse of Fisher's Information, or

$$V(\hat{\theta}) = \mathscr{I}^{-1}(\theta)/T,$$

$$\mathscr{I}(\theta) = -\mathrm{E}\frac{\partial^2 L}{\partial\theta\,\partial\theta'}(\theta)/T. \tag{7}$$

If $\hat{\theta}$ has a limiting normal distribution, and if $\mathscr{I}(\theta)$ is consistently estimated by $\mathscr{I}(\hat{\theta})$, then

$$\xi_W = T(\hat{\theta} - \theta^0)'\mathscr{I}(\hat{\theta})(\hat{\theta} - \theta^0) \tag{8}$$

will have a limiting X^2 distribution with k degrees of freedom when the null hypothesis is true. This is the Wald test based upon Wald's elegant (1943) analysis of the general asymptotic testing problem. It is the asymptotic approximation to the very familiar t and F tests in econometrics.

The likelihood ratio test is based upon the difference between the maximum of the likelihood under the null and under the alternative hypotheses. Under general conditions, the statistic,

$$\xi_{\mathrm{LR}} = -2\big(L(\theta^0, y) - L(\hat{\theta}, y)\big), \tag{9}$$

can be shown to have a limiting X^2 distribution under the null. Perhaps Wilks (1938) was the first to derive this general limiting distribution.

The Lagrange Multiplier test is derived from a constrained maximization principle. Maximizing the log-likelihood subject to the constraint that $\theta = \theta^0$ yields a set of Lagrange Multipliers which measure the shadow price of the constraint. If the price is high, the constraint should be rejected as inconsistent with the data. Letting H be the Lagrangian:

$$H = L(\theta, y) - \lambda'(\theta - \theta^0),$$

the first-order conditions are:

$$\frac{\partial L}{\partial \theta} = \lambda; \qquad \theta = \theta^0,$$

so $\lambda = s(\theta^0, y)$. Thus the test based upon the Lagrange Multipliers by Aitcheson and Silvey (1958) and Silvey (1959) is identical to that based upon the score as originally proposed by Rao (1948). In each case the distribution of the score is easily found under the null since it will have mean zero and variance $\mathscr{I}(\theta^0)T$. Assuming a central limit theorem applies to the scores:

$$\xi_{LM} = s'(\theta^0, y)'\mathscr{I}^{-1}(\theta^0)s(\theta^0, y)/T, \tag{10}$$

will again have a limiting X^2 distribution with k degrees of freedom under the null.

The three principles are based on different statistics which measure the distance between H_0 and H_1. The Wald test is formulated in terms of $\theta^0 - \hat{\theta}$, the LR test in terms of $L(\theta^0) - L(\hat{\theta})$, and the LM test in terms of $s(\theta^0)$. A geometric interpretation of these differences is useful.

With $k = 1$, Figure 3.1 plots the log-likelihood function against θ for a particular realization y.

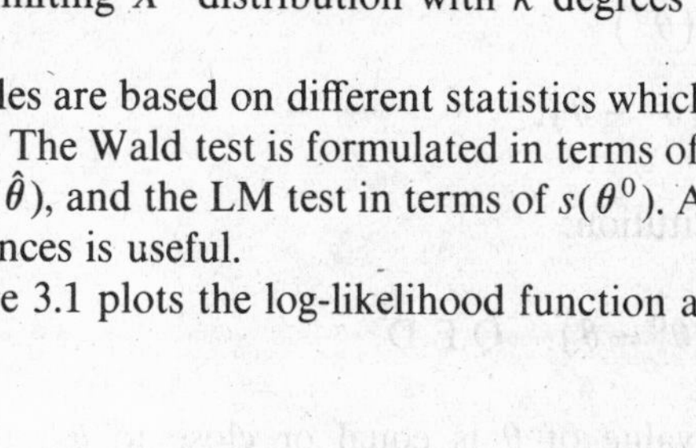

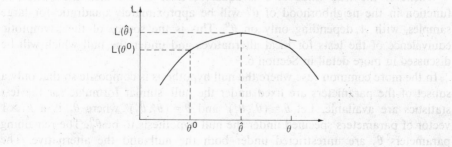

Figure 3.1

The MLE under the alternative is $\hat{\theta}$ and the hypothesized value is θ^0. The Wald test is based upon the horizontal difference between θ^0 and $\hat{\theta}$, the LR test is based upon the vertical difference, and the LM test is based on the slope of the likelihood function at θ^0. Each is a reasonable measure of the distance between H_0 and H_1 and it is not surprising that when L is a smooth curve well approximated by a quadratic, they all give the same test. This is established in Lemma 1.

Lemma 1

If $L = b - 1/2(\theta - \hat{\theta})'A(\theta - \hat{\theta})$ where A is a symmetric positive definite matrix which may depend upon the data and upon known parameters, b is a scalar and $\hat{\theta}$ is a function of the data, then the W, LR and LM tests are identical.

Proof

$$\partial L/\partial\theta = -(\theta - \hat{\theta})'A = s(\theta),$$
$$\partial^2 L/\partial\theta\,\partial\theta' = -A = -T\mathscr{I}.$$

Thus:

$$\xi_{\mathrm{W}} = (\theta^0 - \hat{\theta})'A(\theta^0 - \hat{\theta}),$$
$$\xi_{\mathrm{LM}} = s(\theta^0)'A^{-1}s(\theta^0)$$
$$= (\theta^0 - \hat{\theta})'A(\theta^0 - \hat{\theta}).$$

Finally, by direct substitution:

$$\xi_{\mathrm{LR}} = (\theta^0 - \hat{\theta})'A(\theta^0 - \hat{\theta}). \quad \text{Q.E.D.}$$

Whenever the true value of θ is equal or close to θ^0, then the likelihood function in the neighborhood of θ^0 will be approximately quadratic for large samples, with A depending only on θ^0. This is the source of the asymptotic equivalence of the tests for local alternatives and under the null which will be discussed in more detail in Section 6.

In the more common case where the null hypothesis is composite so that only a subset of the parameters are fixed under the null, similar formulae for the test statistics are available. Let $\theta = (\theta_1', \theta_2')'$ and $\hat{\theta} = (\hat{\theta}_1', \hat{\theta}_2')'$ where θ_1 is a $k_1 \times 1$ vector of parameters specified under the null hypothesis to be θ_1^0. The remaining parameters θ_2 are unrestricted under both the null and the alternative. The maximum likelihood estimate of θ_2 under the null is denoted $\tilde{\theta}_2$ and $\tilde{\theta} = (\theta_1^{0\prime}, \tilde{\theta}_2')'$.

Denote by $\mathscr{I}^{11}$ the partitioned inverse of $\mathscr{I}$ so that:

$$\mathscr{I}^{11^{-1}} = \mathscr{I}_{11} - \mathscr{I}_{12}\mathscr{I}_{22}^{-1}\mathscr{I}_{21}.$$

Then the Wald test is simply:

$$\xi_{\mathrm{W}} = T(\hat{\theta}_1 - \theta_1^0)'\mathscr{I}^{11^{-1}}(\hat{\theta}_1 - \theta_1^0), \tag{11}$$

which has a limiting X^2 distribution with k_1 degrees of freedom when H_0 is true. The LR statistic,

$$\xi_{\mathrm{LR}} = -2(L(\tilde{\theta}, y) - L(\hat{\theta}, y)), \tag{12}$$

has the same limiting distribution. The LM test is again derived from the Lagrangian:

$$H = L(\theta, y) - \lambda'(\theta_1 - \theta_1^0),$$

which has first-order conditions:

$$\frac{\partial L}{\partial \theta_1}(\theta, y) = \lambda,$$

$$\frac{\partial L}{\partial \theta_2}(\theta, y) = 0.$$

Thus:

$$\theta_1 = \theta_1^0,$$

$$\xi_{\mathrm{LM}} = s(\tilde{\theta}, y)'\mathscr{I}^{-1}(\tilde{\theta})s(\tilde{\theta}, y)/T = s_1(\tilde{\theta}, y)'\mathscr{I}^{11}s_1(\tilde{\theta}, y)/T, \tag{13}$$

is the LM statistic which will again have a limiting X^2 distribution with k_1 degrees of freedom under the null. In Lemma 2 it is shown that again for the quadratic likelihood function, all three tests are identical.

Lemma 2

If the likelihood function is given as in Lemma 1 then the tests in (11), (12), and (13) are identical.

Proof

$$\xi_{\mathrm{W}} = (\theta_1^0 - \hat{\theta}_1)'A^{11^{-1}}(\theta_1^0 - \hat{\theta}_1)$$
$$= (\theta_1^0 - \hat{\theta}_1)'(A_{11} - A_{12}A_{22}^{-1}A_{21})(\theta_1^0 - \hat{\theta}_1).$$

For the other two tests, $\tilde{\theta}_2$ must be estimated. This is done simply by setting $S_2(\theta, y) = 0$:

$$\begin{pmatrix} S_1 \\ S_2 \end{pmatrix} = \frac{\partial L}{\partial \theta} = A(\theta - \hat{\theta}) = \begin{bmatrix} A_{11}(\theta_1 - \hat{\theta}_1) + A_{12}(\theta_2 - \hat{\theta}_2) \\ A_{21}(\theta_1 - \hat{\theta}_1) + A_{22}(\theta_2 - \hat{\theta}_2) \end{bmatrix} = 0.$$

So, $S_2 = 0$ implies:

$$\tilde{\theta}_2 - \hat{\theta}_2 = -A_{22}^{-1}A_{21}(\theta_1 - \hat{\theta}_1).$$

The concentrated likelihood function becomes:

$$L = b - \tfrac{1}{2}(\theta_1 - \hat{\theta}_1)'(A_{11} - A_{12}A_{22}^{-1}A_{21})(\theta_1 - \hat{\theta}_1),$$

and hence

$$\xi_{\mathrm{LR}} = (\theta_1^0 - \hat{\theta}_1)(A_{11} - A_{12}A_{22}^{-1}A_{21})(\theta_1^0 - \hat{\theta}_1).$$

Finally, the score is given by:

$$\begin{aligned} S_1(\tilde{\theta}) &= A_{11}(\theta_1^0 - \hat{\theta}_1) + A_{12}(\tilde{\theta}_2 - \hat{\theta}_2) \\ &= (A_{11} - A_{12}A_{22}^{-1}A_{21})(\theta_1^0 - \hat{\theta}_1). \end{aligned}$$

So

$$\xi_{\mathrm{LM}} = (\theta_1^0 - \hat{\theta}_1)'(A_{11} - A_{12}A_{22}^{-1}A_{21})(\theta_1^0 - \hat{\theta}_1). \quad \text{Q.E.D.}$$

Examination of the tests in (11), (12), and (13) indicates that neither the test statistic nor its limiting distribution under the null depends upon the value of the nuisance parameters θ_2. Thus the tests are (asymptotically) similar. It is apparent from the form of the tests as well as the proof of the lemma, that an alternative way to derive the tests is to first concentrate the likelihood function with respect to θ_2 and then apply the test for a simple null directly. This approach makes clear that by construction the tests will not depend upon the true value of the nuisance parameters. If the parameter vector has a joint normal limiting distribution, then the marginal distribution with respect to the parameters of interest will also be normal and the critical region will not depend upon the nuisance parameters either. Under general conditions therefore, the Wald, Likelihood Ratio and Lagrange Multiplier tests will be (asymptotically) similar.

As was described above, each of the tests can be thought of as depending on a statistic which measures deviations between the null and alternative hypotheses, and its distribution when the null is true. For example, the LM test is based upon the score whose limiting distribution is generally normal with variance $(\theta^0)\cdot T$ under the null. However, it is frequently easier to obtain the limiting distribution of the score in some other fashion and base the test on this. If a matrix V can be found so that:

$$T^{-1/2}s(\theta^0, y) \overset{D}{\to} N(0, V)$$

under H_0, then the test is simply:

$$\xi_{\text{LM}} = s'V^{-1}s/T.$$

Under certain non-standard situations V may not equal $\mathscr{I}$ but in general it will. This is the approach taken by Engle (1982) which gives some test statistics very easily in complex problems.

4. Two simple examples

In these two examples, exact tests are available for comparison with the asymptotic tests under consideration.

Consider a set of T independent observations on a Bernoulli random variable which takes on the values:

$$y_t = \begin{cases} 1, & \text{with probability } \theta, \\ 0, & \text{with probability } 1-\theta. \end{cases} \tag{14}$$

The investigator wishes to test $\theta = \theta^0$ against $\theta \neq \theta^0$ for $\theta \in (0,1)$. The mean $\bar{y} = \sum y_t / T$ is a sufficient statistic for this problem and will figure prominently in the solution.

The log-likelihood function is given by:

$$L(\theta, y) = \sum_t (y_t \log\theta + (1-y_t)\log(1-\theta)), \tag{15}$$

with the maximum likelihood estimator, $\hat{\theta} = \bar{y}$. The score is:

$$s(\theta, y) = \frac{1}{\theta(1-\theta)} \sum_t (y_t - \theta).$$

Notice that $y_t - \theta$ is analogous to the "residual" of the fit. The information is:

$$\mathscr{I}(\theta) = \mathrm{E}\left[\frac{T\theta(1-\theta)+(1-2\theta)\Sigma(y_t-\theta)}{\theta^2(1-\theta)^2}\right]\Big/T$$

$$= \frac{1}{\theta(1-\theta)}.$$

The Wald test is given by:

$$\xi_{\mathrm{W}} = T(\theta^0 - \bar{y})^2/\bar{y}(1-\bar{y}). \tag{16}$$

The LM test is:

$$\xi_{\mathrm{LM}} = \left[\frac{\Sigma(y_t-\theta^0)}{\theta^0(1-\theta^0)}\right]^2\frac{\theta^0(1-\theta^0)}{T},$$

which is simply:

$$\xi_{\mathrm{LM}} = T(\theta^0 - \bar{y})^2/\theta^0(1-\theta^0). \tag{17}$$

Both clearly have a limiting chi-square distribution with one degree of freedom. They differ in that the LM test uses an estimate of the variance under the null whereas the Wald uses an estimate under the alternative. When the null is true (or a local alternative) these will have the same probability limit and thus for large samples the tests will be equivalent. If the alternative is not close to the null, then presumably both tests would reject with very high probability for large samples; the asymptotic behavior of tests for non-local alternatives is usually not of particular interest.

The likelihood ratio test statistic is given by:

$$\xi_{\mathrm{LR}} = 2T\{\bar{y}\log\bar{y}/\theta^0 + (1-\bar{y})\log(1-\bar{y})/(1-\theta^0)\}, \tag{18}$$

which has a less obvious limiting distribution and is slightly more awkward to calculate. A two-term Taylor series expansion of the statistic about $\bar{y} = \theta^0$ establishes that under the null the three will have the same distribution.

In each case, the test statistic is based upon the sufficient statistic $\bar{y}$. In fact, in each case the test is a monotonic function of $\bar{y}$ and therefore, the limiting chi squared approximation is not necessary. For each test statistic, the exact critical values can be calculated. Consequently, when the sizes of the tests are equal their critical regions will be identical; they will each reject for large values of $(\bar{y} - \theta^0)^2$.

The notion of how large it should be will be determined from the exact Binomial tables.

The second example is more useful to economists but has a similar result. In the classical linear regression problem, the test statistics are different, however, when corrected to have the same size they are identical for finite samples as well as asymptotically.

Let y^* and x^* be $T \times 1$ and $T \times k$ matrices satisfying:

$$y^*|x^* \sim N(x^*\beta, \sigma^2 I), \tag{19}$$

and consider testing the hypothesis that $R\beta = r$ where R is a $k_1 \times k$ matrix of known constants and r is a $k_1 \times 1$ vector of constants. If R has rank k_1, then the parameters and the data can always be rearranged so that the test is of omitted variable form. That is, (19) can be reparameterized in the notation of (4) as:

$$y|x \sim N(x\theta, \sigma^2 I), \tag{20}$$

where the null hypothesis is $\theta_1 = 0$ and y and x are linear combinations of y^* and x^*. In this particular problem it is just as easy to use (19) as (20); however, in others the latter form will be simpler. The intuitions are easier when the parameters of R and r do not appear explicitly in the test statistics. Furthermore, (20) is most often the way the test is calculated to take advantage of packaged computer programs since it involves running regressions with and without the variables x_1.

For the model in (20) the log-likelihood conditional on x is:

$$L(\theta, y) = k - \frac{T}{2}\log\sigma^2 - \frac{1}{2\sigma^2}(y - x\theta)'(y - x\theta), \tag{21}$$

where k is a constant. If σ^2 were known, Lemmas 1 and 2 would guarantee that the W, LR, and LM tests would be identical. Hence, the important difference between the test statistics will be the estimate of σ^2. The score and information matrix corresponding to the parameters θ are:

$$s(\theta, y) = x'u/\sigma^2; \qquad u = y - x\theta,$$
$$\mathscr{I}_{\theta\theta} = x'x/\sigma^2 T, \tag{22}$$

and the information matrix is block diagonal between θ and σ^2. Notice that the score is proportional to the correlation coefficient between the residuals and the x variables. This is of course zero at $\hat{\theta}$ but not at the estimates under the null, $\tilde{\theta}$.

The three test statistics therefore are:

$$\xi_{\mathrm{W}} = (\theta_1^0 - \hat{\theta}_1)'\left(x_1'x_1 - x_1'x_2(x_2'x_2)^{-1}x_2'x_1\right)(\theta_1^0 - \hat{\theta}_1)/\hat{\sigma}^2, \tag{23}$$

$$\xi_{\mathrm{LM}} = \tilde{u}'x_1\left(x_1'x_1 - x_1'x_2(x_2'x_2)^{-1}x_2'x_1\right)^{-1}x_1'\tilde{u}/\tilde{\sigma}^2, \tag{24}$$

$$\xi_{\mathrm{LR}} = T\log(\tilde{u}'\tilde{u}/\hat{u}'\hat{u}), \tag{25}$$

where $\hat{u} = y - x\hat{\theta}$, $\tilde{u} = y - x\tilde{\theta}$, and $\hat{\sigma}^2 = \hat{u}'\hat{u}/T$, $\tilde{\sigma}^2 = \tilde{u}'\tilde{u}/T$, and x is conformably partitioned as $x = (x_1, x_2)$. From the linear algebra of projections, these can be rewritten as:

$$\xi_{\mathrm{W}} = T(\tilde{u}'\tilde{u} - \hat{u}'\hat{u})/\hat{u}'\hat{u}, \tag{26}$$

$$\xi_{\mathrm{LM}} = T(\tilde{u}'\tilde{u} - \hat{u}'\hat{u})/\tilde{u}'\tilde{u}. \tag{27}$$

This implies that:

$$\xi_{\mathrm{LR}} = T\log(1 + \xi_{\mathrm{W}}/T); \qquad \xi_{\mathrm{LM}} = \xi_{\mathrm{W}}/(1 + \xi_{\mathrm{W}}/T),$$

and that $(T-K)\xi_{\mathrm{W}}/TK_1$ will have an exact $F_{k_1, T-k}$ distribution under the null. As all the test statistics are monotonic functions of the F statistic, then exact tests for each would produce identical critical regions. If, however, the asymptotic distribution is used to determine the critical values, then the tests will differ for finite samples and there may be conflicts between their conclusions. Evans and Savin (1980) calculate the probabilities of such conflicts for the test in (23)–(25) as well as for those modified either by a degree of freedom correction or by an Edgeworth expansion correction. In the latter case, the sizes are nearly correct and the probability of conflict is nearly zero. It is not clear how these conclusions generalize to models for which there are no exact results but similar conclusions might be expected. See Rothenberg (1980) for some evidence for the equivalence of the tests for Edgeworth expansions to powers of $1/T$.

5. The linear hypothesis in generalized least squares models

5.1. *The problem*

In the two preceding examples, there was no reason to appeal to asymptotic approximations for test statistics. However, if the assumptions are relaxed slightly, then the exact tests are no longer available. For example, if the variables were simply assumed contemporaneously uncorrelated with the disturbances as in:

$$y_t|x_t \sim \mathrm{IN}(x_t\beta, \sigma^2), \tag{28}$$

where IN means independent normal, then the likelihood would be identical but the test statistics would not be proportional to an F distributed random variable. Thus, inclusion of lagged dependent variables or other predetermined variables would bring asymptotic criteria to the forefront in choosing a test statistic and any of the three would be reasonable candidates as would the standard F approximations. Similarly, if the distribution of y is not known to be normal, a central limit theorem will be required to find the distribution of the test statistics and therefore only asymptotic tests will be available.

The important case to be discussed in this section is testing a linear hypothesis when the model is a generalized least squares model with unknown parameters in the covariance matrix. Suppose:

$$y|x \sim N(x\beta, \sigma^2\Omega), \qquad \Omega = \Omega(\omega), \tag{29}$$

where ω is a finite estimable parameter vector. The model has been formulated so that the hypothesis to be tested is H_0: $\beta_1 = 0$, where $\beta = (\beta_1', \beta_2')'$ and x is conformally partitioned as $x = (x_1, x_2)$. The collection of parameters is now $\theta = (\beta_1', \beta_2', \sigma^2, \omega')'$.

A large number of econometric problems fit into this framework. In simple linear regression the standard heteroscedasticity and serial correlation covariance matrices have this form. More generally if ARMA processes are assumed for the disturbances or they are fit with spectral methods assuming only a general stationary structure as in Engle (1980), the same analysis will apply. From pooled time series of cross sections, variance component structures often arise which have this form. To an extent which is discussed below, instrumental variables estimation can be described in this framework. Letting X be the matrix of all instruments, $X(X'X)^{-1}X'$ has no unknown parameters but acts like a singular covariance matrix. Because it is an idempotent matrix, its generalized inverse is just the matrix itself, and therefore many of the same results will apply.

For systems of equations, a similar structure is often available. By stacking the dependent variables in a single dependent vector and conformably stacking the independent variables and the coefficient vectors, the covariance matrix of a seemingly unrelated regression problem (SUR) will have a form satisfied by (29). In terms of tensor products this covariance matrix is $\Omega = \Sigma \otimes I$, where Σ is the contemporaneous covariance matrix. Of course more general structures are also appropriate. The three stage least squares estimator also is closely related to this analysis with a covariance matrix $\Omega = \Sigma \otimes X(X'X)^{-1}X'$.

5.2. The test statistics

The likelihood function implied by (29) is given by:

$$L(\theta, y) = k - \frac{T}{2}\log\sigma^2 - \tfrac{1}{2}\log|\Omega| - \frac{1}{2\sigma^2}(y - x\beta)'\Omega^{-1}(y - x\beta). \qquad (30)$$

Under these assumptions it can be shown that the information matrix is block diagonal between the parameters β and (σ^2, ω). Therefore attention can be confined to the β components of the score and information. These are given by:

$$s_{\beta_1}(\theta, y) = x_1'\Omega^{-1}u/\sigma^2, \qquad u = y - x\beta, \qquad (31)$$

$$\mathscr{I}_{\beta\beta}(\theta) = x'\Omega^{-1}x/\sigma^2 T. \qquad (32)$$

Denote the maximum likelihood estimates of the parameters under H_1 by $\hat{\theta} = (\hat{\beta}, \hat{\sigma}^2, \hat{\omega})$ and let $\hat{\Omega} = \Omega(\hat{\omega})$; denote the maximum likelihood estimates of the same parameters under the null as $\tilde{\theta} = (\tilde{\beta}, \tilde{\sigma}^2, \tilde{\omega})$ and let $\tilde{\Omega} = \Omega(\tilde{\omega})$. Further, let $\hat{u} = y - x\hat{\beta}$ and $\tilde{u} = y - x\tilde{\beta}_2$ be residuals under the alternative and the null.

Then substituting into (11), (12), and (13), the test statistics are simply:

$$\xi_{\mathrm{W}} = \hat{\beta}_1'\left(x_1'\hat{\Omega}^{-1}x_1 - x_1'\hat{\Omega}^{-1}x_2\left(x_2'\hat{\Omega}^{-1}x_2\right)^{-1}x_2'\hat{\Omega}^{-1}x_1\right)\hat{\beta}_1/\hat{\sigma}^2, \qquad (33)$$

$$\xi_{\mathrm{LR}} = -2(L(\tilde{\theta}, y) - L(\hat{\theta}, y)), \qquad (34)$$

$$\xi_{\mathrm{LM}} = \tilde{u}'\tilde{\Omega}^{-1}x_1\left(x_1'\tilde{\Omega}^{-1}x_1 - x_1'\tilde{\Omega}^{-1}x_2\left(x_2'\tilde{\Omega}^{-1}x_2\right)^{-1}x_2'\tilde{\Omega}^{-1}x_1\right)^{-1}x_1'\tilde{\Omega}^{-1}\tilde{u}/\tilde{\sigma}^2. \qquad (35)$$

The Wald statistic can be recognized as simply the F or squared t statistic commonly computed by a GLS regression (except for finite sample degree of freedom corrections). This illustrates that for testing one parameter, the square root of these statistics with the appropriate sign would be the best statistic since it would allow one tailed tests if these are desired.

It is well known that the Wald test statistic can be calculated by running two regressions just as in (26). Care must however be taken to use the same metric (estimate of Ω) for both the restricted and the unrestricted regressions. The residuals from the unrestricted regression using $\hat{\Omega}$ as the covariance matrix are the $\hat{u}$, however, the residuals from the restricted regression using $\hat{\Omega}$ are not $\hat{u}$. Let them be denoted u^{01} indicating the model under H^0 with the covariance matrix under H^1. Thus, $u^{01} = y - x_2\beta_2^{01}$ is calculated assuming $\hat{\Omega}$ is a known matrix. The Wald statistic can equivalently be written as:

$$\xi_{\mathrm{W}} = T(u^{01\prime}\hat{\Omega}^{-1}u^{01} - \hat{u}'\hat{\Omega}^{-1}\hat{u})/\hat{u}'\hat{\Omega}^{-1}\hat{u}. \qquad (36)$$

The LM statistic can also be written in several different forms some of which may be particularly convenient. Three different versions will be given below.

Because $\tilde{u}'\tilde{\Omega}^{-1}x_2 = 0$ by the definition of $\tilde{u}$, the LM statistic is more simply written as:

$$\xi_{\mathrm{LM}} = T\tilde{u}'\tilde{\Omega}^{-1}x(x'\tilde{\Omega}^{-1}x)^{-1}x'\tilde{\Omega}^{-1}\tilde{u}/\tilde{u}'\tilde{\Omega}^{-1}\tilde{u}. \tag{37}$$

This can be interpreted as T times the R^2 of a regression where $\tilde{u}$ is the dependent variable, x is the set of independent variables and $\tilde{\Omega}$ is the covariance matrix of the disturbances which is assumed known. From the formula it is clear that this should be the R^2 calculated as the explained sum of squares over the total sum of squares. This is in contrast to the more conventional measure where these sums of squares are about the means. Furthermore, it is clear that the data should first be transformed by a matrix P such that $P'P = \tilde{\Omega}^{-1}$, and then the auxiliary regression and R^2 calculated. As there may be ambiguities in the definition of R^2 when $\Omega \neq I$ and when there is no intercept in the regression, let R_0^2 represent the figure implied by (37). Then:

$$\xi_{\mathrm{LM}} = TR_0^2. \tag{38}$$

In most cases and for most computer packages R_0^2 will be the conventionally measured R^2. In particular when Px includes an intercept under H_0, then $P\hat{u}$ will have a zero mean so that the centered and uncentered sums of squares will be equal. Thus, if the software first transforms the data by P, the R^2 will be R_0^2.

A second way to rewrite the LM statistic is available along the lines of (27). Let u^{10} be the residuals from a regression of y on the unrestricted model using $\tilde{\Omega}$ as the covariance matrix, so that $u^{10} = y - x\beta^{10}$. Then the LM statistic is simply:

$$\xi_{\mathrm{LM}} = T(\tilde{u}'\tilde{\Omega}^{-1}\tilde{u} - u^{10\prime}\tilde{\Omega}^{-1}u^{10})/\tilde{u}'\tilde{\Omega}^{-1}\tilde{u}. \tag{39}$$

A statistic which differs only slightly from the LM statistic comes naturally out of the auxiliary regression. The squared t or F statistics associated with the variables x_1 in the auxillary regressions of $\tilde{u}$ on x using $\tilde{\Omega}$ are of interest. Letting:

$$A = x_1'\tilde{\Omega}^{-1}x_1 - x_1'\tilde{\Omega}^{-1}x_2\left(x_2'\tilde{\Omega}^{-1}x_2\right)^{-1}x_2'\tilde{\Omega}^{-1}x_1,$$

then

$$\beta^{10} = (x'\tilde{\Omega}^{-1}x)^{-1}x'\tilde{\Omega}^{-1}\tilde{u},$$

or the first elements $\beta_1^{10} = A^{-1}x_1'\tilde{\Omega}^{-1}\tilde{u}$. The F statistic aside from degree of freedom corrections is given by:

$$\begin{aligned}\xi'_{LM} &= \beta_1^{10\prime} A \beta_1^{10} / \sigma^{2(10)} \\ &= \tilde{u}'\tilde{\Omega}^{-1} x_1 A^{-1} x_1' \tilde{\Omega}^{-1} \tilde{u} / \sigma^{2(10)},\end{aligned} \tag{40}$$

where $\sigma^{2(10)}$ is the residual variance from this estimation. From (35) it is clear that $\xi_{LM} = \xi'_{LM}$ if $\sigma^{2(10)} = \tilde{\sigma}^2$. The tests will differ when x_1 explains some of $\tilde{u}$, that is, when H_0 is not true. Hence, under the null and local alternatives, these two variances will have the same probability limit and therefore the tests will have the same limiting distribution. Furthermore, adding a linear combination of regressors to both sides of a regression will not change the coefficients or the significance of other regressors. In particular adding $x_2\tilde{\beta}_2$ to both sides of the auxiliary regression converts the dependent variable to y and yet will not change ξ'_{LM}. Hence, the t or F tests obtained from regressing y on x_1 and x_2 using $\tilde{\Omega}$ will be asymptotically equivalent to the LM test.

5.3. *The inequality*

The relationship between the Wald and LM tests in this context is now clearly visible in terms of the choice of Ω to use for the test. The Wald test uses $\hat{\Omega}$ while the LM test uses $\tilde{\Omega}$ and the Likelihood Ratio test uses both. As the properties of the tests differ only for finite samples, frequently computational considerations will determine which to use. The primary computational differences stem from the estimation of Ω which may require non-linear or other iterative procedures. It may further require some specification search over a class of possible disturbance specifications. The issue therefore hinges upon whether $\hat{\Omega}$ or $\tilde{\Omega}$ is already available from previous calculations. If the null hypothesis has already been estimated and the investigator is trying to determine whether an additional variable belongs in the model in the spirit of diagnostic testing, then $\tilde{\Omega}$ is already estimated and the LM test is easier. If on the other hand, the more general model has been estimated, and the test is for a simplification or a test of a theory which predicts the importance of some variable, then $\hat{\Omega}$ is available and the Wald test is easier. In rare cases will the LR test be computationally easier.

The three test statistics differ for finite samples but are asymptotically equivalent. When the critical regions are calculated from the limiting distributions, then there may be conflicts in inference between the tests. The surprising character of this conflict is pointed out by a numerical inequality among the test statistics. It was originally established by Savin (1976) and Berndt and Savin (1977) for special cases of (29) and then by Breusch (1979) in the general case of (29). For any data set y, x, the three test statistics will satisfy the following inequality:

$$\xi_W \geq \xi_{LR} \geq \xi_{LM}. \tag{41}$$

Therefore, whenever the LM test rejects, so will the others and whenever the W fails to reject, so do the others. The inequality, however, has nothing to say about the relative merits of the tests because it applies under the null as well. That is, if the Wald test has a size of 5%, then the LR and LM test will have a size less than 5%. Hence their apparently inferior power performance is simply a result of a more conservative size. When the sizes are corrected to be the same, there is no longer a simple inequality relationship on the powers. As mentioned earlier, both Rothenberg (1979) and Evans and Savin (1982) present results that when the sizes are approximately corrected, the powers are approximately the same.

5.4. *A numerical example*

As an example, consider an equation presented in Engle (1978) which explains employment in Boston's textile industry as a function of the U.S. demand and prices, the stock of fixed factors in Boston and the Boston wage rate. The equation is a reduced form derived from a simple production model with capital as a fixed factor and a constant price elasticity of demand. The variables are specific combinations of logarithms of the original data. Denote the dependent variable by y, and the independent variables by x_1, x_2 and a constant. The hypothesis to be tested is whether a time trend should also be introduced to allow technical progress in the sector. There is substantial serial correlation in the disturbance and several methods of parameterizing it are given in the original paper; however, it will here be assumed to follow a first-order autoregressive process. There are 22 annual observations.

The basic estimate of the relation is:

$$\tilde{y} = \underset{(0.92)}{4.4} + \underset{(2.45)}{0.165x_1} + \underset{(3.11)}{0.669x_2}; \qquad \rho = 0.901, \quad R^2 = 0.339.$$

The estimate is not particularly good but it has the right signs and significant t-statistics. Rho was estimated by searching over the unit interval and the estimate is maximum likelihood.

The residuals from this estimate were then regressed upon the expanded set of regressors, to obtain:

$$\tilde{u} = \underset{(1.90)}{49.2} - \underset{(-1.61)}{0.185x_1} - \underset{(-0.22)}{0.045x_2} - \underset{(1.93)}{0.025 \text{ time}}; \qquad \rho = 0.901, \quad R^2 = 0.171.$$

The same value of rho was imposed upon this estimate. The Lagrange Multiplier statistic is (22) (0.171) = 3.76 which is slightly below the 95% level for $X_1^2(3.84)$ but above the 90% level (2.71) so it rejects at 90% but not 95%. Notice that the t-statistic on time is not significant at 95% but is at the 90% level.

For comparison, the full regression was estimated including a reoptimization of rho. The results were

$$\hat{y} = \underset{(2.26)}{59.9} - \underset{(-0.45)}{0.05x_1} + \underset{(3.18)}{0.611x_2} - \underset{(2.13)}{0.028 \text{ time}}; \qquad \rho = 0.970, \quad R^2 = 0.480.$$

The Wald test involves merely looking at the t-statistic on time; however, the asymptotic formulation would estimate the standard error using 22 degrees of freedom rather than 18. In this case the t-statistic is -2.35 so the test rejects at 95% but not 99%. The Wald statistic $\xi_W = 5.52$ exceeds the 95 point of X_1^2 but not the 99% point (6.63).

In this example the two test statistics give conflicting inference at the 95% level with the Wald statistic rejecting the null hypothesis and the Lagrange Multiplier statistic accepting. However, at both 90% and the 99% level, they agree. The numerical results support the algebraic relationship given above. The benefits from using the Lagrange Multiplier test lie primarily in the avoidance of a recalculation of rho. While this may appear a rather minimal saving for the first-order autoregressive case, it may be substantial for models postulated to have ARMA disturbance processes or general stationary error processes requiring expensive iterative procedures. In establishing the validity of a regression equation, a variety of alternatives may be considered and thus, the computational saving from such a battery of tests will be even more substantial.

5.5. *Instrumental variables*

A closely related set of problems occurs in testing hypotheses in equations or models estimated with instrumental variables methods. The analysis given here concerns only the Wald test in several forms, however, LM versions can be deduced from the results in Engle (1982).

Consider first a single equation in a simultaneous system:

$$y = Y\alpha + x\gamma + \varepsilon = z\beta + \varepsilon, \qquad \varepsilon \sim N(0, \sigma^2 I), \tag{42}$$

and X is a matrix of instrumental variables including x, which is assumed to be uncorrelated with ε but correlated with Y. Limited information maximum likelihood estimation of this model yields asymptotically the same estimates as 2SLS or IV, and hence the standard test statistics are asymptotically equivalent to Wald tests. Letting $G = X(X'X)^{-1}X'$ and H_0; $\beta_1 = 0$ be the hypothesis under test, the standard test statistic is simply:

$$\xi_W = \hat{\beta}_1'\left(z_1'Gz_1 - z_1'Gz_2\left(z_2'Gz_2\right)^{-1} z_2'Gz_1\right)\hat{\beta}_1/\hat{\sigma}^2, \tag{43}$$

where $\hat{\beta}=(z'Gz)^{-1}z'Gy$, $\hat{u}=y-z\hat{\beta}$, $\hat{\sigma}^2=\hat{u}'\hat{u}/T$. This expression is identical to that in (36) except that the estimates of σ^2 are different. In (36) $\hat{\sigma}^2=\hat{u}'\Omega^{-1}\hat{u}/T$ instead of $\hat{u}'\hat{u}/T$. Following the line of reasoning leading to (37), the numerator can be rewritten in terms of the residuals from a restricted regression using the same G matrix. Letting $\tilde{\beta}_2=(z_2'Gz_2)^{-1}z_2'Gy$ and $\tilde{u}=y-z_2\tilde{\beta}_2$, the statistic can be expressed as:

$$\xi_W'=T(\tilde{u}'G\tilde{u}-\hat{u}'G\hat{u})/\hat{u}'\hat{u}. \tag{44}$$

Because G is idempotent, the two sums of squares in the numerator can be calculated by regressing the corresponding residuals on X and looking at the explained sums of squares. Their difference is also available as the difference between the sums of squared residuals from the second stages of the relevant 2SLS regressions.

As long as the instrument list is unchanged from the null to the alternative hypothesis, there is no difficulty formulating this test. If the list does change then the Wald test appropriately uses the list under the alternative. One might suspect that a similar LM test would be available using the more limited set of instruments, however, this is not the case at least in this simple form. When the instruments are different, the LM test can be computed as given in Engle (1979a) but does not have the desired simple form.

In the more general case where (42) represents a stacked set of simultaneous equations the covariance would in general be given by $\Sigma\otimes I$, where Σ is the contemporaneous covariance matrix. The instruments in the stacked system can be formulated as $I\otimes X$ and therefore letting $\hat{\Sigma}$ be the estimated covariance matrix under the alternative, the 3SLS estimator can be written letting $G=\hat{\Sigma}\otimes X(X'X)^{-1}X'$ as:

$$\hat{\beta}=(z'Gz)^{-1}z'Gy.$$

Again, through the equivalence with FIML, the approximate Wald test is:

$$\xi_W'=\hat{\beta}_1'\left(z_1'Gz_1-z_1'Gz_2\left(z_2'Gz_2\right)^{-1}z_2'Gz_1\right)\hat{\beta}_1,$$

which can be reformulated as:

$$=T(\tilde{u}'G\tilde{u}-\hat{u}'G\hat{u}).$$

Notice that $\hat{\sigma}^2$ has disappeared from the test statistic as it is incorporated in G through $\hat{\Sigma}$. Again this difference is equal to the difference between the sums of squared residuals in the restricted and unrestricted third stage of 3SLS.

6. Asymptotic equivalence and optimality of the test statistics

In this section the asymptotic equivalence, the limiting distributions and the asymptotic optimality of the three test statistic will be established under the conditions of Crowder (1976). These rather weak conditions allow some dependence of the observations and do not require that they be identically distributed. Most econometric problems will be encompassed under these assumptions. Although it is widely believed that these tests are optimal in some sense, the discussion in this section is designed to establish their properties under a set of regularity conditions.

The log likelihood function assumed by Crowder allows for general dependence of the random variables and for some types of stochastic or deterministic exogenous variables. Let $Y_0, Y_1, \dots, Y_T$ be $p \times 1$ vectors of random variables which have known conditional probability density functions $f_t(Y_t|\mathscr{F}_{t-1};\theta)$, where $\theta \in \Theta$ an open subset of R^k and $\mathscr{F}_{t-1}$ is the σ field generated by $Y_0, \dots, Y_{t-1}$, the "previous history". The log-likelihood conditional on Y_0 is:

$$L_T(Y;\theta) = \sum_{t=1}^{T} \log f_t(Y_t|\mathscr{F}_{t-1},\theta). \tag{45}$$

In this expression, non-stochastic variables enter through the time subscript on f which allows each random vector to be distributed differently. Stochastic variables which appear in conditioning sets can also be included within this framework if they satisfy the assumptions of weak exogeneity as defined by Engle, Hendry and Richard (1983). Let $Y_t = (y_t, x_t)$, where the parameters of the conditional distribution of y given x, $g_t(y_t|x_t, \mathscr{F}_{t-1}, \theta)$ are of interest. Then expressing the density of x as $h_t(x_t|\mathscr{F}_{t-1}, \phi)$ for some parameters ϕ, the log-likelihood function can be written as:

$$L_T(Y,\theta,\phi) = \sum_{t=1}^{T} \log g_t(y_t|x_t,\mathscr{F}_{t-1},\theta) + \sum_{t=1}^{T} \log h_t(x_t|\mathscr{F}_{t-1},\phi).$$

If ϕ is irrelevant to the analysis, then x_t is weakly exogenous. The information matrix will clearly be block diagonal between θ and ϕ and the MLE of θ will be obtained just by maximizing the first sum with respect to θ. Therefore, if the log-likelihood L_T satisfies Crowder's assumptions, then the conditional log-likelihood,

$$L_T^*(y,x,\theta) = \sum_{t=1}^{T} \log g_t(y_t|x_t,\mathscr{F}_{t-1},\theta),$$

also will. Notice that this result requires only that x be weakly exogenous; it need not be strongly exogenous and can therefore depend upon past values of y.

The GLS models of Section 5 can now also be written in this framework. Letting $P'P = \Omega^{-1}$ for any value of ω, rewrite the model with $y^* = Py$, $x^* = Px$ so that:

$$y^*|x^* \sim N(x^*\beta, \sigma^2 I)$$

The parameters of interest are now β, σ^2 and ω. If the x were fixed constants, then so will be the x^*. If the x were stochastic strongly exogenous variables as implied by (29), then so will be x^*. The density $h(x, \phi)$ will become $h^*(x^*, \phi, \omega)$ but unless there is some strong a priori structure on h, ω will not enter h^*. If the covariance structure is due to serial correlation then rewriting the model conditional on the past will transform it directly into the Crowder framework regardless of whether the model is already dynamic or not.

Based on (45), the score, Hessian and information matrix are defined by:

$$s_T(y,\theta) = \frac{\partial L(y,\theta)}{\partial \theta}, \tag{46}$$

$$H_T(y,\theta) = \frac{\partial^2 L}{\partial \theta\, \partial \theta'}(y,\theta),$$

$$\mathcal{I}_T(\theta) = \frac{1}{T} E s_T(y,\theta) s_T'(y,\theta).$$

Notice that the information matrix depends upon the sample size because the y_t's are not identically distributed.

The essential conditions assumed by Crowder are:

(a) the true θ, θ^*, is an interior point of Θ;
(b) the Hessian matrix is a continuous function of θ in a neighborhood of θ^*;
(c) $\mathcal{I}_T(\theta^*)$ is non-singular;
(d) plim $(\mathcal{I}_T^{-1}(\theta) H_T(y,\theta)/T) = I$ for θ in a neighborhood of θ^*; and
(e) a condition such that no term in y_t dominates the sum to T.

Suppose the hypothesis to be tested is H_0: $\theta = \theta^0$ while the alternative is H_1: $\theta = \theta^T$ where plim $T^{1/2}(\theta^T - \theta^0) = \delta$ for some vector δ.

Under these assumptions the maximum likelihood estimator of θ, $\hat{\theta}$ exists and is consistent with a limiting normal density given by:

$$T^{1/2}\mathcal{I}_T^{1/2}(\theta^*)(\hat{\theta} - \theta^*) \xrightarrow{D} N(0, I) \tag{47}$$

Mean Value Taylor series expansions can be written as:

$$L(\theta, y) = L(\hat{\theta}, y) - \frac{T}{2}(\theta - \hat{\theta})' A_T(\theta, \hat{\theta})(\theta - \hat{\theta}),$$

$$s_T(\theta, y) = -TA_T(\theta, \hat{\theta})(\theta - \hat{\theta}), \tag{48}$$

where $T[A_T(\theta, \hat{\theta})]_{ij} = [H_T(\bar{\theta})]_{ij}$ and $\bar{\theta} \in (\theta, \hat{\theta})$ possibly at different points for different (i, j). From (48) the Likelihood Ratio test is simply:

$$\xi_{LR} = T(\theta^0 - \hat{\theta})' A_T(\theta^0, \hat{\theta})(\theta^0 - \hat{\theta}),$$

and the Wald test is:

$$\xi_W = T(\theta^0 - \hat{\theta})' \mathcal{I}_T(\hat{\theta})(\theta^0 - \hat{\theta}).$$

Thus,

$$\text{plim}|\xi_{LR} - \xi_W| = \text{plim}|T(\theta^0 - \hat{\theta})'(A_T(\theta^0, \hat{\theta}) - \mathcal{I}_T(\hat{\theta}))(\theta^0 - \hat{\theta})|.$$

The plim of the middle terms is zero for $\theta^* = \theta^0$ and for the sequence of local alternatives since again plim $\theta^T = \theta^0$. The terms $T^{1/2}(\hat{\theta} - \theta^0)$ will converge in distribution under both H_0 and H_1 and therefore the product converges in probability to zero under H_0 and H_1. Thus ξ_{LR} and ξ_W have the same limiting distributions. Similarly, from (48) and (10):

$$\begin{aligned}\xi_{LM} &= Ts_T(\theta^0, y)' \mathcal{I}_T(\theta^0)^{-1} s_T(\theta^0, y) \\ &= T(\theta^0 - \hat{\theta})' A_T(\theta^0, \hat{\theta}) \mathcal{I}_T(\theta^0)^{-1} A_T(\theta^0, \hat{\theta})(\theta^0 - \hat{\theta}),\end{aligned}$$

and by the same argument $\text{plim}|\xi_{LR} - \xi_{LM}| = 0$ for H_0 and local alternatives. Thus we have the following theorem:

Theorem 1

Under the assumptions in Crowder (1976), the Wald, Likelihood Ratio and Lagrange Multiplier test statistics have the same limiting distribution when the null hypothesis or local alternative are true.

Another way to describe this result is to rewrite (48) as:

$$L(\theta, y) = L(\hat{\theta}, y) - \frac{T}{2}(\theta - \hat{\theta})' \mathcal{I}_T(\theta^0)(\theta - \hat{\theta}) + O_p(1), \tag{49}$$

where $O_p(1)$ refers to the remainder terms which vanish in probability for H_0 and local alternatives. Thus, asymptotically the likelihood is exactly quadratic and Lemmas 1 and 2 establish that the tests are all the same. Furthermore, (49) establishes that $\hat{\theta}$ is asymptotically sufficient for θ. To see this more clearly, rewrite the joint density of y as:

$$f(y,\theta)=f(y,\hat{\theta})\exp\left[-\tfrac{1}{2}(\theta-\hat{\theta})'\mathscr{I}_T(\theta^0)(\theta-\hat{\theta})\right]+O_p(1)$$

and notice that by the factorization theorem, $\hat{\theta}$ is sufficient for θ as long as y does not enter the exponent which will be true asymptotically.

Finally, because $\hat{\theta}$ has a limiting normal distribution, with a known covariance matrix $\mathscr{I}(\theta^0)^{-1}$, all the testing results for hypotheses on the mean vector of a multivariate normal, now apply asymptotically by considering $\hat{\theta}$ as the data.

To explore the nature of this optimality, suppose that the likelihood function in (49) is exact without the $O_p(1)$ term. Then several results are immediately apparent. If θ is one dimensional, uniformly most powerful (UMP) tests will exist against one sided alternatives and UMP unbiased (UMPU) tests will exist against two sided alternatives.

If $\theta=(\theta_1,\theta_2)$ where θ_1 is a scalar hypothesized to have value θ_1^0 under H_0 but θ_2 are unrestricted, then UMP similar or UMPU tests are available.

When θ_1 is multivariate, an invariance criterion must be added. In testing the hypothesis $\mu=0$ in the canonical model $V\sim N(\mu,I)$, there is a natural invariance with respect to rotations of V. If $\tilde{V}=DV$, where D is an orthogonal matrix, then the testing problem is unchanged so that a test should be invariant to whether V or $\tilde{V}$ are given. Essentially, this invariance says that the test should not depend on which order the V's are in; it should be equally sensitive to deviations in all directions. The maximally invariant statistic in this problem is $\sum V_i^2$ which means that any test which is to be invariant can be based upon this statistic. Under the assumptions of the model, this will be distributed as $X_k^2(\lambda)$ with non-centrality parameter $\lambda=\mu'\mu$. The Neyman–Pearson lemma therefore establishes that the uniformly most powerful invariant test would be based upon a critical region:

$$C=\left\{\sum V_i^2>c\right\}.$$

To rewrite (49) in this form, let $\mathscr{I}_T(\theta^0)^{-1}=P'P$ and $V=P(\hat{\theta}-\theta^0)$. Then the maximal invariant is

$$T(\hat{\theta}-\theta^0)'\mathscr{I}_T(\theta^0)(\hat{\theta}-\theta^0)$$

which is distributed as $X_k^2(\lambda)$ where $\lambda=T\delta'\mathscr{I}_T(\theta^0)\delta$ where $\delta=\theta^1-\theta^0$. The non-centrality parameter depends upon the distance between the null and alternative hypotheses in the metric $\mathscr{I}_T(\theta^0)$.

If the null hypothesis in the canonical model specifies merely H_0: $\mu_1 = 0$, then an additional invariance argument is invoked, namely $\tilde{V}_2' = V_2 + K$, where K is an arbitrary set of constants, and $V' = (V_1', V_2')$. Then the maximal invariant is $V_1'V_1$ which in (49) becomes:

$$\xi = T(\hat{\theta}_1 - \theta_1^0)'(\mathscr{I}_{11} - \mathscr{I}_{12}\mathscr{I}_{22}^{-1}\mathscr{I}_{21})(\hat{\theta}_1 - \theta_1^0). \tag{50}$$

The non-centrality parameter becomes:

$$\lambda = \mu_1'\mu_1 = T\delta_1'(\mathscr{I}_{11} - \mathscr{I}_{12}\mathscr{I}_{22}^{-1}\mathscr{I}_{21})\delta_1. \tag{51}$$

Thus, any test which is invariant can be based on this statistic and a uniformly most powerful invariant test would have a critical region of the form:

$$C = \{\xi \geq c\}.$$

This argument applies directly to the Wald, Likelihood Ratio and LM tests. Asymptotically the remainder term in the likelihood function vanishes for the null hypothesis and for local alternatives. Hence, these tests can be characterized as asymptotically locally most powerful invariant tests. This is the general optimality property of such tests which often will be simply called asymptotic optimality. For further details on these arguments the reader is referred to Cox and Hinckley (1974, chs. 5, 9), Lehmann (1959, chs. 4, 6, 7), and Fergurson (1967, chs. 4, 5).

In finite samples many tests derived from these principles will have stronger properties. For example, if a UMP test exists, a locally most powerful test will be it. Because of the invariance properties of the likelihood function it will automatically generate tests with most invariance properties and all tests will be functions of sufficient statistics.

One further property of Lagrange Multiplier tests is useful as it gives a general optimality result for finite samples. For testing H_0: $\theta = \theta^0$ against a local alternative H_1: $\theta = \theta^0 + \delta$ for δ a vector of small numbers, the Neyman–Pearson lemma shows that the likelihood ratio is a sufficient statistic for the test. The likelihood ratio is:

$$\begin{aligned} e^{\lambda} &= L(\theta^0, y) - L(\theta^0 + \delta, y) \\ &= s(\theta^0, y)'\delta, \end{aligned}$$

for small δ. The best test for local alternatives is therefore based on a critical region:

$$C = \{ s'\delta > c \}.$$

In this case δ chooses a direction. However, if invariance is desired, then the test would be based upon the scores in all directions:

$$C = \left\{ s(\theta^0)' \mathscr{I}_T^{-1}(\theta^0) s(\theta^0) > c \right\},$$

as established above. If an exact value of c can be obtained, the Lagrange Multiplier test will be locally most powerful invariant for finite samples as well as asymptotically. This argument highlights the focus upon the neighborhood of the null hypothesis which is implicit in the LM procedure. King and Hillier (1980) have used this argument to establish this property in a particular case of interest where the exact critical value can be found.

7. The Lagrange Multiplier test as a diagnostic

The most familiar application of hypothesis testing is the comparison of a theory with the data. For some types of departure from the theory which might be of concern the theory may be rejected. The existence of an alternative theory is thus, very important.

A second closely related application is in the comparison of a statistical model with the data. Rarely do we know a priori the exact variables, functional forms and distribution implicit in a particular theory. Thus, there is some requirement for a specification search. At any stage in this search it may be desirable to determine whether an adequate representation of the data has been achieved. Hypothesis testing is a natural way to formulate such a question where the null hypothesis is the statistical model being used and the alternative is a more general specificiation which is being contemplated. A test statistic for this problem is called a *diagnostic* as it checks whether the data are adequately represented by the model. The exact significance of such a test is difficult to ascertain when it is one of a sequence of tests, but it should still be a sufficient statistic for the required inference and conditional on this point in the search, the size is known. In special cases of nested sequential tests, exact asymptotic significance levels can be calculated because the tests are asymptotically independent. For example see Sargan (1980) and Anderson (1971).

Frequently in applied research, the investigator will estimate several models but may not undertake comprehensive testing of the adequacy of his preferred model. Particular types of misspecification are consistently ignored. For example, the use of static models for time series data with the familiar low Durbin–Watson was tolerated for many years although now most applied workers make serial correlation corrections.

However, the next stage in generalization is to relax the "common factors" restriction implicit in serial correlation assumptions [see Hendry and Mizon (1980)] and estimate a dynamic model. Frequently, the economic implications will be very different.

This discussion argues for the presentation of a variety of diagnostics from each regression. Overfitting the model in many different directions allows the investigator to immediately assess the quality and stability of his specification.

The Lagrange Multiplier test is ideal for many of these tests as it is based upon parameters fit under the null which are therefore already available. In particular, the LM test can usually be written in terms of the residuals from the estimate under the null. Thus, it provides a way of checking the residuals for non-randomness. Each alternative considered indicates the particular type of non-randomness which might be expected.

Look for a moment at the LM test for omitted variables described in (37). The test is based upon the R^2 of the regression of the residuals on the included and potentially excluded variables. Thus, the test is based upon the squared partial correlation coefficient between the residuals and the omitted variables. This is a very intuitive way to examine residuals for non-randomness.

In the next sections, the LM test for a variety of types of misspecification will be presented. In Section 8, tests for non-spherical disturbances will be discussed while Section 9 will examine tests for misspecified mean functions including non-linearities, endogeneity, truncation and several other cases.

8. Lagrange Multiplier tests for non-spherical disturbances

A great deal of research has been directed at construction of LM tests for a variety of non-spherical disturbances. In most cases, the null hypothesis is that the disturbances are spherical; however, tests have also been developed for one type of covariance matrix against a more complicated one. In this section we will first discuss tests against various forms of heteroscedasticity as in Breusch and Pagan (1980), Engle (1982) and Godfrey (1978). Then tests against serial correlation as given by Godfrey (1978b, 1979), Breusch (1979), and Breusch and Pagan (1980) are discussed.

Test against other forms of non-spherical disturbances have also been discussed in the literature. For example, Breusch and Pagan (1980) develop a test against variance components structures and Breusch (1979) derives the tests for seemingly unrelated regression models.

8.1. Testing for heteroscedasticity

Following Breusch and Pagan (1980), let the model be specified as:

$$y_t|x_t, z_t \sim \mathrm{IN}(x_t\beta, h(z_t\alpha)) \tag{52}$$

where z_t is a $1\times(p+1)$ vector function of x_t or other variables legitimately taken as given for this analysis. The function h is of known form with first and second derivatives and depends upon an unknown $p+1\times 1$ vector of parameters α. The first element of z is constant with coefficient α_0 so under H_0: $\alpha_1 = \cdots = \alpha_p = 0$, the model is the classical normal regression model. The variance model includes most types of heteroscedasticity as special cases. For example, when

$$h(z_t\alpha) = \mathrm{e}^{z_t\alpha},$$

multiplicative forms are implied, while

$$h(z_t\alpha) = (z_t\alpha)^k$$

gives linear and quadratic cases for $k=1,2$. Special case of this which might be of interest would be:

$$h(z_t,\alpha) = (\alpha_0 + \alpha_1 x_t\beta)^2,$$

$$h(z_t,\alpha) = \exp(\alpha_0 + \alpha_1 x_t\beta),$$

where the variance is related to the mean of y_t.

From applications of the formulae for the LM test given above, Breusch and Pagan derive the LM test. Letting $\theta_1 = (\alpha_1,\dots,\alpha_p)$ and $\partial h/\partial\theta_1|_{\theta_1=0} = \kappa z$, where κ is a scalar, the score is:

$$s(\theta^0, y) = f'z\kappa/\tilde{\sigma}^2,$$

$$\xi_{\mathrm{LM}} = \frac{T}{2} f'z(z'z)^{-1}z'f, \tag{53}$$

where $f_t = \tilde{u}_t^2/\tilde{\sigma}_t^2 - 1$, f and z are matrices with typical rows f_t and z_t and $\tilde{u}$ and $\tilde{\sigma}^2$ are the residuals and variance estimates under the null. This expression is simply one-half the explained sum of squares of a regression of f on z. As pointed out by Engle (1978), plim $f'f/T = 2$ under the null and local alternatives, so an asymptotically equivalent test statistic is TR^2 from this regression. As long as z has an intercept, adding 1 to both sides and multiplying by a constant $\tilde{\sigma}^2$ will not change the R^2, thus, the statistic can be computed by regressing $\tilde{u}^2$ on z and calculating TR^2 of this regression. Koenker (1981) shows that this form is more robust to departures from normality.

The remarkable result of this test however is that κ has vanished. The test will be the same regardless of the form of h. This happens because both the score and the information matrix include only the derivative of h under H_0 and thus the overall shape of h does not matter. As far as the LM test is concerned, the alternative is:

$$h = z_t \alpha \kappa,$$

where κ is a scalar which is obviously irrelevant. This illustrates quite clearly both the strength and the weakness of local tests. One test is optimal for all h much as in the UMP case, however it seems plausible that it suffers from a failure to use the functional form of h.

Does this criticism of the LM test apply to the W and LR tests? In both cases, the parameters α must be estimated by a maximum likelihood procedure and thus the functional form of h will be important. However, the optimality of these tests is only claimed for local alternatives. For non-local alternatives the power function will generally go to one in any case and thus the shape of h is irrelevant from an asymptotic point of view. It remains possible that the finite sample non-local performance of the W and LR tests with the correct functional form for h could be superior to the LM. Against this must be set the possible computational difficulties of W and LR tests which may face convergence problems for some points in the sample space. Some Monte Carlo evidence that the LM test performs well in this type of situation is contained in Godfrey (1981).

Several special cases of this test procedure illustrate the power of the technique. Consider[1] the model $h = \exp(\alpha_0 + \alpha_1 x_t \beta)$, where H_0: $\alpha_1 = 0$. The score as calculated in (53) evaluates all parameters, including β, under the null. Thus, $x_t \tilde{\beta} = \tilde{y}_t$, the fitted values under the null. The heteroscedasticity test can be shown to have the same limiting distribution for $x_t \beta$ as for $x_t \tilde{\beta}$ and therefore it can easily be constructed as TR^2 from $\tilde{u}_t^2$ on a constant and $\tilde{y}_t$. If the model were $h = \exp(\alpha_0 + \alpha_1 (x_t \beta)^2)$ then the regression would be on a constant and $\tilde{y}_t^2$. Thus it is very easy to construct tests for a wide range of, possibly complex, alternatives.

Another interesting example is provided by the Autoregressive Conditional Heteroscedasticity (ARCH) model of Engle (1982). In this case z_t includes lagged squared residuals as well as perhaps other variables. The conditional variance is hypothesized to increase when the residuals increase. In the simplest case:

$$\begin{aligned} h &= \alpha_0 + \alpha_1 \tilde{u}_{t-1}^2 + \cdots + \alpha_p \tilde{u}_{t-p}^2 \\ &= z_t \alpha. \end{aligned}$$

This is really much like that discussed above as $\tilde{u}_{t-1} = y_{t-1} - x_{t-1} \tilde{\beta}$ and both y_{t-1}

[1]Adrian Pagan has suggested and used this model.

and x_{t-1} are legitimately taken as given in the conditional distribution. The test naturally comes out to be a regression of $\tilde{u}_t^2$ on $\tilde{u}_{t-1}^2,\dots,\tilde{u}_{t-p}^2$ and an intercept with the statistic as TR^2 of this regression.

Once a heteroscedasticity correction has been made, it may be useful to test whether it has adequately fixed the problem. Godfrey (1979) postulates the model:

$$\sigma_t^2 = h(z_t\alpha) + g(q_t\gamma), \tag{54}$$

where $g(0)=0$. The null hypothesis is therefore H_0: $\gamma = 0$. Under the null, estimates of $\tilde{\alpha}$ and $\tilde{u} = y_t - x_t\tilde{\beta}$ are obtained, $\tilde{\sigma}_t = h(z_t\tilde{\alpha})$ and the derivative of h at each point $z_t\tilde{\alpha}$ can be calculated as $\tilde{h}'_t$. Of course, if h is linear, this is just a constant. The test is simply again TR^2 of an auxiliary regression. In this case the regression is of:

$$\frac{\tilde{u}_t^2 - \tilde{\sigma}_t^2}{\tilde{\sigma}_t^2} \quad \text{on} \quad \frac{\tilde{h}'_t z_t}{\tilde{\sigma}_t^2} \quad \text{and} \quad \frac{q_t}{\tilde{\sigma}_t^2},$$

and the statistic will have the degrees of freedom of the number of parameters in q_t.

White (1980a) proposes a test for very general forms of heteroscedasticity. His test includes all the alternatives for which the least squares standard errors are biased. The heteroscedastic model includes all the squares and crossproducts of the data. That is, if the original model were $y = \beta_0 + \beta_1 x_1 + \beta_2 x_2 + \varepsilon$, the White test would consider x_1, x_2, x_1^2, x_2^2 and $x_1 x_2$ as determinants of σ^2. The test is as usual formulated as TR^2 of a regression of u^2 on these variables plus an intercept. These are in fact just the regressors which would be used to test for random coefficients as in Breusch and Pagan (1979).

8.2. *Serial correlation*

There is now a vast literature on testing for and estimating models with serial correlation. Tests based on the LM principles are the most recent addition to the econometrician's tool kit and as they are generally very simple, attention will be confined to them.

Suppose:

$$y_t | x_t \sim N(x_t\beta, \sigma_u^2),$$
$$\alpha(L)u_t = \varepsilon, \quad u_t = y_t - x_t\beta, \quad \alpha(L) = 1 - \alpha_1 L - \alpha_2 L^2 - \cdots - \alpha_p L^p, \tag{55}$$

and ε_t is a white noise process. Then it may be of interest to test the hypothesis H_0: $\alpha_1 = \cdots = \alpha_p = 0$. Under H_0, ordinary least squares is maximum likelihood and thus the LM approach is attractive for its simplicity. An alternative formulation of (55) which shows how it fits into Crowder's framework is:

$$y_t|x_t, \psi_{t-1} \sim N(1-\alpha(L))\,y_t + \alpha(L)x_t\beta, \sigma_\varepsilon^2), \tag{56}$$

where ψ_{t-1} is the past information in both y and x. Thus, again under H_0 the regression simplifies to OLS but under the alternative, there are non-linear restrictions. The formulation (56) makes it clear that serial correlation can also be viewed as a restricted model relative to the general dynamic model without the non-linear restrictions. This is the common factor test which is discussed by Hendry and Mizon (1980) and Sargan (1980) and for which Engle (1979a) gives an LM test.

The likelihood function is easily written in terms of (56) and the score is simply:

$$s(y,\theta) = \frac{1}{\sigma^2}U'\tilde{u}, \tag{57}$$

where U has rows $U_t = (\tilde{u}_{t-1}, \tilde{u}_{t-2}, \dots, \tilde{u}_{t-p})$.

From the form of (57) it is clear that the LM test views U_t as an omitted set of variables from the original regression. Thus, as established more rigorously by Godfrey (1978a) and Engle (1979a), the test can be computed by regressing $\tilde{u}_t$ on x_t, U_t and testing TR^2 as a χ^2_p. The argument is essentially that because the score has the form of (31), the test will look like (38). If x_t includes no lagged dependent variables, then $\operatorname{plim} x'U/T = 0$ and the auxiliary regression will be unaffected by leaving out the x's. The test therefore is simply computed by regressing $\tilde{u}_t$ on $\tilde{u}_{t-1}, \dots, \tilde{u}_{t-p}$ and checking TR^2. For $p = 1$, this test is clearly asymptotically equivalent to the Durbin–Watson statistic.

The observation that $U'x$ will have expected value zero when x is an exogenous variable, suggests that in regression models with lagged dependent variables perhaps such products should be set to their expected value which is zero. If this is done systematically, the resulting test is Durbin's (1970) h test, at least for the first order case. Thus the h test uses the a priori structure to set some of the terms of the LM test to zero. One might expect better finite sample performance from this, however, the few Monte Carlo experiments do not show such a difference. Instead, this test performs about equally well when it exists, however, for some points in the sample space, it gives imaginary values. These apparently convey no information about the validity of the null hypothesis and are a result of the approximation of a positive definite matrix by one which is not always so. Because of this fact and the difficulty of generalizing the Durbin test for higher order serial correlation and higher order lags of dependent variables, the LM test is likely to be preferred at least for higher order problems. See Godfrey and Tremayne (1979) for further details.

It would seem attractive to construct a test against moving average disturbances. Thus suppose the model has the form:

$$\begin{aligned} &y_t|x_t \sim N(x_t\beta, \sigma_u^2), \\ &y_t - x_t\beta = u_t, \\ &u_t = \varepsilon_t - \alpha_1\varepsilon_{t-1} - \cdots - \alpha_p\varepsilon_{t-p}, \end{aligned} \tag{58}$$

where ε is again a white noise process. Then $\varepsilon_t = y_t - x_t\beta - \alpha_1\varepsilon_{t-1} - \cdots - \alpha_p\varepsilon_{t-p}$ so the log-likelihood function is proportional to:

$$L = -\sum_{t=1}^{T}(y_t - x_t\beta - \alpha_1\varepsilon_{t-1} - \cdots - \alpha_p\varepsilon_{t-p})^2/2\sigma^2.$$

The score evaluated under the null that $\alpha_1 = \cdots = \alpha_p = 0$ is simply:

$$s(y, \tilde{\theta}) = \tilde{u}'U/\sigma^2,$$

which is identical to that in (57) for the $AR(\rho)$ model. As the null hypothesis is the same, the two tests will be the same. Again, the LM tests for different alternatives turn out to be the same test. For local alternatives, the autoregressive and moving average errors look the same and therefore one test will do for both.

When a serial correlation process has been fit for a particular model, it may still be of interest to test for higher order serial correlation. Godfrey (1978b) supposes that a (p,q) residual model has been fit and that $(p+r,q)$ is to be taken as the alternative not surprisingly, the test against $(p,q+r)$ is identical. Consider here the simplest case where $q=0$. Then the residuals under the null can be written as:

$$\begin{aligned} &\tilde{u}_t = y_t - x_t\tilde{\beta}, \\ &\tilde{\varepsilon}_t = \tilde{u}_t - \tilde{\gamma}_1\tilde{u}_{t-1} - \cdots - \tilde{\gamma}_p\tilde{u}_{t-p}. \end{aligned}$$

The test for $(p+r,0)$ or (p,r) error process can be calculated as TR^2 of the regression of $\tilde{\varepsilon}_t$ on $\tilde{x}_t, \tilde{u}_{t-1},\dots,\tilde{u}_{t-p}, \tilde{\varepsilon}_{t-1},\dots,\tilde{\varepsilon}_{t-r}$, where $\tilde{x}_t = x_t - \tilde{\gamma}_1 x_{t-1} - \cdots - \tilde{\gamma}_p x_{t-p}$. Just as in the heteroscedasticity case the regression is of transformed residuals on transformed data and the omitted variables. Here the new ingredient is the inclusion of $\tilde{u}_{t-1},\dots,\tilde{u}_{t-p}$ in the regression to account for the optimization over γ under the null.

This approach applies directly to diagnostic tests for time series models. Godfrey (1979a), Poskitt and Tremayne (1980), Hosking (1980) and Newbold

(1980) have developed and analyzed tests for a wide range of alternatives. In each case the score depends simply on the residual autocorrelations, however the tests differ from the familiar Box–Pierce–Portmanteau test in the calculation of the critical region. Consequently, the LM tests will have superior properties at least asymptotically for a finite parameterization of the alternative. If the number of parameters under test becomes large with the sample size then the tests become asymptotically equivalent. However, one might suspect that the power properties of tests against low order alternatives might make them the most suitable general purpose diagnostic tools.

When LM tests for serial correlation are derived in a simultaneous equation framework, the statistics are somewhat more complicated and in fact there are several incorrect tests in the literature. The difficulty arises over the differences in instrument lists under the null and alternative models. For a survey of this material plus presentation of several tests, see Breusch and Godfrey (1980). In the standard simultaneous equation model:

$$\begin{aligned} Y_t B + X_t \Gamma &= U_t, \\ U_t &= RU_{t-1} + \varepsilon_t, \end{aligned} \tag{59}$$

where Y and U_t are $1\times G$, X_t is $1\times K$ and R is a square $G\times G$, matrix of autoregressive coefficients, they seek to test H_0: $R=0$ both in the FIML and LIML context. They conclude that if $\tilde{U}_t$ is the set of residuals estimated under the assumption of no serial correlation, then the LM test can be approximated by any standard significance test in the augmented model:

$$Y_t B + X_t \Gamma - R\tilde{U}_{t-1} = \varepsilon_t. \tag{60}$$

Thus comparing the likelihood achieved under (59) and (60) would provide an asymptotically equivalent test to the LM test. As usual, this is just one of many computational techniques.

9. Testing the specification of the mean in several complex models

A common application of LM tests is in econometric situations where the estimation requires iterative procedures to maximize the likelihood function. In this section a variety of situations will be discussed where possibly complex misspecifications of the mean function are tested. LM tests for non-linearities, for common factor dynamics, for weak and strong exogeneity and for omitted variables in discrete choice and truncated dependent variable models are presented below. These illustrate the simplicity of LM tests in complex models and suggest countless other examples.

9.1. Testing for non-linearities

Frequently an empirical relationship derived from economic theory is highly non-linear. This is typically approximated by a linear regression without any test of the validity of the approximation. The LM test generally provides a simple test of such restrictions because it uses estimates only under the null hypothesis. While it is ideal for the case where the model is linear under the null and non-linear under the alternative, the procedures also greatly simplify the calculation when the null is non-linear. Three examples will be presented which show the usefulness of this set of procedures.

If the model is written as:

$$y_t|x_t \sim N\left(g(x_t,\beta),\sigma^2\right),$$

then the score under the null will have the form:

$$s(y,\tilde{\beta}) = \frac{1}{\tilde{\sigma}^2}\sum \tilde{u}_t \frac{\partial g}{\partial \beta}(x_t,\beta)|_0.$$

Thus the derivative of the non-linear relationship evaluated with parameter estimated under the null, can be considered as an omitted variable. The test would be given by the formulations in Section 5.

As an example, consider testing for a liquidity trap in the demand for money. Several studies have examined this hypothesis. Pifer (1969), White (1972) and Eisner (1971) test for a liquidity trap in logarithmic or Box–Cox functional forms while Konstas and Khouja (1969) (K–K) use a linear specification. Most studies find maximum likelihood estimates of the interest rate floor to be about 2% but they differ on whether this figure is significantly different from zero. Pifer says it is not significant, Eisner corrects his likelihood ratio test and says it is, White generalizes the form using a Box–Cox transformation and concludes that it is not different from zero. Recently Breusch and Pagan (1977a) have re-examined the Konstas and Khouja form and using a Lagrange Multiplier test, conclude that the liquidity trap is significant.

Except for minor footnotes in some of the studies, there is no mention of the serial correlation which exists in the models. In re-estimating the Konstas–Khouja model, the Durbin–Watson statistic was found to be 0.3 which is evidence of a severe problem with the specification and that the distribution of all the test statistics may be highly misleading.

The model estimated by K–K is:

$$M = \gamma Y + \beta(r-\alpha)^{-1} + \varepsilon, \tag{61}$$

where M is real money demand, Y is real GNP and r is the interest rate. Perhaps their best results are when $M1$ is used for M and the long-term government bond rate is used for r. The null hypothesis to be tested is $\alpha = 0$. The normal score is proportional to $u'z$ where z, the omitted variable, is the derivative of the right-hand side with respect to α evaluated under the null:

$$z = \left.\frac{\partial g}{\partial \alpha}\right|_0 = \frac{\beta}{r^2}.$$

Therefore, the LM test is a test of whether $1/r^2$ belongs in the regression along with Y and $1/r$.

Breusch and Pagan obtain the statistic $\xi_{LM} = 11.47$ and therefore reject $\alpha = 0$. Including a constant term this becomes 5.92 which is still very significant in the X^2 table. However, correcting for serial correlation in the model under the null changes the results dramatically. A second-order autoregressive model with parameters 1.5295 and -0.5597 was required to whiten the residuals. These parameters are used in an auxiliary regression of the transformed residual on the three transformed right-hand side variables and a constant, to obtain an $R^2 = 0.01096$. This is simply GLS where the covariance parameters are assumed known. Thus, the LM statistic is $\xi_{LM} = 0.515$ which is distributed as X_1^2 if the null is true. As can be seen it is very small suggesting that the liquidity trap is not significantly different from zero.

As a second example, consider testing the hypothesis that the elasticity of substitution of a production function is equal to 1 against the alternative that is constant but not unity. If y is output and x_1 and x_2 are factors of production, the model under the alternative can be written as:

$$\log y = -\frac{\alpha}{\rho}\log\left(\delta x_1^{-\rho} + (1-\delta)x_2^{-\rho}\right) + u. \tag{62}$$

If $\rho = 0$, the elasticity of substitution is one and the model becomes:

$$\log y = \alpha\delta\log x_1 + \alpha(1-\delta)\log x_2 + u.$$

To test the hypothesis $\rho = 0$, it is sufficient to calculate $\partial g/\partial\rho|_{\rho=0}$ and test whether this variable belongs in the regression. In this case

$$\left.\frac{\partial g}{\partial \rho}\right|_{\rho=0} = -\frac{\alpha}{2}\delta(1-\delta)\left(\log\frac{x_1}{x_2}\right)^2$$

which is simply the Kmenta (1967) approximation. Thus the Cobb–Douglas form can be estimated with appropriate heteroscedasticity or serial correlation and the unit elasticity assumption tested with power equal to a likelihood ratio test without ever doing a non-linear regression.

As a third example, Davidson, Hendry, Srba and Yeo (1978) estimate a consumption function for the United Kingdom which pays particular attention to the model dynamics. The equation finally chosen can be expressed as:

$$\Delta_4 c_t = \beta_1 \Delta_4 y_t + \beta_2 \Delta_1 \Delta_4 y_t + \beta_3 (c_{t-4} - y_{t-4}) + \beta_4 \Delta_4 D_t + \beta_5 \dot{p}_t + \beta_6 \Delta_1 \dot{p}_t, \tag{63}$$

where c, y and p are the logs of real consumption, real personal disposable income and the price level, and Δ_i is the ith difference. In a subsequent paper Hendry and Von Ungern-Sternberg (1979) argue that the income series is mismeasured in periods of inflation. The income which accrues from the holdings of financial assets should be measured by the real rate of interest rather than the nominal as is now done. There is a capital loss of $\dot{p}$ times the asset which should be netted out of income. The appropriate log income measure is $y_t^* = \log(Y_t - \alpha \dot{p} L_{t-1})$ where L is liquid assets of the personal sector and α is a scale parameter to reflect the fact that L is not all financial assets.

The previous model corresponds to $\alpha = 0$ and the argument for the respecification of the model rests on the presumption that $\alpha \neq 0$. The LM test can be easily calculated whereas the likelihood ratio and Wald tests require non-linear estimation if not respecification. The derivative of y^* with respect to α evaluated under the null is simply $-pL_{t-1}/Y_t$. Denote this by x_t. The score is proportional to $u'z$, where $z = \tilde{\beta}_1 \Delta_4 x_t + \tilde{\beta}_2 \Delta_1 \Delta_4 x_t - \tilde{\beta}_3 x_{t-4}$, and the betas are evaluated at their estimates under the null. This is now a one degree of freedom test and can be simply performed. The test is significant with a chi squared value of 5. As a one tailed test it is significant at the 2.5% level.

9.2. *Testing for common factor dynamics*

In a standard time series regression framework, there has been much attention given to the testing and estimation of serial correlation patterns in the disturbances. A typical model might have the form:

$$y_t = x_t \beta + u_t, \qquad \rho(L) u_t = \varepsilon_t, \qquad \varepsilon_t \sim \mathrm{IN}(0, \sigma^2), \tag{64}$$

where $\rho(L)$ is an rth order lag polynomial and x_t is a $1 \times k$ row vector which for the moment is assumed to include no lagged exogenous or endogenous variables.

Sargan (1964, 1980) and Hendry and Mizon (1978) have suggested that this is often a strong restriction on a general dynamic model. By multiplying through by $\rho(L)$ the equation can equivalently be written as:

$$\rho(L) y_t = \rho(L) x_t \beta + \varepsilon_t. \tag{65}$$

This model includes a set of non-linear parameter restrictions which essentially reduce the number of free parameters to $k+r$ instead of the full $(k+1)r$ which would be free if the restriction were not imposed. A convenient parameterization of the unrestricted alternative can be given in terms of another matrix of lag polynomials $\theta(L)$ which is a $1\times k$ row vector each element of which is an rth order lag polynomial with zero order lag equal to zero. That is $\theta(0)=0$. The unrestricted model is given by:

$$\rho(L)y_t=\rho(L)x_t\beta+\theta(L)x_t'+\varepsilon_t, \tag{66}$$

which simplifies to the serial correlation case if all elements of θ are zero. Thus, the problem can be parameterized in terms of $z=(x_{-1},\dots,x_{-r})$ as a matrix of kr omitted variables in a model estimated with GLS. The results of Section 5 apply directly. The test is simply TR^2 of $\tilde{\varepsilon}_t$ on $\tilde{\rho}(L)x_t$, z_t and $(\tilde{u}_{t-1},\dots,\tilde{u}_{t-r})$ or equivalently, on $x_t, z_t\ (y_{-1},\dots,y_{-r})$.

Now if x includes lags, the test must be very slightly modified. The matrix z will, in this case, include variables which are already in the model and thus the auxiliary regression will see a data set with perfect multicollinearity. The solution is to eliminate the redundant elements of z as these are not testable in any case. The test statistic will have a correspondingly reduced number of degrees of freedom.

A more complicated case occurs when it is desired to test that the correlation is of order r against the alternative that it is of order $r-1$. Here the standard test procedure breaks down. See Engle (1979a) for a discussion and some suggestions.

9.3. *Testing for exogeneity*

Tests for exogeneity are a source of controversy partly because of the variety of definitions of exogeneity implicit in the formulation of the hypotheses. In this paper the notions of weak and strong exogeneity as formulated by Engle et al. (1983) will be used in the context of linear simultaneous equation systems. In this case weak exogeneity is essentially that the equations defining weakly exogenous variables can be ignored without a loss of information. In textbook cases weakly exogenous variables are predetermined. Strong exogeneity implies, in addition, that the variables in question cannot be forecast by past values of endogenous variables which is the definition implicit in Granger (1969) "non-causality".

Consider a complete simultaneous equation system with G equations and K predetermined variables so that Y, ε, and V are $T\times G$, X is $T\times K$ and the coefficient matrices are conformable. The structural and reduced forms are:

$$YB=X\Gamma+\varepsilon, \qquad \mathrm{E}\varepsilon_t'\varepsilon_t=\Omega, \tag{67}$$

$$Y=X\Pi+V, \tag{68}$$

where ε_t are rows of ε which are independent and the x are weakly exogenous. Partitioning this set of equations into the first and the remaining $G-1$, the structure becomes:

$$y_1 - Y_2\beta = x_1\gamma + \varepsilon_1, \tag{69}$$

$$-y_1\alpha' + Y_2B_2 = X_2\Gamma_2 + \varepsilon_2, \tag{70}$$

where X_2 may be the same as X and

$$B = \begin{pmatrix} 1 & -\alpha' \\ -\beta & B_2 \end{pmatrix}, \qquad \Omega = \begin{pmatrix} \Omega_{11} & \Omega_{12} \\ \Omega_{21} & \Omega_{22} \end{pmatrix}. \tag{71}$$

The hypothesis that Y_2 is weakly exogenous to the first equation in this full information context is simply the condition for a recursive structure:

$$H_0: \alpha = 0, \Omega_{12} = 0, \tag{72}$$

which is a restriction of $2G-2$ parameters.

Several variations on this basic test are implicit in the structure. If the coefficient matrix is known to be triangular, then $\alpha = 0$ is part of the maintained hypothesis and the test becomes simply a test for $\Omega_{12} = 0$. This test is also constructed below; Holly (1979) generalized the result to let the entire B matrix be assumed upper triangular and obtains a test of the diagonality of Ω and Engle (1982a) has further generalized this to block recursive systems. If some of the elements of β are known to be zero, then the testing problem remains the same. In the special case where B_2 is upper triangular between the included and excluded variables of Y_2 and the disturbances are uncorrelated with those of y_1 and the included y_2, then it is only necessary to test that the α's and Ω's of the included elements of y_2 are zero. In effect, the excluded y_2 now form a higher level block of a recursive system and the problem can be defined a priori to exclude them also from y_2. Thus without loss of generality the test in (72) can be used when some components of β take unknown values.

To test (72) with (67) maintained, first construct the normal log likelihood L, apart from some arbitrary constants:

$$L = T\log|B| - \frac{T}{2}\log|\Omega| - \tfrac{1}{2}\sum_{t=1}^{T}\varepsilon_t\Omega^{-1}\varepsilon_t'. \tag{73}$$

Partitioning this as in (71) using the identity $|\Omega| = |\Omega_{22}||\Omega_{11} - \Omega_{12}\Omega_{22}^{-1}\Omega_{21}|$ gives:

$$L = T\log|B_2| + T\log|1 - \alpha' B_2^{-1}\beta| - \frac{T}{2}\log|\Omega_{22}|$$
$$- \frac{T}{2}\log|\Omega_{11} - \Omega_{12}\Omega_{22}^{-1}\Omega_{21}| - \tfrac{1}{2}\sum_t \varepsilon_{1t}\Omega^{11}\varepsilon_{1t}'$$
$$- \tfrac{1}{2}\sum_t \varepsilon_{2t}\Omega^{22}\varepsilon_{2t}' - \sum_t \varepsilon_{1t}\Omega^{12}\varepsilon_{2t}', \tag{74}$$

where the superscripts on Ω indicate the partitioned inverse. Differentiating with respect to α and setting parameters to their values under the null gives the score:

$$\left.\frac{\partial L}{\partial \alpha}\right|_0 = - T\tilde{B}_2^{-1}\tilde{\beta} + \sum_t \tilde{\Omega}^{22}\tilde{U}_{2t}' y_{1t}, \tag{75}$$

where tildes represent estimates under the null and $\tilde{U}_{2t}$ is the row vector of residuals under the null. Recognizing that $\sum_t \hat{\Omega}^{22}\tilde{U}_{2t}'\tilde{U}_{2t}/T = I$, this can be rewritten as:

$$\left.\frac{\partial L}{\partial \alpha}\right|_0 = \sum_t \tilde{\Omega}^{22}\tilde{U}_{2t}'\left(y_{1t} - \tilde{U}_{2t}\tilde{B}_2^{-1}\tilde{\beta}\right) \equiv \sum_t \tilde{\Omega}^{22}\tilde{U}_{2t}'\left(\bar{y}_{1t} + \tilde{u}_{1t}\right), \tag{76}$$

where $\bar{y}_1$ is the reduced form prediction of y_1 which is given in this case as $x_1\tilde{\gamma} + X_2\tilde{\Gamma}_2\tilde{B}_2^{-1}\tilde{\beta}$. Clearly, under the null hypothesis, the score will have expected value zero as it should. Using tensor notation this can be expressed as:

$$s_\alpha = \left(I\otimes(\bar{y}_1 + \tilde{u}_1)\right)'\left(\tilde{\Omega}_{22}^{-1}\otimes I\right)\operatorname{vec}(\tilde{U}_2), \tag{77}$$

which is in the form of omitted variables from a stacked set of regressions with covariance matrix $\tilde{\Omega}_{22}^{-1}\otimes I$. This is a GLS problem which allows calculation of a test for $\alpha = 0$ under the maintained hypothesis that $\Omega_{12} = 0$. Because of the simultaneity, the procedure in Engle (1982a) should be followed.

The other part of the test in (72) is obtained by differentiating with respect to Ω_{12} and evaluating under the null. It is not hard to show that all terms in the derivative vanish except the last. Because $\partial\Omega^{12}/\partial\Omega_{12}|_0 = -\Omega_{11}^{-1}\Omega_{22}^{-1}$ the score can be written as:

$$s_{\Omega_{12}} = \sum_t \tilde{u}_{1t}\tilde{\Omega}_{11}^{-1}\tilde{\Omega}_{22}^{-1}U_{2t}', \tag{78}$$

which can be written in two equivalent forms:

$$s_{\Omega_{12}} = \tilde{\Omega}_{11}^{-1}\tilde{\Omega}_{22}^{-1}\tilde{U}_2'\tilde{u}_1 \tag{79}$$

$$= \tilde{\Omega}_{11}^{-1}(I\otimes\tilde{u}_1)'(\tilde{\Omega}_{22}^{-1}\otimes I)\operatorname{vec}(\tilde{U}_2). \tag{80}$$

Either would be appropriate for testing $\Omega_{12}=0$ when $\alpha=0$ is part of the maintained hypothesis. In (79) the test would be performed in the first equation by considering U_2 as a set of $G-1$ omitted variables. In (80) the test would be performed in the other equations by stacking them and then considering $I\otimes u_1$ as the omitted set of variables. Clearly the former is easier in this case.

To perform the joint test, the two scores must be jointly tested against zero. Here (77) and (80) can easily be combined as they have just the same form. The test becomes a test for two omitted variables, $\bar{y}_1+\tilde{u}_1$ and $\tilde{u}_1$, in each of the remaining $G-1$ equations. Equivalently, $\bar{y}_1$ and $\tilde{u}_1$ can be considered as omitted from these equations.

Engle (1979) shows that this test can be computed as before. If the model is unidentified the test would have no power and if the model is very weakly identified, the test would be likely to have very low power.

In the special case where $G=2$, the test is especially easy to calculate because both equations can be estimated by least squares under the null. Therefore Section 5 can be applied directly.

As an example, the Michigan model of the monetary sector was examined. The equations are reported in Gardner and Hymans (1978). In this model, as in most models of the money market it is assumed that a short term interest rate can be taken as weakly exogenous in an equation for a long-term rate. However, most portfolio theories would argue that all rates are set at the same time as economic agents shift from one asset to another to clear the market.

In this example a test is constructed for the weak exogeneity of the prime rate, *RAAA*, in the 35 year government bond rate equation, *RG*35. The model can be written as:

$$\begin{aligned} RG35 &= \beta\Delta RAAA + x_1\gamma + \varepsilon_1, \\ \Delta RAAA &= \alpha RG35 + x_2\gamma + \varepsilon_2, \end{aligned} \tag{81}$$

where the estimates assume $\alpha=\sigma_{12}=0$, and the x's include a variety of presumably predetermined variables including lagged interest rates. Testing the hypothesis that $\alpha=0$ by considering *RG*35 as an omitted variable is not legitimate as it will be correlated with ε_2. If one does the test anyway, a chi-squared value of 35 is obtained.

The appropriate test of the weak exogeneity of *RG*35 is done by testing u_1 and $RG35-\tilde{\beta}\tilde{u}_2$ as omitted from the second equation where $\tilde{u}_2=\Delta RAAA - x_2\tilde{\gamma}_2$.

This test was calculated by regressing $\tilde{u}_2$ on x_2, $\tilde{u}_1$ and $RG35-\tilde{\beta}\tilde{u}_2$. The resulting $TR^2=1.25$ which is quite small, indicating that the data does not contain evidence against the hypothesis. Careful examination of x_1 and x_2 in this case shows that the identification of the model under the alternative is rather flimsy and therefore the test probably has very little power.

A second class of weak exogeneity tests can be formulated using the same analysis. These might be called limited information tests because it is assumed that there are no overidentifying restrictions available from the second block of equations. In this case equation (70) can be replaced by:

$$Y_2 = X\Pi_2 + \varepsilon_2. \tag{82}$$

Now the definition of weak exogeneity is simply that $\Omega_{12}=0$ because $\alpha=0$ imposes no restrictions on the model. This situation would be expected to occur when the second equation is only very roughly specified.

A very similar situation occurs in the case where Y_2 is possibly measured with error. Suppose Y_2^* is the true unobserved value of Y_2 but one observes $Y_2 = Y_2^* + \eta$. If the equation defining Y_2^* is:

$$Y_2^* = x_2\Gamma_2 + \varepsilon_2,$$

where the assumption that Y_2^* belongs in the first equation implies $\mathrm{E}\varepsilon_1'\varepsilon_2 = 0$, the observable equations become:

$$\begin{aligned} y_1 &= Y_2\beta + x_1\gamma + \varepsilon_1 - \eta\beta, \\ Y_2 &= x_2\Gamma_2 + \varepsilon_2 + \eta. \end{aligned} \tag{83}$$

If there is no measurement error, then the covariance matrix of η will be zero, and $\Omega_{12}=0$. This set up is now just the same as that used by Wu (1973) to test for weak exogeneity of Y_2 when it is known that $\alpha=0$.

The procedure for this test has already been developed. The two forms of the score are given in (79) and (80) and these can be used to test for the presence of U_2 in the first equation. This test is Wu's test and it is also the test derived by Hausman (1979) for this problem. By showing that these are Lagrange Multiplier tests, the asymptotic optimality of the procedures is established when the full set of x_2 is used. Neither Hausman nor Wu could establish this property.

Finally, tests for strong exogeneity can be performed. By definition, strong exogeneity requires weak exogeneity plus the non-predictability of Y_2 from past values of y_1. Partitioning x_2 in (70) into (y_1^0, x_3) where y_1^0 is a matrix with all the relevant lags of y_1, and similarly letting $\Gamma_2 = (\Gamma_{20}, \Gamma_{23})$ the hypothesis of strong exogeneity is:

$$H_0: \alpha = 0, \quad \Omega_{12} = 0, \quad \Gamma_{20} = 0. \tag{84}$$

This can clearly be jointly tested by letting u_1, $\bar{y}_1$ and y_1^0 be the omitted variables from each of the equations. Clearly the weak exogeneity and the Granger non-causality are very separate parts of the hypothesis and can be tested separately. Most often however when Granger causality is being tested on its own, the appropriate model is (82) as overidentifying restrictions are rarely available.

9.4. *Discrete choice and truncated distributions*

In models with discrete or truncated dependent variables, non-linear maximum likelihood estimation procedures are generally employed to estimate the parameters. The estimation techniques are sufficiently complex that model diagnostics are rarely computed and often only a limited number of specifications are tried. This is therefore another case where the LM test is useful. Two examples will be presented: a binary choice model and a self-selectivity model.

In the binary choice model, the outcome is measured by a dependent variable, y, which takes on the value 1 with probability p and 0 with probability $1-p$. For each observation these probabilities are different either because of the nature of the choice or of the chooser. Let $p_t = F(x_t\beta)$, where the function F maps the exogenous characteristics, x_t, into the unit interval. A common source of such functions are cumulative distribution functions such as the normal or the logistic. The log-likelihood of this model is given by

$$L = \sum_t \left(y_t \log p_t + (1-y_t)\log(1-p_t)\right), \qquad p_t = F(x_t\beta). \tag{85}$$

Partitioning the parameter vector and x_t vector conformably into $\beta = (\beta_1', \beta_2')'$, the hypothesis to be tested is H_0: $\beta_1 = 0$. The model has already been estimated using only x_2 as the exogenous variables and it is desired to test whether some other variables were omitted. These estimates under the null will be denoted $\tilde{\beta}_2$ which implies a set of probabilities $\tilde{p}_t$. The score and information matrix of this model are given by:

$$\frac{\partial L}{\partial \beta} = \sum_t \frac{y_t - p_t}{p_t(1-p_t)} f(x_t\beta)x_t', \tag{86}$$

$$\mathscr{I}(\beta) = \mathrm{E}\left(\frac{\partial L}{\partial \beta}\right)\left(\frac{\partial L}{\partial \beta}\right)' = \sum_t \frac{f^2(x_t\beta)}{p_t(1-p_t)} x_t' x_t, \tag{87}$$

where f is the derivative of F. Notice that the score is essentially a function of the "residuals" $y_t - p_t$. Evaluating these test statistics under the null, the LM test statistic is given by:

$$\xi_{\mathrm{LM}} = \left(\frac{\partial L}{\partial \beta}\right)' \mathscr{I}(\tilde{\beta})^{-1} \frac{\partial L}{\partial \beta}$$
$$= \tilde{u}'\tilde{x}(\tilde{x}'\tilde{x})^{-1}\tilde{x}'\tilde{u}, \tag{88}$$

where

$$\tilde{u}_t = (y_t - \tilde{p}_t)/(\tilde{p}_t(1-\tilde{p}_t))^{1/2}, \quad \tilde{x}_t = x_t f(x_{2t}\tilde{\beta})/(\tilde{p}_t(1-\tilde{p}_t))^{1/2},$$

and

$$\tilde{u} = (\tilde{u}_1, \dots, \tilde{u}_T)', \; \tilde{x} = (\tilde{x}_1', \dots, \tilde{x}_T')'.$$

Because $\operatorname{plim} \tilde{u}'\tilde{u}/T = 1$, the statistic is asymptotically equivalent to TR_0^2 of the regression of $\tilde{u}$ on $\tilde{x}$. In the special case of the logit where $p_t = 1/(1+\mathrm{e}^{-x_t\beta})$, $f = \tilde{p}_t(1-\tilde{p}_t)$ and the expressions simplify so that x_t is multiplied by $(\tilde{p}_t(1-\tilde{p}_t))^{1/2}$ rather than being divided by it. For the probit model where F is the cumulative normal, $f = \exp(x_{2t}\tilde{\beta}_2)$ as the factor of proportionality cancels. This test is therefore extremely easy to compute based on estimates of the model under the null.

As a second example, take the self-selectivity model of Hausman and Wise (1977). The sample is truncated based upon the dependent variable. The data come from the negative income tax experiment and when the families reached a sufficiently high income level, they are dropped from the sample. Thus the model can be expressed as:

$$y|x \sim N(x\beta, \sigma^2),$$

but we only have data for $y \leq c$. Thus, the likelihood function is given as the probability density of y divided by the probability of observing this family. The log-likelihood can be expressed in terms of ϕ and Φ which are the Gaussian density and distribution functions respectively as:

$$L = \sum_t \log \phi((y_t - x_t\beta)/\sigma) - \sum \log \Phi((c - x_t\beta)/\sigma). \tag{89}$$

The score is:

$$\frac{\partial L}{\partial \beta} = \frac{1}{\sigma^2}\sum_t \left[y_t - x_t\beta - \sigma\phi\left(\frac{(c-x_t\beta)}{\sigma}\right) \Big/ \Phi\left(\frac{(c-x_t\beta)}{\sigma}\right)\right] x_t'. \tag{90}$$

To estimate this model one sets the score to zero and solves for the parameters. Notice that this implies including another term in the regression which is the ratio of the normal density to its distribution. The inclusion of this ratio, called the Mills ratio, is a distinctive feature of much of the work of self-selectivity. The information matrix can be shown to be:

$$\mathscr{I} = \sum_t x_t'x_t\left(1+(\phi_t/\Phi_t)^2-(\phi_t/\Phi_t)(c-x_t\beta/\sigma)\right), \tag{91}$$

where $\phi_t = \phi((c-x_t\beta)/\sigma)$ and similarly for Φ_t.

To test the hypothesis H_0: $\beta_1 = 0$, denote again the estimates under the null by $\tilde{\beta}, \tilde{\phi}, \tilde{\Phi}$. Let $r_t^2 = 1+(\tilde{\phi}_t/\tilde{\Phi}_t)^2+(\tilde{\phi}_t/\tilde{\Phi}_t)(c-x_t\tilde{\beta}/\tilde{\sigma})$ and define $\tilde{u}_t = (y_t - x_{2_t}\tilde{\beta}_2 + \tilde{\sigma}\tilde{\phi}_t/\tilde{\Phi}_t)/r_t$ and $\tilde{x}_t = x_t r_t$. With $\tilde{u}$ and $\tilde{x}$ being the corresponding vectors and matrices, the LM test statistic is:

$$\xi_{\text{LM}} = \tilde{u}'\tilde{x}(\tilde{x}'\tilde{x})^{-1}\tilde{x}'\tilde{u}. \tag{92}$$

As before, $\operatorname{plim} \tilde{u}'\tilde{u}/T = 1$ so an asymptotically equivalent test statistic is TR_0^2 of the regression of $\tilde{u}$ on $\tilde{x}$. Once again, the test is simply performed by a linear regression on transformed data. All of the components of this transformation such as the Mills ratio, are readily available from the preceding estimation. Thus a variety of complicated model searches and diagnostic tests can easily be carried out even in this complex maximum likelihood framework.

10. Alternative testing procedures

In this section three alternative closely related testing procedures will be briefly explained and the relationship between these methods and ones discussed in this chapter will be highlighted. The three alternatives are Neyman's (1959) $C(\alpha)$ test, Durbin's (1970) general procedure, and Hausman's (1978) specification test.

Throughout this section the parameter vector will be partitioned as $\theta' = (\theta_1', \theta_2')$ and the null hypothesis will be H_0: $\theta_1 = \theta_1^0$. Neyman's test, as exposited by Breusch and Pagan (1980), is a direct generalization of the LM test which allows consistent but inefficient estimation of the parameters θ_2 under the null. Let this estimate be $\bar{\theta}_2$ and let $\bar{\theta} = (\theta_1^0, \bar{\theta}_2')'$. Expanding the score evaluated at θ around the ML estimate $\tilde{\theta}$ gives:

$$\frac{\partial L}{\partial \theta}(\tilde{\bar{\theta}}) = \begin{pmatrix} \partial L/\partial\theta_1(\tilde{\theta}) \\ 0 \end{pmatrix} + \begin{pmatrix} \partial^2 L/\partial\theta_1\,\partial\theta_2'(\bar{\theta})(\tilde{\bar{\theta}}_2 - \tilde{\theta}_2) \\ \partial^2 L/\partial\theta_2\,\partial\theta_2'(\bar{\theta})(\tilde{\bar{\theta}}_2 - \tilde{\theta}_2) \end{pmatrix},$$

where $(\partial L/\partial\theta_2)(\tilde{\theta})=0$. Solving for the desired score:

$$\frac{\partial L}{\partial\theta_1}(\theta)=\frac{\partial L}{\partial\theta_1}(\tilde{\tilde{\theta}})-\frac{\partial^2 L}{\partial\theta_1\partial\theta_2}(\tilde{\theta})\left(\frac{\partial^2 L}{\partial\theta_2\partial\theta_2}(\tilde{\theta})\right)^{-1}\frac{\partial L}{\partial\theta_2}(\tilde{\tilde{\theta}})$$
$$=s_1(\tilde{\tilde{\theta}})-\mathscr{I}_{12}(\tilde{\theta})\mathscr{I}_{22}^{-1}(\tilde{\theta})s_2(\tilde{\tilde{\theta}}). \tag{93}$$

The $C(\alpha)$ test is just the LM test using (93) for the score. This adjustment can be viewed as one step of a Newton–Raphson iteration to find an efficient estimate of θ_2 based upon an initial consistent estimate. In some situations such as the one discussed in Breusch and Pagan, this results in a substantial simplification.

The Durbin (1970) procedure is also based on different estimates of the parameters. He suggests calculating the maximum likelihood estimate of θ_1 assuming $\theta_2=\tilde{\theta}_2$, the ML estimate under the null. Letting this new estimate be $\tilde{\tilde{\theta}}_1$, the test is based upon the difference $\tilde{\tilde{\theta}}_1-\theta_1^0$. Expanding the score with respect to θ_1 about $\tilde{\tilde{\theta}}_1$ holding $\theta_2=\tilde{\theta}_2$ and recognizing that the first term is zero by definition of $\tilde{\tilde{\theta}}_1$ the following relationship is found:

$$\frac{\partial L}{\partial\theta_1}(\tilde{\theta})=-\frac{\partial^2 L}{\partial\theta_1\,\partial\theta_1'}(\theta)(\tilde{\tilde{\theta}}_1-\theta_1^0). \tag{94}$$

Because the Hessian is assumed to be non-singular, any test based upon $\tilde{\tilde{\theta}}_1-\theta_1^0$ will have the same critical region as one based upon the score; thus the two tests are equivalent. In implementation there are of course many asymptotically equivalent forms of the tests, and it is the choice of the asymptotic form of the test which gives rise to the differences between the LM test for serial correlation and Durbin's h test.

The third principle is Hausman's (1978) specification test. The spirit of this test is somewhat different. The parameters of interest are not θ_1 but rather θ_2. The objective is to restrict the parameter space by setting θ_1 to some preassigned values without destroying the consistency of the estimates of θ_2. The test is based upon the difference between the efficient estimates under the null, $\tilde{\theta}_2$, and a consistent but possibly inefficient estimate under the alternative $\hat{\hat{\theta}}_2$. Hausman makes few assumptions about the properties of $\hat{\hat{\theta}}_2$; Hausman and Taylor (1980), however, modify the statement of the result somewhat to use the maximum likelihood estimate under the alternative $\hat{\theta}_2$. For the moment, this interpretation will be used here. Expanding the score around the maximum likelihood estimate and evaluating it at $\tilde{\theta}$ gives:

$$\frac{\partial L}{\partial\theta}(\tilde{\theta})=\frac{\partial^2 L}{\partial\theta\,\partial\theta'}(\theta)(\tilde{\theta}-\hat{\theta}),$$

or

$$\begin{pmatrix} \theta_1^0 - \hat{\theta}_1 \\ \tilde{\theta}_2 - \hat{\theta}_2 \end{pmatrix} = \left(\frac{\partial^2 L}{\partial \theta\, \partial \theta'} \right)^{-1} \begin{pmatrix} \partial L / \partial \theta_1(\tilde{\theta}) \\ 0 \end{pmatrix}. \tag{95}$$

It was shown above that asymptotically optimal tests could be based upon either the score or the difference $(\hat{\theta}_1 - \theta_1^0)$. As these are related by a non-singular transformation which asymptotically is $\mathscr{I}^{11}$, critical regions based on either statistic will be the same. Hausman's difference is based upon $\mathscr{I}^{21}$ times the score asymptotically. If this matrix is non-singular, then the tests will all be asymptotically equivalent. The dimension of $\mathscr{I}^{21}$ is $q \times p$ where p is the number of restrictions and $q = k - p$ is the number of remaining parameters. Thus a necessary condition for this test to be asymptotically equivalent is that $\min(p, q) = p$. A sufficient condition is that $\text{rank}(\mathscr{I}^{21}) = p$. The equivalence requires that there be at least as many parameters unrestricted as restricted. However, parameters which are asymptotically independent of the parameters under test will not count. For example, in a classical linear regression model, the variance and any serial correlation parameters will not count in the number of unrestricted parameters. The reason for the difficulty is that the test is formulated to ignore all information in $\hat{\theta}_1 - \theta_1^0$ even though it frequently would be available from the calculation of $\hat{\theta}_2$.

Hausman and Taylor (1980) in responding to essentially this criticism from Holly (1980) point out that in the case $q < p$, the specification test can be interpreted as an asymptotically optimal test of a different hypothesis. They propose the hypothesis H_0^*: $\mathscr{I}_{22}^{-1}\mathscr{I}_{21}(\theta_1 - \theta_1^0) = 0$ or simply $\mathscr{I}_{21}(\theta_1 - \theta_1^0) = 0$. If H_0^* is true, the bias in θ_2 from restricting $\theta_1 = \theta_1^0$ would asymptotically be zero. The hypothesis H_0^* is explicitly a consistency hypothesis. The Hausman test is one of many asymptotically equivalent ways to test this hypothesis. In fact, the same Wald, LR and LM tests are available as pointed out by Riess (1982). The investigator must however decide which hypothesis he wishes to test, H_0 or H_0^*.

In answering the question of which hypothesis is relevant, it is important to ask why the test is being undertaken in the first place. As the parameters of interest are θ_2, the main purpose of the test is to find a more parsimonious specification, and the advantage of a parsimonious specification is that more efficient estimates of the parameters of interest can be obtained. Thus if consistency were the only concern of the investigator, he would not bother to restrict the model at all. The objective is therefore to improve the efficiency of the estimation by testing and then imposing some restrictions. These restrictions ought, however, to be grounded in an economic hypothesis rather than purely data based as is likely to be the case for H_0^* which simply asserts that the true parameters lie in the column null space of $\mathscr{I}_{21}$.

Finally, if an inefficient estimator $\hat{\hat{\theta}}$ is used in the test, it is unlikely that the results will be as strong as described above. Except in special cases, one would expect the test based upon the MLE to be more powerful than that based upon an inefficient estimator. However, this is an easy problem to correct. Starting from the inefficient estimate, one step of a Newton–Raphson type algorithm will produce asymptotically efficient parameter estimates.

11. Non-standard situations

While many non-standard situations may arise in practice, two will be discussed here. The first considers the properties of the Wald, LM and LR tests when the likelihood function is misspecified. The second looks at the case where the information matrix is singular under the null.

White (1982) and Domowitz and White (1982) have recently examined the problem of inference in maximum likelihood situations where the wrong likelihood has been maximized. These quasi-maximum likelihood estimates may well be consistent, however the standard errors derived from the information matrix are not correct. For example, the disturbances may be assumed to be normally distributed when in fact they are double exponentials. White has proposed generalizations of the Wald and LM test principles which do have the right size and which are asymptotically powerful when the density is correctly assumed. These are derived from the fact that the two expressions for the information matrix are no longer equivalent for QML estimates. The expectation of the outer product of the scores does not equal minus the expectation of the Hessian. Letting L_t be the log-likelihood of the tth observation, White constructs the matrices:

$$A=\frac{1}{T}\frac{\partial^2 L}{\partial\theta\,\partial\theta'};\qquad B=\frac{1}{T}\sum_t\frac{\partial L_t}{\partial\theta}\left(\frac{\partial L_t}{\partial\theta}\right)'\quad\text{and}\quad C=A^{-1}BA^{-1}.$$

Then the "quasi-scores", measured as the derivative of the possibly incorrect likelihood function evaluated under the null, will have a limiting distribution based upon these matrices when the null is true. Letting A^{11} be the first block of the partitioned inverse of A, the limiting covariance of the quasi score is $(A^{11}C_{11}^{-1}A^{11})^{-1}$ so the quasi-LM test is simply:

$$\xi_{\mathrm{LM}}=s'A^{11}C_{11}^{-1}A^{11}s.$$

Notice that if the distribution is correct, then $A=-B$ so that $C=A^{-1}$ and the whole term becomes simply A^{11} as usual. Thus the use of the quasi-LM statistic corrects the size of the test when the distribution is false but gives the asymptotically optimal test when it is true. Except for possible finite sample and computational costs, it appears to be a sensible procedure. Exactly the same correction is made to the Wald test to obtain a quasi Wald test. Because it is the divergence between A and B which creates the situation, White proposes an omnibus test for differences between A and B.

In some situations, an alternative to this approach would be to test for normality directly as well as for other departures from the specification. Jarque and Bera (1980, 1982) propose such a test by taking the Pearson density as the alternative and simultaneously testing for serial correlation, functional form misspecification and heteroscedasticity. This joint test decomposes into independent LM tests because of the block diagonality of the information matrix for this problem.

A second non-standard situation which occurs periodically in practice is when some of the parameters are estimable only when the null hypothesis is false. That is, the information matrix under the null is singular. Two simple examples with rather different conclusions are:

$$y|x_1; x_2 \sim N(\alpha\beta x_1 + \beta x_2, \sigma^2), \qquad H_0: \beta = 0,$$

$$y|x \sim N(\beta x^{\alpha}, \sigma^2), \qquad H_0: \beta = 0.$$

In both cases, the likelihood function can be maximized under both the null and alternative, but the limiting distribution of the likelihood ratio statistic is not clear. Furthermore, conventional Wald and LM tests also have difficulties—the LM will have a parameter which is unidentified under the null which appears in the score, and the Wald will have an unknown limiting distribution. In the first example, it is easy to see that by reparameterizing the model, the null hypothesis becomes a two degree of freedom standard test. In the second example, however, there is no simple solution. Unless the parameter α is given a priori, the tests will have the above-mentioned problems. A solution proposed by Davies (1977) is to obtain the LM test statistic for each value of the unidentified parameter and then base the test on the maximum of these. Any one of these would be chi squared with one degree of freedom, however, the maximum of a set of dependent chi squares would not be chi squared in general. Davies finds a bound for the distribution which gives a test with size less than or equal to the nominal value.

As an example of this, Watson (1982) considers the problem of testing whether a regression coefficient is constant or whether it follows a first order autoregressive process. The model can be expressed as:

$$y_t = x_t\beta_t + z_t\gamma + \varepsilon_t,$$

$$\beta_t = \rho\beta_{t-1} + \eta_t,$$

$$\begin{pmatrix} \varepsilon_t \\ \eta_t \end{pmatrix} \sim N\left(\begin{pmatrix} 0 \\ 0 \end{pmatrix}, \begin{pmatrix} \sigma_\varepsilon^2 & 0 \\ 0 & \sigma_\eta^2 \end{pmatrix}\right).$$

The null hypothesis is that $\sigma_\eta^2 = 0$; this however makes the parameter ρ unidentifiable. The test is constructed by first searching over the possible values of ρ to find the maximum LM test statistic, and then finding the limiting distribution of the test to determine the critical value. A Monte Carlo evaluation of the test showed it to work reasonably well except for values of ρ close to unity when the limiting distribution was well approximated only for quite large samples.

Several other applications of this result occur in econometrics. In factor analytical models, the number of parameters varies with the number of factors so testing the number of factors may involve such a problem. Testing a series for white noise against an $AR(1)$ plus noise again leads to this problem as the parameter in the autoregression is not identified under the null. A closely related problem occurred in testing for common factor dynamics as shown in Engle (1979a). Several others could be illustrated.

12. Conclusion

In a maximum likelihood framework, the Wald, Likelihood Ratio and Lagrange Multiplier tests are a natural trio. They all share the property of being asymptotically locally most powerful invariant tests and in fact all are asymptotically equivalent. However, in practice there are substantial differences in the way the tests look at particular models. Frequently when one is very complex, another will be much simpler. Furthermore, this formulation guides the intuition as to what is testable and how best to formulate a model in order to test it. In terms of forming diagnostic tests, the LM test is frequently computationally convenient as many of the test statistics are already available from the estimation of the null.

The application of these test principles and particularly the LM principle to a wide range of econometric problems is a natural development of the field and it is a development which is proceeding at a very rapid pace. Soon, most of the interesting cases will have been touched in theoretical papers, however, applied work is just beginning to incorporate these techniques and there is a rich future there.

References

Aitcheson, J. and S. D. Silvey (1958), "Maximum Likelihood Estimation of Parameters Subject to Restraints", *Annals of Mathematical Statistics*, 29:813–828.

Anderson, T. W. (1971), *The Statistical Analysis of Time Series*. New York: John Wiley and Sons.

Bera, A. K. and C. M. Jarque (1982), "Model Specification Tests: A Simultaneous Approach", *Journal of Econometrics*, 20:59–82.

Berndt, E. R. and N. E. Savin (1977), "Conflict Among Criteria for Testing Hypotheses in the Multivariate Linear Regression Model", *Econometrica*, 45:1263–1278.

Breusch, T. S. (1978), "Testing for Autocorrelation in Dynamic Linear Models", *Australian Economic Papers*, 17:334–355.

Breusch, T. S. and A. R. Pagan (1979), "A Simple Test for Heteroskedasticity and Random Coefficient Variation", *Econometrica*, 47:1287–1294.

Breusch, T. S. (1979), "Conflict Among Criteria for Testing Hypotheses: Extensions and Comments", *Econometrica*, 47:203–207.

Breusch, T. S. and L. G. Godfrey (1980), "A Review of Recent Work on Testing for Autocorrelation in Dynamic Economic Models", Discussion Paper #8017, University of Southampton.

Breusch, T. S. and A. R. Pagan (1980), "The Lagrange Multiplier Test and Its Applications to Model Specification in Econometrics", *Review of Economic Studies*, 47:239–254.

Cox, D. R. and D. V. Hinckley (1974), *Theoretical Statistics*. London: Chapman and Hall.

Crowder, M. J. (1976), "Maximum Likelihood Estimation for Dependent Observations", *Journal of the Royal Statistical Society, Series B*, 45–53.

Davidson, J. E. H., Hendry, D. F., Srba, F., and S. Yeo (1978), "Econometric Modelling of the Aggregate Time-Series Relationship Between Consumers' Expenditure and Income in the United Kingdom", *Economic Journal*, 88:661–692.

Davies, R. B. (1977), "Hypothesis Testing When a Nuisance Parameter is Present Only Under the Alternative", *Biometrika*, 64:247–254.

Domowitz, I. and H. White (1982), "Misspecified Models with Dependent Observations", *Journal of Econometrics*, 20:35–58.

Durbin, J. (1970), "Testing for Serial Correlation in Least Squares Regression When Some of the Regressors are Lagged Dependent Variables", *Econometrica*, 38:410–421.

Eisner, R. (1971), "Non-linear Estimates of the Liquidity Trap", *Econometrica*, 39:861–864.

Engle, R. F. (1979), "Estimation of the Price Elasticity of Demand Facing Metropolitan Producers", *Journal of Urban Economics*, 6:42–64.

Engle, R. F. (1982), "Autoregression Conditional Heteroskedasticity with Estimates of the Variance of U.K. Inflation", *Econometrica*, 50:987–1007.

Engle, R. F. (1979a), "A General Approach to the Construction of Model Diagnostics Based on the Lagrange Multiplier Principle", U.C.S.D. Discussion Paper 79-43.

Engle, R. F. (1982a), "A General Approach to Lagrange Multiplier Model Diagnostics", *Journal of Econometrics*, 20:83–104.

Engle, R. F. (1980), "Hypothesis Testing in Spectral Regression: the Lagrange Multiplier as a Regression Diagnostic", in: Kmenta and Ramsey, eds., *Criteria for Evaluation of Econometric Models*. New York: Academic Press.

Engle, R. F., D. F. Hendry, and J. F. Richard (1983), "Exogeneity", *Econometrica*, 50:227–304.

Evans, G. B. A. and N. E. Savin (1982), "Conflict Among the Criteria Revisited; The W, LR and LM tests", *Econometrica*, 50:737–748.

Ferguson, T. S. (1967), *Mathematical Statistics*. New York: Academic Press.

Godfrey, L. G. (1978), "Testing for Multiplicative Heteroskedasticity", *Journal of Econometrics*, 8:227–236.

Godfrey, L. G. (1978a), "Testing Against general Autoregressive and Moving Average Error Models When the Regressors Include Lagged Dependent Variables", *Econometrica*, 46:1293–1302.

Godfrey, L. G. (1978b), "Testing for Higher Order Serial Correlation in Regression Equations when the Regressors Include Lagged Dependent Variables", *Econometrica*, 46:1303–1310.

Godfrey, L. G. (1979), "A Diagnostic Check on the Variance Model in Regression Equations with Heteroskedastic Disturbances", unpublished manuscript, University of York.

Godfrey, L. G. (1979a), "Testing the Adequacy of a Time Series Model", *Biometrika*, 66:67–72.

Godfrey, L. G. and A. R. Tremayne (1979), "A Note on Testing for Fourth Order Autocorrelation in Dynamic Quarterly Regression Equations", unpublished manuscript, University of York.

Godfrey, L. G. (1980), "On the Invariance of the Lagrange Multiplier Test with Respect to Certain Changes in the Alternative Hypothesis", *Econometrica*, 49:1443–1456.

Hausman, J. (1978), "Specification Tests in Econometrics", *Econometrica*, 46:1251–1272.

Hausman, J. and D. Wise (1977), "Social Experimentation Truncated Distributions, and Efficient Estimation", *Econometrica*, 45:319–339.

Hausman, J. and W. Taylor (1980), "Comparing Specification Tests and Classical Tests", unpublished manuscript.

Hendry, D. F. and T. von Ungern-Sternberg (1979), "Liquidity and Inflation Effects on Consumers'

Expenditure", in: Angus Deaton, ed., *Festschrift for Richard Stone*. Cambridge: Cambridge University Press.

Hendry, D. F. and G. Mizon (1980), "An Empricial Application and Monte Carlo Analysis of Tests of Dynamic Specification", *Review of Economic Studies*, 47:21–46.

Holly, A. (1982), "A Remark on Hausman's Specification Test," *Econometrica*, v. 50: 749–759.

Hosking, J. R. M. (1980), "Lagrange Multiplier Tests of Time Series Models", *Journal of the Royal Statistical Society B*, 42:170–181.

Jarque, C. and A. K. Bera (1980), "Efficient Tests for Normality, Homoscedasticity, and Serial Independence of Regression Residuals", *Economics Letters*, 6:255–259.

King, M. L. and G. H. Hillier (1980), "A Small Sample Power Property of the Lagrange Multiplier Test", Discussion Paper, Monash University.

Kmenta, J. (1967), "On Estimation of the CES Production Function", *International Economic Review*, 8:180–189.

Koenker, R. (1981), "A Note on Studentizing a Test for Heteroscedasticity", *Journal of Econometrics*, 17:107–112.

Konstas, P. and M. Khouja (1969), "The Keynesian Demand-for-Money Function: Another Look and Some Additional", *Journal of Money Credit and Banking*, 1:765–777.

Lehmann, E. L. (1959), *Testing Statistical Hypotheses*. New York: John Wiley and Sons.

Neyman, J. (1959), "Optimal Asymptotic Tests of Composite Statistical Hypotheses", in (U. Grenander, ed.) *Probability and Statistics*. Stockholm: Almquist and Wiksell, pp. 213–234.

Newbold, P. (1980), "The Equivalence of Two Tests of Time Series Model Adequacy", *Biometrica*, 67:463–465.

Pifer, H. (1969), "A Non-linear Maximum Likelihood Estimate of the Liquidity trap," *Econometrica*, 37:324–332.

Poskitt, D.S. and A.P. Tremayne (1980), "Testing the Specification of a Fitted ARMA Model", *Biometrica*, 67:359–363.

Rao, C. R. (1948), "Large Sample Tests of Statistical Hypothese Concerning Several Parameters with Application to Problems of Estimation", *Proceedings of the Cambridge Philosophical Society*, 44:50–57.

Reiss, P. (1982), "Alternative Interpretations of Hausman's *m* Test", manuscript Yale University.

Rothenberg, T. J. (1980), "Comparing Alternative Asymptotically Equivalent Tests", invited paper presented at World Congress of the Econometric Society, Aix-en-Provence, 1980.

Sargan, J. D. (1964), "Wages and Prices in the United Kingdom: A Study in Econometric Methodology", in (P.E. Hart, G. Mills, J.K. Whitaker, eds.) *Econometric Analysis for National Economic Planning*. London: Butterworths, 1964.

Sargan, J. D. (1980), "Some Tests of Dynamic Specification for a Single Equation", *Econometrica*, 48:879–897.

Savin, N. E. (1976), "Conflicts Among Testing Procedures in a Linear Regression Model with Autoregressive Disturbances", *Econometrica*, 44:1303–1313.

Silvey, D. S. (1959), "The Lagrangean Multiplier Test", *Annals of Mathematical Statistics*, 30:389–407.

Wald, A. (1943), "Tests of Statistical Hypotheses Concerning Several Parameters When the Number of Observations is Large", *Transactions of the American Mathematical Society*, 54:426–482.

Watson, M. (1982). "A Test for Regression Coefficient Stability When a Parameter is Identified Only Under the Alternative", Harvard Discussion Paper 906.

White, H. (1980), "A Heteroskedasticity Consistent Covariance Matrix Estimator and a Direct Test for Heteroskedasticity", *Econometrica*, 48:817–838.

White, H. (1982), "Maximum Likelihood Estimation of Misspecified Models", *Econometrica*, 50:1–26.

White, K. (1972), "Estimation of the Liquidity Trap With a Generalized Functional Form", *Econometrica*, 40:193–199.

Wilks, S. S. (1938), "The Large Sample Distribution of the Likelihood Ratio for Testing Composite Hypotheses", *Annals of Mathematical Statistics*, 9:60–62.

Chapter 14

MULTIPLE HYPOTHESIS TESTING

N. E. SAVIN*

Trinity College, Cambridge

Contents

*This work was supported by National Science Foundation Grant SES 79-12965 at the Institute for Mathematical Studies in the Social Sciences, Stanford University. The assistance of G. B. A. Evans is gratefully acknowledged. I am also indebted to the following people for valuable comments: T. W. Anderson, K. J. Arrow, R. W. Farebrother, P. J. Hammond, D. F. Hendry, D. W. Jorgenson, L. J. Lau, B. J. McCormick, and J. Richmond.

Handbook of Econometrics, Volume II, Edited by Z. Griliches and M.D. Intriligator

1. Introduction

The t and F tests are the most frequently used tests in econometrics. In regression analysis there are two different procedures which can be used to test the hypothesis that all the coefficients are zero. One procedure is to test each coefficient separately with a t test and the other is to test all coefficients jointly using an F test. The investigator usually performs both procedures when analyzing the sample data. The obvious questions are what is the relation between the two procedures and which procedure is better. Scheffé (1953) provided the key to the answers when he proved that the F test is equivalent to carrying out a set of simultaneous t tests. More than 25 years have passed since this result was published and yet the full implications have barely penetrated the econometric literature. Aside from a brief mention in Theil (1971) the Scheffé result has not been discussed in the econometric textbooks; the exceptions appear to be Seber (1977) and Dhrymes (1978). Hence, it is perhaps no surprise there are so few applications of multiple hypothesis testing procedures in empirical econometric research.

This chapter presents a survey of multiple hypothesis testing procedures with an emphasis on those procedures which can be applied in the context of the classical linear regression model. Multiple hypothesis testing is the testing of two or more separate hypotheses simultaneously. For example, suppose we wish to test the hypothesis H: $\beta_1 = \beta_2 = 0$ where β_1 and β_2 are coefficients in a multiple regression. In situations in which we only wish to test whether H is true or not we can use the F test. It is more usual that when H is rejected we want to know whether β_1 or β_2 or both are nonzero. In this situation we have a multiple decision problem and the natural solution is to test the separate hypotheses H_1: $\beta_1 = 0$ and H_2: $\beta_2 = 0$ with a t test. Since H is true if and only if the separate hypotheses H_1: $\beta_1 = 0$ and H_2: $\beta_2 = 0$ are both true, this suggests accepting H if and only if we accept H_1 and H_2. Testing the two hypotheses H_1 and H_2 when we are interested in whether β_1 or β_2 or both are different from zero induces a multiple decision problem in which the four possible decisions are:

d^{00}: H_1 and H_2 are both true,

d^{01}: H_1 is true, H_2 is false,

d^{10}: H_1 is false, H_2 is true,

d^{11}: H_1 and H_2 are both false.

Now suppose that a test of H_1 is defined by the acceptance region A_1 and the rejection region R_1, and similarly for H_2. These two separate tests induce a decision procedure for the four decision problem, this induced procedure being defined by assigning the decision d^{00} to the intersection of A_1 and A_2, d^{01} to the intersection of A_1 and R_2 and so on. This induced procedure accepts H: $\beta_1 = \beta_2 = 0$ if and only if H_1 and H_2 are accepted.

More generally suppose that the hypothesis H is true if and only if the separate hypotheses $H_1, H_2, \ldots$ are true. The induced test accepts H if and only if all the separate hypotheses are accepted. An induced test is either finite or infinite depending on whether there are a finite or infinite number of separate hypotheses. In the case of finite induced tests the exact sampling distributions of the test statistics can be complicated, so that in practice the critical regions of the tests are based on probability inequalities. On the other hand, infinite induced tests are commonly constructed such that the correct critical value can be readily calculated.

Induced tests were developed by Roy (1953), Roy and Bose (1953), Scheffé (1953) and Tukey (1953). Roy referred to induced tests as union–intersection tests. Procedures for constructing simultaneous confidence intervals are closely associated with induced tests and such procedures are often called multiple comparison procedures. Induced tests and their properties are discussed in two papers by Lehmann (1957a, 1957b) and subsequently by Darroch and Silvey (1963) and Seber (1964). A lucid presentation of the union–intersection principle of test construction is given in Morrison (1976). I recommend Scheffé (1959) for a discussion of the contributions of Scheffé and Tukey. A good reference for finite induced tests is Krishnaiah (1979). Miller (1966, 1977) presents an excellent survey of induced tests and simultaneous confidence interval procedures.

The induced tests I will discuss in detail are the Bonferroni test and the Scheffé test. These two induced tests employ the usual t statistics and can always be applied to the classical linear regression model. The Bonferroni test is a finite induced test where the critical value is computed using the well known Bonferroni inequality. While there are inequalities which give a slightly more accurate approximation, the Bonferroni inequality has the advantage that it is very simple to apply. In addition, the Bonferroni test behaves very similarly to finite induced tests based on more accurate approximations. I refer to the F test as the Scheffé test when the F test is used as an infinite induced test. Associated with the Bonferroni and Scheffé tests are the B and S simultaneous confidence intervals, respectively. The Bonferroni test and the B intervals are discussed in Miller (1966) and applications in econometrics are found in Jorgenson and Lau (1975), Christensen, Jorgenson and Lau (1975) and Sargan (1976). The Scheffé test and the S intervals are explained in Scheffé (1959) and the S method is reformulated as the S procedure in Scheffé (1977a). Applications of the Scheffé test and the S intervals in econometrics are given in Jorgenson (1971, 1974) and Jorgenson and Lau (1982). Both the Bonferroni and Scheffé tests are also discussed in Savin (1980).

The organization of the chapter is the following. The relationship between t and F tests is discussed in Section 2. In this section I present a detailed comparison of the acceptance regions of the Bonferroni test and the F test for a special situation. In Section 3 the notion of linear combinations of parameters of primary and secondary interest is introduced. The Bonferroni test is first developed for linear combinations of primary interest and then for linear combinations of secondary interest. The Scheffé test is discussed and the lengths of the B and S intervals are compared. The powers of the Bonferroni test and the Scheffé test are compared in Section 4. The effect of multicollinearity on the power of the tests is also examined. Large sample analogues of the Bonferroni and Scheffé tests can be developed for more complicated models. In Section 5 large sample analogues are derived for a nonlinear regression model. Section 6 presents two empirical applications of the Bonferroni and Scheffé tests.

2. *t* and *F* tests

2.1. The model

Consider the regression model:

$$y = X\beta + u, \tag{2.1}$$

where y is a $T\times 1$ vector of observations on the dependent variable, X is a $T\times k$ nonstochastic matrix of rank k, β is an unknown $k\times 1$ parameter vector and u is a $T\times 1$ vector of random disturbances which is distributed as multivariate normal with mean vector zero and covariance matrix $\sigma^2 I$ where $\sigma^2>0$ is unknown.

Suppose we wish to test the hypothesis:

$$H\colon C\beta - c = \theta = 0, \tag{2.2}$$

where C is a known $q\times k$ matrix of rank $q\le k$ and c is a known $q\times 1$ vector. The minimum variance linear unbiased estimator of θ is:

$$z = Cb - c, \tag{2.3}$$

where $b=(X'X)^{-1}X'y$ is the least squares estimator of β. This estimator is distributed as multivariate normal with mean vector θ and covariance matrix $\sigma^2 V$, where $V = C(X'X)^{-1}C'$. An unbiased estimator of σ^2 is s^2 where $(T-k)s^2 = (y - Xb)'(y - Xb)$.

I will compare the acceptance regions of two tests of H. One test is the F test and the other is a finite induced test based on t tests of the separate hypotheses.

When H is rejected we usually want to know which individual restrictions are responsible for rejection. Hence, I assume that the separate hypotheses are H_i: $\theta_i = 0$, $i = 1,\ldots,q$. It is well known that the F test and the separate t tests can produce conflicting inferences; for example, see Maddala (1977, pp. 122–124). The purpose of comparing the acceptance regions of the two testing procedures is to explain these conflicts.

I first introduce the F test and the finite induced test. Next, I briefly review the distributions and probability inequalities involved in calculating the critical value and significance level of a finite induced test. Then the acceptance regions of the two tests are compared for the case of two restrictions; the exact and Bonferroni critical values are used to perform the finite induced test. Finally, I discuss the effect of a nonsingular linear transformation of the hypothesis H on the acceptance regions of the F test and the finite induced test.

2.2. *Tests*

2.2.1. *F test*

The familiar F statistic is

$$F = \frac{z'V^{-1}z}{qs^2}. \tag{2.4}$$

For an α level F test of H the acceptance region is:

$$F \le F_\alpha(q, T-k), \tag{2.5}$$

where $F_\alpha(q, T-k)$ is the upper α significance point of an F distribution with q and $T-k$ degrees of freedom. The F test of H is equivalent to one derived from the confidence region:

$$(z-\theta)'V^{-1}(z-\theta) \le s^2S^2, \tag{2.6}$$

where $S^2 = qF_\alpha(q, T-k)$. The inequality determines an ellipsoid in the $\theta_1,\ldots,\theta_q$ space with center at z. The probability that this random ellipsoid covers θ is $1-\alpha$. The F test of H accepts H if and only if the ellipsoid covers the origin.

The F test has power against alternatives in all directions. Accordingly, I consider a finite induced test with the same property. It will become apparent the acceptance region of the finite induced test is not the same as the acceptance region of the F test.

2.2.2. *Finite induced test*

Assume the finite induced test of H accepts H if and only if all the separate hypotheses $H_1,\dots,H_q$ are accepted. The t statistic for testing the separate hypothesis H_i: $\theta_i = 0$ is:

$$t_i = \frac{z_i}{\sqrt{s^2 V_{ii}}}, \qquad i = 1,\dots,q, \tag{2.7}$$

where V_{ii} is the ith diagonal element of V. The acceptance region of a δ level equal-tailed test of H_i against the two-sided alternative H_i^*: $\theta_i \neq 0$ is:

$$|t_i| \le t_{\delta/2}(T-k), \qquad i = 1,\dots,q, \tag{2.8}$$

where $t_{\delta/2}(T-k)$ is the upper $\delta/2$ significance point of a t distribution with $T-k$ degrees of freedom.

When all the equal-tailed t tests have the same significance level the acceptance region for an α level finite induced test of H is:

$$|t_i| \le M, \qquad i = 1,\dots,q, \tag{2.9}$$

where the critical value M is such that:

$$P\left[\max(|t_1|,\dots,|t_q|) \le M | H\right] = 1-\alpha. \tag{2.10}$$

In words, this finite induced test rejects H if the largest squared t statistic is greater than the square of the critical value M. The significance level δ of each equal-tailed t test is given by:

$$t_{\delta/2}(T-k) = M. \tag{2.11}$$

The acceptance region of the α level finite induced test is the intersection of the separate acceptance regions (2.9). For this reason Krishnaiah (1979) refers to the above test as the finite intersection test. The acceptance region of the finite induced test is a cube in the $z_1,\dots,z_q$ space with center at the origin and similarly in the $t_1,\dots,t_q$ space.

The finite induced test of H is equivalent to one based on a confidence region. The simultaneous confidence intervals associated with the finite induced test are given by:

$$P\left[z_i - M\sqrt{s^2 V_{ii}} \le \theta_i \le z_i + M\sqrt{s^2 V_{ii}} : i = 1,\dots,q\right] = 1-\alpha. \tag{2.12}$$

I call these intervals M intervals. The intersection of the M intervals is the finite induced confidence region. This region is a cube in the $\theta_1,\dots,\theta_q$ space with center $z_1,\dots,z_q$. The probability that this random cube covers the true parameter point θ is $1-\alpha$. The α level finite induced test accepts H if and only if all the M intervals cover zero, i.e. if and only if the finite induced confidence region covers the origin.

2.3. *Critical values — finite induced test*

To perform an α level finite induced test we need to know the upper α percentage point of the $\max(|t_1|,\dots,|t_p|)$ distribution. The multivariate t and F distributions are briefly reviewed since these distributions are used in the calculation of the exact percentage points. The exact percentage points are difficult to compute except in special cases. In practice inequalities are used to obtain a bound on the probability integral of $\max(|t_1|,\dots,|t_p|)$, when $t_1,\dots,t_p$ have a central multivariate t distribution. Three such inequalities are discussed.

2.3.1. *Multivariate t and F distributions*

Let $x=(x_1,\dots,x_p)'$ be distributed as a multivariate normal with mean vector μ and covariance matrix $\Sigma=\sigma^2\Omega$ where $\Omega=(\rho_{ij})$ is the correlation matrix. Also, let s^2/σ^2 be distributed independently of x as chi-square with n degrees of freedom. In addition, let $t_i=x_i\sqrt{n}/s$, $i=1,\dots,p$. Then the joint distribution of $t_1,\dots,t_p$ is a central or noncentral multivariate t distribution with n degrees according as $\mu=0$ or $\mu\neq 0$. The matrix Ω is referred to as the correlation matrix of the "accompanying" multivariate normal. In the central case, the above distribution was derived by Cornish (1954) and by Dunnett and Sobel (1954) independently. Krishnaiah and Armitage (1965a, 1966) gave the percentage points of the central multivariate t distribution in the equicorrelated case $\rho_{ij}=\rho(i\neq j)$. Tables of $P[\max(t_1,t_2)\le a]$ were computed by Krishnaiah, Armitage and Breiter (1969a). The tables are used for a finite induced test against one-sided alternatives. Such a test is discussed in Section 3.

Krishnaiah (1963, 1964, 1965) has investigated the multivariate F distribution. Let $x_u=(x_{1u},\dots,x_{pu})'$, $u=1,\dots,m$, be m independent random vectors which are distributed as multivariate normal with mean vector μ and covariance matrix $\Sigma=(\sigma_{ij})$. Also let:

$$w_i=\sum_{u=1}^{m}x_{iu}^2,\qquad i=1,\dots,p.$$

The joint distribution of $w_1,\dots,w_p$ is a central or noncentral multivariate chi-square distribution with m degrees of freedom and with Σ as the covariance matrix of the "accompanying" multivariate normal according as $\mu = 0$ or $\mu \neq 0$. Let $F_i = nw_i\sigma_{00}/mw_0\sigma_{ii}$ and let w_0/σ_{00} be distributed independently of $(w_1,\dots,w_p)$ as chi-square with n degrees of freedom. Then the joint distribution of $F_1,\dots,F_p$ is a multivariate F distribution with m and n degrees of freedom with Ω as the correlation matrix of the "accompanying" multivariate normal. When $m = 1$, the multivariate F distribution is equivalent to the multivariate t^2 distribution. Krishnaiah (1964) gave an exact expression for the density of the central multivariate F distribution when Σ is nonsingular. Krishnaiah and Armitage (1965b, 1970) computed the percentage points of the central multivariate F distribution in the equicorrelated case when $m = 1$. Extensive tables of $P[\max(|t_1|,|t_2|) \leq c]$ have been prepared by Krishnaiah, Armitage and Breiter (1969b). Hahn and Hendrickson (1971) gave the square roots of the percentage points of the central multivariate F distribution with 1 and n degrees of freedom in the equicorrelated case. For further details on the multivariate t and F distributions see Johnson and Kotz (1972).

2.3.2. *Probability inequalities*

The well known Bonferroni inequality states that:

$$P(A_1,\dots,A_p) \geq 1 - \sum_{i=1}^{p} P(A_i^c),$$

where A_i is an event and A_i^c its complement. Letting A_i be the event $|t_i| \leq t_{\delta/2}(n)$, $i = 1,\dots,p$, the Bonferroni inequality gives:

$$P\left[\max(|t_1|,\dots,|t_p|) \leq t_{\delta/2}(n)\right] \geq 1 - \delta p, \tag{2.13}$$

i.e. the probability that the point $(t_1,\dots,t_p)$ falls in the cube is $\geq 1 - \delta p$. The probability is $\geq 1 - \alpha$ when the significance level δ is α/p. Tables of the percentage points of the Bonferroni t statistic have been prepared by Dunn (1961) and are reproduced in Miller (1966). A more extensive set of tables has been calculated by Bailey (1977).

Sidák (1967) has proved a general inequality which can be specialized to give a slight improvement over the Bonferroni inequality when both are applicable. The Sidák inequality gives:

$$P\left[\max(|t_1|,\dots,|t_p|) \leq c\right] \geq \prod_{i=1}^{p} P(|t_i| \leq c). \tag{2.14}$$

In words, the probability that a multivariate t vector $(t_1,\dots,t_p)$ with arbitrary correlations falls inside a p-dimensional cube centered at the origin is always at least as large as the corresponding probability for the case where the correlations are zero, i.e. where $x_1,\dots,x_p$ are independent. When the critical value c is $t_{\delta/2}(n)$ the Sidák inequality gives:

$$P\left[\max(|t_1|,\dots,|t_p|)\leq t_{\delta/2}(n)\right]\geq(1-\delta)^p. \tag{2.15}$$

The probability is $\geq 1-\alpha$ when the significance level δ is $1-(1-\alpha)^{1/p}$. The Sidák inequality produces slightly sharper tests or intervals than the Bonferroni inequality because $(1-\delta)^p\geq 1-\delta p$. Games (1977) has prepared tables of the percentage points of the Sidák t statistic. Charts by Moses (1976) may be used to find the appropriate t critical value with either the Bonferroni or Sidák inequality.

In the special case where the correlations are zero, i.e. $\Omega=I$, $\max(|t_1|,\dots,|t_p|)$ has the studentized maximum modulus distribution with parameter p and n degrees of freedom. The upper α percentage point of this distribution is denoted $m(p,n)$. Using a result by Sidák (1967), Hochberg (1974) has proved that:

$$P\left[\max(|t_1|,\dots,t_p|)\leq m_\alpha(p,n)\right]\geq 1-\alpha, \tag{2.16}$$

where Ω is an arbitrary correlation matrix, i.e. $\Omega\neq I$. Stoline and Ury (1979) have shown that if $\delta=1-(1-\alpha)^{1/p}$, then $m_\alpha(p,n)\leq t_{\delta/2}(n)$ with a strict inequality holding when $n=\infty$. This inequality produces a slight improvement over the Sidák inequality. Hahn and Hendrickson (1971) gave tables of the upper percentage points of the studentized maximum modulus distribution. More extensive tables have been prepared by Stoline and Ury (1979).

A finite induced test with significance level exactly equal to α is called an exact finite induced test and the corresponding critical value is called the exact critical value. For a nominal α level test of p separate hypotheses the Bonferroni critical value is $t_{\delta/2}(T-k)$ with $\delta=\alpha/p$, the Sidák critical value is $t_{\delta/2}(T-k)$ with $\delta=1-(1-\alpha)^{1/p}$ and the studentized maximum modulus critical value is $m_\alpha(p,T-k)$. When the exact critical value is approximated by the Bonferroni critical value the finite induced test is called the Bonferroni test. The Sidák test and the studentized maximum modulus test are defined similarly. For the purpose of this paper we use the Bonferroni test since the Bonferroni inequality is familiar and simple to apply. However, the exposition would be essentially unchanged if the Sidák test or the studentized maximum modulus test were used instead of the Bonferroni test.

2.4. *Acceptance regions*

2.4.1. *Case of two restrictions*

The acceptance regions of the F test, the Bonferroni test and the exact finite induced test are now compared for the case of $q=2$ restrictions. It is assumed that σ^2 is known and that

$$V=\frac{1}{1-r^2}\begin{bmatrix} 1 & -r \\ -r & 1 \end{bmatrix}, \qquad |r|<1. \tag{2.17}$$

Christensen (1973) compared the powers of the F test and the Bonferroni test for this case. I will discuss the power comparisons in Section 5.

Since σ^2 is assumed known the t statistics are distributed $N(0,1)$ under the null hypothesis and the F statistic is replaced by the χ^2 statistic. These changes do not change any important features of the tests, at least for the purpose of comparison.

The covariance matrix $\sigma^2 V$ where V is given by (2.17) has a simple interpretation. Consider a model with $K=3$ regressors:

$$y=[eX_1]\beta+u, \tag{2.18}$$

where e is a $T\times 1$ vector of ones, X_1 is $T\times 2$ and $\beta=(\beta_0,\beta_1,\beta_2)'$. Suppose the hypothesis is H: $\beta_1=\beta_2=0$. If both of the columns of X_1 have mean zero and length one, then $\sigma^2 V=\sigma^2(X_1'X_1)^{-1}$, where

$$V^{-1}=\begin{bmatrix} 1 & r \\ r & 1 \end{bmatrix}=X_1'X_1, \tag{2.19}$$

and where r is the correlation between the columns of X_1. In a model with $K>3$ regressors (including an intercept) the covariance matrix of the least squares estimates of the last two regression coefficients is given by $\sigma^2 V$ with V as in (2.17) provided that the last two regressors have mean zero, length one and are orthogonal to the remaining regressors.

Consider the acceptance regions of the tests in the z_1 and z_2 space. The acceptance region of an α level χ^2 test is the elliptical region:

$$z_1^2+2rz_1z_2+z_2^2\le S^2\sigma^2, \tag{2.20}$$

where $S^2=\chi^2_\alpha(2)$ is the upper α significant point of the χ^2 distribution with two degrees of freedom. The acceptance region of a nominal α level Bonferroni test is the square region:

$$|z_i|\le\frac{B\sigma}{\sqrt{1-r^2}}, \qquad i=1,2, \tag{2.21}$$

where $B=t_{\delta/2}(T-k)$ with $\delta=\alpha/2$. This region is a square with sides $2B\sigma/\sqrt{1-r^2}$ and center at the origin. The length of the major axis of the elliptical region (2.20) and the length of the sides of the square become infinite as the absolute value of r tends to one.

It will prove to be more convenient to study the acceptance regions of the tests in the t_1 and t_2 space. The t statistic for testing the separate hypotheses H_i: $\theta_i = 0$ is:

$$t_i = \frac{z_i\sqrt{1-r^2}}{\sigma}, \qquad i=1,2, \tag{2.22}$$

where $\sigma/\sqrt{1-r^2}$ is the standard deviation of z_i and where t_1 and t_2 are $N(0,1)$ variates since σ^2 is known. Dividing both sides of (2.20) by the standard deviation of z_i the acceptance region of the χ^2 test becomes:

$$t_1^2 + 2rt_1t_2 + t_2^2 \le S^2(1-r^2), \tag{2.23}$$

which is an elliptical region in the t_1 and t_2 space. Rewriting the boundary of the elliptical region (2.23) as:

$$(t_2 + rt_1)^2 = (S^2 - t_1^2)(1-r^2), \tag{2.24}$$

we see that the maximum absolute value of t_1 satisfying the equation of the ellipse is S. By symmetry the same is true for the maximum absolute value of t_2. Hence the elliptical region (2.23) is bounded by a square region with sides $2S$ and center at the origin. I refer to this region as the χ^2 box. Dividing (2.21) by the standard deviation of z_i the acceptance region of the Bonferroni test becomes:

$$|t_i| \le B, \qquad i=1,2, \tag{2.25}$$

which is a square region in the t_1 and t_2 space with sides $2B$ and center at the origin. I call this region the Bonferroni box. In this special case $B < S$ so that the Bonferroni box is inside the χ^2 box. The acceptance region of the exact α level finite induced test is a square region which I refer to as the exact box. The exact box is inside the Bonferroni box. The dimensions of the ellipse and the exact box are conditional on r. Since the dimensions of the χ^2 box and the Bonferroni box are independent of r, the dimensions of the ellipse and the exact box remain bounded as the absolute value of r tends to one.

Savin (1980) gives an example of a 0.05 level test of H when $r = 0$. The acceptance region of a 0.05 level χ^2 test of H is:

$$t_1^2 + t_2^2 \le S^2 = 5.991. \tag{2.26}$$

This region is a circle with radius $S = 2.448$ and center at the origin. The acceptance region of a nominal 0.05 level Bonferroni test in the t_1 and t_2 space is a square with sides $2B = 4.482$ since $\delta = 0.05/2$ gives $B = 2.241$. Both the circle and the Bonferroni box are shown in Figure 2.1. When $V = I$ (and σ^2 is known) the t statistics are independent, so the probability that both t tests accept when H is true is $(1-\delta)^2$. If $(1-\delta)^2 = 0.95$, then $\delta = 0.0253$. Hence, for an exact 0.05 level finite induced test the critical value is $M = 2.236$ and the exact box has sides $2M = 4.472$. The difference between the sides of the Bonferroni and the exact box is 0.005. The true significance level of the Bonferroni box is $1-(0.975)^2 = 0.0494$, which is quite close to 0.05.

A comparison of the acceptance regions of the χ^2 test and the finite induced test shows that there are six possible situations:

(1) χ^2 and both t tests reject.
(2) χ^2 and one but not both t tests reject.

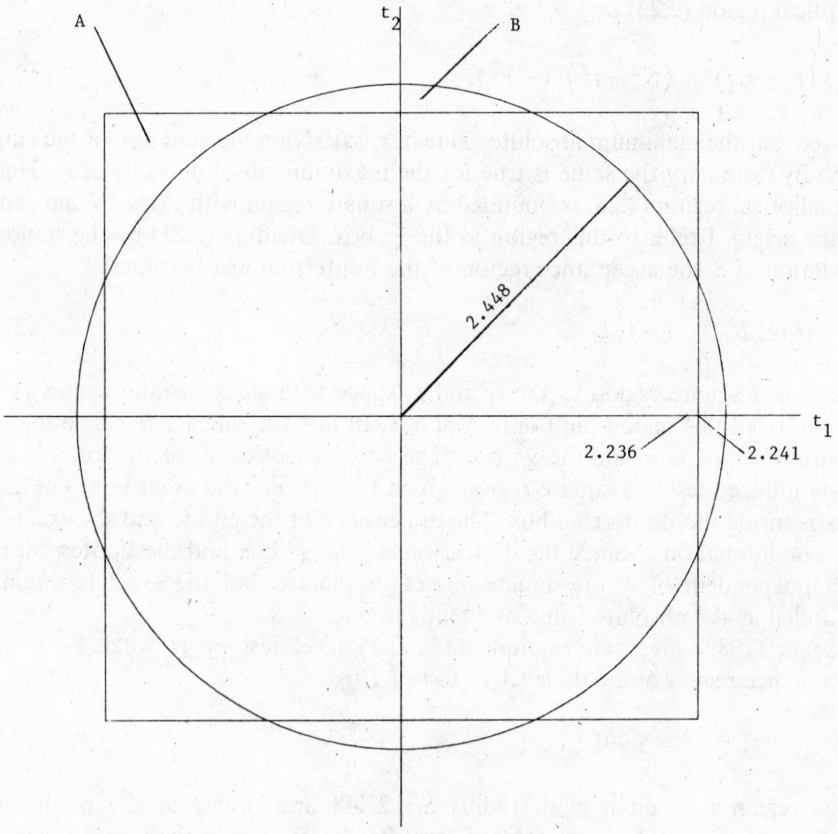

Figure 2.1 The acceptance regions of the Bonferroni and χ^2 tests where the correlation $r = 0$ and the nominal size is $\alpha = 0.05$.

(3) χ^2 test rejects but not the t tests.
(4) Both t tests reject but not χ^2 test.
(5) One, but not both t tests reject, nor χ^2 test.
(6) Neither the t tests nor χ^2 test reject.

Cases 1 and 6 are cases of agreement while the remaining are cases of disagreement. The χ^2 test and the finite induced test can produce conflicting inferences since they use different acceptance regions. These six cases are discussed in the context of the F test and the finite induced test by Geary and Leser (1968) and Maddala (1977, pp. 122–124).

From Figure 2.1 we see that H is accepted by the Bonferroni test and rejected by the χ^2 when A is the point (t_1, t_2) and vice versa when B is the point (t_1, t_2). Case 3 is illustrated by point A and Case 5 by point B. Maddala (1977) remarks that Case 3 occurs often in econometric applications while Case 4 is not commonly observed. Maddala refers to Case 3 as multicollinearity. Figure 2.1 illustrates that Case 3 can occur when $r = 0$, i.e. when the regressors are orthogonal.

Next consider the acceptance regions of the tests when $r \neq 0$. The following discussion is based on the work of Evans and Savin (1980). When r is different from zero the acceptance region of the χ^2 test is an ellipse. The acceptance regions of a 0.05 level χ^2 test in the t_1 and t_2 space are shown in Figure 2.2 for $r = 0.0\ (0.2)\ 1.0$. In Figure 2.2 the inner box is the nominal 0.05 level Bonferroni box and the outer box is the χ^2 box. The ellipse collapses to a line as r increases from zero to one.

Observe that the case where both t tests reject and the χ^2 test accepts (Case 4) cannot be illustrated in Figure 2.1. From Figure 2.2 we see that Case 4 can be illustrated by point C. Clearly, r^2 must be high for Case 4 to occur. Maddala notes that this case is not commonly observed in econometric work.

The true level of significance of the Bonferroni box decreases as r increases in absolute value. The true significance level of a nominal α level Bonferroni box for selected values of α and r are given in Table 2.1. When $\alpha = 0.05$ the true levels are roughly constant for $r < 0.6$. For $r > 0.6$, there is a noticeable decrease in the true level. This suggests that the nominal 0.5 level Bonferroni box is a satisfactory approximation to the exact box for $r < 0.6$. The results are similar when the nominal sizes are $\alpha = 0.10$ and $\alpha = 0.01$.

As noted earlier the χ^2 test and the Bonferroni test can produce conflicting inferences because the tests do not have the same acceptance regions. The probability of conflict is one minus the probability that the tests agree. When H is true the probability that the tests agree and that they conflict are given in Table 2.1 for selected values of α and r. For the case where the nominal size is $\alpha = 0.05$, although the probability of conflict increases as r increases (for $r > 0$), this probability remains quite small, i.e. less than the significance level. This result

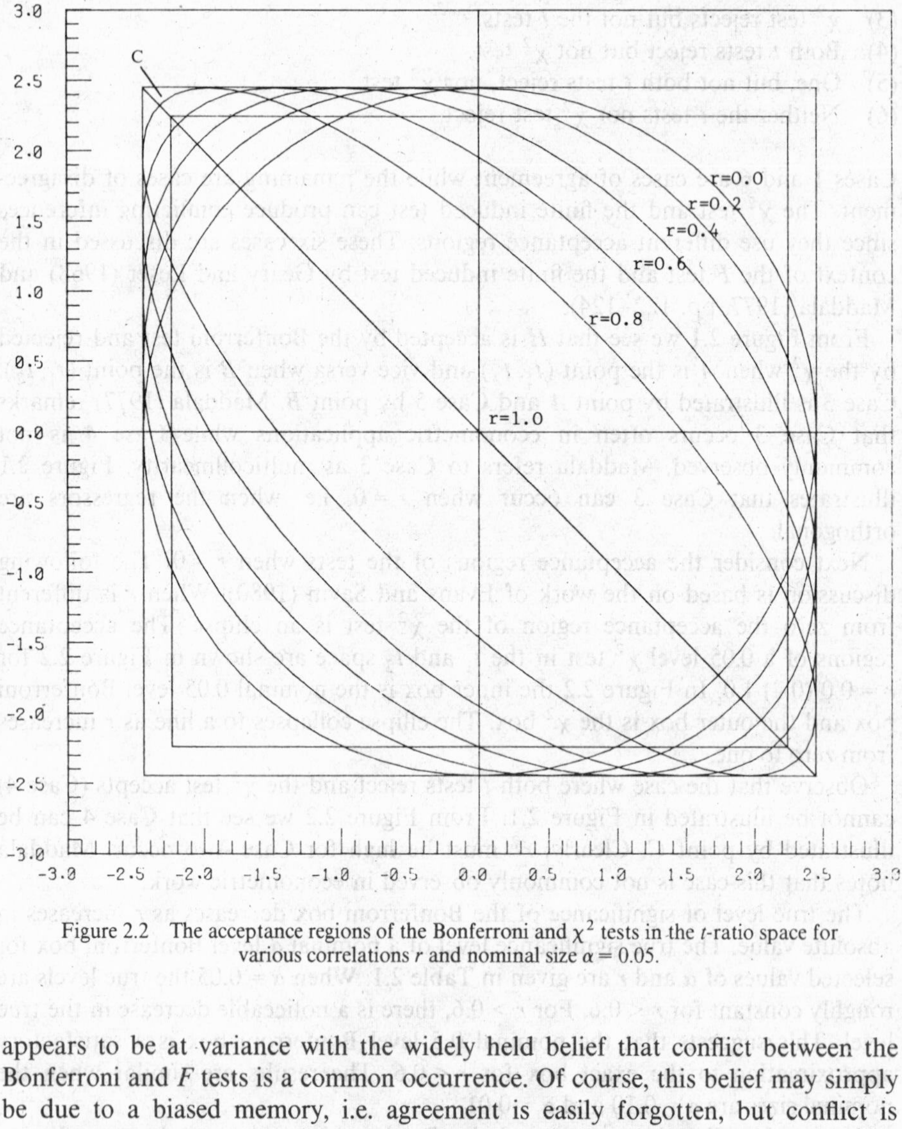

Figure 2.2 The acceptance regions of the Bonferroni and χ^2 tests in the t-ratio space for various correlations r and nominal size $\alpha = 0.05$.

appears to be at variance with the widely held belief that conflict between the Bonferroni and F tests is a common occurrence. Of course, this belief may simply be due to a biased memory, i.e. agreement is easily forgotten, but conflict is remembered. On the other hand, the small probability of conflict may be a special feature of the two parameter case.

Figure 2.2 shows a big decrease in the area of intersection of the two acceptance regions as r increases and hence gives a misleading impression that there is a big decrease in the probability that both tests accept as r increases. In fact, the probability that both tests accept is remarkably constant. The results are similar when the nominal sizes are $\alpha = 0.10$ and $\alpha = 0.01$. As can be seen from

Table 2.1 The Probability of Conflict between the Chi Square and Finite Induced Tests and between the Chi Square and Bonferroni Tests.

Nominal		Finite Induced Test		Bonferroni Test		
Size	r	Agree	Conflict	Agree	Conflict	True Size
0.10	0.0	0.964	0.036	0.965	0.035	0.098
	0.1	0.964	0.036	0.965	0.035	0.097
	0.2	0.963	0.037	0.964	0.036	0.096
	0.3	0.961	0.039	0.962	0.038	0.095
	0.4	0.958	0.042	0.961	0.039	0.093
	0.5	0.954	0.046	0.958	0.042	0.091
	0.6	0.948	0.052	0.955	0.045	0.088
	0.7	0.941	0.059	0.951	0.049	0.083
	0.8	0.934	0.066	0.947	0.053	0.078
	0.9	0.926	0.074	0.942	0.058	0.070
	0.95	0.920	0.080	0.939	0.061	0.065
	0.99	0.913	0.087	0.936	0.064	0.057
	0.9999	0.909	0.091	0.934	0.066	0.051
0.05	0.0	0.978	0.022	0.978	0.022	0.049
	0.1	0.978	0.022	0.978	0.022	0.049
	0.2	0.977	0.023	0.978	0.022	0.049
	0.3	0.976	0.024	0.977	0.023	0.048
	0.4	0.975	0.025	0.976	0.024	0.048
	0.5	0.973	0.027	0.975	0.025	0.046
	0.6	0.971	0.029	0.974	0.026	0.045
	0.7	0.967	0.033	0.972	0.028	0.043
	0.8	0.963	0.037	0.970	0.030	0.040
	0.9	0.959	0.041	0.968	0.032	0.036
	0.95	0.956	0.044	0.966	0.034	0.033
	0.99	0.952	0.048	0.965	0.035	0.029
	0.9999	0.950	0.050	0.964	0.036	0.025
0.01	0.0	0.994	0.006	0.994	0.006	0.010
	0.1	0.994	0.006	0.994	0.006	0.010
	0.2	0.994	0.006	0.994	0.006	0.010
	0.3	0.994	0.006	0.994	0.006	0.010
	0.4	0.994	0.006	0.994	0.006	0.010
	0.5	0.993	0.007	0.994	0.006	0.010
	0.6	0.993	0.007	0.993	0.007	0.009
	0.7	0.992	0.008	0.993	0.007	0.009
	0.8	0.992	0.008	0.993	0.007	0.008
	0.9	0.991	0.009	0.992	0.008	0.008
	0.95	0.990	0.010	0.992	0.008	0.007
	0.99	0.989	0.011	0.992	0.008	0.006
	0.9999	0.989	0.011	0.992	0.008	0.005

Table 2.1 the results are also similar when the Bonferroni box is replaced by the exact box.

2.4.2. *Equivalent hypotheses and invariance*

In this section I discuss the effect of a nonsingular linear transformation of the hypothesis H on the acceptance regions of the F test and the Bonferroni test. Consider the hypothesis:

$$H^*: C^*\beta - c^* = \theta^* = 0, \tag{2.27}$$

where C^* is a known $q^* \times k$ matrix of rank $q^* \leq k$ and c^* is a known $q^* \times k$ vector so that $\theta_1^*, \theta_2^*, \ldots, \theta_{q^*}^*$ are a set of q^* linearly independent functions. The hypotheses H^* and H are equivalent when H^* is true if and only if H is true. Hence, H^* and H are equivalent if the set of β for which $\theta = 0$ is the same as the set for which $\theta^* = 0$.

We now show that H and H^* are equivalent if and only if there exists a nonsingular $q \times q$ matrix A such that $[C^*c^*] = A[Cc]$ and hence $q^* = q$. Our proof follows Scheffé (1959, pp. 31–32]. Suppose first that a $q \times q$ nonsingular matrix A exists such that $[C^*c^*] = A[Cc]$. Then H^* is true implies that $\theta^* = C^*\beta - c^* = A(C\beta - c) = 0$. Thus, $C\beta - c = \theta = 0$ which implies that H is true. Similarly if H is true, then H^* is true.

Suppose next that the equations $C^*\beta = c^*$ have the same solution space as the equations $C\beta = c$. Then the rows of $[C^*c^*]$ span the same space as the rows of $[Cc]$. The q^* rows of C^* are linearly independent and so constitute a basis for this space. Similarly, the q rows of C constitute a basis for the same space. Hence $q^* = q$ and the q rows of C^* must be linear combinations of the q rows of C. Therefore $[C^*c^*] = A[Cc]$, where A is nonsingular since rank $C^* = \text{Rank } C = q$.

If the hypotheses H^* and H are equivalent, the F statistic for testing H^* is the same as the F statistic for testing H. Assume that H^* and H are equivalent. The numerator of the F statistic for testing H^* is

$$\begin{aligned}
&[C^*b - c^*]'\left[C^*(X'X)^{-1}C^{*\prime}\right]^{-1}[C^*b - c^*] \\
&\quad = [Cb - c]'A'(A')^{-1}\left[C(X'X)^{-1}C'\right]^{-1}A^{-1}A[Cb - c] \\
&\quad = [Cb - c]\left[C(X'X)^{-1}C'\right]^{-1}[Cb - c]. \qquad (2.28)
\end{aligned}$$

This is the same as the numerator of the F statistic for testing H, the denominator of the two test statistics being qs^2. Hence the F tests of H^* and H employ the same acceptance region with the result that we accept H^* if and only if we accept H. This can be summarized by saying that the F test has the property that it is invariant to a nonsingular transformation of the hypothesis.

The finite induced test and hence the Bonferroni test does not possess this invariance property. As an example consider the case where $q = 2$ and $\sigma^2 V = I$ which is known. First suppose the hypothesis is H: $\theta_1 = \theta_2 = 0$. Then the acceptance region of the nominal 0.05 level Bonferroni test of H is the intersection of the separate acceptance regions $|z_1| \leq 2.24$ and $|z_2| \leq 2.24$. Now suppose the hypothesis H^* is $\theta_1^* = \theta_1 + \theta_2 = 0$ and $\theta_2^* = \theta_1 - \theta_2 = 0$. The acceptance region of the nominal 0.05 level Bonferroni test of H^* is the intersection of the separate regions $|z_1 + z_2| \leq (2)^{1/2}2.24$ and $|z_1 - z_2| \leq (2)^{1/2}2.24$. The hypotheses H^* and H are equivalent, but the acceptance region for testing H^* is not the same as the region for testing H. Therefore, if the same sample is used to test both hypotheses, H^* may be accepted and H rejected and vice versa.

If all hypotheses equivalent to H are of equal interest we want to accept all these hypotheses if and only if we accept H. In this situation the F test is the natural test. However, hypotheses which are equivalent may not be of equal interest. When this is the case the F test may no longer be an intuitively appealing procedure. Testing linear combinations of the restrictions is discussed in detail in the next section.

3. Induced tests and simultaneous confidence intervals

3.1. *Separate hypotheses*

An important step in the construction of an induced test is the choice of the separate hypotheses. So far, I have only considered separate hypotheses about individual restrictions. In general, the separate hypotheses can be about linear combinations of the restrictions as well as the individual restrictions. This means that there can be many induced tests of H, each test being conditional on a different set of separate hypotheses. The set of separate hypotheses chosen should include those hypotheses which are of economic interest. Economic theory may not be sufficient to determine a unique set of separate hypotheses and hence a unique induced test of H.

Let L be the set of linear combinations ψ such that every ψ in L is of the form $\psi = a'\theta$ where a is any known $q \times 1$ non-null vector. In other words, L is the set of all linear combinations of $\theta_1,\dots,\theta_q$ (excluding the case of $a = 0$). The set L is called a q-dimensional space of functions if the functions $\theta_1,\dots,\theta_q$ are linearly independent, i.e. if rank $C = q$ where C is defined in (2.2).

The investigator may not have an equal interest in all the ψ in L. For example, in economic studies the individual regression coefficients are commonly of most interest. Let G be the set of ψ of primary interest and the complement of G relative to L, denoted by $L - G$, be the set of ψ in L of secondary interest. It is assumed that this twofold partition is fine enough that all ψ in G are of equal interest and similarly for all ψ in $L - G$. Furthermore, it is assumed that G contains q linearly independent combinations ψ.

The set G is either a finite or an infinite set. If G is infinite, then G is either a proper subset of L or equal to L. In the latter case all the ψ in L are of primary interest. All told there are three possible situations: (i) G finite, $L - G$ infinite; (ii) G infinite, $L - G$ infinite; (iii) G infinite, $L - G$ finite. The induced test is referred to as a finite or infinite induced test accordingly as G is finite or infinite.

Let G be a finite set and let ψ_i, $i = 1,\dots,m$, be the linear combinations in G. The finite induced test of

$$H(G): \psi_1 = \cdots = \psi_m = 0 \tag{3.1}$$

accepts $H(G)$ if and only if all the separate hypotheses,

$$H_i: \psi_i = 0, \qquad i = 1,\ldots,m, \tag{3.2}$$

are accepted and rejects $H(G)$ otherwise. Since there are q linearly independent combinations ψ_i, $i = 1,\ldots,q$, in G, the hypotheses $H(G)$ and H: $\theta = 0$ are equivalent and $H(G)$ is true if and only if H is true. Hence, we accept H if all the separate hypotheses H_i, $i = 1,\ldots,m$ are accepted and reject H otherwise. This test procedure is also referred to as the finite induced test of H. Similar remarks apply when G is an infinite set. Since the induced test of H is conditional on the choice of G, it is important that G be selected before analyzing the data.

The set G may be thought of as the set of eligible voters. A linear combination of primary interest votes for (against) H if the corresponding separate hypothesis $H(a)$ is accepted (rejected). A unanimous decision is required for H to be accepted, i.e. all ψ in G must vote for H. Conversely, each ψ in G has the power to veto H. If all ψ in L are of equal interest, then all ψ in L are also in G so there is universal suffrage. On the other hand, the set of eligible voters may have as few as q members. The reason for restricting the right to vote is to prevent the veto power from being exercised by ψ in which we have only a secondary interest.

Instead of having only one class of eligible voters it may be more desirable to have several classes of eligible voters where the weight of each vote depends on the class of the voter. Then the hypothesis H is accepted or rejected depending on the size of the vote. However such voting schemes have not been developed in the statistical literature. In this paper I only discuss the simple voting scheme indicated above.

It is worth remarking that when the number of ψ in G is greater than q the induced test produces decisions which at first sight may appear puzzling. As an example suppose $q = 2$ and that the ψ in G are $\psi_1 = \theta_1$, $\psi_2 = \theta_2$, and $\psi_3 = \theta_1 + \theta_2$. Testing the three separate hypotheses H_i: $\psi_i = 0$, $i = 1,2,3$, induces a decision problem in which one of the eight possible decisions is:

$$H_1 \text{ and } H_2 \text{ are both true,} \quad H_3 \text{ is false.} \tag{3.3}$$

Clearly, when H_1 and H_2 are both known to be true, then H_3 is necessarily true. On the other hand, when testing these three hypotheses it may be quite reasonable to accept that H_1 and H_2 are both true and that H_3 is false. In other words, there is a difference between logical and statistical inference.

3.2. *Finite induced test — ψ of primary interest*

3.2.1. *Exact test*

Suppose that a finite number m of ψ in L are of primary interest. In this case G is a finite set. Let the ψ in G be $\psi_i = a_i'\theta$, $i = 1,\ldots,m$. The t statistic for testing the separate hypothesis $H(a_i)$: $\psi_i = a_i'\theta = 0$ is:

$$t_0(a_i) = \frac{a_i'z}{\sqrt{s^2 a_i' V a_i}} = \frac{\hat{\psi}_i}{\hat{\sigma}_{\hat{\psi}_i}}, \qquad i = 1,\dots,m, \tag{3.4}$$

where $\hat{\psi}_i = a_i'z$ is the minimum variance unbiased estimator of ψ_i and $\hat{\sigma}^2_{\hat{\psi}_i} = s^2 a_i' V a_i$ is an unbiased estimator of its variance where z and V are defined in Section 2.1. For an equal-tailed δ level test of $H(a_i)$ the acceptance region is:

$$|t_0(a_i)| \le t_{\delta/2}(T-k), \qquad i = 1,\dots,m. \tag{3.5}$$

The finite induced test of H accepts H if and only if all the separate hypotheses $H(a_i),\dots,H(a_m)$ are accepted. When all the equal-tailed tests have the same significance level the acceptance region for an α level finite induced test of H is:

$$|t_0(a_i)| \le M, \qquad i = 1,\dots,m, \tag{3.6}$$

where

$$P\left[\max(|t_0(a_1)|,\dots,|t_0(a_m)|) \le M | H\right] = 1 - \alpha. \tag{3.7}$$

The significance level of the separate tests is δ, where $t_{\delta/2}(T-k) = M$. The acceptance region of the finite induced test is the intersection of the separate acceptance regions (3.6). This region is a polyhedron in the $z_1,\dots,z_q$ space and a cube in the $t_0(a_1),\dots,t_0(a_m)$ space.

Simultaneous confidence intervals can be constructed for all ψ in G. The finite induced procedure is based on the following result. The probability is $1-\alpha$ that simultaneously

$$\hat{\psi}_i - M\hat{\sigma}_{\hat{\psi}_i} \le \psi_i \le \hat{\psi}_i + M\hat{\sigma}_{\hat{\psi}_i}, \qquad i = 1,\dots,m. \tag{3.8}$$

I call these intervals M intervals. The intersection of the M intervals is a polyhedron in the θ space with center at z. The α level finite induced test accepts H if and only if all the M intervals (3.8) cover zero, i.e. if and only if the finite induced confidence region covers the origin.

An estimate $\hat{\psi}_i$ of ψ_i is said to be significantly different from zero (*sdfz*) according to the M criterion if the M interval does not cover $\psi_i = 0$, i.e. if $|\hat{\psi}_i| \ge M\hat{\sigma}_{\hat{\psi}_i}$. Hence, H is rejected if and only if the estimate of at least one ψ_i in G is *sdfz* according to the M criterion.

The finite induced test can be tailored to provide high power against certain alternatives. This can be achieved by using t tests which have unequal tails and different significance levels. For example, a finite induced test can be used to test against the alternative H^{**}: $\theta > 0$. The acceptance region of a δ level one-tailed t test against H_i^{**}: $\theta_i > 0$ is:

$$t_i < t_\delta(T - k), \qquad i = i,\dots,q. \tag{3.9}$$

When all the one-tailed t tests have the same significance level the acceptance region for an α level finite induced test of H is

$$t_i < M, \qquad i = 1,\dots,q, \tag{3.10}$$

where

$$P\left[\max(t_1,\dots,t_q) \le M | H\right] = 1 - \alpha. \tag{3.11}$$

The simultaneous confidence intervals associated with the above test procedure are given by:

$$P\left[z_i - M\sqrt{s^2 V_{ii}} \le \theta_i; i = 1,\dots,q\right] = 1 - \alpha. \tag{3.12}$$

A finite induced test against the one-sided alternatives H_i^{**}: $\theta < 0$, $i = 1,\dots,q$, can also be developed. In the remainder of this chapter I only consider two-sided alternatives.

3.2.2. *Bonferroni test*

The Bonferroni test is obtained from the exact test by replacing the exact critical value M by the critical value B given by the Bonferroni inequality. For a nominal α level Bonferroni induced test of H the acceptance region is:

$$|t_0(a_i)| \le B, \qquad i = i,\dots,m, \tag{3.13}$$

where

$$B = t_{\alpha/2m}(T - k). \tag{3.14}$$

The significance level of the separate tests is $\delta = \alpha/m$ and the significance level of the Bonferroni test is $\le \alpha$. The Bonferroni test consists of testing the separate hypotheses using the acceptance region (3.13) where the critical value B is given by (3.14). The acceptance region of the Bonferroni test in the $z_1,\dots, z_q$ space is referred to as the Bonferroni polyhedron and in the $t_0(a_1),\dots, t_0(a_m)$ space as the Bonferroni box. The Bonferroni polyhedron contains the polyhedron of the exact finite induced test and similarly for the Bonferroni box.

The probability is $\geq 1-a$ that simultaneously

$$\hat{\psi}_i - B\hat{\sigma}_{\hat{\psi}_i} \leq \psi_i \leq \hat{\psi}_i + B\hat{\sigma}_{\hat{\psi}_i}, \qquad i=1,\dots,m, \tag{3.15}$$

where these intervals are called B intervals. The B procedure consists in using these B intervals. The Bonferroni test accepts H if and only if all the B intervals cover zero, i.e. if and only if the Bonferroni confidence region covers the origin. An estimate of $\hat{\psi}_i$ of ψ_i is said to be *sdfz* according to the B criterion if the B interval does not cover zero, i.e. $|\hat{\psi}_i| \geq B\hat{\sigma}_{\hat{\psi}_i}$.

The Bonferroni test can be used to illustrate a finite induced test when $m > q$, i.e. the number of separate hypotheses is greater than the number of linear restrictions specified by H. Consider the case where $m=3$, $q=2$, and $\sigma^2 V = I$ which is known. Suppose that the three ψ in G are $\psi_1 = \theta_1$, $\psi_2 = \theta_2$, and $\psi_3 = \theta_1 + \theta_2$ and that tests of the three separate hypotheses H_i: $\psi_i = 0$, $i=1,2,3$, are defined by the three separate acceptance regions:

$$\begin{aligned} &|z_1| \leq 2.39, \qquad |z_2| \leq 2.39, \\ &|z_1 + z_2| \leq (2)^{1/2} 2.39 = 3.380, \end{aligned} \tag{3.16}$$

respectively, where 2.39 is the upper $0.05/2(3) = 0.00833$ significance point of a $N(0,1)$ distribution. The probability is ≥ 0.95 that the Bonferroni test accepts H when H is true.

The acceptance region of the Bonferroni test of H, which is the intersection of the three separate acceptance regions, is shown in Figure 3.1. When A is the point (z_1, z_2) the hypothesis H is rejected and the decision is that H_1 and H_2 are both true and H_3 is false.

For comparison consider the case where $m = q = 2$. The tests of the two separate hypotheses $\psi_1 = \theta_1 = 0$ and $\psi_2 = \theta_2 = 0$ are now defined by the two acceptance regions:

$$|z_1| \leq 2.24, \qquad |z_2| \leq 2.24, \tag{3.17}$$

respectively, where 2.24 is the upper $0.05/2(2) = 0.0125$ significance point of a $N(0,1)$ distribution. The acceptance region of this Bonferroni test of H is the inner square region shown in Figure 3.1. With this region we accept H when A is the point (z_1, z_2). When B is the point (z_1, z_2) the hypothesis H is accepted if ψ_3 is of primary interest and rejected if ψ_3 is of secondary interest. This comparison shows that the Bonferroni test can accept H for one set of ψ of primary interest and reject H for another set.

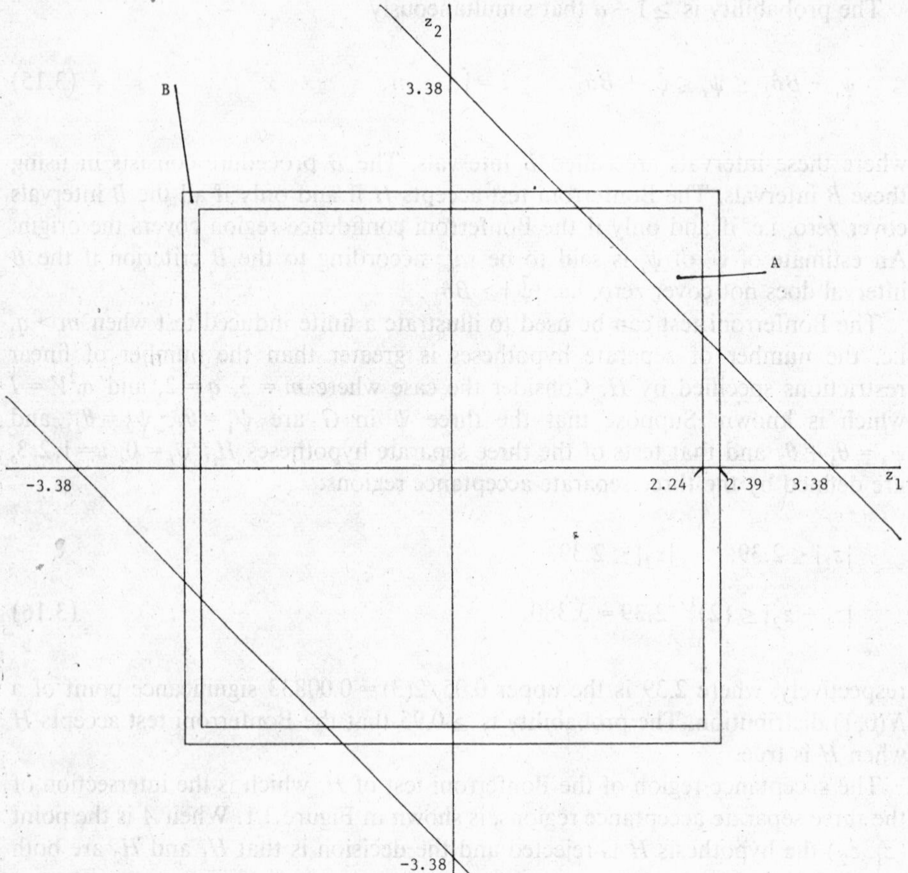

Figure 3.1 Acceptance regions of the Bonferroni test for the cases $m = 2$ and $m = 3$ when $q = 2$ and $\sigma^2 V = I$ which is known. The nominal size is $\alpha = 0.05$.

3.3. *Infinite induced test — Scheffé test*

3.3.1. *Scheffé test*

The Scheffé test is an infinite induced test where all ψ in L are of primary interest. This induced test accepts H if and only if the separate hypothesis,

$$H(a): \psi = a'\theta = 0, \tag{3.18}$$

is accepted for all non-null a. For a δ level equal-tailed test of $H(a)$ the acceptance region is:

$$|t_0(a)| \le t_{\delta/2}(T-k), \tag{3.19}$$

where

$$t_0(a) = \frac{a'z}{\sqrt{s^2 a'Va}} = \frac{\hat{\psi}}{\hat{\sigma}_{\hat{\psi}}}. \tag{3.20}$$

When all the equal-tailed tests have the same significance level the acceptance region for an α level infinite induced test of H is:

$$|t_0(a)| \le S, \quad \text{all non-null } a, \tag{3.21}$$

where

$$P\left[\max_a |t_0(a)| \le S|H\right] = 1-\alpha. \tag{3.22}$$

What is surprising is that the critical value S is given by the relatively simple expression:

$$S = \sqrt{qF_\alpha(q, T-k)}\,. \tag{3.23}$$

The significance level δ of the separate tests is given by $t_{\delta/2}(T-k) = S$.

The acceptance region is the intersection of the separate acceptance regions (3.21) for all non-null a. A remarkable fact is that the acceptance region of an α level Scheffé test of H is the same as the acceptance region of an α level F test of H. As a consequence we start the Scheffé test with an F test of H. If the F test rejects H the next step is to find the separate hypotheses responsible for rejection. The test procedure consists of testing the separate hypotheses using the acceptance region (3.21) where the critical value S is given by (3.23).

The Scheffé test assumes that all ψ in L are of equal interest, i.e. every ψ in L has the power to veto H. When the Scheffé test is used in empirical econometrics we are implicitly assuming that all ψ in L are of equal economic interest. In practice, this assumption is seldom satisfied. As a consequence, if the Scheffé test rejects, the linear combinations which are responsible for rejection may have no economically meaningful interpretation. A solution to the interpretation problem is to use the appropriate finite induced test.

Simultaneous confidence intervals can be constructed for all ψ in L. The probability is $1-\alpha$ that simultaneously for all ψ in L:

$$\hat{\psi} - S\hat{\sigma}_{\hat{\psi}} \le \psi \le \hat{\psi} + S\hat{\sigma}_{\hat{\psi}}, \tag{3.24}$$

where S is given by (3.23). These intervals are called S intervals. In words, the probability is $1-\alpha$ that simultaneously for all ψ in L the S intervals cover ψ. The intersection of the S intervals for all ψ in L is the confidence region (2.6). This is an ellipsoidal region in θ space with center at z.

An estimate $\hat{\psi}$ of ψ is said to be *sdfz* if the S interval does not cover $\psi=0$, i.e. if $|\hat{\psi}|>S\hat{\sigma}_{\hat{\psi}}$. Hence, H is rejected if and only if the estimate of at least one ψ in L is *sdfz* according to the S criterion.

The Scheffé test and the S intervals are based on the following result:

$$P\left[\max_a t^2(a)\le S^2\right]=1-\alpha, \tag{3.25}$$

where $t^2(a)$ is the squared t ratio:

$$t^2(a)=\frac{(\hat{\psi}-\psi)^2}{\hat{\sigma}^2_{\hat{\psi}}}=\frac{\left[a'(z-\theta)\right]^2}{s^2a'Va}, \tag{3.26}$$

and where S is given by (3.23). The result is proved in Scheffé (1959, pp. 69–70). I will now give a simple proof.

Observe that the result is proved by showing that the maximum squared t ratio is distributed as $qF(q,T-k)$. There is no loss in generality in maximizing $t^2(a)$ subject to the normalization $a'Va=1$ since $t^2(a)$ is not affected by a change of scale of the elements of a. Form the Lagrangian:

$$L(a,\lambda)=\left[a'(z-\theta)/s\right]^2-\lambda(a'Va-1), \tag{3.27}$$

where λ is the Lagrange multiplier. Setting the derivative of $L(a,\lambda)$ with respect to a equal to zero gives:

$$\left[(z-\theta)(z-\theta)'-\lambda s^2V\right]a=0. \tag{3.28}$$

Premultiplying (3.28) by a' and dividing by $s^2a'Va$ shows that $\lambda=t^2(a)$. Hence, the determinantal equation:

$$\left|(s^2V)^{-1}(z-\theta)(z-\theta)'-\lambda I\right|=0, \tag{3.29}$$

is solved for the greatest characteristic root λ^*. Since (3.29) has only one non-zero root—the matrix $(z-\theta)(z-\theta)'$ has rank one—the greatest root is:

$$\lambda^*=\operatorname{trace}(s^2V)^{-1}(z-\theta)(z-\theta)'=(z-\theta)'(s^2V)^{-1}(z-\theta), \tag{3.30}$$

which is distributed as $qF(q,T-k)$. The solutions to (3.28), i.e. the characteristic vectors corresponding to λ^*, are proportional to $(s^2V)^{-1}(z-\theta)$ and the characteristic vector satisfying the normalization $a'Va=1$ is $a^*=V^{-1}(z-\theta)/\sqrt{s^2\lambda^*}$.

The Scheffé induced test accepts H if and only if:

$$\max_a t_0^2(a) \le S^2, \tag{3.31}$$

where $t_0^2(a)$ is $t^2(a)$ with $\theta=0$. It follows from (3.30) that:

$$t_0^2(a_0^*) = z'(s^2V)^{-1}z, \tag{3.32}$$

where a_0^* is the vector which maximizes $t_0^2(a)$. Since this t ratio is distributed as $qF(q,T-k)$ when H is true, the α level Scheffé test accepts H if and only if the α level F test accepts H.

When the F test rejects H we want to find which $\hat{\psi}$ are *sdfz*. Since a_0^* can be calculated from (3.30) we can always find at least one $\hat{\psi}$ which is *sdfz*, namely $\hat{\psi}_0=a_0^{*\prime}z$. Unfortunately, computer programs for regression analysis calculate the F statistic, but do not calculate a_0^*.

When the hypothesis H is that all the slope coefficients are zero the components of the a_0^* vector have a simple statistical interpretation. Suppose that the first column of X is a column of ones and let D be the $T\times(k-1)$ matrix of deviations of the regressors (excluding unity) from their means. Since z is simply the least squares estimator of the slope coefficients, $z=(D'D)^{-1}D'y$. Hence $a_0^*=(D'D)z(s^2qF)^{-1/2}=D'y(s^2qF)^{-1/2}$ so that the components of a_0^* are proportional to the sample covariances between the dependent variable and the regressors. If the columns of D are orthogonal, then the components of a_0^* are proportional to the least squares estimates of the slope coefficients, i.e. z. Thus, in the orthogonal case $\hat{\psi}_0$ is proportional to the sum of the squares of the estimates of the slope coefficients.

For an example of the Scheffé test I again turn to the case where $q=2$ and $\sigma^2V=I$ which is known. When $\alpha=0.05$ the test of the separate hypothesis $H(a)$ is defined by the acceptance region:

$$|t_0(a)| = |a'z| \le 2.448, \tag{3.33}$$

where $a'Va=a'a=1$. Thus each separate hypothesis $H(a)$ is tested at the 0.014 level to achieve a 0.05 level separate induced test of H. Geometrically the acceptance region (3.33) is a strip in the z_1 and z_2 space between two parallel lines orthogonal to the vector a, the origin being midway between the lines. The acceptance region or strip for testing the separate hypothesis $H(a)$ is shown in Figure 3.2. The intersection of the separate acceptance regions or strips for all

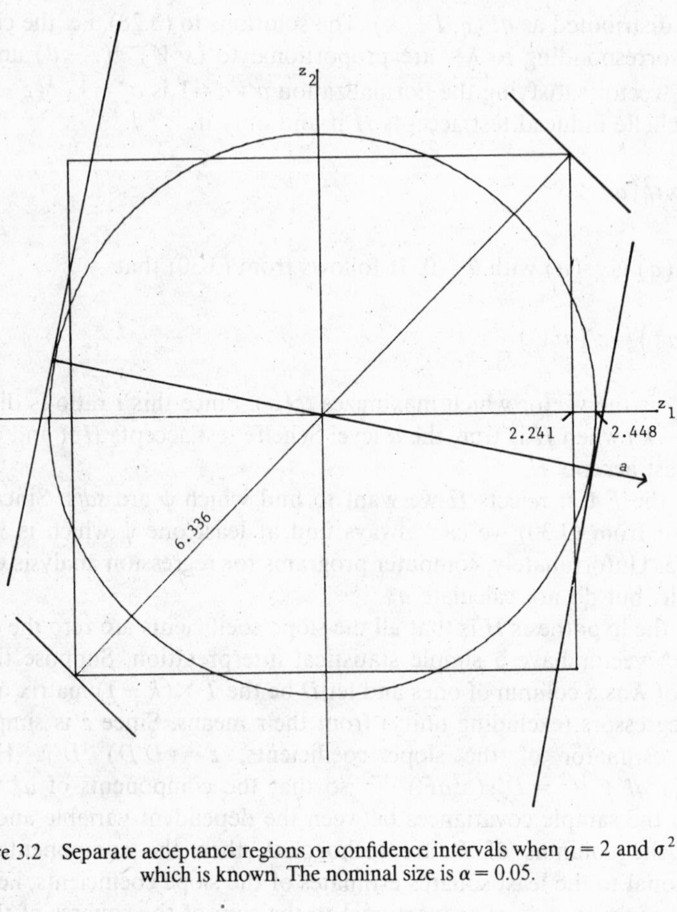

Figure 3.2 Separate acceptance regions or confidence intervals when $q = 2$ and $\sigma^2 V = I$ which is known. The nominal size is $\alpha = 0.05$.

non-null a is the circular region in Figure 3.2. Recall that this circular region is the acceptance region of a 0.05 level χ^2 test of H, i.e. the region shown in Figure 2.1. The square region in Figure 3.2 is the acceptance region of a 0.05 level Bonferroni separate induced test of H when the only ψ in L of primary interest are $\psi_1 = \theta_1$ and $\psi_2 = \theta_2$. As noted earlier these two acceptance regions can produce conflicting inferences and hence the same is true for the Bonferroni and Scheffé separate induced tests of H.

The S interval for $\psi = a'\theta$ is defined by the confidence region:

$$|a'(\theta - z)| \leq 2.448, \tag{3.34}$$

which says that the point θ lies in a strip of θ_1 and θ_2 space between two parallel lines orthogonal to the vector a, the point (z_1, z_2) being midway between the lines. The intersection of the S intervals for all ψ in L is the circular region in Figure 3.2 when it is centered at (z_1, z_2) in the θ_1 and θ_2 space. The S procedure accepts H if and only if all the S intervals cover zero, i.e. if and only if the circular region in Figure 3.2 (interpreted as a 95% confidence region) covers zero.

3.3.2. *An extension*

When the F test rejects, one or more t ratios for individual parameters may be large enough to explain this rejection. As an extension of this result we want to look at F statistics for subsets of the linear restrictions specified by H. If any of these are sufficiently large then we would have found subsets of the restrictions responsible for rejection. To carry out this extension of the S procedure we now present a result due to Gabriel (1964, 1969).

Consider testing the hypothesis:

$$H_*: C_*\beta - c_* = \theta_* = 0, \tag{3.35}$$

where $[C_* c_*]$ consists of any q_* rows of $[Cc]$ defined in (2.2). Let F_* be the F statistic for testing H_* and let $t_0^2(a_*)$ be the squared t ratio for testing:

$$H_*(a_*): a'_*(C_*\beta - c_*) = a'_*\theta_* = 0, \tag{3.36}$$

where a_* is $q_* \times 1$. With no loss of generality we may let $[C_* c_*]$ consist of the last q_* rows of $[Cc]$. From the development (3.23) to (3.26) we find that

$$\max_{a_*} t_0^2(a_*) = q_* F_* = \max_{a \in I} t_0^2(a), \tag{3.37}$$

where I is the set of all non-null a vectors such that the first $q - q_*$ elements are zero. Hence:

$$q_* F_* \leq qF, \tag{3.38}$$

since the constrained maximum of $t_0^2(a)$ is less than or equal to the unconstrained maximum. This establishes that when H is true the probability is $1-\alpha$ that the inequality,

$$q_* F_* \leq qF_\alpha(q, T-k) = S^2, \tag{3.39}$$

is simultaneously satisfied for all hypotheses H_* defined in (3.35) where F_* is the F statistic for testing H_*.

The implication is that using acceptance region (3.39) we can test any number of multivariate hypotheses H_* with the assurance that all will be simultaneously accepted with probability $1-\alpha$ when the hypothesis H is true. The hypotheses H_* may be suggested by the data. When we begin the procedure with an α level F test of H, this is a special case of H_* when $q_* = q$. For further discussion see Scheffé (1977a).

3.3.3. Conditional probability of coverage

The S intervals are usually not calculated when the F test accepts H since none is considered interesting. In light of this practice Olshen (1973, 1977) has argued that we should consider the conditional probability that the S intervals cover the true values given rejection of H. Olshen (1973) has proved that:

$$P\left[(b-\beta)'X'X(b-\beta) < s^2S^2 | b'X'X'b > s^2S^2\right] < 1-\alpha, \tag{3.40}$$

for all β and σ^2 provided $S^2 > 3(T-k)$ and $(T-k) > 2$. This means that under certain mild conditions the conditional probability of coverage is always less than the unconditional probability. Monte Carlo studies show that the conditional probability can be substantially less than the unconditional probability.

A simple example will serve to illustrate the difference between the conditional and unconditional probability of coverage. Let x be an observation from $N(\mu, 1)$. The probability that the nominal 95% confidence interval for μ covers μ given rejection of the hypothesis $\mu = 0$ by a 0.05 level standard normal test is $P(|x-\mu| < 1.96 \,|\, |x| > 1.96)$. For $\mu = 1$ we have $P(|x| > 1.96) = 0.1700$ and $P(|x-\mu| < 1.96,$ $|x| > 1.96) = 0.1435$, so that the conditional probability of coverage is $0.1435/0.1700 = 0.8441$. For $\mu = 4$ the conditional probability is $0.95/0.9793 = 0.9701$. In this example the conditional probability is < 0.95 when $\mu < 3.92$ and > 0.95 when $\mu > 3.92$.

In general the S procedure is not satisfactory if one wants to control the conditional probability of coverage since there is no guarantee that the conditional probability is greater than or equal to the unconditional probability, the latter being the only probability subject to control with the S procedure. Olshen's theorem shows that the unconditional probability can be a very misleading guide to the conditional probability. The S intervals are often criticized for being too wide, but they are two narrow if we want to make the conditional probability at least as great as the unconditional. Thus, if like Olshen we are interested in controlling the conditional probability, then we would want to replace the S procedure with one which controls this probability; see Olshen (1973) for a discussion of some developments along these lines.

Suppose we decide before analyzing the data that we have a multiple decision problem. Then the unconditional probability of coverage is of interest. In this situation the F test is simply the first step in the S procedure. If the F test accepts

H it is not customary to calculate the S intervals since it is known that they all cover zero and if the F test rejects we do not actually calculate all the S intervals since this is not feasible. On the other hand, suppose we do not decide before conducting the F test that we have a multiple decision problem, but decide after the F test rejects that we have such a problem. In this case the conditional probability of coverage is relevant. Of course, we may be interested in both the conditional and the unconditional probabilities. In this paper it has been assumed that we decided to treat the testing problem as a multiple decision problem prior to looking at the data, i.e. that the unconditional probabilities are the focus of attention.

3.4. *Finite induced test — ψ of secondary interest*

Suppose after inspecting the data we wish to make inferences about linear combinations of secondary interest. I now discuss how the finite induced test can be generalized so that inferences can be made about all ψ in L. For this purpose I adopt the general approach of Scheffé (1959, pp. 81–83). Following Scheffé the discussion is in terms of simultaneous confidence intervals.

Let G be a set of ψ in L of primary interest and suppose we have a multiple comparison procedure which gives for each ψ in G an interval:

$$\hat{\psi} - h_{\psi}s \leq \psi \leq \hat{\psi} + h_{\psi}s, \tag{3.41}$$

where h_{ψ} is a constant depending on the vector a but not the unknown θ. The inequality (3.41), which may be written

$$|a'(\theta - z)| \leq h_{\psi}s, \tag{3.42}$$

can be interpreted geometrically to mean that the point θ lies in a strip of the q-dimensional space between two parallel planes orthogonal to the vector a, the point z being midway between the planes. The intersection of these strips for all ψ in G determines a certain convex set C and (3.41) holds for all ψ in G if and only if the point θ lies in C. Thus, the problem of simultaneous confidence interval construction can be approached by starting with a convex set C instead of a set G of ψ in L. For any convex set C we can derive simultaneous confidence intervals for the infinite set of all ψ in L by starting with the relation that the point θ lies in C if and only if it lies between every pair of parallel supporting planes of C.

Let L^* be the set of ψ in L for which $a'Va=1$ and G^* be a set of m linear combinations ψ in L^* of primary interest. This normalization is convenient since the M intervals for all ψ in G^* have length $2Ms$ and the S intervals for all ψ in L^* have length $2Ss$. We now define the confidence set C of the M procedure to be the intersection of the M intervals for all ψ in G^* and the set C of the S procedure to be the intersection of the S intervals for all ψ in L^*. In the M procedure C is a polyhedron and in the S procedure C is the confidence ellipsoid defined by (2.6). When $q=2$ the region C is a polygonal region in the B procedure and an elliptical region in the S procedure. In addition, if $m=2$ and if $\sigma^2V=1$, then C is a square region in the M and B procedures and a circular region in the S procedure, as depicted in Figure 2.1.

Consider the case where the confidence region C is a square with sides $2Ms$. Starting with a square we can derive simultaneous confidence intervals for all ψ in L^*, not just for θ_1 and θ_2. The square has four extreme points which are the four corner points. There are only two pairs of parallel lines of support where each supporting line contains two extreme points. These two pairs of lines define the M intervals for the ψ of primary interest, i.e. θ_1 and θ_2, respectively, and contain all the boundary points of the square. In addition to these two pairs of parallel lines of support, there are an infinite number of pairs of parallel lines of support where each line contains only one extreme point. One such pair is shown in Figure 3.2. This pair defines a simultaneous confidence interval for some ψ of secondary interest. We can derive a simultaneous confidence interval for every ψ of secondary interest by taking into account pairs of supporting lines where each line contains only one extreme point.

A general method for calculating simultaneous confidence intervals is given by Richmond (1982). This method can be used to calculate M intervals for linear combinations of secondary interest. I briefly review this method and present two examples for the case of B intervals.

Let G be a set of a finite number m of linear combinations of primary interest and as before denote the linear combinations in G by $\psi_i = a_i'\theta$, $i=1,2,\dots,m$. Any linear combination in L can be written as $\psi = c_1\psi_1 + c_2\psi_2 + \cdots + c_m\psi_m$, where $c=(c_1,\dots,c_m)'$, i.e. any ψ in L is a linear combination of the ψ in G. The method is based on the following result. The probability is $1-\alpha$ that simultaneously for all ψ in L:

$$\hat{\psi} - M\sum_{i=1}^{m}\sqrt{s^2a_i'Va_i}\,|c_i| \le \psi \le \hat{\psi} + M\sum_{i=1}^{m}\sqrt{s^2a_i'Va_i}\,|c_i|. \tag{3.43}$$

I also call these intervals M intervals. When $c=(0,\dots,0,1,0,\dots,0)'$, the 1 occurring in the ith place, the M interval is for ψ_i, a ψ of primary interest.

This result is a special case of Theorem 2 in Richmond (1982). The result (3.43) is proved by showing that (3.43) is true if and only if:

$$|\hat{\psi}_i - \psi_i| \leq M\sqrt{s^2 a_i' V a_i'}\,, \qquad i=1,\dots,m. \tag{3.44}$$

I will give a sketch of the proof. Suppose (3.44) holds. Multiply both sides of (3.44) by $|c_i|$ and sum over $i=1,\dots,m$. Then:

$$\left|\sum_{i=1}^{m} c_i(\hat{\psi}_i - \psi_i)\right| \leq \sum_{i=1}^{m} |\hat{\psi}_i - \psi_i||c_i| \leq \sum_{i=1}^{m} M\sqrt{s^2 a_i' V a_i}\,|c_i|, \tag{3.45}$$

which is equivalent to (3.43). Conversely, suppose (3.43) holds for all ψ in L. Take $c_i = 1$ and $c_j = 0$, $j=1,\dots,m$, $j \neq i$. Then

$$|\hat{\psi}_i - \psi_i| \leq M\sqrt{s^2 a_i' V a_i}\,, \qquad i=1,\dots,q, \tag{3.46}$$

which completes the proof.

For both examples I assume that $q=2$ and $\sigma^2 V = 1$ which is known. In the first example suppose the $m=2$ linear combinations in G are $\psi_1 = \theta_1$ and $\psi_2 = \theta_2$. Consider the B interval for $\psi = \sqrt{1/2}\,(\psi_1 + \psi_2) = \sqrt{1/2}\,(\theta_1 + \theta_2)$. When $\delta = 0.05/2$ the Bonferroni critical value is $B = 2.24$, so that the length of the B interval is $2(c_1 + c_2)B = 2(2)\sqrt{1/2}\,(2.24) = 6.336$. This is the case shown in Figure 3.2 when the square region is centered at (z_1, z_2) in the θ_1 and θ_2 space, i.e. when the square region is interpreted as a nominal 95% confidence region. In the second example suppose $m=3$ and ψ is of primary interest. When $\delta = 0.05/3$ the Bonferroni critical value is $B = 2.39$ so that the length of the B interval for ψ is $2(2.39) = 4.78$, which is considerably less than when ψ is of secondary interest. This shows that the length of a B interval for a ψ in L can vary considerably depending on whether ψ is of primary or secondary interest. In particular, the length of a B interval for a ψ depends critically on the values of the c_i's.

3.5. *Simultaneous confidence intervals*

In this section I compare the lengths of the finite induced intervals and the S intervals. The lengths are compared for the linear combinations of primary interest and secondary interest. In many cases the B intervals are shorter for the ψ of primary interest. On the other hand, the S intervals are always shorter for at least some ψ of secondary interest.

3.5.1. ψ of primary interest

Consider the set G of linear combinations of primary interest in the finite induced test. The ratio of the length of the M intervals to the length of the S intervals for ψ in G is simply the ratio of M to S. For fixed q the values M and S satisfy the relation:

$$P\left[\max_{a \in I} |t_0(a)| \leq M|H\right] = P\left[\max_{a} |t_0(a)| \leq S|H\right], \tag{3.47}$$

where I is a set of m vectors. Since the restricted maximum is equal to or less than the unrestricted, it follows that $M \leq S$. Hence, the M intervals are shorter than the S intervals for all q and m $(m \geq q)$.

The B intervals can be longer than the S intervals for all ψ in G. Suppose G is fixed. Then S is fixed and from the Bonferroni inequality (2.13) we see that B increases without limit as m increases. Hence, for sufficiently large m the B intervals are longer than the S intervals for all ψ in G. On the other hand, numerical computations show that for sufficiently small m the B intervals are shorter than the S intervals for all ψ in G. The above also holds for intervals based on the Sidák or the studentized maximum modulus inequality. Games (1977) has calculated the maximum number of ψ of primary interest (the number m) such that the intervals based on the Sidák inequality are shorter than the S intervals for all ψ in G.

The effect of varying m (the number of ψ of primary interest) is illustrated by the following examples. Suppose $q = 2$ and $\sigma^2 V = 1$ which is known. If G consists of $m = 4$ linear combinations and if nominally $\alpha = 0.05$, then applying the Bonferroni inequality gives $B = 2.50$. Since $S = 2.448$ the S intervals are shorter than the B intervals for all ψ in G; the ratio of B to S is 1.02. The ratio of the length of the exact finite induced intervals to the S intervals when $m = 2$ and $\alpha = 0.05$ is 0.913 since $M = 2.236$. If instead of calculating the exact 95% finite induced confidence region we use the Bonferroni inequality, then $B = 2.241$ which is also less than S. See Figures 4 and 5 in Miller (1966, pp. 15–16).

In the case where $m = q$ and $\alpha = 0.05$ calculations by Christensen (1973) show that the B intervals are shorter than the S intervals regardless of the size of q. Similar results are reported by Morrison (1976, p. 136) for 95% Bonferroni and Roy–Bose simultaneous confidence intervals on means. The Roy–Bose simultaneous confidence intervals are the same as S intervals in the case of the classical linear normal regression model.

Investigators object to the length of the S intervals. When the Scheffé test rejects, the linear combinations responsible for rejection may be of no economic interest. This may account for the fact that the Scheffé test and the S intervals are not widely used. In theory the solution is to use a procedure where the set G is suitably restricted. In practice it is difficult to construct such a procedure. One approach is to use a finite induced test. The drawback is that to be operational we have to apply approximations based on probability inequalities. As already noted, when m is large relative to q the B intervals are longer than the S intervals and similar results hold for intervals based on the Sidák or studentized maximum modulus inequality. Another approach is to construct an infinite induced test where G is a proper subset of L. No procedure analogous to the S procedure has been developed for this case. It seems that there is no very satisfactory alternative to the S intervals when m is sufficiently large.

3.5.2. ψ *of secondary interest*

When the B intervals are shorter for the ψ of primary interest and the S intervals are shorter for some ψ of secondary interest there is a trade-off between the B procedure and the S procedure. It is instructive to compare the length of the simultaneous confidence intervals derived from the square region with sides $2B = 4.482$ with the intervals derived from the circular region with diameter $2S = 4.895$. The B procedure is the procedure which gives for each ψ in L^* an interval derived from the square region. The B intervals for ψ in L^* include the B intervals for θ_1 and θ_2, which are the ψ of primary interest. The length of the shortest B interval is equal to the length of the side of the square region and the length of the longest B interval is equal to the length of the diagonal which is 6.336. Since the length of the S intervals for all ψ in L^* is 4.895 the S intervals are shorter than the B intervals for some ψ in L^*; in particular, the S interval is shorter for $\psi = \sqrt{1/2}(\theta_1 + \theta_2)$, the B interval for this ψ being the one shown in Figure 3.2.

When G is finite there are a few cases in the one-way lay-out of the analysis of variance where the exact significance level of the induced test of H can be easily calculated. In these cases it is also easy to calculate the probability that simultaneously for all ψ in L the confidence intervals cover the true values. These cases include the generalized Tukey procedure [see Scheffé (1959, theorem 2, p. 74)] where the ψ of primary interest are the pairwise comparisons $(\theta_i - \theta_j)$, $i, j = 1,\dots,q$, $i \neq j$, and the "extended Dunnett procedure" developed by Schaffer (1977), where the ψ of primary interest are the differences $(\theta_1 - \theta_i)$, $i = 2,\dots,q$. Schaffer (1977) found that the Tukey intervals are shorter than the S intervals for the ψ of primary interest in the generalized Tukey procedure and likewise that the Dunnett intervals are shorter than the S intervals for the ψ of primary interest in the extended Dunnett procedure. On the other hand, the S procedure generally gives shorter intervals for the ψ of secondary interest.

Richmond (1982) obtained similar results when extending the Schaffer study to include the case where the ψ of primary interest are taken to be the same as in the extended Dunnett procedure and the intervals are calculated by applying the

Sidák inequality. For further comparisons between Tukey and S intervals see Scheffé (1959, pp. 75–77) and Hochberg and Rodriquez (1977).

4. The power of the Bonferroni and Scheffé tests

4.1. Background

Since the power of the Scheffé test is the same as the power of the F test, it is uniformly most powerful in certain situations. However, it is not uniformly more powerful than the Bonferroni test. An attractive feature of the Bonferroni test is that when it rejects, the linear combinations responsible for rejection are of economic interest. This feature has to be weighed against the power of the test, i.e. the probability that the test rejects H when H is false.

Christensen (1973) and Evans and Savin (1980) have compared the power of the χ^2 Bonferroni tests for the case where $q = 2$, σ^2 is known and V is defined as in (2.17). The acceptance regions of both of these tests have been discussed in Section 2.4. In this Section I review the power of the F test and the results of the Christensen study.

The power of the F test is a function of four parameters: the level of significance α, the numerator and denominator degrees of freedom q and $T - k$, and the noncentrality parameter λ which is given by:

$$\lambda = \theta' V^{-1} \theta / \sigma^2, \tag{4.1}$$

when θ is the true parameter vector. The power of the F test depends on θ and $\sigma^2 V$ only through this single parameter. Therefore it has been feasible to table the power of the F test; for selected cases it can be found from the Pearson and Hartley (1972) charts or the Fox (1956) charts. In addition, the power can be calculated for cases of interest using the procedures due to Imhof (1961) and Tiku (1965). By contrast, little is known about the power of the Bonferroni test and it has proved impracticable to construct tables of the power of the test.

Christensen studied the powers of the 0.05 level χ^2 test and the nominal 0.05 level Bonferroni test along rays in the parameter space. Power calculations by Chrsitensen show that neither test is more powerful against all alternatives. For example, when $r = 0$ the Bonferroni test is more powerful against the alternative $\theta_1 = \theta_2 = 1.585\sigma$. This is not surprising since neither of the acceptance regions contain the other. Despite this, Christensen found that when the absolute value of r was small the power of the two tests was approximately the same regardless of the alternative. However, when the absolute value of r was high the Bonferroni test had very little power against any alternatives considered by Christensen. If only θ_1 or θ_2 is different from zero then the χ^2 test has good power regardless of the value of r. When both θ_1 and θ_2 are different from zero the power of the χ^2 test is mixed. Against some alternatives the power is extremely good—increasing with the absolute value of r. On the other hand, the power against other alternatives decreases badly with increasing absolute value of r. One of the potential explanations for the power of the Bonferroni test is that the actual level of significance of the Bonferroni box decreases as the absolute value of r increases. As noted earlier, for $r = 0$ the actual level is 0.0494 and as the absolute value of r approaches one the actual level approaches 0.025.

4.2. *Power contours*

The behavior of the power function is described by its contours in the parameter space. A power contour is the set of all parameter points θ at which the power is constant. The power contours of the F test can be obtained from the expression for the noncentrality parameter (4.1). This is because the power of the F test is the same at parameter points θ with a given value of the noncentrality parameter. The power of the F test is constant on the surfaces of ellipsoids in the θ space, but the general properties of the power contours of the Bonferroni test are unknown.

Evans and Savin calculate the power contours of the 0.05 level χ^2 test and nominal 0.05 level Bonferroni test in the $(\theta_1\sqrt{1-r^2})/\sigma$ and $(\theta_2\sqrt{1-r^2})/\sigma$ parameter space. The power contours for correlations $r = 0.0, 0.9, 0.99$ at power levels 0.90, 0.95, 0.99 are shown in Figure 4.1(a–c). When $r = 0.0$ [Figure 4.1(a)] the power contours of the χ^2 test are circles with center at the origin while the contours of the Bonferroni test are nearly circular. At a given power level the χ^2 and the Bonferroni power contours are close together. Thus, both tests have similar powers which confirms the results of Christensen. We also see that the contours for a given power level cross so that neither test is uniformly more powerful.

When the correlation is $r = 0.90$ [Figure 4.1(b)] the power contours of the Bonferroni test are not much changed whereas those of the χ^2 test have become narrow ellipses. Hence for a given power level the contours of the two tests are no longer close together. The χ^2 test is more powerful at parameter points in the upper right hand and lower left hand parts of the space and the Bonferroni test at points in the extreme upper left-hand and lower right-hand corners of the space. For $r = 0.99$ [Figure 4.1(c)] we see that the power contours of the Bonferroni test continue to remain much fatter than those of the χ^2 test even when the power is quite close to one. In short, when the correlation r is different from zero the χ^2 test has higher power than the Bonferroni test at most alternatives.

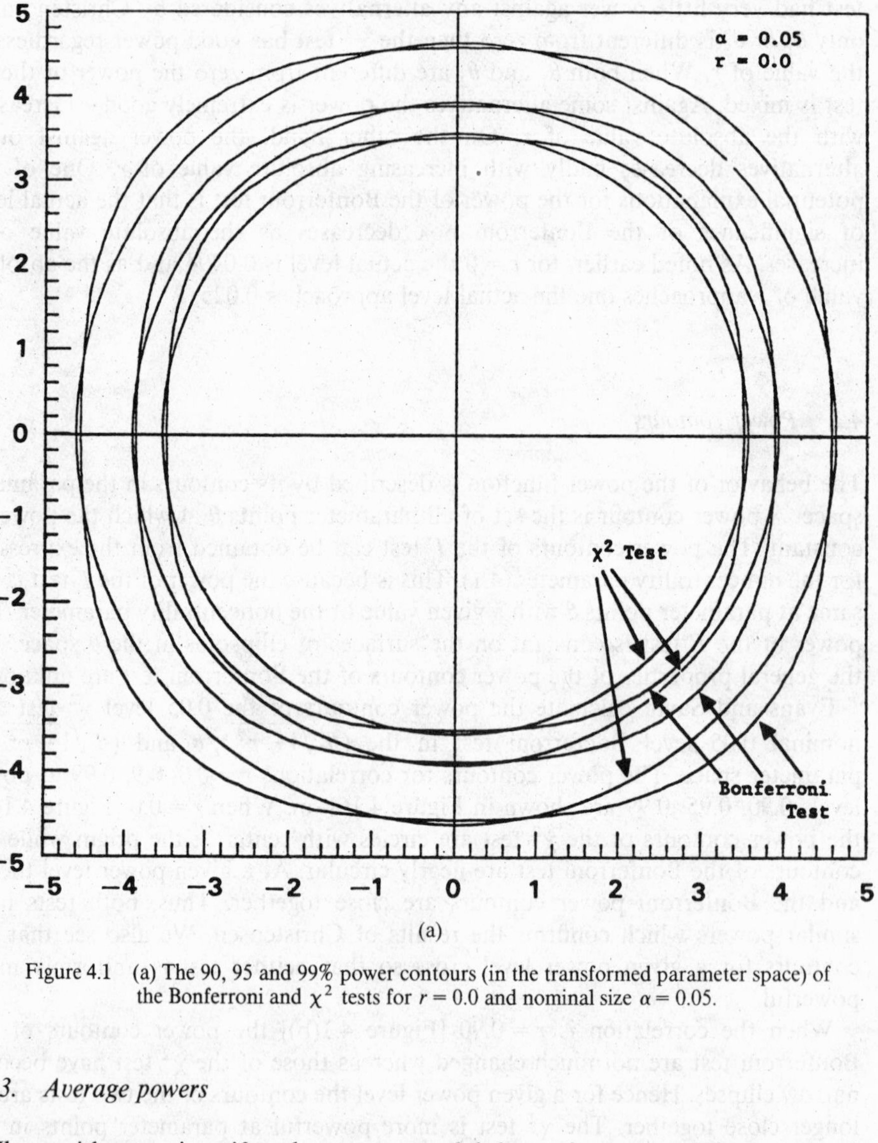

(a)

Figure 4.1 (a) The 90, 95 and 99% power contours (in the transformed parameter space) of the Bonferroni and χ^2 tests for $r = 0.0$ and nominal size $\alpha = 0.05$.

4.3. *Average powers*

When neither test is uniformly more powerful the performance of the tests can be compared on the basis of average power. Since V is a positive definite matrix there exists a nonsingular matrix P such that $P'VP = I$ and $\theta^* = P^{-1}\theta$. Then the noncentrality parameter can be written as:

$$\lambda = \theta^{*\prime}\theta^*/\sigma^2. \tag{4.2}$$

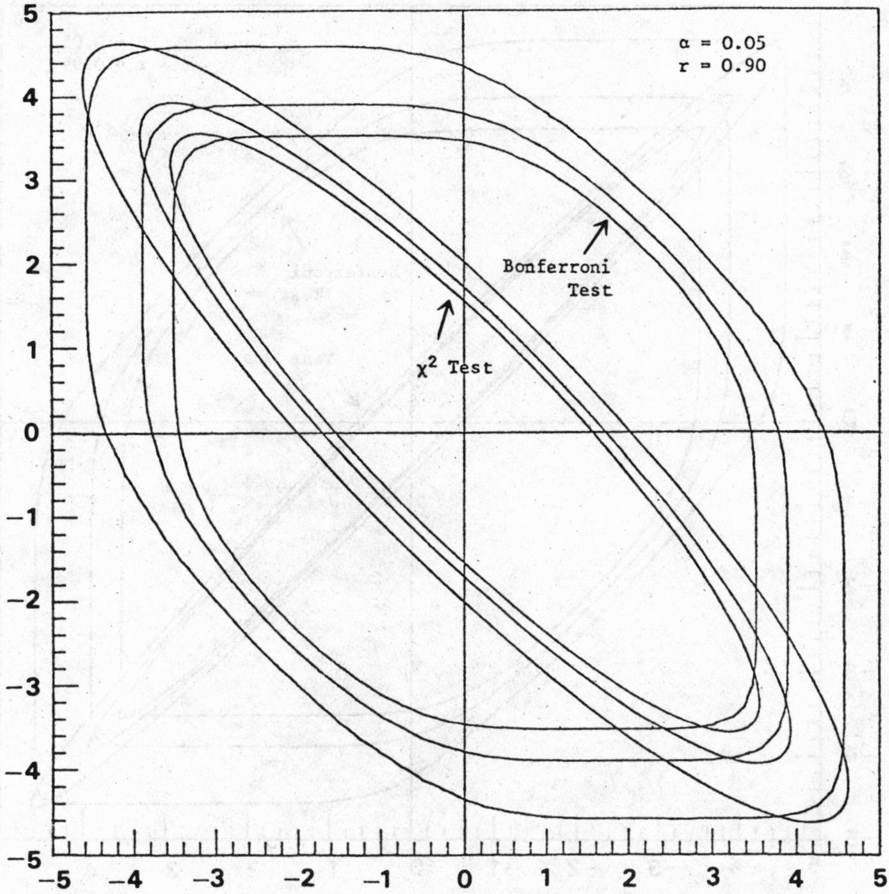

(b)

Figure 4.1 (b) The 90, 95 and 99% power contours (in the transformed parameter space) of the Bonferroni and χ^2 tests for $r = 0.90$ and nominal size $\alpha = 0.05$.

Thus, the power of the F test is constant on the surface of spheres with center at the origin in the θ^* space. In other words, in the transformed space the power of the F test is the same at all alternatives which are the same distance from the null hypothesis (the origin). The F test maximizes the average power on every sphere in the transformed space where the average power is defined with respect to a uniform measure over spheres in this space; see Scheffé (1959, pp. 47–49). Hence the F test is best when we have the same interest in all alternatives which are the same distance from the null in the transformed parameter space.

It may be more natural to suppose that we have an equal interest in all alternatives which are equally distant from the null in the θ parameter space. On this assumption the best test is the one which maximizes the average power on

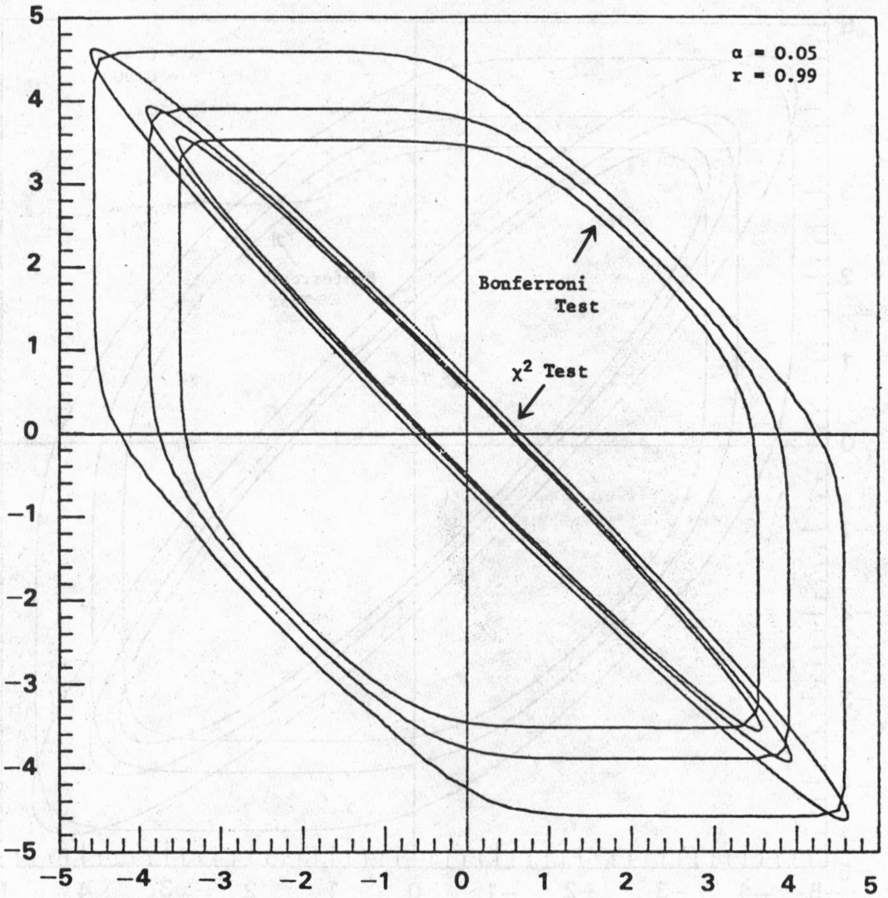

(c)

Figure 4.1 (c) The 90, 95 and 99% power contours (in the transformed parameter space) of the Bonferroni and χ^2 tests for $r = 0.99$ and nominal size $\alpha = 0.05$.

every sphere in the θ parameter space. Evans and Savin (1980) define the average power with respect to a uniform measure over the sphere in the θ space. Using this definition Evans and Savin calculate the average power of an α level χ^2 test, a nominal α level Bonferroni test and an exact α level finite induced test. The results are reported in Table 4.1 for selected values of the radius R of the circle, the correlation r and significance levels α.

When $r = 0$ the average power of both tests is very similar. This is because both tests have very similar power contours in this case, namely circles for the χ^2 test and nearly circles for the Bonferroni test. On the other hand, when r is near one and the radius R of the circle is small the average power of the χ^2 test is markedly

Table 4.1

Average Powers of the Bonferroni(B), Chi-Square(CS) and Exact Finite Induced(E) Tests.

		α=0.10			α=0.05			α=0.01		
r	R	B	CS	E	B	CS	E	B	CS	E
0.0	0.0	0.0975	0.1000	0.1000	0.0494	0.0500	0.0500	0.0100	0.0100	0.0100
	0.5	0.1249	0.1290	0.1278	0.0676	0.0693	0.0683	0.0158	0.0162	0.0158
	1.0	0.2088	0.2177	0.2127	0.1271	0.1327	0.1283	0.0381	0.0404	0.0381
	1.5	0.3468	0.3626	0.3516	0.2365	0.2495	0.2382	0.0897	0.0974	0.0898
	2.0	0.5203	0.5423	0.5254	0.3933	0.4154	0.3954	0.1857	0.2039	0.1858
	2.5	0.6944	0.7182	0.6989	0.5740	0.6028	0.5761	0.3306	0.3634	0.3308
	3.0	0.8347	0.8545	0.8379	0.7419	0.7707	0.7437	0.5081	0.5533	0.5084
	3.5	0.9252	0.9381	0.9270	0.8675	0.8899	0.8687	0.6845	0.7327	0.6847
	4.0	0.9721	0.9786	0.9729	0.9432	0.9567	0.9439	0.8265	0.8666	0.8267
	4.5	0.9915	0.9940	0.9918	0.9799	0.9862	0.9802	0.9194	0.9454	0.9195
	5.0	0.9979	0.9987	0.9980	0.9942	0.9965	0.9943	0.9687	0.9819	0.9687
0.5	0.0	0.0907	0.1000	0.1000	0.0465	0.0500	0.0500	0.0096	0.0100	0.0100
	0.5	0.1177	0.1388	0.1286	0.0642	0.0761	0.0686	0.0153	0.0186	0.0159
	1.0	0.2011	0.2566	0.2159	0.1225	0.1637	0.1293	0.0371	0.0551	0.0382
	1.5	0.3403	0.4377	0.3594	0.2310	0.3204	0.2412	0.0878	0.1447	0.0900
	2.0	0.5183	0.6324	0.5390	0.3890	0.5172	0.4019	0.1826	0.3004	0.1863
	2.5	0.6978	0.7890	0.7159	0.5737	0.6993	0.5870	0.3271	0.4939	0.3322
	3.0	0.8401	0.8903	0.8525	0.7458	0.8313	0.7567	0.5064	0.6732	0.5121
	3.5	0.9285	0.9474	0.9352	0.8722	0.9127	0.8792	0.6862	0.8062	0.6914
	4.0	0.9724	0.9769	0.9753	0.9455	0.9583	0.9490	0.8305	0.8929	0.8343
	4.5	0.9905	0.9909	0.9916	0.9797	0.9819	0.9812	0.9225	0.9453	0.9246
	5.0	0.9970	0.9968	0.9974	0.9931	0.9930	0.9937	0.9694	0.9746	0.9703
0.9	0.0	0.0704	0.1000	0.0995	0.0362	0.0500	0.0500	0.0076	0.0100	0.0100
	0.5	0.0998	0.2538	0.1365	0.0547	0.1640	0.0733	0.0133	0.0575	0.0170
	1.0	0.1855	0.5844	0.2393	0.1130	0.4869	0.1440	0.0344	0.3050	0.0425
	1.5	0.3262	0.7777	0.3978	0.2204	0.7188	0.2673	0.0838	0.5949	0.0997
	2.0	0.5104	0.8655	0.5915	0.3786	0.8250	0.4389	0.1768	0.7404	0.2031
	2.5	0.7055	0.9169	0.7765	0.5698	0.8869	0.6346	0.3198	0.8209	0.3560
	3.0	0.8564	0.9501	0.8962	0.7563	0.9281	0.8083	0.5002	0.8756	0.5417
	3.5	0.9353	0.9716	0.9533	0.8860	0.9563	0.9128	0.6882	0.9158	0.7273
	4.0	0.9714	0.9849	0.9798	0.9492	0.9751	0.9612	0.8429	0.9455	0.8687
	4.5	0.9881	0.9926	0.9918	0.9778	0.9868	0.9834	0.9305	0.9668	0.9421
	5.0	0.9954	0.9967	0.9970	0.9909	0.9936	0.9933	0.9693	0.9812	0.9745

higher than the average power of the Bonferroni test. This is because over a circle of a given radius the average power of the χ^2 test increases as r increases and the average power of the Bonferroni test is virtually constant for all r. As the radius R of the circle increases the average power of the Bonferroni test approaches that of the χ^2 test.

The average power of the exact finite induced test is similar to the average power of the Bonferroni test. For $\alpha = 0.05$ the maximum difference between the average power of the exact test and the Bonferroni test occurs at $r = 0.90$ for a circle of given radius. The average power of the exact test is about 0.065 (11.5%) higher than the average power of the Bonferroni test when the radius is $R = .25$ and 0.027 (3%) higher when the radius is $R = 3.5$. The corresponding figures are

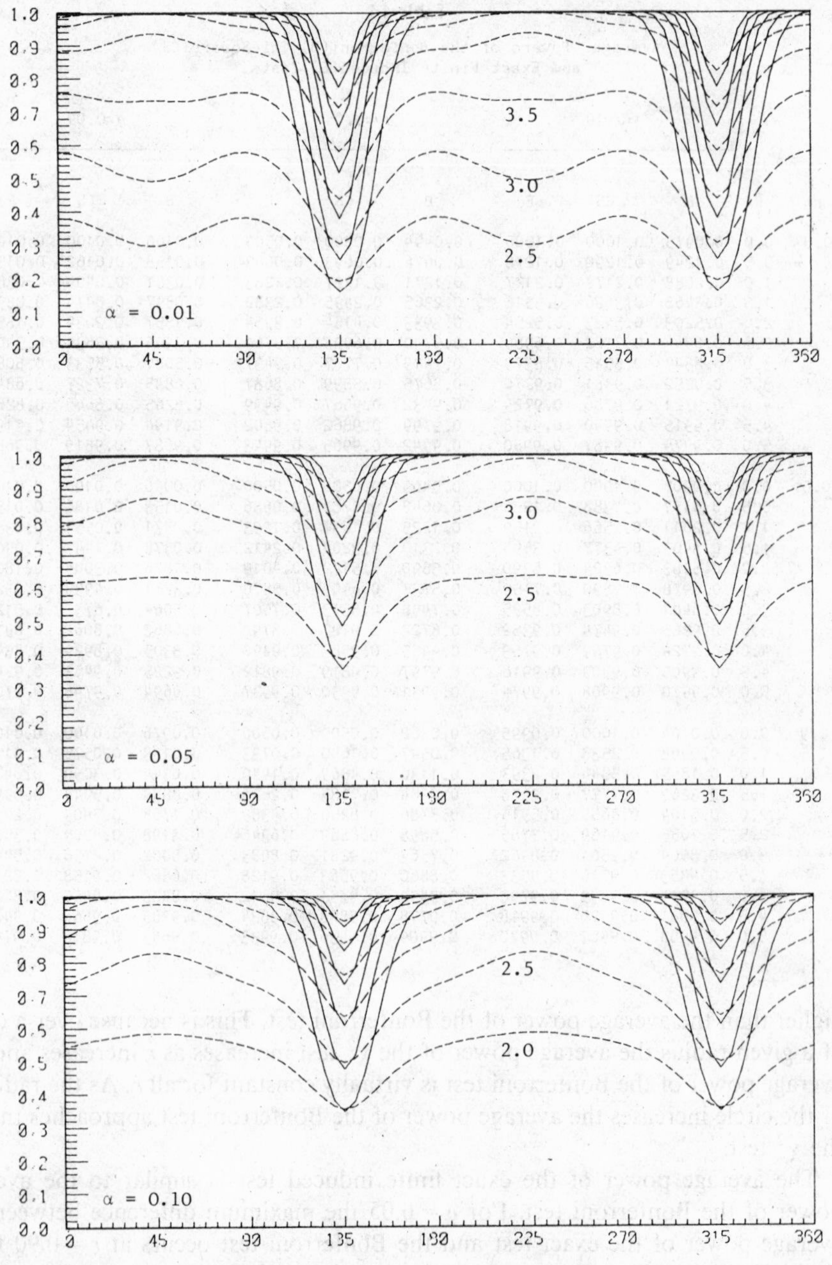

Figure 4.2 The power of the Bonferroni (broken lines) and the χ^2 (full lines) tests at radii $R = 2(0.5)5$ as a function of the direction in degrees. The correlation is $r = 0.9$ and the nominal sizes are $\alpha = 0.10$, 0.05 and 0.01.

somewhat higher if $\alpha = 0.10$ and lower if $\alpha = 0.01$. As a consequence, when the correlation r is near one the exact test is also a poor competitor of the χ^2 test over smaller radius circles.

Evans and Savin have plotted the behavior of the power over the circle for an α level χ^2 test and a nominal α level Bonferroni test. The power over various circles is shown in Figure 4.2 for the case $r = 0.90$ and $\alpha = 0.10$, 0.05 and 0.01. The χ^2 test has excellent power at most points on each circle. The power dips sharply only in the neighborhood of 135 and 315 degrees. The Bonferroni test has better power than the χ^2 test only in the neighborhood of 135 and 315 degrees and even here the power of the Bonferroni test is only marginally better than that of the χ^2 test. The Bonferroni test has more uniform, but substantially lower power over the smaller radius circles. For larger radius circles the power of the Bonferroni test is higher and hence compares more favorably to the χ^2 test. The picture for the exact finite induced test is similar with slightly higher power than the Bonferroni test at all points on the circle.

When the finite induced intervals are shorter than the S intervals for the ψ of primary interest it is common practice to conclude that the finite induced procedure (test) is superior to the S procedure (Scheffé test), for example, see Stoline and Ury (1979). Of course, if the finite induced intervals are shorter for all ψ in L, then the finite induced test is uniformly more powerful. However, the S intervals are generally shorter for some ψ of secondary interest. When the S intervals are shorter for some ψ of secondary interest the Scheffé test may have higher average power. This is clearly demonstrated by the comparison of the average powers of the χ^2 test and the Bonferroni test for the case of $q = 2$ parameters. Hence, it is misleading to conclude that the finite induced test is superior because the finite induced intervals are shorter for the ψ of primary interest. To our knowledge there is no evidence that any of the well known competitors of the Scheffé test have higher average power.

4.4. *The problem of multicollinearity*

The problem of multicollinearity arises when the explanatory variables are correlated, i.e. the columns of the regressor matrix X are not orthogonal. In discussions of the collinearity problem the individual regression coefficients are taken to be the parameters of primary interest. This is a point of crucial importance. A full rank regressor matrix can always be transformed so as to eliminate multicollinearity, but the regression coefficients in the transformed problem may no longer be of primary interest.

Leamer (1979) provides an excellent discussion of the collinearity problem from a Bayesian point of view. He observes (pp. 71–72):

> ...that there is a special problem caused by collinearity. This is the problem of interpreting multi-dimensional evidence. Briefly, collinear data provide

relatively good information about linear combinations of coefficients. The interpretation problem is the problem of deciding how to allocate that information to individual coefficients. This depends on prior information. A solution to the interpretation problem thus involves formalizing and utilizing effectively all prior information. The weak-evidence problem however remains, even when the interpretation problem is solved. The solution to the weak-evidence problem is more and better data. Within the confines of the given data set there is nothing that can be done about weak-evidence.

The interpretation problem can be interpreted as a multiple decision problem where there are q separate hypotheses, each specifying that an individual regression coefficient is equal to zero. In classical inference the finite and infinite induced tests are two approaches to solving the interpretation problem. The finite induced test provides a guaranteed solution to the interpretation problem whereas the infinite induced test has a probability of less than one of providing a solution. Multicollinearity plays an important role because of its effect on the power of the tests. Consider the Christensen two parameter case where the null hypothesis is H: $\beta_1 = \beta_2 = 0$. The correlation $r = 0$ if the relevant two regressors are orthogonal. The Bonferroni and Scheffé tests have similar average power for orthogonal or nearly orthogonal data. As the correlation r increases the average power of the Bonferroni test decreases compared with that of the Scheffé test. This means that for multicollinear data the Bonferroni test solves the interpretation problem at a cost; the cost is lower average power than for the Scheffé test. Hence there is a trade-off between the probability of solving the interpretation problem and the power of the test. The advantage of orthogonal data is that we can always decide which individual regression coefficients are responsible for rejection at a very small sacrifice of average power.

What we want to know is the conditional probability that the Scheffé test solves the interpretation problem given that it has rejected the null hypothesis. The conditional probability that the Scheffé test rejects H_1: $\beta_1 = 0$ or H_2: $\beta_2 = 0$ or both given that it has rejected H is the probability that the point (t_1, t_2) is outside the χ^2 box divided by the probability that is outside the χ^2 ellipse. This conditional probability is calculated for significance levels $\alpha = 0.10, 0.05, 0.01$, correlations $r = 0.0, 0.5, 0.9$ and radii $R = 0.0\ (0.5)\ 4.50$. In the $(\beta_1\sqrt{1-r^2})/\sigma$ and $(\beta_2\sqrt{1-r^2})/\sigma$ parameter space a point can be described by the angle of a ray from the origin to the point and the distance of the point along this ray. Because of the symmetry of the problem the calculations were done for angles between 45 and 135 degrees inclusive. Selected results are reported in Tables 4.2 and 4.3.

The results in Table 4.2 show that on small radius circles the average conditional probability can decrease as the correlation r increases. For example, at $\alpha = 0.05$ and $R = 1.0$ the average conditional probability is 0.637 when $r = 0.0$ and only 0.234 when $r = 0.9$, the decrease being 63%. The decrease is 58.1% when

Table 4.2

Average Conditional Probabilities of rejecting $\beta_1=0$ or $\beta_2=0$ (or both) given that the Chi-Square Test rejects.

		$\alpha=0.10$			$\alpha=0.05$			$\alpha=0.01$		
r	R	ACP	AP	ACS	ACP	AP	ACS	ACP	AP	ACS
0.0	0.0	0.6274	0.0627	0.1000	0.5709	0.0285	0.0500	0.4806	0.0048	0.0100
	0.5	0.6503	0.0839	0.1290	0.5933	0.0411	0.0693	0.5005	0.0081	0.0162
	1.0	0.6966	0.1517	0.2177	0.6369	0.0845	0.1327	0.5364	0.0217	0.0404
	1.5	0.7484	0.2715	0.3626	0.6867	0.1713	0.2495	0.5778	0.0563	0.0974
	2.0	0.8029	0.4354	0.5423	0.7415	0.3081	0.4154	0.6259	0.1276	0.2039
	2.5	0.8576	0.6159	0.7182	0.7999	0.4822	0.6028	0.6811	0.2475	0.3634
	3.0	0.9079	0.7758	0.8545	0.8583	0.6615	0.7707	0.7427	0.4110	0.5533
	3.5	0.9483	0.8897	0.9381	0.9109	0.8106	0.8899	0.8080	0.5920	0.7327
	4.0	0.9757	0.9548	0.9786	0.9519	0.9107	0.9567	0.8711	0.7549	0.8666
	4.5	0.9907	0.9848	0.9940	0.9784	0.9649	0.9862	0.9248	0.8743	0.9454
	5.0	0.9971	0.9958	0.9987	0.9921	0.9886	0.9965	0.9630	0.9455	0.9819
0.5	0.0	0.5881	0.0588	0.1000	0.5422	0.0271	0.0500	0.4672	0.0047	0.0100
	0.5	0.5767	0.0795	0.1390	0.5240	0.0393	0.0761	0.4386	0.0079	0.0186
	1.0	0.5905	0.1461	0.2569	0.5313	0.0817	0.1637	0.4380	0.0212	0.0551
	1.5	0.6373	0.2656	0.4379	0.5700	0.1672	0.3204	0.4640	0.0552	0.1447
	2.0	0.7085	0.4319	0.6326	0.6324	0.3036	0.5172	0.5081	0.1256	0.3004
	2.5	0.7954	0.6171	0.7891	0.7158	0.4798	0.6993	0.5713	0.2445	0.4939
	3.0	0.8810	0.7807	0.8906	0.8107	0.6632	0.8313	0.6559	0.4082	0.6732
	3.5	0.9442	0.8942	0.9476	0.8964	0.8150	0.9127	0.7567	0.5916	0.8062
	4.0	0.9790	0.9566	0.9771	0.9544	0.9143	0.9583	0.8553	0.7576	0.8929
	4.5	0.9934	0.9844	0.9911	0.9835	0.9658	0.9819	0.9298	0.8780	0.9453
	5.0	0.9980	0.9948	0.9968	0.9948	0.9879	0.9930	0.9722	0.9476	0.9746
0.9	0.0	0.4568	0.0457	0.1000	0.4240	0.0212	0.0500	0.3734	0.0037	0.0100
	0.5	0.3074	0.0676	0.2538	0.2581	0.0337	0.1640	0.1905	0.0069	0.0575
	1.0	0.2920	0.1346	0.5844	0.2341	0.0755	0.4869	0.1636	0.0198	0.3050
	1.5	0.3708	0.2533	0.7777	0.2875	0.1594	0.7188	0.1839	0.0528	0.5949
	2.0	0.5104	0.4214	0.8655	0.3985	0.2943	0.8250	0.2414	0.1217	0.7404
	2.5	0.6801	0.6162	0.9169	0.5528	0.4722	0.8869	0.3438	0.2389	0.8209
	3.0	0.8367	0.7943	0.1750	0.7231	0.6651	0.9281	0.4883	0.4016	0.8756
	3.5	0.9318	0.9059	0.9716	0.8670	0.8287	0.9563	0.6544	0.5879	0.9158
	4.0	0.9724	0.9580	0.9849	0.9464	0.9233	0.9751	0.8108	0.7638	0.9455
	4.5	0.9891	0.9819	0.9926	0.9787	0.9660	0.9868	0.9202	0.8898	0.9668
	5.0	0.9960	0.9927	0.9967	0.9918	0.9855	0.9936	0.9696	0.9517	0.9812

ACP Average Conditional Probability of rejecting $\beta_1=0$ or $\beta_2=0$ (or both) given that the chi-square test rejects.

AP Average Probability of rejecting $\beta_1=0$ or $\beta_2=0$ (or both).

ACS Average probability that the Chi-Square test rejects.

$\alpha = 0.10$ and 69.4% when $\alpha = 0.01$. On large radius circles the average conditional probability increases as r increases from $r = 0$, eventually decreasing. Holding the average power of the Scheffé test constant the average conditional probability decreases as the correlation r increases. For instance, when $\alpha = 0.05$ and the average power is roughly 0.45 the average conditional probability falls from about 0.75 to 0.24 as r moves from $r = 0.0$ to $r = 0.9$. For higher power this fall is less

Table 4.3

Conditional Probability(CP) of rejecting $\beta_1=0$ or $\beta_2=0$ (or both), given that the Chi-Square(CS) Test rejects.

R=1.0 α=0.05

r	Angle	CS	CP
0.0	45	0.1327	0.6150
	60	0.1327	0.6260
	75	0.1327	0.6479
	90	0.1327	0.6589
	105	0.1327	0.6479
	120	0.1327	0.6260
	135	0.1327	0.6150
0.5	45	0.2255	0.3660
	60	0.2170	0.3852
	75	0.1939	0.4401
	90	0.1629	0.5220
	105	0.1327	0.6132
	120	0.1112	0.6868
	135	0.1036	0.7153
0.9	45	0.8154	0.1003
	60	0.7879	0.1048
	75	0.6957	0.1200
	90	0.5256	0.1544
	105	0.3114	0.2357
	120	0.1471	0.4269
	135	0.0917	0.6266

dramatic and for sufficiently high power it can reverse. The more detailed results in Table 4.3 show that high power at a given alternative does not insure high conditional probability at that alternative. When the correlation is fixed at $r = 0.9$ there is an inverse relation between the power and the conditional probability even on large radius circles, namely, the higher the power, the lower the conditional probability.

The Bonferroni test solves the interpretation problem whatever the power of the test. But the test is unsatisfactory when the power is low since in this case the test is likely to be misleading. This suggests that we may want to trade off some probability of solving the interpretation problem for some extra power. When the average power of the Bonferroni test is high the average power of the Scheffé test will also be high. In this case the Scheffé test will have a high average conditional probability of solving the interpretation problem. When the Scheffé test has high power but the Bonferroni test has low power, then the sacrifice of power due to using Bonferroni test may be difficult to justify. Therefore the Scheffé test may be more attractive than the Bonferroni test in the presence of multicollinear data. When the average power of the Scheffé test is low then what is needed is more and better data. The weak evidence problem and the low power problem are two sides of the same coin.

5. Large sample induced tests

Large sample analogues of the finite induced tests and the Scheffé test can be constructed for a variety of models. These include single equation and multivariate nonlinear models, linear and nonlinear simultaneous equations models, time series models, and qualitative response models. As an illustration I will briefly discuss large sample analogues of the tests in the context of the standard nonlinear regression model:

$$y_t = f(x_t, \beta_0) + u_t, \qquad t = 1, \ldots, T, \tag{5.1}$$

where y_t is a scalar endogenous variable, x_t is a vector of exogenous variables, β_0 is a $k \times 1$ vector of unknown parameters and the u_t's are unobservable scalar independently identically distributed random variables with mean zero and variance σ_0^2.

The nonlinear least squares estimator, denoted by $\hat{\beta}$, is defined as the value of β that minimizes the sum of squared residuals:

$$S_T(\beta) = \sum_{t=1}^{T} [y_t - f(x_t, \beta)]^2, \tag{5.2}$$

where the β that appears in (5.2) is the argument of the function $f(x_t, \cdot)$. In contrast, β_0 is the true fixed value. The consistency and asymptotic normality of the nonlinear least squares estimator is rigorously proved in Jennrich (1969). Therefore, we have:

$$\sqrt{T}(\hat{\beta} - \beta_0) \to N(0, \sigma_0^2 \Omega^{-1}), \tag{5.3}$$

where

$$\operatorname{plim} \frac{1}{T} \left. \frac{\partial^2 S_T}{\partial \beta \, \partial \beta'} \right|_{\beta^*} = 2\Omega \tag{5.4}$$

is a $k \times k$ matrix and β^* lies between $\hat{\beta}$ and β_0. For a discussion of the assumptions needed to prove (5.3), see Chapter 6 by Amemiya in this *Handbook*.

Amemiya points out that in the process of proving (5.3) we have in effect shown that, asymptotically,

$$\hat{\beta} - \beta_0 \cong (G'G)^{-1} G'u, \tag{5.5}$$

where $G=(\partial f/\partial\beta')_{\beta_0}$, a $T\times k$ matrix. The practical consequence of the approximation (5.5) is that all the results for the linear regression model are asymptotically valid for the nonlinear regression model if G is treated as regressor matrix. In particular, the usual t and F statistics can be used asymptotically. Note that (5.5) holds exactly in the linear case.

As an example consider testing the linear hypothesis:

$$H\colon C\beta - c=\theta=0, \tag{5.6}$$

where C and c are defined as in (2.2). Let:

$$z=C\hat{\beta}-c \tag{5.7}$$

and

$$\hat{V}=C(\hat{G}'\hat{G})^{-1}C', \tag{5.8}$$

where $\hat{G}=(\partial f/\partial\beta')_{\hat{\beta}}$. Then we have asymptotically under the null hypothesis

$$t_i=\frac{z_i}{\sqrt{s^2\hat{V}_{ii}}}\sim t(T-k) \tag{5.9}$$

and

$$F=\frac{z'\hat{V}^{-1}z}{qs^2}\sim F(q,T-k), \tag{5.10}$$

where $s^2=S_T(\hat{\beta})/(T-k)$ and $\hat{V}_{ii}$ is the ith diagonal element of $\hat{V}$.

Suppose that a finite number m of ψ in L are of primary interest. Let the ψ in G be $\psi_i=a_i'\theta$, $i=1,\dots,m$. The usual t statistic for testing the separate hypothesis $H(a_i)$: $\psi_i=a_i'\theta=0$ is:

$$t_0(a_i)=\frac{a_i'z}{\sqrt{s^2a_i\hat{V}a_i}},\qquad i=1,\dots,m. \tag{5.11}$$

The acceptance region of a δ level equal-tailed test of $H(a_i)$ is approximately:

$$|t_0(a_i)|\le t_{\delta/2}(T-k),\qquad i=1,\dots,m. \tag{5.12}$$

The finite induced test accepts H if and only if all the separate hypotheses $H(a_1),\dots,H(a_m)$ are accepted. When all the equal-tailed t tests have the same significance level the acceptance region for an α level Bonferroni test of H is approximately:

$$|t(a_i)| \leq B, \tag{5.13}$$

where $B = t_{\alpha/2m}(T-k)$. The Sidák or studentized maximum modulus critical value can also be used in large samples.

A large sample analogue of the Scheffé test can be developed by using the fact that the maximum of the squared t ratio:

$$t^2(a) = \frac{[a'(z-\theta)]^2}{s^2 a'\hat{V}a}, \tag{5.14}$$

is asymptotically distributed as $qF(q, T-k)$. The proof is essentially the same as the one presented in Section 3.3.1.

Next, consider testing the nonlinear hypothesis:

$$h(\beta) = 0, \tag{5.15}$$

where $h(\beta)$ is a $q \times 1$ vector valued nonlinear function such that $q < k$. If β are the parameters that characterize a concentrated likelihood function $L(\beta)$, where L may or may not be derived from the normal distribution, then the hypothesis (5.15) can be tested using the Wald (W), likelihood ratio (LR), or Lagrange multipler (LM) test. For a discussion of these tests, see Chapter 13 by Engle in this *Handbook*.

When the error vector u is assumed to be normal in the nonlinear regression model (5.1), the three test statistics can be written as

$$\mathrm{W} = h(\hat{\beta})' \left[\left| \frac{\partial h}{\partial \beta'} \right|_{\hat{\beta}} (\hat{G}'\hat{G})^{-1} \left| \frac{\partial h'}{\partial \beta} \right|_{\hat{\beta}} \right]^{-1} h(\hat{\beta}), \tag{5.16}$$

$$\mathrm{LR} = T\left[\log T^{-1}S_T(\tilde{\beta}) - \log T^{-1}S_T(\hat{\beta})\right], \tag{5.17}$$

$$\mathrm{LM} = \frac{[y - f(\tilde{\beta})]'\tilde{G}(\tilde{G}'\tilde{G})^{-1}\tilde{G}'[y - f(\tilde{\beta})]}{S_T(\tilde{\beta})/T}, \tag{5.18}$$

where $\tilde{\beta}$ is the constrained maximum likelihood estimator obtained by maximizing $L(\beta)$ subject to (5.15), and $\tilde{G} = (\partial f/\partial \beta')_{\tilde{\beta}}$. When the hypothesis (5.15) is true all three statistics (5.16), (5.17) and (5.18) are asymptotically distributed as $\chi^2(q)$ if u is normal. In fact, it can be shown that these statistics are asymptotically distributed as $\chi^2(q)$ even if u is not normal. Thus, these statistics can be used to test a nonlinear hypothesis when u is non-normal.

Recall that from any convex set we can derive simultaneous confidence intervals for all ψ in L. This convex set can be the acceptance region of the W, LR or LM tests in large samples. Starting with a finite set G of ψ in L of primary interest the convex set can be defined as the intersection of large sample t intervals for all ψ in G. The t statistics can be based on either the W or the LM principle of test construction. A large sample analogue of the S intervals can be based on the W test of H.

6. Empirical examples

6.1. *Textile example*

Our first empirical illustration is based on the textile example of Theil (1971, p. 103). This example considers an equation of the consumption of textiles in the Netherlands 1923–1939:

$$y = \beta_0 + \beta_1 x_1 + \beta_2 x_2 + u, \tag{6.1}$$

where y = logarithm of textile consumption per capita, x_1 = logarithm of real per capita income and x_2 = logarithm of the relative price of textile goods. The estimated equation is reported by Theil (p. 116) as:

$$y = \underset{(0.31)}{1.37} + \underset{(0.16)}{1.14}\, x_1 - \underset{(0.04)}{0.83}\, x_2, \tag{6.2}$$

where the numbers in parentheses are standard errors.

Theil tests the hypothesis that the income elasticity (β_1) is unity, and that the price elasticity (β_2) is minus unity. This hypothesis is:

$$H\colon C\beta - c = \begin{bmatrix} 0 & 1 & 0 \\ 0 & 0 & 1 \end{bmatrix} \begin{bmatrix} \beta_0 \\ \beta_1 \\ \beta_2 \end{bmatrix} - \begin{bmatrix} 1 \\ -1 \end{bmatrix} = \begin{bmatrix} \theta_1 \\ \theta_2 \end{bmatrix} = \theta = 0. \tag{6.3}$$

The 0.01 level F test rejects H since the value of the F ratio is 11.2 and the upper 1% significance point of an $F(2,14)$ distribution is 6.51.

Consider the Bonferroni test of H where the linear combinations of primary interest are θ_1 and θ_2. The t statistics for testing θ_1 and θ_2 are:

$$t_1 = \frac{z_1}{\sqrt{s^2 V_{11}}} = \frac{0.1430}{0.16} = 0.89 \tag{6.4}$$

and

$$t_2 = \frac{z_2}{\sqrt{s^2 V_{22}}} = \frac{0.1711}{0.04} = 4.28, \tag{6.5}$$

respectively. The nominal 0.01 level Bonferroni test rejects H since $B = t_{\delta/2}(14) = 3.33$ when $\delta = 0.01/2 = 0.005$. Clearly, the separate hypothesis $\beta_2 = -1$ is responsible for the rejection of the Bonferroni test of H. The 0.01 level Scheffé test of H also rejects H since the 0.01 level F test rejects H. In this example the Bonferroni test has roughly the same power contour as the Scheffé test since the correlation between the income and price variables is low, namely about 0.22.

The next step is to calculate simultaneous confidence intervals for θ_1 and θ_2. The B interval for θ_1 is $0.1430 \pm 0.16(3.33)$ and for θ_2 is $0.1711 \pm 0.04(3.33)$ so that the B intervals are $-0.39 \leq \theta_1 \leq 0.68$ and $0.04 \leq \theta_2 \leq 0.30$, respectively. The S interval for θ_1 is $0.1430 \pm 0.16(3.61)$ and for θ_2 is $0.1771 \pm 0.04(3.61)$ since $S = \sqrt{2F_{0.01}(2,14)} = 3.61$. Hence the S intervals are $-0.43 \leq \theta_1 \leq 0.72$ and $0.03 \leq \theta_2 \leq 0.32$, respectively. Note that the S intervals are longer than the B intervals, but not much longer. Both intervals for θ_1 cover zero and both intervals for θ_2 cover only positive values. This suggests that the income elasticity β_1 is unity and that the price elasticity β_2 is greater than minus one. In this example the hypothesis $\beta_2 = -1$ is responsible for the rejection of the Scheffé as well as the Bonferroni test of H. This result also follows from the fact that the absolute value of the t statistic for θ_2 is larger than either B or S. i.e. $|t_2| > B$ and $|t_2| > S$.

The final step is to calculate the normalized a_0 vector:

$$a_0 = \frac{\left[C(X'X)^{-1}C'\right]^{-1}(Cb - c)}{\sqrt{s^2 qF}}, \tag{6.6}$$

where $a_0' V a_0 = 1$. From Theil we have that:

$$s^2\left[C(X'X)^{-1}C\right]^{-1} = \begin{bmatrix} 43.2 & 41.6 \\ 41.6 & 807.0 \end{bmatrix}, \tag{6.7}$$

so that:

$$a_0 = \frac{1}{s(4.733)} \begin{bmatrix} 43.2 & 41.6 \\ 41.6 & 807.0 \end{bmatrix} \begin{bmatrix} 0.1430 \\ -0.1711 \end{bmatrix} = \begin{bmatrix} 207.5 \\ 2247.7 \end{bmatrix}, \tag{6.8}$$

where $s^2 = 0.0001833$. This confirms Theil's conclusions (p. 145) that the specification $\beta_2 = -1$ for the price elasticity is responsible for the F test (Scheffé test) rejecting H, i.e. any linear combination with positive weights and a sufficiently large weight on θ_2 is responsible for rejection.

Suppose in the B procedure that $\psi = \theta_1 - \theta_2$ is of secondary interest. The B interval for ψ is $0.3141 \pm 0.20(3.33)$ or $-0.35 \leq \psi \leq 0.98$. The S interval for ψ is $0.3141 \pm 0.023(3.61)$ or $0.23 \leq \psi \leq 0.40$ so that the S interval is shorter than the B interval. Also notice that $\hat{\psi} = z_1 - z_2$ is *sdfz* according to the S criterion, but not the B criterion. Hence the Scheffé induced test of H is rejected by the separate hypothesis that the income and price elasticities are the same except for sign: $\beta_1 = -\beta_2$. Theil (p. 134) objects to the length of the S intervals for the ψ of primary interest. In fact in the textile example the S intervals give interesting results for both the ψ of primary and secondary interest.

6.2. *Klein's Model I example*

Our second example is based on the unrestricted reduced form equation for consumption expenditures from Klein's Model I of the United States economy 1921–1941:

$$y = \beta_0 + \beta_1 x_1 + \beta_2 x_2 + \beta_3 x_3 + \beta_4 x_4 + \beta_5 x_5 + \beta_6 x_6 + \beta_7 x_7 + u, \tag{6.9}$$

where y = consumption, x_1 = government wage bill, x_2 = indirect taxes, x_3 = government expenditures, x_4 = time (measured as year-1931), x_5 = profits lagged one year, x_6 = end of year capital stock lagged one year, and x_7 = private product lagged one year. For the purpose of this example all regressors are treated as nonstochastic. The data is taken from Theil (1971, p. 456). The estimated equation is:

$$\begin{aligned} y = \underset{(1.90)}{58.3} + \underset{(0.079)}{0.193}\, x_1 - \underset{(-0.871)}{0.366}\, x_2 + \underset{(0.541)}{0.205}\, x_3 + \underset{(0.930)}{0.701}\, x_4 \\ + \underset{(1.49)}{0.748}\, x_5 - \underset{(-1.27)}{0.147}\, x_6 + \underset{(0.842)}{0.230}\, x_7, \end{aligned} \tag{6.10}$$

where now the numbers in parentheses are t ratios. Our estimates of the β's agree with those reported in Goldberger (1964, p. 325). (Note that Goldberger uses $x_1 - x_3$ in place of x_1 so that his estimate of β_1 is $0.19327 - 0.20501 = -0.01174$.)

Consider testing the hypothesis that all the slope coefficients are zero:

$$H: \beta_i = \theta_i = 0, \qquad i = 1, 2, \ldots, 7. \tag{6.11}$$

The slope coefficients are multipliers so we are testing the hypothesis that all the multipliers in the reduced form equation for consumption are zero. The 0.05 level Scheffé test rejects H since the 0.05 level F test overwhelmingly rejects H. The F ratio is 28.2 which is much larger than 2.83, the upper 0.05 significance point of the $F(7,13)$ distribution. Suppose that the linear combinations of primary interest in the Bonferroni test are the slope coefficients: $\psi_i = \theta_i$, $i = 1,2,\ldots,7$. Then the critical t value for a nominal 0.05 level Bonferroni separate induced test of H is $B = t_{\delta/2}(13) = 3.19$, where $\delta = 0.05/7 = 0.00714$. The t ratio with the largest absolute value is the one for lagged profits (β_5). Since this is only 1.49 the Bonferroni test overwhelmingly accepts H. Thus in this example the Scheffé and Bonferroni tests of H produce conflicting inferences.

We now apply the S procedure to find which linear combination of the multipliers led to rejection of the Scheffé test of H. In this example none of the individual multipliers are responsible for rejection since none of the t ratios have an absolute value greater than S. The largest t ratio is 1.49 and $S = \sqrt{7F_{0.05}(7,13)} = 4.45$. To find linear combinations of the multipliers which are responsible for rejection I began by calculating the normalized vector a_0. This vector has components:

$$a_1 = 5.82; \quad a_2 = 4.81; \quad a_3 = 7.37; \quad a_4 = 19.44;$$
$$a_5 = 12.13; \quad a_6 = 14.33; \quad a_7 = 35.84, \qquad (6.12)$$

where these are proportional to the sample covariances between the dependent variable and the regressors. The linear combination (6.12) gives some positive weight to all the multipliers and especially to the multiplier β_7 for lagged private product. Since (6.12) does not seem to have an interesting economic interpretation, I examined a number of other linear combinations. I could not find a linear combination responsible for rejection which was also of economic interest.

In this example the explanatory variables are highly correlated. As a consequence the Bonferroni test can have low average power compared to the Scheffé test. Hence the Bonferroni test may be very misleading. The Scheffé test gives what appears to be a sensible result, but provides little help in deciding which multipliers are nonzero. What is needed is more and better data for a satisfactory solution to the interpretation problem.

References

Bailey, J. R. (1977), "Tables of the Bonferroni t Statistics", *Journal of the American Statistical Association*, 72:469–478.

Christensen, L. R. (1973), "Simultaneous Statistical Inference in the Normal Multiple Linear Regression Model", *Journal of the American Statistical Association*, 68:457–461.

Christensen, L. R., D. W. Jorgenson and L. J. Lau (1975), "Transcendental Logarithmic Utility Function", *American Economic Review*, 65:367–383.

Cornish, E. A. (1954), "The Multivariate Small Sample t-Distribution Associated with a Set of Normal Sample Deviates", *Australian Journal of Physics*, 7:531–542.

Darroch, J. N. and S. D. Silvey (1963), "On Testing More Than One Hypothesis", *Annals of Mathematical Statistics*, 27:555–567.

Dhrymes, P. J. (1978), *Introductory Econometrics*. New York: Springer Verlag.

Dunn, O. J. (1961), "Multiple Comparisons Among Means", *Journal of the American Statistical Association*, 56:52–64.

Dunnett, C. W. (1955), "A Multiple Comparisons Procedure for Comparing Several Treatments with a Control", *Journal of the American Statistical Association*, 50:1096–1121.
Dunnett, C. W. and M. Sobel, (1954), "A Bivariate Generalization of Student's t-Distribution with Tables for Certain Cases", *Biometrika*, 41:153–169.
Evans, G. B. A. and N. E. Savin (1980), "The Powers of the Bonferroni and Scheffé Tests in the Two Parameter Case", Manúscript, Faculty of Economics and Politics, Cambridge.
Fox, M. (1956), "Charts of the Power of the F-Test", *Annals of Mathematical Statistics*, 27:485–494.
Gabriel, K. R. (1964), "A Procedure for Testing the Homogeneity of All Sets of Means in the Analysis of Variance", *Biometrics*, 40:459–477.
Gabriel, K. R. (1969), "Simultaneous Test Procedure—Some Theory of Multiple Comparisons", *Annals of Mathematical Statistics*, 40:224–250.
Games, P. A. (1977), "An Improved Table for Simultaneous Control on g Contrasts", *Journal of the American Statistical Association*, 72:531–534.
Geary, R. C. and C. E. V. Leser (1968), "Significance Tests in Multiple Regression", *The American Statistican*, 22:20–21.
Goldberger, A. S. (1964), *Econometric Theory*. New York: John Wiley & Sons.
Hahn, G. J. and R. W. Hendrickson (1971), "A Table of Percentage Points of The Largest Absolute Value of k Student t Variates and Its Applications", *Biometrika*, 58:323–332.
Hochberg, Y. (1974), "Some Generalization of the T-Method in Simultaneous Inference ", *Journal of Multivariate Analysis*, 4:224–234.
Hochberg, Y. and C. Rodriquez (1977), "Intermediate Simultaneous Inference Procedures", *Journal of the American Statistical Association*, 72:220–225.
Imhof, P. (1961), "Computing the Distribution of Quadratic Forms in Normal Variates", *Biometrika*, 48:419–426.
Jennrich, R. I. (1969), "Asymptotic Properties of Non-linear Least Squares Estimation", *Annals of Mathematical Statistics*, 40:633–643.
Johnson, N. L. and S. Kotz (1972), *Distributions in Statistics: Continuous Multivariate Distributions*. New York: John Wiley & Sons.
Jorgenson, D. W. (1971), "Econometric Studies of Investment Behavior: A Survey", *Journal of Economic Literature*, 9:1111–1147.
Jorgenson, D. W. (1974), "Investment and Production: A Review", Intriligator M. D. and D. A. Kendrick, Eds., *Frontiers in Quantitative Economics, II*. Amsterdam: North Holland.
Jorgenson, D. W. and L. J. Lau (1975), "The Structure of Consumer Preferences", *Annals of Social and Economic Measurement*, 4:49–101.
Jorgenson, D. W. and L. J. Lau (1982), *Transcendental Logarithmic Production Functions*. Amsterdam: North Holland.
Krishnaiah, P. R. (1963), Simultaneous Tests and the Efficiency of Generalized Incomplete Block Designs, ARL 63-174. Wright-Patterson Air Force Base, Ohio.
Krishnaiah, P. R. (1964), Multiple Comparisons Tests in Multivariate Case. ARL 64-124, Wright-Patterson Air Force Base, Ohio.
Krishnaiah, P. R. (1965), "On the Simultaneous ANOVA and MANOVA Tests", *Ann. Inst. Statist. Math.*, 17:35–53.
Krishnaiah, P. R. (1979), "*Some Developments on Simultaneous Test Procedures*", Krishnaiah, P. R. (ed.), Developments in statistics, vol. 2, New York: Academic Press.
Krishnaiah, P. R. and J. V. Armitage (1965a), Percentage Points of the Multivariate t Distribution. ARL 65-199. Wright-Patterson Air Force Base, Ohio.
Krishnaiah, P. R. and J. V. Armitage (1965b), Probability Integrals of the Multivariate F Distribution, With Tables and Applications. ARL 65-236, Wright-Patterson Air Force Base, Ohio.
Krishnaiah, P. R. and J. V. Armitage (1966), "Tables for Multivariate t Distribution", *Sankhya Ser. B*. 28:31–56.
Krishnaiah, P. R. and J. V. Armitage (1970), "*On a Multivariate F Distribution*", in Essays in Probability and Statistics, R. C. Bose et al. eds., Chapel Hill: University of North Carolina Press.
Krishnaiah, P. R., J. V. Armitage and M. C. Breiter (1969a), Tables for the Probability Integrals of the Bivariate t Distribution. ARL 69-060. Wright-Patterson Air Force Base, Ohio.
Krishnaiah, P. R., J. V. Armitage and M. C. Breiter (1969b), Tables for the Bivariate $|t|$ Distribution. ARL 69-0210. Wright-Patterson Air Force Base, Ohio.

Leamer, E. E. (1979), *Specification Searches*. New York: Wiley 1979.
Lehmann, E. L. (1957a), "A Theory of Some Multiple Decision Problems, I", *Annals of Mathematical Statistics*, 28:1–25.
Lehmann, E. L. (1957b), "A Theory of Some Multiple Decision Problems, II", *Annals of Mathematical Statistics*, 28:547–572.
Miller, R. G., Jr. (1966), *Simultaneous Statistical Inference*. New York: McGraw-Hill 1966.
Miller, R. G., Jr. (1977), "Developments in Multiple Comparisons, 1966–1976", *Journal of the American Statistical Association*, 72:779–788.
Morrison, D. F. (1976), *Multivariate Statistical Methods*, 2nd Edition, New York: McGraw-Hill.
Moses, L. E. (1976), "Charts for Finding Upper Percentage Points of Student's t in the Range .01 to .0001", Technical Report No. 24 (5 Rol GM 21215.02), Stanford University.
Olshen, R. A. (1973), "The Conditional Level of the F Test", *Journal American Statistical Association*, 48:692–698.
Olshen, R. A. (1977), "A Note on a Reformulation of the S Method of Multiple Comparison—Comment", *Journal American Statistical Association*, 72:144–146.
Pearson, E. S. and H. O. Hartley (1972), *Biometrika Tables for Statisticians*, *Cambridge*, England: Cambridge University Press.
Richmond, J. (1982), "A General Method for Constructing Simultaneous Confidence Intervals", *Journal of the American Statistical Association*, 77:455–460.
Roy, S. N. (1953), "On a Heuristic Method of Test Construction and Its Uses in Multivariate Analysis", *Annals of Mathematical Statistics*, 24:220–239.
Roy, S. N. and R. C. Bose, (1953), "Simultaneous Confidence Interval Estimation", *Annals of Mathematical Statistics*, 2:415–536.
Sargan, J. D. (1976), "*The Consumer Price Equation in the Post War British Economy: An Exercise in Equation Specification Testing*", mimeo, London School of Economics.
Savin, N. E. (1980), "The Bonferroni and Scheffe Multiple Comparison Procedures", *Review of Economic Studies*, 47:255–273.
Schaffer, J. P. (1977), "Multiple Comparisons Emphasizing Selected Contrasts: An Extension and Generalization of Dunnet's Procedure", *Biometrics*, 33:293–303.
Scheffé, H. (1953), "A Method of Judging All Contrasts in the Analysis of Variance", *Biometrika*, 40:87–104.
Scheffé, H. (1950), *The Analysis of Variance*. New York: John Wiley & Sons.
Scheffé, H. (1977a), "A Note on a Reformulation of the S-Method of Multiple Comparison", *Journal of the American Statistical Association*, 72:143–144.
Scheffé, H. (1977b), "A Note on a Reformulation of the S-Method of Multiple Comparison-Rejoinder", *Journal of the American Statistical Association*, 72:146.
Seber, G. A. F. (1964), "Linear Hypotheses and Induced Tests", *Biometrika*, 51:41–47.
Seber G. A. F. (1977), *Linear Regression Analysis*. New York: Wiley.
Sidák Z. (1967), "Rectangular Confidence Regions for the Means of Multivariate Normal Distributions", *Journal of the American Statistical Association*, 62:626–633.
Stoline, M. R. and N. K. Ury (1979), "Tables of the Studentized Maximum Modulus Distribution and an Application to Multiple Comparisons Among Means", *Technometrics*, 21:87–93.
Theil, H. (1971), *Principles of Econometrics*. New York: John Wiley & Sons.
Tiku, M. (1965), "Laguerre Series Forms of Non-Central Chi-Squared and F Distributions", *Biometrika*, 52:415–427.
Tukey, J. W. (1953), "The Problem of Multiple Comparisons", Princeton University, mimeo.

Chapter 15

APPROXIMATING THE DISTRIBUTIONS OF ECONOMETRIC ESTIMATORS AND TEST STATISTICS

THOMAS J. ROTHENBERG*

University of California, Berkeley

Contents

*I am indebted to Christopher Cavanagh for his help in preparing this survey and to Donald Andrews and James Reeds for comments on an earlier draft. Research support from the National Science Foundation is gratefully acknowledged.

Handbook of Econometrics, Volume II, Edited by Z. Griliches and M.D. Intriligator

1. Introduction

Exact finite-sample probability distributions of estimators and test statistics are available in convenient form only for simple functions of the data and when the likelihood function is completely specified. Often in econometrics these conditions are not satisfied and inference is based on approximations to the sampling distributions. Typically "large sample" methods of approximation based on the central limit theorem are employed. For example, if $\hat{\theta}_n$ is an estimator of a parameter θ based on a sample of size n, it is sometimes possible to find a function $\sigma(\theta)$ such that the distribution of the variable $\sqrt{n}(\hat{\theta}_n - \theta)/\sigma(\theta)$ converges to a standard normal as n tends to infinity. In that case, it is common practice to approximate the distribution of $\hat{\theta}$ by a normal distribution with mean θ and variance $\sigma^2(\theta)/n$. Similar approximations are used for test statistics, although the limiting distribution is often chi-square rather than normal in this context.

These large-sample or asymptotic approximations may be quite accurate even for very small samples. The arithmetic average of independent draws from a rectangular distribution has a bell-shaped distribution for n as low as three. However, it is also easy to construct examples where the asymptotic approximation is poor even when the sample contains hundreds of observations. It is desirable, therefore, to know the conditions under which the asymptotic approximations are reasonable and to have available alternative methods when the asymptotic approximations break down. In what follows we survey some of the basic methods that have been used to approximate distributions in econometrics and describe some typical applications of these methods. Particular emphasis will be placed on "second-order" approximation methods which can be used to compare alternative asymptotically indistinguishable inference procedures.

The subject of our investigation has a long history. Techniques for approximating probability distributions have been studied by mathematical statisticians since the nineteenth century. Indeed, many of the basic methods in current use were developed more than 75 years ago. The transfer of these ideas to econometrics, however, has been very slow; only in the past 15 years has there been substantial progress in improving the approximations used in empirical economics. The reasons for this lag are not hard to fathom. The original work concentrated on one-dimensional statistics based on sums of identically distributed independent random variables. The generalization to multidimensional cases with nonlinearity, dependency, and other complications turns out to involve quite difficult mathematics and nontrivial computation. The advent of more powerful mathematical tools and enormously reduced computation cost in recent years has produced a revolution in the field of statistical approximation. Not only have old methods been applied to more complex problems, a new burst of interest in higher-order asymptotic theory has occurred among mathematical statisticians. With so much recent development both within and without econometrics, this survey must necessarily be incomplete and tentative. It represents a somewhat personal view of the current state of a rapidly changing area of research.

Before turning to the various techniques and applications, it is perhaps useful to raise some general issues concerning the use of approximate distributions in econometrics. First of all, one must decide what one is trying to approximate. In many applications the parameter vector of interest has high dimension. Do we wish to approximate the joint probability distribution of the vector of estimates, or do we wish to approximate each marginal distribution? Is it the cumulative distribution function that needs to be approximated, or is it the density function? Some approaches which lead to good approximations of univariate densities are not convenient for obtaining good approximations of multivariate cumulative distribution functions. In practice the type of approximation method to be employed is strongly influenced by the type of function being approximated. The emphasis in the present survey will be on approximations to univariate distribution functions. It appears that most applications require knowledge of the probability that a scalar random variable lies in some interval. For example, the degree of concentration of an estimator and the power of a test can be measured by such probability statements. Although some discussion of density approximations will be presented, we shall rarely depart from distributions on the real line.

A second issue concerns the approximation of moments. If determining the full probability distribution of a statistic is hard perhaps one can get by with summary values. For many purposes, knowledge of the first few moments of an estimator or test statistic is sufficient. Thus, methods for approximating moments may be just as valuable as methods for approximating distributions. As we shall see, these methods are not unrelated: approximate moments play a key role in developing approximate distribution functions. Hence our survey will cover both topics.

Finally, and perhaps most crucially, there is the issue: What use will be made of the approximation? Generally one can distinguish two distinct reasons for wanting to know the probability distribution of an estimator or test statistic. One reason is that it is needed to make some numerical calculation from the data. For example, one might use the probability distribution to form a confidence interval for a parameter estimate; or one might form a rejection region for a test statistic. An alternative reason for knowing the probability law is that it is needed to evaluate or compare statistical procedures. One might use an estimator's probability distribution to judge whether it was reasonably accurate; or one might use sampling distributions to decide which of two tests is most powerful.

These two different uses of the probability distribution suggest different criteria for judging an approximation. For the former use, we need a computer algorithm that will calculate, quickly and accurately, a number from the actual data. As long as the algorithm is easy to program and does not require too much data as input, it does not matter how complicated or uninterpretable it is. For the latter use, we need more than a number. The probability distribution for an estimator or test statistic generally depends on the unknown parameters and on the values of the exogenous variables. To evaluate statistical procedures we need to know how the key aspects of the distribution (center, dispersion, skewness, etc.) vary with the parameters and the exogenous data. An algorithm which computes the distribution function for any given parameter vector and data set may not be as useful as a simple formula that indicates how the shape of the distribution varies with the parameters. Interpretability, as well as accuracy, is important when comparison of probability distributions is involved.

Since my own interests are concerned with comparing alternative procedures, the present survey emphasizes approximations that yield simple analytic formulae. After reviewing a number of different approaches to approximating distributions in Section 2, the remainder of the chapter concentrates on higher-order asymptotic theory based on the Edgeworth expansion. Although the asymptotic approach rarely leads to the most accurate numerical approximations, it does lead to a powerful theory of optimal estimates and tests. In this context, it is worth recalling the words used by Edgeworth (1917) when discussing the relative merits of alternative approaches to representing empirical data: "I leave it to the impartial statistician to strike the balance between these counterpoised considerations.... I submit, too, that the decision turns partly on the purpose to which representation of statistics is directed. But one of the most difficult questions connected with our investigation is: What is its use?"

2. Alternative approximation methods

2.1. Preliminaries

If we are given the probability distribution of a vector of random variables, we can, in principle, find the distribution of any smooth function of these random variables by multivariate calculus. In fact, however, the mathematics is often too difficult and analytic results are unobtainable. Furthermore, we sometimes wish to learn about certain features of the distribution of a function without specifying completely the exact distribution of the underlying random variables. In this section we discuss a number of alternative methods that can be employed to obtain approximations to the probability distributions of econometric estimators and test statistics under various circumstances.

Although there is a huge statistical literature on the theory and practice of approximating distributions, there are relatively few introductory presentations of this material. The statistics textbook by Bickel and Doksum (1977) gives a very brief survey; the handbook of distributions by Johnson and Kotz (1970) has a more comprehensive discussion. Traditional large-sample theory is developed in Cramer (1946); a detailed treatment is given in Serfling (1980). The extension to asymptotic expansions is presented in Wallace's (1958) excellent (but slightly dated) survey article; some recent developments are discussed in Bickel (1974). For a comprehensive treatment of the subject, however, a major incursion into the textbooks of advanced probability theory and numerical analysis is necessary. For those with the time and patience, chapters 15 and 16 of Feller (1971) and chapters 1, 3, and 4 of Olver (1974) are well worth the effort. In what follows we refer mostly to recent developments in the econometric literature; the bibliographies in the above-mentioned works can give entrée into the statistical literature. The recent survey paper by Phillips (1980) also gives many key references.

The present discussion is intended to be introductory and relatively nontechnical. Unfortunately, given the nature of the subject, considerable notation and formulae are still required. A few notational conventions are described here. Distribution functions will typically be denoted by the capital letters F and G; the corresponding density functions are f and g. The standard univariate normal distribution function is represented by Φ and its density by ϕ. If a p-dimensional random vector X is normally distributed with mean vector μ and covariance matrix Σ, we shall say that X is $N_p(\mu,\Sigma)$; when $p=1$, the subscript will be dropped. The probability of an event will be indicated by $\Pr[\cdot]$. Thus, if X is $N_p(\mu,\Sigma)$ and c is a p-dimensional column vector, $\Pr[c'X\le x]=\Phi[(x-c'\mu)/\sqrt{c'\Sigma c}\,]$, for all real x.

If X is a scalar random variable with distribution function F, its characteristic function is defined as $\psi(t)=\mathrm{E}\exp\{\mathrm{i}tX\}$, where t is real, E represents expectation with respect to the distribution of X, and $\mathrm{i}=\sqrt{-1}$. The function $K(t)=\log\psi(t)$ is called the cumulant function. If X possesses moments up to order r, then $\psi(t)$ is differentiable up to order r; furthermore, the rth moment of X is given by the rth derivative of $\mathrm{i}^{-r}\psi(t)$ evaluated at zero:

$$\mathrm{E}(X^r)=\mathrm{i}^{-r}\psi^{(r)}(0).$$

The rth derivative of $\mathrm{i}^{-r}K(t)$, evaluated at zero, is called the rth cumulant of X and is denoted by:

$$k_r=\mathrm{i}^{-r}K^{(r)}(0).$$

Since the derivatives of $K(t)$ are related to the derivatives of $\psi(t)$, the cumulants are related to the moments. In fact, k_1 is the mean and k_2 is the variance. For a standardized random variable with zero mean and unit variance, k_3 is the third moment and k_4 is the fourth moment less three. For a normal random variable, all cumulants of order greater than two are zero. Hence, these cumulants can be viewed as measures of departure from normality. For further details, one may consult Kendall and Stuart (1969, ch. 3).

Our discussion will concentrate on approximating the cumulative distribution functions for continuous random variables. If the approximating distribution function is differentiable, there will generally be no problem in obtaining an approximate density function. Some approximation methods, however, apply most easily to the density function directly. In that case, numerical integration may be needed to obtain the distribution function if analytic integration is difficult.

2.2. *Curve-fitting*

The simplest way to approximate a distribution is to find a family of curves possessing the right shape and select that member which seems to fit best. If the low-order moments of the true distribution are known, they can be used in the fitting process. If not, Monte Carlo simulations or other information about the true distribution can be employed instead.

Durbin and Watson (1971) describe a number of different approximations to the null distribution of their d statistic for testing serial correlation in regression disturbances. One of the most accurate is the beta approximation proposed by Henshaw (1966). Since d must lie between zero and four and seems to have a unimodal density, it is not unreasonable to think that a linear transformed beta distribution might be a good approximation to the true distribution. Suppose X is a random variable having the beta distribution function:

$$\Pr[X \le x] = \int_0^x \frac{1}{B(p,q)} t^{p-1}(1-t)^{q-1} \mathrm{d}t \equiv G(x; p,q).$$

Then, for constants a and b, the random variable $a + bX$ has moments depending on p, q, a, and b. These moments are easy to express in analytic form. Furthermore, the moments of the Durbin–Watson statistic d are also simple functions of the matrix of regression variables. Equating the first four moments of d to the corresponding moments of $a + bX$, one obtains four equations in the four parameters. For any given matrix of observations on the regressors, these equations give unique solutions, say p^*, q^*, a^*, b^*. Then $\Pr[d \le x]$ can be approximated by $G[(x - a^*)/b^*; p^*, q^*]$. This approximation appears to give third decimal accuracy for a wide range of cases. Theil and Nagar (1961) had earlier proposed a similar approximation, but used approximate rather than actual moments of d. Since these approximate moments do not vary with the matrix of regressors, the Theil–Nagar approximation is independent of the data and can be tabulated once and for all. Unfortunately, the moment approximation is not always accurate and the resulting approximation to the probability distribution is less satisfactory than Henshaw's.

A more sophisticated version of the curve-fitting method is suggested by Phillips (1981). Suppose a statistic X is known to have a density function $f(x)$ that behaves in the tails like the function $s(x)$. For example, if X possess moments only up to order k and takes values everywhere on the real line, $f(x)$ might behave in the tails like a Student density with $k+1$ degrees of freedom. For some small integer r, one might approximate the density function $f(x)$ by a rational function modification of $s(x)$:

$$s(x)\frac{a_0 + a_1 x + \cdots + a_r x^r}{b_0 + b_1 x + \cdots + b_r x^r}, \tag{2.1}$$

where the a_i and b_i are chosen to make the approximation as accurate as possible. Since the function (2.1) does not typically have simple moment formulae (or even possess finite moments), the method of moments is not a useful way to obtain values for the a_i and b_i. But, Monte Carlo experimental data or local power series expansions of the density may be available to help select the parameters. Since (2.1) has $2r+1$ free parameters, it appears that, with a judicious choice for s, this functional form should provide a very accurate approximation to the density function of any econometric statistic. Furthermore, if s is replaced by its integral, a function of the same form as (2.1) could be used to approximate a distribution function.

If considerable information about the true density is available, curve-fitting methods are likely to provide simple and very accurate approximations. Phillips (1981) produces some striking examples. Indeed it is unlikely that any other method will give better numerical results. However, curve-fitting methods are considerably less attractive when the purpose is not quantitative but qualitative. Comparisons of alternative procedures and sensitivity analysis are hindered by the fact that curve-fitting methods do not typically yield a common parametric family. If two statistics are both (approximately) normal, they can be compared by their means and variances. If one statistic is approximately beta and the other approximately normal, comparisons are difficult: the parameters that naturally describe one distribution are not very informative about the other. The very flexibility that makes curve-fitting so accurate also makes it unsuitable for comparisons.

2.3. *Transformations*

Suppose X is a random variable and h is a monotonically increasing function such that $h(X)$ has a distribution function well approximated by G. Since $\Pr[X \le x]$ is the same as $\Pr[h(X) \le h(x)]$, the distribution function for X should be well approximated by $G[h(x)]$. For example, if X has a chi-square distribution with k degrees of freedom, $\sqrt{X} - \sqrt{k}$ has approximately a $N(0, \frac{1}{2})$ distribution when k is large. Hence, one might approximate $\Pr[X \le x]$ by $\Phi[\sqrt{2x} - \sqrt{2k}]$. A better approach, due to Wilson and Hilferty (1931), is to treat $(X/k)^{1/3}$ as $N(1, 2/9k)$ and to approximate $\Pr[X \le x]$ by $\Phi[((x/k)^{1/3} - 1)\sqrt{9k/2}]$.

Fisher's z transformation is another well-known example of this technique. The sample correlation coefficient $\hat{\rho}$ based on random sampling from a bivariate normal distribution is highly skewed if the population coefficient ρ is large in absolute value. However, $z = h(\hat{\rho}) = \log(1 - \hat{\rho})/(1 + \hat{\rho})$ has rather little skewness and is well approximated by a normal random variable with mean $\log(1\text{-}\rho)/(1+\rho)$ and variance n^{-1}. Thus, $\Pr[\hat{\rho} \le x]$ can be approximated by $\Phi[\sqrt{n}\, h(x) - \sqrt{n}\, h(\rho)]$ for moderate sample size n.

Using transformations to approximate distributions is an art. Sometimes, as in the correlation coefficient case, the geometry of the problem suggests the appropriate transformation h. Since $\hat{\rho}$ can be interpreted as the cosine of the angle between two normal random vectors, an inverse trigonometric transformation is suggested. In other cases, arguments based on approximate moments are useful. Suppose $h(X)$ can be expanded in a power series around the point $\mu = \mathrm{E}(X)$:

$$h(X) = h(\mu) + h'(\mu)(X - \mu) + \tfrac{1}{2}h''(\mu)(X - \mu)^2 + \cdots, \tag{2.2}$$

where $X - \mu$ is in some sense small.[1] Then we might act as though:

$$\begin{aligned}
&\mathrm{E}(h) \approx h(\mu) + \tfrac{1}{2}h''(\mu)\mathrm{E}(X - \mu)^2,\\
&\mathrm{Var}(h) \approx [h'(\mu)]^2 \mathrm{Var}(X),\\
&\mathrm{E}(h - \mathrm{E}h)^3 \approx [h'(\mu)]^3 \mathrm{E}(X - \mu)^3 + \tfrac{3}{2}[h'(\mu)]^2 h''(\mu)\left[\mathrm{E}(X - \mu)^4 - \mathrm{Var}^2(X)\right],
\end{aligned}$$

and choose h so that these approximate moments match the moments of the approximating distribution. If the approximating distribution is chosen to be normal, we might require that $\mathrm{Var}(h)$ be a constant independent of μ; or we might want the third moment to be zero. If the moments of X are (approximately) known and the above approximations used, either criterion gives rise to a differential equation in $h(\mu)$. The cube-root transformation for the chi-square random variable can be motivated on the grounds it makes the approximate third moment of h equal to zero. The Fisher transformation for $\hat{\rho}$ stabilizes the approximate variance of h so that it is independent of ρ.

[1]For example, if X is a statistic from a sample of size n, its variance might be proportional to n^{-1}. Expansions like (2.2) are discussed in detail in Section 3 below.

Transformations are discussed in detail by Johnson (1949) and illustrated by numerous examples in Johnson and Kotz (1970). Jenkins (1954) and Quenouille (1948) apply inverse trigonometric transformations to the case of time-series autocorrelation coefficients. The use of transformations in econometrics, however, seems minimal, probably because the method is well developed only for univariate distributions. Nevertheless, as an approach to approximating highly skewed distributions, transformations undoubtedly merit further study.

2.4. Asymptotic expansions

Often it is possible to embed the distribution problem at hand in a sequence of similar problems. If the sequence has a limit which is easy to solve, one might approximate the solution of the original problem by the solution of the limit problem. The sequence of problems is indexed by a parameter which, in many econometric applications, is the sample size n. Suppose, for example, one wishes to approximate the probability distribution of an estimator of a parameter θ based on a sample. We define an infinite sequence $\hat{\theta}_n$ of such estimators, one for each sample size $n = 1, 2, \ldots$, and consider the problem of deriving the distribution of each $\hat{\theta}_n$. Of course, we must also describe the joint probability distribution of the underlying data for each n. Given such a sequence of problems, the asymptotic approach involves three steps: (a) A simple monotonic transformation $T_n = h(\hat{\theta}_n; \theta, n)$ is found so that the distribution of the transformed estimator T_n is not very sensitive to the value n. Since most interesting estimators are centered at the true parameter and have dispersion declining at the rate $n^{-1/2}$, the linear transformation $T_n = \sqrt{n}(\hat{\theta}_n - \theta)$ is often used. (b) An approximation $G_n(x)$ to the distribution function $F_n(x) \equiv \Pr[T_n \leq x]$ is found so that, as n tends to infinity, the error $|G_n(x) - F_n(x)|$ goes to zero. (c) The distribution function for $\hat{\theta}_n$ is approximated using G_n; that is, $\Pr[\hat{\theta}_n \leq t] \equiv \Pr[T_n \leq h(t; \theta, n)]$ is approximated by $G_n[h(t; \theta, n)]$.

For many econometric estimators $\sqrt{n}(\hat{\theta}_n - \theta)$ is asymptotically normal. Hence, using the linear transformation for h, one may choose a normal distribution function for G_n. However, it is possible to develop other approximations. Let $G_n(x)$ be an approximation to the continuous distribution function $F_n(x)$. If, for all x,

$$\lim_{n \to \infty} n^r |F_n(x) - G_n(x)| = 0,$$

we write

$$F_n(x) = G_n(x) + o(n^{-r})$$

and say that G_n is a $o(n^{-r})$ approximation to F_n. (A similar language can be developed for approximating density functions.) The asymptotic distribution is a $o(n^0)$ approximation.

The number r measures the speed at which the approximation error goes to zero as n approaches infinity. Of course, for given sample size n, the value of r does not tell us anything about the goodness of the approximation. If, however, we have chosen the transformation h cleverly so that F_n and G_n vary smoothly with n, the value of r might well be a useful indicator of the approximation error for moderate values of n.

There are two well-known methods for obtaining higher-order approximate distribution functions based on Fourier inversion of the approximate characteristic function. Let $\psi(t) = \mathrm{E}\exp\{\mathrm{i}tT_n\}$ be the characteristic function for T_n and let $K(t) \equiv \log\psi(t)$ be its cumulant function. If ψ is integrable, the density function f_n for T_n can be written as:[2]

$$f_n(x) = \frac{1}{2\pi}\int_{-\infty}^{\infty} \mathrm{e}^{-\mathrm{i}xt}\psi(t)\,\mathrm{d}t = \frac{1}{2\pi}\int_{-\infty}^{\infty} \mathrm{e}^{-\mathrm{i}xt+K(t)}\,\mathrm{d}t. \tag{2.3}$$

Often $K(t)$ can be expanded in a series where the successive terms are increasing powers of $n^{-1/2}$. The integrand can then be approximated by keeping only the first few terms of the series expansion. Integrating term by term, one obtains a series approximation to f_n; further integration yields a series approximation to the distribution function. The Edgeworth approximation (which is obtained by expanding $K(t)$ around $t = 0$) is the simplest and most common method; it does not require complete knowledge of $K(t)$ and can be calculated from the low-order cumulants of T_n. A detailed discussion of the Edgeworth approximation appears in Section 3. The saddlepoint approximation (which is obtained by expanding $K(t)$ around the "saddlepoint" value t^* that maximizes the integrand) is more complex and requires intimate knowledge of the cumulant function. When available, it typically gives more accurate approximations especially in the tails of the distribution. Daniels (1956) and Phillips (1978) have applied the method to some autocorrelation statistics in time series. Unfortunately, knowledge of the cumulant function is rare in econometric applications; the saddlepoint approximation has therefore received little attention to date and will not be emphasized in the present survey.

Wallace (1958) presents an excellent introduction to asymptotic approximations based on expansions of the characteristic function. An exposition with emphasis on multivariate expansions is given by Barndorff-Nielsen and Cox (1979); the comments on this paper by Durbin (1979) are particularly interesting and suggest new applications of the saddlepoint method. In econometrics, T. W.

[2]Cf. Feller (1981, p. 482).

Anderson, P. C. B. Phillips, and J. D. Sargan have been pioneers in applying asymptotic expansions. Some of their work on estimators and test statistics in simultaneous equations models is surveyed in Section 6 below.

The above discussion of asymptotic expansions has focused on estimators, but there is no difficulty in applying the methods to any sample statistic whose cumulant function can be approximated by a power series in $n^{-1/2}$. Furthermore, the parameter which indexes the sequence of problems need not be the sample size. In the context of the simultaneous equations model, Kadane (1971) suggested that it might be more natural to consider a sequence indexed by the error variance. In his "small σ" analysis, the reduced-form error-covariance matrix is written as $\sigma\Omega$; in the sequence, the sample size and the matrix Ω are fixed, but σ approaches zero. Edgeworth and saddlepoint expansions are available as long as one can expand the cumulant function in a power series where successive terms are increasing powers of σ. Anderson (1977) explores this point of view in the context of single-equation estimation in structural models.

2.5. *Ad hoc methods*

Certain statistics permit approximations which take advantage of their special structure. Consider, for example, the ratio of two random variables, say $T = X_1/X_2$. If X_2 takes on only positive values, $\Pr[T \le x] = \Pr[X_1 - xX_2 \le 0]$. If both X_1 and X_2 are sums of independent random variables possessing finite variance, then the distribution of $X_1 - xX_2$ might be approximated by a normal. Defining $\mu_i = \mathrm{E}X_i$ and $\sigma_{ij} = \mathrm{E}(X_i - \mu_i)(X_j - \mu_j)$, we might approximate $\Pr[T \le x]$ by:

$$\Phi\left[(x\mu_2 - \mu_1)\Big/\sqrt{\sigma_{11} - 2\sigma_{12}x + \sigma_{22}x^2}\right].$$

Even if X_2 is not always positive, as long as $\Pr[X_2 \le 0]$ is negligible, the above approximation might be reasonable.

An important example of this situation occurs when X_1 and X_2 are quadratic forms in normal variables. Suppose $X_1 = z'Az$ and $X_2 = z'Bz$, where z is $N_p(0, \Sigma)$. Then, by a rotation of the coordinate system, $X \equiv X_1 - xX_2$ can be written as:

$$X = z'(A - xB)z = \sum_{i=1}^{p} \lambda_i y_i^2,$$

where the λ_i are the characteristic roots of $\Sigma(A - xB)$ and the y_i are independent $N(0,1)$ random variables. If p is moderately large (say, 20 or more) and the λ_i are not too dispersed, a central limit theorem might be invoked and the distribution of X approximated by a normal with mean $\operatorname{tr}\Sigma(A - xB)$ and variance $2\,\mathrm{tr}[\Sigma(A-xB)]^2$. If necessary, Edgeworth or saddlepoint expansions (in powers of $p^{-1/2}$) could be employed to obtain greater accuracy.

In this quadratic case, approximations can be dispensed with entirely. The exact distribution of a weighted sum of independent chi-square random variables can be obtained by one-dimensional numerical integration using the algorithms of Imhof (1961) or Pan Jie-jian (1968). Koerts and Abrahamse (1969) and Phillips (1977a), among others, have used these methods to calculate exact distributions of some time-series statistics. For large p, numerical integration is unnecessary since the Edgeworth approximation to X is likely to be adequate. [Cf. Anderson and Sawa (1979).]

The least-squares estimator of a single coefficient in a linear regression equation can always be written as a ratio. In particular, when one of the regressors is endogenous, its coefficient estimator is the ratio of two quadratic forms in the endogenous variables. Thus ratios occur often in econometrics and their simple structure can easily be exploited. The multivariate generalization $X_2^{-1}X_1$, where X_2 is a random square matrix and X_1 is a random vector, also has a simple structure, but approximation methods for this case seem not to have been explored.

In practice, ad hoc techniques which take advantage of the special structure of the problem are invaluable for developing simple approximations. General methods with universal validity have attractive theoretical features, but are not particularly accurate for any given problem. Approximating distributions is an art involving judgment and common sense, as well as technical skill. The methods discussed in this section are not distinct alternatives. Every approximation involves fitting a curve to a transformed statistic, dropping terms which are judged to be small. In the end, many approaches are merged in an attempt to find a reasonable solution to the problem at hand.

3. Edgeworth approximations

Perhaps the most important and commonly used method to obtain improved approximations to the distributions of estimators and test statistics in econometrics is the Edgeworth expansion. There are a number of reasons for this prominence. First, the method is a natural extension of traditional large-sample techniques based on the central limit theorem. The usual asymptotic approximation is just the leading term in the Edgeworth expansion. Second, since the expansion is based on the normal and chi-square distributions—which are familiar and well tabulated—it is easy to use. Finally, the method can be employed to approximate the distributions of most of the commonly used estimators and test statistics and is very convenient for comparing alternative statistical procedures. Indeed, it is the basis for a general theory of higher-order efficient estimators and tests.

Because of its prominence, the Edgeworth approximation will be described at some length in this section and in the examples which follow. However, it is worth noting at the outset that Edgeworth methods do not lead to particularly accurate approximations. To the contrary, in nearly every application, there exist alternative curve-fitting techniques yielding more satisfactory numerical results. Edgeworth is important, not for its accuracy, but for its general availability and simplicity. Although rarely optimal, it is often quite adequate and leads to a useful, comprehensive approach to second-order comparisons of alternative procedures.

Our discussion of the Edgeworth expansion parallels the traditional approach to asymptotic distribution theory as presented in Theil (1971, ch. 8) or Bishop, Fienberg, and Holland (1975, ch. 14). We first consider the problem of approximating sums of independent random variables. Then we show that the theory also applies to smooth functions of such sample sums. To avoid excessive length and heavy mathematics, our presentation will be quite informal; rigorous proofs and algebraic detail can be found in the literature cited. Although Edgeworth expansions to high order are often available, in practice one rarely goes beyond the first few terms. We shall develop the expansion only up to terms of order n^{-1} and refer to the result as the "second-order" or $o(n^{-1})$ Edgeworth approximation. The extension to higher terms is in principle straightforward, but the algebra quickly becomes extremely tedious.

3.1. *Sums of independent random variables*

Suppose $X_1, X_2, \ldots$ form an infinite sequence of independent random variables with common density function f; each X_i has mean zero, variance one, and possesses moments up to the fourth order. If ψ is the characteristic function associated with f, then the cumulant function $\log\psi$ possesses derivatives up to the fourth order and can be expanded in a neighborhood of the origin as a power series:

$$\log\psi(t) = \tfrac{1}{2}(\mathrm{i}t)^2 + \tfrac{1}{6}k_3(\mathrm{i}t)^3 + \tfrac{1}{24}k_4(\mathrm{i}t)^4 + \cdots, \tag{3.1}$$

where k_r is the rth cumulant of f.

The standardized sum $T_n = \sum X_i/\sqrt{n}$ also has mean zero and variance one; let f_n and ψ_n be its density and characteristic functions. Since

$$\begin{aligned}\log\psi_n(t) &= n\log\psi(t/\sqrt{n})\\ &= \tfrac{1}{2}(\mathrm{i}t)^2 + \frac{1}{6\sqrt{n}}k_3(\mathrm{i}t)^3 + \frac{1}{24n}k_4(\mathrm{i}t)^4 + \cdots,\end{aligned}$$

we observe that the rth cumulant of T_n is simply $k_r n^{1-r/2}$, for $r>2$. Thus, the high-order cumulants are small when n is large. Since the function e^x has the expansion $1+x+\frac{1}{2}x^2+\cdots$, when x is small, $\psi_n(t)$ can be written as a power series in $n^{-1/2}$:

$$\psi_n(t)=\exp\{\log\psi_n(t)\}$$

$$=e^{-1/2t^2}\left[1+\frac{1}{6\sqrt{n}}k_3(\mathrm{i}t)^3+\frac{3k_4(\mathrm{i}t)^4+k_3^2(\mathrm{i}t)^6}{72n}+\cdots\right]. \tag{3.2}$$

The $o(n^{-1})$ Edgeworth approximation to the density function for T_n is obtained by applying the Fourier inversion formula (2.3) and dropping high-order terms. Using the fact that, if f has characteristic function $\psi(t)$, then the rth derivative $f^{(r)}$ has characteristic function $(-\mathrm{i}t)^r\psi(t)$, the inverse Fourier transform is seen to be:

$$f(x)\approx\varphi(x)-\frac{1}{6\sqrt{n}}k_3\varphi^{(3)}(x)+\frac{1}{24n}k_4\varphi^{(4)}(x)+\frac{1}{72n}k_3^2\varphi^{(6)}(x)$$

$$\approx\varphi(x)\left[1+\frac{k_3H_3(x)}{6\sqrt{n}}+\frac{3k_4H_4(x)+k_3^2H_6(x)}{72n}\right], \tag{3.3}$$

where $\varphi^{(r)}$ is the rth derivative of the normal density function φ and H_r is the Hermite polynomial of degree r defined as:

$$H_r(x)=(-1)^r\frac{\varphi^{(r)}(x)}{\varphi(x)}.$$

(By simple calculation, $H_3(x)=x^3-3x$, $H_4(x)=x^4-6x^2+3$, etc.) Integration of (3.3) gives an approximation for the distribution function:

$$F_n(x)\approx\Phi(x)-\varphi(x)\left[\frac{k_3H_2(x)}{6\sqrt{n}}+\frac{3k_4H_3(x)+k_3^2H_5(x)}{72n}\right]. \tag{3.4}$$

This latter formula can be rewritten as:

$$F_n(x)\approx\Phi\left[x-\frac{k_3(x^2-1)}{6\sqrt{n}}+\frac{3k_4(3x-x^3)+2k_3^2(4x^3-7x)}{72n}\right], \tag{3.5}$$

by use of Taylor series expansion and the definition of the Hermite polynomials.

Equations (3.4) and (3.5) are two variants of the $o(n^{-1})$ Edgeworth approximation to the distribution function of T_n; Phillips (1978) refers to them as the Edgeworth-A and Edgeworth-B approximations, respectively. The latter is closely related to the Cornish–Fisher (1937) normalizing expansion for a sample statistic. If (3.5) is written as:

$$\Pr[T_n \le x] \approx \Phi\left[x + \frac{g_1(x)}{\sqrt{n}} + \frac{g_2(x)}{n}\right],$$

then, when n is large enough to ensure that the function in brackets is monotonic for the x values of interest, it can be rewritten as:

$$\Pr\left[T_n + \frac{g_1(T_n)}{\sqrt{n}} + \frac{g_2(T_n)}{n} \le x\right] \approx \Phi[x].$$

Thus, the function inside brackets in (3.5) can be viewed as a transformation h, making $h(T_n)$ approximately normal.

The argument sketched above is, of course, purely formal. There is no guarantee that the remainder terms dropped in the manipulations are really small. However, with a little care, one can indeed prove that the Edgeworth approximations are valid asymptotic expansions. Suppose the power series expansion of ψ_n is carried out to higher order and $G_n^r(x)$ is the analogous expression to (3.4) when terms up to order $n^{-r/2}$ are kept. Then, if the X_i possess moments to order $r+2$ and $|\psi(t)|$ is bounded away from one for large t:

$$\Pr[T_n \le x] = G_n^r(x) + o(n^{-r/2}).$$

A proof can be found in Feller (1971, pp. 538–542). The assumption on the characteristic function rules out discrete random variables like the binomial. Since the distribution of a standardized sum of discrete random variables generally has jumps of height $n^{-1/2}$, it is not surprising that it cannot be closely approximated by a continuous function like (3.4). Edgeworth-type approximations for discrete random variables are developed in Bhattacharya and Rao (1976), but will not be described further in the present survey.

The theory for sums of independent, identically distributed random variables is easily generalized to the case of weighted sums. Furthermore, a certain degree of dependence among the summands can be allowed. Under the same type of regularity conditions needed to guarantee the validity of a central limit theorem, it is possible to show that the distribution functions for standardized sample moments of continuous random variables possess valid Edgeworth expansions as long as higher-order population moments exist.

3.2. A general expansion

Since few econometric estimators or test statistics are simple sums of random variables, these classical asymptotic expansions are not directly applicable. Nevertheless, just as the delta method[3] can be applied to obtain first-order asymptotic distributions for smooth functions of random variables satisfying a central limit theorem, a generalized delta method can be employed for higher-order expansions. With simple modifications, the classical formulae (3.3)–(3.5) are valid for most econometric statistics possessing limiting normal distributions.

Nagar (1959) noted that k-class estimators in simultaneous equations models can be expanded in formal series where the successive terms are increasing powers of $n^{-1/2}$. The expansions are essentially multivariate versions of (2.2). The rth term takes the form $C_r n^{-r/2}$, where C_r is a polynomial in random variables with bounded moments. His approach is to keep the first few terms in the expansion and to calculate the moments of the truncated series. These moments can be interpreted as the moments of a statistic which serves to approximate the estimator. [In some circumstances, these moments can be interpreted as approximations to the actual moments of the estimator; see, for example, Sargan (1974).]

Nagar's approach is quite generally available and can be used to develop higher-order Edgeworth approximations. Most econometric estimators and test statistics, after suitable standardization so that the center and dispersion of the distributions are stabilized, can be expanded in a power series in $n^{-1/2}$ with coefficients that are well behaved random variables. Suppose, for example, T_n is a standardized statistic possessing the stochastic expansion:

$$T_n = X_n + \frac{A_n}{\sqrt{n}} + \frac{B_n}{n} + \frac{R_n}{n\sqrt{n}}, \tag{3.6}$$

where X_n, A_n, and B_n are sequences of random variables with limiting distributions as n tends to infinity. If R_n is stochastically bounded,[4] the limiting distribution of T_n is the same as the limiting distribution of X_n. It is natural to use the information in A_n and B_n to obtain a better approximation to the distribution of T_n. Suppose the limiting distribution of X_n is $N(0,1)$. Let $T' = X_n + A_n n^{-1/2} + B_n n^{-1}$ be the first three terms of the stochastic expansion of T_n. For a large class of cases, T' has finite moments up to high order and its rth cumulant is of order

[3] Suppose a standardized sample mean $X_n = \sqrt{n}(\bar{x}-\mu)/\sigma$ is asymptotically $N(0,1)$ and g is a differentiable function with derivative $b \equiv g'(\mu)$. The delta method exploits the fact that, when n is large, $T_n = \sqrt{n}[g(\bar{x}) - g(\mu)]$ behaves like $b\sigma X_n$; hence T_n is asymptotically $N(0, b^2\sigma^2)$. Cf. Theil (1971, pp. 373–374).

[4] A sequence of random variables Z_n is stochastically bounded if, for every $\varepsilon > 0$, there exists a constant c such that $\Pr[|Z_n| > c] < \varepsilon$, for sufficiently large n. That is, the distribution function does not drift off to infinity. Cf. Feller (1971, p. 247).

$n^{(2-r)/2}$ when r is greater than 2. Furthermore, its mean and variance can be written as:

$$\mathrm{E}(T') = \frac{a}{\sqrt{n}} + o(n^{-1}),$$

$$\mathrm{Var}(T') = 1 + \frac{b}{n} + o(n^{-1}),$$

where a and b depend on the moments of X_n, A_n, and B_n. The restandardized variable,

$$T^* = \frac{T' - \frac{a}{\sqrt{n}}}{\sqrt{1+b/n}},$$

has, to order n^{-1}, zero mean and unit variance. Its third and fourth moments are:

$$\mathrm{E}(T^*)^3 = \frac{c}{\sqrt{n}} + o(n^{-1}),$$

$$\mathrm{E}(T^*)^4 = 3 + \frac{d}{n} + o(n^{-1}),$$

where $c/\sqrt{n}$ is the approximate third cumulant of T' and d/n is the approximate fourth cumulant. Since the cumulants of T' behave like the cumulants of a standardized sum of independent random variables, one is tempted to use the Edgeworth formulae to approximate its distribution. For example, one might approximate $\Pr[T^* \le x]$ by (3.5) with c replacing k_3 and d replacing k_4. Dropping the remainder term and using the fact that

$$\Pr[T' \le x] = \Pr\left[T^* \le \frac{x - a/\sqrt{n}}{\sqrt{1+b/n}}\right]$$

$$= \Pr\left[T^* \le x - \frac{a}{\sqrt{n}} - \frac{bx}{2n} + o(n^{-1})\right],$$

we are led to the Edgeworth-B approximation:

$$\Pr[T_n \le x] \approx \Phi\left[x + \frac{\gamma_1 + \gamma_2 x^2}{6\sqrt{n}} + \frac{\gamma_3 x + \gamma_4 x^3}{72n}\right], \tag{3.7}$$

where

$$\gamma_1 = c - 6a; \qquad \gamma_3 = 9d - 14c^2 - 36b + 24ac,$$

$$\gamma_2 = -c; \qquad \gamma_4 = 8c^2 - 3d.$$

A similar calculation using (3.4) leads to an Edgeworth-A form of the approximation.

Of course, the above discussion in no way constitutes a proof that the approximation (3.7) has an error $o(n^{-1})$. We have dropped the remainder term $R_n/n\sqrt{n}$ without justification; and we have used the Edgeworth formulae despite the fact that T' is not the sum of n independent random variables. With some additional assumptions, however, such a proof can in fact be constructed.

If $|T_n - T'|$ is stochastically of order $o(n^{-1})$, it is reasonable to suppose that the distribution functions for T_n and T' differ by that order. Actually, further assumptions on the tail behavior of R_n are required. Using a simple geometric argument, Sargan and Mikhail (1971) show that, for all x and ε,

$$|\Pr(T_n \leq x) - \Pr(T' \leq x)| \leq \Pr\left[|T_n - T'| > \varepsilon\right] + \Pr\left[|T' - x| < \varepsilon\right].$$

If T' has a bounded density, the last term is of order ε, as ε approaches zero. To show that the difference between the two distribution functions is $o(n^{-1})$ we choose ε to be of that order. Setting $\varepsilon = n^{-3/2}\log^c n$, we find that a sufficient condition for validly ignoring the remainder term is that there exists a positive constant c such that:

$$\Pr\left[|R_n| > \log^c n\right] = o(n^{-1}). \tag{3.8}$$

That is, the tail probability of R_n must be well behaved as n approaches infinity. If R_n is bounded by a polynomial in normal random variables, (3.8) is necessarily satisfied.

To show that T' can be approximated by the Edgeworth formulae, one must make strong assumptions about the sequences X_n, A_n, and B_n. If A_n and B_n are polynomials in variables which, along with X_n, possess valid Edgeworth expansions to order n^{-1}, the results of Chibisov (1980) can be used to prove a validity theorem. The special case where (3.6) comes from the Taylor series expansion of a smooth function $g(p)$, where p is a vector of sample moments, has been studied by Bhattacharya and Ghosh (1978), Phillips (1977b), and Sargan (1975b, 1976). These authors give formal proofs of the validity of the Edgeworth approximation under various assumptions on the function g and the distribution of p. Sargan (1976) gives explicit formulae for the γ_i of (3.7) in terms of the derivatives of the function g and the cumulants of p.

It may be useful to illustrate the approach by a simple example. Suppose $\bar{x}$ and s^2 are the sample mean and (bias adjusted) sample variance based on n independent draws from a $N(\mu, \sigma^2)$ distribution. We shall find the Edgeworth approximation to the distribution of the statistic:

$$T_n = \frac{\sqrt{n}(\bar{x} - \mu)}{s},$$

which, of course, is distributed exactly as Student's t. With $X_n = \sqrt{n}(\bar{x}-\mu)/\sigma$ and $Y_n = \sqrt{n}(s^2-\sigma^2)/\sigma^2$, the statistic can be written as:

$$T_n = \frac{X_n}{\sqrt{1+Y_n/\sqrt{n}}} = X_n - \frac{X_nY_n}{2\sqrt{n}} + \frac{3X_nY_n^2}{8n} + \frac{R_n}{n\sqrt{n}},$$

where the remainder term R_n is stochastically bounded. The random variable X_n is $N(0,1)$; Y_n is independent of X_n with mean zero and variance $2n/(n-1)$. It is easy to verify that T_n satisfies the assumptions of Sargan (1976) and hence can be approximated by a valid Edgeworth expansion. Dropping the remainder term, we find that T' has mean zero and variance $1+2n^{-1}+o(n^{-1})$. Its third cumulant is exactly zero and its fourth cumulant is approximately $6n^{-1}$. Thus, with $a=c=0$, $b=2$, and $d=6$, (3.7) becomes:

$$\Pr[T_n \le x] \approx \Phi\left[x - \frac{x+x^3}{4n}\right], \tag{3.9}$$

which is a well-known approximation to the Student-t distribution function.

There are available a number of alternative algorithms for calculating Edgeworth expansions. The use of (3.7) with Nagar-type approximate moments is often the simplest. Sometimes, however, the moment calculations are tedious and other methods are more convenient. If, for example, the exact characteristic function for T_n is known, it can directly be expanded in a power series without the need to calculate moments. The Edgeworth approximation can be found by Fourier inversion of the first few terms of the series. Anderson and Sawa (1973) employ this method in their paper on the distribution of k-class estimators in simultaneous equations models.

An alternative approach, used by Hodges and Lehmann (1967), Albers (1978), Anderson (1974), and Sargan and Mikhail (1971), exploits the properties of the normal distribution. Suppose the stochastic expansion (3.6) can be written as:

$$T_n = X_n + \frac{A(X_n,Y_n)}{\sqrt{n}} + \frac{B(X_n,Y_n)}{n} + \frac{R_n}{n\sqrt{n}},$$

where R_n satisfies (3.8), X_n is exactly $N(0,1)$, and the vector Y_n is independent of X_n with bounded moments. The functions A and B are assumed to be smooth in both arguments with A' denoting the derivative of A with respect to X_n. Then, conditioning on Y_n and supressing the subscripts, we write:

$$\begin{aligned}\Pr[T' \le x] &= \mathrm{E}_Y \Pr\left[X + \frac{A(X,Y)}{\sqrt{n}} + \frac{B(X,Y)}{n} \le x \,\middle|\, Y\right] \\ &\approx \mathrm{E}_Y \Phi\left[x - \frac{A(x,Y)}{\sqrt{n}} - \frac{B(x,Y)-A(x,Y)A'(x,Y)}{n}\right].\end{aligned}$$

The approximation comes from dropping terms of higher order when inverting the inequality. Taking expectation of the Taylor series expansion of Φ, we obtain the approximation:

$$\Pr[T_n \le x] \approx \Phi\left[x - \mathrm{E}_Y \frac{A(x,Y)}{\sqrt{n}} - \mathrm{E}_Y \frac{B(x,Y) - A(x,Y)A'(x,Y) + \frac{1}{2}x\mathrm{Var}_Y A(x,Y)}{n}\right]. \tag{3.10}$$

Of course, some delicate arguments are needed to show that the error of approximation is $o(n^{-1})$; some conditions on the functions A and B are clearly necessary. Typically, A and B are polynomials and the expectations involved in (3.10) are easy to evaluate. In our Student-t example, we find from elementary calculation $\mathrm{E}_Y(A) = 0$, $\mathrm{E}_Y(B) = 3x/4$, $\mathrm{E}_Y(AA') = x/2$, and $\mathrm{Var}_Y(A) = x^2/2$; hence, we obtain the approximation (3.8) once again.

3.3. *Non-normal expansions*

Edgeworth approximations are not restricted to statistics possessing limiting normal distributions. In the case of multivariate test statistics, the limiting distribution is typically chi-square and asymptotic expansions are based on that distribution. The following general algorithm is developed by Cavanagh (1983). Suppose the sample statistic T_n can be expanded as:

$$T_n = X_n + \frac{A_n}{\sqrt{n}} + \frac{B_n}{n} + \frac{R_n}{n\sqrt{n}},$$

where X_n has, to order n^{-1}, the distribution function F and density function f; the random variables A_n and B_n are stochastically bounded with conditional moments:

$$a(x) = \mathrm{E}(A_n | X_n = x),$$
$$b(x) = \mathrm{E}(B_n | X_n = x),$$
$$v(x) = \mathrm{Var}(A_n | X_n = x),$$

that are smooth functions of x. Define the derivative functions $a' = \mathrm{d}a/\mathrm{d}x$, $v' = \mathrm{d}v/\mathrm{d}x$ and

$$c(x) = \frac{\mathrm{d}\log f(x)}{\mathrm{d}x}.$$

Then, assuming R_n is well behaved and can be ignored, the formal second-order Edgeworth approximation to the distribution of T is given by:

$$\Pr[T_n \le x] \approx F\left[x - \frac{a(x)}{\sqrt{n}} + \frac{2a(x)a'(x) + c(x)v(x) + v'(x) - 2b(x)}{2n}\right]. \tag{3.11}$$

Again, many technical assumptions on the random variables X_n, A_n, and B_n will be needed to prove the validity of the approximation. They seem to be satisfied, however, in actual applications.

For example, suppose z_n is distributed as $N_q(0, I)$ and y_n is a vector, independent of z_n, with zero mean and bounded higher moments. In many hypothesis testing problems the test statistics, under the null hypothesis, posess a stochastic expansion of the form:

$$T_n = z'z + \frac{A(z, y)}{\sqrt{n}} + \frac{B(z, y)}{n} + \frac{R}{n\sqrt{n}},$$

where A is linear in y (for given z) and $A(0, y) = 0$. (Again, the subscript n is dropped to simplify the notation.) Since F in this case is the chi-square distribution function with q degrees of freedom, $c(x) = (q - 2 - x)/2x$. Typically, $a(x) = \mathrm{E}(A|z'z = x) = 0$; $b(x)$ and $v(x)$ are usually homogeneous quadratic functions of x. Thus, using (3.11), we find an approximation of the form:

$$\Pr[T_n \le x] \approx F\left[x + \frac{\beta_1 x + \beta_2 x^2}{n}\right], \tag{3.12}$$

where the β_i are functions of the moments of (z, y). Sargan (1980) gives a detailed derivation for the case where the stochastic expansion arises from a Taylor expansion of a function of moments. Rothenberg (1977, 1981b) analyzes the noncentral case where the mean of z is nonzero and F is the noncentral chi-square distribution.

To summarize, many econometric estimators and test statistics possess, after suitable standardization, stochastic expansions of the form (3.6). It is usually easy to demonstrate that R_n satisfies a regularity condition like (3.8) and the remainder term can be ignored. A formal second-order Edgeworth expansion for the truncated variable T' can be obtained from its moments, using any of the algorithms discussed above. For most econometric applications, the limiting distributions are normal or chi-square and the correction terms A_n and B_n are polynomials in asymptotically normal random variables. Thus the formal Edgeworth approximation is relatively easy to calculate, as we shall demonstrate in later sections. Proofs that the approximation error is indeed $o(n^{-1})$ are much harder. The results of Sargan (1976, 1980) and Phillips (1977b) cover most of the cases met in practice.

4. Second-order comparisons of estimators

In any econometric inference problem, many different ways to estimate the unknown parameters are available. Since the exact sampling distributions are often unknown, choice among the alternative estimators has traditionally been based on asymptotic approximations. Typically, however, there are a number of estimators having the same limiting distributions. In those cases, second-order Edgeworth approximations can be used to distinguish among the asymptotically equivalent procedures. Indeed, a rich and powerful theory of second-order estimation efficiency has developed recently in the statistical literature. Although most of the results concern single-parameter estimation from simple random sampling, the extension of this theory to typical econometric problems is apparent.

Second-order comparisons of estimators based on moments calculated from the first few terms of stochastic expansions have been employed extensively in econometrics after the pioneering work of Nagar (1959). Some recent examples are Amemiya (1980), Fuller (1977), and Taylor (1977). Since the estimators being examined often do not possess finite moments, the status of such comparisons has been questioned by Srinivasan (1970) and others. However, if the calculated expectations are interpreted as the moments of an approximating distribution, it does not seem unreasonable to use them for comparison purposes. In fact, Pfanzagl and Wefelmeyer (1978a) show that most of the general conclusions derivable from second-order moment calculations can be restated in terms of Edgeworth approximations to the quantiles of the probability distributions.

A more serious objection to the econometric work using Nagar-type moment calculations is the lack of strong results. When the alternative estimators have different biases, mean-square-error comparisons typically are inconclusive. No estimator is uniformly best to second order. The comparisons, however, take on new meaning when interpreted in light of the general theory of second-order efficiency. This theory, although initiated over fifty years ago by R. A. Fisher (1925) and explored by C. R. Rao (1961, 1963), has reached maturity only within the past decade. The summary presented here is based on Akahira and Takeuchi (1981), Efron (1975), Ghosh and Subramanyam (1974), and Pfanzagl and Wefelmeyer (1978a, 1979).[5]

[5] Many of these statisticians use the term "second-order" to describe expansions with error $o(n^{-1/2})$ and would refer to our $o(n^{-1})$ Edgeworth approximations as "third-order". Hence, they speak of third-order efficiency.

4.1. General approach

For simplicity, we begin by considering the one-dimensional case under quadratic loss. An unknown parameter θ is to be estimated from observations on a random vector y whose joint probability distribution is $f(y,\theta)$. Under exact sample theory we would evaluate an estimator $\hat{\theta}$ by its mean square error $\mathrm{E}(\hat{\theta}-\theta)^2$. When exact distributions are unavailable, we consider a sequence of estimation problems indexed by the sample size n and use limiting distributions as approximations. For most applications, the commonly proposed estimators converge in probability to the true parameter at rate $n^{-1/2}$ and the standardized estimators are asymptotically normal. These estimators can be evaluated using expectations calculated from the approximating normal distributions. We shall denote such expectations by the symbol E_1.

Suppose $\sqrt{n}(\hat{\theta}-\theta)$ converges to a $N[\mu(\theta),\sigma^2(\theta)]$ distribution where μ and σ^2 are continuous functions of θ in a neighborhood of the true parameter value. Then, first-order mean square error $\mathrm{E}_1(\hat{\theta}-\theta)^2$ is given by $[\mu^2(\theta)+\sigma^2(\theta)]n^{-1}$. We define $\mathscr{S}_1$ to be the set of all such asymptotically normal estimators and consider the problem of finding the best estimator in $\mathscr{S}_1$. Under certain regularity conditions on the density f, the inverse information term can be shown to be a lower bound for the approximate mean square error. That is, for all $\hat{\theta}$ in $\mathscr{S}_1$:

$$n\mathrm{E}_1(\hat{\theta}-\theta)^2 \geq \frac{1}{\lambda(\theta)},$$

where

$$\lambda(\theta) = -\lim \frac{1}{n}\mathrm{E}\frac{\partial^2 \log f(y,\theta)}{\partial\theta^2}$$

is the limiting average information term for f.

An estimator in $\mathscr{S}_1$ whose approximate mean square error attains the lower bound is called asymptotically efficient. Typically, the standardized maximum likelihood estimator $\sqrt{n}(\hat{\theta}_{\mathrm{M}}-\theta)$ converges to a $N[0,\lambda^{-1}(\theta)]$ distribution and hence $\hat{\theta}_{\mathrm{M}}$ is asymptotically efficient. Of course, any other estimator which is asymptotically equivalent to $\hat{\theta}_{\mathrm{M}}$ will share this property; for example, if $\sqrt{n}(\hat{\theta}_{\mathrm{M}}-\hat{\theta})$ converges in probability to zero, then $\hat{\theta}$ and $\hat{\theta}_{\mathrm{M}}$ will be approximated by the same normal distribution and have the same first-order properties. Under suitable smoothness conditions, minimum distance estimators, Bayes estimators from arbitrary smooth priors, and linearized maximum likelihood estimators are all asymptotically efficient. [See, for example, Rothenberg (1973).] It seems natural to compare these estimators using second-order asymptotic approximations.

Let $\hat{\theta}$ be an estimator which, after standardization, possesses an asymptotic expansion of the form:

$$\sqrt{n}(\hat{\theta}-\theta) = X_n + \frac{A_n}{\sqrt{n}} + \frac{B_n}{n} + \frac{R_n}{n\sqrt{n}}, \tag{4.1}$$

where X_n, A_n, and B_n are random variables with bounded moments and limiting distributions as n tends to infinity. Suppose the limiting distribution of X_n is $N[0,\lambda^{-1}(\theta)]$ and A_n, B_n, and R_n are well behaved so that $\hat{\theta}$ is asymptotically efficient and has a distribution which can be approximated by a valid $o(n^{-1})$ Edgeworth expansion. We shall denote by $\mathscr{S}_2$ the set of all such estimators. Expectations calculated from the second-order approximate distributions will be denoted by E_2; thus, $\mathrm{E}_2(\hat{\theta}-\theta)^2$ is the mean square error when the actual distribution of $\hat{\theta}$ is replaced by the $o(n^{-1})$ Edgeworth approximation. These "second-order" moments are equivalent to those obtained by Nagar's technique of term-by-term expectation of the stochastic expansion (4.1).

4.2. *Optimality criteria*

Since the maximum likelihood estimator has minimum (first-order) mean square error in the set $\mathscr{S}_1$, it is natural to ask whether it has minimum second-order mean square error in $\mathscr{S}_2$. The answer, however, is no. If $\hat{\theta}$ is an estimator in $\mathscr{S}_2$ and θ_0 is some constant in the parameter space, then $\hat{\theta}(1-n^{-1})+\theta_0 n^{-1}$ is also in $\mathscr{S}_2$ and has lower mean square error than $\hat{\theta}$ when θ is close to θ_0. Thus, there cannot be a uniformly best estimator in $\mathscr{S}_2$ under the mean square error criterion.

Following the traditional exact theory of optimal inference [for example, Lehmann (1959, ch. 1)], two alternative approaches are available for studying estimators in $\mathscr{S}_2$. We can give up on finding a "best" estimator and simply try to characterize a minimal set of estimators which dominate all others; or we can impose an unbiasedness restriction thus limiting the class of estimators to be considered. The two approaches lead to similar conclusions.

When comparing two estimators in $\mathscr{S}_2$, it seems reasonable to say that $\hat{\theta}_1$ is as good as $\hat{\theta}_2$ if $\mathrm{E}_2(\hat{\theta}_1-\theta)^2 \le \mathrm{E}_2(\hat{\theta}_2-\theta)^2$ for all θ. If the inequality is sometimes strict, we shall say that $\hat{\theta}_2$ is dominated by $\hat{\theta}_1$. When searching for a good estimator, we might reasonably ignore all estimators which are dominated. Furthermore, nothing is lost by excluding estimators which have the same mean square error as ones we are keeping. Suppose $\bar{\mathscr{S}}_2$ is a subset of $\mathscr{S}_2$ such that, for every estimator excluded, there is an estimator included which is as good. Since, in terms of mean square error, one cannot lose by restricting the search to $\bar{\mathscr{S}}_2$, such a set is called *essentially complete*. The characterization of (small) essentially complete classes is, according to one school of thought, the main task of a theory of estimation.

Unfortunately, essentially complete classes are typically very large and include many unreasonable estimators. If one is willing to exclude from consideration all estimators which are biased, a great simplification occurs. Although all the estimators in $\mathscr{S}_2$ are first-order unbiased, they generally are not second-order unbiased. Let $\hat{\theta}_r$ be an estimator in $\mathscr{S}_2$. Its expectation can be written as:

$$\mathrm{E}_2(\hat{\theta}_r) = \theta + \frac{b_r(\theta)}{n} + O(n^{-2}).$$

Although b_r will generally depend on the unknown parameter, it is possible to construct a second-order unbiased estimator by using the estimated bias function. Define

$$\theta_r^* = \hat{\theta}_r - \frac{b_r(\hat{\theta}_r)}{n}$$

to be the bias-adjusted estimator based on $\hat{\theta}_r$. If $b_r(\theta)$ possesses a continuous derivative, the bias-adjusted estimator has a stochastic expansion

$$\sqrt{n}\,(\theta_r^* - \theta) = \left[1 - \frac{b_r'(\theta)}{n}\right]\sqrt{n}\,(\hat{\theta}_r - \theta) - \frac{b_r(\theta)}{\sqrt{n}} + \frac{R^*}{n\sqrt{n}},$$

where b_r' is the derivative of b_r and R^* is a remainder term satisfying the regularity condition (3.8). Thus, θ_r^* is a second-order unbiased estimator in $\mathscr{S}_2$. All estimators in $\mathscr{S}_2$ with smooth bias functions can be adjusted in this way and hence we can construct the subset $\mathscr{S}_2^*$ of all second-order unbiased estimators. If unbiasedness is a compelling property, the search for a good estimator could be restricted to $\mathscr{S}_2^*$.

4.3. *Second-order efficient estimators*

In the larger class $\mathscr{S}_1$ of all (uniformly) asymptotically normal estimators, the maximum likelihood estimator $\hat{\theta}_{\mathrm{M}}$ is first-order minimum variance unbiased; it also, by itself, constitutes an essentially complete class of first-order minimum mean square error estimators. The extension of this result to the set $\mathscr{S}_2$ is the basis for the so-called "second-order efficiency" property of maximum likelihood estimators. Under certain regularity conditions (which take pages to state), it is possible to prove a theorem with the following conclusion:

The bias-adjusted maximum likelihood estimator

$$\theta_{\mathrm{M}}^* = \hat{\theta}_{\mathrm{M}} - \frac{b_{\mathrm{M}}(\hat{\theta}_{\mathrm{M}})}{n} \tag{4.2}$$

has smallest second-order variance among the set $\mathscr{S}_2^*$ of second-order unbiased estimators possessing $o(n^{-1})$ Edgeworth expansions. Furthermore, the class $\bar{\mathscr{S}}_M$ of all estimators of the form:

$$\hat{\theta}_M + \frac{c(\hat{\theta}_M)}{n}, \tag{4.3}$$

where c is any smooth function, is essentially second-order complete in $\mathscr{S}_2$.

For formal statements and proofs of such a theorem under i.i.d. sampling, the reader is directed to Ghosh, Sinha, and Wieand (1980) and Pfanzagl and Wefelmeyer (1978a).

This basic result of second-order estimation theory does *not* say that the maximum likelihood estimator is optimal. Indeed, it does not say anything at all about $\hat{\theta}_M$ itself. If one insists on having an unbiased estimator, then the adjusted MLE θ_M^* is best. Otherwise, the result implies that, in searching for an estimator with low mean square error (based on second-order approximations to sampling distributions), nothing is lost by restricting attention to certain functions of $\hat{\theta}_M$. The choice of an estimator from the class $\bar{\mathscr{S}}_M$ depends on one's trade-off between bias and variance; or, from a Bayesian point of view, on one's prior. Although commonly referred to as a result on second-order *efficiency*, the theorem really says no more than that $\hat{\theta}_M$ is second-order *sufficient* for the estimation problem at hand.

Of course, the second-order optimality properties of the maximum likelihood estimator are shared by many other estimators. If, for $c(\cdot)$ ranging over the set of smooth functions, the class of estimators $\hat{\theta} + c(\hat{\theta})n^{-1}$ is essentially complete in $\mathscr{S}_2$, then $\hat{\theta}$ is said to be second-order efficient. In addition to the MLE, any Bayes estimate calculated from a symmetric loss function and a smooth prior is second-order efficient. Any estimator possessing the same $o(n^{-1})$ Edgeworth expansion as an estimator in $\bar{\mathscr{S}}_M$ is also second-order efficient.

Although the set of second-order efficient estimators is large, it does not include all first-order efficient estimators. Linearized maximum likelihood and minimum distance estimators are generally dominated by functions of Bayes and ML estimators. Indeed, most common procedures used to avoid maximizing the likelihood function in nonlinear models turn out to be second-order inefficient. For example, as pointed out by Akahira and Takeuchi (1981), two-stage and three-stage least squares estimators in overidentified simultaneous equations models with normal errors are dominated by adjusted limited information and full information maximum likelihood estimators.

A full characterization of the set of second-order efficient estimators is difficult. However, it is interesting to note that, although the single iteration method of scoring (linearized maximum likelihood) does not generally lead to second-order efficient estimators, a two-iteration scoring procedure does. Since the line of reasoning is widely used in the literature, a sketch of the argument may be worthwhile. Suppose the logarithmic likelihood function $L(\theta)\equiv \log f(y,\theta)$ possesses well-behaved derivatives so that valid stochastic expansions can be developed. In particular, for $r=1$, 2, and 3, define $L_r = \mathrm{d}^r L(\theta)/\mathrm{d}\theta^r$; we assume L_r/n converges to the constant λ_r as n approaches infinity and that the standardized derivatives $\sqrt{n}(L_r/n-\lambda_r)$ have limiting normal distributions. Let $\hat{\theta}_0$ be some consistent estimator such that $\sqrt{n}(\hat{\theta}_0-\hat{\theta}_{\mathrm{M}})$ has a limiting distribution. Consider the following iterative procedure for generating estimates starting from $\hat{\theta}_0$:

$$\hat{\theta}_{s+1}=\hat{\theta}_s-\frac{L_1(\hat{\theta}_s)}{L_2(\hat{\theta}_s)},\qquad s=0,1,2,\dots.$$

Assuming an interior maximum so that $L_1(\hat{\theta}_{\mathrm{M}})=0$, we can expand by Taylor series around $\hat{\theta}_{\mathrm{M}}$ obtaining:

$$\begin{aligned}\hat{\theta}_{s+1}-\hat{\theta}_{\mathrm{M}}&=\hat{\theta}_s-\hat{\theta}_{\mathrm{M}}-\frac{L_2(\hat{\theta}_{\mathrm{M}})(\hat{\theta}_s-\hat{\theta}_{\mathrm{M}})+\frac{1}{2}L_3(\hat{\theta}_{\mathrm{M}})(\hat{\theta}_s-\hat{\theta}_{\mathrm{M}})^2+\cdots}{L_2(\hat{\theta}_{\mathrm{M}})+L_3(\hat{\theta})(\hat{\theta}_s-\hat{\theta}_{\mathrm{M}})+\cdots}\\&=\frac{L_3(\hat{\theta}_n)(\hat{\theta}_s-\hat{\theta}_{\mathrm{M}})^2}{2L_2(\hat{\theta}_{\mathrm{M}})}+\cdots\\&=\frac{\lambda_3(\theta)}{2\lambda_2(\theta)}(\hat{\theta}_s-\hat{\theta}_{\mathrm{M}})^2+\cdots.\end{aligned}$$

If $\lambda_3(\hat{\theta}_0-\hat{\theta}_{\mathrm{M}})$ is of order $n^{-1/2}$, $\sqrt{n}(\hat{\theta}_1-\hat{\theta}_{\mathrm{M}})$ is of order $n^{-1/2}$ and $\sqrt{n}(\hat{\theta}_2-\hat{\theta}_{\mathrm{M}})$ is of order $n^{-3/2}$. Thus, the second iterate is second-order equivalent to $\hat{\theta}_{\mathrm{M}}$ and the first iterate is not. The first iterate is second-order efficient only if $\lambda_3=0$ or if $\hat{\theta}_0$ is asymptotically efficient.

4.4. *Deficiency*

Second-order inefficient estimators are not necessarily poor, since the efficiency loss may be quite small. It is therefore useful to get an idea of the magnitudes involved. Hodges and Lehmann (1970) propose an interesting measure of second-order inefficiency. Let $\hat{\theta}_1$ and $\hat{\theta}_2$ be two asymptotically efficient estimators in $\mathscr{S}_2$ and consider their bias-adjusted variants

$$\theta_1^*=\hat{\theta}_1-\frac{b_1(\hat{\theta}_1)}{n}\quad\text{and}\quad\theta_2^*=\hat{\theta}_2-\frac{b_2(\hat{\theta}_2)}{n}.$$

Suppose that θ_1^* is second-order optimal and the two bias-adjusted estimators have second-order variances of the form:

$$\mathrm{E}_2(\theta_1^* - \theta)^2 = \frac{1}{n\lambda(\theta)}\left[1 + \frac{B_1(\theta)}{n}\right] + o(n^{-2}),$$

$$\mathrm{E}_2(\theta_2^* - \theta)^2 = \frac{1}{n\lambda(\theta)}\left[1 + \frac{B_2(\theta)}{n}\right] + o(n^{-2}),$$

where the common asymptotic variance is the inverse information term λ^{-1} and $B_2(\theta) \geq B_1(\theta)$ for all θ. The deficiency of θ_2^* is defined to be the additional observations Δ needed, when using θ_2^*, to obtain the same precision as when using θ_1^*. That is, Δ is the solution of:

$$\frac{1}{n}\left[1 + \frac{B_1(\theta)}{n}\right] = \frac{1}{n+\Delta}\left[1 + \frac{B_2(\theta)}{n+\Delta}\right].$$

Solving, we find:

$$\Delta = B_2(\theta) - B_1(\theta) + o(n^{-1}). \tag{4.4}$$

Thus, deficiency is approximately n times the proportional difference in second-order variance.

Although deficiency is defined in terms of the bias-adjusted estimators, it can be calculated from the unadjusted estimators. If $\hat{\theta}_1$ is second-order efficient and $n(\hat{\theta}_2 - \hat{\theta}_1)$ has a limiting distribution with variance V, the deficiency of θ_2^* is $V\lambda$. The proof is based on the fact that $n(\theta_2^* - \theta_1^*)$ must be asymptotically uncorrelated with $\sqrt{n}(\theta_1^* - \theta)$; otherwise, a linear combination of θ_1^* and θ_2^* would have smaller second-order variance than θ_1^*. Hence using $o(n^{-1})$ Edgeworth approximations to compute moments:

$$\begin{aligned}\mathrm{Var}(\theta_2^*) - \mathrm{Var}(\theta_1^*) &= \mathrm{Var}(\theta_2^* - \theta_1^*) \\ &= \frac{1}{n^2}\mathrm{Var}\left[n(\hat{\theta}_2 - \hat{\theta}_1) + b_1(\hat{\theta}_1) - b_2(\hat{\theta}_2)\right] \\ &= \frac{1}{n^2}\mathrm{Var}\left[n(\hat{\theta}_2 - \hat{\theta}_1)\right] + o(n^{-2}).\end{aligned}$$

This result implies that, as far as deficiency is concerned, the key feature which distinguishes the estimators in $\mathscr{S}_2$ is the $n^{-1/2}$ term in their stochastic expansions.

4.5. *Generalizations*

Since the results on second-order efficiency presented in this section have been expressed in terms of Edgeworth expansions, they do not apply to estimation problems for discrete probability distributions. However, the theory can be generalized to include the discrete case. Furthermore, there is no need to restrict ourselves to one-parameter problems under quadratic loss. Pfanzagl and Wefelmeyer (1978a) develop a general theory of multiparameter second-order estimation efficiency for arbitrary symmetric loss functions. When loss is a smooth function of the estimation error, both continuous and discrete distributions are covered. The conclusions are similar to the ones reported here: the set $\bar{\mathscr{S}}_{\mathrm{M}}$ of maximum likelihood estimators adjusted as in (4.3) constitute an essentially complete class of estimators possessing stochastic expansions to order n^{-1}. Bayes and other estimators having the same $o(n^{-1})$ stochastic expansions share this second-order efficiency property. Although general proofs are available only for the case of simple random sampling, it is clear that the results have much broader applicability.

The fact that the second-order optimality properties of maximum likelihood and Bayes estimators hold for arbitrary symmetric loss functions is rather surprising. It suggests that there is no additional information in the third and fourth cumulants that is relevant for the second-order comparison of estimators. In fact, it has been shown by Akahira and Takeuchi (1981) that all well-behaved first-order efficient estimators necessarily have the same skewness and kurtosis to a second order of approximation. The $o(n^{-1})$ Edgeworth expansions for the estimators in $\mathscr{S}_2$ differ only by location and dispersion. Hence, for the purpose of comparing estimators on the basis of second-order approximations to their distributions, nothing is lost by concentrating on the first two moments.

Since the second-order theory of estimation is based on asymptotic expansions, the results can be relied on only to the extent that such expansions give accurate approximations to the true distributions. Clearly, if the tail behavior of estimators is really important, then the second-order comparisons discussed here are unlikely to be useful: there is no reason to believe that Edgeworth-type approximations are very accurate outside the central ninty percent of the distribution. Furthermore, in many cases where two asymptotically efficient estimators are compared, the bias difference is considerably larger than the difference in standard deviations. This suggests that correction for bias may be more important than second-order efficiency considerations when choosing among estimators.

5. Second-order comparisons of tests

5.1. General approach

The theory of second-order efficient tests of hypotheses parallels the theory for point estimation. Again, Edgeworth approximations are used in place of the traditional (first-order) asymptotic approximations to the distributions of sample statistics. We shall consider the case where the probability law for the observed data depends on θ, a q-dimensional vector of parameters which are to be tested, and on ω, a p-dimensional vector of nuisance parameters. The null hypothesis $\theta = \theta_0$ is examined using some test statistic, T, where large values are taken as evidence against the hypothesis. The rejection region $T > t$ is said to have size α if

$$\sup_{\omega} \Pr[T > t] = \alpha,$$

when $\theta = \theta_0$; the constant t is called the critical value for the test. The quality of such a test is measured by its power: the probability that T exceeds the critical value when the null hypothesis is false.

Often the exact distribution of T is not known and the critical value is determined using the asymptotic distribution for large sample size n. Suppose, for example, that under the null hypothesis:

$$\Pr[T \leq x] = F(x) + o(n^0),$$

where the approximate distribution function F does not depend on the unknown parameters. In most applications F turns out to be the chi-square distribution with q degrees of freedom, or, when $q = 1$, simply the standard normal. The asymptotic critical value t_α for a test of size α is the solution to the equation $F(t_\alpha) = 1 - \alpha$. The test which rejects the null hypothesis when $T > t_\alpha$ is asymptotically similar of level α; that is, its type-I error probability, to a first order of approximation, equals α for all values of the nuisance parameter ω.

To approximate the power function using asymptotic methods, some normalization is necessary. The test statistic T typically has a limiting distribution when the null hypothesis is true, but not when the null hypothesis is false. For most problems, the probability distribution for T has a center which moves off to infinity at the rate $\sqrt{n}\,\|\theta - \theta_0\|$ and power approaches unity at an exponential rate as the sample size increases. Two alternative normalization schemes have been proposed in the statistical literature. One approach, developed by Bahadur (1960, 1967) and applied by Geweke (1981a, 1981b) in econometric time-series analysis, employs large deviation theory and measures, in effect, the exponential rate at which power approaches unity. The other approach, due to Pitman and developed by Hodges and Lehmann (1956) and others, considers sequences where the true parameter θ converges to θ_0 and hence examines only local or contiguous alternatives. We shall restrict our attention to this latter approach.

The purpose of the analysis is to compare competing tests on the basis of their abilities to distinguish the hypothesized value θ_0 from alternative values. Of greatest interest are alternatives in the range where power is moderate, say between 0.2 and 0.9. Outside that region, the tests are so good or so poor that comparisons are uninteresting. Therefore, to get good approximations in the central region of the power function, it seems reasonable to treat $\sqrt{n}(\theta - \theta_0)$ as a vector of moderate values when doing the asymptotics. The local approach to approximating power functions finds the limiting distribution of the test statistic T, allowing the true parameter value θ to vary with n so that $\sqrt{n}(\theta - \theta_0)$ is always equal to the constant vector δ. Under such a sequence of local alternatives, the rejection probability approaches a limit:

$$\lim_{n\to\infty} \Pr[T > t_\alpha] = \pi_T(\delta), \tag{5.1}$$

which is called the local power function. Actual power would be approximated by $\pi_T[\sqrt{n}(\theta - \theta_0)]$. The limit (5.1) depends, of course, on θ_0, ω, and t_α in addition to δ; for notational simplicity, these arguments have been suppressed in writing the function π_T.

For many econometric inference problems, there are a number of alternative tests available, all having the same asymptotic properties. For example, as discussed by Engle in Chapter 13 of this Handbook, the Wald, Lagrange multiplier, and likelihood ratio statistics for testing multidimensional hypotheses in smooth parametric models are all asymptotically chi-square under the null hypothesis and have the same limiting noncentral chi-square distribution under sequences of local alternatives. That is, all three tests have the same asymptotic critical value t_α and the same local power functions. Yet, as Berndt and Savin (1977) and Evans and Savin (1982) point out, the small-sample behavior of the tests are sometimes quite different. It seems reasonable, therefore, to develop higher-order asymptotic expansions of the distributions and to use improved approximations when comparing tests.

Suppose the probability distribution function for T can be approximated by an Edgeworth expansion so that, under the null hypothesis:

$$\Pr[T \le t] = F\left[t + \frac{P(t,\theta,\omega)}{\sqrt{n}} + \frac{Q(t,\theta,\omega)}{n}\right] + o(n^{-1}). \tag{5.2}$$

Since F is just the (first-order) limiting distribution, the approximation depends on the unknown parameters only through the functions P and Q. Using these functions and parameter estimates $\hat{\theta}$ and $\hat{\omega}$, a modified critical value,

$$t_T^* = t_\alpha + \frac{g(t_\alpha,\hat{\theta},\hat{\omega})}{\sqrt{n}} + \frac{h(t_\alpha,\hat{\theta},\hat{\omega})}{n}, \tag{5.3}$$

can be calculated so that, when the null hypothesis is true:

$$\Pr\left[T > t_T^*\right] = \alpha + o(n^{-1}),$$

for all values of ω. The critical region $T > t_T^*$ is thus second-order similar of level α. Algorithms for determining the functions g and h from P and Q are given by Cavanagh (1983) and Pfanzagl and Wefelmeyer (1978b). The function g is simply minus P; the function h depends on the method employed for estimating the unknown parameters appearing in g. In the special (but common) case where P is zero, h is simply minus Q.

If the distribution function for T possesses a second-order Edgeworth expansion under a sequence of local alternatives where $\sqrt{n}(\theta - \theta_0)$ is fixed at the value δ as n approaches infinity, the rejection probability can be written as:

$$\Pr\left[T > t_T^*\right] \approx \pi_T(\delta) + \frac{a_T(\delta)}{\sqrt{n}} + \frac{b_T(\delta)}{n} \equiv \pi_T^*(\delta), \tag{5.4}$$

where the approximation error is $o(n^{-1})$. The function π_T^* is the second-order local power function for the size-adjusted test; again, for notational convenience, the dependency on θ_0, ω, and t_α is suppressed. By construction, $\pi_T^*(0) = \alpha$.

Suppose S and T are two alternative asymptotically equivalent test statistics possessing Edgeworth expansions. Since the two tests based on the asymptotic critical value t_α will not usually have the same size to order n^{-1}, it is not very interesting to compare their second-order power functions. It makes more sense to construct the size-adjusted critical values t_S^* and t_T^* and to compare tests with the correct size to order n^{-1}. If $\pi_T^* \geq \pi_S^*$ for all relevant values of δ and ω, then the size-adjusted test with rejection region $T > t_T^*$ is at least as good as the test with rejection region $S > t_S^*$. If the inequality is sometimes strict, then the test based on T dominates the test based on S. A second-order similar test of level α is said to be second-order efficient if it is not dominated by any other such test.

5.2. *Some results when $q = 1$*

When a single hypothesis is being tested, approximate power functions for the traditional test statistics can be written in terms of the cumulative normal distribution function. If, in addition, only one-sided alternatives are contemplated, considerable simplification occurs and a comprehensive theory of second-order optimal tests is available. The pioneering work is by Chibisov (1974) and Pfanzagl (1973); the fundamental paper by Pfanzagl and Wefelmeyer (1978b) contains the main results. More elementary expositions can be found in the survey papers by Pfanzagl (1980) and Rothenberg (1982) and in the application to nonlinear regression by Cavanagh (1981).

Suppose θ is a scalar parameter and the null hypothesis $\theta = \theta_0$ is tested against the alternative $\theta > \theta_0$. As before, ω is a vector of unknown nuisance parameters not involved in the null hypothesis. We consider test statistics whose distributions are asymptotically normal and possess second-order Edgeworth approximations. We reject the null hypothesis for large values of the test statistic.

The Wald, Lagrange multiplier, and likelihood ratio principles lead to asymptotically normal test statistics when the distribution function for the data is well behaved. Let $L(\theta, \omega)$ be the log-likelihood function and let $L_\theta(\theta, \omega)$ be its derivative with respect to θ. Define $(\hat{\theta}, \hat{\omega})$ and $(\theta_0, \hat{\omega}_0)$ to be the unrestricted and restricted maximum likelihood estimates, respectively. If $\sqrt{n}(\hat{\theta} - \theta)$ has a limiting $N[0, \sigma^2(\theta, \omega)]$ distribution, then the Wald statistic for testing $\theta = \theta_0$ is:

$$\mathrm{W} = \frac{\sqrt{n}(\hat{\theta} - \theta_0)}{\sigma(\hat{\theta}, \hat{\omega})}.$$

The Lagrange multiplier statistic is:

$$\mathrm{LM} = \frac{1}{\sqrt{n}} L_\theta(\theta_0, \hat{\omega}_0)\sigma(\theta_0, \hat{\omega});$$

the likelihood ratio statistic is:

$$\mathrm{LR} = \pm\sqrt{2}\left[L(\hat{\theta}, \hat{\omega}) - L(\theta_0, \hat{\omega}_0)\right]^{1/2},$$

where the sign is taken to be the sign of W.

Under suitable smoothness assumptions, all three test statistics have limiting $N(\delta/\sigma, 1)$ distributions under sequences of local alternatives where $\theta = \theta_0 + \delta/\sqrt{n}$. Hence, if one rejects for values of the statistic greater than t_α, the upper α significance point for a standard normal, all three tests have the local power function $\Phi[\delta/\sigma - t_\alpha]$. This function can be shown to be an upper bound for the asymptotic power of any level α test. Thus the three tests are asymptotically efficient.

Using (5.3), any asymptotically efficient test of level α can be adjusted so that it is second-order similar of level α. Let $\mathscr{T}$ be the set of all such size-adjusted tests. Any test in $\mathscr{T}$ has a power function which can be expanded as in (5.4). Since all the tests in $\mathscr{T}$ are asymptotically efficient, the leading term in the power function expansion is given by $\Phi(\delta/\sigma - t_\alpha)$. It can be shown that the next term $a_T(\delta)/\sqrt{n}$ is also independent of the test statistic T. Power differences among the tests in $\mathscr{T}$ are of order n^{-1}. Furthermore, there is an upper bound on the term $b_T(\delta)/n$. For any asymptotically efficient size-adjusted test based on a statistic T, the local power function can be written as:

$$\pi_T^*(\delta) = \pi(\delta) + \frac{a(\delta)}{\sqrt{n}} + \frac{b(\delta)}{n} - \frac{c(\delta)(\delta - \delta_T)^2 + d_T(\delta)}{n} + o(n^{-1}), \tag{5.5}$$

where $c(\delta)$ and $d_T(\delta)$ are non-negative. The functions π, a, b, and c are determined by the likelihood function and the significance level α. To second order, power depends on the test statistic T only through the constant δ_T and the function d_T.

The first three terms of (5.5) constitute the envelope power function:

$$\pi^*(\delta) = \pi(\delta) + \frac{a(\delta)}{\sqrt{n}} + \frac{b(\delta)}{n},$$

which is an upper bound to the (approximate) power of any test in $\mathscr{T}$. A test based on T is second-order efficient if, and only if, $d_T(\delta)$ is identically zero. If $c(\delta)$ is identically zero, all second-order efficient tests have the same power curve $\pi^*(\delta)$ to order n^{-1}. If $c(\delta) > 0$, second-order efficient tests have crossing power curves, each tangent to $\pi^*(\delta)$ at some point δ_T.

In any given one-tailed, one-parameter testing problem, two key questions can be asked. (i) Is $c(\delta)$ identically zero so that all second-order efficient tests have the same power function? (ii) Is the test statistic T second-order efficient and, if so, for what value δ_T is it tangent to the envelope power curve? The classification of problems and test statistics according to the answers to these questions is explored by Pfanzagl and Wefelmeyer (1978b). The value of c depends on the relationship between the first and second derivatives of the log-likelihood function. Generally, tests of mean parameters in a normal linear model have $c = 0$; tests of mean parameters in non-normal and nonlinear models have $c > 0$. Often $c(\delta)$ can be interpreted as a measure of curvature or nonlinearity. Tests based on second-order inefficient estimators have positive d and are dominated. The Lagrange multiplier, likelihood ratio, and Wald tests (based on maximum likelihood estimators) are all second-order efficient. The tangency points for the three tests are $\delta_{LM} = 0$, $\delta_{LR} = \sigma t_\alpha$, and $\delta_W = 2\sigma t_\alpha$. Thus, the LM test dominates all others when power is approximately α; the Wald test dominates all others when power is approximately $1-\alpha$; the LR test dominates at power approximately one-half.

When the alternative hypothesis is $\theta \neq \theta_0$ and $\delta = \sqrt{n}(\theta - \theta_0)$ can be negative as well as positive, the theory of optimal tests is more complicated. If the test statistic T is asymptotically distributed as $N(\delta/\sigma, 1)$, it is natural to reject when T assumes either large positive or large negative values. Using the Edgeworth approximation to the distribution of T, one can find critical values t_1 and t_2 [generally functions of the data as in (5.3)] such that the test which rejects when T lies outside the interval $(-t_1, t_2)$ is second-order similar of level α; that is, when $\theta = \theta_0$:

$$\Pr[-t_1 \leq T \leq t_2] = 1 - \alpha + o(n^{-1}), \tag{5.6}$$

for all ω. Indeed, for any statistic T, there are infinitely many pairs (t_1, t_2) satisfying (5.6), each yielding a different power curve. If, for example, t_2 is considerably greater than t_1, power will be high for negative δ and low for positive δ. Unless some restrictions are placed on the choice of rejection region, no uniformly optimal test is possible.

If the null distribution of T were symmetric about zero, the symmetric rejection region with $t_1 = t_2$ would be a natural choice. However, since the Edgeworth approximation to the distribution of T is generally skewed, the symmetric region is not particularly compelling. Sargan (1975b), in the context of constructing approximate confidence intervals for structural coefficients in the simultaneous equations model, suggested minimizing the expected length $t_1 + t_2$ of the acceptance region. Unfortunately, since the t_i depend on the unknown parameters, he concludes that this criterion is nonoperational and, in the end, recommends the symmetric region.

An alternative approach is to impose the restriction of unbiasedness. This is the basis of much traditional nonasymptotic testing theory [see, for example, Lehmann (1959, ch. 4)] and is commonly employed in estimation theory. A test is said to be unbiased if the probability of rejecting when the null hypothesis is false is always at least as large as the probability of rejecting when the null hypothesis is true. That is, the power function takes its minimum value when $\theta = \theta_0$. In the case $q = 1$, the condition that the test be locally unbiased uniquely determines t_1 and t_2 to second order. Size-adjusted locally unbiased Wald, likelihood ratio, and Lagrange multiplier tests are easily constructed from the Edgeworth expansions. If the curvature measure $c(\delta)$ is zero, the three tests have identical power functions to order n^{-1}; if $c(\delta) \neq 0$, the power functions cross. Again, as in the one-tailed case, the Wald test dominates when power is high and the LR test dominates when power is near one-half. However, the LM test is no longer second-order efficient; when $c(\delta) \neq 0$, one can construct locally unbiased tests having uniformly higher power. Details are given in Cavanagh and Rothenberg (1983).

5.3. *Results for the multiparameter case*

There does not exist a comprehensive theory of second-order optimality of tests when $q > 1$. However, numerous results are available for special cases. Peers (1981), Hayakawa (1975), and others have analyzed multidimensional null hypotheses for arbitrary smooth likelihood functions. These studies have concentrated on power differences of order $n^{-1/2}$. Fujikoshi (1973), Ito (1956, 1960), and Rothenberg (1977, 1981b) have investigated the normal linear model using expansions to order n^{-1}. In this latter case, after adjustment for size, tests based on the Wald, likelihood ratio, and Lagrange multiplier principles (using maximum likelihood estimates) are all second-order efficient. However, their power surfaces generally cross and no one test is uniformly best.

The major findings from these studies do not concern differences in power, but differences in size. When q is large, the adjusted critical values (5.3) based on the Edgeworth approximation often differ dramatically from the asymptotic values. The null distributions of typical test statistics for multidimensional hypotheses are not at all well approximated by a chi-square. This phenomenon has been noted by Evans and Savin (1982), Laitinen (1978), Meisner (1979), and others.

Consider, for example, the multivariate regression model $Y = X\Pi + V$, where Y is an $n \times q$ matrix of observations on q endogenous variables, X is an $n \times K$ matrix of observations on K exogenous variables, and Π is a $K \times q$ matrix of regression coefficients. The n rows of V are i.i.d. vectors distributed as $N_q(0, \Omega)$. Suppose the null hypothesis is that $\Pi'a = 0$, for some K-dimensional vector a. The Wald test statistic is:

$$T = \frac{a'\hat{\Pi}\hat{\Omega}^{-1}\hat{\Pi}'a}{a'(X'X)^{-1}a},$$

where $\hat{\Pi} = (X'X)^{-1}X'Y$ and $\hat{\Omega} = Y'[I - X(X'X)^{-1}X']Y/(n-K)$. Define $m = n - K$. Then it is known that, for $m \geq q$, T is a multiple of an F-statistic. In fact, $T(m-q+1)/mq$ is distributed as F with q and $m-q+1$ degrees of freedom. The usual asymptotic approximation to the distribution of T is a chi-square with q degrees of freedom. The mean of T is actually $mq/(m-q-1)$, whereas the mean of the chi-square is q; even if m is five times q, the error is more than twenty percent. Clearly, very large samples will be needed before the asymptotic approximation is reasonable when many restrictions are being tested.

For most testing problems, the exact distributions and the errors in the asymptotic approximation are unknown. However, it is reasonable to assume that higher-order asymptotic expansions will lead to improved approximations. The second-order Edgeworth approximation typically takes the simple chi-square form (3.12). In multivariate normal linear models (with quite general covariance structures allowing for autocorrelation and heteroscedasticity), the size-adjusted critical values (5.3) for the Wald, Lagrange multiplier, and likelihood ratio tests on the regression coefficients can be written as:

$$t_{\mathrm{W}}^{*} = t_{\alpha}\left[1 + \frac{\gamma_1 + \gamma_0 t_{\alpha}}{n}\right],$$

$$t_{\mathrm{LM}}^{*} = t_{\alpha}\left[1 + \frac{\gamma_2 - \gamma_0 t_{\alpha}}{n}\right],$$

$$t_{\mathrm{LR}}^{*} = t_{\alpha}\left[1 + \frac{\gamma_1 + \gamma_2}{2n}\right],$$

where t_α is the upper α quantile for a chi-square distribution with q degrees of freedom. The coefficients γ_0, γ_1, and γ_2 depend on the particular problem; some examples are given in Rothenberg (1977, 1981b). For this normal regression case, the likelihood ratio has the simplest correction since, to second order, the test statistic is just a multiple of a chi-square. It also turns out to be (approximately) the arithmetic average of the other two statistics. These adjusted critical values appear to be reasonably accurate when q/n is small (say, less than 0.1). Even when q/n is larger, they seem to be an improvement on the asymptotic critical values obtained from the chi-square approximation.

Although the second-order theory of hypothesis testing is not yet fully developed, work completed so far suggests the following conclusions. For moderate sample sizes, the actual significance levels of commonly used tests often differ substantially from the nominal level based on first-order asymptotic approximations. Modified critical regions, calculated from the first few terms of an Edgeworth series expansion of the distribution functions, are available and have significance levels much closer to the nominal level. Unmodified Wald, likelihood ratio, and Lagrange multiplier test statistics, although asymptotically equivalent, often assume very different numerical values. The modified (size-adjusted) tests are less likely to give conflicting results. The formulae for the modified critical regions are relatively simple (for likelihood-ratio tests it typically involves a constant degrees-of-freedom adjustment); their use should be encouraged, especially when the number of restrictions being tested is large.

Once the tests have been modified so that they have the same significance level (to a second order of approximation), it is possible to compare their (approximate) power functions. Here the differences often seem to be small and may possibly be swamped by the approximation error. However, when substantial differences do appear, the second-order theory provides a basis for choice among alternative tests. For example, in nonlinear problems, it seems that the likelihood-ratio test has optimal power characteristics in the interesting central region of the power surface and may therefore be preferable to the Lagrange multiplier and Wald tests.

5.4. *Confidence intervals*

The critical region for a test can serve as the basis for a confidence region for the unknown parameters. Suppose $T(\theta_0)$ is a test statistic for the q-dimensional null hypothesis $\theta = \theta_0$ against the alternative $\theta \neq \theta_0$. Let t^* be a size-adjusted critical value such that, when the null hypothesis is true,

$$\Pr\left[T(\theta_0) > t^*\right] = \alpha + o(n^{-1}),$$

for all values of the nuisance parameters. If $T(\theta_0)$ is defined for all θ_0 in the parameter space, we can form the set $C=\{\theta' : T(\theta') < t^*\}$ of all hypothesized values that are not rejected by the family of tests based on T. By construction the random set C covers the true parameter with probability $1-\alpha+o(n^{-1})$ and hence is a valid confidence region at that level.

A good confidence region covers incorrect parameter values with low probability. Thus, good confidence regions are likely to result from powerful tests. If the test is locally unbiased, then so will be the confidence region: it covers the true θ with higher probability than any false value nearby.

When $q=1$, the results described in Section 5.2 can be applied to construct locally unbiased confidence regions for a scalar parameter. For example, one might use the locally unbiased Lagrange multiplier or likelihood ratio critical region to define a confidence set. Unfortunately, the sets of θ_0 values satisfying

$$-t_1 \le \mathrm{LM}(\theta_0) \le t_2 \quad \text{or} \quad -t_1 \le \mathrm{LR}(\theta_0) \le t_2,$$

are sometimes difficult to determine and need not be intervals. The Wald test, however, always leads to confidence intervals of the form

$$C_{\mathrm{W}} = \left\{ \theta : \hat{\theta} - \frac{t_2}{\sqrt{n}} \sigma(\hat{\theta}, \hat{\omega}) \le \theta \le \hat{\theta} + \frac{t_1}{\sqrt{n}} \sigma(\hat{\theta}, \hat{\omega}) \right\},$$

where t_1 and t_2 are determined by the requirement that the test be locally unbiased and second-order similar of level α. These critical values take the form:

$$t_1 = t + \frac{p(t, \hat{\theta}, \hat{\omega})}{\sqrt{n}} + \frac{q(t, \hat{\theta}, \hat{\omega})}{n},$$

$$t_2 = t - \frac{p(t, \hat{\theta}, \hat{\omega})}{\sqrt{n}} + \frac{q(t, \hat{\theta}, \hat{\omega})}{n},$$

where t is the asymptotic critical value satisfying $1-\Phi(t)=\alpha/2$ and the functions p and q depend on the Edgeworth expansion of the test statistic. Since C_{W} is easy to construct, it would be the natural choice in practice.

6. Some examples

6.1. *Simultaneous equations*

Consider the two-equation model:

$$y_1 = \alpha y_2 + u; \qquad y_2 = z + v, \tag{6.1}$$

where α is an unknown scalar parameter, y_1 and y_2 are n-dimensional vectors of observations on two endogenous variables, and u and v are n-dimensional vectors of unobserved errors. The vector $z = \mathrm{E}y_2$ is unknown, but is assumed to lie in the column space of the observed nonrandom $n \times K$ matrix Z having rank K. The n pairs (u_i, v_i) are independent draws from a bivariate normal distribution with zero means, variances σ_u^2 and σ_v^2, and correlation coefficient ρ.

The first equation in (6.1) represents some structural relationship of interest; the second equation is part of the reduced form and, in the spirit of limited information analysis, is not further specified. Additional exogenous explanatory variables could be introduced in the structural equation without complicating the distribution theory; additional endogenous variables require more work. We shall discuss here only the simple case (6.1). Exact and approximate distribution theory for general simultaneous equations models is covered in detail by Phillips in Chapter 8 of Volume I of this Handbook.[6] Our purpose is merely to illustrate some of the results given in Sections 3–5 above.

The two-stage least squares (2SLS) estimator $\hat{\alpha}$ is defined as:

$$\hat{\alpha} = \frac{y_2'Ny_1}{y_2'Ny_2} = \alpha + \frac{z'u + v'Nu}{z'z + 2z'v + v'Nv},$$

where $N = Z(Z'Z)^{-1}Z'$ is the rank-K idempotent projection matrix for the column space of Z. It will simplify matters if $\hat{\alpha}$ is expressed in terms of a few standardized random variables:

$$X = \frac{z'u}{\sigma_u\sqrt{z'z}}; \qquad Y = \frac{z'v}{\sigma_v\sqrt{z'z}},$$

$$s = \frac{v'Nu}{\sigma_v\sigma_u}; \qquad S = \frac{v'Nv}{\sigma_v^2}.$$

The pair (X, Y) is bivariate normal with zero means, unit variances, and correlation coefficient ρ. The random variable s has mean $K\rho$ and variance $K(1+\rho^2)$; S has mean K and variance $2K$. The standardized two-stage least squares estimator is:

$$d \equiv \frac{\sqrt{z'z}}{\sigma_u}(\hat{\alpha} - \alpha) = \frac{X + (s/\mu)}{1 + (2Y/\mu) + (S/\mu^2)} \tag{6.2}$$

[6] To simplify the analysis, Phillips (and others) transform the coordinate system and study a canonical model where the reduced-form errors are independent. Since the original parameterization is retained here, our formulae must be transformed to agree with theirs.

where $\mu^2 = z'z/\sigma_v^2$ is often called the "concentration" parameter. When μ is large, d behaves like the $N(0,1)$ random variable X. Note that the sample size n affects the distribution of d only through the concentration parameter. For asymptotic analysis it will be convenient to index the sequence of problems by μ rather than the traditional $n^{1/2}$. Although large values of μ would typically be due to a large sample size, other explanations are possible; e.g. a very small value for σ_v^2. Thus, large μ asymptotics can be interpreted as either large n or small σ asymptotics.

Since the two-stage least squares estimator in our model is an elementary function of normal random variables, exact analysis of its distribution is possible. An infinite series representation of its density function is available; see, for example, Sawa (1969). Simple approximations to the density and distribution functions are also easy to obtain. If μ is large (and K is small), the denominator in the representation (6.2) should be close to one with high probability. This suggests developing the stochastic power-series expansion:

$$d = X + \frac{s - 2XY}{\mu} + \frac{4XY^2 - XS - 2Ys}{\mu^2} + \frac{R}{\mu^3}, \tag{6.3}$$

and using the first three terms (denoted by d') to form a second-order Edgeworth approximation. Since R satisfies the regularity condition (3.8), the algorithms given in Section 3 (with μ replacing $n^{1/2}$) are available. To order $o(\mu^{-2})$, the first two moments of d' are:

$$\mathrm{E}(d') = \frac{(K-2)\rho}{\mu}, \qquad \mathrm{Var}(d') = 1 - \frac{(K-4)(1+3\rho^2)+4\rho^2}{\mu^2};$$

the third and fourth cumulants are approximately

$$k_3 = -\frac{6\rho}{\mu}; \qquad k_4 = \frac{12(1+5\rho^2)}{\mu^2}.$$

Substitution into (3.7) yields the Edgeworth-B approximation:

$$\Pr\left[\frac{\sqrt{z'z}}{\sigma_u}(\hat{\alpha}-\alpha) \le x\right] \approx \Phi\left[x + \frac{\rho(x^2+1-K)}{\mu} + \frac{x(K-1)(1-\rho^2)+x^3(3\rho^2-1)}{2\mu^2}\right]. \tag{6.4}$$

This result was obtained by Sargan and Mikhail (1971) and by Anderson and Sawa (1973) using somewhat different methods.

Calculations by Anderson and Sawa (1979) indicate that the Edgeworth approximation is excellent when μ^2 is greater than 50 and K is 3 or less.[7] The approximation seems to be adequate for any μ^2 greater than 10 as long as K/μ is less than unity. When K exceeds μ, however, the approximation often breaks down disastrously. Of course, it is not surprising that problems arise when K/μ is large. The terms s/μ and S/μ^2 are taken to be small compared to X in the stochastic expansion (6.3). When K is the same order as μ, this is untenable and an alternative treatment is required.

The simplest approach is to take advantage of the ratio form of d. Equation (6.2) implies:

$$\begin{aligned}\Pr[d \leq x] &= \Pr\left[X + \frac{s - 2xY}{\mu} - \frac{xS}{\mu^2} \leq x\right] \\ &\equiv \Pr[W(x) \leq x].\end{aligned}$$

Since N is idempotent with rank K, s and S behave like the sum of K independent random variables. When K is large, a central limit theorem can be invoked to justify treating them as approximately normal. Thus, for any value of x and μ, W is the sum of normal and almost normal random variables. The mean $m(x)$ and variance $\sigma^2(x)$ of W can be calculated exactly. Treating W as normal yields the approximation:

$$\Pr[d \leq x] \approx \Phi\left[\frac{x - m(x)}{\sigma(x)}\right].$$

A better approximation could be obtained by calculating the third and fourth cumulants of $(W - m)/\sigma$ and using the Edgeworth approximation to the distribution of W in place of the normal. Some trial calculations indicate that high accuracy is attained when K is 10 or more. Unlike the Edgeworth approximation (6.4) applied directly to d', Edgeworth applied to W improves with increasing K.

Many other methods for approximating the distribution of d are available. Holly and Phillips (1979) derive the saddle-point approximation and conclude that it performs better than (6.4), particularly when K is large. Phillips (1981) experiments with fitting rational functions of the form (2.1) and reports excellent results. In both cases, the density function is approximated analytically and numerical integration is required for the approximation of probabilities. The Edgeworth approximation (6.4) generalizes easily to the case where there are many endogenous explanatory variables; indeed, the present example is based on

[7]Anderson and Sawa actually evaluate the Edgeworth-A form of the approximation, but the general conclusions presumably carry over to (6.4).

the paper by Sargan and Mikhail (1971) which covers the general case. Alternative approximation methods, like the one exploiting the ratio form of d, do not generalize easily. Likewise, Edgeworth approximations can be developed for estimators which do not have simple representations, situations where most other approximating methods are not applicable.

Numerous estimators for α have been proposed as alternatives to two-stage least squares. The second-order theory of estimation described in Section 4 can be employed to compare the competing procedures. Consider, for example, Theil's k-class of estimators:

$$\hat{\alpha}_k = \frac{y_2'(I-kM)y_1}{y_2'(I-kM)y_2},$$

where $M = I - N$. The maximum likelihood estimator is a member of this class when $k_{\mathrm{ML}} - 1$ is given by λ, the smallest root of the determinental equation:

$$|(y_1, y_2)'(N - \lambda M)(y_1, y_2)| = 0.$$

This root possesses a stochastic expansion of the form:

$$\lambda = \frac{u'(N-N_z)u + A_1\mu^{-1} + A_2\mu^{-2} + \cdots}{u'Mu},$$

where $N_z = z(z'z)^{-1}z'$ and the A_i are stochastically bounded as μ tends to infinity.

The standardized k-class estimator can be written as:

$$d_k \equiv \frac{\sqrt{z'z}}{\sigma_u}(\hat{\alpha}_k - \alpha) = \frac{X + (s_k/\mu)}{1 + (2Y/\mu) + (S_k/\mu^2)},$$

where

$$s_k = s + (1-k)\frac{u'Mv}{\sigma_u\sigma_v},$$

$$S_k = S + (1-k)\frac{v'Mv}{\sigma_v^2}.$$

If s_k and S_k are (stochastically) small compared to μ, d_k can be expanded in a power series analogous to (6.3) and its distribution approximated by a second-order Edgeworth expansion. Since $u'Mv$ and $v'Mv$ have means and variances proportional to $n-K$, such expansions are reasonable only if $(n-K)(k-1)$ is small. We shall examine the special class where

$$k = 1 + a\lambda - \frac{b}{n-K}, \tag{6.5}$$

for constants a and b of moderate size (compared to μ). When $a = b = 0$, the estimator is 2SLS; when $a = 1$ and $b = 0$, the estimator is LIML. Thus our subset of k-class estimators includes most of the interesting cases.

The truncated power-series expansion of the standardized k-class estimator is given by:

$$d_k' = X + \frac{s_k - 2XY}{\mu} + \frac{4XY^2 + XS_k - 2Ys_k}{\mu^2};$$

its first two moments, to order μ^{-2}, are:

$$\mathrm{E}(d_k') = \rho\frac{(1-a)(K-1)+b-1}{\mu},$$

$$\mathrm{Var}(d_k') = 1 - \frac{(K-4)(1+3\rho^2)+4\rho^2+2b(1+2\rho^2)-2a(1+a\rho^2)(K-1)}{\mu^2},$$

as long as the sample size n is large compared to the number K.[8] To order $o(\mu^{-2})$, the third and fourth cumulants of d_k' are the same as given above for 2SLS. This implies that all k-class estimators of the form (6.5) have the same skewness and kurtosis to second order. The $o(\mu^{-2})$ Edgeworth approximations differ only with respect to location and dispersion, as implied by the general result stated in Section 4.5.

The mean square error for the standardized estimator d_k' is the sum of the variance and the squared mean. The expression is rather messy but depends on a and b in a very systematic way. Indeed, we have the following striking result first noted by Fuller (1977): the family of k-class estimators with $a = 1$ and $b \geq 4$ is essentially complete to second order. Any other k-class estimator is dominated by a member of that family.

The k-class estimator is unbiased, to second order, if and only if $b = 1 + (a-1)(K-1)$. In that case, the variance becomes:

$$1 + \frac{K(1-\rho^2)+2\rho^2+2\rho^2(K-1)(a-1)^2}{\mu^2}, \tag{6.6}$$

[8]When K/n is not negligible, an additional term is needed in the variance formula; the relative merits of the maximum likelihood estimator are reduced. Cf. Anderson (1977) and Morimune (1981). Sargan (1975a) and Kunitomo (1982) have developed an asymptotic theory for "large models" where K tends to infinity along with the sample size n.

which clearly is minimized when $a = 1$. The estimator with $a = b = 1$ is therefore best unbiased; it is second-order equivalent to the bias-adjusted MLE. The estimator with $a = 0$, $b = 2 - K$ is also approximately unbiased; it is second-order equivalent to the bias-adjusted 2SLS estimator. From (6.6), the deficiency of the bias-adjusted 2SLS estimator compared to the bias-adjusted MLE is:

$$\Delta = n\frac{2(K-1)\rho^2}{\mu^2} = 2(K-1)\rho^2\frac{1-r^2}{r^2},$$

where $r^2 = z'z/(z'z + n\sigma_v^2)$ is the population coefficient of determination for the reduced-form equation. The number of observations needed to compensate for using the second-order inefficient estimator is greatest when K is large and the reduced-form fit is poor; that is, when Z contains many irrelevant or collinear variables.

These results on the relative merits of alternative k-class estimators generalize to the case where the structural equation contains many explanatory variables, both endogenous and exogenous. If y_2 is replaced by the matrix Y_2 and α is interpreted as a vector of unknown parameters, the k-class structural estimator is:

$$\hat{\alpha}_k = \left[Y_2'(I - kM)Y_2\right]^{-1} Y_2'(I - kM)y_1.$$

Let one vector estimator be judged better than another if the difference in their mean square error matrices is negative semidefinite. Then, with λ interpreted as the root of the appropriate determinantal equation, the family of k-class estimators (6.5) with $a = 1$ and $b \geq 4$ is again essentially complete to second order; any other k-class estimator is dominated. The estimator with $a = b = 1$ is again best second-order unbiased. These results can be established by the same methods employed above, with matrix power-series expansions replacing the scalar expansions.

The second-order theory of hypothesis testing can be applied to the simultaneous equations model. The calculation of size-adjusted critical regions and second-order power functions is lengthy and will not be given here. However, the following results can be stated. Tests based on 2SLS estimates are dominated by likelihood based tests as long as $K > 1$. The curvature measure $c(\delta)$ defined in Section 5 is zero for the problem of testing $\alpha = \alpha_0$ in the model (6.1). Hence, Wald, LM, and LR tests, after size correction, are asymptotically equivalent. In more complex, full-information models the curvature measure is nonzero and the power functions for the three tests cross. Detailed results on second-order comparisons of tests in simultaneous equations models are not yet available in print. Some preliminary findings are reported in Cavanagh (1981) and Turkington (1977). Edgeworth approximations to the distribution functions of some test statistics under the null hypothesis are given in Sargan (1975b, 1980).

6.2. Autoregressive models

Suppose the $n+1$ random variables $u_0, u_1, \ldots, u_n$ are distributed normally with zero means and with second moments:

$$\mathrm{E}(u_i u_j) = \sigma^2 \rho^{|i-j|}, \qquad -1 < \rho < 1.$$

The u_i are thus $n+1$ consecutive observations from a stationary first-order autoregressive process. If $u = (u_1, \ldots, u_n)'$ and $u_{-1} = (u_0, \ldots, u_{n-1})'$ are defined as n-dimensional column vectors, the model can be written in regression form:

$$u = \rho u_{-1} + \varepsilon,$$

where ε is $N(0, \sigma_\varepsilon^2 I)$, with $\sigma_\varepsilon^2 = \sigma^2(1-\rho^2)$. The least squares regression coefficient $u'u_{-1}/u'_{-1}u_{-1}$ is often used to estimate ρ. Approximations to its sampling distribution are developed by Phillips (1977a, 1978). We shall consider here the modified estimator:

$$\hat{\rho} = \frac{u'u_{-1}}{u'_{-1}u_{-1} + \left(u_n^2 - u_0^2\right)/2}, \tag{6.7}$$

which treats the end points symmetrically. It has the attractive property of always taking values in the interval $(-1,1)$. Since $\sqrt{n}(\hat{\rho}-\rho)$ has a limiting normal distribution with variance $1-\rho^2$, it is natural to analyze the standardized statistic which can be written as:

$$T = \frac{\sqrt{n}(\hat{\rho}-\rho)}{\sqrt{1-\rho^2}} = \left(X - \frac{\alpha Z}{\sqrt{n}}\right)\left(1 + \frac{Y}{\sqrt{n}}\right)^{-1},$$

where $\alpha = \rho/\sqrt{1-\rho^2}$ and

$$X = \frac{u'_{-1}\varepsilon}{\sigma\sigma_\varepsilon\sqrt{n}}; \qquad Y = \frac{u'_{-1}u_{-1} + \left(u_n^2 - u_0^2\right)/2 - n\sigma^2}{\sigma^2\sqrt{n}}; \qquad Z = \frac{u_n^2 - u_0^2}{2\sigma^2}.$$

Even though X and Y are not sums of independent random variables, they have limiting normal distributions and possess rth order cumulants of order $n^{1-r/2}$ for $r \geq 2$. It seems reasonable to expect that the distribution of T can be approximated by a valid Edgeworth expansion using the techniques developed in Section 3. The truncated power series expansion for T can be written as:

$$T' = X - \frac{XY + \alpha Z}{\sqrt{n}} + \frac{XY^2 + \alpha YZ}{n}.$$

The cumulants of T' are very complicated functions of ρ, but can be approximated to $o(n^{-1})$ as long as ρ is not too close to one. Very tedious calculation yields the approximate first four cumulants:

$$\mathrm{E}(T') = \frac{-2\alpha}{\sqrt{n}}; \qquad \mathrm{Var}(T') = 1 + \frac{7\alpha^2 - 2}{n},$$

$$k_3 = \frac{-6\alpha}{\sqrt{n}}; \qquad k_4 = \frac{6(10\alpha^2 - 1)}{n}.$$

From the general formula (3.7), the Edgeworth-B approximation to the distribution function is:

$$\Pr[T \le t] \approx \Phi\left[t + \frac{\alpha(t^2+1)}{\sqrt{n}} + \frac{t(1+4\alpha^2) + t^3(1+6\alpha^2)}{4n}\right]. \tag{6.8}$$

The Edgeworth approximation is based on the idea that the high-order cumulants of the standardized statistic are small. When $\alpha = 1$ (that is, $\rho = 0.7$), the third cumulant is approximately $-6/\sqrt{n}$ and the fourth cumulant is approximately $54/n$. A high degree of accuracy cannot be expected for sample sizes less than, say, 50. Numerical calculations by Phillips (1977a) and Sheehan (1981) indicate that the Edgeworth approximation is not very satisfactory in small samples when ρ is greater than one half.

The poor performance of the Edgeworth approximation when ρ is large stems from two sources. First, when autocorrelation is high the distribution of $\hat{\rho}$ is very skewed and does not look at all like a normal. Second, the approximations to the cumulants are not accurate when ρ is near one since they drop "end point" terms of the form ρ^n. The former difficulty can be alleviated by considering normalizing transformations which reduce skewness.

A transformation which performs this task and is interesting for its own sake is:

$$T^* = \frac{\frac{n+1}{n}\hat{\rho} - \rho}{\sqrt{(1-\hat{\rho}^2)/(n-1)}}.$$

The numerator is the difference between the bias adjusted estimator and the true parameter value; the denominator is the estimated standard error. The ratio can be interpreted as a modified Wald statistic under the null hypothesis. Since T^* has the power series expansion

$$T^* \approx T + \frac{\alpha(1+T^2)}{\sqrt{n}} + \frac{(1+2\alpha^2)T + (1+3\alpha^2)T^3}{2n},$$

its distribution can be approximated by reverting the series and using (6.8). The Edgeworth-B approximation to the distribution function is given by:

$$\Pr[T^* \le x] \approx \Pr\left[T \le x - \frac{\alpha(1+x^2)}{\sqrt{n}} - \frac{(1-2\alpha^2)x + (1-\alpha^2)x^3}{2n}\right]$$

$$\approx \Phi\left[x - \frac{x+x^3}{4n}\right],$$

which is the Edgeworth approximation to a Student-t distribution. When ρ is large, T^* is not centered at zero and a more complicated bias adjustment is needed. Otherwise, treating T^* as a Student-t random variable with n degrees of freedom seems to give a reasonable approximation for moderate sample sizes.

Many other approaches are available for approximating the distributions of sample statistics from autoregressive models. Anderson (1971, ch. 6) surveys some of the statistical literature. Alternative inverse trigonometric transformations are considered by Jenkins (1954), Quenouille (1948), and Sheehan (1981). Saddle-point methods are employed by Daniels (1956), Durbin (1980), and Phillips (1978). Various curve fitting techniques are also possible. For small values of ρ, all the methods seem reasonably satisfactory. When ρ is large, the general purpose methods seem to break down; ad hoc approaches which take into account the special features of the problem then are necessary.

A slightly more complicated situation occurs when the u_i are unobserved regression errors. Consider the linear model $y = X\beta + u$, where the n-dimensional error vector is autocorrelated as before; the $n \times k$ matrix of regressors are assumed nonrandom. Let $\tilde{\beta} = (X'X)^{-1}X'y$ be the least squares estimator of β and let $\hat{\beta}$ be the maximum likelihood estimator.[9] The two n-dimensional residual vectors are defined as:

$$\tilde{u} = y - X\tilde{\beta} \quad \text{and} \quad \hat{u} = y - X\hat{\beta};$$

their lagged values are:

$$\tilde{u}_{-1} = y_{-1} - X_{-1}\tilde{\beta} \quad \text{and} \quad \hat{u}_{-1} = y_{-1} - X_{-1}\hat{\beta},$$

where y_{-1} and X_{-1} are defined in the obvious way. We assume observations on the variables are available for period 0, but are not used in estimating β; this makes the notation easier and will not affect the results.

Two alternative estimators for ρ are suggested. The residuals $\tilde{u}$ could be used in place of u in (6.7); or the residuals $\hat{u}$ could be used. For purposes of comparison,

[9]Any estimator asymptotically equivalent to the MLE will have the same properties. For example, $\hat{\beta}$ might be the generalized least squares estimator based on some consistent estimator for ρ.

the end-point modifications play no role so we shall consider the two estimators

$$\tilde{\rho} = \frac{\tilde{u}'\tilde{u}_{-1}}{\tilde{u}'_{-1}\tilde{u}_{-1}} \quad \text{and} \quad \hat{\rho} = \frac{\hat{u}'\hat{u}_{-1}}{\hat{u}'_{-1}\hat{u}_{-1}}.$$

Both are asymptotically efficient, but $\hat{\rho}$ dominates to second order since it is based on an efficient estimator of β. Sheehan (1981) calculates the deficiency of the bias-adjusted estimator based on the least-squares residuals. He shows that the deficiency is equal to the asymptotic variance of $n(\tilde{\rho}-\hat{\rho})$ divided by the asymptotic variance of $\sqrt{n}(\hat{\rho}-\rho)$.

The computation of these variances is lengthy, but the basic method can be briefly sketched. By definition, $X'\tilde{u}=0$ and $\tilde{u}=\hat{u}-X(\tilde{\beta}-\hat{\beta})$. Furthermore, the maximum likelihood normal equations imply that $(X-\hat{\rho}X_{-1})'(u-\hat{\rho}u_{-1})\approx 0$. After considerable manipulation of these equations, the standardized difference in estimators can be written, to a first order of approximation, as:

$$n(\tilde{\rho}-\hat{\rho}) = \frac{n}{\tilde{u}'_{-1}\tilde{u}_{-1}}\left[\tilde{u}'\tilde{u}_{-1}-\hat{u}'\hat{u}_{-1}+\hat{\rho}\left(\hat{u}'_{-1}\hat{u}_{-1}-\tilde{u}'_{-1}\tilde{u}_{-1}\right)\right]$$
$$\approx \frac{1}{\rho\sigma^2}(\tilde{\beta}-\hat{\beta})'\left(\rho X'X_{-1}-X'X\right)(\tilde{\beta}-\hat{\beta}).$$

Defining the $K\times K$ matrices:

$$A = X'(X-\rho X_{-1}),$$
$$V = \mathrm{E}(\tilde{\beta}-\hat{\beta})(\tilde{\beta}-\hat{\beta})' \approx (X'X)^{-1}X'\Sigma X(X'X)^{-1}-(X'\Sigma^{-1}X)^{-1},$$

where $\Sigma = \mathrm{E}(uu')$, deficiency has the simple expression:

$$\Delta = \frac{\mathrm{tr}\,AVAV+\mathrm{tr}\,A'VAV}{\rho^2(1-\rho^2)\sigma^4}. \tag{6.9}$$

In the special case where β is a scalar and the single regressor vector x is itself an autoregressive process with autocorrelation coefficient r, the deficiency formula simplifies to:

$$\Delta = \frac{8\rho^2}{1-\rho^2}\left[1+\frac{(r-\rho)^2}{1-r^2}\right]^{-2},$$

which is bounded by $8\rho^2/(1-\rho^2)$. For example, when both u and x have autocorrelation coefficients near 0.7, the use of least squares residuals in place of maximum likelihood residuals is equivalent to throwing away eight observations.

6.3. Generalized least squares

Consider the linear regression model:

$$y = X\beta + u, \tag{6.10}$$

where X is an $n \times K$ matrix of nonrandom regressors and u is an n-dimensional vector of normal random errors with zero mean. The error covariance matrix is written as $\mathrm{E}(uu') = \Omega^{-1}$, where the precision matrix Ω depends on the vector ω of p unknown parameters. For a given estimator $\hat{\omega}$, the estimated precision matrix is $\hat{\Omega} = \Omega(\hat{\omega})$. The generalized least squares (GLS) estimator based on the estimate $\hat{\omega}$ is:

$$\hat{\beta} = (X'\hat{\Omega}X)^{-1}X'\hat{\Omega}y. \tag{6.11}$$

Under suitable regularity conditions on X and Ω, the standardized estimator $\sqrt{n}(\hat{\beta} - \beta)$ is asymptotically normal and possesses a valid second-order Edgeworth expansion as long as $\sqrt{n}(\hat{\omega} - \omega)$ has a limiting distribution. Furthermore, since the asymptotic distribution of $\sqrt{n}(\hat{\beta} - \beta)$ does not depend on which estimator $\hat{\omega}$ is used, all such GLS estimators are first-order efficient. It therefore follows from the general proposition of Akahira and Takeuchi that they all must have the same third and fourth cumulants up to $o(n^{-1})$. Since the estimator using the true ω is exactly normal, we have the surprising result: to a second order of approximation, all GLS estimators based on well-behaved estimates of ω are normally distributed.

It is possible to develop $o(n^{-2})$ expansions for generalized least squares estimators. Suppose $\hat{\omega}$ is an even function of the basic error vector u and has a probability distribution not depending on the parameter β. (The maximum likelihood estimator of ω and all common estimators based on least squares residuals have these properties.) Let c be an arbitrary K-dimensional constant vector. Then, if $c'\hat{\beta}$ has variance σ^2:

$$\Pr\left[\frac{c'(\hat{\beta} - \beta)}{\sigma} \leq x\right] = \Phi\left[x - \frac{x^3 - 3x}{24n^2}a\right] + o(n^{-2}), \tag{6.12}$$

where a/n^2 is the fourth cumulant of $c'(\hat{\beta} - \beta)/\sigma$. The assumption that $\hat{\beta}$ possesses finite moments is not necessary as long as $\hat{\beta}$ has a stochastic expansion with a well-behaved remainder term; σ^2 and a then are the moments of the truncated expansion. The simplicity of the approximation (6.12) results from the following fact: If $\bar{\beta}$ is the GLS estimator using the true Ω and $\hat{\omega}$ satisfies the above-mentioned conditions, then $\hat{\beta} - \bar{\beta}$ is distributed independently of $\bar{\beta}$, is symmetric around zero, and is of order n^{-1}. Details and proofs are given in Rothenberg (1981a).

The variance σ^2 can be approximated to second order using Nagar's technique. Taylor (1977) examines a special case of the GLS model where the errors are independent but heteroscedastic. Phillips (1977c) investigates the seemingly unrelated regression model where Ω has a Kronecker product form. In this latter case, a very simple deficiency result can be stated. Suppose there are G regression equations of the form:

$$y_i = X_i\beta_i + u_i, \qquad i = 1,\dots,G,$$

where each regression has m observations; y_i and u_i are thus m-dimensional vectors and X_i is an $m \times k_i$ matrix of nonrandom regressors. For each observation, the G errors are distributed as $N_G(0,\Sigma)$; the m error vectors for the different observations are mutually independent. The G equations can be written as one giant system of the form (6.10) where $n = Gm$ and $K = \Sigma k_i$. One might wish to compare the GLS estimator $\bar{\beta}$ (which could be used if Σ were known) with the GLS estimator $\hat{\beta}$ based on some asymptotically efficient estimate $\hat{\Sigma}$. (A common choice for $\hat{\sigma}_{ij}$ would be $\hat{u}_i'\hat{u}_j/n$ where $\hat{u}_i$ is the residual vector from an OLS regression of y_i on X_i.) Rothenberg (1981a) shows that the deficiency of $c'\hat{\beta}$ compared to $c'\bar{\beta}$ is bounded by $G+1$ and equals $G-1$ in the special case where the X_i are mutually orthogonal. Although the number of unknown nuisance parameters grows with the square of G, the deficiency grows only linearly.

6.4. *Departures from normality*

All of the above examples concern sampling from normal populations. Indeed, a search of the econometric literature reveals no application of higher-order asymptotic expansions that dispenses with the assumption of normal errors. This is rather odd, since the original intention of Edgeworth in developing his series was to be able to represent non-normal populations.

In principle, there is no reason why second-order approximations need be confined to normal sampling schemes. Although discrete lattice distributions cause some difficulties, valid Edgeworth expansions can be developed for statistics from any continuous population distribution possessing sufficient moments. The basic Edgeworth-B approximation formula (3.7) does not assume normality of the original observations. The normality assumption enters only when the approximate cumulants are computed using Nagar's technique.

In univariate problems, there seems to be no practical difficulty in dropping the normality assumption. The cumulants of the truncated statistic T' will, of course, depend on the higher cumulants of the population error distribution, but the Edgeworth approximation should be computable. In fact, one could conduct interesting studies in robustness by seeing how the approximate distribution of an estimator varies as the error distribution departs from normality.

In multivariate problems, however, things become more complex. The cumulants of T' will depend on all the third and fourth cross cumulants of the errors. Although the calculations can be made, the resulting approximation formula will be very difficult to interpret unless these cross cumulants depend in a simple way on a few parameters. The assumption that the errors are normal can be relaxed if a convenient multivariate distribution can be found to replace it.

7. Conclusions

Approximate distribution theory, like exact distribution theory, derives results from assumptions on the stochastic process generating the data. The quality of the approximation will not be better than the quality of the specifications on which it is based. The models used by econometricians are, at best, crude and rather arbitrary. One would surely not want to rely on a distribution theory unless the conclusions were fairly robust to small changes in the basic assumptions. Since most of the approximation methods discussed here employ information on the first four moments of the data whereas the usual asymptotic theory typically requires information only on the first two moments, some loss in robustness must be expected. However, if a rough idea about the degree of skewness and kurtosis is available, that information often can be exploited to obtain considerably improved approximations to sample statistics.

Clearly, sophisticated approximation theory is most appropriate in situations where the econometrician is able to make correct and detailed assumptions about the process being studied. But the theory may still be quite useful in other contexts. In current practice, applied econometricians occasionally draw incorrect conclusions on the basis of alleged asymptotic properties of their procedures. Even if the specification of the model is incomplete, second-order theory can sometimes prevent such mistakes. For example, in the presence of correlation between regressors and errors in a linear model, the two-stage least squares estimator will be strongly biased if the number of instruments is large and the instruments explain little variation in the regressors. The bias formula derived in section 6.1 under the assumption of normality may be somewhat off if the errors are in fact non-normal. But the general conclusion based on the second-order theory is surely more useful than the assertion that the estimator is consistent and hence the observed estimate should be believed.

In recent years there has developed among econometricians an extraordinary fondness for asymptotic theory. Considerable effort is devoted to showing that some new estimator or test is asymptotically normal and efficient. Of course, asymptotic theory is important in getting some idea of the sampling properties of a statistical procedure. Unfortunately, much bad statistical practice has resulted from confusing the words "asymptotic" and "approximate". The assertion that a standardized estimator is asymptotically normal is a purely mathematical proposition about the limit of a sequence of probability measures under a set of specified assumptions. The assertion that a given estimator is approximately normal suggests that, for the particular problem at hand, the speaker believes that it would be sensible to treat the estimator as though it were really normal. Obviously, neither assertion implies the other.

Accurate and convenient approximations for the distributions of econometric estimators and test statistics are of great value. Sometimes, under certain circumstances, asymptotic arguments lead to good approximations. Often they do not. The same is true of second-order expansions based on Edgeworth or saddlepoint methods. A careful econometrician, armed with a little statistical theory, a modest computer, and a lot of common sense, can always find reasonable approximations for a given inference problem. This survey has touched on some of the statistical theory. The computer and the common sense must be sought elsewhere.

References

Akahira, M. and K. Takeuchi (1981) *Asymptotic Efficiency of Statistical Estimators: Concepts and Higher Order Asymptotic Efficiency*. New York: Springer-Verlag.

Albers, W. (1978) "Testing the Mean of a Normal Population under Dependence", *Annals of Statistics*, 6, 1337–1344.

Amemiya, T. (1980) "The n^{-2}-Order Mean Squared Errors of the Maximum Likelihood and the Minimum Logit Chi-Square Estimator", *Annals of Statistics*, 8, 488–505.

Anderson, T. W. (1971) *The Statistical Analysis of Time Series*. New York: Wiley.

Anderson, T. W. (1974) "An Asymptotic Expansion of the Distribution of the Limited Information Maximum Likelihood Estimate of a Coefficient in a Simultaneous Equation System", *Journal of the American Statistical Association*, 69, 565–573.

Anderson, T. W. (1977) "Asymptotic Expansions of the Distributions of Estimates in Simultaneous Equations for Alternative Parameter Sequences", *Econometrica*, 45, 509–518.

Anderson, T. W. and T. Sawa (1973) "Distributions of Estimates of Coefficients of a Single Equation in a Simultaneous System and Their Asymptotic Expansions", *Econometrica*, 41, 683–714.

Anderson, T. W. and T. Sawa (1979) "Evaluation of the Distribution Function of the Two-Stage Least Squares Estimate", *Econometrica*, 47, 163–182.

Bahadur, R. R. (1960) "Stochastic Comparison of Tests", *Annals of Mathematical Statistics*, 31, 276–295.

Bahadur, R. R. (1967) "Rates of Convergence of Estimates and Test Statistics", *Annals of Mathematical Statistics*, 38, 303–324.

Barndorff-Nielsen, O. and D. R. Cox (1979) "Edgeworth and Saddle-Point Approximations with Statistical Applications", *Journal of the Royal Statistical Society*, B, 41, 279–312.

Berndt, E. and N. E. Savin (1977) "Conflict Among Criteria for Testing Hypotheses in the Multivariate Linear Regression Model", *Econometrica*, 45, 1263–1277.

Bhattacharya, R. N. and J. K. Ghosh (1978) "On the Validity of the Formal Edgeworth Expansion", *Annals of Statistics*, 6, 434–451.

Bhattacharya, R. N. and R. R. Rao (1976) *Normal Approximations and Asymptotic Expansions*. New York: Wiley.

Bickel, P. J. (1974) "Edgeworth Expansions in Nonparametric Statistics", *Annals of Statistics*, 2, 1–20.

Bickel, P. J. and K. A. Doksum (1977) *Mathematical Statistics*. San Francisco: Holden-Day.

Bishop, Y. M. M., S. E. Fienberg, and P. W. Holland (1975) *Discrete Multivariate Analysis: Theory and Practice*. Cambridge, MA: MIT Press.

Cavanagh, C. L. (1981) "Hypothesis Testing in Nonlinear Models", Working Paper, University of California, Berkeley.
Cavanagh, C. L. (1983) "Hypothesis Testing in Models with Discrete Dependent Variables", Ph.D. thesis, University of California, Berkeley.
Cavanagh, C. L. and T. J. Rothenberg (1983) "The Second-Order Inefficiency of the Efficient Score Test", Working Paper, Institute of Business and Economic Research, University of California, Berkeley.
Chibisov, D. M. (1974) "Asymptotic Expansions for Some Asymptotically Optimal Tests", in: J. Hajek, ed., *Proceedings of the Prague Symposium on Asymptotic Statistics*, vol. 2, 37–68.
Chibisov, D. M. (1980) "An Asymptotic Expansion for the Distribution of a Statistic Admitting a Stochastic Expansion", I, *Theory of Probability and its Applications*, 25, 732–744.
Cornish, E. A. and R. A. Fisher (1937) "Moments and Cumulants in the Specification of Distributions", *Review of the International Statistical Institute*, 5, 307–320.
Cramer, H. (1946) *Mathematical Methods of Statistics*, Princeton: Princeton University Press.
Daniels, H. E. (1956) "The Approximate Distribution of Serial Correlation Coefficients", *Biometrika*, 43, 169–185.
Durbin, J. (1979) "Discussion of the Paper by Barndorff-Nielsen and Cox", *Journal of the Royal Statistical Society*, B, 41, 301–302.
Durbin, J. (1980) "The Approximate Distribution of Serial Correlation Coefficients Calculated from Residuals on Fourier Series", *Biometrika*, 67, 335–350.
Durbin, J. and G. Watson (1971) "Serial Correlation in Least Squares Regression", III, *Biometrika*, 58, 1–19.
Edgeworth, F. Y. (1917) "On the Mathematical Representation of Statistical Data", *Journal of the Royal Statistical Society*, 80, 411–437.
Efron, B. (1975) "Defining the Curvature of a Statistical Problem (with Applications to Second-Order Efficiency)", *Annals of Statistics*, 3, 1189–1242.
Evans, G. B. A. and N. E. Savin (1982) "Conflict Among the Criteria Revisited", *Econometrica*, forthcoming.
Feller, W. (1971) *An Introduction to Probability Theory and Its Applications*, vol. 2. New York: Wiley.
Fisher, R. A. (1925) "Theory of Statistical Estimation", *Proceedings of the Cambridge Philosophical Society*, 22, 700–725.
Fujikoshi, Y. (1973) "Asymptotic Formulas for the Distributions of Three Statistics for Multivariate Linear Hypotheses", *Annals of the Institute of Statistical Mathematics*, 25, 423–437.
Fuller, W. A. (1977) "Some Properties of a Modification of the Limited Information Estimator", *Econometrica*, 45, 939–953.
Geweke, J. (1981a) "A Comparison of Tests of the Independence of Two Covariance-Stationary Time Series", *Journal of the American Statistical Association*, 76, 363–373.
Geweke, J. (1981b) "The Approximate Slopes of Econometric Tests", *Econometrica*, 49, 1427–1442.
Ghosh, J. K., B. K. Sinha, and H. S. Wieand (1980) "Second Order Efficiency of the MLE with Respect to any Bowl-Shaped Loss Function", *Annals of Statistics*, 8, 506–521.
Ghosh, J. K. and K. Subramanyam (1974) "Second Order Efficiency of Maximum Likelihood Estimators", *Sankhya*, A, 36, 325–358.
Hayakawa, T. (1975) "The Likelihood Ratio Criteria for a Composite Hypothesis Under a Local Alternative", *Biometrika*, 62, 451–460.
Henshaw, R. C. (1966) "Testing Single-Equation Least-Squares Regression Models for Autocorrelated Disturbances", *Econometrica*, 34, 646–660.
Hodges, J. L. and E. L. Lehmann (1956) "The Efficiency of Some Nonparametric Competitors of the t-Test", *Annals of Mathematical Statistics*, 27, 324–335.
Hodges, J. L. and E. L. Lehmann (1967) Moments of Chi and Powers of t, in: L. LeCam and J. Neyman, eds., *Proceedings of the Fifth Berkeley Symposium on Mathematical Statistics and Probability*, Berkeley: University of California Press.
Hodges, J. L. and E. L. Lehmann (1970) "Deficiency", *Annals of Mathematical Statistics*, 41, 783–801.
Holly, A. and P. C. B. Phillips (1979) "A Saddlepoint Approximation to the Distribution of the k-Class Estimator of a Coefficient in a Simultaneous System", *Econometrica*, 47, 1527–1547.
Imhof, J. P. (1961) "Computing the Distribution of Quadratic Forms in Normal Variables", *Biometrika*, 48, 419–426

Ito, K. (1956) "Asymptotic Formulae for the Distribution of Hotelling's Generalized T^2 Statistic", *Annals of Mathematical Statistics*, 27, 1091–1105.

Ito, K. (1960) "Asymptotic Formulae for the Distribution of Hotelling's Generalized T^2 Statistic", II, *Annals of Mathematical Statistics*, 31, 1148–1153.

Jenkins, G. M. (1954) "An Angular Transformation for the Serial Correlation Coefficient", *Biometrika*, 41, 261–265.

Johnson, N. L. (1949) "Systems of Frequency Curves Generated by Methods of Translation", *Biometrika*, 36, 149–176.

Johnson, N. L. and S. Kotz (1970) *Continuous Univariate Distributions*, vol. 1. New York: Wiley.

Kadane, J. (1971) "Comparison of *k*-Class Estimators when Disturbances are Small", *Econometrica*, 39, 723–739.

Kendall, M. G. and A. Stuart (1969) *The Advanced Theory of Statistics*, vol. 1. London: Griffin.

Koerts, J. and A. P. J. Abrahamse (1969) *On the Theory and Application of the General Linear Model*. Rotterdam: University Press.

Kunitomo, N. (1982) Asymptotic Efficiency and Higher Order Efficiency of the Limited Information Maximum Likelihood Estimator in Large Econometric Models, Technical report 365, Institute for Mathematical Studies in the Social Sciences, Stanford University.

Laitinen, K. (1978) "Why is Demand Homogeneity so Often Rejected"? *Economics Letters*, 1, 187–191.

Lehmann, E. L. (1959) *Testing Statistical Hypotheses*. New York: Wiley.

Meisner, J. F. (1979) "The Sad Fate of the Asymptotic Slutsky Symmetry Test for Large Systems", *Economics Letters*, 2, 231–233.

Morimune, K. (1981) "Asymptotic Expansions of the Distribution of an Improved Limited Information Maximum Likelihood Estimator", *Journal of the American Statistical Association*, 76, 476–478.

Nagar, A. L. (1959) "The Bias and Moment Matrix of the General *k*-Class Estimators of the Parameters in Simultaneous Equations", *Econometrica*, 27, 573–595.

Olver, F. W. J. (1974) *Asymptotics and Special Functions*. New York: Academic Press.

Pan Jie-jian (1968) "Distributions of the Noncircular Serial Correlation Coefficients", *Selected Translations in Mathematical Statistics and Probability*, 7, 281–292.

Peers, H. W. (1971) "Likelihood Ratio and Associated Test Criteria", *Biometrika*, 58, 577–587.

Pfanzagl, J. (1973) "Asymptotically Optimum Estimation and Test Procedures", in: J. Hajek, ed., *Proceedings of the Prague Symposium on Asymptotic Statistics*, vol. 2, 201–272.

Pfanzagl, J. (1980) "Asymptotic Expansions in Parametric Statistical Theory", in: P. R. Krishnaiah, ed., *Developments in Statistics*, vol. 3. New York: Academic Press.

Pfanzagl, J. and W. Wefelmeyer (1978a) "A third-Order Optimum Property of the Maximum Likelihood Estimator", *Journal of Multivariate Analysis*, 8, 1–29.

Pfanzagl, J. and W. Wefelmeyer (1978b) "An Asymptotically Complete Class of Tests", *Zeitschrift fur Wahrscheinlichkeitstheorie*, 45, 49–72.

Pfanzagl, J. and W. Wefelmeyer (1979) Addendum to: "A Third-Order Optimum Property of the Maximum Likelihood Estimator", *Journal of Multivariate Analysis*, 9, 179–182.

Phillips, P. C. B. (1977a) "Approximations to Some Finite Sample Distributions Associated with a First-Order Stochastic Difference Equation", *Econometrica*, 45, 463–485; erratum, 50: 274.

Phillips, P. C. B. (1977b) "A General Theorem in the Theory of Asymptotic Expansions as Approximations to the Finite Sample Distributions of Econometric Estimators", *Econometrica*, 45, 1517–1534.

Phillips, P. C. B. (1977c) "An Approximation to the Finite Sample Distribution of Zellner's Seemingly Unrelated Regression Estimator", *Journal of Econometrics*, 6, 147–164.

Phillips, P. C. B. (1978) "Edgeworth and Saddlepoint Approximations in the First-Order Noncircular Autoregression", *Biometrika*, 65, 91–98.

Phillips, P. C. B. (1980) "Finite Sample Theory and the Distributions of Alternative Estimators of the Marginal Propersity to Consume", *Review of Economic Studies*, 47, 183–224.

Phillips, P. C. B. (1981) A New Approach to Small Sample Theory, Cowles Foundation Discussion Paper, Yale University.

Quenouille, M. H. (1948) "Some Results in the Testing of Serial Correlation Coefficients", *Biometrika*, 35, 261–267.

Rao, C. R. (1961) "Asymptotic Efficiency and Limiting Information", in: *Proceedings of the Fourth*

Berkeley Symposium on Mathematical Statistics and Probability, vol. 1. Berkeley: University of California Press.
Rao, C. R. (1963) "Criteria of Estimation in Large Samples", *Sankhya*, A, 25, 189–206.
Rothenberg, T. J. (1973) *Efficient Estimation with a Priori Information*. New Haven: Yale University Press.
Rothenberg, T. J. (1977) Edgeworth Expansions for Some Test Statistics in Multivariate Regression, Working Paper, University of California, Berkeley.
Rothenberg, T. J. (1981a) Approximate Normality of Generalized Least Squares Estimates, Working Paper, University of California, Berkeley.
Rothenberg, T. J. (1981b) Hypothesis Testing in Linear Models when the Error Covariance Matrix is Nonscalar, Working Paper, University of California, Berkeley.
Rothenberg, T. J. (1982) "Comparing Alternative Asymptotically Equivalent Tests", in: W. Hildenbrand, ed., *Advances in Econometrics*. Cambridge: Cambridge University Press.
Sargan, J. D. (1974) "On the Validity of Nagar's Expansion for the Moments of Econometric Estimators", *Econometrica*, 42, 169–176.
Sargan, J. D. (1975a) "Asymptotic Theory and Large Models", *International Economic Review*, 16, 75–91.
Sargan, J. D. (1975b) "Gram-Charlier Approximations Applied to *t* Ratios of *k*-Class Estimators", *Econometrica*, 43, 327–346.
Sargan, J. D. (1976) "Econometric Estimators and the Edgeworth Approximation", *Econometrica*, 44, 421–448; erratum, 45, 272.
Sargan, J. D. (1980) "Some Approximations to the Distribution of Econometric Criteria which are Asymptotically Distributed as Chi-Squared", *Econometrica*, 48, 1107–1138.
Sargan, J. D. and W. M. Mikhail (1971) "A General Approximation to the Distribution of Instrumental Variables Estimates", *Econometrica*, 39, 131–169.
Sawa, T. (1969) "The Exact Sampling Distributions of Ordinary Least Squares and Two Stage Least Squares Estimates", *Journal of the American Statistical Association*, 64, 923–980.
Serfling, R. J. (1980) *Approximation Theorems of Mathematical Statistics*. New York: Wiley.
Sheehan, D. (1981) Approximating the Distributions of some Time-Series Estimators and Test Statistics, Ph.D. thesis, University of California, Berkeley.
Srinivasan, T. N. (1970) "Approximations to Finite Sample Moments of Estimators Whose Exact Sampling Distributions are Unknown", *Econometrica*, 38, 533–541.
Taylor, W. E. (1977) "Small Sample Properties of a Class of Two Stage Aitken Estimators", *Econometrica*, 45, 497–508.
Theil, H. (1971) *Principles of Econometrics*. New York: Wiley.
Theil, H. and A. L. Nagar (1961) "Testing the Independence of Regression Disturbances", *Journal of the American Statistical Association*, 56, 793–806.
Turkington, D. (1977) Hypothesis Testing in Simultaneous Equations Models, Ph.D. thesis, University of California, Berkeley.
Wallace, D. L. (1958) "Asymptotic Approximations to Distributions", *Annals of Mathematical Statistics*, 29, 635–654.
Wilson, E. B. and M. M. Hilferty (1931) "The Distribution of Chi Square", *Proceedings of the National Academy of Science*, U.S.A., 17, 684–688.

Chapter 16

MONTE CARLO EXPERIMENTATION IN ECONOMETRICS

DAVID F. HENDRY*

Nuffield College, Oxford

Contents

*It is a pleasure to acknowledge my gratitude to Frank Srba for invaluable help in developing the computer programmes NAIVE, CONVAR and DAGER, to the Social Science Research Council for financial support for this research over many years and to Peter Fisk, Robin Harrison, Barry Margolin, Grayham Mizon, Richard Quandt, Jean-Francois Richard, Pravin Trivedi and especially Denis Sargan for much advice on and many suggestions about the underlying work. Julia Campos, Neil Ericsson, Max King, Teun Kloek, Svend Hylleberg and Gene Savin provided helpful comments on an earlier version of this chapter, which was written during a visit to the Center for Operations Research and Econometrics, Louvain-la-Neuve. I am grateful to CORE for making that visit possible, and to the London School of Economics for supporting my work in this area over more than a decade.

Handbook of Econometrics, Volume II, Edited by Z. Griliches and M.D. Intriligator

1. Monte Carlo experimentation

1.1. Introduction

At the outset, it is useful to distinguish Monte Carlo methods from distribution sampling even though their application in econometrics may seem rather similar. The former is a general approach whereby mathematical problems of an analytical nature which prove technically intractable (or their solution involves prohibitively expensive labour costs) can be "solved" by substituting an equivalent stochastic problem and solving the latter. In contrast, distribution sampling is used to evaluate features of a statistical distribution by representing it numerically and drawing observations from that numerical distribution. This last has been used in statistics from an early date and important examples of its application are Student (1908), Yule (1926) and Orcutt and Cochrane (1949) inter alia. Thus, to investigate the distribution of the mean of random samples of T observations from a distribution which was uniform between zero and unity, one could simply draw a large number of samples of that size from (say) a set of one million evenly spaced numbers in the interval [0,1] and plot the resulting distribution. Such a procedure (that is, numerically representing a known distribution and sampling therefrom) is invariably part of a Monte Carlo experiment [the name deriving from Metropolis and Ulam (1949)] but often only a small part. To illustrate a Monte Carlo experiment, consider calculating:

$$\int_a^b f(x)\,\mathrm{d}x = I \quad \text{(say)}, \tag{1}$$

for a complicated function $f(x)$ whose integral is unknown. Introduce a random variable $\nu \in [a, b]$ with a known density $p(\cdot)$ and define $\eta = f(\nu)/p(\nu)$, then:

$$\mathrm{E}(\eta) = \int_a^b \left[\frac{f(x)}{p(x)}\right] p(x)\,\mathrm{d}x = I. \tag{2}$$

Thus, calculating $\mathrm{E}(\eta)$ will also provide I and a "solution" is achieved by *estimating* $\mathrm{E}(\eta)$ [see Sobol' (1974)], highlighting the switch from the initial deterministic problem (evaluate I) to the stochastic equivalent (evaluate the mean of a random variable). Quandt in Chapter 12 of this Handbook discusses the numerical evaluation of integrals in general.

Rather clearly, distribution sampling is involved in (2), but the example also points up important aspects which will be present in later problems. Firstly, $p(\cdot)$

is at choice and hence can be selected to enhance the accuracy with which $\mathrm{E}(\eta)$ is computed for a given cost. Further, the method of estimating $\mathrm{E}(\eta)$ can be chosen as well to improve the accuracy of the computations. One possibility (called direct or naive simulation) is to calculate the mean of a very large random sample of η's, but other estimators could be used (as discussed in a different context below). On the other hand, the disadvantages of switching to the stochastic reformulation can be seen since re-estimation of $\mathrm{E}(\eta)$ is required if either a and/or b is changed and even for fixed values of these parameters, the "solution" is inexact as estimators have sampling distributions: these are the difficulties of *specificity* and *imprecision*, respectively.

An econometric example, which will recur throughout this chapter, further emphasizes the merits and drawbacks of Monte Carlo experimentation. Consider the stationary first-order autoregressive process:[1]

$$y_t = \alpha y_{t-1} + \varepsilon_t, \qquad t = 1,\dots,T, \tag{3}$$

where

$$\varepsilon_t \sim I\mathcal{N}(0,\sigma_\varepsilon^2), \qquad |\alpha| < 1, \tag{4}$$

and

$$y_0 \sim \mathcal{N}(0,\sigma_\varepsilon^2/(1-\alpha^2)). \tag{5}$$

Let $\boldsymbol{\theta} = (\alpha, \sigma_\varepsilon^2)' \in \Theta = \{\boldsymbol{\theta} \,|\, |\alpha| \langle 1, \sigma_\varepsilon^2 \rangle 0\}$ and $T \in \mathscr{T} = [T^0, T^1]$ where $\mathscr{T}$ is preassigned, and let:

$$\hat{\alpha} = \sum_1^T y_t y_{t-1} \Big/ \sum_1^T y_{t-1}^2, \tag{6}$$

then the objective is to calculate $\mathrm{E}(\hat{\alpha})$. Equations (3)–(5) define the *data generation process*; $\Theta \times \mathscr{T}$ defines the *parameter space*; (3) defines the *relationship of interest* (which happens to coincide with one of the equations of the data generation process, but need not do so in general); (6) defines the *econometric technique* to be investigated (econometric estimators are denoted by "$\hat{\ }$" and Monte Carlo estimators by "$\tilde{\ }$" or "$\bar{\ }$"); and, here, calculating $\mathrm{E}(\hat{\alpha})$ is the *objective of the study* (E and $\mathscr{E}$ respectively denote expectations within the "econometric" and "experimental" models to aid clarity).

[1]A known intercept of zero introduces an important element of specificity which affects the finite sample distribution of $\hat{\alpha}$, the limiting distribution for $|\alpha| = 1$ and the usefulness of the limiting distribution as an approximation to the finite sample situation as compared with an estimated, unknown intercept [for an excellent discussion, see Evans and Savin (1981)].

As before, analytical calculation of $E(\hat{\alpha})$ is presumed intractable for the purposes of the illustration [but see, for example, Hurwicz (1950), Kendall (1954), White (1961), Shenton and Johnson (1965), Phillips (1977a) and Sawa (1978)] so that $E(\hat{\alpha})$ has to be estimated. Again, the choice of estimator of $E(\hat{\alpha})$ arises, with some potential distribution of outcomes (imprecision); only estimating $E(\hat{\alpha})$ at a few points in $\Theta \times \mathscr{T}$ is referred to as a "pilot Monte Carlo Study" and can do little more than provide a set of numbers of unknown generality (specificity). Since $E(\hat{\alpha})$ depends on $\boldsymbol{\theta}$ and T, it must be re-estimated as $\boldsymbol{\theta}$ and T vary, but the dependence can be expressed in a *conditional expectations formula*:

$$E(\hat{\alpha}|\boldsymbol{\theta}, T) = G_1(\boldsymbol{\theta}, T), \tag{7}$$

and frequently, the aim of a Monte Carlo study is to evaluate $G_1(\boldsymbol{\theta}, T)$ over $\Theta \times \mathscr{T}$. However, since $E(\hat{\alpha})$ need not vary with all the elements of $(\boldsymbol{\theta}, T)$, it is important to note any *invariance* information; here, $\hat{\alpha}$ is independent of σ_ε^2 which, therefore, is fixed at unity without loss of generality. Also, asymptotic distributional results can help in estimating $E(\hat{\alpha})$ and in checking the experiments conducted; conversely, estimation of $E(\hat{\alpha})$ checks the accuracy of the asymptotic results for $T \in \mathscr{T}$. Thus, we note:

$$\sqrt{T}(\hat{\alpha} - \alpha) \underset{a}{\sim} \mathscr{N}(0, (1-\alpha^2)). \tag{8}$$

It is important to clarify what Monte Carlo can, and cannot, contribute towards evaluating $G_1(\boldsymbol{\theta}, T)$ in (7).

As perusal of recent finite sample distributional results will reveal (see, for example, Phillips, Chapter 8 in this Handbook, and Rothenberg, Chapter 15 in this Handbook), functions such as $G_1(\boldsymbol{\theta}, T)$ tend to be extremely complicated series of sub-functions of $\boldsymbol{\theta}, T$ [for the model (3)–(6), see, for example, the expansions in Shenton and Johnson (1965)]. There is a negligible probability of simulation results establishing approximations to $G_1(\boldsymbol{\theta}, T)$ which are accurate in (say) a Taylor-series sense, such that if terms to $O(T^{-n})$ are included, these approximate the corresponding terms of $G_1(\cdot)$, with the remainder being small relative to retained terms [compare, for example, equations (68) and (69) below]: see White (1980a) for a general analysis of functional form mis-specification. Indeed, draconian simplifications and a large value of N may be necessary to establish results to even $O(T^{-1})$, noting that many asymptotic results are accurate to $O(T^{-1/2})$ anyway. Rather, the objective of establishing "analogues" of $G_1(\cdot)$ [denoted by $H_1(\boldsymbol{\theta}, T)$] is to obviate redoing a Monte Carlo for every new value of $(\boldsymbol{\theta}, T) \in \Theta \times \mathscr{T}$ (which is an expensive approach) by substituting the inexpensive computation of $E(\hat{\alpha}|\boldsymbol{\theta}, T)$ from $H_1(\cdot)$. Consequently, one seeks functions $H_1(\cdot)$ such that over $\Theta \times \mathscr{T}$, the inaccuracy of predictions of $E(\hat{\alpha})$ are of the same order as errors arising from direct estimation of $E(\hat{\alpha})$ by distribution sampling for a prespecified desired accuracy dependent on N (see, for example, Table 6.1 below). In practice, much of the inter-experiment variation observed in Monte Carlo can be accounted for by asymptotic theory [see, for example, Hendry (1973)], and as shown below, often $H_1(\cdot)$ can be so formulated as to coincide with $G_1(\cdot)$ for sufficiently large T.

The approach herein seeks to ensure simulation findings which are at least as accurate as simply numerically evaluating the relevant asymptotic formulae. If the coefficients of $(\boldsymbol{\theta}, T)$ in $G_1(\cdot)$ are denoted by $\boldsymbol{\beta}$, then by construction, $H_1(\cdot)$ depends on a (many → few) reparameterization $\boldsymbol{\gamma} = \boldsymbol{h}(\boldsymbol{\beta})$ defined by orthogonalising excluded effects with respect to included ones, yet ensuring coincidence of $H_1(\cdot)$ and $G_1(\cdot)$ for large enough T. For parsimonious specifications of $\boldsymbol{\gamma}$, simulation based $H_1(\cdot)$ can provide simple yet acceptably accurate formulae for interpreting empirical econometric evidence. Similar considerations apply to other moments, or functions of moments, of econometric techniques.

1.2. *Simulation experiments*

While it is not a universally agreed terminology, it seems reasonable to describe Monte Carlo experiments as "simulation" since they will be conducted by *simulating random processes using random numbers* (with properties analogous to those of the random processes). Thus, for calculating I, one needs random numbers $v_i \in [a, b]$ drawn from a distribution $p(\cdot)$ with $u_i = f(v_i)/p(v_i)$. In the second example, random numbers $e_t \sim I\mathcal{N}(0,1)$ and $y_0 \sim \mathcal{N}(0,(1-\alpha^2)^{-1})$ are required (see Section 3.1 for a brief discussion of random number generators).

The basic naive experiment (which will remain a major component of more "sophisticated" approaches) proceeds as follows. Consider a random sample $(x_1 \dots x_N)$ drawn from the relevant distribution $d(\cdot)$ where $E(x_i) = \mu$ and $E(x_i - \mu)^2 = \sigma^2$; then:

$$\bar{x} = N^{-1} \sum_{i=1}^{N} x_i \quad \text{has } E(\bar{x}) = \mu \quad \text{and} \quad E(\bar{x} - \mu)^2 = \sigma^2/N. \tag{9}$$

This well-known result is applied in many contexts in Monte Carlo, often for $\{x_i\}$ which are very complicated functions of the *original* random variables. Also, for large N, $\bar{x}$ is approximately normally distributed around μ, and if

$$\hat{\sigma}^2 = \frac{1}{N-1} \sum_{i=1}^{N} (x_i - \bar{x})^2, \quad \text{then } E(\hat{\sigma}^2) = \sigma^2. \tag{10}$$

Consequently, unknown $E(\cdot)$ can be estimated using means of simple random samples, with an accuracy which is itself estimable (from $N^{-1}\hat{\sigma}^2$) and which decreases (in terms of the standard error of $\bar{x}$, which has the same units as the $\{x_i\}$) as $\sqrt{N}$ increases, so that "reasonable" accuracy is easy to obtain, whereas high precision is hard to achieve.

Returning to the two examples, the relevant estimators are:

$$\bar{u} = \frac{1}{N}\sum_{i=1}^{N} u_i, \quad \text{with } \mathscr{E}(\bar{u}) = \mathrm{E}(\eta) = I, \tag{11}$$

and

$$\bar{\alpha} = \frac{1}{N}\sum \tilde{\alpha}_i, \quad \text{with } \mathscr{E}(\bar{\alpha}) = \mathrm{E}(\hat{\alpha}), \tag{12}$$

where each $\tilde{\alpha}_i$ is based on an independent set of $(y_0 e_1 \dots e_T)$. Furthermore, letting $\mathrm{E}(\hat{\alpha} - \mathrm{E}(\hat{\alpha}))^2 = V$, then:

$$\bar{V} = \frac{1}{N-1}\sum_{i=1}^{N}(\tilde{\alpha}_i - \bar{\alpha})^2 \quad \text{has } \mathscr{E}(\bar{V}) = V \tag{13}$$

and

$$\mathscr{E}(\bar{\alpha} - \mathrm{E}(\hat{\alpha}))^2 = V/N. \tag{14}$$

Thus, the approximation $\bar{\alpha} \underset{\text{app}}{\sim} \mathscr{N}(\mathrm{E}(\hat{\alpha}), V/N)$ provides a basis for constructing confidence intervals and hypothesis tests about $\mathrm{E}(\hat{\alpha})$.

In what follows, an experiment usually denotes an exercise at *one* point in the space $\Theta \times \mathscr{T}$ (generally replicated N times) with K experiments conducted in total. However, where the context is clear, "an experiment" may also denote the set of K sub-experiments investigating the properties of a single econometric method.

1.3. *Experimentation versus analysis*

The arguments in *favour* of using experimental simulations for studying econometric methods are simply that many problems *are* analytically intractable or analysis thereof is too expensive, and that the relative price of capital to labour has moved sharply and increasingly in favour of capital [see, for example, Summers (1965)]. Generally speaking, compared to a mathematical analysis of a complicated estimator or test procedure, results based on computer experiments are inexpensive and easy to *produce*. As a consequence, a large number of studies has been undertaken [for some surveys thereof, see Teichroew (1965), Smith (1973) and Sowey (1973)].

Certainly, the two *caveats* noted above—namely that simulation experiments can produce imprecise results which are specific to the (often small) range of parameter values investigated—are all too apparent (and are usually cited) in much of the work which has been published. The importance of this issue is that Monte Carlo results are not an end-product in econometrics, econometric theory itself is only an *intermediate* product for the economics profession as a whole, and hence present practice often entails large *consumption costs*. To interpret empirical evidence, users of econometric theory tend to require general but simple formulae which are easily understood and remembered, not vast tabulations of imprecise results which are specific to an unknown extent. Thus, in the second example, accurately estimating $G_1(\boldsymbol{\theta}, T)$ could be useful, whereas simply quoting $\bar{\alpha}$ for a large number of values of α and T (and σ_ε^2?) using small N is less than optimally helpful.

Being an experimental discipline, exactitude and certainty are essentially unobtainable in Monte Carlo so criticisms of specificity and imprecision are serious but not entirely avoidable. Their impact, however, can be reduced and one of the aims of this chapter is to present methods which help to do so. At the same time, an often made and closely related criticism is misplaced and must be rebutted: experimentation can never do *more* than solve the problem for which the analytical solution is desired, so any critique of the choice of data generation process, or estimator, etc. would apply equally to the corresponding analytical derivation and is not a drawback of experimentation per se. Indeed, a powerful advantage of experimental simulations is that much more *general* data generation processes or more complicated techniques can be investigated than can be tackled feasibly by analysis. Thus, in the second example, if it is worthwhile deriving $\mathrm{E}(\hat{\alpha}|\boldsymbol{\theta}, T)$ analytically, then it is equally useful to obtain $G_1(\boldsymbol{\theta}, T)$ by numerical methods. Criticizing the choice of Θ (e.g. the assumption of stationarity) applies equally to analysis and simulation; however, criticizing a simulation for only considering a *few points* in Θ, or inaccurately estimating $G_1(\cdot)$ is a legitimate indictment of the experiments undertaken.

The above discussion points to the potential value of simulation experiments in econometric theory and in investigating finite sample distributions in particular. Below, their role is seen as follows.

Monte Carlo experimentation can efficiently complement analysis to establish numerical-analytical formulae which jointly summarize the experimental findings and known analytical results in order to help interpret empirical evidence and to compute outcomes at other points within the relevant parameter space. The accuracy obtainable depends on the given budget constraint and the relative price of capital to labour, the latter determining the point at which simulation is substituted for analysis, and the former the overall precision of the exercise.

There are several intermediate stages involved in achieving this objective. Firstly, as complete an analysis as feasible of the econometric model should be undertaken (see Section 2). Then, that model should be embedded in a Monte Carlo Model which exploits *all* the information available to the experimenter, and provides an appropriate design for the experiments to be undertaken (see Section 3). Thirdly, simulation specific methods of intra-experiment control should be developed (see Section 4) and combined with covariance techniques for estimating response surfaces between experiments (Section 5). The simple autoregressive model in (3)–(6) is considered throughout as an illustration and in Section 6, results are presented relating to biases, standard errors and power functions of tests. Finally, in Section 7, various loose ends are briefly discussed including applications of simulation techniques to studying estimated econometric systems (see Fair, Chapter 33 in this Handbook) and to the evaluation of integrals [see Quandt, Chapter 12 in this Handbook and Kloek and Van Dijk (1978)]. Three useful background references on Monte Carlo are Goldfeld and Quandt (1972), Kleijnen (1974) and Naylor (1971).

2. The econometric model

2.1. *The data generation process*

The class of processes chosen for investigation defines, and thereby automatically restricts, the *realm of applicability* of the results. Clearly, the class for which the analytical results are desired must be chosen for the simulation! For example, one type of data generation process (DGP) which is often used is the class of stationary, complete, linear, dynamic, simultaneous equations systems with (possibly) autocorrelated errors, or special cases thereof. It is obvious that neither experimentation nor analysis of such processes can produce results applicable to (say) non-stationary or non-linear situations, and if the latter is desired, the DGP must encompass this possibility. Moreover, either or both approaches may be further restricted in the number of equations or parameters or regions of the parameter space to which their results apply.

Denote the parameters of the DGP by $(\boldsymbol{\theta}, T)$ (retaining a separate identity for T because of its fundamental role in finite sample distributions) with the parameter space $\Theta \times \mathscr{T}$. It is important to emphasize that by the nature of computer experimentation, the DGP is *fully known* to the experimenter and in particular the forms of the equations, the numerical values of their parameters and the *actual* values of the random numbers are all known. The use of such information in improving the efficiency of the experiments is discussed below, but its immediate use is that the *correct* likelihood function for the DGP parameters can be specified and the resultant log-likelihood is denoted $L(\boldsymbol{\theta})$. Let

$$\boldsymbol{q}(\boldsymbol{\theta})=\frac{\partial L(\cdot)}{\partial \boldsymbol{\theta}} \tag{15}$$

denote the efficient score for the parameters of the DGP.

The relationship(s) under study need not correspond directly to any equations in the DGP but (obviously) must be derivable from such equations. Let $\boldsymbol{\beta}=\boldsymbol{h}(\boldsymbol{\theta})$ be any requisite one-to-one transformation of $\boldsymbol{\theta}$ and partition $\boldsymbol{\beta}$ into $(\boldsymbol{\beta}_1':\boldsymbol{\beta}_2')'$ such that $\boldsymbol{\beta}_1$ are the parameters of the relationship under study. Then for any given $\boldsymbol{\beta}_2$, say $\boldsymbol{\beta}_2^0$,

$$\left.\frac{\partial L(\cdot)}{\partial \boldsymbol{\beta}_1}\right|_{\boldsymbol{\beta}_2^0}=\boldsymbol{q}_1(\boldsymbol{\beta}_1|\boldsymbol{\beta}_2^0)=\mathbf{0} \tag{16}$$

defines the *estimator generating equation* for $\boldsymbol{\beta}_1$ in that almost all methods for estimating $\boldsymbol{\beta}_1$ are generated by solving $\boldsymbol{q}_1(\cdot)=\mathbf{0}$ (either exactly or approximately depending on the desired estimator and on whether all of the $\boldsymbol{\beta}_1$ are of equal interest) using alternative choices of $\boldsymbol{\beta}_2^0$ [see, for example, Hendry (1976)].

2.2. *Known distributional and invariance results*

Often, it will be known that $\sqrt{T}(\hat{\boldsymbol{\beta}}_1-\boldsymbol{\beta}_1)\underset{a}{\sim}\mathcal{N}(\mathbf{0},\boldsymbol{\Omega})$ and, if incorrect specifications are being considered, a formula for estimating $\operatorname{Var}(\hat{\boldsymbol{\beta}}_1|\boldsymbol{\beta}_2^0)$ (say) $\hat{\boldsymbol{V}}$ will exist with $\operatorname{plim} T\hat{\boldsymbol{V}}=\boldsymbol{V}$ ($=\boldsymbol{\Omega}$ only if the equation is a "correct specification" relative to the DGP under which $\hat{\boldsymbol{V}}$ is obtained). The square roots of diagonal elements of $\hat{\boldsymbol{V}}$ and $T^{-1}\boldsymbol{\Omega}$ are, respectively, referred to as "estimated standard errors" (ESE) and "asymptotic standard errors". Since $\boldsymbol{\theta}$ is known, $\boldsymbol{\beta}_1$, $\boldsymbol{V}$ and $\boldsymbol{\Omega}$ can be evaluated numerically if they are derivable analytically. Clearly, if any results are available to $O(T^{-1})$, or exactly for special cases of the DGP under analysis, such information also should be incorporated in the study.

Denote the unknown cumulative distribution function of $\hat{\boldsymbol{\beta}}_1$ by $\Psi_T(\hat{\boldsymbol{\beta}}_1|\boldsymbol{\theta})$ and its moments (assumed to exist; on which issue see Section 7.1) by $\psi_{Ti}(\hat{\boldsymbol{\beta}}_1|\boldsymbol{\theta})$ where:

$$\mathrm{E}(\hat{\boldsymbol{\beta}}_1)=\psi_{T1}(\hat{\boldsymbol{\beta}}_1|\boldsymbol{\theta})=\boldsymbol{G}_1(\boldsymbol{\theta},T) \tag{17}$$

and

$$\mathrm{E}(\hat{\boldsymbol{\beta}}_1-\psi_{T1})(\hat{\boldsymbol{\beta}}_1-\psi_{T1})'=\psi_{T2}(\cdot)=\boldsymbol{G}_2(\boldsymbol{\theta},T), \tag{18}$$

where the $G_i(\cdot)$ are the conditional expectations *functions* which have to be calculated (or the equivalent thereof for variances such as $\hat{V}$, test powers, etc.). From the above discussion *limiting functional forms* for $G_1(\cdot)$ and $G_2(\cdot)$ (i.e. for large T) are given by $\boldsymbol{\beta}_1$ and $T^{-1}\boldsymbol{\Omega}$, respectively (and by $T^{-1}\boldsymbol{V}$ for $\hat{\boldsymbol{V}}$, etc.).

Frequently, it will be feasible to establish that the $\psi_{T1}(\cdot)$ of relevance do not depend on certain parameters in $\boldsymbol{\theta}$, which may thereby be fixed without loss of generality but with a substantial saving in the cost of the experiments [see $\sigma_\varepsilon^2 = 1$ in the example (3)–(6)]. Such results can be established either by analysis [see, for example, Breusch (1979)] or by "pilot screening" as discussed below in Section 4 when an invariance is anticipated; in both cases, reduction to canonical form is important for clarifying the structure of the analysis [see, for example, Mariano (1982) and Hendry (1979)]. Conversely, it can occur that, unexpectedly, results are more general than claimed because of an invariance in an embedding model [e.g. see King (1980)]. As stressed earlier, other assumptions (such as zero intercepts) may be critical and care is required in establishing invariance, especially in dynamic models.

3. The Monte Carlo model

3.1. Random numbers

The data generation process of the Monte Carlo directly represents that desired for the econometric theory with two important differences. Firstly, the parameters $(\boldsymbol{\theta}, T)$ of the econometric DGP become *design variables* in the experiment and hence the numerical values chosen should be determined by considerations of simulation efficiency, an issue discussed in the following subsection. Secondly, as noted above, the random processes are simulated by random numbers intended to mimic the distributional properties of the former. This does not imply that the random numbers must be generated by an analogue of the random process [although physical devices have been used—see Tocher (1963)]. Rather, whatever method is adopted, the numbers so produced should yield a *valid* answer for the simulation (see the next subsection), the checking of which is one of the advantages of Monte Carlo over pure distribution sampling.

Generally, the basic random numbers in computer experiments have been uniformly distributed values in the unit interval (denoted $n_i \sim R(0,1)$) produced by numerical algorithms such as Multiplicative Congruential Generators (for a more extensive discussion of the numerical aspects of random number generation, see Quandt, Chapter 12 in this Handbook):

$$z_{i+1} = bz_i \pmod r, \qquad i = 0,1,\ldots,m, \tag{19}$$

with $n_i = z_i/r \in [0,1]$. The choices of b and r are important for avoiding autocorrelation, maintaining uniformity and producing the maximum feasible period m and if any study is likely to be dependent on the presence or absence of some feature in the $\{n_i\}$, it is clearly essential to test this on the numbers used in the experiment. The $\{n_i\}$ from (19) are *pseudo-random* in that from knowing the algorithm and the "seed" z_0, they are exactly reproducible but should not be detectably non-random on a relevant series of tests. There is a very large literature on the topic of random number generation, which I will not even attempt to summarise, but useful discussions are provided by Hammersley and Handscomb (1964), Kleijnen (1974), Naylor (1971) and Tocher (1963) inter alia; also, the recent text by Kennedy and Gentle (1980) offers a clear and comprehensive coverage of this issue and Sowey (1972) presents a chronological and classified bibliography.

Other distributions are obtainable from the uniform using the property that:

$$\Pr(n_i \leq k) = P(k) = k \leq 1, \tag{20}$$

so that $P(k)$ and k are interchangeable.

To compute $e_i \sim \Psi(\cdot)$, if $\Psi(\cdot)$ is invertible then $\Psi^{-1}(n_i)$ suffices since:

$$\begin{aligned}\Pr(e_i \leq k) &= \Pr(\Psi(e_i) \leq \Psi(k)) = \Pr(n_i \leq \Psi(k)) \\ &= P(\Psi(k)) = \Psi(k) \text{ as required, if } n_i = \Psi(e_i).\end{aligned} \tag{21}$$

For the exponential distribution, say $\Psi(\cdot) = 1 - \exp(-\mu\varepsilon)$, then $\varepsilon_i = -\mu^{-1}\ln(1 - n_i) \sim \Psi(\cdot)$. However, the Normal distribution does not have an analytical inverse and the two usual methods for generating $e_i \sim I\mathcal{N}(0,1)$ are:

$$\left(\Sigma_1^{12} n_j - 6\right) = e_i \quad \text{(an approximate central limit result)}, \tag{22}$$

or, for bivariate $I\mathcal{N}(\mathbf{0}, \boldsymbol{I})$, the Box–Müller method:

$$(e_i, e_{i+1}) = h_i(\cos 2\pi n_{i+1}, \sin 2\pi n_{i+1}), \tag{23}$$

where $h_i = (-2\ln n_i)^{1/2}$. It is important to use a "good" generator (i.e. one which is well tested and empirically satisfactory) for input to (23), especially if (n_i, n_{i+1}) are successively generated by (19) [see, for example, Neave (1973) and Quandt, Chapter 12 in this Handbook]. Golder (1976) and King (1981) discuss some useful tests on the $\{n_i\}$. Kennedy and Gentle (1980) consider the generation of variates from many useful statistical distributions. Finally, Sylwestrowicz (1981) discusses random number generation on parallel processing machines.

3.2. Efficient simulation: Reliability and validity

A *reliable* experiment is one in which any claimed results are accurately reproducible using a different set $\{n_i\}$ from the same (or an equivalent) generator; a *valid* experiment is one where the claimed results are correct for all points in the space $\Theta \times \mathscr{T}$. An applicable experiment is one in which the assumed DGP is an adequate representation of that required in the equivalent analytical derivation. Reliability is the most easily checked and it is standard practice to quote estimated standard errors of simulation statistics such as $\bar{\psi}_{Ti}$ to indicate the degree of reliability being claimed for these. However, the final products of the type of Monte Carlo being discussed herein are estimates of the conditional expectations functions $G_i(\cdot)$ as in (17) and (18) from $\bar{\psi}_{Ti}$ when the precise functional forms of the $G_i(\cdot)$ are unknown. Consequently, *response surfaces* must be postulated of the general form:

$$\bar{\psi}_{Ti} = H_i(\boldsymbol{\theta}, T) + \boldsymbol{\nu}_{Ti} \quad (i=1,2), \tag{24}$$

where $H_i(\cdot)$ is an approximation[2] to $G_i(\cdot)$ over $\Theta \times \mathscr{T}$. Thus, $\boldsymbol{\nu}_{Ti}$ comprises two components: a "measurement error",

$$\boldsymbol{\nu}_{1Ti} = \left(\psi_{Ti}(\boldsymbol{\theta}) - \bar{\psi}_{Ti}\right), \quad \text{with } \mathscr{E}(\boldsymbol{\nu}_{1Ti}) = \mathbf{0}, \tag{25}$$

and variance, $V_1(\boldsymbol{\nu}_{1Ti})$, estimable from the simulation; and an "approximation error",

$$\boldsymbol{\nu}_{2Ti} = \left(G_i(\boldsymbol{\theta}, T) - H_i(\boldsymbol{\theta}, T)\right), \tag{26}$$

of unknown (but potentially estimable) magnitude. It seems reasonable to assume that the components $\boldsymbol{\nu}_{1Ti}$ and $\boldsymbol{\nu}_{2Ti}$ are independent, but $\boldsymbol{\nu}_{Ti}$ need be neither homoscedastic, nor purely random. The coefficients of the $H_i(\cdot)$ have to be estimated and the net products of the simulation are numerical-analytical expressions of the form $\bar{H}_i(\boldsymbol{\theta}, T)$ [see Section 6; Section 7 briefly considers estimation of $\boldsymbol{\Psi}_T(\cdot)$].

Obtaining $\bar{H}_i(\cdot) \simeq G_i(\cdot)$ for all $(\boldsymbol{\theta}, T) \in \Theta \times \mathscr{T}$ would be an optimal outcome since such results would be both reliable and valid, but to even approximate its attainment requires fulfilling several intermediate steps:

(a) $H_i(\cdot)$ must be a close approximation to $G_i(\cdot)$ over the relevant parameter space so that the error $\boldsymbol{\nu}_{2Ti}$ must be of small magnitude, purely random and have

[2]As discussed above, this is a shorthand for: $H_i(\cdot)$ is the conjectured model of $G_i(\cdot)$ and hence constitutes that reparameterization of the latter which minimizes prediction mean square error over the conducted experiments. Subject to known heteroscedasticity corrections, $H_i(\cdot)$ usually will be a least-squares approximation to $G_i(\cdot)$.

constant variance after appropriate transformation [see White (1980a) for an analysis of functional approximation errors];

(b) $\bar{\psi}_{Ti}$ must be close to $\psi_{Ti}(\cdot)$ and hence $\boldsymbol{\nu}_{1Ti}$ must have similar properties to those noted in (a) for the $\boldsymbol{\nu}_{2Ti}$;

(c) $\bar{\boldsymbol{H}}_i(\cdot)$ must be an "efficient" estimate of $\boldsymbol{H}_i(\cdot)$ so that (after suitable transformations) the $\boldsymbol{\nu}_{Ti}$ have "good" properties also [however, for $\boldsymbol{H}_i(\cdot) \neq \boldsymbol{G}_i(\cdot)$ it may not be possible to achieve this objective];

(d) the experimental design must adequately investigate the parameter space and highlight regions in which $\boldsymbol{G}_i(\boldsymbol{\theta}, T)$ changes most rapidly as a function of its arguments;

(e) the conjectured approximation to $G_i(\cdot)$ has to be tested by predicting outcomes at other points in $\Theta \times \mathscr{T}$ and comparing these with the observed values $\bar{\psi}_{Ti}$, as well as by other relevant diagnostic checks.

Section 4 considers (b) in some detail [overlapping slightly with (a)] and Section 5 deals with (a), (c) and (e); it is beyond the scope of this chapter to adequately discuss (d), but relevant references on experimental design methods are Cochran and Cox (1957), Conlisk (1974) and Naylor (1971), while Myers and Lahoda (1975) consider response surface designs to establish local properties of $\boldsymbol{G}_i(\cdot)$.

3.3. *Embedding the econometric model in the Monte Carlo model*

An overview can be offered as follows. The objective is to compute the $\psi_{Ti}(\boldsymbol{\theta})$ relevant to the econometric theory by making these the population parameters to be estimated from the Monte Carlo. A feasible estimator of any $\psi_{Ti}(\boldsymbol{\theta})$ for a given value of $(\boldsymbol{\theta}, T)$ is provided from distribution sampling by the $\bar{\psi}_{Ti}$ (the means of simple random samples) and the functions generated by varying $\boldsymbol{\theta}$ and/or T [denoted $\boldsymbol{G}_i(\boldsymbol{\theta}, T)$] can be approximated by step-functions (so each experimental outcome is separately reported, equivalent to using a dummy variable of the zero/one variety for every selection of $(\boldsymbol{\theta}, T)$). This is called naive simulation since a vast range of important information is being ignored and no data summarisation of the experimental results is provided.

Rather, as discussed en route above, three sources of information frequently should be exploited to improve efficiency over naive simulation. Firstly, the unique property of *computer* simulation is that the actual values of the DGP parameters $(\boldsymbol{\theta}, T)$ are known as are the exact random numbers $\{n_i\}$ which are generated. The latter is easily used information and almost always provides an efficiency gain (see Sections 4.1 and 4.2), whereas the former plays an indirect role via the second source, namely the known structure of the econometric model. By this is meant that both the population moments of the DGP are computable and the correct form of $\boldsymbol{q}_1(\boldsymbol{\beta}_1 | \boldsymbol{\beta}_2^0)$ is known and this knowledge can be combined to yield *control variates* for the simulation (see Section 4.3). The final source of information comes from the econometric theory again and concerns invariance and (limiting) distributional results; the former reduce the dimensionality of the parameter space needing investigation without losing generality, and the latter provide a useful guide to the formulation of the $\boldsymbol{H}_i(\boldsymbol{\theta}, T)$ by restricting these to coincide with the known form of the corresponding $\boldsymbol{G}_i(\cdot)$ for large T (see Section 5).

Consequently, careful and thorough embedding of the econometric model in the Monte Carlo can yield improved efficiency [sometimes dramatically—see, for example, Hendry and Srba (1977)] and even closer interdependence will emerge in the following sections thereby providing ways of investigating validity as well as further improving reliability.

4. Reducing imprecision: Variance reduction

Variance reduction in the present context entails *intra*-experiment control. The most common techniques are: (a) reusing the known random numbers $\{n_i\}$ (which economises on their generation as well as reducing variability) either directly (4.1) or after transforming (4.2); and (b) developing control variates which ensure variance reductions in pre-specifiable situations [see (4.3)]. Such devices may be used in combination [see, for example, Mikhail (1972), Kleijnen (1974) and Hylleberg (1977)].

4.1. *Common random numbers*

Using the same set $\{n_i\}$ in two situations generally reduces the variability of the *difference* between the estimates in the two situations (although not of the actual estimates). For example, different estimation methods are almost invariably applied to *common* data sets for comparisons. Less usual, but equally useful, the same $\{n_i\}$ also can be used at different points in Θ and/or $\mathscr{T}$ for a single estimator. Thus, "chopping-up" one long realization such as one set of $T = 80$, into two of $T = 40$ and four of $T = 20$ reduces variability between sample size comparisons.

This type of device is generally invaluable in pre-experiment screening for potential invariances. Thus, in examples (3)–(6), estimating $\hat{\alpha}$ with the same $\{n_i\}$ but two different σ_ε^2 values should yield *identical* results. Similarly, for fixed regressors, identical values of $(\hat{\boldsymbol{\beta}} - \boldsymbol{\beta})$ would occur with the same $\{n_i\}$ used at different $\boldsymbol{\beta}$, and so on. However, reusing $\{n_i\}$ across experiments may create non-random $\boldsymbol{\nu}_{1Ti}$.

4.2. *Antithetic random numbers*

Consider two unbiased estimators, $\tilde{\psi}$ and ψ^{+} for an unknown parameter ψ such that the "pooled estimator" $\bar{\psi}=\frac{1}{2}(\tilde{\psi}+\psi^{+})$ has $\mathscr{E}(\bar{\psi})=\psi$ and variance $V(\bar{\psi}) =\frac{1}{4}[V(\tilde{\psi})+V(\psi^{+})+2\,\mathrm{Cov}(\tilde{\psi},\psi^{+})]$. In random replications, $\tilde{\psi}$ and ψ^{+} are based on independent sets $\{n_i\}$ so that $\mathrm{Cov}(\cdot)=0$, $V(\tilde{\psi})=V(\psi^{+})$ and $V(\bar{\psi})=\frac{1}{2}V(\tilde{\psi})$. However, since the $\{n_i\}$ are known, it may be possible to *select matching pairs which offset each other's variability* (i.e. are antithetic) and base $\bar{\psi}$, ψ^{+} on these [see Hammersley and Handscomb (1964) and Kleijnen (1974)]. For example, $\{n_i\}$ and $\{1-{.}_i\}$ are perfectly negatively correlated as are $\{e_t\}\sim I\mathscr{N}(0,\sigma_\varepsilon^2)$ and $\{-e_t\}$. Basing $\tilde{\psi}$ on one and ψ^{+} on the other of an antithetic pair can induce a negative covariance in many cases (see, for example, Mikhail (1972, 1975), Hendry and Trivedi (1972) and Hylleberg (1977)]. In certain respects the effect is equivalent to stratified sampling: $\{n_i\}$ and $\{1-n_i\}$ corresponds to a partition of $R(0,1)$ into $R(0,\frac{1}{2})$ and $R(\frac{1}{2},1)$, while ensuring sampling from each segment, and this idea generalizes to four-way partitions, etc. (with analogous results for normal variates).

Again, antithetic variates can form the basis for invariance determination [see Kakwani (1967)] since if $\tilde{\psi}$ and ψ^{+} are linear in $\{n_i\}$ and $\{1-n_i\}$, respectively, $V(\bar{\psi})=0$ independently of the number of paired replications. In dynamic models, it has proved difficult to locate antithetic transformations which generate negative covariances between *estimators*; in example (3)–(6), basing $\tilde{\psi}$ and ψ^{+} on $(y_0,\{e_t\})$ and $(-y_0,\{-e_t\})$ produces $\tilde{\psi}=\psi^{+}$ and is, therefore, useless. Nevertheless, for stochastic simulation studies of (say) estimated macro-econometric models, carefully chosen antithetic variates may be able to save a considerable expenditure of computer time [see Mariano and Brown (1983) and Calzolari (1979)].

Finally, little work has been done on creating functions of $\{n_i\}$ which improve the efficiency with which moments other than the first are estimated so the next technique seems more promising in econometrics (contrast the conclusions of Kleijnen (1974)).

4.3. *Control variates*

A control variate (CV) is any device designed to reduce intra-experiment variation, by forming a more tractable function of the random numbers than the primary objective of the study. Thus, given $\tilde{\psi}$, create from the same $\{n_i\}$ a ψ^* where $\mathscr{E}(\psi^*)$ is *known* and $\tilde{\psi}$ and ψ^* are positively correlated. Then:

$$\bar{\psi}=\tilde{\psi}-\psi^*+\mathscr{E}(\psi^*)\text{ has }\mathscr{E}(\bar{\psi})=\psi \tag{27}$$

and

$$V(\bar{\psi})=V(\tilde{\psi})+V(\psi^*)-2\,\mathrm{Cov}(\tilde{\psi},\psi^*)\langle V(\tilde{\psi}),\quad\text{if }\mathrm{Cov}(\cdot)\rangle\tfrac{1}{2}V(\psi^*). \tag{28}$$

In much Monte Carlo work, CVs like ψ^* are ad hoc; but it is a major function of the *econometric model to supply estimators from which CVs can be constructed for the Monte Carlo study*. The estimator generating equation $\boldsymbol{q}_1(\boldsymbol{\beta}_1|\boldsymbol{\beta}_2^0)=\boldsymbol{0}$ provides the required solution, since (among other attributes) it defines the class of estimators asymptotically equivalent to $\hat{\boldsymbol{\beta}}_1$ (and hence highly correlated with it for large T). Within the relevant class, choose the most tractable member $\boldsymbol{\beta}_1^*$, seeking a compromise between $\boldsymbol{\beta}_1^*$ behaving similarly to $\hat{\boldsymbol{\beta}}_1$, yet where $\mathrm{E}(\boldsymbol{\beta}_1^*)$ is computable whereas $\mathrm{E}(\hat{\boldsymbol{\beta}}_1)$ is not (compare the analogous problem in choosing Instrumental Variables) [see Hendry and Srba (1977), and for the basis of a general approach based on Edgeworth approximations, see Sargan (1976) and Phillips (1977b)].

For the example in (3)–(6), the DGP of the econometric model is such that $q_1(\alpha|\sigma_\varepsilon^2)=0$ is (asymptotically) equivalent to $(\sum y_{t-1}\varepsilon_t)=0$ and an asymptotically efficient estimator is given by choosing α^* such $\hat{\alpha}=\alpha^*+O_p(1/T)$. To this order:

$$\alpha^*=\alpha+(1-\alpha^2)\sum y_{t-1}\varepsilon_t/T, \tag{29}$$

with $\sqrt{T}(\alpha^*-\alpha)\sim{}_a\mathcal{N}(0,(1-\alpha^2))$ and $\operatorname{plim}\sqrt{T}(\hat{\alpha}-\alpha^*)=0$. Also:

$$\mathrm{E}(\alpha^*)=\alpha \quad\text{and}\quad \mathrm{E}(\alpha^*-\alpha)^2=(1-\alpha^2)/T, \tag{30}$$

so the first two moments are easily calculated. Clearly, α^* requires knowledge of $\boldsymbol{\theta}$ and so is operational *only* in a Monte Carlo setting, but is no less useful for that. In effect, a benefit arises from using α^* as an "intermediary" since:

$$\hat{\alpha}\equiv\alpha^*+(\hat{\alpha}-\alpha^*), \tag{31}$$

splitting the problem of estimating $\mathrm{E}(\hat{\alpha})$ into a part which is *known*, $[\mathrm{E}(\alpha^*)]$, and only simulating the remainder, which is $O_p(1/T)$, whereas $\hat{\alpha}$ is $O_p(1/\sqrt{T})$.

The mapping to the Monte Carlo model is obvious:

$$\tilde{\psi}=\frac{1}{N}\sum\hat{\alpha}_i,\qquad \psi^*=\frac{1}{N}\sum\alpha_i^*\quad\text{and}\quad\bar{\psi}=\alpha+(\tilde{\psi}-\psi^*), \tag{32}$$

so that:

$$\mathscr{E}(\bar{\psi})=\alpha+\mathscr{E}(\tilde{\psi})-\mathscr{E}(\psi^*)=\mathscr{E}(\tilde{\psi})=\mathrm{E}(\hat{\alpha})\quad\text{(as required)}. \tag{33}$$

Since $(\tilde{\psi}-\psi^*)$ is $O_p(1/T\sqrt{N})$ its behaviour is less dependent on the particular random numbers sampled, and $\bar{\psi}$ is increasingly efficient with increasing T, offsetting the rising costs of experimentation.[3]

[3]More accurately, $(\tilde{\psi}-\psi^*)$ is $O_p(1/\sqrt{N}\,T(1-\alpha^2))$: see equations (61) and (64) below.

Similar principles apply to estimating other moments, means of tests, etc. [see Mikhail (1972, 1975) and Hendry and Harrison (1974)].

Furthermore, the *validity* of the experiments now can be checked by testing that the estimated moments of ψ^* do not differ significantly from their known population values. Indeed, we have created a tight specification for determining any ψ, as illustrated for $\mathrm{E}(\hat{\alpha})$ in Table 4.1, where $\overset{\mathscr{E}}{=}$ denotes "equal in expectation", $\rightarrow$ means "helps determine", and $\overset{\mathscr{E}}{\rightarrow}$ implies both $\overset{\mathscr{E}}{=}$ and $\rightarrow$ [see Hendry and Srba (1977)]. The only unknown is ψ, and all other features are checkable: equivalent results hold for second moments, means of estimated variances or tests, etc.

On this basis, it seems possible to reliably and validly estimate $\psi_{\tilde{T}i}(\boldsymbol{\theta})$ thus achieving objective (b) of Section 3.2. Moreover, $V(\tilde{\psi})$ and $V(\bar{\psi})$ will be useful in Section 5 for checking the choice of $\boldsymbol{H}_i(\cdot)$.

The final twist is to note that CVs provide asymptotic approximations to the econometric estimators and have as their finite sample moments, the asymptotic moments of the latter. Consequently CVs allow the *analytical* derivation of moments of estimators which differ by terms of $O_p(1/T)$ from the econometric estimators under study and so, even without a simulation experiment, throw considerable light on the behaviour of the latter *and* the conditions under which asymptotic theory provides an adequate characterisation of finite sample behaviour [see Hendry (1979) and for a correction to the formulae therein; see Maasoumi and Phillips (1982) and Hendry (1982)]. CVs also can be obtained from Nagar Expansions [see Nagar (1959), Hendry and Harrison (1974)], or Taylor Series Expansions [see O'Brien (1979)], and if their exact distributions were known, could help determine $\Psi_T(\cdot)$ directly [see Sargan (1976)].

Moreover, interesting combinations using CV's to accurately estimate moments for Edgeworth approximations are possible for determining significance criteria of tests in dynamic models [see Sargan (1978) and Tse (1979)].

Finally, in the statistics literature, variance reduction methods are often referred to as "swindles" [see, for example, Gross (1973)]. Providing that the costs of the extra labour in deriving, programming, etc. any variance reduction tech-

Table 4.1

	Simulation		Exact		Asymptotic
Econometric estimator	$\tilde{\psi} \overset{\mathscr{E}}{\rightarrow} \bar{\psi}$	$\overset{\mathscr{E}}{\rightarrow}$	$\psi = \mathrm{E}(\hat{\alpha})$	$\leftarrow$	α
	$\nearrow$				$\Vert$
Control variate	ψ^*	$\overset{\mathscr{E}}{=}$	α	$=$	α

nique are more than offset by the efficiency gains, there are no losers (other than computer manufacturers!) so "swindle" is not an accurate description. Generally, CVs are relatively easy to derive and often trivial to compute; in practice, CVs have yielded useful efficiency gains, but by construction they are not helpful in determining differences between asymptotically equivalent estimators (unless expansions to higher order than T^{-1} are used).

5. Reducing specificity: Response surfaces

Following on from Section 3.2, an important component of validity which is open to improvement is the development of approximations to the $G_i(\boldsymbol{\theta}, T)$ valid in $\boldsymbol{\Theta}\times\mathscr{T}$, to help counter one aspect of the criticism of specificity. Various considerations affect the formulation of the $\boldsymbol{H}_i(\boldsymbol{\theta}, T)$ depending on the problem and the moments under study. For simplicity, a scalar case will be discussed and a number of suggestions offered although the literature is not sufficiently advanced to permit a rigorous or comprehensive analysis at this juncture [a sceptical view of the value of simple forms of response surfaces in econometrics is expressed by Maasoumi and Phillips (1982)].

The basic forms proposed below arise from two sets of considerations: firstly, the formulation of $H_i(\cdot)$ functions which satisfy certain restrictions [including coinciding asymptotically with $G_i(\cdot)$]; and secondly, transformations likely to produce homoscedastic response surface residuals when the $G_i(\cdot)$ are known. These are considered in turn below. It is a remarkable feature of much Monte Carlo research that response surface residual variances (after transformation) have an anticipated value of unity if $H_i(\cdot)=G_i(\cdot)$ so that inappropriate functional forms are indicated by estimated residual variances in excess of unity.

5.1. *Response surface formulations*

5.1.1. *First moments of estimators*

Consider the density function of the estimator $\hat{\beta}$ shown in Figure 5.1. This seems a typical situation in econometrics: the parameter of interest θ has a true value θ^0; the econometric estimator $\hat{\beta}$ thereof has a plim of β, an expectation $\mathrm{E}(\hat{\beta})$ and could equal $\hat{\beta}_j$ in any given instance. As θ^0 and T vary over their parameter spaces, most of the variability in $\hat{\beta}$ is due to changing θ^0, some to changing T and some to the vagaries of sampling. Let $I=(\beta-\theta^0)$ denote the inconsistency, $B=(\mathrm{E}(\hat{\beta})-\beta)$ the additional finite sample bias and $(\hat{\beta}_j-\mathrm{E}(\hat{\beta}))$ the sample fluctuation. Since $\mathscr{E}(\bar{\psi}_{T1})=\mathrm{E}(\hat{\beta})$, a useful regressand is $(\bar{\psi}_{T1}-\theta^0)$ when implementing (24). For coherence with the asymptotic result, I should be a regressor, with the term I/T also included as a potential regressor by analogy with results based on Nagar approximations [see Nagar (1959)] and because of its established empirical usefulness [see Hendry and Harrison (1974)]. Let $\phi_{j,1}$ $(j=1,\dots,k_1)$ denote appropriate functions of the design variables in the set of experiments (e.g. $\theta, \theta/T, \theta^2, T^{-1}, T^{-2}$, etc. in this scalar case), chosen on the best available basis, then:

$$H_1(\theta, T) - \theta^0 = \gamma_{01} I + \gamma_{11}/T + \gamma_{21} I/T + \sum_{j=1}^{k_1} \gamma_{j+2,1}\phi_{j1}/T. \tag{34}$$

For $\gamma_0 = 1$ (which should be tested), both $H_1(\cdot)$ and $G_1(\cdot) \to \beta$ as $T \to \infty$. When $I \equiv 0$, the bias $\mathrm{E}(\hat{\beta} - \beta)$ is assumed to be at most $O(T^{-1})$ as in (say) Nagar approximations. Note that, independently of how closely they represent $G_1(\cdot)$, response surfaces such as (34) (after transformation) also provide a useful summarisation of the experimental findings; but as discussed in Section 5.3 below, their validity is open to investigation in any case. The "solutions" of estimated regressions like (34) (for $\gamma_{01} = 1$) yield expressions of the form:

$$\mathrm{E}(\hat{\beta} - \beta) \simeq \left(\hat{\gamma}_{11} + \hat{\gamma}_{21} I + \sum \hat{\gamma}_{j+2,1}\phi_{j1}\right) \Big/ T$$

as the numerical-analytical results approximating the finite sample outcome.

As Nicholls et al. (1975) point out, however, direct estimation of the $\{\gamma_{j,1}\}$ in, for example, (34), will be inefficient and the estimated standard errors will be biased unless appropriate heteroscedasticity corrections (such as those discussed in Section 5.2) are used.

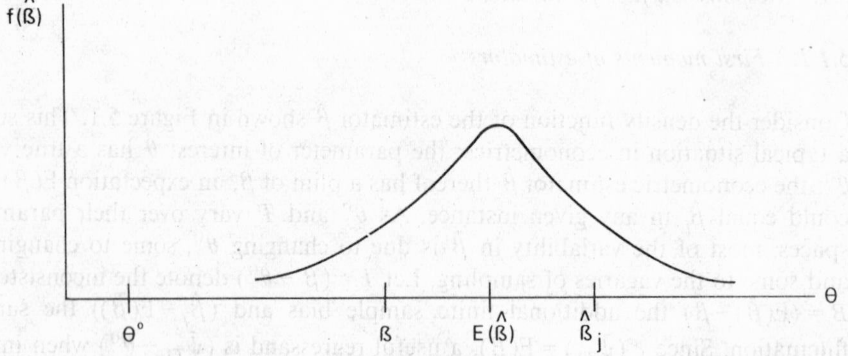

Figure 5.1

5.1.2. Second moments

When analysing $\psi_{T2}(\cdot)$ (where $\hat{\psi}_{T2}^{1/2}$ is referred to as the sampling standard deviation, SSD), the asymptotic variance Ω often is computable to restrict the limiting behaviour of $H_2(\cdot)$. Moreover, a $\log_e(=\ln)$ formulation ensures both positive predictions of ψ_{T2} as well as identical results from $\hat{\psi}_{T2}$ and SSD. Thus, an "obvious" functional form is:

$$\ln H_2(\theta, T) = \gamma_{0,2} \ln \Omega + \gamma_{12} T^{-1} \ln \Omega + \gamma_{2,3}/T + \sum_{j=1}^{k_2} \gamma_{j+2,2}\phi_{j,2}/T, \tag{35}$$

where the $\phi_{j,2}$ may reflect such aspects as degrees of freedom, or the effective sample size, etc. [see (68) below]. Again, for $\gamma_{0,2}=1$, $H_2(\cdot)\to\Omega$ as $T\to\infty$. Similar considerations apply to σ^2 (the equation-error variance in the econometric model) and ESE in correctly specified econometric models. In mis-specified econometric models, however, the role of Ω in (35) is played by V or $\operatorname{plim}\hat{\sigma}^2 = \sigma_p^2$ (say), respectively (see Section 2.2 above). Variants of (35) have proved useful in a number of response surface studies [see, for example, Naylor (1971, ch. 7), Hendry and Srba (1977) and Hendry (1979)].

5.1.3. Test rejection frequencies

Consider a test Z of some hypothesis H_0: $\boldsymbol{\theta} = \boldsymbol{\theta}_0$ such that on H_0, $Z \underset{a}{\sim} \chi^2(l,0)$ where $\chi^2(l,\varphi)$ denotes a non-central chi-square with l degrees of freedom and non-centrality parameter φ. It is assumed that Z is consistent (will reject all fixed false hypotheses with unit probability as $T\to\infty$). The *nominal* and large sample significance level of the test is $\delta = \Pr(\chi^2(l,0)\geq d_1)$, where H_0 is true but is rejected if $Z \geq d_1$. Local alternatives of the form:[4]

$$H_T: \boldsymbol{\theta} = \boldsymbol{\theta}_0 + \boldsymbol{\lambda}/\sqrt{T}, \tag{36}$$

for fixed $\boldsymbol{\lambda}$, are considered, so that on H_T, $Z \underset{a}{\sim} \chi^2(l,\varphi)$ where φ is a scalar function of $\boldsymbol{\theta}$, $\boldsymbol{\theta}_0$ and $\boldsymbol{\lambda}$, independent of T. The large sample power of the test is:

$$\Pr\left(\chi^2(l,\varphi)\geq d_1\right) = P^*(\cdot). \tag{37}$$

[4] Note that H_T varies with $\boldsymbol{\lambda}$ and $\boldsymbol{\theta}_0$ as well as T, and the actual significance level may vary with $\boldsymbol{\theta}$ and T.

The objective of the study is not just to estimate the test rejection frequency at a few points in $\Theta \times \mathscr{T}$ but to determine the power[5] functions:

$$P = \Pr(Z \geq d_1) = G_p(\boldsymbol{\theta}, \boldsymbol{\lambda}, T, \delta). \tag{38}$$

The computation of an integral like P for a fixed value of $(\boldsymbol{\theta}, T, \boldsymbol{\lambda}, \delta)$ is usually based on the random replications estimator $\tilde{P}$ defined as $\sum F_i/N$, where $F_i = 1$ if and only if a reject outcome occurs (and is zero otherwise). Unfortunately, $\tilde{P}$ is usually inaccurate for P unless N is large since:

$$V(\tilde{P}) = P(1-P)/N. \tag{39}$$

To date, few variance reduction methods have proved useful for intra-experiment control [but see Sargan (1976, 1978)]. However, $P^*(\cdot)$ is often obtainable both easily and cheaply and hence can be used as an inter-experiment control in a response surface based on (say):

$$\left(\frac{P}{1-P}\right) = \left(\frac{P^*}{1-P^*}\right)^{\gamma_{03}+\gamma_{13}/T} \exp\left(g\left(\frac{\boldsymbol{\theta}}{T}\right)\right). \tag{40}$$

As discussed in Mizon and Hendry (1980), (40) ensures $P \to P^*$ as $T \to \infty$ for $\gamma_{03} = 1$, and also $0 \leq P \leq 1$.

In the example, to test H_0: $\alpha = \alpha_0$ when $|\hat{\alpha}| < 1$, let

$$Z = T(\hat{\alpha} - \alpha_0)^2/(1-\hat{\alpha}^2) \underset{a}{\sim} \chi^2(1, \varphi), \tag{41}$$

where the asymptotic approximation of $\mathrm{Var}(\hat{\alpha})$ by $(1-\hat{\alpha}^2)/T$ is used for expository purposes [see Evans and Savin (1981)] and $\varphi = \lambda^2/(1-\alpha^2)$ with $\delta = \Pr(\chi^2(1,0) \geq d_1)$. Since $P^*(\alpha, \alpha_0, T, \delta) = \Pr(\chi^2(1, \varphi) \geq d_1)$ when H_T is true, P^* could be computed directly from integrating the non-central χ^2. However [see Kendall and Stuart (1961, ch. 24) and Mizon and Hendry (1980)], it is often convenient to compute $P^*(\cdot)$ instead from a central χ^2 approximation with the same first two moments:

$$P^*(\cdot) \simeq \int_f^{\infty} \mathrm{d}\chi^2(m, 0), \tag{42}$$

where $m = (1+\varphi)^2/(1+2\varphi)$ and $f = md_1/(1+\varphi)$. Using the formula in equation

[5] Estimation of test "power" requires prior calibration of $d_1(T, \boldsymbol{\theta})$ to ensure finite sample significance levels of δ, and most studies simply report rejection frequencies. For the importance of correcting significance levels see, for example, Evans and Savin (1982).

(42), $P^*(\cdot)$ is inexpensive and easy to compute for a wide range of tests, by itself provides considerable insight into factors determining power, is a useful analytical approximation to $P(\cdot)$ for large T, and offers a convenient means of comparing alternative tests [see Hendry (1977) and Mizon and Hendry (1980); also, compare the notion of "approximate slope" comparisons of tests based on φ in Geweke, Chapter 19 in this Handbook]. More accurate approximations to P than P^* could be obtained from Edgeworth expansions as, for example, in Davis (1971).

5.2. *Heteroscedasticity transformations*

In each of the above cases the conjectured response surface functional forms have to be both estimated and tested, and, interestingly, the heteroscedasticity transformations necessary to efficiently achieve the former help provide tests of $H_i(\cdot) = G_i(\cdot)$ over the sampled parameter space.

Consider random sampling certain variates $\{x_i\}$ from a distribution with finite moments given by $\mu'_k = \mathrm{E}(x_i^k)$ for $k = 1,\dots,6$. Let $m'_k = N^{-1}\sum_{i=1}^N x_i^k$, then [see, for example, Kendall and Stuart (1958, ch. 10)]:

$$\mathrm{E}(m'_k) = \mu'_k \quad \text{and} \quad \mathrm{Var}(m'_k) = \frac{1}{N}\left(\mu'_{2k} - (\mu'_k)^2\right). \tag{43}$$

For moments about means, however, exact results are not easily established and instead we use the large sample result [see Cramér (1946, p. 365) and Kendall and Stuart (1958, ch. 10)]:

$$\sqrt{N}\,(m_k - \mu_k) \underset{a}{\sim} \mathcal{N}\left(0, \{\mu_{2k} - 2k\mu_{k-1}\mu_{k+1} - \mu_k^2 + k^2\mu_2\mu_{k-1}^2\}\right), \tag{44}$$

where {} is denoted by ω_k^2 below, $\mu_k = \mathrm{E}(x - \mathrm{E}(x))^k$ with $\mu_0 = 1$, and $m_k = [1/(N-1)]\sum(x_i - m'_1)^k$. In particular:

$$\sqrt{N}\,(m'_1 - \mu'_1)/\sqrt{\mu_2} \underset{a}{\sim} \mathcal{N}(0,1) \tag{43a}$$

and

$$\sqrt{N}\,(m_2 - \mu_2)/\omega_2 \underset{a}{\sim} \mathcal{N}(0,1), \tag{44a}$$

with $\omega_2^2 = \mu_4 - \mu_2^2$. If the $\{x_i\}$ are normally distributed, then $\mu_3 = 0$ and $\mu_4 = 3\mu_2^2$. Note that the $\{x_i\}$ could refer to estimated regression coefficients [in which case (43a) and (44a) relate to ψ_{T1} and ψ_{T2}] or to equation-error variances or estimated standard errors, t-statistics, etc. In most cases, however, ω_k^2 can be estimated directly from the Monte Carlo, and hence (44) can be implemented by replacing ω_k^2 by $\hat{\omega}_k^2$. Examples are reported below in Section 6.

Alternatively, even though the $\{x_i\}$ may be complicated functions of the basic random numbers $\{n_i\}$, often their large-T distributions can be derived and are themselves normal. For example, (8) holds for the model in (3)–(6), and since $\tilde{\psi}_{T1} = N^{-1}\sum\tilde{\alpha}_i$, using (8) and (43a) for $\mu_2 = T^{-1}(1-\alpha^2)$:

$$\sqrt{T\cdot N}\,(\tilde{\psi}_{T1} - \psi_{T1})/\sqrt{1-\alpha^2} \underset{a}{\sim} \mathcal{N}(0,1) \quad \text{(for large } T \text{ and large } N\text{).}^{6} \tag{45}$$

Similarly, for $\tilde{\psi}_{T2}$ using the asymptotic approximation that $\omega_2^2 = 2\mu_2^2$ yields from (44a):

$$\sqrt{N/2}\,(m_2/\mu_2 - 1) \underset{a}{\sim} \mathcal{N}(0,1) \tag{44b}$$

Thus, noting that $(\tilde{\psi}_{T2}/\psi_{T2} - 1)$ is $O_P(N^{-1/2})$, the further approximation that $(\tilde{\psi}_{T2} - \psi_{T2})/\psi_{T2} \simeq \ln(\tilde{\psi}_{T2}/\psi_{T2})$ relates the functional form to (35). In fact, if $x_i \sim \mathcal{N}(\mu_1', \mu_2)$ then $\mathrm{Var}(\ln m_2)$ is independent of μ_2 [see Rao (1952, p. 214)], and from the limiting convergence of $(\chi_l^2 - l)/\sqrt{2l}$ to $\mathcal{N}(0,1)$ [see Johnson and Kotz (1970, ch. 17)], it can be established directly that (44b) holds for large T and large N [see Campos (1980)].

For equation-error variances, σ_ε^2, the $\{x_i\}$ must be interpreted as $\{\hat{\sigma}_{\varepsilon i}^2\}$ so that (43a) applies. However, using the large-T approximation that $\mathrm{E}(\hat{\sigma}_\varepsilon^4) = 3\sigma_\varepsilon^2$ (exact when $\varepsilon_t \sim \mathcal{N}(0, \sigma_\varepsilon^2)$) then:

$$\sqrt{T/2}\,(\hat{\sigma}_\varepsilon^2 - \sigma_\varepsilon^2)/\sigma_\varepsilon^2 \underset{a}{\sim} \mathcal{N}(0,1). \tag{46}$$

Thus, since $\tilde{\sigma}_\varepsilon^2 = N^{-1}\sum_{i=1}^{N}\hat{\sigma}_{\varepsilon i}^2$, and $\ln(\tilde{\sigma}_\varepsilon^2/\sigma_\varepsilon^2) \simeq (\tilde{\sigma}_\varepsilon^2/\sigma_\varepsilon^2 - 1)$:

$$\sqrt{NT/2}\,\ln\left(\tilde{\sigma}_\varepsilon^2/\sigma_\varepsilon^2\right) \underset{a}{\sim} \mathcal{N}(0,1) \quad \text{(for large } T \text{ and large } N\text{).} \tag{46a}$$

If $\mathrm{Var}(\tilde{\sigma}_\varepsilon^2)$ is estimated from the Monte Carlo, then:

$$\frac{\tilde{\sigma}_\varepsilon^2}{\left(\widetilde{\mathrm{Var}}\left(\tilde{\sigma}_\varepsilon^2\right)\right)^{1/2}} \ln\left(\tilde{\sigma}_\varepsilon^2/\sigma_\varepsilon^2\right) \underset{\text{app}}{\sim} \mathcal{N}(0,1). \tag{46b}$$

Note that if a control variate is used in estimating $\tilde{\sigma}_\varepsilon^2$, then (46a) also must be corrected for the efficiency gain.

[6] For $|\alpha|$ near unity, the continuous normalization used by Evans and Savin (1981) may be preferable to $\sqrt{T/(1-\alpha^2)}$; but in practice hardly altered the response surface estimates computed in Section 6 below.

Similar considerations apply for estimated coefficient variances or ESE's when their simulation standard errors (SSE) have been computed, and equations like (46b) are used below in the form:

$$\frac{\sqrt{V/T}}{\text{SSE}}\ln\left(\frac{\text{ESE}}{\sqrt{V/T}}\right)\underset{\text{app}}{\sim}\mathcal{N}(0,1) \tag{47}$$

where $\sqrt{V/T}$ is $1/\sqrt{T}$ times the plim of the estimated standard error ($\sqrt{(1-\alpha^2)/T}$ in the illustration).

Unfortunately, asymptotic approximations to the heteroscedasticity correction factors for ESE are highly model dependent. For example, in a simple regression model with a strongly exogenous, stationary regressor, following Goldberger (1964, ch. 3.8) it can be established that SSE is $O(T^{-1}N^{-1/2})$, whereas for ESE($\hat{\alpha}$) from (3) (for sufficiently large T):

$$\text{SSE} \simeq \sqrt{1+T\alpha^2}\,/T^{3/2}\sqrt{N}\,. \tag{48}$$

Then for $\alpha = 0$, this is of $O(T^{-3/2}N^{-1/2})$ but for (say) $|\alpha| = \frac{1}{2}$ and T large it is close to $(2T\sqrt{N})^{-1}$. For small T, (48) itself is not a good approximation (although closer approximations can be derived from Nagar expansions).

Overall, the asymptotic correction factors have the virtue of simplicity and in practice yield similar results to response surfaces based on $\hat{\omega}_k$. However, they rely on a "double-asymptotic" requirement of large T and large N and require modification by efficiency gain factors for application to results based on control variate estimates. Moreover, as noted, no simple results hold for ESE's and this might also affect other statistics in more complicated data generation processes, so there is a good case for using simulation-estimated ω_k. Nevertheless, as (44) also shows for $k \geq 3$, high order moments are imprecisely estimated. Consequently, below we report various response surfaces based on both forms of correction and also certain descriptive regressions relating the relevant $\hat{\omega}_k$ to their asymptotic counterparts [see equations (64), (66), (67) and (72)].

Finally, for power functions based on $\tilde{P}$, where (39) applies, and a response surface such as (40), after a $\log_e$ transformation, is assumed, then [see Cox (1970, ch. 6)], noting that the Jacobian of the mapping from P to $\mathscr{L}(P) = \ln(P/(1-P))$ is $1/P(1-P)$:

$$J(\mathscr{L}(\tilde{P}) - \mathscr{L}(P)) \underset{a}{\sim} \mathcal{N}(0,1), \tag{49}$$

where $J = (NP(1-P))^{1/2}$ (which is estimable from the simulation). Thus, as remarked at the start of this section, when the relevant $G_i(\cdot)$ is known, each response surface can be formulated to have an anticipated residual variance of unity.

5.3. *Investigating validity*

In addition to the points noted above, each of the conjectured response surfaces entails also that $\gamma_{0j} = 1$ in order to reproduce $G_i(\cdot)$ in large samples and this is potentially testable from the regression estimates. Also, under the null that the error variance (σ^2) should be unity, the residual sum of squares will be distributed as $\chi^2(r,0)$ for r degrees of freedom in the relevant regression, since for correct specifications, $r\hat{\sigma}^2/\sigma^2 \sim \chi^2(r,0)$. Confidence limits for σ^2 for various r have been tabulated [see, for example, Croxton, Cowden and Klein (1968, table L)] but are easily calculated in any case.

As with any regression analysis, the selected response surfaces can be tested by a variety of Lagrange Multiplier based diagnostics (see, for example, Engle, Chapter 13 in this Handbook) of which predictive tests are one of the more important. If K experiments are conducted and K_1 used for selecting and estimating the response surfaces, $K - K_1$ should be used to test the validity of the results to ensure that some credibility beyond mere description attaches to the finally chosen surrogates for $G_i(\cdot)$ [see, for example, Chow (1960)].

Inappropriate choices of $H_i(\cdot)$ could induce either or both of autocorrelation and heteroscedasticity in the residuals. These problems might be detectable directly. The former can be tested by, for example, the Durbin–Watson test when a suitable data ordering exists [as in Mizon and Hendry (1980) or Maasoumi and Phillips (1982)]. A valuable diagnostic for the latter is the general test for functional mis-specification in White (1980b) who also derives a robust estimator of the estimated-parameter variances to support valid, if non-optimal, inference despite heteroscedasticity; both of these statistics are reported below. Discrepancies between the conventional and "robust" coefficient variances are indicative of mis-specification and White (1980a) presents a test based on this notion. Further tests against specific alternatives can be derived following the procedures in Engle (1982).

As noted above, the main advantages of estimated response surfaces over tabulation are their ability to summarize large and non-memorizable quantities of information in simple approximations which in practice do seem able to account for the bulk of inter-experiment variation in simulation outcomes (especially for inconsistent estimators) using formulae known to be correct for sufficiently large values of T. A corresponding disadvantage is that the dependence of the approximation error on the invariants of the data generation process is unknown, but in a well defined parameter space should be estimable for the purposes of predicting outcomes at other points *within* the sampled set [i.e. for experiments which could have been undertaken, but were not, as in Hendry and Harrison (1974)]. Conversely, relative to analytical derivations, the advantages are the use of less restrictive data generating processes than existing techniques can study analytically as well as exploiting the falling relative price of capital to economise on scarce labour resources; whereas the disadvantages are the inherent inexactitude of estimated response surfaces and the absence of a complete understanding of the limitations of the experimentally determined numerical results. As analytical methods improve, the frontier at which simulation is substituted for analysis will move outwards, but is unlikely to obviate any need for efficient Monte Carlo. Equally, simulation based findings seems most helpful when they are tightly circumscribed by analytical results, a point illustrated by the experimental evidence reported in Section 6 [for further discussion, see Hendry (1982)].

6. An application to a simple model

6.1. Formulation of the experiments

To illustrate the application of the experimental approach, we consider the model in (3)–(6) as this highlights the principles involved and indicates what can and cannot be achieved by experimentation. The main objectives of the following experiments—considered as a substantive study—are to:

(a) estimate and test response surfaces for $\psi_{T1} = \mathrm{E}(\hat{\alpha})$, $\psi_{T2} = \mathrm{E}(\hat{\alpha} - \mathrm{E}(\hat{\alpha}))^2$, $\sqrt{V_T} = \mathrm{E}(\hat{V}(\hat{\alpha})^{1/2})$ (= ESE), and $P = \Pr(Z \geq 3.84)$ basing these on the ideas developed in Section 5;

(b) investigate the efficiency gains from using the CVs α^* for $\hat{\alpha}$ and $\sigma_\varepsilon^{*2} = T^{-1}\sum\varepsilon_t^2 \sim T^{-1}\chi^2(T,0)$ for $\hat{\sigma}_\varepsilon^2$ [so that $\mathrm{E}(\sigma_\varepsilon^{*2}) = \sigma_\varepsilon^2$ and $V(\sigma_\varepsilon^{*2}) = 2\sigma_\varepsilon^4/T$];

(c) relate simulation estimates of ω_k to their asymptotic counterparts; and

(d) evaluate the usefulness of asymptotic results as inter-experiment controls.

To recapitulate, the main simulation estimators of the unknown ψ_{Ti}, etc. are given by:

$$\tilde{\psi}_{T1} = N^{-1}\sum\hat{\alpha}_i \quad \text{and} \quad \bar{\psi}_{T1} = \tilde{\psi}_{T1} - \psi_{T1}^* + \alpha \tag{50}$$

(ψ_{Ti}^* are computed as for $\tilde{\psi}_{Ti}$ but with α_i^* replacing $\hat{\alpha}_i$);

$$\tilde{\psi}_{T2} = (N-1)^{-1}\sum(\hat{\alpha}_i - \tilde{\psi}_{T1})^2 \quad \text{and}$$

$$\bar{\psi}_{T2} = \tilde{\psi}_{T2} - \psi_{T2}^* + T^{-1}(1-\alpha^2), \tag{51}$$

$$\widetilde{\mathrm{ESE}} = N^{-1}\sum(\hat{V}(\hat{\alpha}_i))^{1/2}, \tag{52}$$

$$\tilde{\sigma}_\varepsilon^2 = N^{-1}\sum\hat{\sigma}_{\varepsilon i}^2 \quad \text{and} \quad \bar{\sigma}_\varepsilon^2 = \tilde{\sigma}_\varepsilon^2 - N^{-1}\sum\sigma_{\varepsilon i}^{*2} + \sigma_\varepsilon^2, \tag{53}$$

$$\tilde{P} = N^{-1}\sum F_i, \tag{54}$$

where

$$F_i = \begin{cases} 1, & \text{if } Z \geq 3.84 \text{ (for testing } H_0: \alpha = 0\text{)}, \\ 0, & \text{otherwise.} \end{cases}$$

Direct estimation of the cumulative density function of $\hat{\alpha}(\Psi_T(\hat{\alpha}))$ was not an objective of this set of experiments, although it is obviously a legitimate objective in general [see, for example, Orcutt and Winokur (1969)].

The sampling variances of the various simulation estimators were also estimated by the following formulae:

$$\tilde{V}(\tilde{\psi}_{T1}) = N^{-1}\tilde{\psi}_{T2} \quad \text{and} \tag{55}$$

$$\tilde{V}(\bar{\psi}_{T1}) = N^{-1}\left\{(N-1)^{-1}\Sigma(\hat{\alpha}_i - \alpha_i^* - \bar{\psi}_{T1} + \alpha)^2\right\},$$

from which the efficiency gain due to using α^* is given by $\text{EG} = \tilde{V}(\tilde{\psi}_{T1})/\tilde{V}(\bar{\psi}_{T1})$. Next:

$$\tilde{V}(\tilde{\psi}_{T2}) = N^{-1}(\tilde{\mu}_4 - \tilde{\psi}_{T2}^2), \quad \text{where } \tilde{\mu}_4 = (N-1)^{-1}\Sigma(\hat{\alpha}_i - \tilde{\psi}_{T1})^4, \tag{56}$$

$$\tilde{V}(\text{E\tilde{S}E}) = N^{-1}\left\{(N-1)^{-1}\Sigma\left(\hat{V}(\hat{\alpha}_i)^{1/2} - \text{E\tilde{S}E}\right)^2\right\}, \tag{57}$$

$$\tilde{V}(\tilde{\sigma}_\varepsilon^2) = N^{-1}\left\{(N-1)^{-1}\Sigma(\hat{\sigma}_{\varepsilon i}^2 - \tilde{\sigma}_\varepsilon^2)^2\right\}, \tag{58}$$

and

$$\tilde{V}(\bar{\sigma}_\varepsilon^2) = N^{-1}\left\{(N-1)^{-1}\Sigma(\hat{\sigma}_{\varepsilon i}^2 - \sigma_{\varepsilon i}^{*2} - \bar{\sigma}_\varepsilon^2 + \sigma_\varepsilon^2)^2\right\}, \tag{59}$$

with the efficiency gain from σ_ε^{*2} being $\text{SEG} = \tilde{V}(\tilde{\sigma}_\varepsilon^2)/\tilde{V}(\bar{\sigma}_\varepsilon^2)$. Finally, $\tilde{V}(\tilde{P})$ follows from (39), but following Cox (1970, ch. 6) and Mizon and Hendry (1980), (49) is formulated as:

$$(L^*(\tilde{P}) - L(P)) \underset{a}{\sim} \mathcal{N}(0,1), \tag{60}$$

when

$$L^*(\tilde{P}) = \left[\frac{N\tilde{P}(1-\tilde{P})}{(1-N^{-1})}\right]^{1/2} \cdot \ln\left(\frac{\tilde{P} - \xi}{1 - \tilde{P} - \xi}\right),$$

for $\xi = (2N)^{-1}$ and $L(P)$ is similar but replaces the second term by $\ln(P/(1-P))$. Observations with $\tilde{P} = 0$ or 1 are automatically deleted from the regression. We also deleted those for which $(1-P^*)<10^{-5}$, for P^* in (42), when using (40) to approximate the unknown $L(P)$ in (60).

The properties of the experimental design are important for achieving the objectives of the study, and "iterative" designs based on a pilot study followed by intensive searches in regions of θ where the relevant $G_i(\cdot)$ are least "well-behaved" may be needed. For example, it is difficult to design a single experiment which is "good" for estimating (say) both $\psi_{T1}(\cdot)$ and $P(\cdot)$. Here, to "cover" the parameter space, a full factorial design based on $\alpha=\{0,\pm 0.3,\pm 0.6,\pm 0.9\}$ and $T=\{10,20,30,40\}$ was selected, yielding 28 experiments in all with $\sigma_\varepsilon^2=1$ and $N=400$ (so that $\hat{P}$ could be accurate to within 0.0025). It is important to note that the parameter space is now $\{|\alpha|\le 0.9, \sigma_\varepsilon^2=1\}$ and that as $\alpha_0=0$, λ in (36) is implicitly determined by $\sqrt{T}\alpha$ so that $\varphi=T\alpha^2/(1-\alpha^2)$. Six randomly chosen experiments from the 28 were used for predictive testing.[7]

Finally, first order autoregressive processes have been the subject of extensive analysis and experimentation (see inter alia, Bartlett (1946), Hurwicz (1950), Kendall (1954), White (1961), Shenton and Johnson (1965), Copas (1966), Orcutt and Winokur (1969), Phillips (1977a) and Sawa (1978); also Kendall and Stuart (1966, ch. 48) provide a convenient summary of many of the relevant analytical results). Such known analytical results obviously "prejudice" the precise functions chosen to characterise the $G_i(\cdot)$, and where this has been an important influence, it is noted below.

6.2. *Estimating* $\mathrm{E}(\hat{\alpha}|\alpha,T)$

Firstly, the *CV* α^* yielded an average efficiency gain over distribution sampling of 6.4 for trivial extra computational cost. Also, for $|\alpha|\ne 0$, H_0: $\mathrm{E}(\hat{\alpha})=\alpha$ was rejected in every experiment using $\bar{\psi}_{T1}$ but on occasion was not for $|\alpha|=0.3$ using $\tilde{\psi}_{T1}$. The theoretical and simulation moments of α^* matched well, checking the validity of the random numbers used and correlation $(\hat{\alpha},\alpha^*)$ varied from 0.597 to 0.978 as (α,T) went from $(-0.9,10)$ to $(0.0,40)$. Thus, by $T=40$, the asymptotic theory worked fairly well for $|\alpha|\le 0.6$.

Let $T^*=T(1-\alpha^2)$ denote the "effective sample size" [this concept is noted in Sims (1974) and is based on the asymptotic approximations in Hendry (1979)], then EG was described by:

$$\begin{array}{lllll}\ln \mathrm{EG} = - & 2.2 & + \; 1.28\,\ln T^* & + \; 8\;/T & + \; 2.6\,\alpha^2/T^*, \\ & (0.4) & (0.11) & (2) & (0.7) \\ & [0.3] & [0.09] & [1] & [0.4] \end{array}$$

$$R^2=0.93,\quad S=0.21,\quad \eta_1(6)=1.6,\quad \eta_2=1.1, \tag{61}$$

[7]I am grateful to Jan Podivinsky and Frank Srba for assistance in conducting and analysing these experiments.

where

$(\cdot)$ = conventional standard errors,
$[\cdot]$ = heteroscedasticity-consistent standard errors [see White (1980b)],
S = residual standard error,
$\eta_1(k)$ = heteroscedasticity/functional-form mis-specification test based on R_k^2 in the auxiliary regression with k quadratic variables using the form: $R_k^2(T-k-l)/((1-R_k^2)\cdot k)$ for l regressors in, for example, (61), approximately distributed as $F(k, T-k-l)$ under the null [see White (1980a)],
η_2 = Chow (1960) test of parameter constancy, distributed as $F(6, 22-l)$ under the null. This is treated as a Lagrange Multiplier test, and so all regressions quoted are based on the 28 experiments.

From (61), EG increases almost proportionately to T^* (estimating separate coefficients for $\ln T$ and $\ln(1-\alpha^2)$ revealed these to be almost equal magnitude, same sign). Consequently, in experiments with small T^*, CVs like α^* may not yield useful efficiency gains, and conversely, large-T^* is required for asymptotic theory to "work well".

Next, the response surface estimates obtained for $\tilde{\psi}_{T1}$ and $\bar{\psi}_{T1}$ were similar so only the latter are reported. Using the simulation estimated standard errors from (55) (denoted by $S1$) yielded for the simplest bias function:

$$(\bar{\psi}_{T1}-\alpha)/S1 = -\underset{\substack{(0.05)\\ [0.06]}}{1.54}\ \alpha/T\cdot S1,$$

$$R^2 = 0.97, \quad S = 1.67, \quad \eta_1(1) = 9.8, \quad \eta_2 = 0.4. \tag{62}$$

While this accounts for 97% of the simulation variance in $(\tilde{\alpha}-\alpha)/S1$ between experiments, $\eta_1(1)$ rejects homoscedasticity, and the value of S is significantly in excess of unity [$27.S^2$ exceeds the 0.001 critical value of $\chi^2(27,0)$]; this confirms that the diagnostic tests can detect mis-specification. Adding the term α/T^2 yields:

$$(\bar{\psi}_{T1}-\alpha)/S1 = -\underset{\substack{(0.08)\\ [0.07]}}{1.87}\ \alpha/T\cdot S1 + \underset{\substack{(1.2)\\ [1.2]}}{5.6}\ \alpha/T^2\cdot S1,$$

$$R^2 = 0.985, \quad S = 1.26, \quad \eta_1(3) = 1.3, \quad \eta_2 = 0.8. \tag{63}$$

This is obviously a much better approximation (and is "close" to the theoretical result to $O(T^{-2})$ of $-2\alpha(1/T - 2/T^2)$), although S remains significantly larger than unity at the 0.05 level.

Very similar results were obtained on replacing $S1$ by $\xi_1 = (T^* \cdot N \cdot \mathrm{EG})^{-1/2}$ and this is unsurprising given that:

$$\underset{\substack{(0.01)\\[0.01]}}{\ln S1 = -\;0.96} \ln\sqrt{N\cdot T^*} - \underset{\substack{(0.02)\\[0.02]}}{0.58} \ln \mathrm{EG} - \underset{\substack{(0.30)\\[0.21]}}{1.3} /T + \underset{\substack{(0.12)\\[0.09]}}{1.0} \alpha^2/T^*,$$

$$R^2 = 0.991, \quad S = 0.048, \quad \eta_1(8) = 1.1, \quad \eta_2 = 1.5. \tag{64}$$

Thus, while additional finite sample effects can be established, most of the between-experiment variance in $S1$ is attributable to the asymptotic result [note the dependence of EG on the other variables in (64) from (61); also these equations together imply that $S1 \underset{\text{app}}{\sim} O((NT^{*2})^{-1/2})$ as anticipated].

Noting that $T^{-1} - 2T^{-2} \simeq (T+2)^{-1}$, an attempt was made to establish the relevance of α^3/T^3 [based on Shenton and Johnson (1965)]:

$$(\bar{\psi}_{T1} - \alpha)/S1 = -\underset{\substack{(0.05)\\[0.05]}}{1.84}\, \alpha/(T+2)\cdot S1 + \underset{\substack{(11)\\[6]}}{43}\, \alpha^3/T^3 \cdot S1,$$

$$R^2 = 0.989, \quad S = 1.09, \quad \eta_1(2) = 0.7, \quad \eta_2 = 0.9. \tag{65}$$

Since the experimental design actually induced an extremely high correlation between successive odd powers of α, (65) seems a useful approximation to their series expansion:

$$\mathrm{E}(\hat{\alpha} - \alpha) = -\frac{2(T-2)\alpha}{(T+1)^{[2]}} + \frac{12\alpha^3}{(T+5)^{[3]}} + \frac{18(T+8)\alpha^5}{(T+9)^{[4]}} + \frac{24(T+12)(T+10)\alpha^7}{(T+13)^{[5]}} + \cdots,$$

where $T^{[n]} = T(T-2)\ldots(T-2n+2)$. If a larger number of experiments had been conducted, general response surfaces such as (65) might have been estimable directly given (34). The results herein certainly suggest that it can be worthwhile incorporating terms smaller than just $O(T^{-1})$. Finally, replacing $S1$ by ξ_1 in (65) yielded $S = 1.20$ and $\eta_1(2) = 1.3$, so the fit was poorer but not significantly bad, and closely similar coefficient estimates resulted.

Table 6.1 provides some illustrative comparisons of the various regression predictions of biases together with both analytical results and the direct and CV simulation estimates, including one set of values for which experiments were not conducted. Any of the results in (62)–(65) seems adequately accurate for practical

Table 6.1

α	T	(62)	(63)	(65)	(a)[a]	(b)	(c)	$\tilde{\psi}_{T1}$	$\bar{\psi}_{T1}$
0.6,	10	−0.092	−0.079	−0.083	−0.10	−0.091	−0.086	−0.092 (0.013)	−0.087 (0.007)
0.9,	10	−0.139	−0.118	−0.107	−0.15	−0.119	—	−0.102 (0.011)	−0.104 (0.008)
0.6,	30	−0.031	−0.034	−0.034	−0.038	−0.037	−0.036	−0.038 (0.008)	−0.040 (0.003)
0.9,	30	−0.046	−0.051	−0.051	−0.056	−0.053	—	−0.049 (0.006)	−0.052 (0.004)
0.8,	10	−0.123	−0.105	−0.101	−0.133	−0.112	−0.105	—	—

[a](a) $-2\alpha/(T+2)$; (b) to $O(T^{-2})$ and (c) exact, both from Sawa (1978, Table 1a.).

purposes, and the final numerical-analytical summary is given by $\mathrm{E}(\hat{\alpha}-\alpha) \simeq -1.8\alpha/(T+2)+43\alpha^3/T^3$.

6.3. *Estimating $\psi_{T2}(\hat{\alpha})$*

Very similar estimates were produced by $\tilde{\psi}_{T2}$ and $\bar{\psi}_{T2}$, and since variances were estimated only for the former, results are quoted just for these. Firstly, for $\hat{\mu}_4$, since $\mathrm{SSD}=\sqrt{\tilde{\psi}_{T2}}$:

$$\ln\hat{\mu}_4 = \underset{\substack{(0.1)\\ [0.1]}}{3.8}\ \ln\mathrm{SSD} + \underset{\substack{(0.2)\\ [0.2]}}{0.8} + \underset{\substack{(0.3)\\ [0.3]}}{2.4}\ /T^*,$$

$$R^2=0.98, \quad S=0.18, \quad \eta_1(5)=1.1, \quad \eta_2=0.2. \tag{66}$$

Thus, the approximation that $\mu_4 = 3\,\mathrm{SSD}^4$ has some support (note that $\ln 3 = 1.09$).

However, letting $S2=\sqrt{\tilde{V}(\tilde{\psi}_{T2})}$ from (56):

$$\ln\left(\sqrt{\frac{N}{2}}\frac{S2}{\sigma_\alpha^2}\right) = \underset{\substack{(0.3)\\ [0.5]}}{4.2}\ \alpha^2/T^*,$$

$$R^2=0.80, \quad S=0.25, \quad \eta_1(1)=1.0, \quad \eta_2=0.8. \tag{67}$$

Consequently, the asymptotic approximation that $V(\tilde{\psi}_{T2})=2\sigma_\alpha^4/N$ is not very accurate, and this is reflected in the regressions based on (44b):

$$\underset{}{\xi_2 \ln(\tilde{\psi}_{T2}/\sigma_\alpha^2)} = -\underset{\substack{(0.4)\\ [0.3]}}{1.4}\ \xi_2/T + \underset{\substack{(0.4)\\ [0.6]}}{5.2}\ \xi_2\alpha^2/T^* - \underset{\substack{(5)\\ [6]}}{28}\ \xi_2\alpha^2/T^2(1-\alpha^2),$$

$$R^2 = 0.94, \quad S = 1.26, \quad \eta_1(6) = 1.5, \quad \eta_2 = 0.6, \quad \xi_2 = \sqrt{N/2}. \tag{68}$$

Although this regression accounts for much of the variance in $\tilde{\psi}_{T2}$ around σ_α^2 (and almost all of the variance in $\tilde{\psi}_{T2}$ itself), S is significantly in excess of unity. Replacing $\sqrt{\frac{1}{2}N}$ by $\sigma_\alpha^2/S2$ reduced the response surface error variance to unity, but induced so much collinearity between the transformed regressors that sensible individual coefficient estimates could not be obtained. In no case was the unrestricted coefficient of $\ln \sigma_\alpha^2$ significantly different from unity.

By way of comparison, Kendall and Stuart (1966, ch. 48), quote an analytical result which suggests (to $O(T^{-2})$):

$$\ln(\psi_{T2}/\sigma_\alpha^2) \simeq -2/T + 8\alpha^2/T^* - 48\alpha^2/T \cdot T^*. \tag{69}$$

Thus, both (68) and (69) reveal that $\psi_{T2} \simeq \sigma_\alpha^2$ only for large T^*, and the former yields:

$$\mathrm{E}(\hat{\alpha} - \mathrm{E}(\hat{\alpha}))^2 \simeq \left(\frac{1-\alpha^2}{T}\right)\exp\left(\frac{-1.4}{T} + \frac{5\alpha^2}{T(1-\alpha^2)} - \frac{28\alpha^2}{T^2(1-\alpha^2)}\right).$$

6.4. *Estimating $V_T(\hat{\alpha})$*

The response surface based on (47) yielded:

$$\xi_3 \ln\left(\frac{\mathrm{E\tilde{S}E}}{\sigma_\alpha}\right) = -\underset{\substack{(0.03)\\ [0.02]}}{0.09}\ \xi_3(\ln \sigma_\alpha)/T + \underset{\substack{(0.06)\\ [0.09]}}{0.76}\ \xi_3\alpha^2/T^*,$$

$$R^2 = 0.64, \quad S = 1.86, \quad \eta_1(3) = 7.9, \quad \eta_2 = 1.2, \tag{70}$$

where $\xi_3 = \sigma_\alpha/\mathrm{SSE}$. While this explained 99.996% of the variability in $\xi_3 \ln(\mathrm{E\tilde{S}E})$ and the unrestricted coefficient of $\ln \sigma_\alpha$ was 1.002(0.002), both S and $\eta_1(\cdot)$ reject the null of correct functional form. Additional terms rectify this, but at the cost of a cumbersome regression:

$$\xi_3 \ln\left(\frac{\tilde{ESE}}{\sigma_\alpha}\right) = \underset{\substack{(1.2)\\ [0.9]}}{-\ 2.8}\ \xi_3/T + \underset{\substack{(2.1)\\ [1.5]}}{5.5}\ \xi_3/T^{3/2} + \underset{\substack{(0.1)\\ [0.1]}}{1.1}\ \xi_3\alpha^2/T^*$$

$$- \underset{\substack{(1.1)\\ [0.5]}}{7.8}\ \xi_3\alpha^2/T\cdot T^* - \underset{\substack{(0.5)\\ [0.4]}}{1.1}\ \xi_3(\ln\sigma_\alpha)/T,$$

$$R^2 = 0.90, \quad S = 1.05, \quad \eta_1(13) = 0.6, \quad \eta_2 = 1.6. \tag{71}$$

Variables in common with (68) have the same signs of coefficients and (when expressed in terms of estimated variances) similar magnitudes.

The heteroscedasticity correction in (48) was reasonably accurate and regression yielded:

$$\ln \text{SSE} = \underset{\substack{(0.03)\\ [0.04]}}{-\ 1.4}\ \ln T + \underset{\substack{(0.02)\\ [0.02]}}{0.39}\ \ln(1+T\alpha^2) - \underset{\substack{(0.10)\\ [0.11]}}{3.16},$$

$$R^2 = 0.987, \quad S = 0.087, \quad \eta_1(5) = 1.7, \quad \eta_2 = 0.9. \tag{72}$$

($\ln\sqrt{N} = 3.0$), but no response surfaces for ESE were based on the approximation that $\sqrt{N}\cdot T\cdot \text{SSE} \simeq (T^{-1} + \alpha^2)^{0.4}$ as suggested by (72).

6.5. *Estimating* $E(\hat{\sigma}_\varepsilon^2\,|\,\alpha, T)$

All estimates based on $\bar{\sigma}_\varepsilon^2$ had very small standard errors and were close to $\sigma_\varepsilon^2 = 1$. The average efficiency gain, SEG, was over 25 (i.e. equivalent to the accuracy of 10,000 random replications) and was described by:

$$\ln\sqrt{\text{SEG}} = \underset{\substack{(0.10)\\ [0.09]}}{0.19} + \underset{\substack{(0.03)\\ [0.03]}}{0.46}\ \ln T,$$

$$R^2 = 0.89, \quad S = 0.086, \quad \eta_1(2) = 0.1, \quad \eta_2 = 0.4, \tag{73}$$

so that SEG $\simeq T$.

Moreover, using (58):

$$\ln\sqrt{\tilde{V}(\tilde{\sigma}_\varepsilon^2)} = \underset{\substack{(0.1)\\ [0.1]}}{-\ 0.1} - \underset{\substack{(0.03)\\ [0.03]}}{0.97}\ \ln\sqrt{NT/2},$$

$$R^2 = 0.977, \quad S = 0.041, \quad \eta_1(2) = 3.3, \quad \eta_2 = 0.5, \tag{74}$$

so that $\tilde{V}(\tilde{\sigma}^2) \simeq 2/NT$ and from (73), $\tilde{V}(\bar{\sigma}_\varepsilon^2) \simeq 2/NT^2$.

These results are again consistent with CV theory, although $\eta_1(\cdot)$ may indicate an inappropriate functional form; note that a more stringent check of the simulation could have been conducted by testing the within-experiment distribution of $T\cdot\sigma_\varepsilon^{*2}$ against $\chi^2(T,0)$.

6.6. Estimating $P(Z|\alpha,T,\delta)$

The experimental design was such that $\hat{P}=1$ occurred in 13 experiments out of the 28, but unlike the results in Mizon and Hendry (1980), there was no systematic tendency for $\hat{P}$ to underestimate P^* when $P^*>\frac{1}{2}$ (consistent with their conjecture that this was an artifact due to reusing the random numbers). A simple response surface based on (60) yielded:

$$\left(L^*(\tilde{P})-L(P^*)\right)=\underset{\substack{(2)\\ [2]}}{17}\,\phi\tilde{J}/T^2,$$

$$R^2=0.73,\quad S=1.57,\quad \eta_1(1)=0.01,\quad \eta_2(4,10)=0.5, \tag{75}$$

where $\tilde{J}=[N\tilde{P}(1-\tilde{P})/(1-N^{-1})]^{1/2}$. The terms T^{-1} and $L(P^*)/T$ were insignificant if added to (75), and the unrestricted coefficient of $L(P^*)$ was not significantly different from unity. When $\alpha=0$, the rejection frequencies were:

T	10	20	30	40	mean,
$\tilde{P}$	0.053	0.058	0.048	0.045	0.051,

all of which are close to the nominal significance level of $\delta=0.05$. Moreover, $(1+\phi)$ accounted for over 99.9% of the between-experiment variance in the mean of Z, consistent with $\mathrm{E}(\chi^2(1,\phi))=1+\phi$. Thus, although S in (75) is significantly in excess of unity, a reasonable summary power function is:

$$\left(\frac{P}{1-P}\right)=\left(\frac{P^*}{1-P^*}\right)\exp\left[\frac{17\phi}{T^2}\right]. \tag{76}$$

Finally, the rejection frequency $\tilde{P}_\alpha$ for the true value of α [i.e. $\alpha_0=\alpha$ in (41) so $\phi=0$] was investigated:

$$\tilde{P}_\alpha=\underset{\substack{(0.003)\\ [0.003]}}{0.050}+\underset{\substack{(0.016)\\ [0.008]}}{0.024}\,/T^*,$$

$$R^2=0.08,\quad S=0.011,\quad \eta_1(1)=1.0,\quad \eta_2=0.8, \tag{77}$$

so that $\alpha_0 = \alpha$ is indeed rejected around 5% of the time at all values of $\alpha(\sqrt{P(1-P)/N} = 0.011$ when $P = 0.05$, $N = 400$). Overall, the results in (61)–(77) highlight what Monte Carlo can achieve (e.g. simple numerical-analytical formulae) and what it cannot (e.g. provide little insight into what happens as $\alpha \to 1$; compare Phillips (1977a)). It is not a complete substitute for analysis, but may be a helpful partner, and is often a cost-effective solution which need not entail high consumption costs if adequate summarization is provided.

7. Some loose ends

7.1. Non-existent moments

There are many "respectable" estimators which have no finite moments in small samples (e.g. LIML and FIML) yet which have been investigated by Monte Carlo methods. Possible approaches to this "tail area" problem are:

(a) pre-define an "acceptable" region for $\hat{\boldsymbol{\beta}}$ and discard outliers;
(b) use non-parametric statistics [like medians, etc.; see Summers (1965)];
(c) investigate the existence of moments by varying N (and possibly $\boldsymbol{\theta}$);
(d) report only $\psi_T(\cdot)$; and
(e) only derive the CV, and do not do the simulation.

Sargan (1982) has investigated the outcome which is likely to occur if conventional simulation means are used to estimate non-existent moments (with and without CVs) and found that N could be chosen as a function of T such that the Monte Carlo provided reasonable estimates of the Nagar approximations to the moments (which in turn help in understanding the Edgeworth Approximation to the distribution function). Even so, some truncation bounds for deleting outliers seemed better than using none, supporting (a); no bounds could produce rather unreliable results, and non-parametric statistics (b) in effect operate by "discounting" discrepant results. The natural alternative is direct estimation of $\psi_T(\boldsymbol{\theta})$.

In low-dimensional problems, *numerical tabulation* of $\Psi_T(\cdot)$ for very large N can be useful [see Orcutt and Winokur (1969)] but otherwise, the *function* has to be estimated. Sargan (1976) considers using CVs to improve the accuracy of estimating $\Psi_T(\cdot)$, but this requires that the exact distribution function of the CV is known, and Basmann et al. (1974) test various hypotheses about forms of $\Psi_T(\cdot)$ in specific models. Improved simulation methods in this area would be of great value, but at present it is rarely feasible to attempt estimation of distribution functions which depend on many parameters.

7.2. Evaluating integrals

*CV*s for test powers would be a useful advance [closely related to estimating $\Psi_T(\cdot)$]. These can be derived in certain static models, but their use depends on knowing $\Psi_T(\beta_1^*)$, not just its moments, and so test-power *CV*s are difficult to obtain in dynamic models. Experiments in which significance levels rather than local alternatives were changed also would be interesting and helpful in understanding the behaviour of tests.

Returning to the example in equations (1), (2) and (11), some cross-fertilization of ideas may prove fruitful. Kloek and van Dijk (1978) discuss Monte Carlo integration for economic estimation, and demonstrate its feasibility using importance functions. Also, Van Dijk and Kloek (1980) discuss the choice of importance function and implement nine-dimensional integration. However, on the one hand, $p(\nu)$ also might be of use in estimating integrals corresponding to test powers even though the density function is unknown (e.g. by generating *CV*s which are exactly distributed as the importance function which in turn is chosen to be the asymptotic distribution of the test). On the other hand, naive estimators such as $\bar{u}$ in (11) surely could be improved upon by using some functions of the $\{v_i\}$ as a *CV*: e.g. calculating $\tilde{u}$ from $\bar{u}$ and $f(\{v_i\})$ so as to correct for chance departures of $\bar{v}$ from $\mathrm{E}(\nu)=\int_a^b xp(x)\mathrm{d}x$ which will in general be known (although this ad hoc suggestion may not guarantee efficiency gains).

A further problem which is equivalent to computing an integral is estimating the mean stochastic simulation path of a non-linear econometric system. Here, antithetic variates switching $\{\varepsilon_t\}\sim I\mathcal{N}(\mathbf{O},\mathbf{I})$ to $\{-\varepsilon_t\}$ and creating $\omega_t=\mathbf{K}\varepsilon_t\sim I\mathcal{N}(\mathbf{O},\mathbf{\Sigma})$ from $\mathbf{\Sigma}=\mathbf{KK}'$ seem to be of use. The efficiency gains depend on the extent of the non-linearity and the relative "explanatory" power of the strongly exogenous variables compared to the endogenous dynamics, varying from infinite efficiency for linear, static systems to zero for closed, dynamic models with squared errors [see Fair, Chapter 33 in this Handbook, and Mariano and Brown (1983); and for an application, Calzolari (1979)].

Much work remains to be done on determining factors which influence such simulation efficiency (e.g. dependence of the data on such features as the sign and/or scale of the errors) and hence on deriving appropriate antithetic selections. Recently, Calzolari and Sterbenz (1981) have derived control variates from local linearization of non-linear systems and find very large efficiency gains over straightforward random replications for the Klein–Goldberger model.

Manifestly, other applications are legion since very many problems in econometrics are equivalent to computing integrals which in turn can be estimated by averages, and hence are susceptible to efficiency improvements.

And notwithstanding all the above arguments, when only a couple of points in $\Theta\times\mathcal{T}$ are believed to be of *empirical* relevance, naive simulation "pilot" studies remain an easy and inexpensive means of learning about finite sample properties in complicated models or methods.

References

Bartlett, M. S. (1946) "On the Theoretical Specification and Sampling Properties of Autocorrelated Time Series", *Journal of the Royal Statistical Society*, B, 8, 27–41.
Basmann, R. L., D. H. Richardson and R. J. Rohr (1974) "Finite Sample Distributions Associated with Stochastic Difference Equations—Some Experimental Evidence", *Econometrica*, 42, 825–840.
Breusch, T. S. (1980) "Useful Invariance Results for Generalised Regression Models", *Journal of Econometrics*, 13, 327–340.
Calzolari, G. (1979) "Antithetic Variates to Estimate the Simulation Bias in Non-linear Models", *Economics Letters*, 4, 323–328.
Calzolari, G. and F. Sterbenz (1981) "Efficient Computation of Reduced Form Variances in Nonlinear Econometric Models", IBM, Pisa, mimeo.
Campos, J. (1980) "The Form of Response Surface for a Simulation Standard Error in Monte Carlo Studies", unpublished paper, London School of Economics.
Chow, G. C. (1960) "Tests of Equality Between Sets of Coefficients in Two Linear Regressions", *Econometrica*, 28, 591–605.
Cochran, W. G. and G. M. Cox (1957) *Experimental Designs*. New York: John Wiley and Sons.
Conlisk, J. (1974) "Optimal Response Surface Design in Monte Carlo Sampling Experiments", *Annals of Economic and Social Measurement*, 3, 463–473.
Copas, J. B. (1966) "Monte Carlo Results for Estimation in a Stable Markov Time Series", *Journal of the Royal Statistical Society*, A, 129, 110–116.
Cox, D. R. (1970) *Analysis of Binary Data*. London: Chapman and Hall.
Cramér, H. (1946) *Mathematical Methods of Statistics*. Uppsala: Almqvist and Wicksells.
Croxton, F. E., D. J. Cowden and S. Klein (1968) *Applied General Statistics*. London: Sir Isaac Pitman and Sons Ltd., 3rd edn.
Davis, A. W. (1971) "Percentile Approximations for a Class of Likelihood Ration Criteria", *Biometrika*, 58, 349–356.
Engle, R. F. (1982) "A General Approach to Lagrange Multiplier Model Diagnostics", *Journal of Econometrics*, 20, 83–104.
Evans, G. B. A. and N. E. Savin (1981) "Testing for Unit Roots: I", *Econometrica*, 49, 753–779.
Evans, G. B. A. and N. E. Savin (1982) "Conflict Among the Criteria Revisited; the W, LR and LM Tests", *Econometrica*, 50, 737–748.
Goldberger, A. S. (1964) *Econometric Theory*. New York: John Wiley and Sons.
Golder, E. R. (1976) "Algorithm AS98: The Spectral Test for the Evaluation of Congruential Pseudo-Random Generators", *Applied Statistics*, 25, 173–180.
Goldfeld, S. M. and R. E. Quandt (1972) *Nonlinear Methods in Econometrics*. Amsterdam: North-Holland.
Gross, A. M. (1973) "A Monte Carlo Swindle for Estimators of Location", *Journal of the Royal Statistical Society*, C, 22, 347–353.
Hammersley, J. M. and D. C. Handscomb (1964) *Monte Carlo Methods*. London: Metheun.
Hendry, D. F. (1973) "On Asymptotic Theory and Finite Sample Experiments", *Economica*, 160, 210–217.
Hendry, D. F. (1976) "The Structure of Simultaneous Equations Estimators", *Journal of Econometrics*, 4, 51–88.
Hendry, D. F. (1977) "On the Time Series Approach to Econometric Model Building", in: C. A. Sims, Ed., *New Methods in Business Cycle Research*. Federal Reserve Bank of Minneapolis, 183–208.
Hendry, D. F. (1979) "The Behaviour of Inconsistent Instrumental Variables Estimators in Dynamic Systems with Autocorrelated Errors", *Journal of Econometrics*, 9, 295–314.

Hendry, D. F. (1982) "A Reply to Professors Maasoumi and Phillips", *Journal of Econometrics*, 19, 203–213.

Hendry, D. F. and R. W. Harrison (1974) "Monte Carlo Methodology and the Finite Sample Behaviour of Ordinary and Two-Stage Least Squares", *Journal of Econometrics*, 2, 151–174.

Hendry, D. F. and F. Srba (1977) "The Properties of Autoregressive Instrumental Variables Estimators in Dynamic Systems", *Econometrica*, 45, 969–990.

Hendry, D. F. and P. K. Trivedi (1972) "Maximum Likelihood Estimation of Difference Equations with Moving Average Errors: A Simulation Study", *The Review of Economic Studies*, 39, 117–145.

Hurwicz, L. (1950) "Least Squares Bias in Time Series", in: T. C. Koopmans (Ed.): *Statistical Inference in Dynamic Economic Models*. Cowles Commission Monograph 10; New York: John Wiley and Sons, ch. 15.

Hylleberg, S. (1977) "A Comparative Study of Finite Sample Properties of Band Spectrum Regression Estimators", *Journal of Econometrics*, 5, 167–182.

Johnson, N. L. and S. Kotz (1970) *Continuous Univariate Distributions-1; Distributions in Statistics*. New York: John Wiley and Sons.

Kakwani, N. C. (1967) "The Unbiasedness of Zellner's Seemingly Unrelated Regression Equations Estimator", *Journal of the American Statistical Association*, 62, 141–142.

Kendall, M. G. (1954) "Note on Bias in the Estimation of Autocorrelation", *Biometrika*, 41, 403–404.

Kendall, M. G. and A. Stuart (1958, 1961, 1966) *The Advanced Theory of Statistics*. Vols. 1–3. New York: Charles Griffen.

Kennedy, W. J. Jr. and J. E. Gentle (1980) *Statistical Computing*. New York: Marcel Dekker, Inc.

King, M. L. (1980) "Small Sample Properties of Econometric Estimators and Tests Assuming Elliptically Symmetric Disturbances", Paper presented to the Fourth World Congress of the Econometric Society, France.

King, M. L. "A Note on the Burroughs B6700 Pseudo-Random Number Generator", *New Zealand Statistician* (forthcoming).

Kleijnen, J. P. C. (1974) *Statistical Techniques in Simulation*. New York: Marcel Dekker Inc.

Kloek, T. and H. K. Van Dijk (1978) "Bayesian Estimates of Equation System Parameters: An Application of Integration by Monte Carlo", *Econometrica*, 46, 1–19.

Maasoumi, E. and P. C. B. Phillips (1982) "On the Behaviour of Inconsistent Instrumental Variable Estimators", *Journal of Econometrics*, 19, 183–201.

Mariano, R. S. (1982) "Analytical Small-Sample Distribution Theory in Econometrics: The Simultaneous-Equations Case", *International Economic Review*, 23, 503–533.

Mariano, R. S. and B. W. Brown (1983) "Asymptotic Behaviour of Predictors in a Nonlinear Simultaneous System", *International Economic Review*, 24, 523–536.

Metropolis, N. and S. Ulam (1949) "The Monte Carlo Method", *Journal of the American Statistical Association*, 44, 335–341.

Mikhail, W. M. (1972) "Simulating the Small Sample Properties of Econometric Estimators", *Journal of the American Statistical Association*, 67, 620–624.

Mikhail, W. M. (1975) "A Comparative Monte Carlo Study of the Properties of Econometric Estimators", *Journal of the American Statistical Association*, 70, 91–104.

Mizon, G. E. and D. F. Hendry (1980) "An Empirical Application and Monte Carlo Analysis of Tests of Dynamic Specification", *Review of Economic Studies*, 47, 21–45.

Myers, R. H. and S. J. Lahoda (1975) "A Generalisation of the Response Surface Mean Square Error Criterion with a Specific Application to the Slope", *Technometrics*, 17, 481–486.

Nagar, A. L. (1959) "The Bias and Moment Matrix of the General *k*-Class Estimators of the Parameters in Simultaneous Equations", *Econometrica*, 27, 575–595.

Naylor, T. H. (1971) *Computer Simulation Experiments with Models of Economic Systems*. New York: John Wiley and Sons, 1971.

Neave, H. R. (1973) "On Using the Box–Müller Transformation with Multiplicative Congruential Pseudo-Random Number Generators", *Applied Statistics*, 22, 92–97.

Nicholls, D. F., A. R. Pagan and R. D. Terrell (1975) "The Estimation and Use of Models with Moving Average Disturbance Terms: A Survey", *International Economic Review*, 16, 113–134.

O'Brien, R. J. (1979) "The Sensitivity of Econometric Estimators to Data Perturbations: II. Instrumental Variables", Unpublished Paper, Southampton University.

Orcutt, G. H. and D. Cochrane (1949) "A Sampling Study of the Merits of Autoregressive and Reduced Form Transformations in Regression Analysis", *Journal of the American Statistical Association*, 44, 356–372.
Orcutt, G. H. and H. S. Winokur (1969) "First Order Autoregression: Inference Estimation and Prediction", *Econometrica*, 37, 1–14.
Phillips, P. C. B. (1977) "Approximations to Some Finite Sample Distributions Associated with a First Order Stochastic Difference Equation", *Econometrica*, 45, 463–485.
Phillips, P. C. B. (1977) "A General Theorem in the Theory of Asymptotic Expansions as Approximations to Finite Sample Distributions of Econometric Estimators", *Econometrica*, 45, 1517–1534.
Phillips, P. C. B. (1980) "Finite Sample Theory and the Distributions of Alternative Estimators of the Marginal Propensity to Consume", *The Review of Economic Studies*, 47, 183–224.
Rao, C. R. (1952) *Advanced Statistical Methods in Biometric Research*. New York: John Wiley and Sons.
Sargan, J. D. (1976) "Econometric Estimators and the Edgeworth Approximation", *Econometrica*, 44, 421–448.
Sargan, J. D. (1978) "The Estimation of Edgeworth Approximations by Monte Carlo Methods", Unpublished Paper, London School of Economics.
Sargan, J. D. (1982) "On Monte Carlo Estimates of Moments That are Infinite", *Advances in Econometrics*, 1, 267–299.
Sawa, T. (1978) "The Exact Moments of the Least Squares Estimator for the Autoregressive Model", *Journal of Econometrics*, 8, 159–172.
Shenton, L. R. and W. L. Johnson (1965) "Moments of a Serial Correlation Coefficient", *Journal of the Royal Statistical Society*, B, 27, 308–320.
Sims, C. A. (1974) "Distributed Lags", in: M. D. Intriligator and D. A. Kendrick, eds., *Frontiers of Quantitative Economics, Vol. II*. Amsterdam: North-Holland, ch. 5.
Smith, V. K. (1973) *Monte Carlo Methods*. London: D. C. Heath.
Sobol', I. M. (1974) *The Monte Carlo Method*. Popular Lectures in Mathematics, London: University of Chicago Press, Ltd.
Sowey, E. R. (1972) "A Chronological and Classified Bibliography on Random Number Generation and Testing", *International Statistical Review*, 40, 355–371.
Sowey, E. R. (1973) "A Classified Bibliography of Monte Carlo Studies in Econometrics", *Journal of Econometrics*, 1, 377–395.
Student (1908) "On the Probable Error of a Mean", *Biometrika*, 6, 1–25.
Summers, R. (1965) "A Capital Intensive Approach to the Small Sample Properties of Various Simultaneous Equations Estimators", *Econometrica*, 33, 1–41.
Sylwestrowicz, J. D. (1981) "Applications of the ICL Distributed Array Processor in Econometric Computations", *ICL Technical Journal*, 280–286.
Teichroew, D. (1965) "A History of Distribution Sampling Prior to the Era of the Computer and Its Relevance to Simulation", *Journal of the American Statistical Association*, 60, 27–49.
Tocher, K. D. (1963) *The Art of Simulation*. London: English Universities Press.
Tse, Y. K. (1979) "Finite Sample Approximations to the Distribution of the Autoregressive Coefficients in a First Order Stochastic Difference Equation with Exogenous Variables", Unpublished Paper, London School of Economics.
Van Dijk, H. K. and T. Kloek (1980) "Further Experience in Bayesian Analysis using Monte Carlo Integration", *Journal of Econometrics*, 14, 307–328.
White, H. (1980) "Using Least Squares to Approximate Unknown Regression Functions", *International Economic Review*, 21, 149–170.
White, H. (1980) "A Heteroskedastic-Consistent Covariance Matrix Estimator and a Direct Test for Heteroskedasticity", *Econometrica*, 48, 817–838.
White, J. S. (1961) "Asymptotic Expansions for the Mean and Variance of the Serial Correlation Coefficient", *Biometrika*, 48, 85–95.
Yule, G. U. (1926) "Why Do We Sometimes Get Nonsense-Correlations Between Time-Series?—A Study in Sampling and the Nature of Time-Series", *Journal of the Royal Statistical Society*, 89, 1–64.

PART 5

TIME SERIES TOPICS

Chapter 17

TIME SERIES AND SPECTRAL METHODS IN ECONOMETRICS

C. W. J. GRANGER and MARK W. WATSON

Contents

Handbook of Econometrics, Volume II, Edited by Z. Griliches and M.D. Intriligator
© Elsevier Science Publishers BV, 1984

1. Introduction

A discrete time series is here defined as a vector x_t of observations made at regularly spaced time points $t=1,2,\ldots,n$. These series arise in many fields, including oceanography, meterology, medicine, geophysics, as well as in economics, finance and management. There have been many methods of analysis proposed for such data and the methods are usually applicable to series from any field. For many years economists and particularly econometricans behaved as though either they did not realize that much of their data was in the form of time series or they did not view this fact as being important. Thus, there existed two alternative strategies or approaches to the analysis of economic data (excluding cross-sectional data from this discussion), which can be called the time series and the classical econometric approaches. The time series approach was based on experience from many fields, but that of the econometrician was viewed as applicable only to economic data, which displayed a great deal of simultaneous or contemporaneous interrelationships. Some influences from the time series domain penetrated that of the classical econometrician, such as how to deal with trends and seasonal components, Durbin–Watson statistics and first-order serial correlation, but there was little influence in the other direction. In the last ten years, this state of affairs has changed dramatically, with time series ideas becoming more mainstream and the procedures developed by econometricians being considered more carefully by the time series analysts. The building of large-scale models, worries about efficient estimation, the growing popularity of rational expectations theory and the consequent interest in optimum forecasts and the discussion of causality testing have greatly helped in bringing the two approaches together, with obvious benefits to both sides.

In Section 2 the methodology of time series is discussed and Section 3 focuses on the theory of forecasting. Section 4 emphasizes the links between the classical econometric and time series approaches while Section 5 briefly discusses the question of differencing of data, as an illustration of the alternative approaches taken in the past. Section 6 considers seasonal adjustment of data and Section 7 discusses some applications of time series methods to economic data.

2. Methodology of time series analysis

A discrete time series consists of a sequence of observations x_t taken at equi-spaced time intervals, examples being annual automobile production, monthly unemployment, weekly readings on the prime interest rate and daily (closing) stock market prices. x_t may be a vector. Underlying these observations will be a theoretical stochastic process X_t which can, of course, be fully characterized by a (possibly countable-infinitely dimensioned) distribution function. The initial and basic objective of time series analysis is to use the observed series x_t to help characterize or describe the unobserved theoretical sequence of random variables X_t. The similarity between this and the ideas of sample and population in classical statistics is obvious. However, the involvement of time in our sequences and the fact, or assumed fact, that time flows in a single direction does add a special structure to time-series data and it is imperative that this extra structure be fully utilized. When standing at time t, it is important to ask how will the next value of the series be generated. The general answer is to consider the conditional distribution of x_{t+1} given x_{t-j}, $j \geq 0$, and then to say that x_{t+1} will be drawn from this distribution. However, a rather different kind of generating function is usually envisaged in which the x_{t+1} is given by:

$$x_{t+1} = (\text{function of } \tilde{x}_t) + e_{t+1}, \tag{2.1}$$

where

$$\tilde{x}_t = (x_t, x_{t-1}, \dots)$$

and the parameters of the distribution of e_{t+1} other than the mean, can depend on x_{t-j}, $j \geq 0$. It is usually overly ambitious to consider the whole distribution of e_{t+1} and, at most, the variance is considered unless e_{t+1}, or a simple transformation of it, is assumed to be normally distributed. An obviously important class of models occurs when the function in (2.1) is linear, so that:

$$x_{t+1} = \sum_{j=0}^{\infty} \alpha_{j,t} x_{t-j} + e_{t+1}. \tag{2.2}$$

For linear models, an appropriate set of characterizing statistics are the first and second moments of the process, that is the mean:

$$\mathrm{E}[X_t] = \mu_t;$$

the variance:

$$\mathrm{E}\left[(X_t - \mu_t)^2\right] = \sigma_t^2;$$

and the covariances:

$$\mathrm{E}\left[(X_t - \mu_t)(X_{t-s} - \mu_{t-s})\right] = \lambda_{t,s},$$

assuming that these quantities exist.

Given a finite amount of data and a single realization, which is the usual case in practice with economic data, it is fairly clear that one cannot estimate these quantities without imposing some further structure. A case which provides a good base situation is when the process is stationary. A process is said to be second-order stationary if the mean and variance, μ and σ^2, do not vary with time and the covariances, λ_s, depend only on the time interval between X_t and X_{t-s} rather than on time itself. A general definition of stationarity has that any group of x's, and the same group shifted by a finite time interval, have identical joint distributions. In terms of the generating function (2.1), x_t will be stationary if the form and parameters of the function do not vary through time. For the linear form (2.2) a sufficient set of conditions are that the parameters of the distribution of ε_t are time invariant and the parameters $\alpha_{j,t}$ are both time invariant and are such that the difference equation:

$$X_{t+1} = \sum_{j=0}^{\infty} \alpha_j X_{t-j},$$

is stable. An assumption of stationarity is not made because it is believed to be realistic, but because a number of important results derive from the assumption and these results can then be studied as the stationarity assumption is relaxed in useful ways.

If x_t is a univariate, stochastic process, its linear properties can be studied from knowledge of its mean, which is henceforth assumed known and to be zero, variance σ^2 and the autocovariances λ_s, or equivalently the autocorrelations $\rho_s = \lambda_s/\sigma^2$. Given a single realization x_t, $t = 1,\dots, n$, consistent estimates of these quantities are easily found provided that the process is ergodic, which essentially means that as n increases the amount of useful information about the process continually increases. (An example of a non-ergodic process is $X_t = a\cos(bt)$ where a is a random variable with finite mean.) Although these quantities, particularly the autocorrelations, do characterize the linear properties of the process, they are not always easy to interpret or to use, if, for example, one is interested in forecasting. For many purposes there is greater interest in the generating process, or at least approximations to it. Ideally, one should be able to look at the correlogram, which is the plot of ρ_s against s, decide which is the appropriate model, estimate this model and then use it. To do this, one naturally first requires a list, or menu of possible and interesting models. There is actually no shortage of time series models, but in the stationary case just a few models are of particular importance.

The most fundamental process, called white noise, consists of an uncorrelated sequence with zero mean, that is ε_t such that $\mathrm{E}[\varepsilon_t] = 0$, $\mathrm{var}(\varepsilon_t) < \infty$ and $\mathrm{corr}(\varepsilon_t, \varepsilon_{t-s}) = 0$, all $s \neq 0$. The process can be called pure white noise if ε_t and

ε_{t-s} are independent for $s \neq 0$. Clearly a pure white-noise process cannot be forecast from its own past, and a white noise cannot be forecast linearly, in each case the optimal forecast is the mean of the process. If one's objective when performing an analysis is to find a univariate model that produces optimum linear forecasts, it is clear that this objective has been reached if a linear transformation of x_t can be found that reduces the series to white noise, and this is why the white noise process is so basic. It can be shown that any univariate stationary process can, in theory at least, be reduced uniquely to some white-noise series by linear transformation. If non-linear or multivariate processes are considered there may not be a unique transformation.

A class of generating processes, or models, that are currently very popular are the mixed autoregressive moving averages given by:

$$x_t = \sum_{j=1}^{p} a_j x_{t-j} + \sum_{j=0}^{q} b_j \varepsilon_{t-j}, \qquad b_0 = 1,$$

where ε_t is white noise. In terms of the extremely useful backward shift operator, B, where

$$B^k x_t = x_{t-k},$$

these ARMA (p,q) models can be expressed as:

$$a(B)x_t = b(B)\varepsilon_t,$$

where

$$a(B) = 1 - \sum_{j=1}^{p} a_j B^j$$

and

$$b(B) = \sum_{j=0}^{q} b_j B^j, \qquad b_0 = 1.$$

If $q = 0$, one has an autoregressive, AR(p), model and if $p = 0$ the model is a moving average, denoted MA(q). The ε_t's are, of course not directly observable, but a model is said to be invertible if the original ε_t can be re-constructed from the observed x_t. Given a long enough series for x_t, the models are invertible if the roots of the equation $b(z) = 0$ all lie outside the unit circle.

Consider now the AR(1) model:

$$x_t = \alpha x_{t-1} + \varepsilon_t.$$

This simple difference equation has the solution:

$$x_t = \sum_{j=0}^{t+n} \alpha^j \varepsilon_{t-j},$$

if the process started up at time $t = -n$. If ε_t has zero mean and variance σ^2, then clearly the variance of x_t is:

$$\operatorname{var}(x_t) = \left[\frac{1-\alpha^{2[n+t+1]}}{1-\alpha^2}\right]\sigma^2,$$

and x_t has mean zero. If now the starting up time is moved into the distant past, the variance of x_t tends to $1/(1-\alpha^2)$ if $|\alpha|<1$, but increases exponentially and explodes if $|\alpha|>1$. A borderline case, known as a random walk when $\alpha=1$, has $\operatorname{var} x_t = (t+n+1)\sigma^2$. It is clear that if $|\alpha|\geq 1$, x_t will have infinite variance. More generally, if all of the roots of $a(z)=0$ lie outside the unit circle and the process started in the distant past, the series will be stationary, if any roots lie inside the unit circle the series will be explosive. If d roots lie on the unit circle and all others outside one has an integrated process. Suppose that x_t is generated by

$$(1-B)^d a(B) x_t = b(B)\varepsilon_t,$$

where $a(B)$ is a polynomial of order p with all roots outside the unit circle and $b(B)$ is a polynomial of order q, then x_t is said to be an integrated autoregressive–moving average series, denoted $x_t \sim \text{ARIMA}(p,d,q)$ by Box and Jenkins (1976) who introduced and successfully marketed these models. It should be noted that the result of differencing x_t d times is a series $y_t = (1-B)^d x_t$, which is ARMA(p,q) and stationary. Although, when $d>0$ and x_t is not stationary, then these models are only a rather simple subset of the class of all non-stationary series. There has been a rather unfortunate confusion in the literature recently about distinguishing between integrated and general non-stationary processes. These terms have, incorrectly, been used as synonyms.

One reason for the popularity of the ARMA models derives from Wold's theorem, which states that if x_t is a stationary series it can be represented as the sum of two components, x_{1t} and x_{2t}, where x_{1t} is deterministic (i.e. $x_{1,t+k}$, $k>0$, can be forecast without any error by a linear combination of $x_{1,t-j}$, $j>0$) and x_{2t} has an MA(q) representation where q may be infinite. As an infinite series can frequently be well approximated by a rational function, the MA(∞) process may be adequately approximated by an ARMA(p,q) process with finite p and q. The ARIMA(p,d,q) models give the analyst a class of linear time series processes that are general enough to provide a good approximation to the true model, but are still sufficiently uncomplicated so that they can be analyzed. How this is done is discussed later in this section.

Many other models have been considered. The most venerable considers a series as being the sum of a number of distinct components called trend, long waves, business cycles of various periods, seasonal and a comparatively unimportant and undistinguished residual. Many economic series have a tendency to steadily grow, with only occasional lapses, and so may be considered to contain a trend in mean. Originally such trends were usually represented by some simple function of time, but currently it is more common to try to pick up these trends by using integrated models with non-zero means after differencing. Neither technique seems to be completely successful in fully describing real trends, and a "causal" procedure, which attempts to explain the trend by movements in some other series—such as population or price—may prove to be better. The position that economic data contains deterministic, strictly periodic cycles is not currently a popular one, with the exception of the seasonal which is discussed in Section 5. The ARIMA models can adequately represent the observed long swings or business cycles observed in real economics, although, naturally, these components can be better explained in a multivariate context.

The decomposition of economic time series into unobserved components (e.g. permanent and transitory, or, "trend" and seasonal components) can be accomplished by signal extraction methods. These methods are discussed in detail in Nerlove, Grether and Carvalho (1979). In Section 6 we show how the Kalman filter can be used for this purpose.

A certain amount of consideration has been given to both non-stationary and non-linear models in recent years, but completely practical procedures are not usually available and the importance of such models has yet to be convincingly demonstrated in economics. The non-stationary models considered include the ARIMA models with time-varying parameters, the time variation being either deterministic, following a simple AR(1) process or being driven by some other observed series. Kalman filter techniques seem to be a natural approach with such models and a useful test for time-varying autoregressive parameters has been constructed by Watson and Engle (1980).

Estimation and prediction in models with time varying autoregressive parameters generated by an independent autoregressive process is a straightforward application of the techniques discussed by Chow in Chapter 20 of this Handbook. Stochastically varying moving average coefficients are more difficult to handle. Any stochastic variation in the coefficients yields a model which is not invertible as it is impossible to completely unscramble the shocks to the coefficients from the disturbance. In the moving average model this introduces a non-linear relationship between the unobservables, the disturbances and the coefficients. The Kalman filter cannot be used directly. It is possible to linearize the model and use an extended Kalman filter as Chow does in Chapter 20 for the simultaneous equation model. The properties of the coefficient estimates and forecasts derived from this method are not yet established.

Useful classes of non-linear models are more difficult to construct, but a class with some potential is discussed in Granger and Andersen (1978). These are the bilinear models, an example being:

$$x_t = \alpha x_{t-1} + \beta x_{t-2}\varepsilon_{t-1} + \varepsilon_t.$$

When $\alpha = 0$, this particular model has the interesting property that the autocorrelations ρ_s all vanish for $s \neq 0$, and so appears, in this sense, to be similar to white noise. Thus, in this case x_t cannot be forecast linearly from its own past, but it can usually be very well forecast from its own past non-linearly. Conditions for stationarity and invertibility are known for some bilinear models, but it is not yet known if they can be used to model the types of non-linearity that can be expected to occur in real economic data.

Priestly (1980) introduces a state-dependent model which in its general form encompasses the bilinear model and several other non-linear models. The restricted and conceivably practical form of the model is a mix of the bilinear and stochastically time varying coefficient models.

Engle (1982) has proposed a model which he calls autoregressive conditional heteroscedastic (ARCH) in which the disturbances, ε_t, have a variance which is unconditionally constant, but conditional on past data may change, so that:

$$\mathrm{E}[\varepsilon_{t+1}^2] = \sigma^2,$$

but

$$\mathrm{E}[\varepsilon_{t+1}^2 | x_t, x_{t-1}, \ldots, x_1] = h(x_t, x_{t-1}, \ldots, x_1) = h_{t+1}.$$

As will be shown in the next section, ε_{t+1} is just the one step ahead forecast error x_{t+1}. The ARCH model postulates that x_{t+1} will sometimes be relatively easy to forecast from x_t, i.e. $h_{t+1} < \sigma^2$, while at other times it may be relatively difficult. This seems an attractive model for economic data.

One of the basic tools of the time series analyst is the correlogram, which is the plot of the (estimated) autocorrelations ρ_s against the lag s. In theory, the shape of this plot can help discriminate between competing linear models. It is usual practice in time series analysis to initially try to identify from summaries of the data one or just a few models that might have generated the data. This initial guess at model specification is now called the identification stage and decisions are usually made just from evidence from the data rather than from some preconceived ideas, or theories, about the form of the true underlying generating process. As an example, if a process is ARMA (p, q) with $p > 0$, then $\rho_s = \theta^s$ for s

large, with $|\theta|<1$, but if $p=0$, $\rho_s=0$ for $s\geq q+1$ so that the shape of the correlogram can, theoretically, help one decide if $p>0$ and, if not, to choose the value of q. A second diagram, which is proposed by Box and Jenkins to help with identification is the partial correlogram, being the plot of $a_{s,s}$ against s, where $a_{k,k}$ is the estimated coefficient of x_{t-k} when an kth order AR model is fitted. If $q>0$, this diagram also declines as θ^s for s large, but if $q=0$, then $a_{s,s}=0$ for $s\geq p+1$. Thus, the pair of diagrams, the correlogram and the partial correlogram, can, hopefully, greatly help in deciding which models are appropriate. In this process, Box and Jenkins suggest that the number of parameters used, $p+q$, should be kept to a minimum—which they call the principal of parsimony—so that estimation properties remain satisfactory. The value of this suggestion has not been fully tested.

The Box and Jenkins procedure for identifying the orders p and q of the ARMA(p,q) model is rather complicated and is not easily conducted, even by those experienced in the technique. This is particularly true for the mixed model, when neither p nor q vanishes. Even for the pure AR or MA models difficulties are often encountered and identification is expensive because it necessitates decision making by a specially trained statistician. A variety of other identification procedures have been suggested to overcome these difficulties. The best known of these is the Akaike information criteria (AIC) in which if, for example, an AR(k) model is considered using a data set of size N resulting in an estimated residual variance $\hat{\sigma}_k^2$, then one defines

$$AIC(k)=\log\hat{\sigma}_k^2+2k/N.$$

By choosing k so that this quantity is minimized, an order for the AR model is selected. Hannan and Quinn (1979) have shown that this criteria provides upward-biased estimates of the order of the model, and that minimization of the criterion:

$$\phi_k=\log\hat{\sigma}_k^2+N^{-1}2kc\log\log N,\qquad c>1,$$

provides better, and strongly consistent estimates of this order.

Although c is arbitrary, a value $c=1$ appears to work well according to evidence of a simulation. So for instance, if $N=100$ an AR(4) model would be prefered to an AR(5) model if the increase in $\hat{\sigma}^2$ is less than 2% using AIC and less than 3% using ϕ. These procedures can be generalized to deal also with mixed ARMA(p,q) models. (A critical discussion on the use of information criteria in model selection can be found in Chapter 5 of the Handbook.) Another partly automated method has been proposed by Gray, Kelly and McIntire (1978) which is particularly useful with the mixed model. Although the method lacks intuitive appeal, examples of its use indicate that it has promise. As these, and other, automated methods become generally available, the original Box–Jenkins procedures will probably be used only as secondary checks on models derived. There is also a possibility that these methods can be used in the multiple series case, but presently they are inclined to result in very non-parsimonious models.

The identification stage of time series modeling is preceded by making an estimate of d, in the ARIMA(p, d, q) model. If $d > 0$, the correlogram declines very slowly—and theoretically not at all—so the original series is differenced sufficiently often so that such a very smooth correlogram does not occur. In practice, it is fairly rare for a value of d other than zero or one to be found with economic data. The importance and relevance of differencing will be discussed further in Section 5. Once these initial estimates of p, d and q have been obtained in the identification stage of analysis, the various parameters in the model are estimated and finally various diagnostic checks applied to the model to see if it adequately represents the data.

Estimation is generally carried out using maximum likelihood or approximate maximum likelihood methods. If we assume the ε's are normally distributed with mean zero and variance (conditional on past data) σ_ε^2, the likelihood function is proportional to:

$$(\sigma_\varepsilon^2)^{-T/2} f(\beta)\exp[-S(\beta, X_T)/-2\sigma_\varepsilon^2],$$

where β contains the parameters in $a(B)$ and $b(B)$ and now $X_T = (x_1, x_2, \dots, x_T)'$. Analytic expressions for $f(\beta)$ and $S(\beta, X_T)$ can be found in Newbold (1974).

One of three methods, all with the same asymptotic properties, is generally used to estimate the parameters. The first is the exact maximum likelihood method, and Ansley (1979) proposes a useful transformation of the data when this method is used. The second method, sometimes called exact least squares, neglects the term $f(\beta)$, which does not depend on the data, and minimizes $S(\beta, X_T)$. The method is called exact least squares since $S(\beta, X_T)$ can be written as:

$$\sum_{t=-\infty}^{T} = \tilde{\varepsilon}_t^2,$$

where $\tilde{\varepsilon}_t = \mathrm{E}[\varepsilon_t | X_T, \beta]$. Box and Jenkins (1976) suggest approximating this by "back-forecasting" (a finite number of) the pre-sample values of ε. The third and simplest approach, called conditional least squares, is the same as exact least squares except pre-sample values of the disturbances are set equal to their unconditional expected values.

Monte Carlo evidence [see Newbold and Ansley (1979)] suggests that the exact maximum likelihood method is generally superior to the least squares methods. Conditional least squares performs particularly poorly when the roots of the MA polynomial, $b(z)$, are near the unit circle.

Once the model has been estimated diagnostic checks are carried out to test the adequacy of the model. Most of the procedures in one way or another test the residuals for lack of serial correlation. Since diagnostic tests are carried out after estimation Lagrange Multiplier tests are usually the simplest to carry out (see Chapter 12 of this Handbook). For the exact form of several of the tests used the reader is referred to Hosking (1980). Higher moments of the residuals should also be checked for lack of serial correlation as these tests may detect non-linearities or ARCH behavior.

The use of ARIMA models and the three stages of analysis, identification, estimation and diagnostic testing are due to Box and Jenkins (1976), and these models have proved to be relatively very successful in forecasting compared to other univariate, linear, time-invariant models, and also often when compared to more general models. The models have been extended to allow for seasonal effects, which will be discussed in Section 6.

A very different type of analysis is known as spectral analysis of time series. This is based on the pair of theorems [see, for instance, Anderson (1971, sections 7.3 and 7.4)] that the autocorrelation sequence ρ_s of a discrete-time stationary series, x_t has a Fourier transformation representation:

$$\rho_s = \int_{\pi}^{-\pi} e^{i\omega s} dS(\omega),$$

where $S(\omega)$ has the properties of a distribution function, and the spectral representation for x_t:

$$x_t = \int_{\pi}^{-\pi} e^{it\omega} dz(\omega),$$

where

$$\begin{aligned} E[dz(\omega)\overline{dz}(\lambda)] &= 0, & \omega \neq \lambda, \\ &= \sigma^2 dS(\omega), & \omega = \lambda, \end{aligned}$$

where $\sigma^2 = \mathrm{var}(x_t)$. When x_t contains no purely cyclical components $dS(\omega)$ can be replaced by $s(\omega)d\omega$, where $s(\omega)$ is known as the spectral function and is given by:

$$s(\omega) = \frac{1}{2\pi} \sum_{\text{all } s} (\rho_s e^{-is\omega}).$$

The spectral representation for x_t can be interpreted as saying that x_t is the sum of an uncountably infinite number of random components, each associated with a particular frequency, and with each pair of components being uncorrelated. The variance of the component with frequencies in the range $(\omega, \omega+d\omega)$ is $\sigma^2 s(\omega)d\omega$ and the sum (actually integral) of all these variances is σ^2, the variance of the original series. This property can obviously be used to measure the relative importance of the frequency components. Small, or low, frequencies correspond to long periods, as frequency $= 2\pi$ (period)$^{-1}$, and thus to long swings or cycles in the economy if x_t is a macro-variable. High frequencies, near π, correspond to short oscillations in the series. In one sense, spectral analysis or frequency-domain analysis gives no more information than the more conventional time-domain analysis described earlier, as there is a unique one-to-one relationship between the set of autocorrelations ρ_s, $s=1,2,\dots$, and the spectral function $s(\omega)$. However, the two techniques do allow different types of interpretation to be achieved and for each there are situations where they are clearly superior. Thus, for example, if one is interested in detecting cycles or near cycles in one's data, spectral analysis is obviously appropriate.

If x_t is a stationary series and a second series is formed from it by a linear transformation of the form:

$$y_t = \sum_{j=0}^{m} g_j x_{t-j},$$

then their respective spectral representations are easily seen to be:

$$y_t = \int_{-\pi}^{\pi} e^{it\omega} g(\omega)\, dz(\omega),$$

if

$$x_t = \int_{-\pi}^{\pi} e^{it\omega} dz(\omega),$$

where

$$g(\omega) = \sum_{j=0}^{m} g_j z^j, \qquad z = e^{-i\omega}.$$

By considering the autocovariance sequence of y, it follows immediately that the spectrum of y_t is $g(\omega)\bar{g}(\omega)s_x(\omega)$ where $s_x(\omega)$ is the spectrum of x_t and $\bar{g}$ is the complex conjugate of g. y_t is known as a (one-sided) filter of x_t and the effect on a series of the application of a filter is easily determined in the frequency domain.

A zero-mean, white-noise series ε_t with variance of σ_ε^2 has spectrum $s_\varepsilon(\omega) = \sigma_\varepsilon^2/(2\pi)$, so that the spectrum of a white noise is flat, meaning that all frequency components are present and contribute equal proportions to the total variance. Considering a series x_t generated by an ARMA(p, q) process as a filtered version of ε_t, that is:

$$a_p(B)x_t = b_q(B)\varepsilon_t,$$

or

$$x_t = a_p^{-1}(B)b_q(B)\varepsilon_t,$$

it follows that the spectrum of x_t is:

$$\frac{b_q(\omega)\bar{b}_q(\omega)\sigma_\varepsilon^2}{a_p(\omega)\bar{a}_p(\omega)^{2\pi}}.$$

Some applications of spectral analysis in econometrics will be discussed in Section 7. Potentially, the more important applications do not involve just single series, but occur when two or more series are being considered. A pair of series, x_t, y_t, that are individually stationary are (second-order) jointly stationary, if all cross correlations $\rho_s^{xy} = \operatorname{corr}(x_t y_{t-s})$ are time invariant. In terms of their spectral representations, it is necessary that:

$$\begin{aligned} \mathrm{E}\left[\mathrm{d}z_x(\omega)\overline{\mathrm{d}z_y}(\lambda)\right] &= 0, & \omega \neq \lambda, \\ &= \operatorname{cr}(\omega)\mathrm{d}\omega, & \omega = \lambda, \end{aligned}$$

where x_t and y_t have spectral representations:

$$x_t = \int_{-\pi}^{\pi} \mathrm{e}^{it\omega}\mathrm{d}z_x(\omega)$$

and

$$y_t = \int_{-\pi}^{\pi} \mathrm{e}^{it\omega}\mathrm{d}z_y(\omega).$$

$\operatorname{cr}(\omega)$ is known as the cross spectrum and is, in general, a complex valued quantity. Interpretation is easier in terms of three derived functions, the phase $\phi(\omega)$, the coherence $C(\omega)$, and the gain $R_{xy}(\omega)$ given by:

$$\phi(\omega) = \tan^{-1}\left[\frac{\text{imaginary part of } \operatorname{cr}(\omega)}{\text{real part of } \operatorname{cr}(\omega)}\right],$$

$$C(\omega) = \frac{|\operatorname{cr}(\omega)|^2}{s_x(\omega)s_y(\omega)},$$

$$R_{xy}(\omega) = \frac{|\operatorname{cr}(\omega)|}{s_y(\omega)}.$$

When the two series are related in a simple fashion:

$$x_t = ay_{t-k} + v_t,$$

where v_t is a stationary series uncorrelated with y_{t-s}, all s, the phase diagram takes the form:

$$\phi(\omega) = k\omega.$$

This is true whether k is an integer or not, so a plot of the estimate of $\phi(\omega)$ against ω will give an estimate of the lag k in this simple model. Models relating x_t and y_t involving more complicated structures do not lead to such easily interpreted phase diagrams, this being particularly true for two-way causal relationships. The coherence function measures the square of the correlation between corresponding frequency components of the two series and is always important. For instance, it might be found that two series are highly interrelated at low frequencies ("in the long rum") but not at high frequencies ("in the short run") and this could have interesting econometric implications. The gain can be interpreted as the regression coefficient of the ω-frequency component of x on the corresponding component of y.

The extension of spectral techniques to analyze more than two series is much less well developed, although partial cross spectra can be easily determined but have been little used.

Spectral estimation has generated a considerable literature and only the rudiments will be discussed here. Since the spectral density function is given by:

$$s(\omega) = \frac{1}{2\pi} \sum_{j=-\infty}^{\infty} \rho_j e^{-ij\omega}.$$

A natural estimator is its sample counterpart:

$$\hat{s}(\omega) = \frac{1}{2\pi} \sum_{j=-T+1}^{T+1} \hat{\rho}_j e^{-ij\omega}.$$

This estimator has the desirable property of being asymptotically unbiased but also has the undesirable properties of being inconsistent and producing a rather "choppy" graph when plotted against frequency even when $s(\omega)$ is smooth. This last property follows from the fact that $\hat{s}(\omega_1)$ and $\hat{s}(\omega_2)$ will be asymptotically uncorrelated for $\omega_1 \neq \omega_2$.

To alleviate these problems $\hat{s}(\omega)$ is usually smoothed to produce an estimator $\hat{s}_k(\omega)$ given by:

$$\hat{s}_k(\omega) = \int_{-\pi}^{\pi} k(\lambda)\hat{s}(\omega - \lambda)\,d\lambda.$$

The weighting function $k(\lambda)$ is called the spectral window. It is symmetric about

ω and most of its mass is concentrated around this frequency. Specific forms for spectral windows are given in the references below.

Since $\hat{s}_k(\omega)$ is a weighted averaged of $\hat{s}(\lambda)$ for λ near ω large changes in the spectrum near ω cause a large bias in $\hat{s}_k(\omega)$. These spillover effects are called leakage, and will be less of a problem the flatter the spectrum. To avoid leakage series are often "prewhitened" prior to spectral estimation and the spectrum is then "recolored". A series is prewhitened by applying a filter to the series to produce another series which is more nearly white noise, i.e. has a flatter spectrum than the original series. So, for example, x_t might be filtered to produce a new series y_t as:

$$y_t = \phi(B)x_t.$$

The filter $\phi(B)$ may be chosen from a low order autoregression or an ARMA model. Once the spectrum of y_t has been estimated, the spectrum of x_t can be recovered by recoloring, that is:

$$\hat{s}_x(\omega) = |\phi(\omega)|^{-2}\hat{s}_y(\omega).$$

The details of spectral estimation and the properties of the estimators can be found in the books by Anderson (1971), Fishman (1969), and Koopmans (1974). There are many computer packages for carrying out spectral and cross-spectral estimation. For the length of time series generally encountered in economics computation costs are trivial.

If in the spectral representation,

$$x_t = \int_{-\pi}^{\pi} e^{it\omega}\, dz(\omega),$$

the random amplitudes $dz(\omega)$ are not orthogonal, so that

$$E[dz(\omega)\overline{dz}(\lambda)] = d^2F(\omega, \lambda),$$

which is not necessarily zero when $\omega \neq \lambda$, a very general class of non-stationary processes result, known as harmonizable processes. They have recently been discussed and applied to economic data by Joyeux (1979).

3. Theory of forecasting[1]

In applied economics as well as many other sciences much of the work on time series analysis has been motivated by the desire to generate reliable forecasts of future events. Many theoretical models in economics now assume that agents in

[1] This section relies heavily on Granger and Newbold (1977).

the economy optimally or "rationally" forecast future events and take actions based on these forecasts. This section will be devoted to discussing certain aspects of forecasting methodology and forecast evaluation.

Let X_t be a discrete time stochastic process, and suppose that we are at time n (n = now) and seek a forecast of X_{n+h} (h = hence). Anything that can be said about X_{n+h} at time n will obviously be based on some information set available at time n, which will be denoted by I_n. As an example, a univariate forecast might use the information set:

$$I_n' = (x_t, -\infty < t \leq n; \text{model}),$$

where by "model" we mean the process generating the data. Any information set containing the past and present of the variable being forecast will be called a proper information set.

Everything that can be inferred about X_{n+h} given the information set I_n is contained in the conditional distribution of X_{n+h} given I_n. Typically it is too ambitious a task to completely characterize the entire distribution, and the forecaster must settle for a confidence band for X_{n+h}, or a single value, called a point forecast.

To derive an optimal point forecast a criterion is needed, and one can be introduced using the concept of a cost function. Agents engage in forecasting presumably because knowledge about the future aids them in deciding which actions to take today. An accurate forecast will lead to an appropriate action and an inaccurate forecast to an inappropriate action. An investor, for example, will forecast the future price of an asset to decide whether to purchase the asset today or to sell the asset "short". An accurate forecast implies a profit for the investor and an inaccurate forecast implies a loss. A cost function measures the loss associated with a forecast error. If we define the forecast of X_{n+h} based on information set I_n as $f^x_{n,h}(I_n)$, then the forecast error will be:

$$e^x_{n,h}(I_n) = X_{n+h} - f^x_{n,h}(I_n). \tag{3.1}$$

The cost associated with this error can be denoted as $c(e^x_{n,h}(I_n))$. (For notational convenience we will often suppress the subscripts, superscripts, and information set when they are easily inferred from the context.) A natural criterion for judging a forecast is the expected cost of the forecast error.

The most commonly used cost function is the quadratic:

$$C(e) = ae^2,$$

where a is some positive constant. This cost function is certainly not appropriate in all situations—it is symmetric for example. However, it proves to be the most tractable since standard least squares results can be applied. Many results obtained from the quadratic cost function carry over to other cost functions with only minor modification. For a discussion of more general cost functions the reader is referred to Granger (1969) and Granger and Newbold (1977).

Standard theory shows that the forecast which minimizes the expected squared forecast error is:

$$f_{n,h} = \mathrm{E}(X_{n+h}/I_n).$$

Calculating the expected value of the conditional distribution may be difficult or impossible in many cases, since as mentioned earlier the distribution may be unknown. Attention has therefore focused on forecasts which minimize the mean square forecast error and which are linear in the data contained in I_n. Except for a brief mention of non-linear forecasts at the end of this section, we will concern ourselves only with linear forecasts.

We will first derive the optimal linear forecast of X_{n+h} for the quadratic cost function using the information set I_n' introduced above. We will assume that X_t is covariance stationary and strictly non-deterministic. The deterministic component of the series can, by definition, be forecast without error from I_n so there is no loss in generality in the last assumption. For integrated processes, X_t is the appropriate differenced version of the original series. Since the infinite past of X_t is never available the information set I_n' is rather artificial. In many cases, however, the backward memory of the X_t process [see Granger and Newbold (1977)] is such that the forecasts from I_n' and

$$I_n'' = (x_t, t = 0,1,\dots,n;\ \text{model}).$$

differ little or not at all.

The optimal forecast for the quadratic cost function is just the minimum mean square error forecast. The linear minimum mean square error forecasts from the information set I_n' will be of the form:

$$f_{n,h} = \sum_{i=0}^{\infty} c_i x_{n-i} = c(B)x_n,$$

where $c(B)$ minimizes:

$$\mathrm{E}\left[(x_{n+h} - c(B)x_n)^2\right]$$

From Wold's theorem x_t has a moving average representation:

$$x_t = b(B)\varepsilon_t, \tag{3.2}$$

where ε_t is white noise. If we define:

$$w(B)=b(B)c(B),$$

and we assume that $b(B)$ is invertible, the problem reduces to finding $w(B)$ which minimizes:

$$\mathrm{E}\left[\left(x_{n+h}-w(B)\varepsilon_n\right)^2\right].$$

It is then straightforward to show [Granger and Newbold (1977, p. 121)] that the equations which characterize $w(B)$ are:

$$w_i=b_{i+h}, \qquad i=0,1,\dots.$$

A compact way of writing this is:

$$w(B)=\left[\frac{b(B)}{B^h}\right]_+,$$

where "+" means ignore all negative powers of B. The linear mean square error forecast can then be written as:

$$f_{n,h}=\left[\frac{b(B)}{B^h}\right]_+\varepsilon_n,$$

or:

$$f_{n,h}=\left[\frac{b(B)}{B^h}\right]_+\frac{1}{b(B)}x_n. \tag{3.3}$$

Substituting (3.2) and (3.3) into (3.1) shows that the forecast error will be:

$$e_{n,h}=\sum_{i=0}^{h-1}b_i\varepsilon_{n+h-i},$$

so that the h step forecast errors are generated by a moving average process of order $h-1$. The one step ahead forecast error is just ε_{n+1} which is white noise. Furthermore, x_{n+h} can be decomposed as:

$$x_{n+h}=f_{n,h}+e_{n,h},$$

where $f_{n,h}$ and $e_{n,h}$ are uncorrelated. The variance of the forecast will therefore be bounded above by the variance of the series.

The formulae given above for the optimal univariate forecast may look rather imposing, but simple recursions can easily be derived. Note, for instance, that:

$$\begin{aligned} f_{n,h} &= f_{n-1,h+1} + b_n \varepsilon_n \\ &= f_{n-1,h+1} + b_n (X_n - f_{n-1,1}), \end{aligned}$$

so that forecasts of X_{n+h} can easily be updated as more data becomes available. A very simple method is also available for ARMA models. Suppose that x_t is ARMA(p,q) so that:

$$x_{n+h} = a_1 x_{n+h-1} + \cdots + a_p x_{n+h-p} + \varepsilon_{n+h} - b_1 \varepsilon_{n+h-1} - \cdots - b_q \varepsilon_{n+h-q}.$$

$f_{n,h}$ can be formed by replacing the terms on the right-hand side of the equation by their known or optimal forecast values. The optimal forecast for ε_{n+k} is, of course, zero for $k \geq 0$.

While univariate forecasting methods have proved to be quite useful (and popular) the dynamic interaction of economic time series suggests that there may be substantial gains from using wider information sets. Consider the forecast of x_{n+h} from the information set:

$$I_n''' = \{ (x_t, y_t'), -\infty < t \leq n; \text{model} \}.$$

where y is a vector of other variables. If we assume that (X_t, Y_t') is a covariance stationary process, then an extension of Wold's theorem allows us to write:

$$z_t \equiv \begin{bmatrix} x_t \\ y_t \end{bmatrix} = \begin{bmatrix} a_{11}(B) & a_{12}(B) \\ a_{21}(B) & a_{22}(B) \end{bmatrix} \xi_t,$$

where ξ is vector of white noise with contemporaneous matrix Σ, so that $A(0) = I$. The linear mean square error forecast will be of the form:

$$f_{n,h}^x (I_n''') = Q(B) z_n,$$

where $Q(B)$ minimizes:

$$\mathrm{E}[X_{n+h} - Q(B) z_n]^2.$$

If the matrix polynomial, $A(B)$ is invertible it can be shown that:

$$Q(B) = \left[\frac{a_1(B)}{B^h} \right]_+ [A(B)]^{-1},$$

where $a_1(B)$ is the first row of $A(B)$.

Once again, the forecast errors, $e_{n,h}(I_n''')$ will follow a moving average process of order $h-1$. Furthermore, it must be the case that:

$$\operatorname{var}\left(e_{n,h}\left(I_n'\right)\right) \geq \operatorname{var}\left(e_{n,h}\left(I_n'''\right)\right),$$

since adding more variables to the information set cannot increase the forecast error variance.

These optimal forecasting results have been used to derive variance bounds implied by a certain class of rational expectations models. [The discussion below is based on Singleton (1981); see also Shiller (1981) and LeRoy and Porter (1981).] The models under consideration postulate a relationship of the form:

$$P_n = \sum_{i=0}^{k} \delta_i f_{n,i}^x\left(I_n'''\right), \tag{3.4}$$

where the forecasts are linear minimum mean square error. In some models P_n could represent a long-term interest rate and X_n a short-term rate, while in others P_n represents an asset price and X_n is the value of services produced by the asset over the time interval.

If we define

$$P_n^* = \sum_{i=0}^{k} \delta_i x_{n+i}$$

and

$$\tilde{P}_n = \sum_{i=0}^{k} \delta_i f_{n,i}^x\left(I_n'\right),$$

where $f_{n,i}^x(I_n')$ is the linear mean square error forecast, then:

$$P_n^* = P_n + \eta_n = \tilde{P}_n + \nu_n,$$

where

$$\eta_n = \sum_{i=1}^{k} \delta_i e_{n,i}\left(I_n'''\right)$$

and

$$\nu_n = \sum_{i=1}^{k} \delta_i e_{n,i}\left(I_n'\right).$$

Since P_n and $\tilde{P}_n$ are linear combinations of optimal forecasts:

$$\mathrm{E}[P_n\eta_n] = \mathrm{E}[\tilde{P}_n\nu_n] = 0,$$

which implies:

$$\sigma_{P*}^2 = \sigma_P^2 + \sigma_\eta^2 = \sigma_{\tilde{P}}^2 + \sigma_\nu^2.$$

Furthermore, since I_n' is a subset of I_n''':

$$\sigma_\eta^2 \le \sigma_\nu^2,$$

which leaves us with the inequality

$$\sigma_{P*}^2 > \sigma_P^2 \ge \sigma_{\tilde{P}}^2.$$

The variances σ_{P*}^2 and $\sigma_{\tilde{P}}^2$ are then the bounds for the variance of the observed series. If σ_P^2 falls outside of these bounds the model (3.4) must be rejected. The first two variances can be calculated from the available data in a straightforward manner. Singleton proposes a method for estimating the last variance, derives the asymptotic distribution of these estimators and proposes a test based on this asymptotic distribution.

The discussion thus far has dealt only with optimal forecasts. It is often the case that a researcher has at his disposal forecasts from disparate information sets, none of which may be optimal. These forecasts could be ranked according to mean square error and the best one chosen, but there may be gains from using a combination of the forecasts. This was first noted by Bates and Granger (1969) and independently by Nelson (1972) and has been applied in a number of research papers [see, for example, Theil and Feibig (1980)].

To fix notation, consider one step ahead forecasts of x_{N+1}, denoted $f^1, f^2, \ldots, f^m$, with corresponding errors $e^1, e^2, \ldots, e^m$. Since bias in a forecast is easily remedied we will assume that all of the forecasts are unbiased. An optimal linear combined forecast is:

$$f^c = \sum_{i=1}^{m} a_i f^i,$$

where the a_i's are chosen to minimize:

$$\mathrm{E}(x_{n+1} - f^c)^2.$$

If the mean of X is not zero the resulting combined forecast will be unbiased only if:

$$\sum_{i=1}^{m} a_i = 1.$$

The papers by Bates and Granger and Nelson derive the weights subject to this constraints. This is just a constrained least squares problem.

Granger and Ramanathan (1981) point out that the constraint will generally be binding and so a lower mean square root error combined forecast is available. As an example suppose that x_t is generated by:

$$x_t = y_{t-1} + z_{t-1} + \eta_t,$$

where y_t, z_t, and η_t are independent white noise. If I_n^1 contains only past and present y and I_n^2 contains only past and present z, the optimal forecasts are:

$$f^1 = y_{n-1},$$
$$f^2 = z_{n-1}$$

and

$$f^c = f^1 + f^2.$$

The combined forecast has a mean square error equal to σ_η^2. Imposing the constraint yields:

$$\tilde{f}^c = a_1 f_1 + a_2 f_2,$$

where

$$a_1 = \frac{\sigma_y^2}{\sigma_y^2 + \sigma_z^2}; \quad a_2 = \frac{\sigma_z^2}{\sigma_y^2 + \sigma_z^2},$$

and the mean square error of $\tilde{f}^c$ is:

$$\sigma_\eta^2 + \frac{\sigma_y^2 \sigma_z^2}{\sigma_y^2 + \sigma_z^2} \geq \sigma_\eta^2.$$

When the weights are unconstrained the combined forecast will generally be biased. This is easily remedied. One merely expands the list of available forecasts to include the mean of X. There is no need to impose the constraint as it will be satisfied by the unconstrained least squares solution, for the same reson that if a constant is included in an OLS regression the residuals will sum to zero.

Evaluation of forecast performance is by no means a clear-cut procedure. The discussion of optimal forecasts does however suggest some properties which are easily checked. The optimal linear forecast of X_{n+h} based on the information set I_n is the projection of X_{n+h} on the data I_n. This implies that the forecast error, $e_{n,h}$, is orthogonal to any linear combination of variables in the information set. Forecast errors can then be regressed on linear combinations of data in the information set and the estimated coefficients can be tested to see if they are significantly different from zero. Care must be taken in carrying out these tests. We showed earlier that the optimal h-step forecast errors from a proper information set followed a moving average process of order $h-1$, and therefore even under the null the residuals in this regression will not be white noise for h larger than 1. One step ahead forecast errors from proper information sets should be white noise and this is an easy property to check. The variance bounds derived above also suggest a weak test. The variance of the forecast should be less than the variance of the series being forecast.

When more than one forecast of the same quantity is available additional tests can be constructed. Forecasts can be ranked on a mean square error criterion and the best chosen. More demanding tests can also be constructed. If f is the optimal forecast from an information set I_n^f, and g is a forecast from an additional information set I_n^g, which is a subset of I_n^f, then the forecast error from f will be uncorrelated with g. A regression of the forecast error, e_f, on g should yield a coefficient which is not significantly different from zero. Equivalently, if the optimal combined forecast using f and g is formed the weights on f and g should not be significantly different from one and zero, respectively. Tests similar to these have been constructed to evaluate the forecasting performance of macro models and are briefly discussed in Section 7. A thorough discussion of these tests and others is contained in Granger and Newbold (1977, ch. 8).

We have largely been concerned in this section with linear forecasts; however, even for covariance stationary processes considerable gains can occur from considering nonlinear forecasts. Consider for example a special case of the bilinear model introduced in Section 2:

$$x_t = \beta \varepsilon_{t-1} x_{t-2} + \varepsilon_t,$$

where ε_t is white noise. The process will be covariance stationary if $\beta^2\sigma_\varepsilon^2 < 1$ [Granger and Andersen (1978, p. 40)]. Since the lagged autocovariances are all zero, it follows that the optimal univariate linear one step ahead forecast of X_{n+1} is zero. The forecast mean square error is then:

$$\sigma_x^2 = \frac{\sigma_\varepsilon^2}{1-\beta^2\sigma_\varepsilon^2}.$$

The optimal non-linear one step ahead forecast is $\beta\varepsilon_{t-1}x_{t-2}$ which will have an expected mean square forecast error σ_ε^2.

Identification of complicated bilinear models is a difficult procedure, but the book by Granger and Andersen suggests methods which seem practical for simple models. Their procedure is to examine the autocorrelations of the squares of the residuals from linear time series models. Many nonlinear models have linear approximations with serially correlated squared residuals. If the squared residuals appear to be serially correlated it is not clear which non-linear models should be considered as alternatives. A further discussion of non-linear forecasting and forecasting non-linear transformations of the data can be found in Granger and Newbold (1976, 1977) and in Priestley (1980).

4. Multiple time series and econometric models

Econometric models (for time series data) and multiple time series models both attempt to describe or at least approximate the dynamic relationship between the variables under consideration. As mentioned in the first section the approaches taken in building these two types of models have historically been quite different. To facilitate the comparison of these approaches it is useful to introduce a variety of multiple time series representations.

Let $\boldsymbol{Z}_t$ be an $N\times 1$ vector stationary time series. Then an extension of Wold's theorem [Hannan (1970)] allows us to write:

$$\boldsymbol{Z}_t = \boldsymbol{c}(B)\boldsymbol{\varepsilon}_t,$$

where $\boldsymbol{c}(B)$ is an $N\times N$ matrix of (possibly infinite degree) polynomials in the backward shift operator and $\boldsymbol{\varepsilon}_t$ is an $N\times 1$ vector white noise, that is:

$$\boldsymbol{\varepsilon}_t = (\varepsilon_{1t}, \varepsilon_{2t}, \dots, \varepsilon_{Nt}),$$

with

$$\mathrm{E}[\boldsymbol{\varepsilon}_t] = 0$$

and

$$\mathrm{E}[\boldsymbol{\varepsilon}_t\boldsymbol{\varepsilon}'_s] = \delta_{st}\Sigma,$$

where δ is the Kronecker delta.

As was the case with the univariate model, it may be true that $\boldsymbol{c}(B)$ can be represented, or at least well approximated, by the rational function:

$$\boldsymbol{c}(B) = \boldsymbol{a}^{-1}(B)\boldsymbol{b}(B), \tag{4.1}$$

where both $\boldsymbol{a}(B)$ and $\boldsymbol{b}(B)$ are $N\times N$ matrices of finite order polynomials in B. We will assume that these matrices are of full rank, so that their inverses exist. When (4.1) is satisfied, $\boldsymbol{Z}_t$ is said to follow a vector ARMA or VARMA process of order (P,Q). P and Q are now $N\times N$ matrices with p_{ij} equal to the order of the polynomial $a_{ij}(B)$ and q_{ij} equal to the order of the polynomial $b_{ij}(B)$. The generating process for $\boldsymbol{Z}_t$ can then be written as:

$$\boldsymbol{a}(B)\boldsymbol{Z}_t=\boldsymbol{b}(B)\boldsymbol{\varepsilon}_t. \tag{4.2}$$

The AR side of (4.2) states that each component of $\boldsymbol{Z}_t$ is at least partially explained by its own past and the present and past of the other components. The whole model then states that when the lag operator $\boldsymbol{a}(B)$ is applied to $\boldsymbol{Z}_t$, the resulting vector time series is such that its autocovariances and cross covariances can be represented by the multivariate moving average model $\boldsymbol{b}(B)\boldsymbol{\varepsilon}_t$. It should be noted that the variables which are observed are the components of $\boldsymbol{Z}_t$ and that the disturbances, $\boldsymbol{\varepsilon}_t$, are at best estimated from the model, provided that the moving average part is invertible. Invertibility is satisfied in the multivariate model if $\boldsymbol{b}^{-1}(B)$ exists.

The representation (4.2) is by no means unique and normalizations must be imposed if the parameters are to be identified in the econometric sense. One source of under-identification comes from the contemporaneous relationship or causality of the data. The elements $\boldsymbol{Z}_t$ will be contemporaneously related if any of the off-diagonal elements of $\boldsymbol{a}(0)$, $\boldsymbol{b}(0)$, or Σ are non-zero. Clearly, there will be no way to tell these apart given only data on $\boldsymbol{Z}$. A common normalization sets $\boldsymbol{a}(0)=\boldsymbol{b}(0)=I$ and leaves Σ unrestricted. Others are, of course, possible. Sims (1980) for example uses the recursive form of the model for his vector autoregressions in which $a(0)$ is lower triangular and Σ is diagonal. This is a useful form for forecasting and for the vector AR model implies that the parameters can be efficiently estimated by ordinary least squares. Sufficient conditions for parameter identification in VARMA models are given in Hannan (1969).

As $\boldsymbol{a}(B)$ is assumed of full rank, (4.2) may also be written as:

$$\boldsymbol{Z}_t=\boldsymbol{a}^{-1}(B)\boldsymbol{b}(B)\boldsymbol{\varepsilon}_t.$$

If $\boldsymbol{a}^*(B)$ is the adjoint matrix associated with $\boldsymbol{a}(B)$ and $|\boldsymbol{a}(B)|$ is the determinant of this matrix. This results in the equivalent model:

$$|\boldsymbol{a}(B)|\boldsymbol{Z}_t=\boldsymbol{a}^*(B)\boldsymbol{b}(B)\boldsymbol{\varepsilon}_t,$$

and the jth equation of this system is:

$$|\boldsymbol{a}(B)|Z_{jt}=\boldsymbol{\alpha}_j(B)\boldsymbol{\varepsilon}_t,\qquad j=1,\dots,N,$$

where $\boldsymbol{\alpha}_j(B)$ is the jth row of $\boldsymbol{a}^*(B)\boldsymbol{b}(B)$. If no cancellation of factors of the form $(1-\beta B)$ from both sides of these equations occurs, it follows that all the single series ARMA (p,q) models for the components of $\boldsymbol{Z}_t$ will have identical AR parts, and further that p and q will be very large if the number of components is large. As neither of these features is actually observed, this suggests that considerable cancellations do occur or that the present single series modeling techniques tend to choose models that are too simple. Zellner and Palm (1976) and Protheo and Wallis (1976) have suggested that the common AR property can be utilized to indicate relevant constraints on the form of the matrix $\boldsymbol{a}(B)$ in the full model (4.2), but the technique has been applied only to small systems so far. A possible limitation to this technique can be seen by noting that the $\boldsymbol{Z}_{jt}$'s could all be univariate white noises, but still be related through a model of the form (4.2), although this model will be constrained so that $|a(B)|$ and the moving average process implied by $\boldsymbol{\alpha}_j(B)\boldsymbol{\varepsilon}_t$ are equal for all j. Such constraints are not easily used in practice.

Time series identification, that is the choice of p and q, for VARMA models is a difficult task and completely satisfactory methods are not yet available. Tiao et al. (1979) suggest a method similar to univariate methods of Box and Jenkins which is practical for AR or MA models. Mixed models are substantially more difficult. A procedure for bivariate models is proposed in Granger and Newbold (1977). A computer package, Tiao et al. (1979), is available for estimating small scale (up to five series) VARMA models.

A model more familiar to traditional econometricians is achieved by using the partition:

$$\begin{bmatrix} \boldsymbol{a}_{11} & \boldsymbol{a}_{12} \\ \boldsymbol{a}_{21} & \boldsymbol{a}_{22} \end{bmatrix} \begin{bmatrix} \boldsymbol{y}_t \\ \boldsymbol{x}_t \end{bmatrix} = \begin{bmatrix} \boldsymbol{b}_{11} & \boldsymbol{b}_{12} \\ \boldsymbol{b}_{21} & \boldsymbol{b}_{22} \end{bmatrix} \begin{bmatrix} \boldsymbol{\varepsilon}_{1t} \\ \boldsymbol{\varepsilon}_{2t} \end{bmatrix},$$

where the lag operators have not been shown for notational convenience. If it is now assumed that $\boldsymbol{a}_{21} \equiv 0$, $\boldsymbol{b}_{12} \equiv 0$, and $\boldsymbol{b}_{21} \equiv 0$, one obtains the two sets of equations:

$$\boldsymbol{a}_{11}(B)\boldsymbol{y}_t + \boldsymbol{a}_{12}(B)\boldsymbol{x}_t = \boldsymbol{b}_{11}(B)\boldsymbol{\varepsilon}_{1t} \tag{4.3}$$

and

$$\boldsymbol{a}_{22}(B)\boldsymbol{x}_t = \boldsymbol{b}_{22}(B)\boldsymbol{\varepsilon}_{2t}. \tag{4.4}$$

If, furthermore, there are no contemporaneous correlations between the components of the white-noise vector $\boldsymbol{\varepsilon}_{1t}$ and the white noise vector $\boldsymbol{\varepsilon}_{2t}$, the $\boldsymbol{Z}_t$ is decomposed into $\boldsymbol{x}_t$ and $\boldsymbol{y}_t$, where the components of x_t are called exogenous. The question of how exogeneity should be defined and tested is discussed in Chapter 18, on causality, in this Handbook. Alternative definitions of exogeneity can be found in Engle, Hendry, and Richard (1981). The correct division of variables into these two classes is clearly important for forecasting, as well as other purposes. Equations (4.3) and (4.4) provide the link between times series and econometric models. Equation (4.3) can be viewed as the structural form of a dynamic simultaneous equation model, while (4.4) describes the evolution of the exogenous variables. Traditionally, the existence of the subsystem (4.4) is not considered, as the exogenous variables are said to be "generated outside of the system." In the time series literature, systems such as (4.3) are now being called ARMAX systems, for autoregressive-moving average with exogenous variables.

Although the structural form (4.3) is of fundamental importance, some other derived models are also of interest. Denote $\boldsymbol{a}_{110}(B)=\boldsymbol{a}_{11}(B)-\boldsymbol{a}_{11}(0)$, then (4.3) may be written either as:

$$y_t=-\boldsymbol{a}_{11}^{-1}(0)\boldsymbol{a}_{110}(B)y_t+\boldsymbol{a}_{11}^{-1}(0)\boldsymbol{a}_{12}(B)\boldsymbol{x}_t+\boldsymbol{a}_{11}^{-1}(0)\boldsymbol{b}_{11}(B)\varepsilon_{1t},$$

which is known as the reduced form, or as:

$$y_t=-\boldsymbol{a}_{11}^{-1}(B)\boldsymbol{a}_{12}(B)\boldsymbol{x}_t+\boldsymbol{a}_{11}^{-1}(B)\boldsymbol{b}_{11}(B)\boldsymbol{\eta}_{1t}, \tag{4.5}$$

which has been called the final form, a multidimensional rational-distributed lag model, or of a unidirectional transfer-function form. In the reduced form, endogenous variables are explained by "predetermined variables"—that is, exogenous and lagged endogenous variables—whereas in the final form y_t appears to be explained by just the exogenous variables. If parameter values are known, or have been estimated, both the reduced form and the final form can be used to produce forecasts. The reduced form used the information set $I_n^{(1)}$: $[x_{n-j}, y_{n-j}, j\geq 0]$, plus forecasts of exogenous variables and the final form appears to use just $I_n^{(2)}$: $[x_{n-j}, j\geq 0]$, plus exogenous variable forecasts. However, as is easily seen from (4.5), the use of $I_n^{(2)}$ will generally produce forecasts with errors that are not white noise. These forecasts can then be improved by modeling the residuals, but to do this earlier values of the residuals are required and to know this earlier values of y_t are needed, so that effectively one ends up using $I_n^{(1)}$. As situations are rare in which past values of exogenous variables are available, but not the past values of endogenous variables, the proper information set $I_n^{(1)}$ is the appropriate one in most cases.

Traditionally, econometricians have viewed their task as specifying and estimating the model (4.3) while ignoring (4.4). The time series analyst, on the other hand, would identify and estimate both (4.3) and (4.4). To the econometrician, the parameters of (4.3) were thought to be the most important as these presumably contained the sought after information about the working of the economy. These parameters could then be subjected to hypothesis tests, etc. Time series analysts, being primarily interested in forecasting and not economic theory, required both (4.3) and (4.4) for their purpose. Lucas (1976) showed that the parameters of (4.3) were in general not the parameters of economic interest. He persuasively argued that the important economic parameters could not be deduced without knowledge of the process generating the exogenous variables. The main point of Lucas is that the parameters of (4.3) are not structural at all. They will in general be functions of underlying structural parameters and the parameters of (4.4). The Lucas critique has spawned a new class of econometric models in which the time series properties of the exogenous variables play a crucial role. Examples can be found in Wallis (1980) and Sargent (1981).

Other clear differences between the time series and classical econometric approaches are the size of the information sets used and the intensity with which they are analyzed. Time series models often involve just a few series, but a wide variety of different lag structures are considered. Classically, econometric models involved very large numbers of series, a model of 400 equations now being classified as moderate in size, but are sparse in that most variables do not enter most equations. To the time series analyst's eyes, econometric models involve remarkably few lags. It has been said that when a time series analyst is unhappy with his model, he adds further lagged terms, but an unhappy econometrician is inclined to add further equations. One reason why econometricians rely heavily on an economic theory is that they have so many variables, but usually with rather small amounts of data, so that it would be impossible to consider a wide enough variety of models to be able to get anywhere near the true model. The use of the theory severely limits the number of alternative model specifications that need to be considered. Thus, the theory effectively greatly expands the available data set, but the difficulty is that if an incorrect theory is imposed an incorrect model specification results.

A further use of time series analysis in econometric model building is based on the precept that one man's errors may be another man's data. Thus, the residuals from an econometric model can be analyzed using time series methods to check for model mis-specification. Calling the procedure TSAR, for time series analysis of residuals, Ashley and Granger (1979) looked at the residuals from the St. Louis Federal Reserve Bank Model. Some of the individual residual series were found not to be white-noise and so could be forecast from their own past, and some residuals could be forecast from other residuals, suggesting missing variables, model mis-specification and inefficient estimation. The classification of some variables as exogenous was also found to be questionable.

5. Differencing and integrated models

An example of differences in attitudes between time series analysts and the classical econometricians concerns the question of whether the levels or changes of economic variables should be modeled. If one has a properly specified model in levels, then there will correspond an equally properly specified model in changes. Forecasting from either will lead to identical results, for example, by noting that next level equals next change plus current level. However, if it is possible that the model is mis-specified, which is certainly a sensible viewpoint to take, there can be advantages in using differenced data rather than levels. The occurrence of spurious relationships between independent variables has been known for a long time and was documented again, using theory and simulation, by Granger and Newbold (1974). There it was shown, for example, that if x_t and y_t were each ARIMA (p, d, q), with $d=1$, but independent, then regressions of the form:

$$x_t = \alpha + \beta y_{t-k} + \varepsilon_t,$$

when estimated by ordinary least squares would frequently show apparently significant β and R^2 values. The problem can be seen by considering the null hypothesis, $\beta = 0$, which implies $\varepsilon_t = x_t - \alpha$. This shows that ε_t is serially correlated under the null so that standard t-tests based on ordinary least squares are not appropriate. Estimation methods which assume ε_t is AR(1) improve matters, but do not totally remove the problem, as spurious relationships can still occur. Clearly, if a sufficiently general model is allowed for the errors, the problem is less likely to occur, but if the dependent variable x_t has infinite variance, as occurs when $d=1$, but the model for ε_t only allows finite variance, then spurious relationships are often found. If all series involved are differenced, the residual need not be white noise, so that ordinary least squares is not efficient, but now at least the change series and the residual all have finite variance. Plosser and Schwert (1977, 1978) have shown that, in a sense, over-differencing is less dangerous than under-differencing and have provided illustrations using real data of spurious relationships and the effects of differencing. Using differenced data is not, of course, a general panacea and, as Plosser and Schwert state "the real issue is not differencing, but an appropriate appreciation of the role of the error term in regression". As some econometricians were traditionally rather casual about the error specification, to the eyes of a time series analyst, until recently the possibility that apparently significant relationships were spurious or weaker than they appear remained.

Despite these results, some econometricians have been reluctant to build models other than in levels or have rejected the idea of differencing all variables. Partly this is because they feel more comfortable in specifying models in levels from their understanding of economic theory and also because differencing may not always seem appropriate, particularly when non-linear terms are present or if a change in one variable is to be explained by the difference between the levels of two other variables. Another reason for this reluctance is that econometricians have become used to extremely high R^2, or corrected R^2, values when explaining levels, but R^2 often falls to modest, or even embarrassingly low values, when changes in a variable are explained. Partly this is due to the removal of spurious relationships, but is largely due to the fact that a very smooth, high-momentum variable, such as many levels, are very well explained from past values of this variable, but this is no longer true with the highly variable change series. An extreme case is stock market prices, the levels following a random walk and the changes being white noise, or very nearly so. Econometricians have also been worried that differencing may greatly reduce or even largely remove the very important low-frequency component, corresponding to the long-swings and the business cycle. This can certainly occur if one over-differences, but should not be a problem if the correct amount of differencing occurs to reduce the series to an ARMA generated sequence. Differencing may also exacerbate errors in variables problems, but the presence of errors in variables can often be tested, and these tests can be carried out on the differences as well as the levels. There has also been some debate about the usefulness of differencing by time series analysts. It has been pointed out that if a series has a mean, then this mean cannot be reconstructed from the differenced series, but this would not be so if the difference operator $(1-B)$ is replaced by $(1-\alpha B)$ with α near, but less than, one. The obvious response is that an ARIMA series need not possess a mean.

A way of generalizing this discussion in a potentially useful fashion follows by noting that differencing a series d times means that the spectrum of the series is multiplied by:

$$|1-z|^{2d},$$

where

$$z=\varepsilon^{i\omega}.$$

If a series x_t has a spectrum of the form:

$$|1-z|^{-2d}f(\omega),$$

where $f(\omega)$ is the spectrum of a stationary ARMA series, it will be said to be integrated of order d, and denoted $x_t \sim I(d)$. Note that x_t needs to be differenced d times to become stationary ARMA. As just defined, d need not be an integer and one can talk of fractional differencing a series if a filter of the form $a(B)=(1-B)^d$ is applied to it. It has been shown that integrated series, with non-integer d, arise from the aggregation of dynamic microvariables and from large dynamic systems [see Granger (1980a)].

When $d \geq \frac{1}{2}$, x_t will have infinite variance and if $d<\frac{1}{2}$, the series has finite variance. An integrated series with $d \geq \frac{1}{2}$ will be inclined to be identified by standard Box–Jenkins techniques as requiring differencing. Note that if also

$d<1$, the differencing will produce a series whose spectrum is zero at zero frequency. Thus, the time series analysts will, in a sense, be correct in requiring differencing to remove infinite variance, but the econometricians' worries about losing their critical low-frequency components are well founded. The proper procedure is, of course, to fractionally difference, provided that the correct value of d is known. The best way to estimate d has yet to be determined, as has the importance and actual occurrence of integrated series with non-integer d.

Possible use of fractional integrated models, if they occur in practice, is in long-run forecasting. It can easily be shown that if the MA(∞) model corresponding to $x_t \sim I(d)$ is considered, then the coefficients will decline in the form:

$$b_j \sim Aj^{d-1},$$

whereas a stationary ARMA(p,q) model, with infinite p and q, will have coefficients declining at least exponentially, i.e.

$$b_j \sim A\theta^j, \qquad |\theta|<1.$$

This "long-memory" property can be utilized to improve long-run forecasts in a simple fashion, once d is known or has been reliably estimated.

6. Seasonal adjustment

Many important economic series show a consistent tendency to be relatively high in one part of the year and low in another part, examples being unemployment, retail sales, exports, and money supply. It is fairly uncontroversial to say that a series contains seasonal variation if its spectrum shows peaks, that is extra power, at the seasonal frequencies, which are:

$$2\pi\frac{j}{12}, \qquad j=1,\dots,6,$$

for monthly series. For some series, the seasonal component is an important one, in that the seasonal frequencies contribute a major part of the total variance. For reasons that are not always clearly stated, many econometricians feel that if the seasonal component is reduced, or removed, analysis of the remaining components becomes easier. Presumably, the seasonal part is considered to be economically unimportant or easily understood, but that leaving it in the series confuses the analysis of the more important low-frequency business cycle components. By "seasonal adjustment" is meant any procedure that is designed to remove, or reduce, the seasonal component. The problem of how best to design seasonal adjustment procedures is a very old one and it has generated a considerable literature. Although much progress has been made the problem can hardly be classified as solved. Two excellent recent references are the extensive collection of papers and discussions edited by Zellner (1979) and the survey by Pierce (1980).

Much of the discussion of seasonal adjustment begins with the additive decomposition of an observed series y_t into two unobserved components:

$$y_t = n_t + s_t,$$

where s_t is strongly seasonal—so that its spectrum is virtually nothing except peaks at the seasonal frequencies and n_t is non-seasonal. For this model, "seasonal adjustment" is any procedure which yields an estimate of the non-seasonal component. If this estimate is based on an information set which contains only the past, present, and possibly future values of y_t, the method is called auto-adjustment. A procedure based on a wider information set, called causal adjustment, will be discussed at the end of this section. Most of the literature on seasonal adjustment concerns auto-adjustment procedures and these are by far the most widely used methods. Consequently, much of our discussion will be devoted to these methods.

Early methods of seasonal adjustment relied on the additive decomposition above, and assumed that s_t followed a periodic deterministic process, an example for monthly data being:

$$s_t = \sum_{i=1}^{12} \alpha_i D_{ti},$$

where the D_{ti}'s are a set of monthly dummy variables or sine and cosine terms. The non-seasonal component was assumed to be composed of a "trend" and "irregular" component. These components were approximated by a polynomial in t and white noise. The seasonal component in this model can be estimated using standard regression techniques. Subtracting this estimate from the observed series yields an estimate of the non-seasonal component. This method and its statistical properties are discussed in Jorgenson (1964, 1967).

The causes of seasonal fluctuations, e.g. weather, and the inspection of estimated spectra for economic time series suggest that the deterministic model for s_t is a poor one. A popular approach is to assume that each component is stochastic and generated by an ARMA model. (The possible need to difference the series can be handled, but introduces further complications that will not be discussed here. More details can be found in the references given above.) Thus, we can write:

$$a_n(B) n_t = b_n(B) \eta_t$$

and

$$a_s(B) s_t = b_s(B) \varepsilon_t,$$

where η_t and ε_t are independent white noise:

$$a_n(B) = 1 - a_1^n B - a_2^n B^2 - ,\dots, - a_{p_n}^n B^{p_n},$$
$$b_n(B) = 1 - b_1^n B - b_2^n B^2 - ,\dots, - b_{q_n}^n B^{q_n},$$

and $a_s(B)$ and $b_s(B)$ are similarly defined. The polynomials are such that s_t is strongly seasonal, so that

$$s_s(\omega) = \frac{b_s(z)\bar{b}_s(z)}{a_s(z)\bar{a}_s(z)} \frac{\sigma_\varepsilon^2}{2\pi}, \qquad z = \mathrm{e}^{-\mathrm{i}\omega},$$

has most of its power concentrated around the seasonal frequency and n_t is non-seasonal. The implied model for y_t is:

$$a(B)y_t = b(B)e_t,$$

where $a(B) = a_s(B)a_n(B)$ if $a_s(B)$ and $a_n(B)$ have no common roots, and $b(B)e_t$ is a moving average having the same autocovariances as $a_n(B)b_s(B)\varepsilon_t + a_s(B)b_n(B)\eta_t$.

Since only the sum of n_t and s_t is observed it is impossible to deduce the values of the components if both σ_ε^2 and σ_η^2 are non-zero. We will denote the seasonal adjustment error at time t by:

$$\sigma_t = n_t - \hat{n}_t = \hat{s}_t - s_t,$$

where $\hat{s}_t$ and $\hat{n}_t$ are the estimated values of the components. The linear estimate of n_t which minimizes the mean square seasonal adjustment error is the projection of n_t on the available data (conditional expected value if y_t is normal). If an entire realization of y_t is available the optimal linear estimate of the seasonally adjusted series is then:

$$\hat{n}_t = P(n_t | y_k, -\infty < k < \infty) = V(B)y_t,$$

where P is the projection operator and [Weiner (1949), Whittle (1963), Grether and Nerlove (1970)]:

$$V(z) = \frac{\text{spectrum of } n_t}{\text{spectrum of } y_t},$$

where $z = \mathrm{e}^{-\mathrm{i}\omega}$.

Several properties of the optimal linear estimate follow immediately. First, $\hat{n}_t$ is obtained from a time invariant linear filter applied to y_t, so that the coherence between $\hat{n}_t$ and y_t is one. Second, the filter is symmetric, $v_j = v_{-j}$, implying that the phase between y_t and $\hat{n}_t$ is zero. Finally, the spectrum of $\hat{n}_t$ is

$$s_{\hat{n}}(\omega) = |V(z)|^2 s_y(\omega)$$
$$= \frac{s_n(\omega)}{s_y(\omega)/s_n(\omega)},$$

and since

$$s_y(\omega) = s_n(\omega) + s_s(\omega),$$
$$s_{\hat{n}}(\omega) = \frac{s_n(\omega)}{1 + s_s(\omega)/s_n(\omega)},$$

so that

$$s_{\hat{n}}(\omega) \leq s_n(\omega).$$

The spectrum of $\hat{n}_t$ will be substantially less than the spectrum of n_t over those frequencies where the spectrum of s_t is large relative to the spectrum of n_t. Since this occurs at the seasonal frequencies the spectrum of the adjusted series will contain "dips" at these frequencies. Equivalently, the adjusted series will have negative autocorrelations at the seasonal lags. The "optimal" procedure will tend to "overadjust" for seasonality.

This optimal filter cannot be used for obvious reasons. The parameters of the model and hence the elements of $V(B)$ are rarely known, and a complete realization of y_t is never available. Since the process is stationary $v_j = v_{-j} \approx 0$ for large j implying that the last problem is most serious near the beginning and end of the sample.

Pagan (1975) and Engle (1979) overcome this problem through the use of the Kalman filter and smoother. The Kalman filter produces linear minimum mean square error estimates of n_t using observed data up through time t. The smoother optimally updates these estimates as data beyond time t becomes available. (The Kalman filter and smoother are discussed in detail in Chapter 20 of this Handbook.) To implement the filter the model is written in state space form. Although moving average terms can easily be handled [see Harvey and Phillips (1979)] it is notationally convenient to assume that $b_s(B) = b_n(B) = 1$. With this assumption the model can be written as:

$$y_t = \begin{bmatrix} 1'_{p_n} & 1'_{p_s} \end{bmatrix} \begin{bmatrix} \tilde{n}_t \\ \tilde{s}_t \end{bmatrix},$$

$$\begin{bmatrix} \tilde{n}_t \\ \tilde{s}_t \end{bmatrix} = \begin{bmatrix} \phi_n & 0 \\ 0 & \phi_s \end{bmatrix} \begin{bmatrix} \tilde{n}_{t-1} \\ \tilde{s}_{t-1} \end{bmatrix} + \begin{bmatrix} 1_{p_n} & 0 \\ 0 & 1_{p_s} \end{bmatrix} \begin{bmatrix} n_t \\ \varepsilon_t \end{bmatrix},$$

where 1_k is a k vector with one as its first element and all other elements zero:

$$\tilde{n}_t' = \left[n_t, n_{t-1}, \dots, n_{t-p_n+1}\right],$$

$$\tilde{s}_t' = \left[s_t, s_{t-1}, \dots, s_{t-p_s}\right],$$

and

$$\phi_k = \left[\begin{array}{c|c} a_1^k \; a_2^k \dots a_{p_k-1}^k & a_{p_k} \\ \hline & 0 \\ I_{(p_k-1)} & 0 \\ & \vdots \\ & 0 \end{array}\right], \quad \text{for } k = n \text{ or } s.$$

As Engle (1979) notes, this formulation has several advantages. Computationally it is easier to implement than the Weiner filter, which requires a factorization of the spectral density of y [see Nerlove, Grether and Carvalho (1979)]. The model is also more general as a slight modification will allow weakly exogenous variables to appear as explanatory variables for n_t and s_t. Models with deterministic components can easily be handled. The filter also insures that the revisions made in n_t at time $t+k$ follow a (time varying) moving average process of order $k-1$. This follows since the revision will be a (time varying) linear function of $e_{t+1}, e_{t+2}, \dots, e_{t+k}$.

The filter does require a value of the mean and variance of n_0 and s_0 to begin the recursions. In the case under consideration these components are covariance stationary and the correct starting values are just the unconditional means and variances. For non-stationary models the initial values can be estimated as nuisance parameters, as described in Rosenberg (1973) or Engle and Watson (1981b).

Since the parameters of the model are rarely known, they will generally need to be estimated prior to the adjustment process. If ε_t and η_t are assumed to be normally distributed, the parameters can be estimated using the maximum likelihood methods discussed in Chapter 20 of this Handbook. The scoring algorithm presented in Engle and Watson (1981a) and the EM algorithm discussed in Engle and Watson (1981b) have been successfully used in similar models.

There are of course many ways to additively decompose y_t into two uncorrelated components. The parameters of the model will not in general be identified. Identification can sometimes be achieved by assuming specific forms for the processes as in Engle (1979), or by finding a representation which minimizes the variance of the seasonal component as in Pierce (1979).

Some of the other approaches to seasonal adjustment rely on models which have parameters varying in a seasonal manner, such as the cyclo-stationary models investigated by Parzen and Pagano (1978), while others have the amplitude of the seasonal changing with the size of other components, such as the multiplicative and the harmonizable models. Havenner and Swamy (1981) propose a model similar to the deterministic model discussed above, but they allow the regression coefficients to vary stochastically. When some of these models are employed the concept of seasonal adjustment can become rather confused.

The most widely used program for seasonal adjustment is the Census Bureau's X-11. The program consists primarily of a set of symmetric linear filters applied to the data, but also has features which correct for the number of trading days and "extreme" values. For recent data the symmetric filter is inappropriate and special "end weights" are used. Young (1968) presents a symmetric linear filter which approximates the filter used by X-11, and Cleveland and Tiao (1976) present models for which X-11 is approximately optimal. Details on the characteristics of X-11 can be found in Shiskin, Young, and Musgrave (1967) and Kupier (1979). A discussion of the models for X-11 is presented in the survey paper by Pierce.

In practice, the use of seasonally adjusted data can lead to considerable modeling problems. Many techniques, including X-11, will usually insert "over-adjustment problems", such as the above mentioned negative autocorrelations at seasonal frequencies and the relationships between pairs of series can be considerably disturbed, as various studies have indicated. Partly this is due to the use of robust techniques, which attempt to reduce the relevance of outliers. When actual outliers occur, these methods are valuable, but if over-used, as in X-11, the resulting non-linearities that are introduced can have serious consequences for modeling relationships, for parameter estimation, for causality testing and for forecasting.

Godfrey and Karreman (1967) present evidence that the methods of adjustment often used in practice will have no unfortunate effects on low-frequency components (that is components with frequencies lower than the seasonal frequency), but that all other components are badly affected, even non-seasonal higher-frequency components. The original components with frequencies higher than the seasonal frequencies are partly replaced with variables uncorrelated with them, so that coherences between the original non-seasonal components and the corresponding components of the adjusted series are reduced. This suggests that modeling pairs of seasonally adjusted series can lead to difficulties, and Newbold (1981) presents convincing evidence that this does occur. Wallis (1974, 1979) and Sims (1974) have discussed this problem in detail. Their conclusions suggest that in general it is preferable to use seasonally unadjusted data and explicitly model the seasonality.

The question of how to evaluate a seasonal adjustment procedure is not an easy one, partly because the seasonal and non-seasonal components introduced above are not clearly distinguished. A white-noise series has a flat spectrum and, thus, has some power at seasonal frequencies. The seasonal component is thought of as giving extra power at seasonal frequencies, over and above that provided by the non-seasonal component. However, this statement does not provide enough information to ensure a unique decomposition of a given series into seasonal and non-seasonal components. A similar criterion applies to the simple criterion that a series, after adjustment, has no peaks remaining in its spectrum. A clearer criterion is to require that the variance of the seasonal component, or a suitable transformation of it, should be minimized. This criterion can be characterized in either time or frequency domains and in a sense removes no more than necessary to achieve no seasonality. When one knows the correct model, or a reasonable approximation to it, such a criterion can be used to provide a good seasonal adjustment procedure. However, if the assumed model does not approximate the true world, an inappropriate adjustment may occur.

To evaluate an adjustment procedure, it has been suggested that spectral techniques are the most appropriate and that, (a) the adjusted series should have neither peaks nor dips (over adjustment) at seasonal frequencies, and (b) if the adjustment procedure is applied to a non-seasonal series, the cross spectrum between the original and the adjusted series should have a coherence near one and a phase near zero at all frequencies. Although these appear to be sensible criteria, as shown above the "optimal" adjustment method mentioned earlier will not obey them, producing dips in the spectrum at seasonal frequencies or, equivalently, negative autocorrelation at seasonal lags. This merely means that a pair of "sensible" criteria are inconsistent, but it does leave the choice of proper criteria for the selection and evaluation of techniques for further consideration.

The methods discussed above have all been "auto-adjustment," in that just the observed series x_t has been utilized. As one must expect the seasonal components to be, at least partially, the results of various causal variables a sounder approach would be to seasonally adjust in a multivariate context. Thus, if the weather causes the seasonality in Chicago house construction, it should be natural for econometricians to model this relationship. The effects of a severe winter, for example, are then directly allowed for rather than being considered as some vague, unexplained outlier. Of course, it is by no means easy to correctly model the required relationships, particularly as the series involved will all be strongly seasonal and the use of causal adjustment procedures would be far too expensive for the government to use on all of the series that are said to need adjustment. Nevertheless, if an econometrician is anxious to produce a really sound model, it is advisable to use unadjusted, raw data and to build seasonal causal terms in the model. However, even then the data may still need application of a seasonal adjustment procedure as some causes could be unobservable, but if one does it oneself at least the methods used is under one's own control and need produce less unpleasant surprises than the use of an "off-the-shelf" technique. Further discussion of these points may be found in the papers by Granger and Engle in the volume edited by Zellner mentioned above.

7. Applications

In this section a few examples of the way in which time series techniques have been applied to economic data will be briefly discussed. It would be virtually impossible to survey all of the applications that exist. Two applications that will not be discussed, although they are currently very much in vogue, are testing for causality and the use of Kalman filter techniques for investigating time-varying parameter models, as these are described in Chapters 18 and 20 of this Handbook. Additional applications using frequency domain techniques can be found in Granger and Engle (1981).

The most obvious, and oldest, application is to model a single series to provide what are termed "naive" forecasts against which the forecasts from a full-scale econometric model can be compared. Of course, the comparison is not strictly fair, as the econometric model uses a much larger information set, and also has the "advantage" of being based on an economic theory, but, nevertheless, econometricians have behaved as though they believe that such naive models are worthy forecasting opponents. In fact, the econometric models have found it difficult to beat the time-series forecasts, an example being Cooper (1972), who used only AR(4) models. More recently, the econometric models have performed relatively better, although a more stringent criterion suggested in Granger and Newbold (1977, ch. 8), involving the combination of forecasts, would still probably suggest that there is still room for considerable improvement by the econometric models. It will be interesting to continue to compare forecasts from the two types of model, as each is certainly improving through time.

More natural comparisons are between econometric models and multivariate time series, although the best way to specify the latter is still uncertain. Some examples are the papers by Zellner and Palm (1974), Sargent (1981) and Taylor (1979). No complete comparison of relative forecasting abilities is available at this time. Multivariate time series techniques can also be used to measure the importance, in terms of improved forecasting ability, of adding further variables to the model. An obvious example is to ask how useful is anticipation data. The technique used is the same as that developed for causality testing, as discussed in Chapter 18 of this Handbook. The results are sometimes rather surprising, such as the weak relationships found between some financial series by Pierce (1977). Neftci (1979) investigated the usefulness of the NBER leading indicator for forecasting the index of industrial production (IIP). He modeled IIP in terms of its own lags and the added various leading indicators to the model. Using post-sample forecasts, he found that the leading indicators did not improve forecasts for "normal" times, but did help during the recession year of 1974. The results thus agree with the NBER claims about the usefulness of this indicator series at turning points, but nothing more. Auerbach (1981) studied the usefulness of the leading indicator series in predicting changes in both IIP and the adult civilian unemployment rate. Based on both in-sample fit and forecasting performance he found the leading indicator series useful, but his in-sample results suggest that it may be possible to choose better (possibly time varying) weights for the components of the leading indicator series.

The ARCH model introduced in Section 2 has been used in a number of applications. Engle (1980, 1982) has shown that there are significant ARCH effects in U.S. and U.K. inflation data, and Engle and Kraft (1981) derive conditional multiperiod forecast variances from an autoregressive model where the disturbance follows an ARCH process. Robbins (1981) estimates a model in which the conditional variance of excess returns for short rates affects the liquidity premium for long rates. Engle, Granger and Kraft (1981) use a multivariate ARCH model to compute optimal time varying weights for forecasts of inflation from two competing models.

The obvious applications of univariate spectral analysis are to investigate the presence or not of cycles in data. Thus, for example, Hatanaka and Howrey (1969) looked for evidence of long swings or long cycles in the economy, by asking if there were peaks in the spectrum corresponding to such cycles. The results were inconclusive, because very long series would be required to find significant peaks, particularly against the "typical spectral shape" background, corresponding to the high power at low frequencies found with ARIMA (p, d, q) models, $d > 0$, which we often observed for the levels of economic macro variables. A related application is to compare the estimated spectral shape with that suggested by some theory. For example, the random-walk theory of stock market prices suggests that price changes should be white noise and thus have a flat spectrum. Granger and Morgenstern (1970) found evidence that was generally in favor of the hypothesis, although a very slight evidence for a seasonal in price changes was occasionally observed. Estimated spectra of a wide range of economic series give no evidence of strict cycles except for the seasonal component. Howrey (1972) calculated the spectra of major variables implied by the Wharton model and compared them to the typical spectral shape, and generally found the econometric model did produce the correct spectral shape.

The power spectrum is obviously useful in consideration of the seasonal, both to find out if a series contains a seasonal component, to measure its strength and also to investigate the effects of seasonal adjustment. One of the very first applications of frequency domains techniques to economic data was by Nerlove

(1964) investigating these aspects of seasonality. He also used the spectrum to define the seasonal component in a similar way to that used in Section 6. He gave clear indication that seasonal adjustment could disrupt the data in an unfortunate manner with the follow-up study by Godfrey and Karreman (1967) providing further illustrations of this problem.

The first application of cross-spectral analysis in economics were by Nerlove (1964) on seasonals and by Hatanaka in Granger and Hatanaka (1964), who considered the leads and strength of the relationship between the NBER leading indicators and the level of the economy. Hatanaka found some coherence at low frequencies, but the leads observed in the phase diagram were less than found by the NBER using less sophisticated methods. A later investigation of leading indicators by Hymans (1973) also used spectral methods. The results threw some doubts on the usefulness of several of the components of the index of leading indicators and using the observed coherence values an alternative weighted index was proposed, which would seem to be superior to that now in use. Most subsequent applications of cross-spectral analysis try simply to measure the extent to which pairs of series are related and whether or not there is evidence for a simple lag. Examples may be found in Labys and Granger (1970). When there is a feedback relationship between the variables, the lag structure cannot be determined, and so difficulties in interpretation frequently occur.

The Fourier transform of a stationary series allows one to look at the different frequency components of the series, at least to some extent. This idea was used in Granger and Hatanaka (1964) to test for stationarity by considering the possibility of the amplitude of the frequency components varying through time. By isolating frequency components in a group of series, the possibility of the relationships between the series varying with frequency can be analyzed. Calling the technique band spectrum regression, Engle (1974) considered a simple time-domain regression, transformed it into the frequency domain and then used a test similar to the Chow test for structure stability, to see if relationships were frequency dependent. The method is an obvious generalization of the familiar decomposition into "permanent" and "transitory" components and has similar interpretational advantages. In Engle (1978) the technique was applied to a variety of wage and price series and it was found, for example, that "the effect on prices of a low-frequency change in wages is much greater than the effect of a high-frequency change".

Spectral techniques have also been used recently by Sargent and Sims (1977), Geweke (1975, 1977), and Singleton (1980) to search for unobserved variables or factors, in a group of series, such as a common "business cycle factor" in a group of macro variables or a "national factor" in a group of regional employment series. The model is a dynamic generalization of the factor analysis model typically applied to cross-section data and postulates that all of the dynamic interrelationships between the series can be accounted for by a small number of common factors. In the exploratory version of the model, which is useful for determining the number of common factors, standard estimation techniques adapted for complex arithmetic can be applied. Rather than applying these techniques to a covariance matrix, as in the cross-section case, they are applied to the spectral density matrix, frequency by frequency. When there are constraints on the model, as in confirmatory factor analysis, estimation is more difficult as constraints must be imposed across frequency bands. Often these constraints are more easily imposed in the time domain, and Engle and Watson (1981b) discuss time domain estimation and hypothesis testing methods.

8. Conclusion

Because of the way econometrics has been developing in recent years, the distinction between time series methods and the rest of econometrics has become much less clear. It seems very likely that this will continue and the tendency is already being reflected in modern textbooks such as Maddala (1977). It is nevertheless true that many econometricians do not appreciate the theoretical results and techniques available in the time series field, and so a list of some of the textbooks in this field is provided. The first four books concentrate on the frequency domain, and the others are general in coverage or deal just with the time domain (in each group, the books are approximately in order of increasing mathematical sophistication): Granger and Hatanaka (1964), Bloomfield (1976), Koopmans (1974), Priestly (1981), Granger (1980c), Nelson (1973), Box and Jenkins (1976), Granger and Newbold (1977), Fuller (1976), Anderson (1971), Brillinger (1975), and Hannan (1970).

References

Anderson, T. W. (1971) *The Statistical Analysis of Time Series*. New York: Wiley.

Ansley, C. (1979) "An Algorithm for the Exact Likelihood of a Mixed Autoregressive-Moving Average Process", *Biometrika*, 66, 59–65.

Ashley, R. A. and C. W. J. Granger (1979) "Time-Series Analysis of Residuals from the St. Louis Model", *Journal of Macroeconomics*, 1, 373–394.

Auerbach, A. J. (1981) "The Index of Leading Indicators: 'Measurement Without Theory', Twenty-Five Years Later", Harvard Institute of Economic Research, Discussion Paper 841.

Bates, J. W. and C. W. J. Granger (1969) "The Combination of Forecasts", *Operations Research Quarterly*, 20, 451–468.

Bloomfield, P. (1976) *Fourier Analysis of Time Series*. New York: Wiley.

Box, G. E. P. and G. M. Jenkins (1976) *Time Series, Forecasting and Control*, Holden Day, San Francisco, revised edition.

Brillinger, D. R. (1975) *Time Series Data Analysis and Theory*. New York: Holt, Rinehart and Winston.

Cleveland, W. P. and G. C. Taio (1976) "Decomposition of Seasonal Time Series: A Model for the Census X-11 Program", *Journal of the American Statistical Association*, 71, 581–587.

Cooper, R. L. (1972) "The Predictive Performance of Quarterly Econometric Models of the United States" in *Econometric Models of Cyclical Behavior*, ed. by B. G. Hickman, Columbia University Press.

Engle, R. F. (1976) "Band Spectrum Regression Inter". *International Economic Review*, 15, 1–11.

Engle, R. F. (1978) "Testing Price Equations for Stability Across Spectral Frequency Bands", *Econometrica*, 46, 869–882.

Engle, R. F. (1979) "Estimating Structural Models of Seasonality", in Zellner (1979).

Engle, R. F. (1980) "Estimates of the Variance of Inflation Based Upon the ARCH Model", University of California, San Diego, Discussion Paper.

Engle, R. F. and D. Kraft (1981) "Multiperiod Forecast Error Variances of Inflation Estimated from ARCH Models", paper presented to the Conference on Applied Time Series Analysis of Economic Data, Washington, D.C., October, 1981.

Engle, R. F., C. W. J. Granger, and D. Kraft (1981) "Combining Competing Forecasts of Inflation Using a Bivariate ARCH Model", University of California, San Diego, mimeo.

Engle, R. F. and M. W. Watson (1981a) "A One-Factor Model of Metropolitan Wage Rates", *Journal of the American Statistical Association*, 76 (December).

Engle, R. F. and M. W. Watson (1981b) "A Time Domain Approach to Dynamic MIMIC and Factor Models" (revised), Harvard University, mimeo.

Engle, R. F., D. F. Hendry, and J. F. Richard (1981) "Exogeneity", University of California, San Diego Discussion Paper 81-1 (revised).

Engle, R. F. (1982) "Autoregressive Conditional Heteroscedasticity With Estimates of the Variance of Inflationary Expectations", forthcoming in *Econometrica*.

Fuller, W. A. (1976) *Introduction to Statistical Time Series*. New York: Wiley.

Geweke, J. (1975) "Employment Turnover and Wage Dynamics in U.S. Manufacturing", Unpublished Ph.D. Dissertation, University of Minnesota.

Geweke, J. (1977), "The Dynamic Factor Analysis of Economic Time Series", in: D. J. Aigner and A. S. Goldberger, eds., *Latent Variables in Socio-Economic Models*, North-Holland, Amsterdam, ch. 19.

Godfrey, M. D. and H. Kareman (1976) "A Spectrum Analysis of Seasonal Adjustment", *Essays in Mathematical Economics in Honor of Oskar Morgenstern*, ed. M. Shubik, Princeton University Press.

Granger, C. W. J. (1969) "Prediction with a Generalized Cost of Error Function", *Operations Research Quarterly*, 20, 199–207.

Granger, C. W. J. (1980a) "Long-Memory Relationships and The Aggregation of Dynamic Models", *Journal of Econometrics*.

Granger, C. W. J. (1980b) "Some Properties of Time-Series Data and Their Use in Econometric Model Specification", *Annals of Applied Econometrics* (supplement to *Journal of Econometrics*).

Granger, C. W. J. (1980c) *Forecasting in Business and Economics*. New York: Academic Press.

Granger, C. W. J. and A. P. Andersen (1979) *An Introduction to Bilinear Time Series Models*. Gottingen: Vandenhoeck and Ruprecht.

Granger, C. W. J. and M. Hatanaka (1964) *Spectral Analysis of Economic Time Series*. Princeton University Press.

Granger, C. W. J. and P. Newbold (1974) "Spurious Regressions in Econometric", *Journal of Econometrics*, 26, 1045–1066.

Granger, C. W. J. and P. Newbold (1976) "Forecasting Transformed Series", *Journal of the Royal Statistical Society B*, 38, 189–203.

Granger, C. W. J. and P. Newbold (1977) *Forecasting Economic Time Series*. New York: Academic Press.

Granger, C. W. J. and O. Morgenstern (1970) *Predictability of Stock Market Prices*. Lexington: Heath-Lexington Book.

Granger, C. W. J. and R. Ramanathan (1981) "On the Combining of Forecasts", University of California, San Diego, mimeo.

Gray, Kelley and McIntire (1978) "A New Approach to ARMA Modeling", *Communications in Statistics, B*.

Hannan, E. J. (1969) "The Identification of Vector Mixed Autoregressive Moving Average Systems", *Biometrika*, 56, 223–225. .

Hannan, E. J. (1970) *Multiple Time Series*. New York: Wiley.

Hannan, E. J. and B. G. Quinn (1979) "The Determination of the Order of an Autoregression", *Journal of the Royal Statistical Society B*, 1, 190–195.
Harvey, A. C. and G. D. A. Phillips (1979) "Maximum Likelihood Estimation of Regression Models with Autoregressive-Moving Average Disturbances", *Biometrika*, 66, 49–58.
Hatanaka, M. and E. P. Howrey (1969) "Low-Frequency Variations in Economic Time Series", *Kyklos*, 22, 752–766.
Havenner, A. and P. A. V. B. Swamy (1981) "A Random Coefficient Approach to Seasonal Adjustment", *Journal of Econometrics*.
Hosking, J. R. M. (1980) "Lagrange-Multiplier Tests of Time-Series Models", *Journal of the Royal Statistical Society, B*, 42, no. 2, 170–181.
Howrey, E. P. (1972) "Dynamic Properties of a Condensed Version of the Wharton Model", *Econometric Models of Cyclical Behavior*, Vol. II, ed. B. Hickman, Columbia University Press, 601–663.
Hymans, S. H. (1973) "On the Use of Leading Indicators to Predict Cyclical Turning Points", *Brookings Papers on Economic Activity*, no. 2.
Jorgenson, D. W. (1964) "Minimum Variance, Linear, Unbiased Seasonal Adjustment of Economic Time Series", *Journal of the American Statistical Association*, 59, 681–687.
Jorgenson, D. W. (1967) "Seasonal Adjustment of Data for Econometric Analysis", *Journal of the American Statistical Association*, 62, 137–140.
Joyeux, R. (1979) "Harmonizable Processes in Economics", Ph.D. Thesis, Department of Economics, University of California, San Diego.
Koopmans, L. H. (1974) *The Spectral Analysis of Time Series*. New York: Academic Press.
Kupier, J. (1979) "A Survey of Comparative Analysis of Various Methods of Seasonal Adjustment", in Zellner (1979).
Labys, W. C. and C. W. J. Granger (1970) *Speculation, Hedging and Forecasts of Commodity Prices*. Lexington: Heath-Lexington Books.
LeRoy, S. F. and R. D. Porter (1981) "The Present-Value Relation: Tests Based on Implied Variance Bounds", *Econometrica*, 49, 555–574.
Lucas, R. E., Jr. (1976) "Econometric Policy Evaluation: A Critique", *The Phillips Curve and the Labor Market* (K. Brunner and A. Meltzer, ed.) Vol. 1 of Carnegie-Rochester Conferences in Public Policy.
Maddala, G. S. (1977) *Econometrics*. New York: McGraw-Hill.
Neftci, S. N. (1979) "Lead-Lag Relationships and Prediction of Economic Time Series", *Econometrica*, 47, 101–114.
Nelson, C. R. (1972) "The Prediction Performance of the FRB-MIT-PENN Model of the U.S. Economy", *American Economic Review*, December.
Nelson, C. R. (1973) *Applied Time-Series Analysis for Managerial Forecasting*. San Francisco: Holden Day.
Nerlove, M. (1964) "Spectral Analysis of Seasonal Adjustment Procedures", *Econometric*, 32, 241–286.
Nerlove, M., D. M. Grether, and J. L. Carvalho (1979) *Analysis of Economic Time Series, A Synthesis*. New York: Academic Press.
Newbold, P. (1974) "The Exact Likelihood Functions for a Mixed Autoregressive-Moving Average Process", *Biometrika* 61, 423–426.
Newbold, P. (1981) "A Note on Modelling Seasonally Adjusted Data", *Journal of Time Series Analysis*.
Newbold, P. and C. Ansley (1979) "Small Sample Behavior of Some Precedures Used in Time Series Model Building and Forecasting", mimeo.
Pagan, A. R. (1975) "A Note on the Extraction of Components from Time Series", *Econometrica*, 43, 163–168.
Parzen, E. and M. Pagano (1978) "An Approach to Modelling Seasonally Stationary Time Series", *Journal of Econometrics*, 9 (*Annals of Applied Econometrics*, 1979-1), 137–154.
Pierce, D. A. (1977) "Relationships and the Lack Thereof Between Economic Time Series, with Special Reference to Money and Interest Rates", *Journal of the American Statistical Society*, 72, 11–21.
Pierce, D. A. (1979) "Seasonal Adjustment When Both Deterministic and Stochastic Seasonality are Present", in Zellner (1979).

Pierce, D. A. (1980) "Recent Developments in Seasonal Adjustment", *Proceedings of the IMS Special Time Series Meeting on Time Series*.
Plosser, C. I. and G. W. Schwert (1977) "Estimation of a Non-Invertible Moving Average Process: The Case of Over Differencing", *Journal of Econometrics*, 5, 199–224.
Priestly, M. B. (1980) "State-Dependent Models: A General Approach to Non-Linear Time Series Analysis", *Journal of Time Series Analysis*, 1, 45–71.
Priestly, M. B. (1981) *Spectral Analysis and Time Series*. New York: Academic Press.
Prothero, D. L. and K. F. Wallis (1976) "Modelling Macroeconomic Time Series", *Journal of the Royal Statistical Society A*, 139, 468–500.
Robbins, R. (1981), Unpublished Ph.D. Dissertation, University of California, San Diego.
Rosenberg, B. (1973) "The Analysis of a Cross-Section of Time Series by Stochastically Convergent Parameter Regression", *Annals of Economic and Social Measurement*, 2, 399–428.
Sargent, T. J. (1981) "Interpreting Economic Time Series", *Journal of Political Economy*, 89, 213–248.
Sargent, T. J. and C. A. Sims (1977) "Business Cycle Modelling Without Pretending to Have Too Much A Priori Economic Theory", *New Methods in Business Cycle Research: Proceedings from a Conference*, Federal Reserve Bank of Minneapolis.
Shiller, R. J. (1981), "Do Stock Prices Move Too Much to be Justified by Subsequent Changes in Dividends?" *American Economic Review*, June.
Shiskin, J., A. H. Young, and J. C. Musgrave (1967) "The X-11 Variant of the Census Method-II Seasonal Adjustment Program", Technical Paper No. 15, U.S. Bureau of the Census.
Singleton, K. J. (1980) "A Latent Time Series Model of the Cyclical Behavior of Interest Rates", *International Economic Review*, 21.
Singleton, K. J. (1980) "Expectations Models of the Term Structure and Implied Variance Bounds", *Journal of Political Economy*, 88.
Sims, C. A. (1974) "Seasonality in Regression", *Journal of the American Statistical Association*, 69, 618–626.
Sims, C. A. (1980) "Macroeconomics and Reality", *Econometric*, 48, 1–48.
Taylor, J. B. (1979) "Estimation and Control of a Macroeconomic Model with Rational Expectations", *Econometrica*, 47, 1267–1286.
Tiao, G. C., G. E. P. Box, G. B. Hudak, W. R. Bell, and I. Chang (1979) "An Introduction to Applied Multiple Time Series Analysis", University of Wisconsin, mimeo.
Wallis, K. F. (1974) "Seasonal Adjustment and Relations Between Variables", *Journal of the American Statistical Association*, 69, 18–31.
Wallis, K. F. (1979) "Seasonal Adjustment and Multiple Time Series", in Zellner (1979).
Watson, M. W. and R. F. Engle (1980) "Testing for Varying Regression Coefficients When a Parameter is Unidentified Under the Null", unpublished manuscript, University of California, San Diego, July, 1980.
Weiner, N. (1949) *The Extrapolation, Interpolation, and Smoothing of Stationary Time Series*. Boston: M.I.T. Press.
Whittle, P. (1963) *Prediction and Regulation*. London: English Universities Press.
Young, A. H. (1968) "Linear Approximations to the Census and BLS Seasonal Adjustment Methods", *Journal of the American Statistical Association*, 63, 445–457.
Zellner, A. (1979) "Seasonal Analysis of Economic Time Series", *U.S. Department of Commerce, Bureau of the Census, Economic Research Report*, ER-1.
Zellner, A. and F. Palm (1976) "Time-Series Analysis and Simultaneous Equation Econometric Models", *Journal of Econometrics*, 2, 17–54.

Chapter 18

DYNAMIC SPECIFICATION

DAVID F. HENDRY
Nuffield College, Oxford

ADRIAN R. PAGAN
Australian National University

J. DENIS SARGAN*
London School of Economics

Contents

*We are grateful to Jean-Francois Richard for his helpful advice, and to James Davidson, Rob Engle, Clive Granger, Andrew Harvey, Svend Hylleberg and Timo Teräsvirta for comments on an earlier draft. Financial support from the Social Science Research Council to the Programme in Methodology, Inference and Modelling at the London School of Economics and from the Centre for Operations Research and Econometrics at the Catholic University of Louvain-la-Neuve is gratefully acknowledged.

Handbook of Econometrics, Volume II, Edited by Z. Griliches and M.D. Intriligator

1. Introduction

Dynamic specification denotes the problem of appropriately matching the lag reactions of a postulated theoretical model to the autocorrelation structure of the associated observed time-series data. As such, the issue is inseparable from that of stochastic specification if the finally chosen model is to have a purely random error process as its basic "innovation", and throughout this chapter, dynamic and stochastic specification will be treated together. In many empirical studies, most other econometric "difficulties" are present jointly with those of dynamic specification but to make progress they will be assumed absent for much of the discussion.

A number of surveys of dynamic models and distributed lags already exist [see, inter alia, Griliches (1967), Wallis (1969), Nerlove (1972), Sims (1974), Maddala (1977), Thomas (1977) and Zellner (1979)], while Dhrymes (1971) treats the probability theory underlying many of the proposed estimators. Nevertheless, the subject-matter has advanced rapidly and offers an opportunity for critically examining the main themes and integrating previously disparate developments. However, we do not consider in detail: (a) Bayesian methods [see Drèze and Richard in Chapter 9 of this Handbook for background and Guthrie (1975), Mouchart and Orsi (1976) and Richard (1977) for recent studies]; (b) frequency domain approaches [see, in particular, Granger and Watson in Chapter 17 of this Handbook, Sims (1974), Espasa (1977) and Engle (1976)]; nor (c) theoretical work on adjustment costs as discussed, for example, by Nerlove (1972). Although theories of intertemporal optimising behaviour by economic agents are continuing to develop, this aspect of the specification problem is not stressed below since, following several of the earlier surveys, we consider that as yet economic theory provides relatively little prior information about lag structures. As a slight caricature, *economic-theory based models* require strong ceteris paribus assumptions (which need not be applicable to the relevant data generation process) and take the form of *inclusion* information such as $y = f(z)$ where z is a vector on which y is claimed to depend. While knowledge that z may be relevant is obviously valuable, it is usually unclear whether z may in *practice* be treated as "exogenous" and whether other variables are *irrelevant* or are simply assumed constant for analytical convenience (yet these distinctions are important for empirical modelling).

By way of contrast, *statistical-theory based models* begin by considering the joint density of the observables and seek to characterise the processes whereby the data were generated. Thus, the focus is on means of *simplifying* the analysis to allow valid inference from sub-models. Throughout the chapter we will maintain this distinction between the (unknown) Data Generation Process, and the econometric model postulated to characterise it, viewing "modelling" as an attempt to match the two. Consequently, both aspects of economic and statistical theory require simultaneous development. All possible observables cannot be considered from the outset, so that economic theory restrictions on the analysis are essential; and while the data are the result of economic behaviour, the actual statistical properties of the observables corresponding to y and z are also obviously relevant to correctly analysing their empirical relationship. In a nutshell, measurement without theory is as valueless as the converse is non-operational.[1] Given the paucity of dynamic theory and the small sample sizes presently available for most time series of interest, as against the manifest complexity of the data processes, all sources of information have to be utilised.

Any attempt to resolve the issue of dynamic specification first involves developing the relevant concepts, models and methods, i.e. the deductive aspect of statistical analysis, prior to formulating inference techniques. In an effort to reduce confusion we have deliberately restricted the analysis to a particular class of stationary models, considered only likelihood based statistical methods and have developed a typology for interpreting and interrelating dynamic equations. Many of our assumptions undoubtedly could be greatly weakened without altering, for example, asymptotic distributions, but the resulting generality does not seem worth the cost in complexity for present purposes. In a number of cases, however, we comment parenthetically on the problems arising when a sub-set of parameters changes. Nevertheless, it is difficult to offer a framework which is at once simple, unambiguous, and encompasses a comprehensive range of phenomena yet allows "economic theory" to play a substantive role without begging questions as to the validity of that "theory", the very testing of which may be a primary objective of the analysis.

Prior to the formal analysis it seems useful to illustrate by means of a relatively simple example why dynamic specification raises such difficult practical problems. Consider a consumption–income $(C-Y)$ relationship for quarterly data given by:

$$\Delta_4 \ln C_t = \delta_0 + \delta_1 \Delta_4 \ln Y_t^{\mathrm{n}} + \delta_2 \Delta_4 \ln C_{t-1} + \delta_3 \ln(C/Y^{\mathrm{n}})_{t-4} + \varepsilon_t, \tag{1}$$

where $\Delta_4 x_t = (x_t - x_{t-4})$, ln is logarithm to the base e, ε_t is *assumed* to be white noise and Y_t^{n} is "normal" income, such that:

$$\ln Y_t^{\mathrm{n}} = 0.1 \sum_{i=0}^{3} (4-i) \ln Y_{t-i}. \tag{2}$$

[1]This is a very old point, but bears repetition: "all induction is blind, so long as the deduction of causal connections is left out of account; and all deduction is barren so long as it does not start from observation" [taken from J. N. Keynes (1890, p. 164)]. Also, it has long been seen as essential to treat economic theory as a "working 'first approximation to reality' in statistical investigations", e.g. see Persons (1925).

The unrestricted distributed lag relationship between $\ln C_t$ and $\ln Y_t$ has the form:

$$\ln C_t = \delta_0 + \sum_{i=1}^{m} (\alpha_i \ln C_{t-i-1} + \beta_i \ln Y_{t-i}) + \varepsilon_t. \tag{3}$$

When $\boldsymbol{\delta}' = (0, 0.5, 0.25, -0.2)$ (but this is unknown) (3) has coefficients:

j	0	1	2	3	4	5	6	7	
$\ln C_{t-j}$	-1	0.25	0	0	0.80	-0.25	0	0	(4)
$\ln Y_{t-j}$	0.2	0.15	0.10	0.05	-0.12	-0.09	-0.06	-0.03	

Under appropriate conditions on Y_t, estimation of the unknown value of $\boldsymbol{\delta}$ (or of $\delta_0, \boldsymbol{\alpha}, \boldsymbol{\beta}$) is straightforward, so this aspect will not be emphasised below. However, the formulation in (1)–(4) hides many difficulties experienced in practice and the various sections of this chapter tackle these as follows.

Firstly, (1) is a single relationship between two series (C_t, Y_t), and is, at best, only a part of the data generation process (denoted DGP). Furthermore, the validity of the representation depends on the properties of Y_t. Thus, Section 2.1 investigates conditional sub-models, their derivation from the DGP, the formulation of the DGP itself, and the resulting behaviour of $\{\varepsilon_t\}$ (whose properties cannot be arbitrarily chosen at convenience, since by construction, ε_t contains everything not otherwise explicitly in the equation). To establish notation and approach, estimation, inference and diagnostic testing are briefly discussed in Section 2.2, followed in Section 2.3 by a more detailed analysis of the interpretation of equations like (1). However, dynamic models have many representations which are equivalent when no tight specification of the properties of $\{\varepsilon_t\}$ is available (Section 2.4) and this compounds the difficulty of selecting equations from data when important features [such as m in (3), say] are not known a priori. Nevertheless, the class of models needing consideration sometimes can be delimited on the basis of theoretical arguments and Section 2.5 discusses this aspect. For example, (1) describes a relatively simple situation in which agents make annual decisions, marginally adjusting expenditure as a short distributed lag of changes in "normal" income and a "disequilibrium" feedback to ensure a constant static equilibrium ratio of C to Y (or Y^n). This model constrains the values in (3) to satisfy $1 - \sum \alpha_i = \sum \beta_i$ (inter alia) although appropriate converse reformulations of (3) as in (1) are rarely provided by economic theory alone.

Since (3) has a complicated pattern of lagged responses [with eleven non-zero coefficients in (4)] unrestricted estimation is inefficient and may yield very imprecise estimates of the underlying coefficients (especially if m is also estimated from the data). Consequently, the properties of restricted dynamic models representing economic data series are important in guiding parsimonious yet useful characterisations of the DGP and Section 2.6 offers a typology of many commonly used choices. For example, (1) is an "error correction" model (see also Section 4.2) and, as shown in (4), negative effects of lagged Y on C may be correctly signed if interpreted as arising from "differences" in (1). Note, also, that long lags in (3) (e.g. $m = 7$) need not entail slow reactions in (1), [e.g. from (4) the median lag of Y^{n} on C_t is one-quarter]. The typology attempts to bring coherence to a disparate and voluminous literature.

This is also used as a framework for structuring the more detailed analyses of finite distributed lag models in Section 3 and other dynamic formulations in Section 4 (which include partial adjustment models, rational distributed lags and error correction mechanisms). Moreover, the typology encompasses an important class of error autocorrelation processes (due to common factors in the lag polynomials), clarifying the dynamic-stochastic link and leading naturally to an investigation of stochastic specification in Section 5.

While the bulk of the chapter relates to one equation sub-models to clarify the issues involved, the results are viewed in the context of the general DGP and so form an integral component of system dynamic specification. However, multi-dimensionality also introduces new issues and these are considered in Section 6, together with the generalised concepts and models pertinent to systems or sub-models thereof.

Since the chapter is already long, we do not focus explicitly on the role of expectations in determining dynamic reactions. Thus, on one interpretation, our analysis applies to derived equations which, if expectations are important, confound the various sources of lags [see Sargent (1981)]. An alternative interpretation is that by emphasising the econometric aspects of time-series modelling, the analysis applies howsoever the model is obtained and seeks to be relatively neutral as to the economic theory content [see, for example, Hendry and Richard (1982)].

2. Data generation processes

2.1. *Conditional models*

Let $\boldsymbol{x}_t$ denote a vector of n observable random variables, $\boldsymbol{X}_0$ the matrix of initial conditions, where $\boldsymbol{X}_t^1 = (\boldsymbol{x}_1 \dots \boldsymbol{x}_t)'$ and $\boldsymbol{X}_t = (\boldsymbol{X}_0' \boldsymbol{X}_t^{1\prime})'$. For a sample of size T, let $D(\boldsymbol{X}_T^1 | \boldsymbol{X}_0, \boldsymbol{\theta})$ be the joint data density function where $\boldsymbol{\theta} \in \Theta$ is an identifiable vector of unknown parameters in the interior of a finite dimensional parameter space Θ. Throughout, the analysis is conducted conditionally on $\boldsymbol{\theta}$ and $\boldsymbol{X}_0$, and the likelihood function is denoted by $\mathscr{L}(\boldsymbol{\theta}; \boldsymbol{X}_T^1)$. The joint data density is sequentially factorised into:

$$D(X_T^1 | X_0, \theta) = \prod_{t=1}^{T} D(x_t | X_{t-1}, \theta). \tag{5}$$

It is assumed that the conditional density functions in (5) have the common functional form:

$$x_t | X_{t-1} \sim \mathcal{N}(\mu_t, \Omega), \quad \text{with } \mu_t = E(x_t | X_{t-1}, \theta), \tag{6}$$

where $x_t - \mu_t = v_t$ is an "innovation", and by construction:

$$E\left(v_t x'_{t-j}\right) = \mathbf{0}, \quad \forall_j \geq 1, \quad \text{so that } E\left(v_t v'_{t-j}\right) = \mathbf{0}, \quad \forall_j \geq 1.$$

Implicitly, we are ignoring important issues of aggregation (over agents, space, time, goods, etc.) and marginalisation (with respect to all other variables than those in x_t) by assuming that (5) is an adequate statistical representation for a DGP. Hopefully, this conflation of the concepts of DGP and Model, due to deliberate exclusion of other difficulties, will not prove confusing. Concerning the economic behaviour determining x_t, we suppose economic agents to form *contingent plans* based on limited information [see Bentzel and Hansen (1955) and Richard (1980)]. Such plans define *behavioural relationships* which could correspond to optimising behaviour given expectations about likely future events, allow for adaptive responses and/or include mechanisms for correcting previous mistakes. To express these in terms of x_t will require marginalising with respect to all unobservables. Thus, assuming linearity (after suitable data transformations) and a fixed finite lag length (m) yields the model:

$$\mu_t = \sum_{i=1}^{m} \pi_i x_{t-i}. \tag{7}$$

In (7) the value of m is usually unknown but in practice must be small relative to T. The corresponding "structural" representation is given by:

$$Bx_t + \sum_{i=1}^{m} C_i x_{t-i} = \varepsilon_t, \tag{8}$$

with $\varepsilon_t = Bv_t$ and $B\pi_i + C_i = \mathbf{0}$, where B and $\{C_i\}$ are well defined functions of θ and B is of rank n $\forall \theta \in \Theta$ [strictly, the model need not be complete, in that (6) need only comprise $g \leq n$ equations to be well defined: see Richard (1979)].

From (5)–(8), $\varepsilon_t \sim I\mathcal{N}(\mathbf{0}, \Sigma)$ where $\Sigma = B\Omega B'$, but as will be seen below, this class of processes does not thereby exclude autocorrelated error representations.

Also, while not considered below, the model could be generalised to include, for example, Autoregressive-Conditional Heteroscedasticity [Engle (1982)].

Direct estimation of $\{\pi_i\}$ is generally infeasible [see, however, Section 6.3 and Sargent and Sims (1977)] and in any case still involves important assumptions concerning parameter constancy, the choices of n and m and the constituent components of $\boldsymbol{x}_t$. Generally, econometricians have been more interested in conditional sub-models suggested by economic theory and hence we partition $\boldsymbol{x}_t'$ into $(y_t'z_t')$ and factorise the data densities $D(\boldsymbol{x}_t|\boldsymbol{X}_{t-1},\boldsymbol{\theta})$ and likelihood function correspondingly as:

$$D(\boldsymbol{x}_t|\boldsymbol{X}_{t-1},\boldsymbol{\theta}) = D_1(y_t|z_t,\boldsymbol{X}_{t-1},\boldsymbol{\phi}_1)D_2(z_t|\boldsymbol{X}_{t-1},\boldsymbol{\phi}_2),$$

where $(\boldsymbol{\phi}_1,\boldsymbol{\phi}_2)$ is an appropriate reparameterisation of $\boldsymbol{\theta}$, and:

$$\mathscr{L}(\boldsymbol{\theta};\boldsymbol{X}_T^1|\boldsymbol{X}_0) = \prod_{t=1}^{T}\mathscr{L}_1(\boldsymbol{\phi}_1;y_t|z_t,\boldsymbol{X}_{t-1})\prod_{t=1}^{T}\mathscr{L}_2(\boldsymbol{\phi}_2;z_t|\boldsymbol{X}_{t-1}). \tag{9}$$

Certain parameters, denoted ψ, will be of interest in any given application either because of their "invariance" to particular interventions or their relevance to policy, or testing hypotheses suggested by the associated theory etc. If ψ is a function of $\boldsymbol{\phi}_1$ alone, and $\boldsymbol{\phi}_1$ and $\boldsymbol{\phi}_2$ are variation free, then z_t is *weakly exogenous* for ψ and fully efficient inference is possible from the partial likelihood $\mathscr{L}_1(\cdot)$ [see Koopmans (1950), Richard (1980), Florens and Mouchart (1980), Engle et al. (1983) and Geweke in Chapter 19 of this Handbook]. Thus, the model for z_t does not have to be specified, making the analysis more robust, more comprehensible, and less costly, *hence facilitating model selection* when the precise specification of (8) is not given a priori. Indeed, the practice whereby $\mathscr{L}_1(\cdot)$ is specified in most econometric analyses generally involves many implicit weak exogeneity assertions and often proceeds by specifying the conditional model *alone* leaving $\mathscr{L}_2(\cdot)$ to be whatever is required to "complete" $\mathscr{L}(\cdot)$ in (9). That ψ can be estimated efficiently from analysing only the conditional sub-model, does *not* entail that z_t is *predetermined* in:

$$\boldsymbol{B}_{11}y_t + \boldsymbol{B}_{12}z_t + \sum \boldsymbol{C}_{1i}\boldsymbol{x}_{t-i} = \boldsymbol{\varepsilon}_{1t} \tag{10}$$

(using an obvious notation for the partition of $\boldsymbol{B}$ and $\{\boldsymbol{C}_i\}$), merely that the model for z_t does not require joint estimation with (10).

If in addition to being weakly exogenous for ψ, the following holds for z_t:

$$D_2(z_t|\boldsymbol{X}_{t-1},\boldsymbol{\phi}_2) \equiv D_2(z_t|\boldsymbol{Z}_{t-1},\boldsymbol{Y}_0,\boldsymbol{\phi}_2) \qquad (t=1,\dots,T), \tag{11}$$

so that lagged y's are uninformative about z_t given $\boldsymbol{Z}_{t-1}$, and hence $\boldsymbol{y}$ does not

Granger cause z [see Granger (1969), Sims (1977) and Geweke in Chapter 19 of this Handbook], then z_t is said to be *strongly exogenous* for ψ. Note that the initial choice of $\boldsymbol{x}_t$ in effect required an assertion of strong exogeneity of $\boldsymbol{x}_t$ for the parameters of other potentially relevant (economic) variables. Also, as shown in subsection 2.6, paragraph (g), if (11) does not hold, so that y does Granger cause z, then care is required in analysing model formulations which have autocorrelated errors since z will also Granger cause such errors.

The remainder of this chapter focusses on dynamic specification in models like (10) since these encompass many of the equation forms and systems (with a "linearity in variables" caveat) occurring in empirical research. For example, the system:

$$\boldsymbol{B}^*\boldsymbol{x}_t + \sum_{i=1}^{m^*} \boldsymbol{C}_i^*\boldsymbol{x}_{t-i} = \boldsymbol{u}_t, \quad \text{where } \boldsymbol{u}_t = \sum_{i=1}^{r^*} \boldsymbol{R}_i^*\boldsymbol{u}_{t-i} + \boldsymbol{\varepsilon}_t, \tag{8*}$$

with $m^* + r^* = m$, can be re-expressed as (8) with non-linear relationships between the parameters. However, unique factorisation of the $\{\boldsymbol{\pi}_i\}$ into $(\boldsymbol{B}_1^*\{\boldsymbol{C}_i^*\}\{\boldsymbol{R}_i^*\})$ requires further restrictions on $\{\boldsymbol{R}_i^*\}$ such as block diagonality and/or *strong* exogeneity information [see Sargan (1961) and Sections 5 and 6.1].

2.2. *Estimation, inference and diagnostic testing*

Since *specific* techniques of estimation, inference and diagnostic testing will not be emphasised below [for a discussion of many estimation methods, see Dhrymes (1971), Zellner (1979) and Hendry and Richard (1983)] a brief overview seems useful notwithstanding the general discussions provided in other chapters. At a slight risk of confusion with the lag operator notation introduced below, we denote $\log_e$ of the relevant partial likelihood from (9) by:[2]

$$L(\psi) = \sum_{t=1}^{T} L(\psi; y_t | z_t, \boldsymbol{X}_{t-1}). \tag{12}$$

In (12), ψ is considered as an argument of $L(\cdot)$, when z_t is weakly exogenous and (8) is the data generation process. Let:

$$\boldsymbol{q}(\psi) = \frac{\partial L(\cdot)}{\partial \psi} \quad \text{and} \quad \boldsymbol{Q}(\psi) = \frac{\partial \boldsymbol{q}(\cdot)}{\partial \psi'}. \tag{13}$$

[2]Strictly, (12) relates to $\boldsymbol{\phi}_1$ but ψ is used for notational simplicity; $L(\cdot)$ can be considered as the reparameterised concentrated likelihood if desired.

The general high dimensionality of ψ forces summarisation in terms of maximum likelihood estimators (denoted MLEs), or appropriate approximations thereto, and under suitable regularity conditions [most of which are satisfied here granted (6)]—see, for example, Crowder (1976)—MLEs will be "well behaved". In particular if the roots of

$$\left| I - \sum_{i=1}^{m} \pi_i g^i \right| = 0 \tag{14}$$

(a polynomial in g of order no greater than nm) are all outside the unit circle, then when $\hat{\psi}$ is the MLE of ψ:

$$\sqrt{T}(\hat{\psi} - \psi) \underset{a}{\sim} \mathcal{N}(O, V_\psi), \quad \text{where } V_\psi = -\operatorname{plim} T \cdot Q(\hat{\psi})^{-1}, \tag{15}$$

and is positive definite. Note that $\hat{\psi}$ is given by $q(\hat{\psi}) = 0$ [with $Q(\hat{\psi})$ negative definite] and numerical techniques for computing $\hat{\psi}$ are discussed in Dent (1980) and in Quandt in Chapter 12 of this Handbook. Phillips (1980) reviews much of the literature on exact and approximate finite sample distributions of relevant estimators. If (8) is not the DGP, a more complicated expression for V_ψ is required although asymptotic normality still generally results [see, for example, Domowitz and White (1982)].

Note that $q(\psi) = 0$ can be used as an estimator generating equation for most of the models in the class defined by (10) when not all elements of ψ are of equal interest [see Hausman (1975) and Hendry (1976)].

To test hypotheses of the general form H_0: $F(\psi) = 0$, where $F(\cdot)$ has continuous first derivatives at ψ and imposes r restrictions on $\psi = (\psi_1 \dots \psi_k)'$, three principles can be used [see Engle in Chapter 13 of this Handbook] namely: (a) a Wald-test, denoted W [see Wald (1943)]; (b) the Maximised Likelihood Ratio, LR [see, for example, Cox and Hinkley (1974, ch. 9)]; and (c) Lagrange Multiplier, LM [see Aitchison and Silvey (1960), Breusch and Pagan (1980) and Engle (1982)]. Since (a) and (c) are respectively computable under the maintained and null hypotheses alone, they are relatively more useful as their associated parameter sets are more easily estimated. Also, whereas (b) requires estimation of both restricted and unrestricted models, this is anyway often necessary given the outcome of either W or LM tests. Because of their relationship to the unrestricted and restricted versions of a model, W and LM tests frequently relate respectively to tests of specification and mis-specification [see Mizon (1977b)], that is, within and outside initial working hypotheses. Thus, [see Sargan (1980c)] Wald forms apply to common factor tests, whereas LM forms are useful as diagnostic checks for residual autocorrelation. Nevertheless, *both* require specification of the "maintained" model.

Formally, when (8) is the DGP, $E\boldsymbol{q}(\boldsymbol{\psi})=\mathbf{0}$ and $E\boldsymbol{Q}(\boldsymbol{\psi})=-\boldsymbol{I}(\boldsymbol{\psi})$, with $T^{-1/2}\boldsymbol{q}(\boldsymbol{\psi})\underset{a}{\sim}\mathcal{N}(\mathbf{0},\overline{\mathscr{I}}(\boldsymbol{\psi}))$, where $\overline{\mathscr{I}}(\cdot)=\operatorname{plim} T^{-1}\mathscr{I}(\cdot)=\boldsymbol{V}_\psi^{-1}$. Then we have:

(a) From (15), on H_0: $\boldsymbol{F}(\boldsymbol{\psi})=\mathbf{0}$:

$$\sqrt{T}\boldsymbol{F}(\hat{\boldsymbol{\psi}})\underset{a}{\sim}\mathcal{N}\left(0,\boldsymbol{J}'\boldsymbol{V}_\psi\boldsymbol{J}\right)=\mathcal{N}(\mathbf{0},\boldsymbol{V}_{\boldsymbol{F}}),\tag{16}$$

where $\boldsymbol{J}=\partial\boldsymbol{F}(\cdot)/\partial\boldsymbol{\psi}$. Let $\hat{\boldsymbol{J}}$ and $\hat{\boldsymbol{V}}_{\boldsymbol{F}}$ denote evaluation at $\hat{\boldsymbol{\psi}}$, then on H_0:

$$\mathrm{W}_F=T\boldsymbol{F}(\hat{\boldsymbol{\psi}})'\hat{\boldsymbol{V}}_{\boldsymbol{F}}^{-1}\boldsymbol{F}(\hat{\boldsymbol{\psi}})\underset{a}{\sim}\chi^2_r\qquad\left(\hat{\boldsymbol{V}}_{\boldsymbol{F}}=\hat{\boldsymbol{J}}'\overline{\mathscr{I}}(\hat{\boldsymbol{\psi}})^{-1}\hat{\boldsymbol{J}}\right).\tag{17}$$

Furthermore if W_a and W_b are two such Wald criteria based upon two sets of constraints such that those for W_a are obtained by adding constraints to those characterising W_b, then:

$$(\mathrm{W}_a-\mathrm{W}_b)\underset{a}{\sim}\chi^2_{r_a-r_b},\quad\text{independently of }\mathrm{W}_b\underset{a}{\sim}\chi^2_{r_b}.\tag{18}$$

Such an approach adapts well to commencing from a fairly unconstrained model and testing a sequence of nested restrictions of the form $\boldsymbol{F}_i(\boldsymbol{\psi})=\mathbf{0}$, $i=1,2,\dots$, where $r_i>r_{i-1}$ and rejecting $\boldsymbol{F}_j(\cdot)$ entails rejecting $\boldsymbol{F}_l(\cdot)$, $l>j$. This occurs, for example, in a "contracting search" (see Leamer in Chapter 5 of this Handbook), and hence W is useful in testing dynamic specification [see Anderson (1971, p. 42), Sargan (1980c), Mizon (1977a) and Section 5].

(b) Let $\tilde{\boldsymbol{\psi}}$ denote the MLE of $\boldsymbol{\psi}$ subject to $\boldsymbol{F}(\boldsymbol{\psi})=\mathbf{0}$, then:

$$\mathrm{LR}_F=2[L(\hat{\boldsymbol{\psi}})-L(\tilde{\boldsymbol{\psi}})]\underset{a}{\sim}\chi^2_r,\quad\text{if } H_0\text{ is true.}\tag{19}$$

(c) Since $\tilde{\boldsymbol{\psi}}$ is obtained from the Lagrangian expression:

$$L(\boldsymbol{\psi})+\boldsymbol{\lambda}'\boldsymbol{F}(\boldsymbol{\psi}),\quad\text{using } \boldsymbol{q}(\boldsymbol{\psi})+\boldsymbol{J}\boldsymbol{\lambda}=\mathbf{0},\tag{20}$$

then, when H_0 is true:

$$\mathrm{LM}_F=T\boldsymbol{q}(\tilde{\boldsymbol{\psi}})'\overline{\mathscr{I}}(\tilde{\boldsymbol{\psi}})^{-1}\boldsymbol{q}(\tilde{\boldsymbol{\psi}})=T\tilde{\boldsymbol{\lambda}}'\tilde{\boldsymbol{J}}'\overline{\mathscr{I}}(\tilde{\boldsymbol{\psi}})^{-1}\tilde{\boldsymbol{J}}\tilde{\boldsymbol{\lambda}}\underset{a}{\sim}\chi^2_r,\tag{21}$$

and hence the test is also known as the "efficient score" test [see Rao (1965)]. Note that $\boldsymbol{q}(\hat{\boldsymbol{\psi}})\equiv\mathbf{0}$, whereas $\boldsymbol{F}(\tilde{\boldsymbol{\psi}})\equiv\mathbf{0}$, the converses not holding. Also (17), (19) and (21) show the three tests to be asymptotically equivalent both under H_0 and under the sequence of local alternatives H_T: $\boldsymbol{F}(\boldsymbol{\psi})=T^{-1/2}\boldsymbol{\delta}$ (for constant $\boldsymbol{\delta}$). All three tests are non-central χ^2_r with non-centrality parameter $\boldsymbol{\delta}'\boldsymbol{V}_{\boldsymbol{F}}^{-1}\boldsymbol{\delta}$ and are, therefore, consistent against any *fixed* alternative (i.e. $T^{-1/2}\boldsymbol{\delta}$ constant).[3] As yet, little is known about their various finite sample properties [but see Berndt and Savin (1977), Mizon and Hendry (1980) and Evans and Savin (1982)].

It must be stressed that rejecting H_0 by any of the tests provides evidence only against the validity of the restrictions and does not necessarily "support" the alternative against which the test might originally have been derived. Also, careful consideration of significance levels is required when sequences of tests are used. Finally, generalisations of some of the test forms are feasible to allow for (8) not being the DGP [see Domowitz and White (1982)].

2.3. *Interpreting conditional models*

For simplicity of exposition and to highlight some well-known but important issues we consider a single equation variant of (10) with only one lag namely:

$$y_t = \beta_1 z_t + \boldsymbol{\beta}_2' \boldsymbol{x}_{t-1} + e_t. \tag{22}$$

There are (at least) four distinct interpretations of (22) as follows [see for example, Richard (1980) and Wold (1959)].

(a) Equation (22) is a *regression equation* with parameters *defined by*:

$$E(y_t | z_t, \boldsymbol{x}_{t-1}) = \beta_1 z_t + \boldsymbol{\beta}_2' \boldsymbol{x}_{t-1}, \tag{23}$$

where $e_t = y_t - E(y_t|\cdot)$ so that $E(z_t e_t) = 0$, and $E(\boldsymbol{x}_{t-1} e_t) = \mathbf{0}$. When (23) holds, $\boldsymbol{\beta} = (\beta_1 \boldsymbol{\beta}_2')'$ minimises the variance of e.

Whether $\boldsymbol{\beta}$ is or is not of interest depends on its relationship to ψ and the properties of z_t (e.g. $\boldsymbol{\beta}$ is clearly of interest if ψ is a function of $\boldsymbol{\beta}$ and z_t is weakly exogenous for $\boldsymbol{\beta}$).

(b) Equation (22) is a *linear least-squares approximation* to some dynamic relationship linking y and z, chosen on the *criterion* that e_t is purely random and uncorrelated with $(z_t, \boldsymbol{x}_{t-1})$. The usefulness of such approximations depends partly on the objectives of the study (e.g. short-term forecasting) and partly on the properties of the actual data generation process (e.g. the degree of non-linearity in $y = f(z)$, and the extent of joint dependence of y_t and z_t): see White (1980).

(c) Equation (22) is a *structural relationship* [see, for example, Marschak (1953)] in that $\boldsymbol{\beta}$ is a constant with respect to changes in the data process of z_t (at least for the relevant sample period) and the equation is *basic* in the sense of Bentzel and Hansen (1955). Then (22) directly characterises how agents form plans in

[3] For boundary points of θ, the situation is more complicated and seems to favour the use of the LM principle—see Engle in Chapter 13 of this Handbook. Godfrey and Wickens (1982) discuss locally equivalent models.

terms of observables and consequently $\boldsymbol{\beta}$ is of interest. In economics such equations would be conceived as deriving from *autonomous* behavioural relations with structurally-invariant parameters [see Frisch (1938), Haavelmo (1944), Hurwicz (1962) and Sims (1977)]. The last interpretation is:

(d) Equation (22) is derived from the *behavioural relationship*:

$$E(y_t|\boldsymbol{X}_{t-1}) = \gamma_1 E(z_t|\boldsymbol{X}_{t-1}) + \boldsymbol{\gamma}_2'\boldsymbol{x}_{t-1}. \tag{24}$$

If

$$\varepsilon_{2t} = z_t - E(z_t|\boldsymbol{X}_{t-1}), \tag{25}$$

then e_t is the composite: $e_t = (\varepsilon_{1t} - \gamma_1\varepsilon_{2t})$ so that $E(e_t\varepsilon_{2t}) \neq 0$ in general and depends on γ_1.

More generally, if $E(z_t|\boldsymbol{X}_{t-1})$ is a non-constant function of $\boldsymbol{X}_{t-1}$, $\boldsymbol{\beta}$ need not be structurally invariant, and if incorrect weak exogeneity assumptions are made about z_t, then *estimates* of $\boldsymbol{\gamma}$ need not be constant when the data process of z_t alters.

That the four "interpretations" are distinct is easily seen by considering a data density with a non-linear regression function $[(a) \neq (b)]$ which does not coincide with a non-linear behavioural plan $[(a) \neq (d), (b) \neq (d)]$ in which the presence of $E(z_t|\boldsymbol{X}_{t-1})$ inextricably combines $\boldsymbol{\phi}_1$ and $\boldsymbol{\phi}_2$, thereby losing structurality for all changes in $\boldsymbol{\phi}_2$ [i.e. (c) does not occur]. Nevertheless, in stationary linear models with normally distributed errors, the four cases "look alike".

Of course, structural invariance is only interesting in a non-constant world and entails that in practice, the four cases will behave differently if $\boldsymbol{\phi}_2$ changes. Moreover, even if there exists some structural relationship linking y and z, failing to specify the model thereof in such a way that its coefficients and $\boldsymbol{\phi}_2$ are variation free can induce a loss of structurality in the estimated equation to interventions affecting $\boldsymbol{\phi}_2$. This point is important in dynamic specification as demonstrated in the following sub-section.

2.4. *The status of an equation*

Any given dynamic model can be written in a large number of equivalent forms *when no tight specification is provided for the error term*. The following example illustrates the issues involved:

Suppose there existed a well-articulated, dynamic but non-stochastic economic theory (of a supply/demand form) embodied in the model:

$$Q_t = \alpha_1 Q_{t-1} + \alpha_2 I_t + \alpha_3 P_t + v_{1t}, \tag{26}$$

$$P_t = \alpha_4 P_{t-1} + \alpha_5 C_t + \alpha_6 Q_{t-1} + v_{2t}, \tag{27}$$

where Q_t, P_t, I_t and C_t are quantity, price, income and cost, respectively, but the properties of v_{it} are not easily prespecified given the lack of a method for relating *decision* time periods to observation intervals (see Bergstrom in Chapter 20 of this Handbook for a discussion of continuous time estimation and discrete approximations). It is assumed below that (C_t, I_t) is weakly, but not strongly, exogenous for $\{\alpha_i\}$, and that (26) and (27) do in fact correspond "reasonably" to basic structural behavioural relationships, in the sense just discussed.

Firstly, consider (26); eliminating lagged Q's yields an alternative dynamic relation linking Q to I and P in a distributed lag:

$$Q_t = \sum_{i=0}^{\infty} (a_{2i}I_{t-i} + a_{3i}P_{t-i}) + u_{1t}, \tag{28}$$

where $a_{ji} = \alpha_1^i \alpha_j$ $(j = 2,3)$. Alternatively, eliminating P_t from (26) using (27) yields the reduced form:

$$Q_t = \pi_1 Q_{t-1} + \pi_2 I_t + \pi_3 C_t + \pi_4 P_{t-1} + e_{1t}, \tag{29}$$

which in turn has a distributed lag representation like (28), but including $\{C_{t-j}|j \geq 0\}$ and excluding P_t. Further, (27) can be used to eliminate all values of P_{t-j} from equations determining Q_t to yield:

$$Q_t = \beta_1 Q_{t-1} + \beta_2 Q_{t-2} + \beta_3 I_t + \beta_4 I_{t-1} + \beta_5 C_t + w_{1t}, \tag{30}$$

transformable to the distributed lag:

$$Q_t = \sum_{i=0}^{\infty} (b_{3i}I_{t-i} + b_{4i}C_{t-i}) + \eta_{1t} \tag{31}$$

(where the expressions for b_{ji} as functions of α_k are complicated), which is similar to (28) but with $\{C_{t-i}\}$ in place of $\{P_{t-i}\}$.

Manifestly, the error processes of the various transformations usually will have quite different autocorrelation properties and we have:

$$\begin{aligned} u_{1t} &= \alpha_1 u_{1t-1} + v_{1t}, \\ e_{1t} &= v_{1t} + \alpha_3 v_{2t}, \\ w_{1t} &= e_{1t} - \alpha_4 v_{1t-1}, \\ \eta_{1t} &= \beta_1 \eta_{1t-1} + \beta_2 \eta_{1t-2} + w_{1t}. \end{aligned}$$

Almost all of these errors are likely to be autocorrelated, with correlograms that may not be easy to characterise simply and adequately, emphasising the link of dynamic to stochastic specification.

In the illustration, all of the "distributed lag" representations are *solved* versions of (26)+(27) and if estimated unrestrictedly (but after truncating the lag length!) would produce very inefficient estimates (and hence inefficient forecasts etc.). Consequently, before estimating any postulated formulation, it seems important to have some cogent justifications for it, albeit informal ones in the present state of the art: simply asserting a given equation and "treating symptoms of residual autocorrelation" need not produce a useful model.

Indeed, the situation in practice is far worse than that sketched above because of two additional factors: mis-specification and approximation. By the former, is meant the possibility (certainty?) that important influences on y_t have been excluded in defining the model and that such variables are not independent of the included variables. By the latter, is meant the converse of the analysis from (26)+(27) to (31) namely that *theory* postulates a general lag relationship between Q_t and its determinants I_t, C_t as in (31) (say) and to reduce the number of parameters in b_{3i} and b_{4i} various restrictions are imposed. Of course, a similar analysis applies to all forms derived from (27) with P_t as the regressand. Moreover, "combinations" of any of the derived equations might be postulated by an investigator. For an early discussion, see Haavelmo (1944).

For example, consider the case where C_t is omitted from the analysis of (26)+(27) when a "good" time-series *description* of C_t is given by:

$$d_1(L)C_t = d_2(L)Q_t + d_3(L)I_t + d_4(L)P_t + \zeta_t, \tag{32}$$

where $d_i(L)$ are polynomials in the lag operator L, $L^k x_t = x_{t-k}$, and ζ_t is "white noise", independent of Q, P and I. Eliminating C_t from the analysis now generates a different succession of lag relationships corresponding to (28)–(31). In turn, each of these can be "adequately" approximated by other lag models, especially if full allowance is made for residual autocorrelation. Nevertheless, should the stochastic properties of the data generation process of any "exogenous" variable change [such as C_t in (32)], equations based on eliminating that variable will manifest a "structural change" even if the initial structural model (26)+(27) is unaltered. For this reason, the issue of the validity of alternative approximations to lag forms assumes a central role in modelling dynamic processes. A variety of possible approximations are discussed in Section 3, and in an attempt to provide a framework, Section 2.6 outlines a typology of single equation dynamic models. First, we note a few quasi-theoretical interpretations for distributed lag models.

2.5. *Quasi-theoretical bases for dynamic models*

Firstly, equations with lagged dependent variables arise naturally in situations where there are types of *adjustment costs* like transactions costs, search costs, optimisation costs, etc. and/or where agents react only slowly to changes in their environment due to habit, inertia or lags in perceiving changes and so on. Thus economic agents may attach monetary or utility costs to instantaneous alteration of instruments to fully achieve plans. Even when there are no adjustment costs, slow reactions are likely because of the uncertainty engendered by the future and the lack of perfect capital and futures markets. Although formal modelling of such costs is still badly developed—Nerlove (1972) and Sims (1974) provide references and discussion—it appears that what optimal rules there are prove to be extraordinarily complex and, given the fact that only aggregates are observed, such theory would seem to be only a weak source of prior information. In fact it is not impossible that distributed lags between aggregate variables reflect the distribution of agents through the population. For example, if agents react with fixed time delays but the distribution of the length of time delays across agents is geometric, the aggregate lag distribution observed would be of the Koyck form. In the same way that Houthakker (1956) derived an aggregate Cobb–Douglas production function from individual units with fixed capital/labour ratios, some insight might be obtained for the format of aggregate distributed lags from similar exercises [see, for example, Trivedi (1982)].

However, it seems likely that many agents use simple adaptive decision rules rather than optimal ones although, as Day (1967) and Ginsburgh and Waelbroeck (1977) have shown, these have the capability of solving quite complex optimization problems. A further example of the potential role of these adaptive "rules of thumb" arises from the monetarists' contention that disequilibria in money balances provide *signals* to agents that their expenditure plans are out of equilibrium [e.g. Jonson (1977)] and that simple rules based on these signals may be adopted as the costs are low and information value high. Stock-flow links also tend to generate models with lagged dependent variables.

In any case, state-variable feedback solutions of optimization problems often have alternative representations in terms of servo-mechanisms of a form familiar to control engineers, and it has been argued that simple control rules of the type discussed by Phillips (1954, 1957) may be more robust to mis-specification of the objective function and/or the underlying economic process [see Salmon and Young (1979) and Salmon (1979)]. For quadratic cost functions, linear decision rules result and can be expressed in terms of proportional, derivative and integral control mechanisms. This approach can be used for deriving dynamic econometric equations [see, for example, Hendry and Anderson (1977)], an issue discussed more extensively below. Since such adaptive rules seem likely solutions of many decision problems [see, for example, Marschak (1953)] lagged dependent variables will commonly occur in economic relationships. Thus, one should *not* automatically interpret (say) "rational lag" models such as (26) as *approximations* to "distributed lag" models like (28); often the latter will be the solved form, and it makes a great deal of difference to the structurality of the relationship and the properties of the error term whether an equation is a solved variant or a direct representation.

Next, finite distributed lags also arise naturally in some situations such as order–delivery relationships, or from aggregation over agents, etc. and often some knowledge is available about properties of the lag coefficients (such as their sum being unity or about the "smoothness" of the distribution graph). An important distinction in this context is between imposing restrictions on the *model* such that (say) only steady-state behaviour is constrained, and imposing restrictions on the *data* (i.e. constraints binding at *all* points in time). This issue is discussed at greater length in Davidson et al. (1978), and noted again in Section 2.6, paragraph (h).

Thirdly, unobservable expectations about future outcomes are frequently modelled as depending on past information about variables included in the model, whose current values influence y_t. Eliminating such expectations also generates more or less complicated distributed lags which can be approximated in various ways although as noted in Section 2.3, paragraph (d), changes in the processes generating the expectations can involve a loss of structurality [see, for example, Lucas (1976)]. Indeed, this problem occurs on omitting observables also, and although the conventional interpretation is that estimates suffer from "omitted variables bias" we prefer to consider omissions in terms of eliminating (the orthogonalised component of) the corresponding variable with associated transformations induced on the original parameters. If all the data processes are stationary, elimination would seem to be of little consequence other than necessitating a reinterpretation of coefficients, but this does not apply if the processes are subject to intervention.

Finally, observed variables often are treated as being composed of "systematic" and "error" components in which case a lag polynomial of the form $d(L) = \sum_{i=0}^{m} d_i L^i$ can be interpreted as a "filter" such that $d(L)z_t = z_t^*$ represents a systematic component of z_t, and $z_t - z_t^* = w_t$ is the error component. If y_t responds to z_t^* according to some theory, but the $\{d_i\}$ are unknown, then a finite distributed lag would be a natural formulation to estimate [see, for example, Godley and Nordhaus (1972) and Sargan (1980b) for an application to models of full-cost pricing]. Conversely, other models assert that y_t only responds to w_t [see, for example, Barro (1978)] and hence restrict the coefficients of z_t and z_t^* to be equal magnitude, opposite sign.

As should be clear from the earlier discussion but merits emphasis, any decomposition of an observable into (say) "systematic" and "white noise" components depends on the choice of information set: white noise on one information set can be predictable using another. For example:

$$V_{1t} = \sum_{0}^{m} \gamma_j \nu_{jt-j} \tag{33}$$

is white noise if each of the independent ν_{jt-j} is, but is predictable apart from

$\gamma_0 \nu_{0t}$ using linear combinations of lagged variables corresponding to the $\{\nu_{jt-j}\}$. Thus, there is an inherent lack of uniqueness in using white noise residuals as a criterion for data coherency, although non-random residuals do indicate data "incoherency" [see Granger (1981) and Davidson and Hendry (1981) for a more extensive discussion]. In practice, it is possible to estimate all of the relationships derivable from the postulated data generation process and check for mutual consistency through mis-specification analyses of parameter values, residual autocorrelation, error variances and parameter constancy [see Davidson et al. (1978)]. This notion is similar in principle to that underlying "non-nested" tests [see Pesaran and Deaton (1978)] whereby a correct model should be capable of predicting the residual variance of an incorrect model and any failure to do so demonstrates that the first model is not the data generation process [see, for example, Bean (1981)]. Thus, *ability to account for previous empirical findings is a more demanding criterion of model selection than simply having "data coherency"*: that is, greater power is achieved by adopting a more general information set than simply lagged values of variables already in the equation [for a more extensive discussion, see Hendry and Richard (1982)].

Moreover, as has been well known for many years,[4] testing for predictive failure when data correlations alter is a strong test of a model since in modern terminology (excluding chance offsetting biases) it indirectly but jointly tests structurality, weak exogeneity and appropriate marginalisation (which includes thereby both dynamic and stochastic aspects of specification). A well-tested model with white-noise residuals and constant parameters (over various sub-samples), which encompasses previous empirical results and is consonant with a pre-specified economic theory seems to offer a useful approximation to the data generation process.

2.6. *A typology of single dynamic equations*

In single equation form, models like (22) from the class defined in (6) and (7) are called Autoregressive-Distributed lag equations and have the general expression:

$$d_0(L)y_t = \sum_{j=1}^{k} d_j(L)z_{jt} + \varepsilon_{1t}, \tag{34}$$

where $d_i(L)$ is a polynomial in L of degree m_i. Thus, (34) can be denoted

[4] See, for example, Marget's (1929) review of Morgenstern's book on the methodology of economic forecasting.

$\mathrm{AD}(m_0, m_1, \ldots, m_k)$ although information on zero coefficients in the $d_i(L)$ is lost thereby. The *class* has $\{\varepsilon_{1t}\}$ white noise *by definition* so not all possible data processes can be described parsimoniously by a member of the AD($\cdot$) class; for example, moving-average errors (which lead to a "more general" class called ARMAX—see Section 4) are formally excluded but as discussed below, this raises no real issues of principle. In particular, AD(1,1) is given by:

$$y_t = \beta_1 z_t + \beta_2 z_{t-1} + \beta_3 y_{t-1} + \varepsilon_{1t}, \tag{35}$$

which for present purposes is assumed to be a structural behavioural relationship wherein z_t is weakly exogenous for the parameter of interest $\boldsymbol{\beta}' = (\beta_1 \beta_2 \beta_3)$, with the error $\varepsilon_{1t} \sim I\mathcal{N}(0, \sigma_{11})$. Since all models have an error variance, (35) is referred to for convenience as a *three*-parameter model. Although it is a very restrictive equation, rather surprisingly *AD*(1,1) *actually encompasses schematic representatives of nine distinct types of dynamic model as further special cases*. This provides a convenient pedagogical framework for analysing the properties of most of the important dynamic equations used in empirical research, highlighting their respective strengths and weaknesses, thereby, we hope, bringing some coherence to a diverse and voluminous literature.

Table 2.1 summarises the various kinds of model subsumed by AD(1,1). Each model is only briefly discussed; cases (a)–(d) are accorded more space in this subsection since Sections 3, 4 and 5, respectively, consider in greater detail case (e), cases (f), (h) and (i), and case (g).

The nine models describe very different lag shapes and long-run responses of y to x, have different advantages and drawbacks as descriptions of economic time series, are differentially affected by various mis-specifications and prompt generalisations which induce different research avenues and strategies. Clearly (a)–(d) are one-parameter whereas (e)–(i) are two-parameter models and on the assumptions stated above, all but (g) are estimable by ordinary least squares [whereas (g) involves iterative least squares]. Each case can be interpreted as a model "in its own right" or as derived from (or an approximation to) (35) and these approaches will be developed in the discussion.

The generalisations of each "type" in terms of increased numbers of lags and/or distinct regressor variables naturally resemble each other more than do the special cases chosen to highlight their specific properties, although major differences from (34) persist in most cases. The exclusion restrictions necessary to obtain various specialisations from (34) [in particular, (36)–(40) and (44)] seem difficult to justify in general. Although there may sometimes exist relevant theoretical arguments supporting a specific form, it is almost always worth testing whatever model is selected against the general unrestricted equation to help gain protection from major mis-specifications.

Table 2.1

Type of model	Equation	Restrictions on (35)	Generalisation $(\sum_0^n)$
(a) Static regression	(36) $y_t = \beta_1 z_t + e_t$	$\beta_2 = \beta_3 = 0$	$y_t = \sum \beta_j z_{j,t} + e_t$
(b) Univariate time series	(37) $y_t = \beta_3 y_{t-1} + e_t$	$\beta_1 = \beta_2 = 0$	$y_t = \sum \beta_{3j} y_{t-j-1} + e_t$
(c) Differenced data/ growth rate	(38) $\Delta y_t = \beta_1 \Delta z_t + e_t$	$\beta_3 = 1, \beta_2 = -\beta_1$	$\Delta y_t = \sum \beta_{1j} \Delta z_{j,t} + \sum \beta_{3j} \Delta y_{t-j-1} + e_t$
(d) Leading indicator	(39) $y_t = \beta_2 z_{t-1} + e_t$	$\beta_1 = \beta_3 = 0$	$y_t = \sum\sum \beta_{1jk} z_{j,t-k-1} + e_t$
(e) Distributed lag	(40) $y_t = \beta_1 z_t + \beta_2 z_{t-1} + e_t$	$\beta_3 = 0$	$y_t = \sum\sum \beta_{1jk} z_{j,t-k} + e_t$
(f) Partial adjustment	(41) $y_t = \beta_1 z_t + \beta_3 y_{t-1} + e_t$	$\beta_2 = 0$	$y_t = \sum \beta_{1j} z_{j,t} + \sum \beta_{3j} y_{t-j-1} + e_t$
(g) Common factor (autoregressive error)	(42) $\begin{cases} y_t = \beta_1 z_t + u_t \\ u_t = \beta_3 u_{t-1} + e_t \end{cases}$	$\beta_2 = -\beta_1 \beta_3$	$\begin{cases} y_t = \sum\sum \beta_{1j} z_{j,t-k} + \sum \beta_{3j} y_{t-j-1} + u_t \\ u_t = \sum \rho_l u_{t-l-1} + e_t \end{cases}$
(h) Error correction	(43) $\Delta y_t = \beta_1 \Delta z_t + (1-\beta_3)(z-y)_{t-1} + e_t$	$\sum \beta_i = 1$	$\Delta y_t = \left\{ \sum\sum \beta_{1jk} \Delta z_{j,t-k} + \sum \gamma_j \Delta y_{t-j-1} \right\} + \sum\sum \lambda_l (z_l - y)_{t-k-1} + e_t$
(i) Reduced form/ dead start	(44) $y_t = \beta_2 z_{t-1} + \beta_3 y_{t-1} + e_t$	$\beta_1 = 0$	$y_t = \sum \beta_{2jk} z_{j,t-k-1} + \sum \beta_{3j} y_{t-j-1} + e_t$

(a) *Static regression* models of the general form:

$$y_t = \sum_j \beta_j z_{jt} + e_t, \tag{45}$$

rarely provide useful approximations to time-series data processes [but see Hansen (1982)]. This occurs both because of the "spurious regressions" problem induced by the observations being highly serially correlated [see Yule (1926) and Granger and Newbold (1974)] with associated problems of residual autocorrelation and uninterpretable values of R^2, and because the assertion that (45) is structural with z_t weakly exogenous for $\boldsymbol{\beta}$ has not proved viable in practice. While equilibrium economic theories correctly focus on interdependence and often entail equations such as $y = f(z)$ where linearity seems reasonable, imposing (45) on *data* restricts short-run and long-run responses of y to z to be identical and instantaneous. It seems preferable simply to require that the dynamic model *reproduces* $y = f(z)$ under equilibrium assumptions; this restricts the class of model but not the range of dynamic responses [see (h)]. Finally, for forecasting y_{t+j} (45) requires a prior forecast of z_{t+j} so lagged information is needed at some stage and seems an unwarranted exclusion from behavioural equations.

(b) In contrast, *univariate time-series* models focus only on dynamics but often serve as useful data-descriptive tools especially if selected on the criterion of white-noise residuals [see Box and Jenkins (1970)]. A general stationary form is the autoregressive moving average (ARMA) process:

$$\gamma(L) y_t = \delta(L) e_t, \tag{46}$$

where $\gamma(L)$ and $\delta(L)$ are polynomials of order m_0, m_1 (with no redundant factors), and (46) is denoted ARMA(m_0, m_1) with (37) being ARMA (1,0). Equations like (37) can be suggested by economic theory and, for example, efficient-market and rational expectations models often have $\beta_3 = 1$ [see, for example, Hall (1978) and Frenkel (1981)], but for the most part ARMA models tend to be derived rather than autonomous. Indeed, every variable in (7) has an ARMA representation[5] [see, for example, Zellner and Palm (1974) and Wallis (1977)] but such reformulations need not be structural and must have larger variances. Thus, econometric models which do not *fit* better than univariate time-series processes have at least mis-specified dynamics, and if they do not *forecast* "better"[6] must be highly suspect for policy analysis [see, inter alia, Prothero and Wallis (1976)].

[5]Implicitly, therefore, our formulation excludes deterministic factors, such as seasonal dummies, but could be generalised to incorporate these without undue difficulty.

[6]It is difficult to define "better" here since sample data may yield a large variance for an effect which is believed important for policy, but produces inefficient forecasts. A minimal criterion is that the econometric model should not experience predictive failure when the ARMA model does not.

In principle, all members of our typology have generalisations with moving-average errors, which anyway are likely to arise in practice from marginalising with respect to autoregressive or Granger-causal variables, or from measurement errors, continuous time approximations etc. However, detailed consideration of the enormous literature on models with moving average errors is precluded by space limitations (see, Section 4.1 for relevant references). In many cases, MA errors can be quite well approximated by autoregressive processes [see, for example, Sims (1977, p. 194)] which are considered under (g) below, and it seems difficult to discriminate in practice between autoregressive and moving-average approximations to autocorrelated residuals [see, for example, Hendry and Trivedi (1972)].

(c) *Differenced data* models resemble (a) but after transformation of the observations y_t, z_t to $(y_t - y_{t-1}) = \Delta y_t$ and Δz_t. The filter $\Delta = (1 - L)$ is commonly applied on the grounds of "achieving stationarity", to circumvent awkward inference problems in ARMA models [see Box and Jenkins (1970), Phillips (1977), Fuller (1976), Evans and Savin (1981) and Harvey (1981)] or to avoid "spurious regressions" criticisms. Although the equilibrium equation that $y = \beta_1 z$ implies $\Delta y = \beta_1 \Delta z$, differencing fundamentally alters the properties of the error process. Thus, even if y is proportional to z in equilibrium, the solution of (38) is indeterminate and the estimated magnitude of β_1 from (38) is restricted by the relative variances of Δy_t to Δz_t. A well-known example is the problem of reconciling a low marginal with a high and constant average propensity to consume [see Davidson et al. (1978) and compare Wall et al. (1975) and Pierce (1977)]. In any case, there are other means of inducing stationarity, such as using ratios, which may be more consonant with the economic formulation of the problem.

(d) *Leading indicator* equations like (39) attempt to exploit directly differing latencies of response (usually relative to business cycles) wherein, for example, variables like employment in capital goods industries may "reliably lead" GNP. However, unless such equations have some "causal" or behavioural basis, β_2 need not be constant and unreliable forecasts will result so econometric models which *indirectly* incorporate such effects have tended to supercede leading indicator modelling [see, inter alia, Koopmans (1947) and Kendall (1973)].

(e) As discussed in Section 2.4, *distributed lags* can arise either from structural/behavioural models or as implications of other dynamic relationships. Empirically, equations of the form:

$$y_t = \alpha(L) z_t + e_t, \tag{47}$$

where $\alpha(L)$ is a polynomial of order m_1 frequently manifest substantial residual autocorrelation [see, inter alia, many of the AD$(0, m_1, \dots, m_k)$ equations in Hickman (1972) or, for example, new housing "starts–completions" relationships in Waelbroeck (1976)]. Thus, whether or not z_t is *strongly* exogenous becomes important for the detection and estimation of the residual autocorrelation. "Eliminating" autocorrelation by fitting autoregressive errors imposes "common factor restrictions" whose validity is often dubious and merits testing [see (g) and Section 5], and even after removing a first order autoregressive error, the equation may yet remain prey to the "spurious regressions" problem [see Granger and Newbold (1977)]. Moreover, collinearity between successive lagged z's has generated a large literature attempting to resolve the profligate parameterisations of unrestricted estimation (and the associated large standard errors) by subjecting the $\{\alpha_j\}$ to various "a priori constraints". Since relatively short "distributed lags" also occur regularly in other AD$(\cdot)$ models, and there have been important recent technical developments, the finite distributed lag literature is surveyed in Section 3.

(f) *Partial adjustment* models are one of the most common empirical species and have their basis in optimization of quadratic cost functions where there are adjustment costs [see Eisner and Strotz (1963) and Holt et al. (1960)]. *Invalid* exclusion of z_{t-1} can have important repercussions since the shape of the distributed lag relationship derived from (41) is highly skewed with a large mean lag when β_3 is large even though that derived from (35) need not be for the same numerical value of β_3: this may be part of the explanation for apparent "slow speeds of adjustment" in estimated versions of (41) or generalisations thereof (see, especially, studies of aggregate consumers' expenditure and the demand for money in the United Kingdom). Moreover, many derivations of "partial adjustment" equations like (41) entail that e_t is autocorrelated [see, for example, Maddala (1977, ch. 9), Kennan (1979) and Muellbauer (1979)] so that OLS estimates are inconsistent for the β_i [see Malinvaud (1966)], have inconsistently estimated standard errors, and residual autocorrelation tests like the Durbin–Watson (DW) statistic are invalid [see Griliches (1961) and Durbin (1970)]. However, appropriate Lagrange multiplier tests can be constructed [see Godfrey (1978) and Breusch and Pagan (1980)]. Finally, generalised members of this class such as:

$$\gamma(L)y_t = \sum_{i=1}^{K} \delta_i z_{it} + e_t, \tag{48}$$

have unfortunate parameterisations since "levels" variables in economics tend to be highly intercorrelated.

(g) *Common factor* representations correspond 1–1 to autoregressive error models and most clearly demonstrate the dynamic-stochastic specification link in terms of "equation dynamics" versus "error dynamics" [see Sargan (1964, 1980c), Hendry and Mizon (1978) and Mizon and Hendry (1980)]. To illustrate the principles involved, reconsider (35) written in lag operator notation (with $\beta_1 \neq 0$):

$$(1-\beta_3 L)y_t = \beta_1(1+(\beta_2/\beta_1)L)z_t + e_t, \tag{35*}$$

where both lag polynomials have been normalised. Under the condition:

$$-\beta_3 = \beta_2/\beta_1 \quad \text{or} \quad \beta_1\beta_3+\beta_2 = 0, \tag{49}$$

the lag polynomials coincide and constitute a *common factor* of $(1-\beta_3 L)$. Dividing both sides of (35*) by $(1-\beta_3 L)$ yields:

$$y_t = \beta_1\left(\frac{1+(\beta_2/\beta_1)L}{1-\beta_3 L}\right)z_t + \frac{e_t}{1-\beta_3 L} = \beta_1 z_t + u_t, \tag{50}$$

where

$$u_t = \beta_3 u_{t-1} + e_t. \tag{51}$$

Consequently, the equations:

$$\begin{matrix} y_t = \beta_1 z_t + u_t \\ u_t = \beta_3 u_{t-1} + e_t \end{matrix} \quad \begin{bmatrix} \text{AD}(0,0) \\ \text{AD}(1) \end{bmatrix} \tag{52}$$

uniquely imply and are uniquely implied by:

$$y_t = \beta_1 z_t + \beta_3 y_{t-1} - \beta_1\beta_3 z_{t-1} + e_t \quad [\text{AD}(1,1)]. \tag{53}$$

Usually, $|\beta_3| < 1$ is required; note that (52) can also be written as:

$$y_t^+ = \beta_1 z_t^+ + e_t, \tag{54}$$

where $z_t^+ = (z_t - \beta_3 z_{t-1})$ is a "quasi-difference" and the operator $(1-\beta_3 L)$ "eliminates" the error autocorrelation.

This example highlights two important features of the AD($\cdot$) class. Firstly, despite formulating the class as one with white-noise error, it does *not* exclude autoregressive error processes. Secondly, such errors produce a *restricted* case of the class and hence the assumption of an autoregressive error *form* is testable against a less restricted member of the AD($\cdot$) class. More general cases and the implementation of appropriate tests of common factor restrictions are discussed in Section 5.

The equivalence of autoregressive *errors* and common factor dynamics has on occasion been misinterpreted to mean that autocorrelated *residuals* imply common factor dynamics. There are many reasons for the existence of autocorrelated residuals including: omitted variables, incorrect choice of functional form, measurement errors in lagged variables, and moving-average error processes as well as autoregressive errors. Consequently, for example, a low value of a Durbin–Watson statistic does *not* uniquely imply that the errors are a first-order autoregression and automatically "eliminating" residual autocorrelation by assuming an AD(1) process for the error can yield very misleading results.

Indeed, the order of testing is incorrect in any procedure which tests for autoregressive errors by *assuming* the existence of a common factor representation of the model: the validity of (49) should be tested before assuming (52) and attempting to test therein H_b: $\beta_3 = 0$. In terms of commencing from (35), if and only if H_a: $\beta_2 + \beta_1\beta_3 = 0$ is true will the equation have a representation like (52) and so only if H_a is *not* rejected can one proceed to test H_b: $\beta_3 = 0$. If H_b is tested alone, conditional on the belief that (49) holds, then failure to reject $\beta_3 = 0$ does not imply that $y_t = \beta_1 z_t + e_t$ (a common mistake in applied work) nor does *rejection* of H_b imply that the equations in (52) are valid. It is sensible to test H_a first since only if a common factor exists is it meaningful to test the hypothesis that its root is zero. While (52) is easily interpreted as an approximation to some more complicated model with the error autocorrelation simply acting as a "catch all" for omitted variables, unobservables, etc. a *full behavioural* interpretation is more difficult. Formally, on the one hand, $E(y_t|X_{t-1}) = \beta_1 z_t + \beta_3 u_{t-1}$ and hence agents adjust to this shifting "optimum" with a purely random error. However, if the $\{u_t\}$ process is viewed as being autonomous then the first equation of (52) entails an immediate and complete adjustment of y to changes in z, but if agents are perturbed above (below) this "equilibrium" they will stay above (below) for some time and do not adjust to remove the discrepancy. Thus, (52) also characterises a "good/bad fortune" model with persistence of the chanced-upon state in an equilibrium world. While these paradigms have some applications, they seem likely to be rarer than the present frequency of use of common factor models would suggest, supporting the need to *test* autoregressive error restrictions before imposition. The final interpretation of (53) noted in Section 5 serves to reinforce this statement.

Despite these possible interpretations, unless y does *not* Granger cause z, then z Granger causes u. If so, then regressing y_t on z_t when $\{u_t\}$ is autocorrelated will yield an inconsistent estimate of β_1, and the residual autocorrelation coefficient will be inconsistent for β_3. Any "two-step" estimator of (β_1, β_3) commencing from *these* initial values will be inconsistent, even though: (a) there are no *explicit* lagged variables in (52) and (b) *fully iterated* maximum likelihood estimators are consistent and fully efficient when z_t is weakly exogenous for $\boldsymbol{\beta}$ [see Hendry (1976) for a survey of estimators in common factor equations]. Finally, it is worth emphasising that under the additional constraint that $\beta_3 = 1$, model (c) is a common factor formulation.

(h) *Error correction models* such as (43) are a natural reparameterisation of AD$(\cdot)$ equations when:

$$\left(\sum \beta_i - 1\right) = \delta = 0. \tag{55}$$

If $\beta_3 \neq 1$, the steady-state solution of (43) for $\Delta z = g = \Delta y$ is:

$$y = (1-\beta_1)g/(1-\beta_3) + z = k(g) + z, \tag{56}$$

and hence $y = z$ in static equilibrium, or $Y = K(g)Z$ (more generally) when y and z are $\ln Y$ and $\ln Z$, respectively [see Sargan (1964) and Hendry (1980)]. Thus, (55) implements *long-run proportionality* or homogeneity and ensures that the dynamic equation reproduces in an equilibrium context the associated equilibrium theory. Moreover, H_0: $\delta = 0$ is easily tested, since (35) can be rewritten as:

$$\Delta y_t = \beta_1 \Delta z_t + (1-\beta_3)(z-y)_{t-1} + \delta z_{t-1} + e_t, \tag{57}$$

which anyway offers the convenient interpretation that agents marginally adjust y_t from y_{t-1} in response to *changes* in z_t (β_1 being the short-run effect), the previous *disequilibrium* $(z-y)_{t-1}$ ($(1-\beta_3)$ being the "feedback" coefficient) and the previous *level* z_{t-1} (which is irrelevant under proportionality). Since many economic theories have proportional forms in static equilibrium, error correction models might be expected to occur frequently. Indeed, an important property of (43) is that when $\delta = 0$, (57) coincides with (43) and *all of the other models in this typology become special cases of* (43). Thus, given $\delta = 0$ a modelling exercise which commenced from (43) even when one of the *other* types represented the actual data generation process would involve no mis-specification and which other special case was correct would be readily detectable from the values of the parameters in (43) given in Table 2.2. The converse does *not* hold: fitting any of (a)–(g) when (h) is true but Table 2.2 restrictions are invalid, induces mis-specifications, the precise form of which could be deduced by an investigator who used (h). Thus, when $\delta = 0$, error correction is essentially a necessary and sufficient model form and it is this property which explains the considerable practical success of error correction formulations in encompassing and reconciling diverse empirical estimates in many subject areas [see, inter alia, Henry et al. (1976), Bean (1977), Hendry and Anderson (1977), Davidson et al. (1978), Cuthbertson (1980),

Table 2.2

(a) $\beta_1 = 1 - \beta_3 = 1$	(b) $\beta_1 = 1 - \beta_3 = 0$	(c) $1 - \beta_3 = 0$
(d) $\beta_1 = \beta_3 = 0$	(e) $\beta_3 = 0$	(f) $\beta_1 = 1 - \beta_3$
(g) $\beta_1 = 1$	(i) $\beta_1 = 0$	

Hendry (1980) and Davis (1982)]. In an interesting way, therefore, (43) nests "levels" and "differences" formulations and, for example, offers one account of why a small value of β_1 in (c) is compatible with proportionality in the long run, illustrating the interpretation difficulties deriving from imposing "differencing filters".

(i) Equation (44) could constitute either the *reduced form* of (35) on eliminating z_t [assuming its process to be AD(1,1) also, or a special case thereof] or a "deadstart" model in its own right. For example, if $z_t = \lambda z_{t-1} + \varepsilon_{2t}$ and (35) is the behavioural equation, (44) is also " valid" with parameters:

$$y_t = (\beta_2 + \beta_1 \lambda) z_{t-1} + \beta_3 y_{t-1} + e_t, \qquad e_t \sim I\mathcal{N}(0, \sigma_{11} + \beta_1^2 \sigma_{22}), \tag{58}$$

but is no longer structural for changes in λ, and λ is required for estimating $\boldsymbol{\beta}$. Indeed if $\delta = 0$ in (55), (58) will not exhibit proportionality unless $\beta_1(1-\lambda) = 0$. Also, $\beta_2 + \beta_1 \lambda < 0$ does not exclude $y = z$ in equilibrium, although this interpretation will only be noticed if (y_t, z_t) are *jointly* modelled.

Conversely, if (44) is structural because of an inherent lag before z affects y, then it is a *partial adjustment type* of model, and other types have deadstart variants in this sense.

The discussions in Sections 3, 4 and 5, respectively, concern the general forms of (e); (f), (h) and (i); and (g), plus certain models excluded above, with some overlap since distributed lags often have autocorrelated errors, and other dynamic models usually embody short distributed lags. Since generalisations can blur important distinctions, the preceding typology is offered as a clarifying framework.

3. Finite distributed lags

3.1. A statement of the problem

A finite distributed-lag relationship has the form:

$$y_t = \sum_{i=1}^{n} W_i(L) z_{it} + u_t, \tag{59}$$

where

$$W_i(L) = \sum_{j=m_i^0}^{m_i} w_{ij} L^j, \tag{60}$$

and is a member of the AD$(0, m_1, \ldots, m_n)$ class. For ease of exposition and notation, attention is centered on a bivariate case, namely AD(0, m) denoted by:

$$y_t = \sum_{j=m^0}^{m} w_j z_{t-j} + u_t = W(L) z_t + u_t, \tag{61}$$

where $\{z_t\}$ is to be treated as "given" for estimating $w = (w_{m^0}, \dots, w_m)'$, and u_t is a "disturbance term". It is assumed that sufficient conditions are placed upon $\{u_t\}$ and $\{z_t\}$ so that OLS estimators of w are consistent and asymptotically normal [e.g. that (8) is the data generation process and is a stable dynamic system with w defined by $E(y_t | \mathbf{Z}_{t-m^0})$].

Several important and interdependent difficulties hamper progress. Firstly, there is the issue of the status of (61), namely whether it is basic or derived and whether or not it is structural, behavioural, etc. or just an assumed approximation to some more complicated lag relationship between y and z (see Sections 2.3 and 2.4). Unless explicitly stated otherwise, the following discussion assumes that (61) is structural, that $u_t \sim I\mathcal{N}(0, \sigma_u^2)$ and that z_t is weakly exogenous for w. These assumptions are only justifiable on a pedagogic basis and are unrealistic for many economics data series; however, most of the technical results discussed below would apply to short distributed lags in a more general dynamic equation. Secondly, $W(L)$ is a polynomial of the same degree as the lag length and for highly intercorrelated $\{z_{t-j}\}$, unrestricted estimates of w generally will not be well determined. Conversely, it might be anticipated that a lower order polynomial, of degree $k < m$ say, over the same lag length might suffice, and hence one might seek to estimate the $\{w_j\}$ subject to such restrictions. Section 3.2 considers some possible sets of restrictions whereas Section 3.4 discusses methods for "weakening" lag weight restrictions ("variable lag weights" wherein the $\{w_j\}$ are dependent on economic variables which change over time, are considered in Section 3.6).

However, k, m^0 and m are usually unknown and have to be chosen jointly, and this issue is investigated in Section 3.3 together with an evaluation of some of the consequences of incorrect specifications. Further, given that formulations like (61) are the correct specification, many alternative estimators of the parameters have been proposed and the properties of certain of these are discussed in Section 3.5 and related to Sections 3.2 and 3.4.

Frequently, equations like (61) are observed to manifest serious residual autocorrelation and Section 3.6 briefly considers this issue as well as some alternative specifications which might facilitate model selection.

3.2. *Exact restrictions on lag weights*

If (61) is the correct specification and in its initial form $W(1) = \sum w_i = h$ (say) then working with $\{h^{-1} w_i\}$ produces a lag weight distribution which sums to unity. It ssumed below that such rescaling has occurred so that $W(1) = 1$, although it is *not* assumed that this is necessarily *imposed* as a restriction for purposes of estimation. It should be noted at the outset that all non-stochastic static equilibrium solutions of (61) take the simple form: $y = hz$ and the importance of this is evaluated in (3.6). Moreover, provided all of the w_i are non-negative, they are analogous to discrete probabilities and derived "moments" such as the mean and/or median lag (denoted μ and η, respectively), variance of the lag distribution etc. are well defined [for example, see Griliches (1967) and Dhrymes (1971)]:

$$\mu = \sum_{i=m^0}^{m} i w_i \tag{62}$$

and η is an integer such that:

$$\sum_{i=0}^{\eta-1} w_i < \tfrac{1}{2} \leq \sum_{i=0}^{\eta} w_i .$$

Nevertheless, even assuming (61) is structural, economic theory is not usually specific about various important features of the $W(\cdot)$ polynomial, including its "shape" (i.e. multimodality, degree, etc.), starting point m^0, and lag length m. For the present, we take $m^0 = 0$ and m to be known, and first consider the issue of the "shape" of $W(\cdot)$ as a function of $k < m$.

If little information exists on what might constitute likely decision rules, or if (say) the relationship is an order/delivery one, so that (61) is a reasonable specification but m is large, some restrictions *may* need to be placed on $\{w_i\}$ to obtain "plausible" estimates. However, as Sims (1974) and Schmidt and Waud (1973) argue, this should not be done without first estimating w unrestrictedly. From such results, putative restrictions can be tested. Unrestricted estimates can provide a surprising amount of information, notwithstanding prior beliefs that "collinearity" would preclude sensible results from such a profligate parameterisation. Even so, some simplification is usually feasible and a wide range of possible forms of restrictions has been proposed including arithmetic, inverted "υ", geometric, Pascal, gamma, low order polynomial and rational [see, for example, the discussion in Maddala (1977)]. Of these, the two most popular are the low order polynomial distributed lag [denoted PDL; see Almon (1965)]:

$$w_j = \sum_{i=0}^{k} \gamma_i j^i, \qquad j = 0,\dots,m, \tag{63}$$

and the rational distributed lag [denoted RDL; see Jorgenson (1966)]:

$$W(L) = A(L)/B(L): A(L) = \sum_{j=0}^{p} a_j L^j, B(L) = 1 - \sum_{j=1}^{q} b_j L^j. \tag{64}$$

These are denoted PDL(m,k) and RDL(p,q) respectively. If $k=m$ then the $\{w_j\}$ are unrestricted and $\{\gamma_i\}$ is simply a one–one reparameterisation. Also, if $A(L)$ and $B(L)$ are defined to exclude *redundant common factors*, then RDLs cannot be finite[7] but:

(a) as shown in Pagan (1978), PDL restrictions can be implemented via an RDL model denoted the finite RDL, with $B(L)=(1-L)^{k+1}$ and $p=k$; and

(b) RDLs can provide close approximations to PDLs as in:

$$\begin{aligned}W(L)&=(0.50+0.30L+0.15L^2+0.05L^3)\simeq\tfrac{1}{2}(1-0.5L)^{-1}\\&=(0.50+0.25L+0.13L^2+0.06L^3+0.03L^4\ldots).\end{aligned}\tag{65}$$

Indeed, early treatments of RDL and PDL methods regarded them as ways of approximating unknown functions to any desired degree of accuracy, but as Sims (1972) demonstrated, an approximation to a distribution which worked quite well in one sense could be terrible in other respects. Thus, solved values from $A(L)/B(L)$ could be uniformly close to $W(L)$ yet (say) the implied mean lag could be "infinitely" wrong. In (65), for example, the actual mean lag is 0.75 while that of the illustrative approximating distribution is 1.0 (i.e. 33% larger). Lütkepohl (1980) presents conditions which ensure accurate estimation of both μ and the long-run response (also see Sections 4.3 and 5 below).

Rather than follow the "approximations" idea, it seems more useful instead to focus attention on the nature of the constraints being imposed upon the lag coefficients by any parametric assumptions, especially since the consequences of invalid restrictions are well understood and are capable of analytical treatment. For the remainder of this section, only PDL(m,k) is considered, RDL models being the subject of Section 4. Schmidt and Mann (1977) proposed combining PDL and RDL in the LaGuerre distribution, but Burt (1980) argued that this just yields a particular RDL form.

The restrictions in (63) can be written in matrix form as:

$$\boldsymbol{w}=\boldsymbol{J}\boldsymbol{\gamma}\quad\text{or}\quad\left(\boldsymbol{I}-\boldsymbol{J}(\boldsymbol{J}'\boldsymbol{J})^{-1}\boldsymbol{J}'\right)\boldsymbol{w}=\boldsymbol{R}\boldsymbol{w}=\boldsymbol{0},\tag{66}$$

where $\boldsymbol{J}$ is an $(m+1)\times(k+1)$ Vandermonde matrix and rank $(\boldsymbol{R})=m-k$. Perhaps the most useful parameterisation follows from Shiller's (1973) observation that the $(k+1)$th differences of a kth order polynomial are zero and hence the linear restrictions in (66) for PDL(m,k) imply that:

$$(1-L)^{k+1}w_j=0,\qquad j=k+1,\ldots,m.$$

[7]For example, $(1-(\lambda L)^{m+1})/(1-\lambda L)=\sum_{i=0}^{m}(\lambda L)^i$ is finite since $A(L)$ and $B(L)$ have the factor $(1-\lambda L)$ in common (and have unidentifiable coefficients unless specified to have a common factor).

Thus, $\boldsymbol{R}$ is a differencing matrix such that $\boldsymbol{RJ}=\boldsymbol{0}$. Expressing (61) in matrix form:

$$\boldsymbol{y}=\boldsymbol{Zw}+\boldsymbol{u} \quad \text{or} \quad \boldsymbol{y}=(\boldsymbol{ZJ})\boldsymbol{\gamma}+\boldsymbol{u}=\boldsymbol{Z}^*\boldsymbol{\gamma}+\boldsymbol{u}, \tag{67}$$

shows that $\boldsymbol{\gamma}$ can be estimated either directly or via the first expression in (67) subject to $\boldsymbol{Rw}=\boldsymbol{0}$.

Indeed, (63) can be reparameterised as desired by replacing j^i by any other polynomial $\Psi_i(j)=\sum_{l=0}^{i}\psi_{il}j^l$ and appropriate choices of $\Psi_i(j)$ can facilitate the testing of various restrictions, selecting parsimonious models and/or computational accuracy [see Robinson (1970), Pagano and Hartley (1981), Trivedi and Pagan (1979) and Sargan (1980b)]. In particular, orthogonal polynomials could be used if computational accuracy was likely to be a problem.

A rather different reparameterisation is in terms of the moments of $\boldsymbol{w}$ [see Hatanaka and Wallace (1980), Burdick and Wallace (1976) and Silver and Wallace (1980) who argue that the precision of estimating lag moments from economic data falls as the order of the moments rises]. As shown by Yeo (1978), the converse of the Almon transformation is involved, since, when $m=k$:

$$\boldsymbol{y}=\boldsymbol{ZJ}^{-1}\boldsymbol{Jw}+\boldsymbol{u}=(\boldsymbol{ZJ}^{-1})\boldsymbol{\phi}+\boldsymbol{u} \tag{68}$$

yields $\boldsymbol{\phi}$ as the moments of $\boldsymbol{w}$ (assuming $w_i\geq 0, \forall_i$). Moreover, from analytical expressions for $\boldsymbol{J}^{-1}$, Yeo establishes that $(\boldsymbol{ZJ}^{-1})$ involves linear combinations of powers of differences of z_t's (i.e. $\sum\lambda_i\Delta^i z_t$) and that the parameters ψ_j of the equation:

$$y_t=\sum_{j=0}^{k}\psi_j\Delta^j z_t+u_t \tag{69}$$

are the factorial moments (so that $\psi_0=\phi_0$ and $\psi_1=\phi_1$).

When z is highly autoregressive, z_t will not be highly correlated with $\Delta^i z_t$ for $j\geq 1$ so that ψ_0 will often be well determined. Finally, (67)–(69) allow intermatching of prior information about $\boldsymbol{w}$ and $\boldsymbol{\phi}$ or $\boldsymbol{\psi}$.

The formulation in (66), and the stochastic equivalent $\boldsymbol{Rw}=\boldsymbol{\varepsilon}\sim\text{i.d.}(\boldsymbol{0},\sigma_\varepsilon^2\boldsymbol{I})$, both correspond to "smoothness" restrictions on how rapidly the lag weights change. Sims (1974) doubted the appropriateness of such constraints in many models, although this is potentially testable. Since the case $k=m$ is unrestricted in terms of *polynomial* restrictions (but, for example, imposes an exact polynomial response of y_t to lagged z's with a constant mean lag, etc.), the larger k the less can be the conflict with sample information—but the smaller the efficiency gain if k is chosen too large. Nevertheless, it must be stressed that in addition to all the other assumptions characterising (61), low order k in PDL(m,k) approximating large m entails strong smoothness restrictions.

3.3. *Choosing lag length and lag shape*

Once m and k are specified, the PDL(m,k) model is easily estimated by unrestricted least squares; consequently, most research in this area has been devoted either to the determination of m and k or the analysis of the properties of the PDL estimator when any choice is incorrect. Such research implicitly accepts the proposition that there is a "true" lag length and polynomial degree—to be denoted by (m^*,k^*)—and this stance is probably best thought of as one in which the "true" model, if known and subtracted from the data, would yield only a white noise error.

In such an orientation it is not asserted that any model can fully capture reality, but only that what is left is not capable of being predicted in any systematic way, and this viewpoint (which is an important element in data analysis) is adopted below. For the remainder of this section, m^0 is taken to be known as zero; lack of this knowledge would further complicate both the analysis and any applications thereof.

The various combinations of (m,k) and their relationships to (m^*,k^*) are summarized in Figure 3.1 using a six-fold partition of (m,k) space. Each element of the partition is examined separately in what follows, as the performance of the PDL estimator varies correspondingly.

A $(m=m^*, k\geq k^*)$
On this half line, m is fixed at m^* and k varies above k^*, which is to be determined.

It is well-known [see Dhrymes (1971), Frost (1975) and Godfrey and Poskitt (1975)] that *decreasing* k results in an *increasing* number of linear restrictions upon the model and this is easily seen by noting that the number of differencing restrictions is $(m-k)$, which increases as k decreases. Furthermore [since $(1-L)^k w_j = (1-L)^{k-k_1}(1-L)^{k_1} w_j = 0$ if $(1-L)^{k_1} w_j = 0$], when the coefficients lie on a (k_1-1)th degree polynomial they also lie on a $(k-1)$th order polynomial for $k_1 \leq k$. Hence, the sequence of hypotheses that the polynomial is of degree

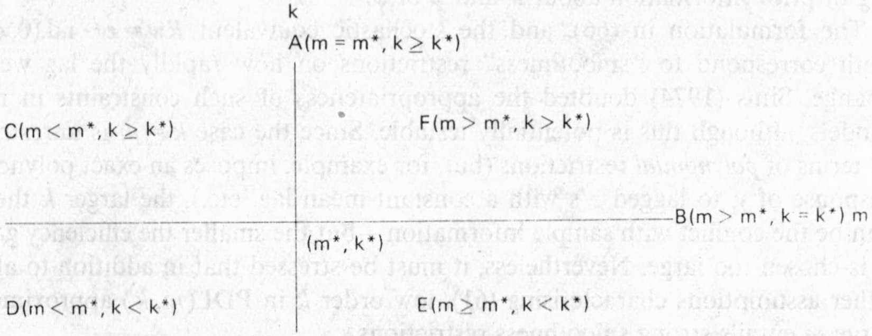

Figure 3.1

$k-1, k-2, k-3, \ldots, 1$, is ordered and nested [see Mizon (1977b)] and Godfrey and Poskitt (1975) selected the optimal polynomial degree by applying Anderson's method for determining the order of polynomial in polynomial regression [see Anderson (1971, p. 42)]. The main complication with this procedure is that the significance level changes at each step (for any given nominal level) and the formula for computing this is given in Mizon (1977b) [some observations on efficient ways of computing this test are available in Pagano and Hartley (1981) and Sargan (1980b)].

When $m = k$, $\boldsymbol{J}$ is non-singular in (66) so either $\boldsymbol{w}$ or $\boldsymbol{\gamma}$ can be estimated directly, with Wald tests used for the nested sequence; those based on $\boldsymbol{\gamma}$ appear to have better numerical properties than those using $\boldsymbol{w}$ [see Trivedi and Pagan (1979)].

B ($m > m^*, k = k^*$)

The next stage considers the converse of known k^* and investigates the selection of m. From (66), it might seem that increasing m simply increases the number of restrictions but as noted in Thomas (1977) [by reference to an unpublished paper of Yeo (1976)] the sum of squared residuals may either increase or decrease in moving from PDL(m, k) to PDL$(m+1, k)$. This arises because *different* parameter vectors are involved while the same order of polynomial is being imposed.

This situation (k^* known, m^* unknown) has been analysed informally by Schmidt and Waud (1973) and more formally by Teräsvirta (1976), Frost (1975), Carter, Nagar and Kirkham (1975) and Trivedi and Pagan (1979). Teräsvirta suggests that overstating m leads to biased estimates of the coefficients while Frost says (p. 68): "Overstating the length of the lag, given the correct degree of polynomial, causes a bias. This bias eventually disappears as k increases." Support for these propositions seems to come from Cargill and Meyer's (1974) Monte Carlo study, but they are only a statement of *necessary* conditions for the existence of a bias. As proved in Trivedi and Pagan (1979), the sufficient condition is that stated by Schmidt and Waud (1973); *the sufficient condition for a bias in the PDL estimator is that the lag length be overstated by more than the degree of approximating polynomial.* For example, if $k = 1$, and m is chosen as $m^* + 1$, it is possible to give an interpretation to the resulting restriction, namely it is an *endpoint restriction* appropriate to a PDL$(m^*, 1)$ model. Although it has been appreciated since the work of Trivedi (1970a) that the imposition of endpoint restrictions should not be done lightly, there are no grounds for excluding them from any analysis a priori, and *if valid*, no bias need result from $m^* + k \geq m > m^*$.

To reconcile this with Teräsvirta's theory it is clear from his equation (5) (p. 1318) that the bias is zero if $\boldsymbol{R}_1\boldsymbol{\beta}_1 = \boldsymbol{0}$ and no reasons are given there for believing that this cannot be the case. Examples can be constructed in which biases will and will not be found and the former occurred in Cargill and Meyer's work: these biases do not translate into a general principle, but reflect the design of the experiments. As an aside, for fixed $\{z_t\}$ there is no need to resort to imprecise and specific *direct* simulation experiments to study mis-specifications in PDLs. This is an area where controlled experiments could yield accurate and fairly general answers from such techniques as control variables, antithetic variates and response surfaces (see Hendry in Chapter 16 of this Handbook). For example, using antithetic variates for u_t yields exact simulation answers for biases. In general, if:

$$y = G(Z)\boldsymbol{\Theta} + \boldsymbol{u} \text{ and } G(\cdot) \text{ is any constant function of fixed } Z, \tag{70}$$

with $\boldsymbol{u}$ distributed symmetrically according to $f(\boldsymbol{u})$, and:

$$(\hat{\boldsymbol{\Theta}} - \boldsymbol{\Theta}) = (G'G)^{-1}G'\boldsymbol{u}, \tag{71}$$

then $f(\boldsymbol{u}) = f(-\boldsymbol{u})$, whereas $(\hat{\boldsymbol{\Theta}} - \boldsymbol{\Theta})$ switches sign as $\boldsymbol{u}$ does. Consequently, if, for example, $E(\boldsymbol{u}) = \mathbf{0}$ and (70) is correctly specified, simulation estimates of (71) always average to zero over $(\boldsymbol{u}, -\boldsymbol{u})$ proving unbiasedness in *two* replications [see Hendry and Trivedi (1972)].

Is it possible to design tests to select an optimal m if k^* is known? Carter, Nagar and Kirkham (1975) propose a method of estimating the "bias" caused by overspecifying m and argue for a strategy of overspecifying m, computing the "bias" and reducing m if a large "bias" is obtained. This is an interesting suggestion but, as noted above, the bias may be zero even if m is incorrect. Sargan (1980b) points out that the models PDL($m+1, k$) and PDL(m, k) are non-nested and that two separate decisions need to be made, for which "t"-tests can be constructed:

(i) Is there a longer lag, i.e. does $w_{m+1} = 0$?

(ii) Does the coefficient w_{m+1} lie on the kth order polynomial?

To test the first, form a regression with the PDL(m, k) variables and z_{m+1} in the regression. The second can be constructed from the covariance matrix $\hat{w}$ that is a by-product of the regression. If both (i) and (ii) are rejected, then a more general specification is required.

A possible difficulty with this proposal is that for a valid test of $w_{m+1} = 0$, the estimator under the *alternative* hypothesis must be unbiased, but a bias is certain if the lag length is overstated by more than the polynomial degree. Accordingly, to implement (i) and (ii) above, it is important to have good prior information on the true lag length, at least within k^* periods. Thus, the first step of the analysis should be to select an optimal polynomial order for a sufficiently long lag length in which case (ii) is accepted and a test is required for the validity of an endpoint restriction;[8] if accepted, these two steps can be conducted sequentially till an appropriate termination criterion is satisfied.

While this procedure at least yields an ordered sequence of hypotheses, its statistical properties remain to be investigated, and to be compared to alternative

[8]We owe this point to George Mailath.

approaches. There are some further comments that need to be made. Firstly, what happens if PDL$(m, m-1)$ is rejected at the first step? As this could be due either to the true lag being m and unrestricted or to the need for a shorter lag, the unrestricted estimates PDL(m, m) allow a test of this. Secondly, an ordered sequence (even when nested) merely yields control of Type I errors and says nothing about power (necessarily). Thus, gross overstatement of the lag length causing a large number of tests in the sequence almost certainly results in low power (e.g. for $m^* = 4$, $k^* = 2$ choosing $m \geq 8$ initially could result in a lengthy test sequence). However, Bewley (1979b), in a comparison of various methods to select m and k, examined one approach similar to that described above, finding it to have good power. Notice that the unrestricted estimation of the distributed lag parameters is an important part of the strategy, because it is the comparison of a restricted with an unrestricted model that enables a check on the validity of the first set of restrictions imposed; once these are accepted it is possible to continue with the restricted model as the new alternative.

C, D$(m < m^*, k \gtreqless k^*)$

Partitioning $\boldsymbol{Z}$ as $\boldsymbol{Z} = [\boldsymbol{Z}_1 \boldsymbol{Z}_2]$ with $\boldsymbol{Z}_1 \sim T \times (m+1)$ and $\boldsymbol{Z}_2 \sim T \times (m^* - m)$, $\boldsymbol{w}$ as $(\boldsymbol{w}_1' \boldsymbol{w}_2')'$ and $\boldsymbol{R}$ as $(\boldsymbol{R}_1 \boldsymbol{R}_2)$ (the last two conformable with $\boldsymbol{Z}$), the PDL estimator of $\boldsymbol{w}_1$—the underspecified lag distribution—is from (67):

$$\hat{\boldsymbol{w}}_1^R = \hat{\boldsymbol{w}}_1^u - \left(\boldsymbol{Z}_1'\boldsymbol{Z}_1\right)^{-1}\boldsymbol{R}_1'\left(\boldsymbol{R}_1\left(\boldsymbol{Z}_1'\boldsymbol{Z}_1\right)^{-1}\boldsymbol{R}_1'\right)^{-1}\boldsymbol{R}_1\hat{\boldsymbol{w}}_1^u, \tag{72}$$

where $\hat{\boldsymbol{w}}_1^u$ is the unrestricted OLS estimator of $\boldsymbol{w}_1$ in the model $\boldsymbol{y} = \boldsymbol{Z}_1\boldsymbol{w}_1 + \boldsymbol{v}$. From standard regression theory $\hat{\boldsymbol{w}}_1^u$ is biased whenever $\boldsymbol{Z}_1'\boldsymbol{Z}_2 \neq \boldsymbol{0}$, or whenever the included and excluded regressors are correlated. The existence of this bias will induce a bias in the PDL estimator, so that a sufficient condition for the PDL estimator to be biased whenever the lag length is understated is that the regressors be correlated. However, it is not a necessary condition as $\hat{\boldsymbol{w}}_1^R$ is biased whenever $\boldsymbol{R}_1\boldsymbol{w}_1 \neq \boldsymbol{0}$ even if $\hat{\boldsymbol{w}}_1^u$ is unbiased. For $k < k^*$, $\boldsymbol{R}\boldsymbol{w}_1$ never equals zero and understatement of the polynomial order therefore results in a bias. Furthermore, the condition noted above in the analysis of the half-line B applies in reverse; *understating* the lag length by more than the degree of approximating polynomial leads to a bias.

E$(m \geq m^*, k < k^*)$

The restrictions are incorrect and the PDL estimator must be biased.

F$(m > m^*, k > k^*)$

In contrast to the cases above which have been extensively analysed in the literature, little attention has been paid to the quadrant F despite its potential relevance. With any PDL(m, k) combination there are $(m-k)$ independent homogeneous restrictions upon the $(m+1)$ w_j coefficients, with m^*+1 of these coefficients, $w_0, \dots, w_{m^*}$, lying upon a k^*th order polynomial by assumption and being linked by m^*-k^* homogenous differencing restrictions. Because of this latter characteristic, $w_{k^*+1}, \dots, w_{m^*}$ can be expressed as linear functions of $w_0, \dots, w_{k^*}$, thereby reducing the number of coefficients involved in the $(m-k)$ restrictions from $(m+1)$ to $(m+1)-(m^*-k^*)$. Now two cases need to be distinguished, according to whether the assumed polynomial order is less than or equal to the true lag length, and the bias situation in each instance is recorded in the following two propositions:

Proposition 1

When $k < m^*$, the PDL(m, k) estimator is certainly biased if $m - m^* > k^*$.

Proposition 2

When $k \geq m^*$, the PDL(m, k) estimator is certainly biased if $m - k > k^*$.

Proofs of these and other propositions presented below are provided in Hendry and Pagan (1980).

Propositions 1 and 2 indicate that the effects of an incorrect choice of polynomial and lag order are complex. Frost's conjecture cited in the analysis of B is borne out, but there may not be a monotonic decline in bias; until $k \geq m^*$ the possibility of bias is *independent* of the assumed polynomial order. Certainly the analysis reveals that the choice of m and k cannot be done arbitrarily, and that indifference to the selection of these parameters is likely to produce biased estimators. Careful preliminary thought about the likely values of m^* and k^* is therefore of some importance to any investigation.

To complete this sub-section, we briefly review other proposals for selecting m^* and k^*. Because PDL(m_1, k_1) and PDL(m_2, k_2) models are generally non-nested, many of the methods advocated for the selection of one model as best out of a range of models might be applied [these are surveyed in Amemiya (1980)]. The only evidence of the utility of such an approach is to be found in Frost (1975), where m and k are chosen in a simulation experiment by maximizing $\bar{R}^2$ [as recommended by Schmidt and Waud (1973)]. There it is found that a substantial upward bias in the lag length results, an outcome perhaps not unexpected given the well-known propensity of $\bar{R}^2$ to augment a regression model according to whether the t-statistic of the augmenting variable exceeds unity or not. Teräsvirta (1980a), noting that the expected residual variance of a model is the sum of two terms—the true variance and a quadratic form involving the bias induced by an incorrect model—showed that the bias in Frost's experiments was very small once a quadratic polynomial was selected, causing little difference between the expected residual variances of different models. Consequently, the design of the experiments plays a large part in Frost's conclusions. It may be that other criteria which can be expressed in the form $g(T)(1-R^2)$, where $g(\cdot)$ is a known function [examples being Akaike (1972), Mallows (1973), Deaton (1972a) and Amemiya (1980)] would be superior to $\bar{R}^2$ as a model selection device, and it would be of interest to compute their theoretical performance for Frost's study. Nevertheless, both Sawa (1978) and Sawyer (1980) have produced analyses suggesting that none of these criteria is likely to be entirely satisfactory [see also Geweke and Meese (1981) for a related analysis].

Harper (1977) proposed to choose m and k using the various model mis-specification tests in Ramsey (1969). The rationale is that an incorrect specification could lead to a disturbance with a non-zero mean. This contention is correct whenever the lag length and polynomial degree is understated, but in other circumstances need not be valid.

A final technique for selecting m and k has been provided by Teräsvirta (1980a, 1980b). This is based upon the risk of an estimator $\hat{\boldsymbol{\beta}}$ of $\boldsymbol{\beta}$ [where $\boldsymbol{\beta}$ would be $\boldsymbol{w}$ for a PDL$(\cdot)$ like (61)]:

$$r(\boldsymbol{\beta},\hat{\boldsymbol{\beta}})=E(\hat{\boldsymbol{\beta}}-\boldsymbol{\beta})'\boldsymbol{Q}(\hat{\boldsymbol{\beta}}-\boldsymbol{\beta}), \tag{73}$$

where $\boldsymbol{Q}$ is a positive definite matrix, frequently taken to be $\boldsymbol{Q}=\boldsymbol{I}$ or $\boldsymbol{Q}=\boldsymbol{Z}'\boldsymbol{Z}$. From Judge and Bock (Chapter 10 in this Handbook, eq. (3.7)) a PDL(m,k) estimator of $\boldsymbol{\beta}$ exhibits lower risk than OLS when $\boldsymbol{Q}=\boldsymbol{I}$ if and only if:

$$\sigma_u^{-2}\boldsymbol{\beta}'\boldsymbol{R}'\left(\boldsymbol{R}(\boldsymbol{Z}'\boldsymbol{Z})^{-1}\boldsymbol{R}'\right)^{-1}\boldsymbol{R}\boldsymbol{\beta}\le 1, \tag{74a}$$

or when $Q=\boldsymbol{Z}'\boldsymbol{Z}$ if and only if [eq. (3.8)]:

$$\sigma_u^{-2}\boldsymbol{\beta}'\boldsymbol{R}'\left(\boldsymbol{R}(\boldsymbol{Z}'\boldsymbol{Z})^{-1}\boldsymbol{R}'\right)^{-1}\boldsymbol{R}\boldsymbol{\beta}\le m-k. \tag{74b}$$

Replacing $\boldsymbol{\beta}$ and σ_u^2 by their OLS estimators in (74), test statistics that the conditions are satisfied can be constructed using a non-central F distribution. A disadvantage of this rule is that it applies strictly to a comparison of any particular PDL(m,k) estimator and OLS, but does not provide a way of comparing different PDL estimators; ideally, a sequential approach analogous to Anderson's discussed above is needed. Another problem arises when the lag length is overspecified. Teräsvirta shows that the right-hand side of (74b) would then be $m-k-p_2$, where p_2 is the degree of overspecification of the lag length. As p_2 is unknown, it is not entirely clear how to perform the test of even one (m,k) combination against OLS. Teräsvirta (1980b), utilizing Almon's original investment equation as a test example, discusses these difficulties, and more details can be found therein.

3.4. *Weaker restrictions on lag weights*

The essence of PDL procedures is the imposition of linear deterministic constraints upon the parameters and there have been a number of suggestions for widening the class or allowing some variation in the rigidity of the restrictions. Thus, Hamlen and Hamlen (1978) assume that $\boldsymbol{w} = \boldsymbol{A}\boldsymbol{\gamma}$, where $\boldsymbol{A}$ is a matrix of cosine and sine terms, while a more general proposal was made by Corradi and Gambetta (1974), Poirier (1975) and Corradi (1977) to allow the lag distribution to be a spline function. Each of these methods is motivated by the "close approximation" idea but are capable of being translated into a set of linear restrictions upon $\boldsymbol{w}$ [see Poirier (1975) for an example from the spline lag]. The spline lag proposal comes close to the PDL one as the idea is to have piecewise polynomials and the restrictions are a combination of differencing ones and others representing join points (or knots). In both cases, however, users should present an F-statistic on the validity of the restrictions—arguments from numerical analysis on the closeness of approximation of trigonometric functions and "natural cubic splines" are scarcely convincing, however suggestive they might be. Although the spline lag proposal does not yet seem to have had widespread application, attention might be paid to the use of a variety of differencing restrictions upon any one set of parameters. For example, if the location of the mode was important and it was believed that this lay between four and eight lags, low order differencing might be applied for lags up to four and after eight, and very high order differencing restrictions between four and eight. Thus, one could constrain the distribution in the regions where it matters least and to leave it relatively free where it is likely to be changing shape most rapidly. There is no compelling reason why an investigator must retain the same type of linear restrictions throughout the entire region of the lag distribution.

Shiller (1973, 1980) has made two proposals that involve stochastic differencing restrictions—if one wishes to view his approach in a classical rather than Bayesian framework as was done by Taylor (1974)—and a good account of these has been given by Maddala (1977). The first of Shiller's methods has $(1-L)^k w_j = \varepsilon_j$, where $\varepsilon_j \sim \text{i.i.d.}(0, \sigma_w^2)$. One might interpret this as implying that w_j is *random across the lag distribution*, where the mean $\bar{w}_j$ lies on a kth order polynomial and the error $w_j - \bar{w}_j$ is autocorrelated. Of course if one wanted to press this random coefficients interpretation it would make more sense to have $w_j = \bar{w}_j + \varepsilon_j$ as in Maddala's "Bayesian Almon" estimator (p. 385). In keeping with the randomness idea it would be possible to allow the coefficients to be *random across time* as in Ullah and Raj (1979), even though this latter assumption does not "break" collinearity in the same way as Shiller's estimator does, and seems of dubious value unless one suspects some structural change. Shiller's second suggestion is to use $(1-L)^k \log w_j = \varepsilon_j$. Mouchart and Orsi (1976) also discuss alternative parameterisations and associated prior distributions.

Shiller terms his estimators "smoothness priors" (SP) and a number of applications of the first estimator have appeared, including Gersovitz and Mackinnon (1978) and Trivedi, Lee and Yeo (1979). Both of these exercises are related to Pesando's (1972) idea that distributed lag coefficients may vary seasonally and SP estimators are suited to this context where there are a very large number of parameters to be estimated. For a detailed analysis, see Hylleberg (1981).

It is perhaps of interest to analyse the SP estimator in the same way as the PDL estimator. Because of the differencing restrictions underlying the SP estimator, in what follows it is convenient to refer to a correct choice of k as one involving a "true" polynomial order. Furthermore, the assumption used previously of the existence of a "true" model holds again; this time there being an added dimension in the variance parameters $\boldsymbol{\Sigma} = \operatorname{diag}\{\sigma_j^2\}$. Under this set of conditions, the SP estimator of $\boldsymbol{\beta}$, $\tilde{\boldsymbol{\beta}}_{\mathrm{SP}}$ is [Shiller (1973)]:[9]

$$\tilde{\boldsymbol{\beta}}_{\mathrm{SP}} = (\boldsymbol{Z}'\boldsymbol{Z} + \boldsymbol{R}'\boldsymbol{\Sigma}^{-1}\boldsymbol{R})^{-1}\boldsymbol{Z}'\boldsymbol{y}. \tag{75}$$

From (75), $\tilde{\boldsymbol{\beta}}_{\mathrm{SP}}$ is a biased estimator of $\boldsymbol{\beta}$ but, with the standard assumption $\lim_{T\to\infty} T^{-1}\boldsymbol{Z}'\boldsymbol{Z} > \boldsymbol{0}$, $\tilde{\boldsymbol{\beta}}_{\mathrm{SP}}$ is a consistent estimator of $\boldsymbol{\beta}$ provided $\hat{\boldsymbol{\beta}}_{\mathrm{OLS}}$ is. Accordingly, under-specification of the lag length will result in $\tilde{\boldsymbol{\beta}}_{\mathrm{SP}}$ being inconsistent.

To obtain some appreciation of the consequences of over-specification of the lag length or mis-specification of the polynomial order, it is necessary to set up a benchmark. Because $\tilde{\boldsymbol{\beta}}_{\mathrm{SP}}$ is biased, this can no longer be the true value $\boldsymbol{\beta}$, and for the purpose of enabling a direct comparison with the PDL estimator it is convenient to assess performance relative to the expected value of $\tilde{\boldsymbol{\beta}}_{\mathrm{SP}}$ if $\boldsymbol{R}$ and $\boldsymbol{\Sigma}$ were known. This is only one way of effecting a comparison—for example, the impact upon risk used by Trivedi and Lee (1979) in their discussion of the ridge estimator would be another—but the present approach enables a sharper contrast with the material in Section 3.3 above.

So as to focus attention upon the parameters (m, k) only, $\boldsymbol{\Sigma}$ is taken to be known in the propositions below, and it is only $\boldsymbol{R}$ that is mis-specified at $\overline{\boldsymbol{R}}$. Then:

$$E(\tilde{\boldsymbol{\beta}}_{\mathrm{SP}}) = (\boldsymbol{Z}'\boldsymbol{Z} + \boldsymbol{R}'\boldsymbol{\Sigma}^{-1}\boldsymbol{R})^{-1}E(\boldsymbol{Z}'\boldsymbol{y}), \tag{76a}$$

$$E(\overline{\boldsymbol{\beta}}_{\mathrm{SP}}) = (\boldsymbol{Z}'\boldsymbol{Z} + \overline{\boldsymbol{R}}'\boldsymbol{\Sigma}^{-1}\overline{\boldsymbol{R}})^{-1}E(\boldsymbol{Z}'\boldsymbol{y}), \tag{76b}$$

and $E(\overline{\boldsymbol{\beta}}_{\mathrm{SP}}) \neq E(\tilde{\boldsymbol{\beta}}_{\mathrm{SP}})$ unless $\overline{\boldsymbol{R}}'\boldsymbol{\Sigma}^{-1}\overline{\boldsymbol{R}} = \boldsymbol{R}'\boldsymbol{\Sigma}^{-1}\boldsymbol{R}$.

Propositions 3 and 4 then record the effects of particular incorrect choices of m and k:

Proposition 3

Overstatement of the lag length with a correct polynomial degree need not induce a difference in $E(\overline{\boldsymbol{\beta}}_{\mathrm{SP}})$ and $E(\tilde{\boldsymbol{\beta}}_{\mathrm{SP}})$.

[9]Assuming that the covariance matrix of $\boldsymbol{u}$ in (67) is $\boldsymbol{I}$. The more general assumption that it is $\sigma_u^2\boldsymbol{I}$ (to be used in a moment) would require $\boldsymbol{\Sigma}$ to be defined as the variance ratios $\sigma_u^{-2}\sigma_j^2$ in (78). Note that we use $\boldsymbol{\beta}$ rather than $\boldsymbol{w}$ for results which are not specific to PDLs, as with (75) for general $\boldsymbol{R}$ and $\boldsymbol{\Sigma}$.

Proposition 4

Whenever the polynomial order is under or overstated with a correct lag length, $E(\bar{\beta}_{\mathrm{SP}})$ is different from $E(\tilde{\beta}_{\mathrm{SP}})$.

These propositions reveal that the SP estimator provides an interesting contrast to the PDL estimator, being insensitive to over-specification of the lag length but not to an over-specified polynomial order. This last result is a surprising one, and its source seems to be that:

$$(1-L)^k\beta_j = \varepsilon_j \Rightarrow (1-L)^{k+1}\beta_j = (1-L)\varepsilon_j,$$

and the autocorrelation induced by an over-specified polynomial is ignored in the construction of $\tilde{\beta}_{\mathrm{SP}}$.

As the analysis above demonstrates, the choice of $\boldsymbol{\Sigma}$ is an important one. Shiller's original treatment was a Bayesian one and $\boldsymbol{\Sigma}$ represented the variance of a prior distribution. Because $\boldsymbol{\Sigma}$ depends upon the units of measurement of $\boldsymbol{Z}$, his second specification involving the logs of β_j has greater appeal as better prior information is likely to be available concerning the percentage changes in β_j. If a Bayesian treatment of this model is desired, the choice of prior is clearly critical, and the papers by Mouchart and Orsi (1976) and Trivedi and Lee (1981) contain extensive examinations of this issue.

A more classically oriented approach derives from the special case when $\boldsymbol{\Sigma} = \sigma_\beta^2 \boldsymbol{I}$. The SP estimator becomes:

$$\tilde{\beta}_{\mathrm{SP}} = (\boldsymbol{Z}'\boldsymbol{Z} + h\cdot \boldsymbol{R}'\boldsymbol{R})^{-1}\boldsymbol{Z}'\boldsymbol{y}, \tag{77}$$

where $h = \sigma_\beta^{-2}\sigma_u^2$, and the formal similarity of (77) to the ridge regression estimator has prompted a number of authors—Hill and Johnson (1975), Maddala (1977) and Ullah and Raj (1979)—to utilize the principles from that literature to select h (also see Section 3.5).

Finally, mention should be made of Teräsvirta (1980a, 1980b) who proposed to select h and k by reference to the risk for different combinations of these parameters in a similar fashion to the methodology described in Section 3.3). Fomby (1979) is another to select h according to the mean square error of estimators.

3.5. *Alternative estimators*

The treatment so far has effectively assumed that lack of bias was an appropriate way to classify different estimators and that the unrestricted estimates would be selected in preference to any restricted estimator if the restrictions were invalid.

Such a position is by no means universally accepted, and there have been advocates for imposing restrictions, even if invalid, if this reduces a specified loss function. For example, Amemiya and Morimune (1974) selected an optimal polynomial order by minimizing a particular loss function and Trivedi and Pagan (1979) used this loss function to compare various restricted estimators. Essentially the argument for such procedures is that of "good forecasting" but there is another tradition of biased estimation that aims at "breaking" the collinearity between the lagged values of z_t that may be the cause of badly determined unrestricted estimates. As there have been a few applications of these ideas to the estimation of distributed lag models, we propose to make some comments upon the direction and utility of this research.

We first focus on the question of whether restricted and other "improved" estimators (e.g. Stein–James), do in fact yield substantial reductions in a loss function relative to the unrestricted estimator (OLS). Yancey and Judge (1976, p. 286) have ably summed up the importance of this question: "...there has been no rush in econometrics to abandon maximum likelihood estimators... Possibly one reason for the reluctance to change estimators may be uncertainty relative to the magnitude of the risk gains from changing estimation rules."

Conventionally, the loss function has been taken as in (73) for an estimator $\hat{\boldsymbol{\beta}}$ of $\boldsymbol{\beta}$ and $\boldsymbol{Q}$ positive definite. There has, however, been little agreement over $\boldsymbol{Q}$. Schmidt and Sickles (1976) set $\boldsymbol{Q} = \boldsymbol{Z}'\boldsymbol{Z}$, while in Aigner and Judge (1977), $\boldsymbol{Q} = \boldsymbol{I}$ and $\boldsymbol{Z}'\boldsymbol{Z}$ were selected. Strawderman (1978) notes that: "The former case seems appropriate when an error of any given magnitude is equally serious to all coordinates, while the latter case corresponds to the usually fully invariant situation" (p. 626) and shows that adaptive ridge estimators would be minimax if $\boldsymbol{Q} = (\boldsymbol{Z}'\boldsymbol{Z})^2$. Probably $\boldsymbol{Q} = \boldsymbol{Z}'\boldsymbol{Z}$ is interesting if the goal is forecasting as Amemiya and Morimune (1974) stress the relationship of this loss function to the conditional mean square prediction error. The relative risks of different estimators when $\boldsymbol{Q} = \boldsymbol{I}$ or $\boldsymbol{Z}'\boldsymbol{Z}$ feature inequalities involving the eigenvalues of $(\boldsymbol{Z}'\boldsymbol{Z})^{-1}\boldsymbol{R}'[\boldsymbol{R}(\boldsymbol{Z}'\boldsymbol{Z})^{-1}\boldsymbol{R}']^{-1}\boldsymbol{R}(\boldsymbol{Z}'\boldsymbol{Z})^{-1}$. It does not seem possible to say much about these inequalities without specifying $\boldsymbol{R}$ and $\boldsymbol{\beta}$. When $\boldsymbol{R} = \boldsymbol{I}$, Aigner and Judge have pointed out that it is the eigenvalues of $(\boldsymbol{Z}'\boldsymbol{Z})$ which are required, and Trivedi (1978) exploited this result to show that the risk reductions obtained with the Stein–James estimator on the imposition of false restrictions *decreased* with the degree of autocorrelation of $\{z_t\}$. The poor performance of the Stein–James estimator in the presence of collinear data has also been observed by Aigner and Judge (1977).

Ridge regression has already been mentioned in the context of the "smoothness priors" estimator of Section 3.4 and at times has been put forth as a direct estimator of (61). As the literature on ridge techniques is vast, our comments pertain only to those features that have been of concern to investigators estimating distributed lag models.

Foremost amongst these has been the determination of the parameter $h = \sigma_u^2/\sigma_w^2$ in the estimator:

$$\tilde{w} = (Z'Z + h\cdot D)^{-1}Z'y, \tag{78}$$

where $D = R'R$ in the case of "smoothness priors". In practice, h has frequently been determined in an iterative way [see, for example, Maddala (1977) and Spencer (1976)], based on sample information.

The question that arises with such data-based priors is whether there are implicit restrictions being placed upon the estimated coefficients. Some analysis of these schemes seems necessary as Spencer found that his iterated Shiller estimator converged to the Almon estimator, i.e. $\hat{\sigma}_w^2 = 0$, and there have been other reports that ridge estimators have a tendency to produce rectangular distributions. For example, Maddala (1977) says: "...the Hoerl and Kennard method and the Lindley–Smith method are not too promising for distributed lag estimation". As shown in Hendry and Pagan (1980): *the iterated Shiller estimator has a tendency to converge to the Almon estimator*. Whether it will terminate at the Almon estimator or not depends upon the existence of a local minimum to the function of which the iterative rules are the derivatives [denoted by $S(w, \sigma_w^2)$] since $S(\cdot)$ has a global minimum at $\sigma_w^2 = 0$. Although it is hard to decide on the likelihood of $\sigma_w^2 \neq 0$ on theoretical grounds, nevertheless, one might conjecture that as the data become more collinear, the iterated Shiller estimator will converge to the Almon estimator. This occurs because, with collinear data, large variations in $\hat{w}$ result in only small changes in the residual sum of squares and it is this term which must rise to offset the reduction in $S(\cdot)$ caused by falling σ_w^2. It would also seem that, as the lag length m increases, this tendency would be intensified. Some theoretical work on this question is available in Trivedi, Lee and Yeo (1979).

A similar situation exists if the Lindley–Smith scheme which sets $\hat{\sigma}_w^2 = (m+1)^{-1}\Sigma(\hat{w}_j - \bar{\hat{w}})^2$ is adopted, as the analysis can be repeated to show that the global minimum occurs as $\hat{\sigma}_w^2 \to 0$, i.e. where $\hat{w}_j$ are equal $\forall_j$. This indicates that iterating with this scheme tends to produce rectangular lag distributions *regardless* of the "true" lag distribution. Again, this is only a tendency, but it is disturbing that the possibility exists that a lag distribution can be obtained that simply reflects the way in which the "prior" was constructed and which may bear no resemblance to whatever prior knowledge does exist. Thus care is needed in the application of these estimators and more analytical work is necessary before they become widely used.

The above problems apply only to those estimators that choose σ_w^2 in some data-based way and not when σ_w^2 is selected on a priori grounds. Even then, one must have some misgivings about shrinkage estimators that are supposed to be "robust" and to produce "reasonable" answers in any situation, irrespective of the true model and however badly specified the approximation. This is a major problem in distributed lags where m is unknown, as few diagnostic tests have yet been developed for the detection of specification errors for "shrinkage" estimators and, until tests are better developed, one must be sceptical of their general application. There appear to have been a large number of estimators proposed within the ridge class—see Vinod (1978)—but the amount of work done on their implications and range of applications seems quite small. The "classical" estimators discussed in preceding sections have been subject to intensive analysis and we would be loathe to discard them for fashionable estimators derivative from other fields.

3.6. *Reformulations to facilitate model selection*

The many difficulties noted above for choosing m^* and k^* even when it is known that $m^0 = 0$ and m^* is finite are in practice exacerbated by the failure of $\{\hat{u}_t\}$ to be white noise and the dubiety of asserting that z_t is strongly exogenous for w; yet the joint failure of these entails that $\hat{w}$ will be inconsistent and that, for example, DW statistics have an incorrect significance level (although LM tests for residual autocorrelation remain valid). Moreover, as shown in Section 2.6, paragraph (g), "correcting" for residual autocorrelation by (say) Cochrane–Orcutt or other autoregressive processes involves untested common factor assumptions, the invalidity of which would throw into question the very assumption that m^* is finite (see Section 2.4).

When y and z are inherently positive, and the static equilibrium postulate is $y = hz$, then $\ln y = \ln h + \ln z$ is an admissible transform and suggests an error correction rather than a distributed lag approach since the latter is a "special case" of the former in a unit elasticity world [see Section 2.6, paragraph (h)]. Moreover, the error process need no longer be truncated to ensure y_t, $z_t > 0$ and even for $h = 1$, additive "modifiers" (i.e. additional variables) do not produce logical inconsistencies [which they would in (61) unless restricted to vanish in equilibrium]. Such considerations become increasingly important in formulations where the $\{w_j\}$ depend on economic variables [as in Tinsley (1967) and Trivedi (1970b)]: these models pose no insurmountable estimation problems, but raise awkward selection issues when so many features have to be jointly chosen from the data.

Finally, as noted in Section 3.2 above, even within the PDL class reformulations of the polynomials can greatly economise on parameters; the suggestion in Sargan (1980b) of using $\Psi_i(j) = (m+1-j)^i$ so that:

$$\sum_{j=0}^{m} w_j z_{t-j} = \sum_{i=0}^{k} \gamma_i^* \left\{ \sum_{j=0}^{m} (m+1-j)^i z_{t-j} \right\} = \sum_{i=0}^{k} \gamma_i^* z_{i,t}^+ \tag{79}$$

can reduce the required number of $\{\gamma_i^*\}$ well below $\{\gamma_i\}$ (or conversely) depending on the lag shape. For example:

$$w_j = (1-0.1j)^2, \qquad j = 0,\dots,9, \tag{80}$$

only involves *one* γ^* (i.e. γ_2^*) but *three* γ's. *Short* distributed lags in general dynamic equations often can be parameterised along these lines [see, for example, Hendry and Ungern-Sternberg (1981)].

4. Infinite distributed lags

4.1. Rational distributed lags

Almost all individual estimated equations in macro-econometric systems have been members of the general class of Autoregressive-Moving Average Models with "Explanatory" variables, denoted by ARMAX$(\cdot)$ and written as:

$$\alpha_0(L)y_t = \sum_{j=1}^{n} \alpha_j(L)z_{jt} + \alpha_{n+1}(L)e_t, \qquad e_t \sim I\mathcal{N}(0,\sigma_e^2) \tag{81}$$

where:

$$\alpha_i(L) = \sum_{j=0}^{m_i} \alpha_{ij}L^j, \qquad \alpha_{0,0} \equiv \alpha_{n+1,0} \equiv 1, \tag{82}$$

and there are *no* polynomial factors common to *all* the $\alpha_j(L)$. Then (81) is said to be ARMAX $(m_0, m_1, \dots, m_n, m_{n+1})$ [generalising the AD$(\cdot)$ notation with the last argument showing the order of the moving average error process]. The $\{z_{jt}\}$ in (81) are not restricted to be "exogenous" in the sense defined in Section 2, and could be endogenous, weakly or strongly exogenous or lagged values of variables endogenous elsewhere in the systems, and might be linear or nonlinear transformations of the original (raw) data series. However, it is assumed that the parameters of (81) are identifiable and constant over any relevant time period.

The formulation in (81) can be expressed equivalently as:

$$y_t = \sum_{i=1}^{n} \frac{\gamma_i(L)}{\delta_i(L)} z_{it} + \frac{\phi(L)}{\rho(L)} e_t, \tag{83}$$

where all common factors have been cancelled in the ratios of polynomials. An important special case of (83) is where $\phi(L)=\rho(L)$ [i.e. $\alpha_0(L)=\alpha_{n+1}(L)$ in (81) which we call the Rational Distributed Lag (RDL)]:

$$y_t=\sum_{i=1}^{n}\frac{\gamma_i(L)}{\delta_i(L)}z_{it}+e_t=\sum_{i=1}^{n}w_i(L)z_{it}+e_t, \tag{84}$$

and like the AD($\cdot$) class, RDL is *defined here* to have white-noise disturbances relative to its information set. As discussed in Section 3 above, (84) generalises (59) to infinite lag responses. Thus, ARMAX($\cdot$) is RDL with ARMA($\cdot$) errors or AD($\cdot$) with MA($\cdot$) errors, and if any denominator polynomial is of non-zero order, some of the derived lag distributions are infinite. Relative to the class defined by (81) the parameter spaces of AD($\cdot$) and RDL($\cdot$) models constitute a set of measure zero in the general parameter space. In practical terms, however, all of the models in this chapter constitute more or less crude first approximations to complicated underlying economic processes, and for high order lag polynomials, provide rather similar data descriptions. Indeed, if all of the roots of the $\delta_i(L)$ $(i=1,\dots,n)$, $\rho(L)$ and $\phi(L)$ polynomials in (83) lie outside the unit circle, by expanding the inverses of these polynomials as power series, a wide range of alternative approximations can be generated (extending the analysis in Section 2.3 above). But selecting equations *purely* on the basis of "goodness of approximation" is of little comfort if the resulting model does not correspond to either a behavioural or a structural relationship, and as stressed below derived parameters (such as mean lags, long-run outcomes, etc.) can differ greatly between "similar" approximations.

Consequently, the choice of model *class* relevant to empirical research does not seem to us to be an issue of principle, but a matter of whether: (a) the formulation is coherent with available theory and/or prior information concerning structural/behavioural relationships; (b) the parameterisation is parsimonious with easily understood properties; and (c) the equation is easily manipulated, estimated (when its form is known) and selected (when the exact orders of all the lag polynomials, relevant regressors, etc. are not known a priori). These criteria may conflict since simple, easily estimated equations may not provide the most parsimonious representations or may be non-structural, etc. Moreover, if the unknown data generation process takes one form (e.g. an error correction AD(1,1)) but an encompassing model is investigated (say, ARMAX(1,1,1)), then parsimony cannot be claimed even if a "minimal representation" of the dynamics is selected. For example, (43) becomes:

$$y_t=\frac{(\beta_1+(1-\beta_3-\beta_1)L)}{1-\beta_3L}z_t+\frac{e_t}{1-\beta_3L}=\frac{(\gamma_{10}+\gamma_{11}L)}{1-\delta_{11}L}z_t+\frac{e_t}{1-\rho_1L}, \tag{85}$$

which necessitates four rather than two parameters in the absence of knowledge that $\delta_{11} = \rho_1$ and $\gamma_{10} + \gamma_{11} = 1 - \delta_{11}$, the imposition of which restrictions depends on the relevant behavioural theory. Conversely, an inadequate dynamic–stochastic representation entails inconsistency of parameter estimates and a loss of structural invariance, so both data coherency and theory validity are necessary, and such considerations must take precedence over arguments concerning approximation accuracy, generality of class, etc.

An important consequence for econometric analysis (as against data description) is that closely similar dynamic model specifications can entail rather different behavioural implications. To isolate some of the differences, consider the three simplest cases of partial adjustment (PA), error correction (ECM) and RDL, with one strongly exogenous variable $\{z_t\}$, each model defined to have white noise disturbances relative to its information set:

$$\Delta y_t = \gamma(\beta z_t - y_{t-1}) + u_t \quad \text{(PA)}, \tag{86}$$

$$\Delta y_t = \alpha \Delta z_t + \gamma(\beta z_{t-1} - y_{t-1}) + v_t \quad \text{(ECM)}, \tag{87}$$

$$y_t = (1-(1-\gamma)L)^{-1}\gamma\beta z_t + e_t \quad \text{(RDL)}. \tag{88}$$

The three models have the same non-stochastic, static equilibrium solution, namely:

$$y = \beta z = y^e \quad \text{(say)}, \tag{89}$$

and so could be interpreted as alternative implementations of a common theory. Expressed in ECM form, however, (86) and (88) are:

$$\Delta y_t = \gamma \Delta y_t^e + \gamma(y_{t-1}^e - y_{t-1}) + u_t \quad \text{(PA)}, \tag{90}$$

$$\Delta y_t^* = \gamma \Delta y_t^e + \gamma(y_{t-1}^e - y_{t-1}^*) \quad \text{(RDL)}, \tag{91}$$

where $y_t = y_t^* + e_t$. Thus, both (86) and (88) constrain the response to changes in y^e and to past disequilibria to be the same, a strong specification which may well be at variance with observed behaviour [compare the arguments for the "optimal partial adjustment" model in Friedman (1976)]. Also, the disequilibria in the PA/ECM models are measured differently from those of the RDL in that the latter are relative to y_{t-1}^* rather than y_{t-1}. Accordingly, an RDL formulation is appropriate to behaviour wherein agents ignore the impact of past disturbances on the measured data, concentrating instead upon the "permanent" component y_{t-1}^* so that disturbances in any period are *not* transmitted into future behaviour unlike in PA/ECM models.

Which formulation of the impact on plans of past disturbances is most appropriate to any particular situation must be an empirical matter, although in general the truth probably lies at neither extreme since adjustments to pure shocks are likely to differ from responses to past plans; and equation disturbances are anyway composites of measurement errors and *all* mis-specifications as well as shocks. Since the RDL form in (88) generalises easily to:

$$y_t = (1-(1-\gamma)L)^{-1}(\alpha\Delta z_t + \beta\gamma z_{t-1}) + e_t, \tag{92}$$

which still has (89) as its static solution but corresponds to:

$$\Delta y_t^* = \alpha\Delta z_t + \gamma(\beta z_{t-1} - y_{t-1}^*) \quad \text{(with } \alpha \text{ unrestricted)}, \tag{93}$$

the real distinction between AD($\cdot$) and RDL lies in their respective stochastic specifications. Yet investigators alter error assumptions for convenience without always acknowledging the consequential changes entailed in *behavioural* assumptions.

With the conventional practice of "allowing for autocorrelated residuals", distinctions between model types become hopelessly blurred since disturbances in ARMAX($\cdot$) models are transmitted k periods into the future if $\phi(L)/\rho(L)$ is of degree k in L (and hence k is infinite if $\rho(L)$ is not of degree zero).

The literature on ARMAX models and all their special cases is vast and it is quite beyond the scope of this chapter to even reference the main relevant papers, let alone adequately survey the results [see, among many others, Anderson (1980), Aigner (1971), Nicholls et al. (1975), Harvey and Phillips (1977), Osborn (1976), Palm and Zellner (1980), Wallis (1977), Zellner (1979), Harvey (1981, Section 7.3) and Davidson (1981) and the references therein]. When all z_{it} are strongly exogenous in (83) separate estimation of $\gamma_i(\cdot)/\delta_i(\cdot)$ and $\phi(\cdot)/\rho(\cdot)$ is possible [see Pesaran (1981), who also derives several LM-based residual diagnostic tests]. However, this last result is not valid if any of the z_i are Granger caused by y in the model information set, nor will conventionally estimated standard errors provide a useful basis for model selection until the residuals are white noise. The general issue of stochastic specification is considered in Section 5 below.

4.2. General error correction mechanisms

There is a close relationship between error correction formulations and "servomechanism" control rules [see Phillips (1954, 1957)]. Hendry and Ungern-Sternberg (1981) interpret α and γ in (87) as parameters of "derivative" and "proportional" feedback controls, introducing the additional interpretation of stock variables in flow equations as "integral controls". Also, Nickell (1980) derives the ECM as the optimal decision rule for an infinite horizon quadratic optimization problem when the "exogenous" variables are neither static nor random walk processes and Salmon (1979) demonstrates that state-variable feedback rules can be reparameterised in servomechanism (and hence, if appropriate, in ECM) form. Thus, the ECM specification is compatible with "forward looking" as well as "servomechanistic" behaviour, and since many static-equilibrium economic theories yield proportionality or homogeneity results (or are transformable thereto), this model form has a potentially large range of applications.

Suppose a given static theory to entail (in logs) that:

$$y=\lambda_0+\lambda_1 z_1+(1-\lambda_1)z_2+\lambda_2 z_3, \tag{94}$$

and no theory-based dynamic speficiation is available. Then the following model at least ensures consistency with (94) in static equilibrium:

$$\begin{aligned}\Delta y_t=\beta_0&+\sum_{i=0}^{m_1}\beta_{1i}\Delta z_{1t-i}+\sum_{i=0}^{m_2}\beta_{2i}\Delta z_{2t-i}+\sum_{i=0}^{m_3}\beta_{3i}\Delta z_{3t-i}\\&+\sum_{i=1}^{m_4}\beta_{4i}\Delta y_{t-i}+\gamma_1(y-z_1)_{t-k_1}+\gamma_2(z_1-z_2)_{t-k_2}+\gamma_3 z_{3t-k_3}+e_t.\end{aligned} \tag{95}$$

Such a formulation has a number of useful features. Firstly, the proportionality restriction is easily tested by adding y_{t-k_4} as a separate regressor, and non-rejection entails that (94) is the static solution of (95) for $\gamma_1 \neq 0$. Generally, low values of the m_i suffice to make e_t white noise and the resulting short distributed lags usually can be adequately represented by one or two Almon polynomial functions, so that the final parameterisation is relatively parsimonious [see, for example, Hendry (1980)]. Also, the k_i are often unity (or four for quarterly—seasonally unadjusted—data); the parameterisation is frequently fairly orthogonal (certainly more so than the levels of variables); and despite the "common" lagged dependent variable coefficient [i.e. $(1+\gamma_1)$] the formulation allows for very different lag distributions of y with respect to each z_i. Moreover, using Δy_t as the dependent variable helps circumvent the most basic "spurious" regressions problem without losing long-run information from using differenced data only [compare, for example, Pierce (1977)]. Also, using Δz_{jt-i} as regressors shows that "levels representations" (of y_t on z_{jt-i}) will have *negative* coefficients at some lag lengths but this does *not* preclude all the *solved distributed lag weights from being positive*. Furthermore, if (95) is a good data description when (94) is a useful equilibrium assertion, then omitting the feedback variables $(y-z_1)_{t-k_1}$ and $(z_1-z_2)_{t-k_2}$ need not produce detectable residual autocorrelation, so that a model in differenced data *alone* might seem acceptable on a "white-noise residual" criterion although it violates homogeneity [see, for example, Davidson et al.

(1978) and as a possible example, Silver and Wallace (1980)]. Finally, in practice, ECMs have successfully reconciled disparate empirical evidence in many areas, as discussed in Section 2.6, paragraph (h).

On a steady-state growth path, the solution of (95) entails that λ_0 in (94) depends on the growth rates of the z_i, a feature which has been criticised by Currie (1981). This issue is closely related to the existence of short-run (apparent) trade-offs (since sequences of above or below average values of Δz_i's will lower or raise the ratios of y to the z_i's in levels), and hence to the "Lucas critique" of (1976) concerning the non-invariance of certain econometric equations to changes in policy rules. Also, Salmon and Wallis (1982) discuss the need for the input variables over the estimation period to "stimulate" responses relevant to later behaviour if structurality is to be retained when policy alters the time profile of some z_{it}'s as well as emphasising the need to correctly allocate dynamic responses to expectation formation and behavioural responses. On both issues, again see Haavelmo (1944).

Constant-parameter linear models are only locally useful and *adaptive* processes in which the β_{ji} (say) depend on other functions (e.g. higher order differences) of the data merit consideration, so that "trade-offs" in effect disappear if they entail exploiting information which actually ceases to be neglected *when* it becomes relevant. Sometimes, such models can be reparameterised as linear in parameters with non-linear variables acting as modifiers when they are non-constant. Also, note that the restriction of ECMs to cases in which y has a unit elasticity response to one variable (or a combination of variables) is not essential since "logit" feedbacks with variable elasticities which eventually converge to unity are easily introduced [see, for example, Hendry and Richard (1983)]; other recent discussions are Salmon (1982), Kloek (1982) and Patterson and Ryding (1982).

We have not discussed partial adjustment models extensively since there are already excellent textbook treatments, but it is interesting that ECM is equivalent to partial adjustment of $(y-z)$ to Δz in (87) (not of y to z unless $\alpha=\gamma\beta$). Thus, on the one hand, care is required in formulating to which variable the PA principle is applied, and on the other hand the equivalence reveals that the ECM in (87) is most heavily dampening of discrepancies from equilibrium due to once-for-all impulses in z_t (so Δz_t goes $\dots,0,\delta,-\delta,0,\dots$), than of permanent changes in the level of z_t, and least for changes in the growth rate of z_t (although integral corrections and higher order derivative responses help mitigate the last two). In the case $\beta=1$, $\alpha\neq\gamma$ in (87) if the data generation process is ECM but this is approximated by a PA model, the impact effect of z on y is generally underestimated although the derived mean lag need not be overestimated since the coefficient of y_{t-1} can be downward biased. Specifically, rewriting (87) (for $\beta=1$) as:

$$y_t=\gamma z_t+(\alpha-\gamma)\Delta z_t+(1-\gamma)y_{t-1}+v_t, \tag{96}$$

when z_t is highly autoregressive, the impact effect will be estimated for PA at around γ (rather than α) and the feedback coefficient at around $(1-\gamma)$, whereas if Δz_t is sufficiently negatively correlated with y_{t-1}, the mean lag will be underestimated. This issue conveniently leads to the general topic of derived statistics in AD$(\cdot)$ models.

4.3. *Derived statistics*

Given the general equation (81), there are many derived statistics of interest including long-run responses, roots of the lag polynomials, summary statistics for the solved lag distributions, etc. and approximate or asymptotic standard errors of these can be calculated in many cases (subject to various regularity conditions). The general problem is given $\hat{\boldsymbol{\theta}} \underset{\text{app}}{\sim} \mathcal{N}(\boldsymbol{\theta}, \boldsymbol{V})$ for a sufficiently large sample size T, to compute $\boldsymbol{f}(\hat{\boldsymbol{\theta}}) \underset{\text{app}}{\sim} \mathcal{N}(\boldsymbol{f}(\boldsymbol{\theta}), \boldsymbol{\Omega})$ where, to first order, $\boldsymbol{\Omega} = \boldsymbol{JVJ}'$ and $\boldsymbol{J} = \partial \boldsymbol{f}(\cdot)/\partial\boldsymbol{\theta}'$ [which, if necessary, can be computed numerically as in Sargan (1980c)]. Of course, normality could be a poor approximation when $\boldsymbol{f}(\boldsymbol{\theta})$ corresponds to (say), a latent root or the mean lag [see, for example, Griliches (1967) who discusses asymmetrical confidence intervals], but in the absence of better approximations it seems more useful to quote the relevant values of $\boldsymbol{f}(\hat{\boldsymbol{\theta}})$ and $\hat{\boldsymbol{\Omega}}$ than provide no summaries at all. However, the mean lag can be a misleading statistic for lag distributions that are highly asymmetrical and is meaningless if the derived lag weights are not all of the same sign. For many distributions, it could be more useful to quote some of the fractiles rather than the first two moments (e.g. the median lag and the time taken for nine-tenths of the response to be completed): as an illustration, when $\beta = 1$ in (87), $\alpha = 0.5$ and $\gamma = 0.05$ yields a mean lag of 10 periods yet has a median lag of one period and 70% of the adjustment has taken place by the mean lag (but 90% adjustment takes 31 periods!). Changing γ to 0.1 halves the mean lag but does not alter the median lag or the percentage response at the mean lag, while reducing the number of periods at which 90% response is reached to 15. For skew distributions there seems little substitute to presenting several fractiles (or some measure of the skewness).

At first sight it may seem surprising that derived estimates of long-run responses might have large standard errors given that the typical spectral shape of economic variables has much of the power near the origin (i.e. in low frequency components)—see Granger (1966). There is no paradox here, however, since highly autoregressive series also have primarily low frequency components yet may provide little long-run information about relations between variables. Alternatively expressed, the long-run of (81) for $n = 1$ is $y = [\alpha_1(1)/\alpha_0(1)]z = Hz$, and if $\alpha_0(L)$ has a root close to unity, estimates of H can fluctuate wildly for seemingly small changes in $\{\hat{\alpha}_{0j}\}$. Thus, valid theoretical information about H

can be of immense value in empirical analysis and, for example, if $H=1$, switching from unrestricted estimation of (84) to (87) can substantially reduce parameter standard errors (and hence forecast error variances). Conversely, for highly autoregressive series much of the sample variability may be due to the dynamics and until this is partialled-out, a misleading picture of the economic inter-relationships may emerge (not just from "spurious" regressions, but also the converse of attenuating important dependencies). For econometric research, there seems little alternative to careful specification of the dynamics—and hence of the "error term" as discussed in Section 5. Note that reparameterisations of the original formulation (81) can allow direct estimation of the long-run response and/or mean lag, etc. as in Bewley (1979a).

5. Stochastic specification

If hazardous inference is to be avoided, it is crucial that the stochastic error generation processes are correctly specified. There is no a priori procedure guaranteeing this: the correct specification can only be decided ex post by using appropriate tests. As noted above, the simple rational lag model:

$$y_t = \beta(L)^{-1}\alpha(L)z_t, \tag{97}$$

where y_t and z_t are scalar endogenous/exogenous variables [as in Dhrymes, Klein and Steiglitz (1970) or Dhrymes (1971)] has the alternative AD$(\cdot)$ form:

$$\beta(L)y_t = \alpha(L)z_t. \tag{98}$$

If it is assumed in either case that there is an additive white-noise error, then (98) can be estimated by OLS or using instrumental variables non-iteratively, whereas (97) requires a non-linear iterative procedure. If the DGP is (98) then the estimation of (97) will produce inconsistent estimates and/or standard errors and vice versa.

When the z_{it} in (81) include both endogenous and exogenous variables, it is convenient to ignore the distinction between the two sets of variables and write x_{it} for the ith variable, specifying the most general ARMA structure in the form of (83) as:

$$\sum_{i=0}^{n} \frac{\gamma_i(L)}{\delta_i(L)}x_{it} = \frac{\phi(L)}{\rho(L)}e_t. \tag{99}$$

Alternatively, defining $\delta_i^* = \prod_{j \neq i} \delta_j$, (81) can be expressed as:

$$\sum_i \rho(L)\gamma_i(L)\delta_i^*(L)x_{it} = \left(\prod_j \delta_j(L)\right)\phi(L)e_t. \tag{100}$$

There is considerable difficulty in testing for the maximal lags in the $\gamma_i(L)$, $\delta_i(L)$, $\phi(L)$ and $\rho(L)$. The simplest possibility is to set all the lags at the largest feasible values, and then to use a sequence of Wald tests to consider whether the maximal lags can be reduced. Even so, the tests required are not simply nested and if maximum lags up to, say, four are specified, then eight parameters per variable are to be estimated which can only be done if the sample size is large. A particular problem with the formulation in (100) is that each x_{it} has applied to it a set of lag polynomial operators which give a large total lag, so effectively reducing the available sample size. On the other hand, if (99) is used, then the latent roots of the $\delta_i(L)$ must be kept away from the unit circle since truncation of the power series corresponding to $(\delta_i(L))^{-1}$ will give very poor approximations if a root is near the unit circle. This problem only arises in estimating some parameter sets so that (99) may give sensible results in some cases, but it suggests that simpler models with fewer adjustable parameters (and less likelihood of such difficulties) may be preferred. One possibility is to assume $\delta_i(L) = \delta(L)\forall_i$:

$$\sum_i \rho(L)\gamma_i(L)x_{it} = \delta(L)\phi(L)e_t. \tag{101}$$

Note that the maximum lags on the x_{it} variables have been considerably reduced when written in this form, and if the same degree $\gamma_i(L)$ is considered as before the number of parameters in the model has been roughly halved. Of course, it is not possible to identify the parameters $\delta(L)$ and $\phi(L)$ separately, so a model of this form can be written:

$$\sum_i \gamma_i(L)x_{it} = \eta_t, \tag{102}$$

$$\rho(L)\eta_t = \phi^*(L)e_t, \tag{103}$$

with the gloss that (102) is a structural equation (which may have the advantages of simplifying the interpretation of the structural dynamics and easing the imposition of restrictions implied by price homogeneity of the economic variables, say, as well as any identifying restrictions) whereas (103) is the ARMA process generating the errors on the structural equation.

Further alternative simplifications are: (a) $\rho(L) = 1$; or (b) $\phi^*(L) = 1$; (a) has the advantage that the equation can be estimated by standard computer programs. If the maximum feasible lag is introduced in the $\gamma_i(L)$ and $\phi^*(L)$, then

Wald tests can be used to decide whether these lags can be reduced. Two problems have been found to arise using this methodology. One is that if the degrees of all the $\gamma_i(L)$, $\phi^*(L)$ are over-stated, then the equation is badly identified and iterative estimation procedures converge very slowly. Conversely, by setting the maximal lags too low there is the problem that there is a tendency for $\phi^*(L)$ to have latent roots biased toward the unit circle [and indeed a non-zero probability in finite samples that a latent root will be found on the unit circle see Kang (1973) and Sargan and Bhargava (1983)]. This may be avoided by sequentially testing (starting with the smallest expected values of the maximum lag) using Box–Pierce type portmanteau autocorrelation statistics or Lagrange Multiplier tests, but such a strategy involves re-estimation if lags are set initially at too low levels.

Assumption (b), that $\phi^*(L)=1$, avoids some of these difficulties, since setting the degree of $\rho(L)$ too large does not lead to lack of identification, and although it may lead to multicollinearity, the asymptotic t-ratios and asymptotic error variance matrices of the resulting estimators still give valid tests. A suggested technique (the COMFAC procedure) for setting provisional lags for the $\gamma_i(L)$ is to write (b) as [see Sargan (1980c)]:

$$\sum_i \psi_i(L)x_{it} = \phi^*(L)e_t. \tag{104}$$

Then for this case or with the special form $\phi^*(L)=1$, the equation is estimated with no restrictions on $\psi_i(L)$, the maximum lags being determined by using the usual t-ratio significance test, or are fixed by taking the maximum lags which are feasible. Different lags on different variables are allowed, based upon significance tests or upon a priori considerations. Then a set of Wald tests can be used to test whether the $\psi_i(L)$ satisfy equations of the form:

$$\psi_i(L) = \rho(L)\gamma_i(L), \qquad i=0,\dots,n. \tag{105}$$

We also write $\psi(L)=\rho(L)\gamma(L)$, where without the suffixes $\psi(L)$ and $\gamma(L)$ denote vectors of lag polynomials (see Section 2.6, paragraph (g), for an exposition).

Equation (105) states that all the $\psi_i(L)$ contain a scalar factor polynomial $\rho(L)$ of degree r, and a set of constraints on the coefficients ψ_{ij} can be calculated which ensure that the common factor exists for each r. If the maximum lag on $\psi_i(L)$ is f_i, then the $\psi_i(L)$, $i=0,\dots,n$, have $n+\sum_{i=0}^{n} f_i$ unknown coefficients, assuming that one of the zero order coefficients is standardised as one, $\rho(L)$ has r unknown coefficients, and the $\gamma_i(L)$ have $\sum_{i=0}^{n}(f_i-r)+n$ unknown coefficients. It follows that there are implicitly nr constraints. For computing purposes, let $m=n+1$ and take summations over $i=1,\dots,m$, renumbering $\{x_{it}\}$ accordingly.

The procedure used to define the constraints is an algorithm, which is equivalent to the "single-division" algorithm used to compute the values of a set of related determinants. The criterion which is a necessary and sufficient condition (except in cases where pivots in the single division algorithm are zero, which occur with probability zero in matrices derived from sample estimates) is that a certain matrix has appropriate rank. The matrix is specified by first defining $N = m(f_1 + 1) - \sum_{j=1}^{m} f_j$, where it is assumed that $f_1 \geq f_2 \geq, \dots, \geq f_m$. Then k is defined as the smallest integer satisfying $(m-1)k \geq f_1 + m - r - N$.

The simplest case [case 1 of Sargan (1980c)] is when $N < f_1 + m - r$, or $k > 0$. For a discussion of the case where $k \leq 0$, see Sargan (1980c).

We define $d_i = f_1 - f_i$, $i = 1, \dots, m$, and then the matrix $\boldsymbol{\Psi}_0$ of N rows, using the notation $\boldsymbol{\psi}_i$ to mean a row vector of the elements $(\psi_{i0}, \psi_{i1}, \dots, \psi_{if_i})$, $i = 1, \dots, m$, is given by:

$$\boldsymbol{\Psi}_0' = \begin{bmatrix} & & 0 & \vdots & \mathbf{0}'_{d_2} & \boldsymbol{\psi}'_3 & \vdots\ \vdots\ \vdots & \mathbf{0}'_{d_m} \\ & \boldsymbol{\psi}'_2 & \boldsymbol{\psi}'_2 & \vdots & & & & \\ \boldsymbol{\psi}'_1 & & & \vdots & & & & \\ & \mathbf{0}'_{d_2} & \mathbf{0}'_{(d_2-1)} & \vdots & \boldsymbol{\psi}'_2 & \mathbf{0}'_{d_3} & & \boldsymbol{\psi}'_m \end{bmatrix}. \tag{106}$$

Here $\mathbf{0}_s$ is used to mean a row vector with s zeros. $\boldsymbol{\Psi}_0$ can be divided vertically into sub-matrices; each sub-matrix has $(d_i + 1)$ rows, and each row of the ith sub-matrix contains a vector $\boldsymbol{\psi}_i$ and d_i zeros. The number of zeros on the left of the row increases in each successive row by one. $\boldsymbol{\Psi}_0$ has $f_1 + 1$ columns. Define also:

$$\boldsymbol{\Psi}_a = \begin{bmatrix} & \boldsymbol{\psi}_1 \\ \mathbf{0}_{d_2} & \boldsymbol{\psi}_2 \\ \mathbf{0}_{d_3} & \boldsymbol{\psi}_3 \\ \cdots & \cdots \\ \cdots & \cdots \\ \mathbf{0}_{d_m} & \boldsymbol{\psi}_m \end{bmatrix}$$

which has m rows and $f_1 + 1$ columns, and then:

$$\boldsymbol{\Psi}_{(k)} = \begin{pmatrix} \boldsymbol{\Psi}_0 & \mathbf{0}_{Nk} & \\ \mathbf{0}_{m1} & \boldsymbol{\Psi}_a & \mathbf{0}_{m(k-1)} \\ \mathbf{0}_{m2} & \boldsymbol{\Psi}_a & \mathbf{0}_{m(k-2)} \\ \cdots & \cdots & \cdots \\ \mathbf{0}_{mk} & \boldsymbol{\Psi}_a & \end{pmatrix}$$

where $\mathbf{0}_{pq}$ is a zero matrix of order $p \times q$.

The condition that is required is that $\Psi_{(k)}$ should be of rank $f_1+k+1-r$. In fact there are too many constraints here if we consider equating determinants of degree $(f_1+k+2-r)$ to zero, but it can be shown that it is sufficient to consider the matrix obtained from $\Psi_{(k)}$ by taking only its first $(f_1+k+m-r)$ rows, and we then obtain $(m-1)r$ constraints by taking a suitable set of $(m-1)r$ determinants which ensure that the reduced matrix is of rank $(f_1+k+1-r)$. A suitable sub-routine (COMFAC) has been written to compute the required determinants given the vectors ψ_i, $i=1,\dots,m$. This can be attached to any single equation estimation computer program to give the appropriate test statistics [see Sargan (1980a) and Mizon and Hendry (1980)]. Let $\bar{r}$ be the true value. If a value of r is specified which is less than $\bar{r}$, then theoretically the asymptotic distribution of the Wald test for the constraints for this value of r is no longer the usual χ^2 distribution, and in practice it has been found that in finite samples the statistic takes much smaller values than would be expected if the regularity conditions which lead to the usual χ^2 approximation were satisfied.

Note that for different values of r we have a set of nested constraints on the coefficients of the $\psi_i(L)$. We need to choose the optimal r, and a sequence of Wald test criteria have the advantage that they can be computed from a single set of estimates of the unconstrained model. Following the pattern discussed towards the end of Section 2.2, Wald test criteria can be computed for a set of increasing values of r, and the corresponding differences $w_r - w_{(r-1)}$ are used as a basis for the choice of the appropriate r.

Alternatively, asymptotically equivalent tests can be made by using likelihood ratio tests, or tests of Durbin/Lagrange Multiplier type. Both methods have their difficulties. The likelihood ratio test is upset by the existence of multiple maxima of the likelihood function if r is specified to be lower than $\bar{r}$ since then we can write: $\rho(L)=\rho_1(L)\rho_2(L)$, where $\rho_1(L)$ is a polynomial of rank r, containing any r of the $\bar{r}$ roots of the polynomial $\rho(L)$. Let:

$$\gamma_1(L)=\rho_2(L)\gamma(L),$$

then

$$\psi(L)=\rho_1(L)\gamma_1(L).$$

This gives a valid factorisation of $\psi(L)$ into a scalar factor of degree r, and a set of function $\gamma_{1i}(L)$ of degree $\bar{f}_i+\bar{r}-r$. Note that there are as many ways of specifying $\gamma_1(L)$ and $\rho_1(L)$ as the number of ways of splitting $\rho(L)$ up into two real factors $\rho_1(L)$ and $\rho_2(L)$ of appropriate degrees. Thus if all these roots of $\rho(L)$ are real there are $\bar{r}!/((\bar{r}-r)!r!)$ different ways of stating an equation with $\rho_1(L)$ and $\gamma_1(L)$ of the given form. From the discussion in Sargan (1975) it follows that a maximum likelihood estimator of $\gamma_1(L)$, $\rho_1(L)$ will find local maxima corresponding to each of these alternative parameterisations with a high probability for large T. If the estimated model is taken which corresponds to the global maximum, then minus the corresponding value of the log-likelihood function is asymptotically distributed as half the maximum of a set of related χ^2 statistics. The differences of the log-likelihood functions which are used to discriminate between alternative values of r then do not have the usual asymptotic distributions. Although the directions of the biases can often be sorted out as in Sargan and Mehta (1983), the magnitude of the bias is not easy to establish, and this suggests that the Wald test is a better alternative. The use of the Durbin three-part-division test discussed in Sargan and Mehta is also somewhat arbitrary in the choice of the exact form of the test and as with the likelihood ratio test involves a sequence of non-linear computer optimisations to decide between all possible values of r.

A similar type of test of the Lagrange multiplier form also can be used, but suffers from the same disadvantages as the three-part-division test. It may be, however, that there is an additional consideration when using the Wald test. If a value of r is specified which is less than the true $\bar{r}$, it has been found in practice that the Wald test for r against $r=0$ will be biased downwards. Denote this criterion by w_r and the corresponding criterion for $\bar{r}$ against $r=0$ by $w_{\bar{r}}$. This latter criterion is asymptotically distributed as χ^2 of $n\bar{r}$ degrees of freedom. The difference $(w_{\bar{r}}-w_r)$ is not distributed asymptotically as a χ^2 of degrees of freedom $n(\bar{r}-r)$ (as would be expected if the conditions for the Wald criteria to have the usual χ^2 asymptotic distribution were satisfied). Thus the use of a sequence of Wald tests each obtained by taking differences of two successive Wald criteria for r against $r-1$ will lead to difference test criteria which are biased. However, when the biases are allowed for, then if, for $r<\bar{r}$ the Wald criteria are below what would be expected for an asymptotic χ^2 of nr degrees of freedom, and if for $r=\bar{r}$ the Wald criterion is not significant considered as an asymptotic χ^2, and for $r=\bar{r}+1$ it is significantly larger than its asymptotic χ^2 confidence limit, then this confirms $\bar{r}$ as the true value.

The equation tested below was part of a three equation model for wage–price inflation in the United Kingdom reported in Sargan (1980a). All the variables are logarithms of the corresponding economic variables as follows: w_t is the official weekly wage rates index, p_t is the consumption price deflator for the official estimates of quarterly real consumption, a_t is the corresponding official estimate of average weekly earnings, S_t is the moving average of the working days lost through strikes in the three years up to and including the current quarter. The equation also included a constant term, three quarterly seasonal dummies, and a linear trend. The coefficients of these variables will not be reported in the following tables. The sample ran from 1953 $Q1$ to 1973 $Q4$. The basic form of equation is illustrated by the OLS estimates:

$$w_t-w_{t-1}+\gamma_1(w_{t-1}-p_{t-1})+\gamma_2(a_{t-1}-w_{t-1})+\gamma_3 S_{t-1}=\eta_t. \qquad (107)$$

Money wage rates move in reaction to real wage rates, the ratio of earnings to wage rates, and a "pushfulness" variable obtained from a strike variable.

The Wald criteria are obtained by taking this as $\gamma'(L)x_t = \eta_t$ and considering autoregressive equations of order up to $r = 4$. Thus, $\psi'(L)x_t = \varepsilon_t$ was estimated by OLS including up to fourth order lags in all the variables of (107). The Wald criteria are of determinental type.

In Table 5.1 the Wald criteria for $r = 1,2,3,4$, are given in column 2 with the appropriate degrees of freedom in column 3. In column 4 the successive differences are given of the Wald criteria. Note that if it were not for the problems raised above it would be appropriate to assume that all of these are asymptotically distributed as independent χ^2 of three degrees of freedom provided $r \geq \bar{r}$.

The Wald criteria for $r = 1$, 2 and 3 are all biased downwards in the sense of being smaller than would be expected if distributed as χ^2 of degrees of freedom 3, 6 and 9, respectively. On the other hand, the $r = 4$ Wald criterion is in line with the assumption that it is distributed as χ^2 of degrees of freedom of 12. The differences of the successive criteria given in column 4 of Table 5.1 have the consequent pattern of biases that those for $r = 1,2,3$, are all biased downwards whereas that for $r = 4$ is above the 1% confidence limit for a χ^2 of 3 D.F. However the pattern of the biases is clear, and confirms that the true value of r is 4. Some simulation evidence on the rejection frequencies of COMFAC tests in simple models is presented in Mizon and Hendry (1980) together with a further empirical application.

Finally, while common factor models can closely approximate the fit of other dynamic processes, derived moments can be very inaccurate. In the AD(1,1) case, for example, if the data generation process is as given in Section 2.6, paragraph (h), but the postulated model is as given in Section 2.6, paragraph (g), the common factor restriction is nearly satisfied if either $\beta_1 = 1$ or β_3 is small [specifically, if $\beta_3(1-\beta_1) \simeq 0$]. Nevertheless, the long-run response and mean lag in paragraph (g) are estimated as $(\beta_1, 0)$ rather than $[1,(1-\beta_1)/\beta_3]$, potentially distorting both the magnitude and the timing of the impact of z on y. This arises because paragraph (g) can be written in ECM form as:

$$\Delta y_t = \beta_1 \Delta z_t + (\beta_3 - 1)(y_{t-1} - \beta_1 z_{t-1}) + e_t, \tag{108}$$

Table 5.1
Wald criteria

r	Wald criteria	D.F.	Differences	D.F.
1	0.00	3	0.00	3
2	0.84	6	0.84	3
3	2.14	9	1.30	3
4	14.53	12	12.39	3

and hence the measure of disequilibrium is $(y-\beta_1 z)_{t-1}$ rather than $(y-z)_{t-1}$. Correspondingly, (108) enforces a static solution equal to the impact effect, hence the mean lag of zero. Since invalidly imposing common factor dynamics can produce misleading results, it cannot be justified as an automatic "solution" to residual autocorrelation even where the final residuals end up being "white noise".

It is clear that theories which legitimately restricted the data analysis to one model *type* would be of great value. Thus, even non-stochastic static equilibrium results can be useful if they constrain the model and are not simply imposed on the data.

6. Dynamic specification in multi-equation models

6.1. *Identification with autoregressive errors*

The problems that arise in multi-equation models are very similar to those discussed in earlier sections: to introduce suitable lag structures which represent correctly our a priori economic intuitions about the behaviour of the variables in the long and the short period, but which are not limited by an over-simplistic specification of the lags in the system nor made over-complex by the confusion of the basic dynamics of the economy with the stochastic processes generating the errors in the system.

Consider this latter problem first. Suppose that in lag operator notation we write the structural equations in the form:

$$\boldsymbol{A}(L)\boldsymbol{x}_t = \boldsymbol{B}(L)\boldsymbol{y}_t + \boldsymbol{C}(L)\boldsymbol{z}_t = \boldsymbol{u}_t, \qquad t = 1,\dots,T, \tag{109}$$

where $\boldsymbol{A}(L) = (\boldsymbol{B}(L), \boldsymbol{C}(L))$ is a matrix of polynomials in the lag operator L, with specified maximum lags on each variable, $\boldsymbol{x}_t$ is a vector of observed variables, made up of n endogenous variables $\boldsymbol{y}_t$, and m strongly exogenous variables $\boldsymbol{z}_t$, and $\boldsymbol{u}_t$ is the vector of errors on the structural equations, all in period t. $\boldsymbol{B}(L)$ is a square matrix such that $\boldsymbol{B}_0$ (the zero lag coefficient matrix) is non-singular. Suppose now that the $\boldsymbol{u}_t$ are generated by an ARMA process of the form:

$$\boldsymbol{R}(L)\boldsymbol{u}_t = \boldsymbol{S}(L)\boldsymbol{e}_t, \tag{110}$$

where $\boldsymbol{R}(L)$ and $\boldsymbol{S}(L)$ are square matrix lag polynomials of degree r and s respectively, and $\boldsymbol{R}_0 = \boldsymbol{S}_0 = \boldsymbol{I}_n$. Our general econometric methodology first requires us to discuss identification for such models. We can find sufficient conditions for identification by formulating the problem as follows. Eliminating

$\boldsymbol{u}_t$ between equations (109) and (110) we obtain:

$$\boldsymbol{R}(L)\boldsymbol{A}(L)\boldsymbol{x}_t = \boldsymbol{S}(L)\boldsymbol{e}_t. \tag{111}$$

Writing this in the form:

$$\boldsymbol{\Psi}(L)\boldsymbol{x}_t = \boldsymbol{S}(L)\boldsymbol{e}_t, \tag{112}$$

where

$$\boldsymbol{\Psi}(L) = \boldsymbol{R}(L)\boldsymbol{A}(L), \tag{113}$$

consider conditions which ensure that the factorisation is unique, for a given $\boldsymbol{\Psi}(L)$ with a given maximum lag on each variable. Clearly, if $\boldsymbol{A}(L)$ and $\boldsymbol{R}(L)$ satisfy (113), then $\boldsymbol{HA}(L)$ and $\boldsymbol{HR}(L)\boldsymbol{H}^{-1}$ satisfy:

$$\boldsymbol{H\Psi}(L) = \left(\boldsymbol{HR}(L)\boldsymbol{H}^{-1}\right)\left(\boldsymbol{HA}(L)\right),$$

and if there are no prior restrictions on the covariance matrix of $\boldsymbol{e}_t$, then if we write $\boldsymbol{A}^*(L) = \boldsymbol{HA}(L)$, $\boldsymbol{R}^*(L) = \boldsymbol{HR}(L)\boldsymbol{H}^{-1}$, $\boldsymbol{S}^*(L) = \boldsymbol{HS}(L)\boldsymbol{H}^{-1}$, and $\boldsymbol{e}^* = \boldsymbol{He}_t$, then the model consisting of equations (109) and (110) with stars on the lag matrices is observationaly equivalent to (111). Conditions similar to those discussed by Hsiao in Chapter 4 of this Handbook are necessary for identification. Sufficient conditions for identification are: (a) that equation (112) is identified when $\boldsymbol{\Psi}_0$ is of the form $\boldsymbol{\Psi}_0 = (\boldsymbol{I} : \boldsymbol{\Psi}_{02})$, and the only constraints specify the minimum lag on each variable. Sufficient conditions for this are those given by Hannan (1970) discussed by Kohn (1979) and Hsiao in Chapter 4 of this Handbook. (b) Conditions which ensure that there is a unique factorisation for (113) subject to the same maximal lag conditions, and $\boldsymbol{B}_0 = \boldsymbol{I}$. (c) Standard conditions for identification, which ensure that linear or non-linear constraints on the coefficients of $A(L)$ are only satisfied if $\boldsymbol{H} = \boldsymbol{I}$, discussed by Hsiao.

However, he does not deal with conditions of type (b), and these will be discussed briefly here. Necessary and sufficient conditions for identification are given in Sargan (1978a), when only the maximum lags on the variables are specified. The conditions depend on the presence or absence of latent roots of the $\boldsymbol{A}(L)$ polynomial. $\boldsymbol{A}(L)$ has a latent root λ, if for some non-zero vector $\boldsymbol{h}$:

$$\boldsymbol{h}'(\boldsymbol{A}(\lambda)) = \boldsymbol{0}'. \tag{114}$$

A necessary condition for there to be more than one solution is that (114) is satisfied for some λ and $\boldsymbol{h}$. (The article referred to above gives a slightly different formulation which makes it easier to discuss cases where $A(z)$ has an infinite latent root.) This condition is also sufficient, provided a factorisation condition is satisfied which can be taken to have a prior probability of unity.

A necessary condition that the model is not locally identified is that $\boldsymbol{A}(z)$ and $\boldsymbol{R}(z)$ have a latent root λ in common, in the sense that for some non-zero vector $\boldsymbol{h}$, (114) is satisfied, and for some non-zero vector $\boldsymbol{k}$:

$$\boldsymbol{R}(\lambda)\boldsymbol{k} = \boldsymbol{0}.$$

This is a sufficient condition that the Jacobian (first order derivative) conditions for identification are not satisfied. But even if the Jacobian is not full rank, it does not follow that the model is not locally identified. This is discussed in the above article.

The estimation of the model has two stages. The first is to decide on the various lags on the different variables, and on the autoregressive and moving average processes. For this suitable test procedures are required and will be discussed in the next section.

Given the specification of these maximum lags then parameter estimation can proceed using maximum likelihood procedures, or procedures asymptotically equivalent to these. For a complete model, if a numerical optimisation program which does not require analytic derivatives of the likelihood function is used to optimise the likelihood function, such as a conjugate gradient procedure or one using numerical differentiation, it is no more difficult to fit a model of the form (111) than a less restricted model of form (112), since all that is required as an addition to a program for producing maximum likelihood estimates of (112) is a sub-routine for computing the coefficients of $\boldsymbol{\Psi}(L)$ as functions of the unconstrained elements of $\boldsymbol{A}(L)$ and $\boldsymbol{R}(L)$.

It can be argued that since, in using ARMA models for the generation of the errors in econometric models, we are merely making use of convenient approximations, there might be considerable advantages (at least in the stage of making preliminary estimates of the model to provisionally settle its economic specification) in using a model with a fairly high order autoregressive specification and a zero order moving average specification. In practice the time to compute moving average specifications can be large when the latent roots of the moving average matrix polynomials tend to move towards the unit circle, and the convergence properties of autoregressive specifications may be much better. Hendry (1976) contains a discussion of estimators for $\boldsymbol{S}(L) = \boldsymbol{0}$ which are asymptotically equivalent to maximum likelihood estimators for models of this type but which may be lower in computing requirements.

For "incomplete" models it may be necessary to modify the model, before it is feasible to estimate it. The simplest way of defining the modified model is to retain both equations (109) and (110), but now allow $\boldsymbol{B}(L)$ to be a rectangular matrix. Thus, it is assumed that the errors on the incomplete model are generated by an ARMA model, which involves only the errors on the set of equations to be estimated. Note that starting from a complete set of equations whose errors are generated by an ARMA model, by eliminating the errors of the equations whose coefficients are not to be estimated, it is possible to obtain a higher order ARMA process generating the errors on the equations to be estimated. Thus the current formulation is of some generality. One method of estimating the incomplete system is to use a set of instrumental variables. These can be chosen rather arbitrarily initially, but as the specification is refined, a set can be chosen which is efficient if the model is linear in the variables. Generalising to the case where the $\boldsymbol{A}(L)$ coefficients depend in a general non-linear way on a set of p parameters forming a vector $\boldsymbol{\Theta}$, the estimators can be regarded as minimising a criterion function of the form:

$$\det\left(\hat{\boldsymbol{\Omega}}^{-1}(\boldsymbol{E}'\boldsymbol{Z}^{+})(\boldsymbol{Z}^{+\prime}\boldsymbol{Z}^{+})^{-1}(\boldsymbol{Z}^{+\prime}\boldsymbol{E})\right), \tag{115}$$

where $\boldsymbol{E}$ is the matrix of white-noise errors or "innovations" in the ARMA process, and $\hat{\boldsymbol{\Omega}}$ is some preliminary consistent estimate of the variance matrix of $\boldsymbol{e}_t$. $\boldsymbol{Z}^{+}$ is the matrix of instrumental variables, which may include lagged values of the predetermined variables. If the $\boldsymbol{A}(L)$ coefficients considered as functions of $\boldsymbol{\Theta}$ have continuous first order derivatives in some neighbourhood of the true value $\overline{\boldsymbol{\Theta}}$, the instrumental variables estimates will be as efficient as the corresponding limited information maximum likelihood estimates if it is possible to express the expectations of $(\partial \boldsymbol{A}(L)/\partial\Theta_i)\boldsymbol{x}_t$, conditional on all lagged values of $\boldsymbol{y}_t$, as linear functions of the z_{jt}^{+} for all j, and for all i. This result follows from the discussion of Hausman (1975), and in the case of a purely autoregressive specification is most easily satisfied by using as instrumental variables the current values of $\boldsymbol{z}_t$, and the lagged values of $\boldsymbol{x}_t$ up to and including the rth order lag. When the ARMA model contains a moving average process, it is difficult to produce estimates of the conditional expectations from an incomplete model, but if the latent roots of the moving average process are not too close to the unit circle there may be a comparatively small loss of efficiency in using $\boldsymbol{x}_{t-s}$ up to some maximum s^*, which is such that the total number of instrumental variables is not more than a fixed proportion (say 40%) of the sample size. With such a set of instrumental variables an iterative minimisation of (115) is possible, by computing $\boldsymbol{u}_t = \boldsymbol{A}(L)\boldsymbol{x}_t$:

$$\boldsymbol{e}_t = \boldsymbol{S}(L)^{-1}\boldsymbol{R}(L)\boldsymbol{u}_t, \tag{116}$$

recursively for given values of the parameters, starting from the values $\boldsymbol{e}_0 = \boldsymbol{e}_{-1} = \boldsymbol{e}_{-2} = \boldsymbol{e}_{-3}\cdots = \boldsymbol{e}_{-(s-1)} = \boldsymbol{0}$. This procedure may not be optimal in a model with no exogenous variables, where end corrections corresponding to $\boldsymbol{u}_t$ being a stationary time series might give better results, but in a model with an autoregressive side there seems no simple alternative to the crude assumptions for $\boldsymbol{e}_t$ listed above. The recursive generation of $\boldsymbol{e}_t$, $t \geq 1$, uses the equation (116) in the form:

$$\boldsymbol{e}_t = [\boldsymbol{I} - \boldsymbol{S}(L)]\boldsymbol{e}_t + \boldsymbol{R}(L)\boldsymbol{u}_t, \qquad t = 1, \ldots, T,$$

noting that $\boldsymbol{I} - \boldsymbol{S}(L)$ has a zero order coefficient matrix equal to zero. Recent discussions of estimators for models with vector moving average error processes include Osborn (1977), Anderson (1980), Reinsel (1979) and Palm and Zellner (1980).

6.2. *Reduced form, final form and dynamic multipliers*

From (109) it is of some interest to discuss the behaviour of y_t in response to changes in the z_t, particularly when some of the z_t may be regarded as government controlled variables which can be changed independently so as to affect the level of the y_t variables. The standard reduced form of the model can be written:

$$y_t = -\boldsymbol{B}_0^{-1}\boldsymbol{B}^*(L)Ly_t - \boldsymbol{B}_0\boldsymbol{C}(L)z_t + \boldsymbol{B}_0^{-1}\boldsymbol{u}_t,$$

where

$$\boldsymbol{B}(L) = \boldsymbol{B}_0 + L\boldsymbol{B}^*(L),$$

and $\boldsymbol{B}^*(L)$ has a degree one less than that of $\boldsymbol{B}(L)$. This equation is useful for directly simulating the impact of a change in z_t. Two types of dynamic multiplier can be distinguished, (i) the impact multiplier, (ii) the cumulative multiplier. The first considers the impact of a unit change in an element of z_t in time period t on all subsequent values of y_t, the second considers the change in y_s, $s \geq t$, if an element of z_τ is changed by one unit for all $\tau \geq t$. Since the second multiplier is obtained from the impact multiplier by summation for all $\tau \leq s$, only the impact multiplier will be considered here. Suppose that we wish to consider the impact multipliers for some subset of elements of z_t, which we form into a vector z_t^*, and denote the corresponding rows of $\boldsymbol{C}(L)$ by $\boldsymbol{C}^*(L)$. Then clearly if we denote the change in z_t^* by $\boldsymbol{D}z_t^*$, the corresponding endogenous-variable changes $\boldsymbol{D}y_\tau$, $\tau \geq t$, will be obtained by solving the equation:

$$\boldsymbol{B}(L)\boldsymbol{D}y_\tau = -\boldsymbol{C}^*(L)\boldsymbol{D}z_\tau^*, \qquad \tau = t, t+1, \ldots,$$

where $\boldsymbol{D}z_\tau^* = \boldsymbol{0}$, if $\tau \neq t$. If we write for the solution:

$$\boldsymbol{D}y_\tau = \boldsymbol{\Pi}(L)\boldsymbol{D}z_\tau^*,$$

then the coefficients $\boldsymbol{\Pi}_s$ give the impact multipliers in period $\tau = t + s$, of the change in z_t^* in period t.

Formally, we may write:

$$\boldsymbol{\Pi}(L) = -(\boldsymbol{B}(L))^{-1}\boldsymbol{C}^*(L),$$

but a more practical computing procedure is to solve sequentially the equations:

$$\boldsymbol{B}(L)\boldsymbol{\Pi}(L) = -\boldsymbol{C}^*(L), \quad \text{for } \boldsymbol{\Pi}_s, s = 0,1,\dots. \tag{117}$$

In fact it is better to use the reduced form, and if we write:

$$\boldsymbol{P}_1(L) = \boldsymbol{B}_0^{-1}\boldsymbol{B}(L) = \sum_{i=0}^{k_1} \boldsymbol{P}_{1i}L^i,$$

$$\boldsymbol{P}_2(L) = \boldsymbol{B}_0^{-1}\boldsymbol{C}^*(L) = \sum_{i=0}^{k_2} \boldsymbol{P}_{2i}L^i,$$

then the equations (117) are equivalent to:

$$\sum_{i=1}^{j} \boldsymbol{P}_{1(j-i)}\boldsymbol{\Pi}_i = \boldsymbol{P}_{2j}, \qquad j = 0,\dots,\infty, \tag{118}$$

where

$$\boldsymbol{P}_{ji} = 0, \text{ if } i > k_j \quad (j = 1,2).$$

These can be solved for $\boldsymbol{\Pi}_j$ sequentially noting that in the jth equation the matrix coefficient of $\boldsymbol{\Pi}_j$ is $\boldsymbol{P}_{10} = \boldsymbol{I}$. Asymptotic standard errors for the $\boldsymbol{\Pi}_j$ can be computed in the usual way, expressing them as functions of the $\boldsymbol{B}(L)$ and $\boldsymbol{C}(L)$, and using implicit differentiation to obtain the first derivatives from (117) [see, for example, Theil and Boot (1962), Goldberger et al. (1961), and Brissimis and Gill (1978)].

The final equations of Tinbergen [see Goldberger (1959)] are obtained by multiplying equation (109) by adj $\boldsymbol{B}(L)$ where this is the adjoint matrix of $\boldsymbol{B}(L)$ considered as a matrix polynomial. Since

$$\text{adj}\,\boldsymbol{B}(L)\cdot\boldsymbol{B}(L) = \det\boldsymbol{B}(L)\cdot\boldsymbol{I},$$

we can then write:

$$[\det\boldsymbol{B}(L)]\,\boldsymbol{y}_t = -\,\text{adj}\,\boldsymbol{B}(L)\cdot\boldsymbol{C}(L)\boldsymbol{z}_t + \text{adj}\,\boldsymbol{B}(L)\boldsymbol{u}_t. \tag{119}$$

A possible method of testing models which is particularly appropriate for comparison with ARIMA statistical time series models [used, for example, by Zellner and Palm (1974) and Wallis (1977)] is to estimate a model of the form (119) first neglecting the constraints that every endogenous variable has the same scalar lag polynomial on the left-hand side of (119). Thus, unconstrained ARMA explanations of each y_{it} in terms of lagged z_t are estimated by single equation ARMA maximum likelihood estimation. Then tests are made to check that the coefficients of the lag polynomials applied to the y_{it}, $i=1,\dots,n$, are all the same. There are severe difficulties in doing this successfully. First, if there are more than two endogenous variables, and more than one lag on each endogenous variable in (109), then $\det(\boldsymbol{B}(L))$ and $\operatorname{adj}\boldsymbol{B}(L)\cdot\boldsymbol{C}(L)$ are both of at least the fifth degree in L, and in models which are at all realistically treated as complete econometric models the degree must be much larger than this. This of course requires a large sample before asymptotic theory can be a good approximation, since each equation to be estimated will contain a large number of variables of various lags. If the total number of lags on the variables in the final equation form (119) is determined by significance tests then there is an obvious probability that the subsequent tests will reject the constraints that all the y_{it} are subject to the same lag operator. Indeed, there is no reason why the unconstrained estimates of the longer lagged coefficients should be significantly different from zero. The true values of these coefficients can be expected to be small if the model is stable since the higher order coefficients contain the products of many latent roots all less than one in modulus. Thus, it would be better to allow the maximum lag to be determined by feasibility. Even then, the size of model may have to be small to estimate unconstrainedly the set of equations of the final form. Finally, there are many implicit restrictions on the coefficients of $\operatorname{adj}\boldsymbol{B}(\cdot)$ which it is difficult to put in explicit form. Since unless the right-hand-side polynomials satisfy these implicit constraints, the constraints that all the left-hand side polynomials are the same is of little interest, it appears that starting from the unconstrained final equation is not really an adequate way of testing the specification of realistic econometric models. Moreover, parameter constancy in derived equations like (119) relies on all the equations in (109) being structurally invariant.

If the z_t are regarded as generated by ARMA processes, so that:

$$\boldsymbol{D}(L)z_t=\boldsymbol{F}(L)\boldsymbol{\varepsilon}_t, \tag{120}$$

where $\boldsymbol{\varepsilon}_t$ is a white-noise series, then we can eliminate z_t from (119) using (120) to give:

$$\boldsymbol{B}(L)y_t=-\boldsymbol{C}(L)\boldsymbol{D}(L)^{-1}\boldsymbol{F}(L)\boldsymbol{\varepsilon}_t+\boldsymbol{u}_t$$

or

$$(\det\boldsymbol{D}(L))\boldsymbol{B}(L)y_t=-\boldsymbol{C}(L)(\operatorname{adj}\boldsymbol{D}(L))\boldsymbol{F}(L)\boldsymbol{\varepsilon}_t+(\det\boldsymbol{D}(L))\boldsymbol{u}_t. \tag{121}$$

The error term on (121) contains a complicated moving average of the $\boldsymbol{\varepsilon}_t$, and if $\boldsymbol{u}_t$ itself is generated by a general ARMA stochastic model then the stochastic specification is even more complicated. Assuming for simplicity that $\boldsymbol{u}_t = \boldsymbol{S}(L)\boldsymbol{e}_t$, where $\boldsymbol{e}_t$ is white noise, there is a corresponding final form:

$$(\det \boldsymbol{D}(L))(\det \boldsymbol{B}(L))\, y_t = -(\operatorname{adj} \boldsymbol{B}(L))\boldsymbol{C}(L)(\operatorname{adj} \boldsymbol{D}(L))\boldsymbol{F}(L)\boldsymbol{\varepsilon}_t + (\operatorname{adj} \boldsymbol{B}(L))(\det \boldsymbol{D}(L))\boldsymbol{S}(L)\boldsymbol{e}_t. \tag{122}$$

Equation (122) gives separate ARMA-type representation for each element of y_t. Note that the autoregressive polynomial, $(\det \boldsymbol{D}(L))(\det \boldsymbol{B}(L))$, will generally be the same for each y_{it} (the exception occurs if a recursive structure can be set up by paritioning y_t into sub-sections). For a given y_{it}, the right-hand side of (122) also can be represented as a single moving average process, the maximum lag of which is the same as the maximum lag in the terms of the right-hand side of (122). Note, however, that this new representation neglects the detailed correlation structure of the different components of the right-hand side of (122) and so loses a great deal of information which is contained in the specification (122). Thus, using the individual ARMA equations to forecast y_{it} would give less accurate forecasts than using the detailed model (122), and the use of original model (109) to forecast should also give more accurate forecasts than (122). With a correctly specified system, this should be true for an estimated model. And in estimating the model it will be found that greater efficiency in the estimation of the coefficients of (122) is obtained by first estimating (109) and (120) taking account of any appropriate constraints and then substituting the resulting estimates of $\boldsymbol{A}(L)$, $\boldsymbol{S}(L)$, $\boldsymbol{D}(L)$ and $\boldsymbol{F}(L)$ into equation (122) to give ARMA equations for the individual y_{it}. For an example of some relevant applied work see Prothero and Wallis (1976), and for a different emphasis, Zellner (1979). Also, Trivedi (1975) compares ARMA with econometric models for inventories.

The suggested alternative for testing the specification is that the original model or its reduced from is estimated using the maximal feasible lags. Then constraints reducing the order of the lags in each reduced form equation are tested using asymptotic t- or F-ratio tests (Wald test) or by using likelihood ratio tests [see Sargan (1978b), and for an example, Hendry and Anderson (1977)].

6.3. *Unconstrained autoregressive modelling*

Alternatively, one may consider, following Sargent and Sims (1977), the possibility of an autoregressive representation for the economy in which the distinction between endogenous and exogenous variables is ignored. In an effort to estimate the dynamics of the system with no a priori constraints, the equations are written in the form:

$$y_t = P(L)y_{t-1} + e_t, \tag{123}$$

where $P(L)$ is an unconstrained matrix lag polynomial of maximum lag q and e_t is a white-noise vector. This can be regarded as a linear approximation to an autoregressive representation of the stochastic model generating y_t if the y_t are stationary time series [see Hannan (1970)] with the e_t being approximations to the innovations in the y_t. If the y_t are non-stationary but Δy_t are stationary, then a set of equations of the form (123) may still be a good approximation but unit latent roots should occur in the latent roots equation for the system. However, there may be problems if we consider (123) as an approximation to an ARMA system of the form:

$$S(L)^{-1}B(L)y_t = e_t, \tag{124}$$

if $S(L)$ has roots close to the unit circle. In particular, if the true system is of ARMA form in a set of endogenous variables y_t^*, and the equations are misspecified by taking a sub-set of variables y_t which includes the first differences of the corresponding variables y_t^*, then corresponding differences of white noise will occur in the errors of the ARMA model for the y_t. Thus, over-differencing the variables will invalidate the Wold autoregressive representation and the corresponding finite autoregressive representation will not hold and Sims has tended to work with levels variables in consequence. With the length of sample available for estimating the equations by multiple regression, it is necessary to work with a relatively small model and to restrict q, rather drastically. Sargent and Sims (1977) also considered models which contain index variables, which in effect introduce non-linear restrictions in the coefficients $P(L)$ by requiring each y_{it} to depend upon past values of k index variables $k < n$, where n is the number of variables y_t. It is to be expected that when the number of regression equations is allowed to increase to be of the same order as T, the estimated coefficients become sensitive to changes in the variables. Sims interprets his results by considering the effect of an innovation in a particular variable on later values of the other variables, but the interpretation is complicated by the correlation between the contemporaneous innovations on the different variables [see, for example, Sims (1980)]. Additionally, marginalising with respect to other elements of y_t^* than those retained in y_t will produce an ARMA form as in (124) unless none of the excluded variates Granger causes the y_t, and as in all derived representations, strong assumptions are necessary to ensure parameter constancy.

6.4. Alternative forms of disequilibrium model

The simplest model of disequilibrium is that of Samuelson–Tobin in which the tendency of economic variables to their equilibrium values is modelled by introducing an equilibrium static equation for each endogenous variable:

$$y_t^* = y_e(z_t),$$

where this is a vector of n endogenous variables' equilibrium values expressed as functions of a set of m exogenous variables. Writing $\Delta y_t = y_t - y_{t-1}$ the Samuelson–Tobin model is [see, for example, Samuelson (1947)]:

$$\Delta y_t = D(y_t^* - y_{t-1}) + u_t, \tag{125}$$

where D is an arbitrary square matrix. The simplest special case is where D is a positive diagonal matrix, with every diagonal element satisfying:

$$0 < d_{ii} < 1.$$

This type of model can be regarded as derived from an optimal control problem where the function whose expected value is to be minimised is:

$$\sum_{t=1}^{S} \left[(\Delta y_t')W\Delta y_t + (y_t^* - y_t)'M(y_t^* - y_t)\right]. \tag{126}$$

As $S \to \infty$, the optimal control solution when $E(y_{t+s}^* | Y_t^*) = y_t^*$ is:

$$\Delta y_t = D(y_t^* - y_{t-1}),$$

where if $K = I - D$, then:

$$K + K^{-1} = 2I + W^{-1}M,$$

or if we write:

$$K^* = W^{1/2}KW^{-1/2},$$
$$K^* + K^{*-1} = 2I + W^{-1/2}MW^{-1/2}.$$

Now if W and M are both positive definite the matrix on the right-hand side is positive definite such that every latent root is real and greater than 2. K^* can then clearly be chosen to be symmetric, so that every root is real, and if λ_k is a root of K^* and λ_m a corresponding root of $W^{-1/2}MW^{-1/2}$, then:

$$\lambda_k + \lambda_k^{-1} = 2 + \lambda_m.$$

We can pick $\boldsymbol{K}^*$ such that λ_k satisfies:

$$0<\lambda_k<1,$$

provided $\boldsymbol{M}$ is non-singular, and $\boldsymbol{K}=\boldsymbol{W}^{-1/2}\boldsymbol{K}^*\boldsymbol{W}^{1/2}$, and has the same set of latent roots.

This choice of $\boldsymbol{K}$ is appropriate, since unstable solutions cannot be optimal. Then for $\boldsymbol{D}=\boldsymbol{I}-\boldsymbol{K}$, we have that $\lambda_d=1-\lambda_k$, and so:

$$0<\lambda_d<1. \tag{127}$$

Note that if $\boldsymbol{W}$ is a diagonal matrix then $\boldsymbol{D}$ is symmetric. Of course without prior knowledge of $\boldsymbol{W}$ and $\boldsymbol{M}$ it is difficult to specify $\boldsymbol{D}$, and even the constraints that $\boldsymbol{D}$ has real latent roots satisfying (127) are difficult to enforce.

The generalisation of the model (125) to more complicated time lags is obvious, but perhaps rather arbitrary. Using the lag operator notation a general formulation would be:

$$\boldsymbol{C}(L)\Delta \boldsymbol{y}_t=\boldsymbol{D}(L)(\boldsymbol{y}_t^*-\boldsymbol{y}_{t-1})+\boldsymbol{u}_t, \tag{128}$$

where $\boldsymbol{C}(L)$ and $\boldsymbol{D}(L)$ are matrix polynomials of any order. However, if $\boldsymbol{y}_t^*$ is being written as an arbitrary linear function of current and lagged values of a set of exogenous variables, then (128) may contain some redundancy and in any case a useful simplification may be obtained by considering either of the following special cases:

$$\Delta \boldsymbol{y}_t=\boldsymbol{D}(L)(\boldsymbol{y}_t^*-\boldsymbol{y}_{t-1})+\boldsymbol{u}_t \tag{129}$$

or

$$\boldsymbol{C}(L)\Delta \boldsymbol{y}_t=\boldsymbol{D}(\boldsymbol{y}_t^*-\boldsymbol{y}_{t-1})+\boldsymbol{u}_t. \tag{130}$$

If no attempt is made to put constraints on the $\boldsymbol{C}(L)$ and $\boldsymbol{D}$ matrices in (130), a further transformation can be considered by using this form with the restriction that $\boldsymbol{D}$ is diagonal and $\boldsymbol{C}_0$ [the zero order coefficient matrix in $\boldsymbol{C}(L)$] has its diagonal elements equal to one or perhaps better that $\boldsymbol{D}=\boldsymbol{I}$ and $\boldsymbol{C}_0$ is unrestricted. In specifying $\boldsymbol{y}_t^*=\boldsymbol{y}_e(z_t)$ when there are linear restrictions on these equilibrium functions each of which affects only one element of $\boldsymbol{y}_e(z_t)$, such forms (rather than the more usual assumption that $\boldsymbol{C}_0=\boldsymbol{I}$) have the advantage that the corresponding restriction affects only one equation of (130).

If there are restrictions on $\boldsymbol{C}(L)$ and $\boldsymbol{D}$, then an equivalent model with an arbitrary lag on $(\boldsymbol{y}_t^*-\boldsymbol{y}_{t-1})$ may make a better formulation if the rather ad hoc economic considerations to be considered below are applied. It is less easy to formulate an optimal control approach which will give significant restrictions on $\boldsymbol{C}(L)$ and $\boldsymbol{D}$. Thus, for example, if we write:

$$\tilde{y}_t = \begin{pmatrix} y_t \\ y_{t-1} \\ \vdots \\ y_{t-f} \end{pmatrix}, \quad y_t^+ = \begin{pmatrix} y_t - y_t^* \\ y_{t-1} - y_{t-1}^* \\ \vdots \\ y_{t-f} - y_{t-f}^* \end{pmatrix},$$

then a loss function of the form

$$\sum_{t=1}^{S} \left(y_t^{+\prime} \boldsymbol{M}^+ y_t^+ \right) + \sum_{t=1}^{S} (\Delta \tilde{y}_t)' \boldsymbol{W}^+ (\Delta \tilde{y}_t)$$

leads to optimal control equations of the form (128) (with the property that if y_t^* is held constant the adjustment equations are stable) but further restrictions, along the lines that the latent roots are all real, are not appropriate, since the results for the first order case are dependent on the special separability assumption for the loss function used in the first order case. [For a discussion of optimal control closed loop paths see Chow (1975).]

A possibility of some importance, which prevents the normalisation $\boldsymbol{D} = \boldsymbol{I}$, is that $\boldsymbol{D}$ is in fact singular. This arises particularly where there are identities corresponding to exact equations satisfied by the whole set of endogenous variables, or where some variables react so quickly during the unit time period, that the general equation of type (130) becomes inappropriate. If the partial equilibrium for this variable is stable, and is attained within the unit time period, then a static equation, or a dynamic equation obtained by differencing the static equation, is introduced into the model.

This possibility can be found in applied studies from various fields, for example, in models of wage–price inflation, as in Sargan (1980a) or Laidler and Cross (1976), models of entrepreneurial behaviour, as in Brechling (1973) or Nadiri and Rosen (1969), or models of consumer behaviour, as in Phlips (1978) or Deaton (1972b), or in models of portfolio choice, as in Tobin (1969).

Somewhat similar specialisations in the form of the general adjustment equations occur where there are buffer variables, such as cash in portfolio choice models, or inventories in some models of firm behaviour, or hours worked in models of labour demand. Buffer variables in the short period are regarded as absorbing the consequences of disequilibrium. Here if a sufficiently short time period is employed it may be appropriate to assume that the change in the buffer variable is determined by an appropriate overall identity.

If it is known how agents formulate $E(y_{t+s}^*|Y_t^*)$, and a loss function of the form (126) is appropriate, then a useful theory-based dynamic specification can be derived. For example, if y_t^* is generated by: $\Delta y_t^* = A\Delta y_{t-1}^* + V_t$ then the solution linear decision rule is:

$$\Delta y_t = D_1 \Delta y_t^* + D_2(y_t^* - y_{t-1}) + u_t, \tag{131}$$

which is a system error correction form [see Section 2.6, Hendry and Anderson (1977) and Nickell (1980)]. In (131), D_1 and D_2 depend on A, W and M such that $D_1 = 0$ and $D_2 = D$ in (125) if $A = 0$, and an intercept in the y_t^* equation would produce an intercept in (131) (so that the decision rules would depend on the growth rate of y_t^* in a log-linear model). Similarly, a rational expectations assumption in models with strongly exogenous variables provides parameter restrictions [see Wallis (1980), noting that the vector of first derivatives of the likelihood function provides an estimator generating equation for the model class, in the sense of Section 2.2, suggesting fully efficient, computationally cheap estimators and highlighting the drawbacks of "fixed point" methods.] Nevertheless, stringent diagnostic testing of models must remain an essential component of any empirical approach to dynamic specification [see Hendry (1974) and Sargan (1980a)].

Finally, where the economic model is set up in a form which makes it depend on a discrete decision period, and the unit time period is inappropriate, the use of continuous time period models, and the discrete time period approximations to them discussed by Bergstrom in Chapter 20 of this Handbook may considerably improve the dynamic specifications of the model whenever there are valid a priori restrictions on the continuous time model.

References

Aigner, D. J. (1971) "A Compendium on Estimation of the Autoregressive Moving Average Model from Time Series Data", *International Economic Review*, 12, 348–369.

Aigner, D. J. and G. G. Judge (1977) "Application of Pre-Test and Stein Estimators to Economic Data", *Econometrica*, 45, 1279–1288.

Aitchison, J. and S. D. Silvey (1960) "Maximum Likelihood Estimation and Associated Tests of Significance", *Journal of the Royal Statistical Society*, Series B, 22, 154–171.

Akaike, H. (1972) "Information Theory and an Extension of the Maximum Likelihood Principle", *Proceedings, 2nd International Symposium on Information Theory*, 267–281.

Almon, S. (1965) "The Distributed Lag between Capital Appropriations and Expenditures", *Econometrica*, 33, 178–196.

Amemiya, T. (1980) "Selection of Regressors", *International Economic Review*, 21, 331–354.

Amemiya, T. and K. Morimune (1974) "Selecting the Optimal Order of Polynomial in the Almon Distributed Lag", *Review of Economics and Statistics*, 56, 378–386.

Anderson, T. W. (1971) *The Statistical Analysis of Time Series*. New York: John Wiley.

Anderson, T. W. (1980) "Maximum Likelihood Estimation for Vector Autoregressive Moving Average Models", pp. 49–59 in D. R. Brillinger and G. C. Tiao (eds.): *New Directions in Time Series*, Institute of Mathematical Statistics.

Barro, R. J. (1978) "Unanticipated Money, Output and the Price Level in United States", *Journal of Political Economy*, 86, 549–580.
Bean, C. R. (1977) "More Consumer Expenditure Equations", Paper AP (77) 35, H. M. Treasury, London.
Bean, C. R. (1981) "An Econometric Model of Investment in the United Kingdom", *Economic Journal*, 91, 106–121.
Bentzel, R. and B. Hansen, (1955) "On Recursiveness and Interdependency in Economic Models", *Review of Economic Studies*, 22, 153–168.
Berndt, E. R. and N. E. Savin (1977) "Conflict Among Criteria for Testing Hypotheses in the Multivariate Linear Regression Model", *Econometrica*, 45, 1263–1278.
Bewley, R. (1979a) "The Direct Estimation of the Equilibrium Response in a Linear Dynamic Model", *Economic Letters*, 3, 357–362.
Bewley, R. (1979b) "On Searching for the Lag Length and Polynomial Degree in the Almon Lag Method", Discussion Paper No. 42, University of New South Wales.
Box, G. E. P. and G. M. Jenkins (1970), *Time Series Analysis, Forecasting and Control*, San Francisco: Holden-Day.
Brechling, F. (1973) *Investment and Employment Decisions*, The University Press, Manchester.
Brissimis, S. N. and L. Gill (1978) "On the Asymptotic Distribution of Impact and Interim Multipliers", *Econometrica*, 46, 463–469.
Breusch, T. S. and A. R. Pagan (1980) "The Lagrange Multiplier Test and its Applications to Model Specification in Econometrics", *Review of Economic Studies*, 47, 239–253.
Burdick, D. A. and D. Wallace (1976) "A Theorem on an Inequality of Two Quadratic Forms and an Application to Distributed Lags", *International Economic Review*, 17, 769–772.
Burt, O. R. (1980) "Schmidt's La Guerre Lag is a Pascal Rational Lag", University of Kentucky, mimeo.
Cargill, T. F. and R. A. Meyer (1974) "Some Time and Frequency Domain Distributed Lag Estimators: A Comparative Monte Carlo Study", *Econometrica*, 42, 1031–1044.
Carter, R. A. L., A. L. Nagar and P. G. Kirkham (1975) "The Estimation of Misspecified Polynomial Distributed Lag Models", Report No. 7525, Dept. of Economics, University of Western Ontario.
Chow, G. C. (1975) *Analysis and Control of Dynamic Economic Systems*, New York: Wiley.
Corradi, C. (1977) "Smooth Distributed Lag Estimators and Smoothing Spline Functions in Hilbert Spaces", *Journal of Econometrics*, 5, 211–220.
Corradi, C. and G. Gambetta (1974) "A Numerical Method for Estimating Distributed Lag Models", *Proc. I.F.I.P. Congress*, (North-Holland, Amsterdam), 638–641.
Cox, D. R. and D. V. Hinkley (1974) *Theoretical Statistics*, London: Chapman and Hall.
Crowder, M. J. (1976) "Maximum Likelihood Estimation for Dependent Observations", *Journal of the Royal Statistical Society*, B, 38, 45–53.
Currie, D. (1981) "Some Long Run Features of Dynamic Time Series Models", *Economic Journal*, 91, 704–715.
Cuthbertson, K. (1980) "The Determination of Consumer Durables Expenditure: An Exercise in Applied Econometric Analysis", Unpublished Paper, National Institute for Economic and Social Research, London.
Davidson, J. E. H. (1981) "Problems with the Estimation of Moving Average Processes", *Journal of Econometrics*, 16, 295–310.
Davidson, J. E. H., D. F. Hendry, F. Srba and S. Yeo (1978) "Econometric Modelling of the Aggregate Time-Series Relationship Between Consumers' Expenditure and Income in the United Kingdom", *Economic Journal*, 88, 661–692.
Davidson, J. E. H. and D. F. Hendry (1981) "Interpreting Econometric Evidence: Consumers' Expenditure in the United Kingdom", *European Economic Review*, 16, 177–192.
Davis, E. P. (1982) "The Consumption Function in Macroeconomic Models: A Comparative Study", Bank of England Discussion Paper No. 1, Technical Series.
Day, R. H. (1967) "Technological Change and the Sharecropper", *American Economic Review*, 57, 427–449.
Deaton, A. S. (1972a) "Letter to the Editor", *American Statistician*, 26, 63.
Deaton, A. S. (1972b) "The Estimation and Testing of Systems of Demand Equations", *European Economic Review*, 3, 390–411.

Dent, W. T. (ed.), (1980) "Computation in Econometric Models", *Journal of Econometrics*, 12, No. 1.
Dhrymes, P. (1971) *Distributed Lags: Problems of Estimation and Formulation*, San Francisco: Holden Day.
Dhrymes, P. J., L. R. Klein, and K. Steiglitz (1970) "The Estimation of Distributed Lags", *International Economic Review*, 11, 235–250.
Durbin, J. (1970) "Testing for Serial Correlation in Least-Squares Regression when Some of the Regressors are Lagged Dependent Variables", *Econometrica*, 38, 410–421.
Eisner, R. and R. H. Strotz (1963) "Determinants of Business Investment", pp. 60–138 in *Commission on Money and Credit: Impacts of Monetary Policy*, Englewood Cliffs, N.J.: Prentice-Hall.
Engle, R. F. (1976) "Interpreting Spectral Analyses in Terms of Time-Domain Models" *Annals of Economic and Social Measurement*, 5, 89–109.
Engle, R. F. (1982) "A General Approach to Lagrange Multiplier Model Diagnostics", *Annals of Applied Econometrics*, 20, 83–104.
Engle, R. F. (1982) "Autoregressive Conditional Heteroscedasticity with Estimates of the Variances of United Kingdom Inflations", *Econometrica*, 50, 987–1008.
Engle, R. F., D. F. Hendry, and J-F. Richard (1983) "Exogeneity", *Econometrica*, 51, 277–304.
Espasa, A. (1977) *The Spectral Maximum Likelihood Estimation of Econometric Models with Stationary Errors*, Göttingen: Vanderhoeck und Ruprecht.
Evans, G. B. A. and N. E. Savin (1981) "Testing for Unit Roots", *Econometrica*, 49, 753–779.
Evans, G. B. A. and N. E. Savin (1982) "Conflict Among the Criteria Revisited: The W, LR and LM Tests", *Econometrica*, 50, 737–748.
Florens, J-P. and M. Mouchart (1980) "Initial and Sequential Reduction of Bayesian Experiments", CORE Discussion Paper 8015, Université Catholique de Louvain, Louvain-la-Neuve, Belgium.
Fomby, T. B. (1979) "MSE Evaluation of Shiller's Smoothness Priors", *International Economic Review*, 20, 203–215.
Frenkel, J. (1981) "Flexible Exchange Rates, Prices and the Role of "News": Lessons from the 1970's", *Journal of Political Economy*, 89, 665–705.
Friedman, B. M. (1976) "Substitution and Expectation Effects on Long-Term Borrowing Behaviour and Long-Term Interest Rates", Discussion Paper No. 495, Harvard University.
Frisch, R. (1938) "Statistical versus Theoretical Relations in Economic Macro-Dynamics", Business Cycle Conference, Cambridge, England.
Frost, P. A. (1975) "Some Properties of the Almon Lag Technique when one Searches for Degree of Polynomial and Lag", *Journal of the American Statistical Association*, 70, 606–612.
Fuller, W. A. (1976) *Introduction to Statistical Time-Series*. New York: John Wiley and Sons.
Gersovitz, M. and J. G. Mackinnon (1978) "Seasonality in Regression: An Application of Smoothness Priors", *Journal of the American Statistical Association*, 73, 264–273.
Geweke, J. F. and R. Meese (1981) "Estimating Regression Models of Finite But Unknown Order", *International Economic Review*, 22, 55–70.
Ginsburgh, V. and J. Waelbroeck (1976) "Computational Experience with a Large General Equilibrium Model", *Computing Equilibria: How and Why*, ed. by J. Los and M. W. Los, (North Holland, Amsterdam), 257–269.
Godfrey, L. G. (1978) "Testing Against General Autoregressive and Moving Average Error Models when the Regressors Include Lagged Dependent Variables", *Econometrica*, 46, 1293–1301.
Godfrey, L. G. and D. S. Poskitt (1975) "Testing the Restrictions of the Almon Technique", *Journal of the American Statistical Association*, 70, 105–108.
Godfrey, L. G. and M. R. Wickens (1982) "Tests of Misspecification using Locally Equivalent Alternative Models", Ch. 6 in G. C. Chow and P. Corsi (eds.), *Evaluating the Reliability of Macro-Economic Models*. New York: John Wiley and Sons.
Godley, W. A. H. and W. D. Nordhaus (1972) "Pricing in the Trade Cycle", *The Economic Journal*, 82, 853–882.
Goldberger, A. S. (1959) *Impact Multipliers and Dynamic Properties of the Klein–Goldberger Model*, Amsterdam: North Holland.
Goldberger, A. S., A. L. Nagar, and H. S. Odeh (1961) "The Covariance Matrices of Reduced Form Coefficients and of Forecasts for a Structural Econometric Model", *Econometrica*, 29, 556–573.
Granger, C. W. J. (1966) "The Typical Spectral Shape of an Economic Variable, *Econometrica*, 34, 150–161.

Granger, C. W. J. (1969) "Investigating Causal Relations by Econometric Models and Cross-Spectral Methods", *Econometrica*, 37, 424–438.

Granger, C. W. J. (1981) "Forecasting White Noise", Discussion Paper 80-31, University of California, San Diego.

Granger, C. W. J. and P. Newbold (1974) "Spurious Regressions in Econometrics", *Journal of Econometrics*, 2, 111–120.

Granger, C. W. J. and P. Newbold (1977) "The Time Series Approach to Econometric Model Building", Chapter 1, in C. A. Sims (ed.) op. cit.

Griliches, Z. (1961) "A Note on Serial Correlation Bias in Estimates of Distributed Lags", *Econometrica*, 29, 65–73.

Griliches, Z. (1967) "Distributed Lags: A Survey", *Econometrica*, 35, 16–49.

Guthrie, R. S. (1976) "A Note on the Bayesian Estimation of Solow's Distributed Lag Model", *Journal of Econometrics*, 4, 295–300.

Haavelmo, T. (1944) "The Probability Approach in Econometrics" Supplement to *Econometrica*, 12.

Hall, R. E. (1978) "Stochastic Implications of the Life Cycle-Permanent Income Hypothesis: Theory and Evidence", *Journal of Political Economy*, 86, 971–987.

Hamlen, S. S. and W. A. Hamlen (1978) "Harmonic Alternatives to the Almon Polynomial Technique", *Journal of Econometrics*, 7, 57–66.

Hannan, E. J. (1970) *Multiple Time Series*, New York: Wiley.

Hansen, L. P. (1982) "Large Sample Properties of Generalised Method of Moments Estimators", *Econometrica*, 50, 1029–1054.

Harper, C. P. (1977) "Testing for the Existence of a Lagged Relationship within Almon's Method", *Review of Economics and Statistics*, LIX, 204–210.

Harvey, A. C. (1981) *Econometric Analysis of Time Series*. London: Phillip Allan.

Harvey, A. C. and G. D. A. Phillips (1979) "The Estimation of Regression Models with Autoregressive-Moving Average Disturbances", *Biometrika*, 66, 49–58.

Hatanaka, M. and T. D. Wallace (1980) "Multicollinearity and the Estimation of Low Order Moments in Stable Lag Distributions", pp. 322–337 in J. Kmenta and J. B. Ramsay (eds.), *Evaluation of Econometric Models*. New York: Academic Press.

Hausman, J. A. (1975) "An Instrumental Variable Approach to Full Information Estimators for Linear and Certain Non-Linear Econometric Models", *Econometrica*, 43, 727–753.

Hendry, D. F. (1974) "Stochastic Specification in an Aggregate Demand Model of the United Kingdom", *Econometrica*, 42, 559–578.

Hendry, D. F. (1976) "The Structure of Simultaneous Equations Estimators", *Journal of Econometrics*, 4, 51–88.

Hendry, D. F. (1979) "The Behaviour of Inconsistent Instrumental Variables Estimators in Dynamic Systems with Autocorrelated Errors", *Journal of Econometrics*, 9, 295–314.

Hendry, D. F. (1980) "Predictive Failure and Econometric Modelling in Macro-Economics: The Transactions Demand for Money", Chapter 9 in P. Ormerod (ed.) *Economic Modelling*. London: Heinemann Educational Books.

Hendry, D. F. and G. J. Anderson (1977) "Testing Dynamic Specification in Small Simultaneous Systems: An Application to a Model of Building Society Behaviour in the United Kingdom", *Frontiers in Quantitative Economics*. Vol. IIIA (ed. M. D. Intriligator). Amsterdam: North Holland.

Hendry, D. F. and G. F. Mizon (1978) "Serial Correlation as a Convenient Simplification, not a Nuisance: A Comment on a Study of the Demand for Money by the Bank of England", *The Economic Journal* 88, 549–563.

Hendry, D. F. and A. R. Pagan (1980) "A Survey of Recent Research in Distributed Lags", unpublished paper, Australian National University.

Hendry, D. F. and J-F. Richard (1982) "On the Formulation of Empirical Models in Dynamic Econometrics", *Journal of Econometrics*, 20, 3–34.

Hendry, D. F. and J-F. Richard (1983) "The Econometric Analysis of Economic Time Series", *International Statistical Review*, 51, 111–163.

Hendry, D. F. and P. K. Trivedi (1972) "Maximum Likelihood Estimation of Difference Equations with Moving Average Errors: A Simulation Study", *The Review of Economic Studies*, 39, 117–145.

Hendry, D. F. and T. von Ungern-Sternberg (1981) "Liquidity and Inflation Effects on Consumers' Behaviour", Chapter 9, in A. S. Deaton (ed.), *Essays in the Theory and Measurement of Consumers'*

Behaviour, Cambridge University Press.
Henry, S. G. B., M. C. Sawyer, and P. Smith (1976) "Models of Inflation in the United Kingdom", *National Institute Economic Review*, 76/3, 60–71.
Hickman, B. G. (1972) *Econometric Models of Cyclical Behaviour*, New York: Columbia University Press.
Hill, R. C. and S. R. Johnson (1975) "Distributed Lag Estimators Derived from Smoothness Priors: A Comment", Unpublished paper, University of Missouri-Columbia.
Holt, C., F. Modigliani, J. Muth and H. Simon (1960), *Planning Production, Inventories and Work Force*. Englewood Cliffs: Prentice-Hall.
Houthakker, H. S. (1956) "The Pareto Distribution and the Cobb-Douglas Production Function in Activity Analysis", *Review of Economic Studies*, 23, 27–31.
Hurwicz, L. (1962) "On the Structural Form of Interdependent Systems", In *Logic, Methodology and the Philosophy of Science*, ed. by E. Nagel, et al. Palo Alto: Stanford University Press.
Hylleberg, S. (1981) "Seasonality in Regressions. The Time Varying Parameter Model", NBER/NSF Conference Paper, San Diego, California.
Jonson, P. D. (1976) "Money, Prices and Output: An Integrative Essay", *Kredit und Kapital*, 84, 979–1012.
Jorgenson, D. W. (1966) "Rational Distributed Lag Functions", *Econometrica*, 34, 135–149.
Judge, G. G. and M. E. Bock (1978) *The Statistical Implications of Pre-Test and Stein-Rule Estimators in Econometrics*. Amsterdam: North Holland.
Kang, K. M. (1973) "A Comparison of Estimators for Moving Average Processes", Report of Australian National Bureau of Statistics, Canberra.
Kendall, M. G. (1973) *Time-Series*. London: Charles Griffen and Co.
Kendall, M. G. and A. Stuart (1961) *The Advanced Theory of Statistics*. Vol. 2, New York: Charles Griffen and Co.
Kennan, J. (1979) "The Estimation of Partial Adjustment Models with Rational Expectations", *Econometrica*, 47, 1441–1455.
Keynes, J. N. (1890) *The Scope and Method of Political Economy*. London: MacMillan and Co.
Kloek, T. (1982) "Dynamic Adjustment when the Target is Nonstationary", unpublished paper, Erasmus University, Rotterdam.
Kohn, R. (1979) "Identification Results for Armax Structures", *Econometrica*, 47, 1295–1304.
Koopmans, T. C. (1947) "Measurement without Theory", *Review of Economics and Statistics*, 29, 161–179.
Koopmans, T. C. (1950) "When is an Equation System Complete for Statistical Purposes?" *Statistical Inference in Dynamic Economic Models* (T. C. Koopmans, ed.), Ch. 17, New York: John Wiley.
Laidler, D. and R. Cross (1976) "Inflation, Excess Demand in Fixed Exchange Rate Open Economies", in *Inflation in the World Economy*, edited by M. Parkin and G. Zis, Manchester University Press.
Lucas, R. E. (1976) "Econometric Policy Evaluation: A Critique", in *The Phillips Curve and Labor Markets* (K. Brunner and A. H. Meltzer, eds.), pp. 19–46. Amsterdam: North-Holland (Carnegie-Rochester Conference Series on Public Policy, vol. 1).
Lütkepohl, H. (1980) "Approximation of Arbitrary Distributed Lag Structures by a Modified Polynomial Lag: An Extension", *Journal of the American Statistical Association*, 75, 428–430.
Maddala, G. S. (1977) *Econometrics*. New York: McGraw Hill.
Malinvaud, E. (1966) *Statistical Methods of Econometrics*. Amsterdam: North-Holland.
Mallows, C. L. (1973) "Some Comments on C_P", *Technometrics*, 15, 661–675.
Marget, A. W. (1929) "Morgenstern on the Methodology of Economic Forecasting", *Journal of Political Economy*, 37, 312–339.
Marschak, J. (1953) "Economic Measurements for Policy and Prediction" in W. C. Hood and T. C. Koopmans (eds.): *Studies in Econometric Method*. New Haven: Yale University Press.
Mizon, G. E. (1977a) "Model Selection Procedures", Chapter 4 in Artis, M. J. and Nobay, A. R. (eds.) *Studies in Modern Economic Analysis*, Oxford: Basil Blackwell.
Mizon, G. E. (1977b) "Inferential Procedures in Nonlinear Models: An Application in a UK Industrial Cross Section Study of Factor Substitution and Returns to Scale", *Econometrica*, 45, 1221–1242.
Mizon, G. E. and D. F. Hendry, (1980) "An Empirical Application and Monte Carlo Analysis of Tests of Dynamic Specification, *Review of Economic Studies*, 47, 21–46.

Mouchart, M. and R. Orsi (1976) "Polynomial Approximation of Distributed Lags and Linear Restrictions: A Bayesian Approach", *Empirical Economics*, 1, 129–152.
Muellbauer, J. (1979) "Are Employment Decisions Based on Rational Expectations?" Unpublished paper, Birkbeck College, London.
Nadiri, M. I. and S. Rosen (1969) "Interrelated Factor Demand Functions", *American Economic Review*, 59, 457–471.
Nerlove, M. (1972) "On Lags in Economic Behaviour", *Econometrica*, 40, 221–252.
Nicholls, D. F., A. R. Pagan and R. D. Terrell (1975) "The Estimation and Use of Models with Moving Average Disturbance Terms: A Survey", *International Economic Review*, 16, 113–134.
Nickell, S. (1980) "Error Correction, Partial Adjustment and All That: An Expository Note", unpublished paper, London School of Economics.
Osborn, D. R. (1976) "Maximum Likelihood Estimation of Moving Average Processes", *Annals of Economic and Social Measurement*, 5, 75–87.
Osborn, D. R. (1977) "Exact and Approximate Maximum Likelihood Estimators for Vector Moving Average Processes", *Journal of the Royal Statistical Society*, B39, 114–118.
Pagan, A. R. and D. F. Nicholls (1976) "Exact Maximum Likelihood Estimation of Regression Models with Finite Order Moving Average Errors", *Review of Economic Studies*, 43, 383–387.
Pagan, A. R. (1978) "Rational and Polynomial Lags: The Finite Connection", *Journal of Econometrics*, 8, 247–254.
Pagano, M. and M. J. Hartley (1981) "On Fitting Distributed Lag Models Subject to Polynomial Restrictions", *Journal of Econometrics*, 16, 171–198.
Patterson, K. D. and J. Ryding (1982) "Deriving and Testing Rate of Growth and Higher Order Growth Effects in Dynamic Economic Models", Bank of England Discussion Paper 21.
Palm, F. and A. Zellner (1980) "Large Sample Estimation and Testing Procedures for Dynamic Equation Systems", *Journal of Econometrics*, 12, 251–284.
Persons, W. M. (1925) "Statistics and Economic Theory", *Review of Economic Statistics*, 7, 179–197.
Pesando, J. E. (1972) "Seasonal Variability in Distributed Lag Models", *Journal of the American Statistical Association*, 71, 835–841.
Pesaran, M. H. (1981) "Diagnostic Testing and Exact Maximum Likelihood Estimation of Dynamic Models", Chapter 3 in E. G. Charatsis (ed.), *Proceedings of the Econometric Society European Meeting 1979: Selected Econometric Papers in Memory of Stefan Valavanis*, North-Holland, 63–87.
Pesaran, M. H. and A. S. Deaton (1978) "Testing Non-nested Non-linear Regression Models", *Econometrica*, 46, 677–694.
Phillips, A. W. (1954) "Stabilisation Policy in a Closed Economy", *The Economic Journal*, 64, 290–323.
Phillips, A. W. (1957) "Stabilization Policy and The Time Form of Lagged Responses", *The Economic Journal*, 67, 265–277.
Phillips, P. C. B. (1977) "Approximations to Some Finite Sample Distributions Associated with a First Order Stochastic Difference Equation", *Econometrica*, 45, 463–485.
Phillips, P. C. B. (1980) "Finite Sample Theory and the Distributions of Alternative Estimators of the Marginal Propensity to Consume", *Review of Economic Studies*, 47, 183–224.
Phlips, L. (1978) *Applied Consumption Analysis*, North Holland Publishing Co., Amsterdam.
Pierce, D. A. (1977) "Relationships—and the lack thereof—Between Economic Time Series, with Special Reference to Money and Interest Rates", *Journal of the American Statistical Association*, 72, 11–22.
Poirier, D. J. (1975) "Spline Lags: Why the Almon Lag has Gone to Pieces" in D. J. Poirier, *The Econometrics of Structural Change*, North Holland, Amsterdam.
Prothero, D. L. and K. F. Wallis (1976) "Modelling Macro-Economic Time Series" (with Discussion), *Journal of the Royal Statistical Society*, Series A, 139, 486–500.
Ramsey, J. (1969) "Tests for Specification Errors in Classical Linear Least Squares Regression Analysis", *Journal of the Royal Statistical Society*, Series B, 350–371.
Rao, C. R. (1965) *Linear Statistical Inference and Its Applications*. New York: John Wiley.
Reinsel, G. (1979) "Maximum Likelihood Estimation of Stochastic Linear Difference Equations with Autoregressive Moving Average Errors", *Econometrica*, 47, 129–152.
Richard, J.-F. (1977) "Bayesian Analysis of the Regression Model When the Disturbances are Generated by an Autoregressive Process", *New Developments in the Applications of Bayesian*

Methods, ed. by A. Aykac, and C. Brumat, (North Holland, Amsterdam), 185–209.

Richard, J.-F. (1979) "Exogeneity, Inference and Prediction in So-Called Incomplete Dynamic Simultaneous Equation Models", CORE Discussion Paper 7922, Université Catholique de Louvain, Louvain-la-Neuve, Belgium.

Richard, J.-F. (1980) "Models with Several Regimes and Changes in Exogeneity", *The Review of Economic Studies*, 47, 1–20.

Robinson, S. (1970) "Polynomial Approximations of Distributed Lag Structures", Discussion Paper No. 1, London School of Economics.

Salmon, M. (1979) "Notes on Modelling Optimising Behaviour in the Absence of an 'Optimal' Theory", unpublished paper, University of Warwick.

Salmon, M. (1982) "Error Correction Mechanisms", *Economic Journal*, 92, 615–629.

Salmon, M. and K. F. Wallis (1982) "Model Validation and Forecast Comparisons: Theoretical and Practical Considerations", Ch. 12 in G. C. Chow and P. Corsi (eds.), *Evaluating the Reliability of Macro-Economic Models*. New York: John Wiley and Sons.

Salmon, M. and P. C. Young (1978) "Control Methods for Quantitative Economic Policy", in S. Holly, B. Rustem and M. Zarrop (eds.): *Optimal Control for Econometric Models: An Approach to Economic Policy Formation*. London: Macmillan.

Samuelson, P. A. (1947) *Foundations of Economic Analysis*, Harvard University Press.

Sargan, J. D. (1961) "The Maximum Likelihood Estimation of Economic Relationships with Autoregressive Residuals", *Econometrica*, 29, 414–426.

Sargan, J. D. (1964) "Wages and Prices in the United Kingdom: A Study in Econometric Methodology", in P. E. Hart, G. Mills and J. K. Whitaker (eds.) *Econometric Analysis for National Economic Planning* (Colston Papers No. 16) London, Butterworths.

Sargan, J. D. (1975) "The Identification and Estimation of Sets of Simultaneous Stochastic Equations", L.S.E. Econometrics Programme Discussion Paper A1.

Sargan, J. D. (1978a) "Identification in Models with Autoregressive Errors", L.S.E. Econometrics Programme Discussion Paper A.17.

Sargan, J. D. (1978b) "Dynamic Specification for Models with Autoregressive Errors. Vector Autoregressive Case", L.S.E. Econometrics Programme Working Paper.

Sargan, J. D. (1980a) "A Model of Wage-Price Inflation", *Review of Economic Studies*, 47, 97–112.

Sargan, J. D. (1980b) "The Consumer Price Equation in the Post War British Economy. An Exercise in Equation Specification Testing", *The Review of Economic Studies*, 47, 113–135.

Sargan, J. D. (1980c) "Some Tests of Dynamic Specification for a Single Equation", *Econometrica*, 48, 879–897.

Sargan, J. D. and A. Bhargava (1983) "Maximum Likelihood Estimation of Regression Models with First Order Moving Average Errors when the Root Lies on the Unit Circle", *Econometrica*, 51, 153–174.

Sargan, J. D. and F. Mehta (1983) "A Generalisation of the Durbin Significance Test and its Application to Dynamic Specification", *Econometrica*, 51, 1551–1567.

Sargent, T. J. (1981) "Interpreting Economic Time Series", *Journal of Political Economy*, 89, 213–248.

Sargent, T. J. and C. A. Sims (1977) "Business Cycle Modelling Without Pretending to Have Too Much a Priori Economic Theory", in *New Methods of Business Cycle Research*, edited by C. A. Sims, Federal Reserve Bank of Minneapolis.

Sawa, T. (1978) "Information Criteria for Discriminating Among Alternative Regression Models", *Econometrica*, 46, 1273–1291.

Sawyer, K. R. (1980) *The Theory of Econometric Model Selection*, unpublished doctoral dissertation, Australian National University.

Schmidt, P. (1974) "A Modification of the Almon Distributed Lag", *Journal of the American Statistical Association*, 69, 679–681.

Schmidt, P. and R. N. Waud (1973) "Almon Lag Technique and the Monetary Versus Fiscal Policy Debate", *Journal of the American Statistical Association*, 68, 11–19.

Schmidt, P. and R. Sickles (1975) "On the Efficiency of the Almon Lag Technique", *International Economic Review*, 16, 792–795.

Schmidt, P. and W. R. Mann (1977) "A Note on the Approximation of Arbitrary Distributed Lag Structures by a Modified Almon Lag", *Journal of the American Statistical Association*, 72, 442–443.

Shiller, R. J. (1973) "A Distributed Lag Estimator Derived from Smoothness Priors", *Econometrica*, 41, 775–788.
Shiller, R. J. (1980) "Distributed Lag Estimators Based on Linear Coefficient Restrictions and Bayesian Generalisations of These Estimators", *I.H.S. Journal*, 4, 163–180.
Silver, J. L. and T. D. Wallace (1980) "The Lag Relationship Between Wholesale and Consumer Prices: An Application of the Hatanaka-Wallace Procedure", *Journal of Econometrics*, 12, 375–388.
Sims, C. A. (1972) "The Role of Approximate Prior Restrictions in Distributed Lag Estimation", *Journal of the American Statistical Association*, 67, 169–175.
Sims, C. A. (1974) "Distributed Lags", Ch. 5, in M. D. Intriligator and D. A. Kendrick (eds.) *Frontiers of Quantitative Economics*. vol. 2, Amsterdam: North-Holland.
Sims, C. A. (ed.) (1977) *New Methods in Business Cycle Research*, Federal Reserve Bank of Minneapolis.
Sims, C. A. (1980) "Macroeconomics and Reality", *Econometrica*, 48, 1–48.
Spencer, G. (1976) "A Comparison of Some Alternative Estimators for Distributed Lag Models", unpublished M.Sc. thesis, London School of Economics.
Strawderman, W. E. (1978) "Minimax Adaptive Generalized Ridge Regression" *Journal of the American Statistical Association*, 73, 623–627.
Taylor, W. E. (1974) "Smoothness Priors and Stochastic Prior Restrictions in Distributed Lag Estimation", *International Economic Review*, 15, 803–804.
Teräsvirta, T. (1976) "A Note on Bias in the Almon Distributed Lag Estimator", *Econometrica*, 44, 1317–1321.
Teräsvirta, T. (1980a) "The Polynomial Distributed Lag Revisited", *Empirical Economics*, 5, 69–81.
Teräsvirta, T. (1980b) "Linear Restrictions in Misspecified Linear Models and Polynomial Distributed Lag Estimation", Research Report No. 16, Dept. of Statistics, University of Helsinki.
Theil, H. and J. C. G. Boot (1962) "The Final Form of Econometric Equation Systems", *Review of International Statistical Institute*, 30, 162–170.
Thomas, J. J. (1977) "Some Problems in the Use of Almon's Technique in the Estimation of Distributed Lags", *Empirical Economics*, 2, 175–193.
Tinsley, P. A. (1967) "An Application of Variable Weight Distributed Lags", *Journal of the American Statistical Association*, 62, 1277–1290.
Tobin, J. (1969) "A General Equilibrium Approach to Monetary Theory", *Journal of Money, Credit and Banking*, 1, 15–29.
Trivedi, P. K. (1970a) "A Note on the Application of Almon's Method of Calculating Distributed Lag Coefficients", *Metroeconomica*, 22, 281–286.
Trivedi, P. K. (1970b) "The Relation Between the Order-Delivery Lag and the Rate of Capacity Utilisation in the Engineering Industry in the United Kingdom, 1958–1967", *Economica*, 37, 54–67.
Trivedi, P. K. (1975) "Time Series Analysis Versus Structural Models: A Case Study of Canadian Manufacturing Behavior", *International Economic Review*, 16, 587–608.
Trivedi, P. K. (1978) "Estimation of a Distributed Lag Model Under Quadratic Loss", *Econometrica*, 46, 1181–1192.
Trivedi, P. K. and A. R. Pagan (1979) "Polynomial Distributed Lags: A Unified Treatment", *Economic Studies Quarterly*, XXX, 37–49.
Trivedi, P. K., B. M. S. Lee and J. S. Yeo (1979) "On Using Ridge-Type Estimators for a Distributed Lag Model", unpublished paper, Australian National University.
Trivedi, P. K. and B. M. S. Lee (1979) "The Effects of Model Misspecification on the Properties of the Simple Ridge Estimator", Discussion Paper 8009, University of Southampton.
Trivedi, P. K. and B. M. S. Lee (1981) "Seasonal Variability in a Distributed Lag Model", *Review of Economic Studies*, 48, 497–505.
Trivedi, P. K. (1982) "Distributed Lags, Aggregation and Compounding: A Suggested Interpretation", unpublished paper, Australian National University.
Ullah, A. and B. Raj (1979) "A Distributed Lag Estimator Derived from Shiller's Smoothness Priors", *Economic Letters*, 2, 219–223.
Vinod, H. D. (1978) "A Survey of Ridge Regression and Related Techniques for Improvements Over Ordinary Least Squares", *Review of Economics and Statistics*, 60, 121–131.
Waelbroeck, J. K. (1976) (ed.), *The Models of Project Link*. Amsterdam: North-Holland.

Wald, A. (1943) "Tests of Statistical Hypotheses Concerning Several Parameters When the Number of Observations is Large", *Transactions of the American Mathematical Society*, 54, 426–482.
Wall, K. D. (1976) "FIML Estimation of Rational Distributed Lag Structural Form Models", *Annals of Economic and Social Measurement*, 5, 53–63.
Wall, K. D., A. J. Preston, J. W. Bray and M. H. Peston (1975) "Estimates of a Simple Control Model of the U.K. Economy", Chapter 14 in G. A. Renton (ed.), *Modelling the Economy*. London: Heinemann Educational Books.
Wallis, K. F. (1969) "Some Recent Developments in Applied Econometrics: Dynamic Models and Simultaneous Equation Systems", *Journal of Economic Literature*, 7, 771–796.
Wallis, K. F. (1977) "Multiple Time Series Analysis and the Final Form of Econometric Models", *Econometrica*, 45, 1481–1498.
Wallis, K. F. (1980) "Econometric Implications of the Rational Expectations Hypothesis", *Econometrica*, 48, 49–73.
White, H. (1980) "Non-linear Regression on Cross-Section Data", *Econometrica*, 48, 721–746.
Wold, H. (1959) "Ends and Means in Econometric Model Building", pp. 355–434 in U. Grenander (ed.) *Probability and Statistics*. New York: John Wiley and Sons.
Yancey, T. A. and G. G. Judge (1976) "A Monte Carlo Comparison of Traditional and Stein-Rule Estimators Under Squared Error Loss", *Journal of Econometrics*, 4, 285–294.
Yeo, J. S. (1976) "Testing the Length of the Almon Lag Using the Anderson Procedure", London School of Economics, mimeo.
Yeo, J. S. (1978) "Multicollinearity and Distributed Lags", unpublished paper, London School of Economics.
Yule, G. U. (1926) "Why do we Sometimes get Nonsense-Correlations Between Time-Series?—A Study in Sampling and the Nature of Time-Series", *Journal of the Royal Statistical Society*, 89, 1–64.
Zellner, A. and F. I. Palm (1974) "Time Series Analysis and Simultaneous Equation Econometric Models", *Journal of Econometrics*, 2, 17–54.
Zellner, A. (1979) "Statistical Analysis of Econometric Models", *Journal of the American Statistical Association*, 74, 628–643.

Chapter 19

INFERENCE AND CAUSALITY IN ECONOMIC TIME SERIES MODELS

JOHN GEWEKE

Duke University

Contents

Handbook of Econometrics, Volume II, Edited by Z. Griliches and M.D. Intriligator

1. Introduction

Many econometricians are apt to be uncomfortable when thinking about the concept "causality" (in part, because they usually do so under some duress). On the one hand, the concept is a primitive notion which is indispensable when thinking about economic phenomena, econometric models, and the relation between the two. On the other, the idea is notoriously difficult to formalize, as casual reading in the philosophy of science will attest. In this chapter we shall be concerned with a particular formalization that has proved useful in empirical work: hence the juxtaposition of "causality" and "inference". It also bears close relation to notions of strictly exogenous and predetermined variables, which have considerable operational significance in statistical inference, and to the concepts of causal orderings and realizability which are important in model construction in econometrics and engineering, respectively.

Our concept of causality was introduced to economists by C. W. J. Granger [Granger (1963, 1969)], who built on earlier work by Wiener (1956). We shall refer to the concept as Wiener–Granger causality. It applies to relations among time series. Let $X = \{x_t, t \text{ real}\}$ and $Y = \{y_t, t \text{ real}\}$ be two time series, and let X_t and Y_t denote their entire histories up to and including time t: $X_t = \{x_{t-s}, s \geq 0\}$, $Y_t = \{y_{t-s}, s \geq 0\}$. Let U_t denote all information accumulated as of time t, and suppose that $X_s \subseteq U_t$ if and only if $s \leq t$, and $Y_s \subseteq U_t$ if and only if $s \leq t$. If we are better able to predict x_t using U_{t-1} than we are using $U_{t-1} - Y_{t-1}$, then Y *causes* X. If we are better able to "predict" x_t using $U_{t-1} \cup y_t$ than we are using U_{t-1}, then Y *causes* X *instantaneously*.[1]

Since Wiener–Granger causality is defined in terms of predictability, it cannot be an acceptable definition of causation for most philosophers of science [Bunge (1959, ch. 12)]. We do not take up that argument in this chapter. Rather, we concentrate on the operational usefulness of the definition in the construction, estimation, and application of econometric models. In Section 2, for example, we consider the logical relationships among Wiener–Granger causality, Simon's (1952) definition of causal ordering, the engineer's criterion of realizability [e.g. Zemanian (1972)], and the concept of structure set forth by Hurwicz (1962).

Although Wiener–Granger causality is an empirical rather than a logical or ontological concept, it must be made much more specific before propositions like

[1]Granger's (1963, 1969) definitions assume that the time series are stationary, predictors are linear least-squares projections, and mean-square error is the criterion for comparison of forecasts. While these assumptions are convenient to make when conducting empirical tests of the proposition that causality of a certain type is absent, they are not *sui generis* and therefore have not been imposed here.

"Y does not cause X" can be refuted, even in principle. One must always specify the set of "all information" assumed in the definition since Y may cause X for some sets but not others. One must also have a criterion for the comparison of predictors, and the validity of propositions like "Y does not cause X" can be assessed only for restricted classes of predictors and distribution functions. In Section 3 we take up the case, frequently assumed in application, in which $U_t = X_t \cup Y_t$, predictors are linear, and the time series are jointly wide sense stationary, purely nondeterministic, and have autoregressive representations.

In Sections 4 and 5 we move on to issues of statistical inference. In Section 4 it is shown that unidirectional causality from X to Y (i.e. Y does not cause X, and X may or may not cause Y) is logically equivalent to the existence of simultaneous equation models with X exogenous. It is also shown that unidirectional causality from X to Y is not equivalent to the assertion that X is predetermined in a particular behavioral relationship whose parameters are to be estimated. In Section 5 we take up the narrower problem of testing the proposition that Y does not cause X under the assumptions made in Section 3.

Section 6 is devoted to some of the problems which arise in testing the proposition of unidirectional causality using actual economic time series, due to the fact that these series need not satisfy the ideal assumptions made in Sections 3 and 5. We concentrate on parameterization problems, processes which are nonautoregressive or have deterministic components or are nonstationary, and inference about many variables. The reader who is only interested in the mechanics of testing hypotheses about unidirectional causality can skip Sections 2 and 4, and read Sections 3, 5, and 6 in order. The material in Sections 2 and 4, however, is essential in the interpretation of the results of those tests.

2. Causality

Whether or not Wiener–Granger causality is consistent with formal definitions of causality offered by philosophers of science is an open question. In most definitions, "cause" is similar in meaning to "force" or "produce" [e.g. Blalock (1961, pp. 9–10)], which are clearly not synonymous with "predict". Perhaps the definition closest to Wiener–Granger causality is Feigl's in which "causation is defined in terms of *predictability according to a law*" [Feigl (1953, p. 408)]. It has been argued [Zellner (1979)] that statistical "laws" of the type embodied in Wiener–Granger causality are not admissible, as opposed to those of economic theory. Wiener–Granger causality is therefore "devoid of subject matter considerations, including subject matter theory, and thus is in conflict with others' definitions, including Feigl's, that do mention both predictability and laws" [Zellner (1979, p. 51)]. Bunge (1959, p. 30), on the other hand, argues forcefully

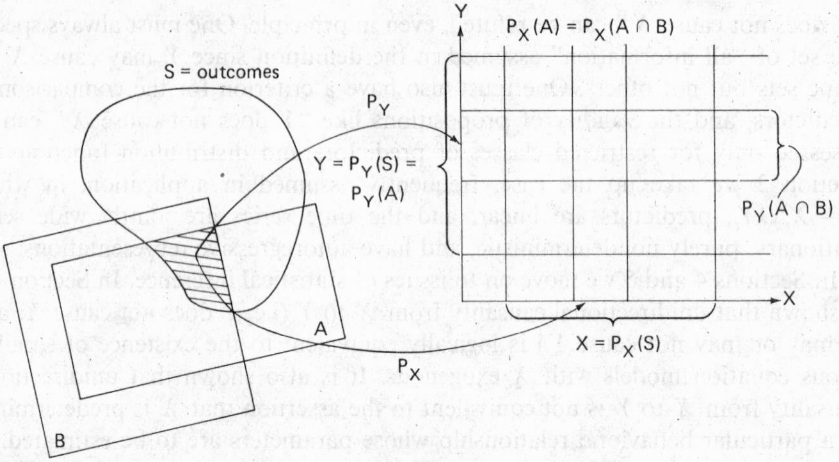

Figure 2.1

against a distinction between statistical and other kinds of laws: "The claim that statistical laws, in contrast to other kinds of scientific law, are incomplete, hence provisional, is largely a matter of metascientific inertia.... In contemporary science and technology, and even in everyday life, we often ask questions that simply cannot be answered on any individual or dynamical laws, questions requiring a statistical approach and analysis."

The *usefulness* of the concept of Wiener–Granger causality in the conceptualization, construction, estimation and manipulation of econometric models is independent of its consistency or inconsistency with formal definitions. To evaluate its usefulness, we review and formalize some operational concepts implicit in econometric modelling.[2]

A definition of *causal ordering* in any econometric model (as opposed to the real world) was proposed by Simon (1952). Suppose S is a space of possible outcomes, and that the model imposes two sets of restrictions, A and B, on these outcomes. The entire model imposes the restriction $A \cap B$ on S. Suppose that S is mapped into two spaces, X and Y, by P_X and P_Y, respectively. Then the ordered pair of restrictions (A, B) implies a causal ordering from X to Y if A restricts X (if at all) but not Y, and B restricts Y (if at all) without further restricting X. Formally we have the following:

Definition

The ordered pair (A, B) of restrictions on S determines a *causal ordering from X to Y* if and only if $P_Y(A) = Y$ and $P_X(A \cap B) = P_X(A)$.

[2] Much (but not all) of what follows in this section may be found in Sims (1977a).

A geometric interpretation of this definition is provided in Figure 2.1. Some examples may also be helpful. Perhaps the simplest one which can be constructed is the following. Let $S = \{(x, y) \in \mathbb{R}^2\}$, and consider the restrictions:.

$$x = a \quad \text{"A"}$$
$$y + bx = c \quad \text{"B"}$$

on S. Let P_X map S into the x coordinate and let P_Y map S into the y coordinate. Then (A, B) determines a causal ordering from X to Y because A determines x without affecting y, while B together with A determines y without further restricting x. The causal ordering is a property of the model, not a property of the restrictions on S to which the model happens to give rise: clearly, there are many pairs of restrictions (C, D) such that $P_X(C \cap D) = P_X(A \cap B) = a$ and $P_Y(C \cap D) = P_Y(A \cap B) = c - ba$, and in fact one of these establishes a causal ordering from Y to X.

As a second example, let S be the family of pairs of random variables (x, y) with bivariate normal distribution. Consider the restrictions:

$$x = u_1 \sim N(\mu_1, \sigma_1^2) \quad \text{"A"}$$
$$y + bx = u_2 \sim N(\mu_2, \sigma_2^2) \quad \text{"B"}$$

on S, where u_1 and u_2 are independent. Suppose P_X and P_Y map S into the marginal distributions for x and y, respectively. Then (A, B) determines a causal ordering from the marginal for x to the marginal for y. The model consisting of A, B, and the stipulation that u_1 and u_2 are independent is the simplest example of a recursive model [Strotz and Wold (1960)]. As Basmann (1965) has pointed out, any outcome in S can be described by such a model—again, the causal ordering is a property of the model, not of the outcome.

Causal orderings, or recursive models, are intended to be more than just descriptive devices. Inherent in such models is the notion that if A is changed, the outcome will still be $A \cap B$, with B unchanged. Once the possibility of changing the first restriction in the ordered pair is granted, it makes a great deal of difference which causal ordering is inherent in the model: different models describe different sets of restrictions on S arising from manipulation of the first restriction. Hence attention is focused on B. We formalize the notion that B is unchanged when A is manipulated as follows.

Definition

The set $B \subseteq S$ *accepts* X *as input* if for any $A \subset S$ which constraints only X (i.e. $P_X^{-1}(P_X(A)) = A$), (A, B) determines a causal ordering from X to Y.

In econometric modelling, the notion that B should accept X as input is so entrenched and natural that it is common to think of B as the model itself, with little or no attention given to the set A which restricts the admissible inputs for the model, although these restrictions may be very important. Conventional manipulation of an econometric model for policy or predictive purposes assumes that the manipulated variables are accepted as input by the model.

In many applications X and Y are time series, as they were in the notation of Section 1. Consider the simple case in which X and Y are univariate, normally distributed, jointly stationary time series, and S is the family of bivariate, normally distributed, jointly stationary time series. Suppose that the restriction A is:

$$A(L)x_t = v_t,$$

where $A(L)$ is one-sided (i.e. involves only non-negative powers of the lag operator L) and has all roots outside the unit circle; and $V = \{v_t, t \text{ real}\}$ is a serially uncorrelated, normally distributed, stationary time series. Let the restriction B be:

$$B(L)y_t + C(L)x_t = w_t,$$

where $B(L)$ has no roots on the unit circle, both $B(L)$ and $C(L)$ may be two-sided (i.e. involve negative powers of the lag operator L) and $W = \{w_t, t \text{ real}\}$ is a serially uncorrelated normally distributed, stationary time series independent of U. Since A implies $x_t = A(L)^{-1}u_t$, it establishes the first time series without restricting the second, while "B" implies

$$y_t = -B(L)^{-1}C(L)x_t + B(L)^{-1}v_t, \tag{2.1}$$

which establishes the second without changing the first. Hence, the model establishes a causal ordering from X to Y, and if for any normally distributed, jointly stationary X the outcome of the model satisfies (2.1), then B accepts X as input. Such a model might or might not be interesting for purposes of manipulation, however. In general, y_t will be a function of past, current, and future X, which is undesirable if B is supposed to describe the relation between actual inputs and outputs; the restriction that $B(L)$ and $C(L)$ be one-sided and that $B(L)$ have no roots inside the unit circle would obviate this difficulty.

The notion that future inputs should not be involved in the determination of present outputs is known in the engineering literature as realizability [Zemanian (1972)], and we can formalize it in our notation as follows.

Definition

The set $B \subseteq S$ is *realizable with time series X as input* if B accepts X as input, and $P_{X_t}(A_1) = P_{X_t}(A_2)$ implies $P_{Y_r}(A_1 \cap B) = P_{Y_r}(A_2 \cap B)$ for all $A_1 \subseteq S$ and $A_2 \subseteq S$ which constrain only X, and all $t \geq r$.

If B accepts X as input but is not realizable, then a specification of inputs up to time t will not restrict outputs, but once outputs up to time t are restricted by B, then further restrictions on inputs—those occurring after time t—are implied. This is clearly an undesirable characteristic of any model which purports to treat time in a realistic fashion.

The concepts of causal ordering, inputs, and realizability pertain to models. One can establish whether models possess these properties without reference to the phenomena which the models are supposed to describe. Of course, our interest in these models stems from the possibility that they do indeed describe actual phenomena. Hurwicz (1962) attributes the characteristic *structural* to models which meet this criterion.

Definition

The set $B \subseteq S$ is *structural* for inputs X if B accepts X as input, and when any set $C \subseteq X$ is implemented, then $P_Y(P_X^{-1}(C) \cap B)$ is true.

Notice that the use of the word "structural" here is not the same as its use in the parlance of simultaneous equation models. The sets of "structural", "reduced form" and "final form" equations are either all structural or not structural in the sense of the foregoing definition, depending on whether or not the model depicts actual phenomena.

This definition incorporates two terms which shall remain primitive: "implemented" and "true". Whether or not $P_Y(P_X^{-1}(C) \cap B)$ is true for a *given* C is a question to which statistical inference can be addressed; at most, we can hope to attach a posterior probability to the truth of this statement. We can never know whether $P_Y(P_X^{-1}(C) \cap B)$ is true for any C: one can never prove that a model is structural, although by implementing one or more sets C serious doubts could be cast on the assertion. Since the definition allows *any* set $C \subseteq X$ to be implemented, those implementing inputs in real time are permitted to change their plans. It seems implausible that the current outputs of an actual system should depend on future inputs as yet undetermined. We formalize this idea as follows.

Axiom of causality

$B \subseteq S$ is structural for inputs X only if B is realizable with X as input.

The axiom of causality is a formalization of the idea that the future cannot cause the past, an idea which appears to be uniformly accepted in the philosophy of science despite differences about the relations between antecedence and causality. For example, Blalock (1964, p. 10) finds this condition indispensable:

"Since the forcing or producing idea is not contained in the notion of temporal sequences, as just noted, our conception of causality should not depend on temporal sequences, except for the impossibility of an effect preceding its cause." Bunge argues that the condition is universally satisfied:

> Even relativity admits the reversal of time series of physically disconnected events but excludes the reversal of causal connections, that is, it denies that effects can arise before they have been produced...events whose order of succession is reversible cannot be causally connected with one another; at most they may have a common origin...To conclude, a condition for causality to hold is that C [the cause] be previous to or at most simultaneous with E [the event] (relative to a given reference system) [Bunge (1959, p. 67)].

It is important to note that the converse of the axiom of causality is the *post hoc ergo propter hoc* fallacy. The fallaciousness of the converse follows from the fact that there are many $B_j \subseteq S$ which are realizable with X as input, but for which $P_Y(P_X^{-1}(C) \cap B_j) \neq P_Y(P_X^{-1}(C) \cap B_k)$ when $j \neq k$ for some choices of C. For C which have actually been implemented, B_j and B_k may of course produce identical outputs in spite of their logical inconsistency: one cannot establish that a restriction is structural through statistical inference, even to a specified level of *a posteriori* probability.[3] It may seem curious to provide the name "axiom of causality" to a statement which nowhere mentions the word "cause". The name is chosen because of Sims' (1972) result that (in our language, and with appropriate restrictions on classes of time series and predictors) B is realizable with X as input if and only if in B Wiener–Granger causality is unidirectional from X to Y. To develop this result we shall be quite specific about the structure of the time series X and Y.

3. Causal orderings and their implications

In any empirical application the concept of Wiener–Granger causality must be formulated more narrowly than it is in Granger's definitions. The relevant universe of information must be specified, and the class of predictors to be considered must be limited. If formal, classical hypothesis testing is contemplated, then the question of whether or not Y is causing X must be made to depend on the values of parameters which are few in number relative to the number of observations at hand. The determination of the relevant universe of information rests primarily on *a priori* considerations from economic theory, in much the

[3]An extended discussion of specific pitfalls encountered in using a finding that a restriction B which is realizable with X as input is in agreement with the data, to buttress a claim that B is structural, is provided by Sims (1977).

same way that the specification of which variables should enter a behavioral equation or system of equations does. Empirical studies which examine questions of Wiener–Granger causality differ greatly in the care with which the universe of information is chosen; in many instances, it is suggested by earlier work on substantively similar issues which did not address questions of causality. However, virtually all of these studies consider only predictors which are linear either in levels or logarithms. This choice is due mainly to the analytical convenience of the linearity specification, as it is elsewhere in econometric theory. In the present case it is especially attractive because only linear predictors are necessarily time invariant when time series are assumed to be wide sense stationary, the least restrictive class of time series for which a rich and useful theory of prediction is available. In this section we will discuss the portions of this theory essential for developing the testable implications of Wiener–Granger causality. Considerations of testing and inference are left to Section 5.

3.1. *A canonical form for wide sense stationary multiple time series*

We focus our attention on a wide sense stationary, purely non-deterministic time series z_t: $m \times 1$. By wide sense stationary, it is meant that the mean of z_t exists and does not depend on t, and for all t and s $\mathrm{cov}(z_t, z_{t+s})$ exists and depends on s but not t. By purely non-deterministic, it is meant that the correlation of z_{t+p} and z_t vanishes as p increases so that in the limit the best linear forecast of z_{t+p} conditional on $\{z_{t-s}, s>0\}$ is the unconditional mean of z_{t+p}, which for convenience we take to be $\mathbf{0}$. It is presumed that the relevant universe of information at time t consists of $Z_t = \{z_{t-s}, s>0\}$. These assumptions restrict the universe of information which might be considered, but they are no more severe than those usually made in standard linear or simultaneous equation models for the purposes of developing an asymptotic theory of inference.

We further suppose that there exists a moving average representation for z_t:

$$z_t = \sum_{s=0}^{\infty} A_s \boldsymbol{\varepsilon}_{t-s}, \qquad E(\boldsymbol{\varepsilon}_t) = \mathbf{0}, \qquad \mathrm{var}(\boldsymbol{\varepsilon}_t) = \Upsilon. \tag{3.1}$$

In the moving average representation, all roots of the generating function $\sum_{s=0}^{\infty} A_s z^s$ have modulus not less than unity, the coefficients satisfy the square summability condition $\sum_{s=0}^{\infty} \|A_s\|^2 < \infty$,[4] and the vector $\boldsymbol{\varepsilon}_t$ is serially uncorrelated [Wold (1938)]. The existence of the moving average representation is important to

[4]For any square complex matrix C, $\|C\|$ denotes the square root of the largest eigenvalue of $C'C$, and $|C|$ denotes the square root of the determinant of $C'C$.

us for two reasons. First, it is equivalent to the existence of the spectral density matrix $S_z(\lambda)$ of z_t at almost all frequencies $\lambda \in [-\pi, \pi]$ [Doob (1953, pp. 499–500)]. Second, it provides a lower bound on the mean square error of one-step ahead minimum mean square error linear forecasts, which is:

$$|\Upsilon| = \exp\left(\frac{1}{2\pi}\int_{-\pi}^{\pi} \ln|S_z(\lambda)|\,d\lambda\right) > 0. \tag{3.2}$$

The condition $|\Upsilon| > 0$ is equivalent to our assumption that Z is strictly non-deterministic [Rozanov (1967, p. 72)].

Whether this lower bound can be realized depends on whether the relation in (3.1) can be inverted so that z_t becomes a linear function of Z_{t-1} and ε_t,

$$z_t = \sum_{s=1}^{\infty} B_s z_{t-s} + \varepsilon_t. \tag{3.3}$$

A sufficient condition for invertibility is that there exist a constant $c \geq 1$ such that for almost all λ:

$$c^{-1} I_n \circledleq S_z(\lambda) \circledleq c I_n \tag{3.4}$$

[Rozanov (1967, pp. 77–78)], which we henceforth assume.[5] This assumption is nontrivial, because processes like $z_t = \varepsilon_t + \varepsilon_{t-1}$ are excluded. The requirement (3.4) that the spectral density matrix be bounded uniformly away from zero almost everywhere in $[-\pi, \pi]$ is more restrictive than (3.2). On the other hand (3.4) is less restrictive than the assumption that Z is a moving average, autoregressive process of finite order with invertible moving average and autoregressive parts, which is sometimes taken as the point of departure in the study of multiple time series.

Suppose now that z_t has been partitioned into $k \times 1$ and $l \times 1$ subvectors x_t and y_t, $z_t' = (x_t', y_t')$, reflecting an interest in causal relationships between X and Y. Adopt a corresponding partition of $S_z(\lambda)$:

$$S_z(\lambda) = \begin{bmatrix} S_x(\lambda) & S_{xy}(\lambda) \\ S_{yx}(\lambda) & S_y(\lambda) \end{bmatrix}.$$

From (3.4) X and Y each possess autoregressive representations, which we denote:

$$x_t = \sum_{s=1}^{\infty} E_{1s} x_{t-s} + u_{1t}, \quad \operatorname{var}(u_{1t}) = \Sigma_1, \tag{3.5}$$

[5] $A \circledleq B$ indicates that $B - A$ is positive semidefinite; $A \circledless B$ indicates that $B - A$ is positive semidefinite and not null.

and

$$y_t = \sum_{s=1}^{\infty} G_{1s} x_{t-s} + v_{1t}, \quad \text{var}(v_{1t}) = T_1, \tag{3.6}$$

respectively. The disturbance u_{1t} is the one-step-ahead error when x_t is forecast from its own past alone, and similarly for v_{1t} and y_t. These disturbance vectors are each serially uncorrelated, but may be correlated with each other contemporaneously and at various leads and lags. Since u_{1t} is uncorrelated with all X_{t-1}, (3.5) denotes the linear projection of x_t on its own past, and likewise (3.6) denotes the linear projection of y_t on past Y_{t-1}.

The linear projection of x_t on X_{t-1} and Y_{t-1}, and of y_t on X_{t-1} and Y_{t-1} is given by (3.3), which we partition:

$$x_t = \sum_{s=1}^{\infty} E_{2s} x_{t-s} + \sum_{s=1}^{\infty} F_{2s} y_{t-s} + u_{2t}, \quad \text{var}(u_{2t}) = \Sigma_2, \tag{3.7}$$

$$y_t = \sum_{s=1}^{\infty} G_{2s} y_{t-s} + \sum_{s=1}^{\infty} H_{2s} x_{t-s} + v_{2t}, \quad \text{var}(v_{2t}) = T_2. \tag{3.8}$$

The disturbance vectors u_{2t} and v_{2t} are each serially uncorrelated, but since each is uncorrelated with X_{t-1} and Y_{t-1}, they can be correlated with each other only contemporaneously. We shall find the partition:

$$\Upsilon = \text{var}\begin{pmatrix} u_{2t} \\ v_{2t} \end{pmatrix} = \begin{bmatrix} \Sigma_2 & C \\ C' & T_2 \end{bmatrix}$$

useful.

If the system (3.7)–(3.8) is premultiplied by the matrix

$$\begin{bmatrix} I_k & -CT_2^{-1} \\ -C'\Sigma_2^{-1} & I_l \end{bmatrix},$$

then in the first k equations of the new system x_t is a linear function of X_{t-1}, Y_t, and a disturbance $u_{2t} - CT_2^{-1}v_{2t}$. Since the disturbance is uncorrelated with v_{2t} it is uncorrelated with y_t—as well as X_{t-1} and Y_{t-1}. Hence, the linear projection of x_t on X_{t-1} and Y_t,

$$x_t = \sum_{s=1}^{\infty} E_{3s} x_{t-s} + \sum_{s=0}^{\infty} F_{3s} y_{t-s} + u_{3t}, \quad \text{var}(u_{3t}) = \Sigma_3, \tag{3.9}$$

is provided by the first k equations of the new system. Similarly, the existence of the linear projection of $\boldsymbol{y}_t$ on Y_{t-1} and X_t,

$$\boldsymbol{y}_t = \sum_{s=1}^{\infty} G_{3s}\boldsymbol{y}_{t-s} + \sum_{s=0}^{\infty} H_{3s}\boldsymbol{x}_{t-s} + \boldsymbol{v}_{3t}, \quad \mathrm{var}(\boldsymbol{v}_{3t}) = T_3, \tag{3.10}$$

follows from the last l equations.

We finally consider the linear projections of $\boldsymbol{x}_t$ on X_{t-1} and Y, and $\boldsymbol{y}_t$ on Y_{t-1} and X. Let $\tilde{D}(\lambda) = S_{xy}(\lambda)S_y(\lambda)^{-1}$, for all $\lambda \in [-\pi, \pi]$ for which the terms are defined and (3.4) is true. Because of (3.4), the inverse Fourier transform,

$$D_s = \frac{1}{2\pi}\int_{-\pi}^{\pi} \tilde{D}(\lambda)e^{i\lambda s}\mathrm{d}\lambda,$$

of $\tilde{D}(\lambda)$ satisfies the condition $\sum_{s=0}^{\infty}\|D_s\|^2 < \infty$. From the spectral representation of Z it is evident that $\boldsymbol{w}_t = \boldsymbol{x}_t - \sum_{s=-\infty}^{\infty} D_s\boldsymbol{y}_{t-s}$ is uncorrelated with all $\boldsymbol{y}_s$, and that

$$\boldsymbol{x}_t = \sum_{s=-\infty}^{\infty} D_s\boldsymbol{y}_{t-s} + \boldsymbol{w}_t \tag{3.11}$$

therefore provides the linear projection of $\boldsymbol{x}_t$ on Y. Since $S_w(\lambda) = S_x(\lambda) - S_{xy}(\lambda)S_y(\lambda)^{-1}S_{yx}(\lambda)$ consists of the first k rows and columns of $S_z(\lambda)^{-1}$, $c^{-1}I_k \circledleq S_w(\lambda) \circledleq cI_k$ for almost all λ. Hence, $\boldsymbol{w}_t$ possesses an autoregressive representation, which we write:

$$\boldsymbol{w}_t = \sum_{s=1}^{\infty} B_s\boldsymbol{w}_{t-s} + \boldsymbol{u}_{4t}. \tag{3.12}$$

Consequently,

$$\boldsymbol{x}_t = \sum_{r=1}^{\infty} B_r\boldsymbol{x}_{t-r} - \sum_{r=0}^{\infty} B_r \sum_{s=-\infty}^{\infty} D_s\boldsymbol{y}_{t-s-r} + \boldsymbol{u}_{4t}, \tag{3.13}$$

where $B_0 = -I_k$. Grouping terms, (3.13) may be written:

$$\boldsymbol{x}_t = \sum_{s=1}^{\infty} E_{4s}\boldsymbol{x}_{t-s} + \sum_{s=-\infty}^{\infty} F_{4s}\boldsymbol{y}_{t-s} + \boldsymbol{u}_{4t}, \quad \mathrm{var}(\boldsymbol{u}_{4t}) = \Sigma_4, \tag{3.14}$$

where $E_{4s} = B_s$ and $F_{4s} = \sum_{r=0}^{\infty} B_r D_{s-r}$. Since $\boldsymbol{u}_{4t}$ is a linear function of W_t, it is uncorrelated with Y; and since X_{t-1} is a linear function of Y and W_{t-1}, $\boldsymbol{u}_{4t}$ is

Table 3.1
A canonical form for the wide sense stationary time series $z_t' = (x_t', y_t')$.

$$x_t = \sum_{s=1}^{\infty} E_{1s} x_{t-s} + u_{1t} \quad (3.5) \qquad y_t = \sum_{s=1}^{\infty} G_{1s} y_{t-s} + v_{1t} \quad (3.6)$$

$$x_t = \sum_{s=1}^{\infty} E_{2s} x_{t-s} + \sum_{s=1}^{\infty} F_{2s} y_{t-s} + u_{2t} \quad (3.7) \qquad y_t = \sum_{s=1}^{\infty} G_{2s} y_{t-s} + \sum_{s=1}^{\infty} H_{2s} x_{t-s} + v_{2t} \quad (3.8)$$

$$x_t = \sum_{s=1}^{\infty} E_{3s} x_{t-s} + \sum_{s=0}^{\infty} F_{3s} y_{t-s} + u_{3t} \quad (3.9) \qquad y_t = \sum_{s=1}^{\infty} G_{3s} y_{t-s} + \sum_{s=0}^{\infty} H_{3s} x_{t-s} + v_{3t} \quad (3.10)$$

$$x_t = \sum_{s=1}^{\infty} E_{4s} x_{t-s} + \sum_{s=-\infty}^{\infty} F_{4s} y_{t-s} + u_{4t} \quad (3.14) \qquad y_t = \sum_{s=1}^{\infty} G_{4s} y_{t-s} + \sum_{s=-\infty}^{\infty} H_{4s} x_{t-s} + v_{4t} \quad (3.15)$$

$$\operatorname{var}[u_{jt}] = \Sigma_j \qquad \operatorname{cov}(u_{2t}, v_{2t}) = C \qquad \operatorname{var}[v_{jt}] = T_j$$

$$\Upsilon = \begin{bmatrix} \Sigma_2 & C \\ C' & T_2 \end{bmatrix}$$

uncorrelated with X_{t-1}. Hence, (3.14) provides the linear projection of x_t on X_{t-1} and all Y, $\sum_{s=1}^{\infty} \|E_{4s}\|^2 < \infty$ and $\sum_{s=-\infty}^{\infty} \|F_{4s}\|^2 < \infty$. The same argument may be used to demonstrate that the linear projection of y_t on Y_{t-1} and X_t,

$$y_t = \sum_{s=1}^{\infty} G_{4s} y_{t-s} + \sum_{s=-\infty}^{\infty} H_{4s} x_{t-s} + v_{4t}, \quad \operatorname{var}(v_{4t}) = T_4, \tag{3.15}$$

exists and all coefficients are square summable.

3.2. *The implications of unidirectional causality*[6]

If the universe of information at time t is Z_t, all predictors are linear, and the criterion for the comparison of forecasts is mean square error, then the Wiener–Granger definition of causality may be stated in terms of the parameters of the canonical form displayed in Table 3.1, whose existence was just demonstrated. For example, Y causes X if, and only if, $F_{2s} \not\equiv 0$; equivalent statements are $\Sigma_1 \textcircled{>} \Sigma_2$ and $|\Sigma_1| > |\Sigma_2|$. Since $\Sigma_1 \textcircled{\geq} \Sigma_2$ in any case, Y does not cause X if and only if $|\Sigma_1| = |\Sigma_2|$. Define the *measure of linear feedback from Y to X*:

$$F_{Y \to X} \equiv \ln(|\Sigma_1| / |\Sigma_2|).$$

The statement "Y does not cause X" is equivalent to $F_{Y \to X} = 0$. Symmetrically, X does not cause Y if, and only if, the *measure of linear feedback from X to Y*,

$$F_{X \to Y} \equiv \ln(|T_1| / |T_2|),$$

is zero.

[6]Most of the material in this subsection may be found in Geweke (1982a).

The existence of instantaneous causality between X and Y amounts to non-zero partial correlation between $\boldsymbol{x}_t$ and $\boldsymbol{y}_t$ conditional on $X_{t-1} \cup Y_{t-1}$, or equivalently a non-zero *measure of instantaneous linear feedback*:

$$F_{X \cdot Y} \equiv \ln(|\Sigma_2|/|\Sigma_3|).$$

A concept closely related to the notion of linear feedback is that of linear dependence. From (3.11), X and Y are linearly independent if and only if $D_s \equiv 0$, which from (3.14) is equivalent to $F_{4s} \equiv 0$. Hence, X and Y are linearly independent if and only if $|\Sigma_1| = |\Sigma_4|$, which suggests the *measure of linear dependence*:

$$F_{X,Y} = \ln(|\Sigma_1|/|\Sigma_4|).^7$$

Since instantaneous linear feedback and linear dependence are notions in which the roles of X and Y are symmetric (unlike linear feedback), $F_{X \cdot Y}$ and $F_{X,Y}$ could have been expressed in terms of the T_j's rather than the Σ_j's. The following result shows that the alternative definitions would be equivalent, and demonstrates some other relationships among the parameters of the canonical form in Table 3.1.

Theorem

In the canonical form in Table 3.1,

(i) $F_{X,Y} = \ln(|\Sigma_1|/|\Sigma_4|) = \ln(|T_1|/|T_4|)$.
(ii) $F_{Y \to X} = \ln(|\Sigma_1|/|\Sigma_2|) = \ln(|T_3|/|T_4|)$.
(iii) $F_{X \to Y} = \ln(|\Sigma_3|/|\Sigma_4|) = \ln(|T_1|/|T_2|)$.
(iv) $F_{X \cdot Y} = \ln(|\Sigma_2|/|\Sigma_3|) = \ln(|T_2|/|T_3|) = \ln(|T_2| \cdot |\Sigma_2|/|\Upsilon|)$.

Proof

(i) Since $\boldsymbol{u}_{4t}$ is the disturbance in the autoregressive representation (3.12) of $\boldsymbol{w}_t$,

$$\begin{aligned}
\ln|\Sigma_4| &= \frac{1}{2\pi}\int_{-\pi}^{\pi} \ln|S_w(\lambda)|\,d\lambda \\
&= \frac{1}{2\pi}\int_{-\pi}^{\pi} \ln|S_x(\lambda) - S_{xy}(\lambda)S_y(\lambda)^{-1}S_{yx}(\lambda)|\,d\lambda \\
&= \frac{1}{2\pi}\int_{-\pi}^{\pi} (\ln|S_z(\lambda)| - \ln|S_y(\lambda)|)\,d\lambda. \\
&= \ln(|\Upsilon|/|T_1|).
\end{aligned} \tag{3.16}$$

[7]In the case $k = l = 1$, our measure of linear dependence is the same as the *measure of information* per unit time contained in X about Y and vice versa, proposed by Gel'fand and Yaglom (1959). In the case $l = 1$, $F_{X \to Y} + F_{X \cdot Y} = -\ln(1 - R^2{}_*)$ and $F_{X,Y} = -\ln(1 - R^2{}_*(\pm))$, $R^2{}_*$ and $R_*(\pm)$ being proposed by Pierce (1979). In the case in which there is no instantaneous causality, Granger (1963) proposed that $1 - |\Sigma_2|/|\Sigma_1|$ be defined as the "strength of causality $Y \Rightarrow X$" and $1 - |T_2|/|T_1|$ be defined as the "strength of causality $X \Rightarrow Y$."

Hence,

$$\ln(|\Sigma_1|/|\Sigma_4|) = \ln(|\Sigma_1|\cdot|T_1|/|\Upsilon|),$$

and by an argument symmetric in X and Y:

$$\ln(|T_1|/|T_4|) - \ln(|\Sigma_1|\cdot|T_1|/|\Upsilon|).$$

(ii) By construction of (3.9) $\Sigma_3 = \Sigma_2 - CT_2^{-1}C'$, so $|\Sigma_3|\cdot|T_2| = |\Upsilon|$. Combining this result with $|\Sigma_4|\cdot|T_1| = |\Upsilon|$ from (3.16), (ii) is obtained.

(iii) Follows by symmetry with (ii).

(iv) Follows from $|\Sigma_3|\cdot|T_2| = |\Upsilon|$ and the symmetry of the right-hand side of that equation in X and Y.

We have seen that the measures $F_{X\cdot Y}$ and $F_{X,Y}$ preserve the notions of symmetry inherent in the concepts of instantaneous causality and dependence, in the case where relations are constrained to be linear and the metric of comparison is mean square error. Since

$$F_{X,Y} = F_{Y\to X} + F_{X\cdot Y} + F_{X\to Y},$$

linear dependence can be decomposed additively into three kinds of linear feedback. Absence of a particular causal ordering is equivalent to one of these three types of feedback being zero. As we shall see in Section 5, the relations in this theorem provide a basis for tests of null hypotheses which assert the absence of one or more causal orderings.

It is a short step from this theorem to Sims' (1972) result that Y does not cause X if, and only if, in the linear projection of Y on future, current and past X coefficients on future X are zero. The statement "Y does not cause X" is equivalent to $\Sigma_1 = \Sigma_2$ and $T_3 = T_4$, which is in turn equivalent to $H_{3s} \equiv H_{4s}$. From our derivation of (3.14) from (3.11), coefficients on $X - X_t$ in (3.15) are zero if and only if the coefficients on $X - X_t$ in the projection of y_t on X are zero. This implication provides yet another basis for tests of the null hypothesis that Y does not cause X.

3.3. *Extensions*

The concept of Wiener–Granger causality has recently been discussed in contexts less restrictive than the one presented here. The assumption that the multiple time series of interest is stationary and purely non-deterministic can be relaxed, attention need not be confined to linear relations, and characterizations of bidirectional causality have been offered. We briefly review the most important developments in each of these areas, providing citations but not proofs.

The extension to the case in which Z may be non-stationary and have deterministic components is relatively straightforward, so long as only linear relations are of interest. The definition of Wiener–Granger causality given in Section 1 remains pertinent with the understanding that only linear predictors are considered. If Z is non-stationary, then the linear predictors are in general not time invariant, as was the case in this section. Hosoya (1977) has shown that if Y does not cause X, then the difference between y_t and its projection on X_t is orthogonal to x_{t+p} $(p \geq 1)$. The latter condition is the one offered by Sims (1972) under the assumptions of stationary and pure non-determinism, and is the natural analogue of the condition $\ln(|T_3|/|T_4|)=0$. If Z contains deterministic components, then the condition that Y does not cause X implies that these components are linear functions of the deterministic part of X, plus a residual term which is uncorrelated with X at all leads and lags [Hosoya (1977)].

When we widen our attention to include possibly non-linear relations, more subtle issues arise. Consider again the condition that Y does not cause X. Corresponding natural extensions of the conditions we developed for the linear, stationary, purely non-deterministic case are:

(1) X_{t+1} is independent of Y_t conditional on X_t for all t, for the restriction $F_{2s} \equiv 0$ in (3.7);

(2) y_t is independent of $x_{t+1}, x_{t+2}, \dots$ conditional on X_t for all t, for the restriction that the linear projections of y_t on X and on X_t be identical; and

(3) y_t is independent of $x_{t+1}, x_{t+2}, \dots$ conditional on X_t and Y_{t-1} for all t, for the restriction $H_{3s} \equiv H_{4s}$ in (3.10) and (3.15).

Chamberlain (1982) has shown that under a weak regularity condition (analogous to $\sum_{s=0}^{\infty}\|A_s\|^2<\infty$ introduced in Section 3.1) conditions (1) and (3) are equivalent, just as their analogues were in the linear case. However, (1) or (3) implies (2), but not conversely: the natural extension of Sims' (1972) result is not true. Further discussion of these points is provided in Chamberlain's paper and in the related work of Florens and Mouchart (1982).

When causality is unidirectional it is natural to seek to quantify its importance and provide summary characteristics of the effect of the uncaused on the caused series. When causality is bidirectional – as is perhaps the rule – these goals become even more pressing. The measures of linear feedback provide one practical answer to this question, since they are easy to estimate. Two more elaborate suggestions have been made, both motivated by problems in the interpretation of macroeconomic aggregate time series.

Sims (1980) renormalizes the moving average representation (3.1) in recursive form

$$z_t = \sum_{s=0}^{\infty} A_s^* \varepsilon_{t-s}^*, \tag{3.17}$$

with A_0^* lower triangular and $\Sigma^* = \text{var}(\varepsilon_t^*)$ diagonal. [The renormalization can be computed from (3.1) by exploiting the Choleski decomposition $\Sigma = MM'$, with M lower triangular. Denote $L = \text{diag}(M)$. Then $LM^{-1}z_t = \sum_{s=0}^{\infty} LM^{-1}A_s\varepsilon_{t-s}$; $LM^{-1}A_0 = LM^{-1}$ is lower triangular with units on the main diagonal and $\text{var}(LM^{-1}\varepsilon_t) = LL'$.] If we let $\sigma_i^2 = \text{var}(\varepsilon_{it}^*)$ and $[A_s^*]_{ij} = a_{sij}^*$, it follows from the diagonality of Σ^* that the m-step-ahead forecast error for z_{jt} is $\sigma^2(j,m) \equiv \sum_{i=1}^{n}\sigma_i^2\sum_{s=0}^{m-1}a_{sij}^{*2}$. The function $\sigma_i^2\sum_{s=0}^{m-1}a_{sij}^{*2}/\sigma^2(j,m)$ provides a measure of the relative contribution of the disturbance corresponding to z_i in (3.17) to the m-step-ahead forecast error in z_j. This measure is somewhat similar to the measures of feedback discussed previously; when $m = 1$ and Σ is diagonal, there is a simple arithmetic relationship between them. An important advantage of this decomposition is that for large m it isolates relative contributions to movements in the variables which are, intuitively, "persistent". An important disadvantage, however, is that the measures depend on the ordering of the variables through the renormalization of (3.1).

Geweke (1982a) has shown that the measures of feedback $F_{Y\to X}$ and $F_{X\to Y}$ may be decomposed by frequency. Subject to some side conditions which as a practical matter are weak, there exist non-negative bonded functions $f_{Y\to X}(\lambda)$ and $f_{X\to Y}(\lambda)$ such that $F_{Y\to X} = (1/2\pi)\int_{-\pi}^{\pi} f_{Y\to X}(\lambda)\,\mathrm{d}\lambda$ and $F_{X\to Y} = (1/2\pi)\int_{-\pi}^{\pi} f_{X\to Y}(\lambda)\,\mathrm{d}\lambda$. The measures of feedback are thus decomposed into measures of feedback by frequency which correspond intuitively to the "long run" (low frequencies, small λ) and "short run" (high frequencies, large λ). In the case of low frequencies, this relationship has been formalized in terms of the implications of comparative statics models for time series [Geweke (1982b)].

4. Causality and exogeneity

The condition that Y not cause X, in the sense defined in Section 1, is very closely related to the condition that X be strictly exogenous in a stochastic model. The two are so closely related that tests of the hypothesis that Y does not cause X are often termed "exogeneity tests" in the literature [Sims (1977), Geweke (1978)]. The strict exogeneity of X is in turn invoked in inference in a wide variety of situations, for example the use of instrumental variables in the presence of serially correlated disturbances. The advantage of the strict exogeneity assumption is that there is often no loss in limiting one's attention to distributions conditional on strictly exogenous X, and this limitation usually results in considerable simplification of problems of statistical inference. As we shall soon see, however, the condition that Y not cause X is not *equivalent* to the strict exogeneity of X. All that can be said is that if X is strictly exogenous in the complete dynamic simultaneous equation model, then Y does not cause X, where Y is endogenous in that model. This means that tests for the absence of a Wiener–Granger causal ordering can be used to refute the strict exogeneity specification in a certain class of stochastic models, but never to establish it. In addition, there are many circumstances in which nothing is lost by undertaking statistical inference conditional on a subset of variables which are not strictly exogenous—the best known being that in which there are predetermined variables in the complete dynamic simultaneous equation model. Unidirectional causality is therefore neither a necessary nor a sufficient condition for inference to proceed conditional on a subset of variables.

To establish these ideas, specific terminology is required. We begin by adopting a definition due to Koopmans and Hood (1953, pp. 117–120), as set forth by Christ (1966, p. 156).[8]

Definition

A *strictly exogenous variable* in a stochastic model is a variable whose value in each period is statistically independent of the values of all the random disturbances in the model in all periods.

Examples of strictly exogenous variables are provided in complete, dynamic simultaneous equation models in which all variables are normally distributed:[9]

$$\begin{aligned} &B(L)\boldsymbol{y}_t + \Gamma(L)\boldsymbol{x}_t = \boldsymbol{u}_t; \\ &A(L)\boldsymbol{u}_t = \boldsymbol{\varepsilon}_t; \\ &\operatorname{cov}(\boldsymbol{\varepsilon}_t, \boldsymbol{y}_{t-s}) = 0, \qquad s > 0; \\ &\operatorname{cov}(\boldsymbol{\varepsilon}_t, \boldsymbol{x}_{t-s}) = 0, \quad \text{all } s; \end{aligned} \tag{4.1}$$

Roots of $|B(L)|$ and $|A(L)|$ have modulus greater than 1.

This model is similar to Koopmans' (1950) and those discussed in most econometrics texts, except that serially correlated disturbances and possibly infinite lag lengths are allowed. The equation $A(L)^{-1}B(L)\boldsymbol{y}_t + A(L)^{-1}\Gamma(L)\boldsymbol{x}_t = \boldsymbol{\varepsilon}_t$ corresponds to (3.10) in the canonical form derived in Section 3, and since $\boldsymbol{\varepsilon}_t$ is uncorrelated with X, it corresponds to (3.15) as well. Hence, $F_{Y \to X} = 0$: Y does not cause X. In view of our discussion in Section 2 and the fact that the complete dynamic simultaneous equation model is usually perceived as a structure which accepts X as input, this implication is not surprising.

If Y does not cause X then there exists *a* complete dynamic simultaneous equation model with Y endogenous and X strictly exogenous, in the sense that

[8]We use the term "strictly exogenous" where Christ used "exogenous" in order to distinguish this concept from weak exogeneity, to be introduced shortly.

[9]The strong assumption of normality is made because of the strong condition of independence in our definition of strict exogeneity. As a practical matter, quasi-maximum likelihood methods are usually used, and the independence condition can then be modified to specify absence of correlation.

there exist systems of equations formally similar to (4.1)—e.g. (3.10) and (3.15). However, none of these systems need satisfy the overidentifying restrictions in the model of interest, and in this case X is not strictly exogenous. For example, consider the case of (4.1) in which there are no exogenous variables:

$$B(L)y_t = \begin{bmatrix} \underset{l_1\times l_1}{B_{11}(L)} & \underset{l_1\times l_2}{B_{12}(L)} \\ \underset{l_2\times l_1}{B_{21}(L)} & \underset{l_2\times l_2}{B_{22}(L)} \end{bmatrix} \begin{pmatrix} \underset{l_1\times l_1}{y_{1t}} \\ \underset{l_2\times 1}{y_{2t}} \end{pmatrix} = \begin{pmatrix} \underset{l_1\times 1}{\varepsilon_{1t}} \\ \underset{l_2\times 1}{\varepsilon_{2t}} \end{pmatrix}. \tag{4.2}$$

The autoregressive representation of Y is $B(0)^{-1}B(L)y_t = B(0)^{-1}\varepsilon_t$. If the elements in the first l_1 rows and last l_2 columns of $B(0)^{-1}B(L)$ turn out to be zeroes, then Y_2 does not cause Y_1. For instance, this will always be the case if only contemporaneous values of Y_2 enter the system (4.2), i.e. $B_{12}(L) = B_{12}(0)$ and $B_{22}(L) = B_{22}(0)$.

If a simultaneous equation model of the form (4.1) is hypothesized, then a finding that Y causes X is grounds for rejection of the model. A finding that Y does not cause X supports the model, in the sense that a refutable implication of the model fails to be rejected, but in the same sense the finding supports all models with the same lists of exogenous and endogenous variables.

The outcome of a test of the hypothesis that Y does not cause X is in general not equivalent to answering the question of whether X or a subset of X can be regarded as fixed for purposes of inference. There are many examples in which this assumption is justified by *a priori* information, and yet Y causes X. One of the most familiar is the special case of (4.2) in which $B_{12}(0) = 0$ and $\mathrm{cov}(\varepsilon_{1t}, \varepsilon_{2t}) = 0$: the system is block recursive, and if all restrictions consist of exclusions of variables then full information maximum likelihood estimates of the parameters in $B_{21}(L)$ and $B_{22}(L)$ coincide with maximum likelihood estimates in the last m_2 equations computed under the assumption that $Y_{1t-1} = \{y_{1t-s}, s > 0\}$ is fixed. Unless $B_{12}(L) = 0$, Y causes X. Evidently, the variables in Y_{1t-1} are predetermined, according to the generally accepted definitions of Koopmans and Hood (1953, pp. 117–121) and Christ (1966, p. 227).

Definition

A variable z_t is *predetermined* at time t if all its current and past values $Z_t = \{z_{t-s}, s \geq 0\}$ are independent of the vector of current disturbances in the model, $\boldsymbol{u}_t$, and these disturbances are serially independent.

In the foregoing examples the predetermined and strictly exogenous variables could be treated as fixed for purposes of inference because they satisfied the condition of weak exogeneity [Engle, Hendry and Richard (1982)].

Definition

Let $\bar{X}=(x_1,\ldots,x_n)'$ and $\bar{Y}=(y_1,\ldots,y_n)'$ be $n\times r$ and $n\times s$ matrices of observations on X and Y. Suppose the likelihood function $L(\theta;\bar{X},\bar{Y})$ can be reparameterized by $\lambda=F(\theta)$, where F is a one-to-one transformation; $\lambda'=(\lambda_1',\lambda_2')$, $(\lambda_1,\lambda_2)\in\Lambda_1\times\Lambda_2$; and the investigator's loss function depends on the parameters of interest λ_1 but not on the nuisance parameters λ_2. Then $\bar{X}$ is *weakly exogenous* if

$$L(\lambda_1,\lambda_2;\bar{X};\bar{Y})=L_1(\lambda_1;\bar{Y}\bar{X})\cdot L_2(\lambda_2;\bar{X}). \tag{4.3}$$

Weak exogeneity is a sufficient condition for treating $\bar{X}$ as fixed for purposes of statistical inference without loss. The joint likelihood function may always be factored as the product of a likelihood in X and a likelihood in Y conditional on X, but the parameters of interest need not be confined to the latter function, as (4.3) stipulates. Whether or not $\bar{X}$ is weakly exogenous depends on the investigator's loss function.

Consider the case in which X is strictly exogenous in a complete model. We denote the joint distribution of the disturbances $\bar{U}=(u_1,\ldots,u_n)'$ in the model by $f_u^n(\bar{U};\lambda_1)$ and the joint distribution of the exogenous variables $\bar{X}=x_1,\ldots,x_n$ in the model by $f_X^n(\bar{X};\lambda_2)$. The model itself is $M_n(\bar{Y},\bar{X};\lambda_1)=\bar{U}$ and the condition of completeness is that the Jacobian of transformation $J(\lambda_1)$ from $\bar{U}$ to $\bar{Y}$ be non-zero for all $\lambda_1\in\Lambda_1$. Since X is strictly exogenous, the joint distribution of $\bar{U}$ and $\bar{X}$ is $f_U^n(\bar{U};\lambda_1)f_X^n(\bar{X};\lambda_2)$, and the joint distribution of $\bar{Y}$ and $\bar{X}$ is $f_U^n(M_n(\bar{Y},\bar{X};\lambda_1);\lambda_1)J(\lambda_1)f_X^n(\bar{X};\lambda_1)$. Consequently, the distribution of $\bar{Y}$ conditional on $\bar{X}$ is $f_U^n(M(Y,X;\lambda_1))J(\lambda_1)$. The condition (4.3) for weak exogeneity will be met *if* the investigator's loss function depends only on λ_1 and there are no cross restrictions on λ_1 and λ_2. The former condition will be met if the investigator cares about the model, and the latter may or may not be met. In many simultaneous equations models the distribution of $\bar{X}$ is of no consequence. An important class of exceptions is provided by rational expectations models, in which behavioral parameters are generally linked to the distribution of exogenous variables; efficient estimation then requires that we work with $L(\theta;\bar{X},\bar{Y})$.

In many applications predetermined variables will also be weakly exogenous, but this again need not be the case. Let $f_u^t(u_t;\lambda_1)$ be the distribution of the tth disturbance; by virtue of serial independence, $f_U^n(\bar{U};\lambda_1)=\prod_{t=1}^n f_u^t(u_t;\lambda_1)$. Let the distribution of the vector of predetermined variables in period t, conditional on their past history, be denoted $f_x^t(x_t|X_{t-1};\lambda_2)$. (Some of the predetermined variables may in fact be exogenous, of course.) Since the disturbances are serially independent it is convenient to write the model for period t, $M_t(y_t,x_t;\lambda_1)=u_t$. The Jacobian of transformation from u_t to y_t, $J_t(\lambda_1)$, is assumed to be non-zero.

Let $Z = X_0, \ldots, X_{T-1}$. Then:

$$L(\lambda_1, \lambda_2; \bar{X}, \bar{Y}|Z) = \prod_{t=1}^{n} f_u^t(M_t(y_t, x_t; \lambda_1); \lambda_1) J_t(\lambda_1) f_x^t(x_t|X_{t-1}; \lambda_2)$$

and

$$L(\lambda_1, \lambda_2; \bar{Y}|\bar{X}, Z) = \prod_{t=1}^{n} f_u^t(M_t(y_t, x_t; \lambda_1); \lambda_1) J_t(\lambda_1).$$

The question of whether or not there are cross-restrictions on λ_1 and λ_2 again arises in the case of predetermined variables. For example, in the block recursive version of (4.2) in which $B_{12}(0) = 0$ and $\mathrm{cov}(\varepsilon_{1t}, \varepsilon_{2t}) = 0$, λ_2 consists of the parameters in $B_{11}(0)$ and $\mathrm{var}(\varepsilon_{1t})$: so long as there is no functional dependence between these and the parameters in $B_{21}(0)$, $B_{22}(0)$, and $\mathrm{var}(\varepsilon_{2t})$, y_{1t} will be weakly exogenous for purposes of estimating the latter parameters.

The assertion that $\bar{X}$ is weakly exogenous always relies to some degree on *a priori* assumptions. For example, Y_{1t} is weakly exogenous in (4.2) if $B_{12}(0) = 0$ and $\mathrm{cov}(\varepsilon_{1t}, \varepsilon_{2t}) = 0$, but this assumption is by itself not refutable as Basmann (1965) pointed out years ago. As another example, weak exogeneity of $\bar{X}$ follows from strict exogeneity of X in (4.1) if all the parameters in the distribution of X are nuisance parameters. Strict exogeneity is refutable, because it implies that Y does not cause X, but it is not equivalent to a verifiable restriction on the joint distribution of Y and X. In each example, weak exogeneity can only be tested together with other restrictions on the model. Indeed, this will *always* be the case, because (4.3) is a condition on parameters, not a restriction on joint distributions. One can always construct a "model" for which (4.3) will be true. The fact that the constructed model might not be of substantive interest corresponds precisely to the notion that any consistent test of weak exogeneity must be a joint test of weak exogeneity and other hypotheses.

The specification of weak exogeneity is attractive because it simplifies statistical inference. To the extent that this specification is grounded in economic theory it is more attractive, and if that theory implies refutable restrictions it is more attractive still. The strict exogeneity assumption in the complete dynamic simultaneous equation model is an example of such a specification. It often arises naturally from the assumption that the model in question has, historically, taken X as input [e.g. Koopmans' (1953, pp. 31–32) discussion of "weather" in econometric models]. It implies the refutable restriction that the endogenous variables do not cause the exogenous variables. Specifications of weak exogeneity which do not rely on strict exogeneity require assumptions about the serial correlation of disturbances which are usually more difficult to ground in economic theory. For example, if a variable is predetermined but not strictly exogenous, then there always exists a pattern of serial correlation for which the assumption that the variable is weakly exogenous will lead to inconsistent estimates. Assumptions about serial correlation in the latter case should therefore be tested just as unidirectional causality should be tested when the weak exogeneity specification rests on strict exogeneity. In both cases, weak exogeneity will still rely on *a priori* assumptions; no set of econometric tests will substitute for careful formulation of the economic model.

5. Inference[10]

Since the appearance of Sims' (1972) seminal paper, causal orderings among many economic time series have been investigated. The empirical literature has been surveyed by Pierce (1977) and Pierce and Haugh (1977). Virtually all empirical studies have been conducted under the assumptions introduced in Section 3: time series are wide sense stationary, purely non-deterministic with autoregressive representation, the relevant universe of information at t is Y_t and X_t, predictors are linear, and the criterion for comparison of forecasts is mean square error. A wide array of tests has been used. In this section we will describe and compare those tests which conceivably allow inference in large samples—i.e. those for which the probability of Type I error can, under suitable conditions, be approximated arbitrarily well as sample size increases. The development of a theory of inference is complicated in a non-trivial way by the fact that expression of all possible relations between wide sense stationary time series requires an infinite number of parameters, as illustrated in the canonical form derived in Section 3. This problem is not insurmountable, but considerable progress on the "parameterization problem" is required before a rigorous and useful theory of inference for time series is available; as we proceed, we shall take note of the major *lacunae*.

5.1. *Alternative tests*

Suppose that Y and X are two vector time series which satisfy the assumptions of Section 3. We find it necessary to make the additional assumption that Y and X are linear processes [Hannan (1970, p. 209)], which is equivalent to the specification that the disturbances $\boldsymbol{u}_{jt}$ and $\boldsymbol{v}_{jt}$ in Table 3.1 are serially independent, and not merely serially uncorrelated. Consider the problem of testing the null hypothesis that Y does not cause X. From the Theorem of Section 3, this may be done by testing (3.5) as a restriction on (3.7) or (3.10) as a restriction on (3.15). We shall refer to tests based on the first restriction as "Granger tests", since the restriction emerges immediately from Granger's (1969) definition, and to tests

[10]Much, but not all, of the material in this section is drawn from Geweke, Meese and Dent (1983).

based on the second restriction as "Sims tests" since that restriction was first noted, in a slightly different form, by Sims (1972). Suppose it were known *a priori* that $E_{1s} = E_{2s} = 0$ for $s > p$ and $F_{2s} = 0$ for $s > q$. One could then calculate ordinary least squares estimates E_{1s}, E_{2s}, and F_{2s} of the parameters of the equations:

$$\boldsymbol{x}_t = \sum_{s=1}^{p} E_{1s}\boldsymbol{x}_{t-s} + \boldsymbol{u}_{1t}, \tag{5.1}$$

$$\boldsymbol{x}_t = \sum_{s=1}^{p} E_{2s}\boldsymbol{x}_{t-s} + \sum_{s=1}^{q} F_{2s}\boldsymbol{y}_{t-s} + \boldsymbol{u}_{2t}, \tag{5.2}$$

and form estimates $\hat{\boldsymbol{\Sigma}}_j = \sum_{t=1}^{n} \hat{\boldsymbol{u}}_{jt}\hat{\boldsymbol{u}}'_{jt}/n$, where $\hat{\boldsymbol{u}}_{jt}$ denotes the vector of ordinary least squares residuals corresponding to the disturbance vector $\boldsymbol{u}_{jt}$ and n denotes sample size. From the usual central limit theorems for autoregressions [e.g. Theil (1970, pp. 411–413)] the asymptotic distribution of the test statistic:

$$T_n^{\mathrm{GW}} = n\left(\operatorname{tr}\left(\hat{\Sigma}_1\hat{\Sigma}_2^{-1}\right) - k\right), \tag{5.3}$$

under the null hypothesis is $\chi^2(klq)$. The superscript "G" in (5.3) denotes the Granger test, while W refers to the fact that this is the Wald (1943) variant of that test. The Lagrange multiplier,

$$T_n^{\mathrm{GL}} = n\left(k - \operatorname{tr}\left(\hat{\Sigma}_2\hat{\Sigma}_1^{-1}\right)\right), \tag{5.4}$$

and likelihood ratio,

$$T_n^{\mathrm{GR}} = n\ln\left(|\hat{\Sigma}_1|/|\hat{\Sigma}_2|\right), \tag{5.5}$$

variants of this test statistic will have the same asymptotic distribution under the null hypothesis. (See Chapter 13 of this Handbook.)

There is in fact rarely any reason to suppose that lag lengths in autoregressions are finite, or more generally to suppose that any particular parameterization (e.g. the autoregressive-moving average) is completely adequate. Our problem cannot be cast into the assumptions of the classical theory of inference which assume that coefficients can be made to depend in a known way on a finite number of unknown parameters. It is more similar to the one encountered in non-parametric inference about spectra (Chapter 17 of this Handbook), which can be resolved by estimating more autocovariances or using narrower spectral windows as sample size increases. In this procedure the number of parameters implicitly estimated increases, but decreases relative to sample size, as sample size increases. A similar strategy may be used in estimating autoregressions like those in Table 3.1. For example, p and q in (5.1) and (5.2) can be made functions of n such that $p(n)\to\infty$ and $q(n)\to\infty$ as $n\to\infty$, but $p(n)/n\to 0$ and $q(n)/n\to 0$. Since the coefficients in Table 3.1 are all square summable the equations there can all be approximated arbitrarily well in mean square by autoregressions of finite order. Therefore there exist rates of increase $p(n)$ and $q(n)$ such that consistent estimates of all coefficients in (3.5) and (3.7) can be obtained by estimating (5.1) and (5.2) by ordinary least squares.[11] Sufficient conditions on lag length for consistency have yet to be derived, however. A more pressing, unresolved problem is whether there exist rates of increase such that the limiting distributions of (5.3)–(5.5) under the null are $\chi^2(klq)$. In what follows, *we shall assume that such rates of increase do exist*, an assumption which is uncomfortable but which does not appear to be contradicted by sampling experiments which have been conducted. More bothersome is the practical problem of choosing the number of parameters to estimate in a given circumstance, to which we shall return in the next section.

We may test (3.10) as a restriction on (3.15) by estimating the equations,

$$y_t = \sum_{s=1}^{p} G_{3s} y_{t-s} + \sum_{s=0}^{q} H_{3s} x_{t-s} + v_{3t} \tag{5.6}$$

and

$$y_t = \sum_{s=1}^{p} G_{4s} y_{t-s} + \sum_{s=-r}^{q} H_{4s} x_{t-s} + v_{4t}, \tag{5.7}$$

by ordinary least squares, p, q, and r being allowed to increase with n. For suitable rates of increase:

$$T_n^{\mathrm{SW}} = n\left(\operatorname{tr}\left(\hat{T}_3\hat{T}_4^{-1}\right) - l\right), \tag{5.8}$$

$$T_n^{\mathrm{SL}} = n\left(l - \operatorname{tr}\left(\hat{T}_4\hat{T}_3^{-1}\right)\right) \tag{5.9}$$

and

$$T_n^{\mathrm{SR}} = n\ln\left(|\hat{T}_3|/|\hat{T}_4|\right) \tag{5.10}$$

are all distributed as $\chi^2(klr)$ in large samples under the null hypothesis. In finite sample, there is no numerical relation between the Granger test conducted using (5.1) and (5.2) and the Sims test conducted using (5.6) and (5.7); the null may well be rejected using one test but not the other.

Other tests about Wiener–Granger causal orderings, instantaneous causality, and combinations of the two may be undertaken in rather obvious fashion using the canonical form of Section 3 and truncating lag lengths. For example, a test of

[11]For an elaboration, see Section 6.1 of this chapter.

the hypothesis of no instantaneous causality can be based on ordinary least squares residuals from (5.2) and those from:

$$x_t = \sum_{s=1}^{p} E_{3s} x_{t-s} + \sum_{s=0}^{q} F_{3s} y_{t-s} + u_{3t}. \tag{5.11}$$

If $p = q$ the Wald, likelihood ratio and Lagrange multiplier test statistics will be identical to those which would have been obtained using the ordinary least squares residuals from:

$$y_t = \sum_{s=1}^{p} G_{2s} y_{t-s} + \sum_{s=1}^{q} H_{2s} x_{t-s} + v_{2t}, \tag{5.12}$$

and those from (5.4). A test of the hypothesis that Y does not cause X and there is no instantaneous causality could be based on a comparison of (5.1) and (5.11) or (5.12) and (5.7). The measures of feedback $F_{X\to Y}$, $F_{Y\to X}$, $F_{X\cdot Y}$, and $F_{X,Y}$ introduced in Section 3 can be estimated consistently using the $\hat{\Sigma}_j$ and $\hat{T}_j$.

Most of the empirical studies reported in the literature do not use any of these tests. Instead, they follow Sims (1972), who based inference directly on the projection of y_t on X_t,

$$y_t = \sum_{s=0}^{\infty} D_{1s} x_{t-s} + w_{1t}, \tag{5.13}$$

and y_t on X,

$$y_t = \sum_{s=-\infty}^{\infty} D_{2s} x_{t-s} + w_{2t}. \tag{5.14}$$

As we saw in Section 3, under the null hypothesis that Y does not cause X the two projections are the same. We also saw that the disturbance terms in (5.13) and (5.14) will be serially correlated. In particular, the disturbance term in (5.14) has autoregressive representation,

$$w_{2t} = \sum_{s=1}^{\infty} G_{4s} w_{2t-s} + v_{4t},$$

and under the null hypothesis the disturbance in (5.13) will of course have the same autoregressive representation. These tests can cope with the infinite length of lag distributions in the way just described, replacing (5.13) and (5.14) with:

$$y_t = \sum_{s=0}^{q} D_{1s} x_{t-s} + w_{1t} \tag{5.15}$$

and

$$y_t = \sum_{s=-r}^{q} D_{2s} x_{t-s} + w_{2t}, \tag{5.16}$$

respectively, but they also have to deal with serial correlation in w_{1t} and w_{2t}. In earlier studies [e.g. Sims (1972), Barth and Benet (1974)], this was often done by asserting a particular pattern of serial correlation in the disturbances, but in more recent work [e.g. Sims (1974), Neftci (1978)] a consistent estimate of the unknown pattern of serial correlation has been used in lieu of the actual pattern.

In general, corrections for serial correlation proceed as follows. Let it be assumed that $\Omega_n^1 = \text{var}(w_1) = \Omega_n^1(\alpha_1)$ and $\Omega_n^2 = \text{var}(w_2) = \Omega_n^2(\alpha_2)$, where $w_j = (w'_{j1}, \dots, w'_{jn})'$. The functions $\Omega_n^1(\cdot)$ and $\Omega_n^2(\cdot)$ are known, but α_1 and α_2 are unknown $s_1 \times 1$ and $s_2 \times 1$ vectors, respectively. [There is no presumption that $\Omega_n^1 = \Omega_n^2$, $\Omega_n^1(\cdot) = \Omega_n^2(\cdot)$, or $\alpha_1 = \alpha_2$.] Let $(\hat{D}'_1, \hat{\alpha}'_1)'$ denote the vector of maximum likelihood estimates of the $kl(q+1)+s_1$ unknown parameters in (5.15), and let $(\hat{D}'_2, \hat{\alpha}'_2)'$ denote the vector of maximum likelihood estimates of the $kl(q+r+1)+s_2$ unknown parameters in (5.16). Let $\hat{w}_1$ be the vector of residuals corresponding to $\hat{D}_1$, and let $\hat{w}_2$ be the vector of residuals corresponding to $\hat{D}_2$. Define also the estimator $\hat{D}^*_{1s}$ of the D_{1s}, which is the maximum likelihood estimator of the D_{1s} assuming $\Omega_n^1 = \Omega_n^2(\hat{\alpha}_2)$, and the estimator $\hat{D}^*_{2s}$ of the D_{2s}, which is the maximum likelihood estimator of the D_{2s} assuming $\Omega_n^2 = \Omega_n^1(\hat{\alpha}_1)$. Let $\hat{w}_1^*$ denote the vector of residuals corresponding to the $\hat{D}^*_{1s}$ and let $\hat{w}_2^*$ denote the vector of residuals corresponding to the $\hat{D}^*_{2s}$. Then Wald and Lagrange multiplier test statistics are

$$T_n^{(\text{SW})} = \hat{w}_1^{*\prime}\left(\Omega_n^2(\hat{\alpha}_2)\right)^{-1}\hat{w}_1^* - \hat{w}_2'\left(\Omega_n^2(\hat{\alpha}_2)\right)^{-1}\hat{w}_2 \tag{5.17}$$

and

$$T_n^{(\text{SL})} = \hat{w}_1'\left(\Omega_n^1(\hat{\alpha}_1)\right)^{-1}\hat{w}_1 - \hat{w}_2^{*\prime}\left(\Omega_n^1(\hat{\alpha}_1)\right)^{-1}\hat{w}_2^*, \tag{5.18}$$

respectively. Under the null hypothesis, the limiting distribution of each of these statistics is $\chi^2(klr)$. Since s_1 and s_2 are not known to be finite, their values must be increased as n increases as is the case with q and r.

There are several methods for parameterizing the variance matrix Ω_n of a wide sense stationary disturbance. The Hannan efficient method [Hannan (1963)] exploits the approximation $\Omega_n \doteq F_n S_n F_n'$ due to Grenander and Szego (1954), where F_n is the Fourier matrix with typical $((j,k)\text{th})$ element $\exp(2\pi i\, j(k-1)/n)/\sqrt{n}$ and S_n is a diagonal matrix whose jth diagonal element is the spectral density of the disturbance evaluated at the frequency $2\pi(j-1)/n$. In practice, a consistent estimate of the spectral density is constructed in the conventional way from ordinary least squares residuals [those from (5.15) for the Lagrange multiplier variant and those from (5.16) for the Wald variant]. In

Amemiya's procedure [Amemiya (1973)] the autoregressive representation of the disturbance is approximated by an autoregressive process of finite order, say $\boldsymbol{w}_{jt} = \sum_{s=1}^{s_j-1} A_j \boldsymbol{w}_{jt-s} + \boldsymbol{\varepsilon}_t$, which is estimated from ordinary least square residuals. The equations (5.15) and (5.16) are then premultiplied by $(1-\sum_{s=1}^{s_j-1}\hat{A}_j L^j)$ before estimation by ordinary least squares. A conventional, Wald "F" Test statistic for the restriction (5.15) on (5.16) is then asymptotically equivalent to (5.17). Strictly speaking, Wald and Lagrange multiplier variants of neither procedure lead to (5.17) or (5.18), because maximum likelihood estimates of the Ω_n^j are not computed. However, the procedures amount to the first step of iterative maximum likelihood procedures along the lines suggested by Oberhoffer and Kmenta (1974), and asymptotically the first step is equivalent to the full, maximum likelihood procedure.

Although tests based on (5.13) and (5.14) have been the most popular in the empirical literature, it is evident that they demand more computation than the other tests. Other methods, requiring even more computation, have been suggested and occasionally used. Pierce and Haugh (1977) propose that autoregressive moving average models for X and the ordinary least squares residuals of (5.15) be constructed, and the test of Haugh (1976) for their independence be applied. Under the null hypothesis X and the disturbances of (5.13) are independent, and Haugh's test statistic has a limiting $\chi^2(klr)$ distribution. Pierce (1977) has proposed a method involving the construction of autoregressive moving average models for X and Y and a modification of Haugh's statistic, but this method does not lead to test statistics whose asymptotic distribution under the null hypothesis is known [Sims (1977b)].

5.2. Comparison of tests

All the test statistics just discussed have the same limiting distribution under the null hypothesis. The adequacy of the limiting distribution in finite samples need not be the same for all the statistics, however, and thus far nothing has been said about their relative power. Sampling experiments can be used to address the first question, but without some sort of paradigm these experiments cannot be used to investigate the question of relative power. A convenient paradigm is the criterion of approximate slope, introduced by Bahadur (1960) and discussed in its application to time series by Geweke (1981).

Suppose that test i rejects the null in favor of the alternative in a sample of size n if the statistic T_n^i exceeds a critical value. Suppose further, as is the case here, that the limiting distribution of T_n^i under the null is chi-square. The approximate slope of test i is then the almost sure limit of T_n^i/n, which we shall denote $T^i(\boldsymbol{\theta})$. The set $\boldsymbol{\theta}$ consists of all unknown parameters of the population distribution, perhaps countably infinite in number, and the approximate slope in general depends on the values of the elements of $\boldsymbol{\theta}$. For all $\boldsymbol{\theta}$ which satisfy the null hypothesis $T^i(\boldsymbol{\theta})=0$, and for most test statistics (all of ours), $T^i(\boldsymbol{\theta})>0$ for all other $\boldsymbol{\theta}$. The approximate slopes of different tests are related to their comparative behavior under the alternative in the following way [Geweke (1981)]. Let $n^i(t^*,\beta;\boldsymbol{\theta})$ denote the minimum number of observations required to insure that the probability that the test statistic T_n^i exceeds a specified critical point t^* is at least $1-\beta$. Then $\lim_{t^*\to\infty} n^1(t^*,\beta;\boldsymbol{\theta})/n^2(t^*,\beta;\boldsymbol{\theta})=T^2(\boldsymbol{\theta})/T^1(\boldsymbol{\theta})$: the ratio of the number of observations required to reject the alternative as t^* is increased without bound (or equivalently, the asymptotic significance level of the test is reduced toward zero) is inversely proportional to the ratio of their approximate slopes. Similarly, if $t^i(n,\beta;\boldsymbol{\theta})$ indicates the largest non-rejection region (equivalently, smallest significance level) possible in a sample of size n if power $1-\beta$ is to be maintained against alternative $\boldsymbol{\theta}$, then $\lim_{n\to\infty} t^1(n,\beta;\boldsymbol{\theta})/t^2(n,\beta;\boldsymbol{\theta})=T^1(\boldsymbol{\theta})/T^2(\boldsymbol{\theta})$.

The criterion of approximate slope is a useful paradigm because it suggests that when the null is false and the number of observations is large, we might observe that tests with greater approximate slopes reject the null more often than tests with smaller approximate slopes, and those with the same approximate slopes reject the null with the same frequency. Differences in approximate slopes may be a factor in test choice if the number of observations required is not so large that the asymptotic significance levels involved are minute. Approximate slope is not a normative concept. For example, the statistic for the test with the larger approximate slope might exhibit slower convergence to the limiting distribution under the null than the statistic for the test with the smaller approximate slope. Indeed, if exact distributions under the null became known, it might turn out that if the critical points for the two tests were modified appropriately then the test with larger approximate slope would require more observations to assure given power against a given alternative than would the test with smaller approximate slope.

It is fairly straightforward to derive the approximate slopes of the alternative tests of the hypothesis that Y does not cause X. From (5.3)–(5.5) the approximate slopes of the tests based on the limiting distribution of T_n^{GW}, T_n^{GL}, and T_n^{GR} are:

$$T^{\mathrm{GW}}=\operatorname{tr}\left(\Sigma_1\Sigma_2^{-1}\right)-k,\qquad T^{\mathrm{GL}}=k-\operatorname{tr}\left(\Sigma_2\Sigma_1^{-1}\right)$$

$$\text{and}\quad T^{\mathrm{GR}}=\ln(|\Sigma_1|/|\Sigma_2|),$$

respectively. Likewise the approximate slopes of the Sims tests not involving serial correlation corrections are:

$$T^{\mathrm{SW}}=\operatorname{tr}\left(T_3T_4^{-1}\right)-l,\qquad T^{\mathrm{SL}}=l-\operatorname{tr}\left(T_4T_3^{-1}\right)\quad\text{and}\quad T^{\mathrm{SR}}=\ln(|T_3|/|T_4|).$$

Derivation of the approximate slopes of $T_n^{(\mathrm{SW})}$, $T_n^{(\mathrm{SL})}$, and $T_n^{(\mathrm{SR})}$ from (5.17)–(5.18) is rather tedious [see Geweke, Meese, and Dent (1983)].

It is evident that $T^{\mathrm{GW}} \geq T^{\mathrm{GR}} \geq T^{\mathrm{GL}}$, which is not surprising in view of relations between Wald, likelihood ratio and Lagrange multiplier statistics which often prevail in finite sample [Berndt and Savin (1977)]. Let $\lambda_1, \dots, \lambda_k$ denote the roots of $|\Sigma_1 - \lambda\Sigma_2| = 0$ and let $\Lambda = \mathrm{diag}(\lambda_1, \dots, \lambda_k)$; observe that $\lambda_j \geq 1$, $j = 1, \dots, k$. There exists non-singular F such that $\Sigma_1 = F'\Lambda F$ and $\Sigma_2 = F'F$ [Anderson (1958, p. 343)]. Substituting in the expressions for T^{GW}, T^{GL} and T^{GR}:

$$T^{\mathrm{GW}} = \sum_{j=1}^{k} \lambda_j - k, \qquad T^{\mathrm{GL}} = k - \sum_{j=1}^{k} \lambda_j^{-1}, \qquad T^{\mathrm{GR}} = \sum_{j=1}^{k} \ln(\lambda_j).$$

From Jensen's inequality $T^{\mathrm{GW}} \geq T^{\mathrm{GR}} \geq T^{\mathrm{GL}}$. This set of inequalities is strict unless all the approximate slopes are zero, i.e. the null hypothesis is true. Similarly $T^{\mathrm{SW}} \geq T^{\mathrm{SR}} \geq T^{\mathrm{SL}}$. From the Theorem of Section 3, $T^{\mathrm{GR}} = T^{\mathrm{SR}}$. If $k = l = 1$ this implies $T^{\mathrm{GW}} = T^{\mathrm{SW}}$ and $T^{\mathrm{GL}} = T^{\mathrm{SL}}$, but when $k > 1$ or $l > 1$ the relative sizes of T^{GW} and T^{SW}, and of T^{GL} and T^{SL}, are indeterminate. Geweke, Meese, and Dent (1983) show that $T^{(\mathrm{SW})} \geq T^{\mathrm{SW}}$, but this is the only determinate relation involving the approximate slope of a Sims test requiring correction for serial correction.

We therefore expect that when sample size is large and the null hypothesis of unidirectional causality is false, Wald tests will reject more frequently than likelihood ratio tests which will reject more frequently than Lagrange multiplier tests. A Wald test requiring correction for serial correlation should reject more frequently than one which does not. If X and Y are univariate, then Wald tests requiring a serial correlation correction will reject more frequently than any of the other tests, and the rejection frequencies of T^{SW} and T^{GW} should be the same as should the rejection frequencies of T^{SL} and T^{GL}.

To be of much practical use this asymptotic theory must be supplemented with sampling experiments. Several have been conducted [Geweke, Meese and Dent (1983), Guilkey and Salemi (1982), Nelson and Schwert (1982)]. The design of part of the experiment described in the first study is set forth in Table 5.1. Data for X and Y were generated to mimic quarterly economic time series, and the null hypotheses Y does not cause X (true) and X does not cause Y (false) were tested. Parameters were chosen so that an interesting frequency of rejections of the latter hypothesis should be obtained. Based on the results of Wald (1943) the rejection frequency should be 90% for $T^{(\mathrm{SW})}$ for tests conducted at the 10% significance level when $r = 4$, and less for the other tests (e.g. 76% for T^{SW} and T^{GW} when four parameters are restricted under the null and test size is 5%). For tests of the null that Y does not cause X the parameterization used in (5.1) for the Granger tests is exact, while that used in (5.6) and (5.15) for the Sims tests is exact only in Model B. For tests of the null that X does not cause Y (5.2) is exact when $p \geq 5$,

Table 5.1[a]
Design of a sampling experiment

Model A:	$y_t = 1.0 + 1.5y_{t-1} - 0.5625y_{t-2} + 0.215x_{t-1} + v_{4t}$	
	$x_t = 1.0 + 0.8x_{t-1} + u_{2t}$	
	$v_{4t} \sim N(0,1)$; $u_{2t} \sim N(0,1)$	
Model B:	$y_t = 1.0 + 0.258x_{t-1} + 0.172x_{t-2} - 0.086x_{t-3} + w_{2t}$	
	$w_t = 1.5w_{t-1} - 0.5625w_{t-2} + v_{4t}$	
	$x_t = 1.0 + 0.8x_{t-1} + w_{2t}$	
	$v_{4t} \sim N(0,1)$; $u_{2t} \sim N(0,1)$	
	Approximate slopes	
	Model A	Model B
$T^{\mathrm{GL}} = T^{\mathrm{SL}}$	0.0985	0.0989
$T^{\mathrm{GR}} = T^{\mathrm{SR}}$	0.1037	0.1041
$T^{\mathrm{GW}} = T^{\mathrm{SW}}$	0.1093	0.1098
$T^{(\mathrm{SW})}$	0.1240	0.1125
$T^{(\mathrm{SL})}$	0.1103	0.0954
$T^{(\mathrm{SL})}$(prefiltered)	0.0711	0.1016

[a]*Source*: Geweke, Meese and Dent (1983, table 4.1).

but (5.7) and (5.16) are never exact. In all instances in which parameterizations are inexact, however, the contributions of the omitted variables to variance in the dependent variable is very small.

The outcome of the sampling experiment is presented in Table 5.2, for tests that Y does not cause X, and in Table 5.3, for tests that X does not cause Y. The sampling distribution of the test statistics for Sims tests requiring a serial correlation correction is very unsatisfactory when the null is true. If no prefilter is applied, rejection frequencies range from 1.5 to 6.4 times their asymptotic values. When the prefilter $(1-0.75L)^2$ is used, the sampling distribution of these test statistics is about as good as that of the others. In the present case this prefilter removes all serial correlation in the disturbances of the regression equation (5.13). With actual data one cannot be sure of doing this, and the sensitivity of these test results to prefiltering, combined with the greater computational burden which they impose, argues against their use. By contrast, the sampling distribution of the Granger and Sims test which use lagged dependent variables appears much closer to the limiting distribution. Overall, Wald variants reject somewhat too often and Lagrange variants somewhat too infrequently, but departures from limiting frequencies are not often significant (given that only 100 replications of data were generated).

The rejection frequencies when the null is false, presented in Table 5.3, accord very well with what one would expect given the approximate slopes in Table 5.1 and the rejection frequencies anticipated in the design of the experiment. Rejection frequencies for T^{GW} and T^{SW} are about the same; those for T^{GL} and T^{SL} are each lower but similar to each other. Rejection frequencies are greatest for the Sims tests requiring a serial correlation correction, but their distribution under the

Table 5.2[a]
Outcome of sampling experiment

				Number of rejections[b] in 100 replications when null is true			
	Parameterization			Model *A*		Model *B*	
Test[c]	p	q	r	5% level	10% level	5% level	10% level
T^{GW}	4	4		7	14	7	16
	8	4		5	19	11	19
	12	4		9	16	11	22
	12	12		8	11	22	36
T^{GL}	4	4		4	11	4	7
	8	4		3	11	5	10
	12	4		5	13	4	9
	12	12		0	2	1	2
T^{SW}	4	4	4	6	10	3	7
	4	8	4	7	13	4	6
	4	12	4	9	13	3	8
	4	12	12	6	8	3	5
T^{SL}	4	4	4	4	9	1	3
	4	8	4	3	11	1	4
	4	12	4	4	12	3	4
	4	12	12	0	6	1	3
$T^{(\mathrm{SW})}(F)$		4	4	6	15	12	14
		8	4	7	12	12	21
		12	4	8	17	15	24
		12	12	9	15	26	35
$T^{(\mathrm{SW})}(U)$		4	4	26	33	32	38
		8	4	24	32	32	41
		12	4	21	28	31	38
		12	12	26	31	30	34
$T^{(\mathrm{SL})}(F)$		4	4	2	8	4	6
		8	4	5	7	3	9
		12	4	4	9	4	12
		12	12	2	4	2	5
$T^{(\mathrm{SL})}(U)$		4	4	23	30	18	26
		8	4	20	26	9	21
		12	4	18	27	8	13
		12	12	9	17	8	15

[a]*Source*: Geweke, Meese and Dent (1983, table 5.2).
[b]The appropriate F, rather than chi-square, distribution was used.
[c]For tests requiring correction for serial correlation, the Hannan efficient method of estimation was used. The spectral density of the disturbances was estimated using an inverted "V" spectral window with a base of 19 ordinates, applied to the 100 residual periodogram ordinates. For those tests with the suffix (F), data were initially prefiltered by $(1-0.75L)^2$, which flattens the spectral density of the disturbance w_{2t}. For those with (U), no prefilter was applied.

Table 5.3[a]
Outcome of sampling experiment

	Number of rejections in 100 replications[b] when null is false						
	Parameterization			Model A		Model B	
Test[c]	p	q	r	5% level	10% level	5% level	10% level
T^{GW}	4	4		59	76	69	79
	8	4		63	71	64	73
	12	4		61	70	64	76
	12	12		36	50	39	59
T^{GL}	4	4		55	67	59	74
	8	4		54	68	58	70
	12	4		55	69	58	70
	12	12		15	24	8	24
T^{SW}	4	4	4	60	70	71	79
	4	8	4	64	76	73	77
	4	12	4	58	68	71	76
	4	12	12	41	56	49	61
T^{SL}	4	4	4	51	65	63	74
	4	8	4	49	64	65	75
	4	12	4	50	64	65	72
	4	12	12	32	37	25	44
$T^{(\text{SW})}(F)$		4	4	90	96	87	93
		8	4	88	94	87	91
		12	4	85	94	85	90
		12	12	81	86	62	79
$T^{(\text{SL})}(F)$		4	4	82	92	78	88
		8	4	78	89	78	89
		12	4	77	86	74	84
		12	12	32	52	17	33

[a]*Source*: Geweke, Meese and Dent (1983, table 5.4).
[b]The appropriate F, rather than chi-square, distribution was used.
[c]For tests requiring correction for serial correlation, the Hannan efficient method of estimation was used. The spectral density of the disturbances was estimated using an inverted "V" spectral window with a base of 19 ordinates, applied to the 100 residual periodogram ordinates. For those tests with the suffix (F), data were initially prefiltered by $(1-0.75L)^2$, which flattens the spectral density of the disturbance w_{2t}.

null is less reliable; in view of the results in Table 5.2, rejection frequencies for those tests which use unfiltered data have not been presented in Table 5.3.

These results are corroborated in the rest of the Geweke, Meese and Dent study, as well as in Guilkey and Salemi, and Nelson and Schwert. Guilkey and Salemi have, in addition, studied the case in which the chosen parameterization omits important contributions of the explanatory variables; as one would expect, this biases tests of the null toward rejection and diminishes power under the alternative. The consensus is that inference can be undertaken with greatest reliability and computational ease employing either (5.1) as a restriction on (5.2) or (5.15) as a restriction on (5.16), and using either the Lagrange or Wald variant of the conventional "F". There are a host of practical considerations in the application of either procedure, like the choices of p, q, and r, and the treatment of apparent non-stationarity. We shall deal briefly with a few of these.

6. Some practical problems for further research

As is the case in many branches of econometrics, in the study of causal orderings and feedback in time series development of reliable and practical methods of inference has lagged well behind work on the population theory. In this final section we discuss those unanswered questions which presently pose the most serious impediments to empirical work. We concentrate on those problems specific to economic time series for which conventional econometric methods are often unsuited, to the exclusion of those matters (e.g. omitted variables and aggregation over time) which are discussed elsewhere in this volume and whose application to the problems at hand is fairly straightforward.

6.1. *The parameterization problem and asymptotic distribution theory*

In our discussion of inference, we explicitly assumed that the question of whether or not Y causes X cannot be reduced to one of inference about a finite number of unknown parameters. Although this is the only defensible assumption about parameterization which can be made in applied work,[12] it leads to some complications. In the comparison of test performance under the alternative based on approximate slopes, it was necessary to assume that the dimension of the parameter space was expanded in such a way that parameter estimates are strongly consistent. For example, in the Wald variant of the Granger test of the hypothesis that Y does not cause X, p and q must be expanded at a rate which guarantees that $\hat{\Sigma}_1 \overset{\text{a.s.}}{\rightarrow} \Sigma_1$ and $\hat{\Sigma}_2 \overset{\text{a.s.}}{\rightarrow} \Sigma_2$. Such rates must exist, because for purely non-deterministic processes with autoregressive representation $\lim_{p\to\infty} \hat{\text{var}}(\boldsymbol{x}_t|\boldsymbol{x}_{t-1},\dots,\boldsymbol{x}_{t-p}) = \Sigma_1$ and $\lim_{p,q\to\infty} \hat{\text{var}}(\boldsymbol{x}_t|\boldsymbol{x}_{t-1},\dots,\boldsymbol{x}_{t-p}; \boldsymbol{y}_{t-1},\dots,\boldsymbol{y}_{t-q}) = \Sigma_2$.[13] The residuals of linear least squares regression of $\boldsymbol{x}_t$ on $(\boldsymbol{x}_{t-1},\dots,\boldsymbol{x}_{t-p})$ can be used in the usual way to yield a consistent estimate of $\hat{\text{var}}(\boldsymbol{x}_t|\boldsymbol{x}_{t-1},\dots,\boldsymbol{x}_{t-p})$ and those of $\boldsymbol{x}_t$ on $(\boldsymbol{x}_{t-1},\dots,\boldsymbol{x}_{t-p}; \boldsymbol{y}_{t-1},\dots,\boldsymbol{y}_{t-q})$ to yield a strongly consistent estimate of $\hat{\text{var}}(\boldsymbol{x}_t|\boldsymbol{x}_{t-1},\dots,\boldsymbol{x}_{t-p}; \boldsymbol{y}_{t-1},\dots,\boldsymbol{y}_{t-q})$. Hence, there

[12] Unless one is willing to test the hypothesis that Y does not cause X maintaining hypotheses in addition to the assumptions made in Section 3.

[13] $\hat{E}(x|A)$ denotes the linear, least-squares projection of x on the random variables A, and $\hat{\text{var}}(x|A) = \text{var}(x - \hat{E}(x|A))$.

must exist an upper bound on the rate of expansion of p and q with n such that if p and q increase without limit as n grows but satisfy the upper bound constraint on rate, estimates of $\hat{\text{var}}(x_t x_{t-1},\ldots,x_{t-p})$ and $\hat{\text{var}}(x_t|x_{t-1},\ldots,x_{t-p};y_{t-1},\ldots,y_{t-q})$ will be strongly consistent for Σ_1 and Σ_2, respectively.[14] The problem of prescribing which values of p and q should be used in a particular situation is, of course, more difficult.

In our discussion of test performance under the null, it was necessary to make the much stronger assumption that the dimension of the parameter space can be expanded with n in such a way that the distributions of the test statistics approach those which would obtain were the true parameterization finite and known. In the context of our example of the Wald variant of the Granger test of the hypothesis that Y does not cause X, this is equivalent to saying that p and q grow with n slowly enough that strongly consistent estimates of Σ_1 and Σ_2 are achieved, but rapidly enough that their squared bias becomes negligible relative to their variance. It is not intuitively clear that such rates of expansion must exist, and there has been little work on this problem. Sampling study results suggest that as a practical matter this problem need not be overwhelming; witness the behavior of the Granger tests and the Sims tests incorporating lagged dependent variables, under the null hypothesis. On the other hand, the poor performance of the Sims tests involving a correction for serial correlation, under the null hypothesis, may be due in large measure to this difficulty.

A variety of operational solutions of the problem, "how to choose lag length", has appeared in the literature. Most of these solutions [Akaike (1974), Amemiya (1980), Mallows (1973), Parzen (1977)] emerge from the objective of choosing lag length to minimize the mean square error of prediction. For example, Parzen suggests that in a multivariate autoregression for Z, that lag length which minimizes the values of

$$\text{trace}\left\{\frac{m}{n}\sum_{j=1}^{p}\left(\hat{\hat{\Sigma}}_j\right)^{-1}-\hat{\hat{\Sigma}}_p^{-1}\right\} \tag{6.1}$$

be chosen, where m is the number of variables in Z, p is lag length, and $\hat{\hat{\Sigma}}_j=\sum_{t=1}^{n}\hat{\varepsilon}_t\hat{\varepsilon}_t'/(n-jm)$, where $\hat{\varepsilon}_t$ is the vector of ordinary least squares residuals in the linear regression of z_t on $z_{t-1},\ldots,z_{t-j}$. Choice of p in the other solutions is based on the minimization of different functions, but the value of p chosen will

[14] Upper bounds of this type have been derived by Robinson (1978) for a problem which is formally similar, that of estimating the coefficients in the population linear projection of Y on X consistently by ordinary least squares. The bounds require that lag lengths be $O(n^{(\nu-1)/\nu})$, where ν is a constant between 0 and $-1/2$ which depends on higher moments of X, and that the rates of increase in the number of past and number of future X not be too disparate.

usually be the same [Geweke and Meese (1981)]. On the other hand, Schwarz (1978) has shown that for a wide variety of priors which place positive prior probability on all finite lag lengths (but none on infinite lag length), the posterior estimate of lag length in large samples will be that which minimizes the value of

$$\ln|\hat{\Sigma}_p| + m^2 p \ln(n)/n, \tag{6.2}$$

when Z is Gaussian. Lag lengths chosen using (6.2) will generally be shorter than those chosen using (6.1). These solutions to the problem of choosing lag length are in many respects convenient, but it must be emphasized that they were not designed as the first step in a regression strategy for estimating coefficients in an autoregression whose length may be infinite. Neither analytic work nor sampling studies have addressed the properties of inference conditional on lag length chosen by either method, although the first one has been used in empirical work [Hsiao (1979a, 1979b)].

6.2. *Non-autoregressive processes*

A time series may be wide sense stationary and purely non-deterministic, and fail to have autoregressive representation because one or more of the roots of the Laurent expansion of its moving average representation lie on the unit circle. A non-autoregressive process does not possess an optimal linear, one step ahead predictor with coefficients which converge in mean square. The representations derived in Section 3 fail to exist for such a series, and if one attempts inference about Wiener–Granger causality using non-autoregressive processes, then misleading results are likely to emerge. A simple example provides some indication of what can go wrong.

Consider first two time series with joint moving average representation

$$x_t = \varepsilon_t + \rho\varepsilon_{t-1}, \qquad y_t = \varepsilon_{t-1} + \eta_t, \tag{6.3}$$

where $(\varepsilon_t, \eta_t)'$ is the bivariate innovation for the $(x_t, y_t)'$ process, $|\rho| < 1$, $E(\varepsilon_t) = E(\eta_t) = \text{cov}(\varepsilon_t, \eta_t) = 0$ and $\text{var}(\varepsilon_t) = \text{var}(\eta_t) = 1$. In the population, $\text{var}(x_t | x_{t-1}, \dots, x_{t-k}) = 1 + \rho^2 - \boldsymbol{m}_k' M_k^{-1} \boldsymbol{m}_k$, where

$$\underset{k\times 1}{\boldsymbol{m}_k} = \begin{pmatrix} \rho \\ 0 \\ \vdots \\ 0 \\ 0 \end{pmatrix}, \qquad \underset{k\times k}{M_k} = \begin{bmatrix} 1+\rho^2 & \rho & & & \\ \rho & 1+\rho^2 & \cdot & & \text{-0-} \\ & \cdot & \ddots & & \\ & & \ddots & 1+\rho^2 & \rho \\ \text{-0-} & & & \rho & 1+\rho^2 \end{bmatrix}.$$

Using a result of Nerlove, Grether and Carvalho (1979, p. 419) for the exact inverse of a tridiagonal matrix, the entry in the first row and column of M_k^{-1} is $(1-\rho^{2k})/(1-\rho^{2(k+1)})$, from which

$$\hat{\mathrm{var}}(x_t|x_{t-1},\dots,x_{t-k}) = \left(1-\rho^{2(k+2)}\right)/\left(1-\rho^{2(k+1)}\right).^{15}$$

Suppose that k lagged values of y_t are also used to predict x_t. In the population,

$$\begin{aligned}&\hat{\mathrm{var}}(x_t|x_{t-1},\dots,x_{t-k};y_{t-1},\dots,y_{t-k})\\ &=1+\rho^2-\left(\boldsymbol{m}_k'\;\boldsymbol{0}'\right)\begin{bmatrix} M_k & B_k\\ B_k' & 2I_k\end{bmatrix}^{-1}\begin{pmatrix}\boldsymbol{m}_k\\ \boldsymbol{0}\end{pmatrix},\end{aligned}$$

where

$$\underset{k\times k}{B_k}=\begin{bmatrix}\rho & 0 & & \text{-0-}\\ 1 & \ddots & & \\ \vdots & \cdot & \rho & \cdot \\ \cdot & \cdot & \ddots & \cdot 0\\ & \cdot & \ddots & \rho\\ & \text{-0-} & \cdot & \\ & & 1 & \rho\end{bmatrix}.$$

The element in the first row and column of

$$\begin{bmatrix} M_k & B_i\\ B_k' & 2I_k\end{bmatrix}^{-1}$$

is the same as that in the first row and column of the inverse of:[16]

$$M_k-0.5B_kB_k'=\begin{bmatrix}1+.5\rho^2 & 0.5\rho & & & \text{-0-}\\ 0.5\rho & 0.5(1+\rho^2) & \ddots & & \\ & \ddots & \ddots & \ddots & \\ \text{-0-} & & \ddots & 0.5(1+\rho^2) & 0.5\rho\\ & & & 0.5\rho & 0.5(1+\rho^2)\end{bmatrix}.$$

[15] For an alternative derivation see Whittle (1983, p. 75).

[16] We use the standard result [e.g. Theil (1971, pp. 17–19)] on the inverse of a partitioned matrix.

Partitioning out the first row and column of this matrix, the desired element is $(1+0.5\rho^2-0.5\boldsymbol{m}'_{k-1}M_{k-1}^{-1}\boldsymbol{m}_{k-1})^{-1}$. Hence:

$$\begin{aligned}\hat{\operatorname{var}}&(x_t|x_{t-1},\dots,x_{t-k};\, y_{t-1},\dots,y_{t-k})\\ &=1+\rho^2-\rho^2\left(1+0.5\rho^2-0.5\boldsymbol{m}'_{k-1}M_{k-1}^{-1}\boldsymbol{m}_{k-1}\right)^{-1}\\ &=1+\rho^2-\left(\rho^2\left(1-\rho^{2k}\right)\right)/\left(1-0.5\rho^{2k}-0.5\rho^{2(k+1)}\right)\\ &=\left(1-0.5\rho^{2k}-0.5\rho^{2(k+2)}\right)/\left(1-0.5\rho^{2k}-0.5\rho^{2(k+1)}\right). \qquad (6.4)\end{aligned}$$

Clearly k must grow with n, or else the hypothesis that Y does not cause X—which is true—will be rejected in large samples. If k grows at a rate such that the asymptotic results of Wald (1943) on the distribution of test statistics under the alternative are valid, then in large samples the Wald variant of the Granger test will have a limiting non-central chi square distribution with non-centrality parameter:

$$\left[\frac{\left(1-\rho^{2(k+2)}\right)}{\left(1-\rho^{2(k+1)}\right)}\cdot\frac{\left(1-0.5\rho^{2k}-0.5\rho^{2(k+1)}\right)}{\left(1-0.5\rho^{2k}-0.5\rho^{2(k+2)}\right)}-1\right]n$$

$$=\frac{0.5\rho^{2(k+1)}\left(1-\rho^2\right)\left(1-\rho^{2k}\right)n}{\left(1-\rho^{2(k+1)}\right)\left(1-0.5\rho^{2k}-0.5\rho^{2(k+2)}\right)}.$$

So long as $k/\ln(n)$ increases without bound, the non-centrality parameter will vanish asymptotically—a necessary and sufficient condition for the test statistic to have a valid asymptotic distribution under the null, given our assumptions. In particular, this will occur if $k=[n^\alpha]$, $0<\alpha<0.5$.

Suppose now that $\rho=1$ in (6.3): the process (x_t, y_t) is then stationary but has no autoregressive representation. Nerlove, Grether and Carvalho's result does not apply when $\rho=1$; let c_k denote the element in the first row and column of M_k^{-1}. Partitioning out the first row and column of M_k:

$$c_k=\left(2-\boldsymbol{m}'_{k-1}M_{k-1}^{-1}\boldsymbol{m}_{k-1}\right)^{-1}=\left(2-c_{k-1}\right)^{-1}.$$

By mathematical induction $c_k=k/(k+1)$. Hence:

$$\hat{\operatorname{var}}(x_t|x_{t-1},\dots,x_{t-k})=2-\boldsymbol{m}'_kM_k^{-1}\boldsymbol{m}_k=2-c_{k+1}^{-1}=(k+2)/(k+1).$$

Substituting in (6.4):

$$\hat{\operatorname{var}}(x_t|x_{t-1},\dots,x_{t-k},y_{t-1},\dots,y_{t-k})=(2k+2)/(2k+1).$$

The non-centrality parameter of the limiting distribution of the Wald variant of the Granger test in this case is:

$$\left[\frac{(k+2)}{(k+1)}\cdot\frac{(2k+1)}{2(k+1)}-1\right]n=\frac{kn}{2(k+1)^2}.$$

The non-centrality parameter will increase without bound unless $\lim_{n\to\infty}(n/k)=0$, a condition which cannot be met.

It therefore appears plausible that when $|\rho|<1$, Granger tests may be asymptotically unbiased if the number of parameters estimated increases suitably with sample size; in particular, $k=n^{\alpha}$, $0<\alpha<0.5$ may be a suitable rule. It appears equally implausible that such rules are likely to exist when $|\rho|=1$; in particular, Granger tests are likely to be biased against the null. This result is not peculiar to the Granger test: $S_x(\pi)=0$ causes equally troublesome problems in Sims tests of the hypothesis of absence of Wiener–Granger causality from Y to X.

6.3. *Deterministic processes*

From Wold's decomposition theorem [Wold (1938)] we know that a stationary time series may contain deterministic components. Although the definition of Wiener–Granger causality can be extended to encompass processes which are not purely non-deterministic, inference about directions of causality for such series is apt to be treacherous unless the deterministic process is known up to a finite number of unknown parameters. The difficulty is that to the extent the two series in question are mutually influenced by the same deterministic components, the influence may be perceived as causal. A simple example will illustrate this point.

Let $x_t=x_{1t}+x_{2t}$, $y_t=y_{1t}+y_{2t}$, $x_{1t}=y_{1t}=\cos(\pi t/2)$, and suppose x_{2t} and y_{2t} are independent white noises with zero means and unit variances. The bivariate process $(x_t, y_t)'$ may be interpreted as quarterly white noise contaminated by deterministic seasonal influence. Following Hosoya (1977), X does not cause Y and Y does not cause X because $x_{1t}=y_{1t}$ is in the Hilbert space generated by either X_{t-1} or Y_{t-1}. It is clear from the symmetry of the situation, however, that given only finite subsets of X_{t-1} and Y_{t-1}, the subsets taken together will permit a better separation of the deterministic and non-deterministic components of the series than either alone, and hence a better prediction of x_t and y_t. For instance, the linear projection of x_t on $(x_{t-1},\dots,x_{t-4k})$ is $\sum_{j=1}^{k}(x_{t-4j}-x_{t+2-4j})/(2(k+1))$, $\hat{\operatorname{var}}(x_t|x_{t-1},\dots,x_{t-k})=(2k+3)/(2k+2)$. The linear projection of x_t on $4k$ lagged values of x_t and y_t is $\sum_{j=1}^{k}(x_{t-4j}+y_{t-4j}-x_{t+2-4j}-y_{t+2-4j})/4(k+1)$, $\hat{\operatorname{var}}(x_t|x_{t-1},\dots,x_{t-4k};y_{t-1},\dots,y_{t-4k})=(4k+3)/(4k+2)$. Following the argument used in the example for non-autoregressive processes, if the limiting distribution of the Wald variant of the Granger tests is (possibly non-central) chi square, the non-centrality parameter is $2kn/(2k+2)(4k+3)$. This parameter could vanish only if the number of parameters estimated increased faster than the number of observations, which is impossible.

This result is consistent with the intuition that tests for the absence of Wiener–Granger causality are apt to be biased against the null when deterministic components are involved. It is equally clear from the example that if the determinism can be reduced to dependence on a few unknown parameters—here, the coefficients on perhaps four seasonal dummies—then the results of Sections 3 and 5 apply to the processes conditional on deterministic influences.

6.4. *Non-stationary processes*

Although in principle non-stationarity can take on many forms, most non-stationarity which we suspect[17] in economic time series is of certain specific types. If the non-stationarity arises from deterministic influences on mean or variance and their functional forms are known, the methods of Section 5 can be modified directly to accommodate these influences. For example, trend terms may be incorporated in estimated equations to allow for conditional means which change with time, and means and variances which increase with time may in some cases be eliminated by the logarithmic transformation and incorporation of a linear trend.

Non-stationarity need not be deterministic, and in fact many economic time series appear to be well described as processes with autoregressive representations which have Laurent expansions with one or more roots on the unit circle.[18] The asymptotic distribution theory for the estimates of autoregressions of stationary series does not apply directly in such cases. For example, in the case of the first-order autoregression $x_t = \rho x_{t-1} + \varepsilon_t$, $|\rho| > 1$, the least squares estimator of ρ has a limiting distribution that is non-normal if ε_t is non-normal [Rao (1961)] and a variance that is $O(1/n)$. If $|\rho| = 1$, the limiting distribution of the least squares estimator is not symmetric about ρ even when the distribution of ε_t is normal [Anderson (1959)] and the variance of the limiting distribution is again $O(1/n)$. The limiting distributions in these cases reflect the fact that information about the

[17]One can never demonstrate non-stationarity with a finite sample: e.g. apparent trends can be ascribed to powerful, low frequency components in a purely non-deterministic, stationary series. As a practical matter, however, inference with only asymptotic justification is apt to be misleading in such cases.

[18]Witness the popularity of the autoregressive integrated moving average models proposed by Box and Jenkins (1970) in which autoregressive representations often incorporate factors of the form $(1-L)$, or $(1-L^s)$ when there are s observations per year.

unstable or explosive roots of these processes is accumulated at a more rapid rate than is information about the roots of stationary processes, since the variance of the regressors grows in the former case but remains constant in the latter. This intuition is reinforced by a useful result of Fuller (1976) who considers an autoregression of finite order, $y_t = \sum_{s=1}^{p} c_s y_{t-s} + \varepsilon_t$ with one root of $\sum_{s=1}^{p} c_s L^s$ equal to unity and all others outside the unit circle. In the transformed equation $y_t = \theta_1 y_{t-1} + \sum_{i=2}^{p} \theta_i (y_{t+1-i} - y_{t-i}) + \varepsilon_t$ the ordinary least squares estimates of $\theta_2, \dots, \theta_p$ have the same limiting distribution as they would if all roots of the expansion were outside the unit circle, and the ordinary least squares estimate of θ_1 has the same limiting distribution as in the first order autoregressive equation with unit root. Hence, in the original equation inference about $c_2, \dots, c_p$ (but not c_1) may proceed in the usual way.

Whether or not Fuller's result can be extended to multiple roots and vector processes is an important problem for future research. If this extension is possible, then the methods of Section 5 involving the estimation of vector autoregressions can be applied directly to processes which are non-stationary because of roots on the unit circle in the expansions of their autoregressive representations. For example, if the process Z has autoregressive representation $[(1-L) \otimes I_m] A(L) z_t = \varepsilon_t$ with all roots of $|A(L)|$ outside the unit circle, tests of unidirectional causality between subsets of Z can proceed as in the stationary case. An attraction of this procedure is its conservatism: it allows non-stationarity, rather than requires it as is the case if one works with first differences.

6.5. *Multivariate methods*

While our discussion has been cast in terms of Wiener–Granger causal orderings between vector time series, in most empirical work causal relations between univariate time series are considered. In many instances relations between vector series would be more realistic and interesting than those between univariate series. The chief impediment to the study of such relations is undoubtedly the number of parameters required to describe them, which in the case of the vector autoregressive parameterization increases as the square of the number of series, whereas available degrees of freedom increases linearly. The orders of magnitude are similar for other parameterizations. Practical methods for the study of vector time series are only beginning to be developed, and the outstanding questions are numerous. All that will be done here is to mention those which seem most important for inference about causal orderings.

The first question is whether or not inference about causal orderings is adversely affected by the size of the system, in the sense that for vector time series the power of tests is substantially less than for univariate series. Any answer to this question is a generalization about relationships among observed economic time series; confidence intervals for measures of feedback constructed by Geweke (1982a) suggest that for samples consisting of 60–80 quarterly observations and lags of length six feedback must be roughly twice as great for a pair of bivariate series as for a pair of univariate series if unidirectional causality is to be rejected. The large number of parameters required for pairs of vector time series thus appears potentially important in the assessment of feedback and tests of unidirectional causal orderings.

Two kinds of approaches to the large parameter problem are being taken. The first is to reduce the number of parameters, appealing to "principles of parsimony" and using exact restrictions. For example, the autoregressive representation may be parameterized as a mixed moving average autoregression of finite order, with the form of the moving average and autoregressive components specified by inspection of various diagnostic statistics [e.g. Wallis, (1978)]. This is the extension to vector time series of the methods of Box and Jenkins (1970). There are two serious complications which emerge when going from univariate to multivariate series: estimation, in particular by exact maximum likelihood methods, is difficult [Osborn (1977)]; and the restrictions specified subjectively are quite large in number and not suggested in nearly so clear a fashion from the diagnostic statistics as in the univariate case. At least one alternative set of exact restrictions has been suggested for vector autoregressions: Sargent and Sims (1977) experiment with the restriction that in the finite order autoregressive model $A(L)z_t = \varepsilon_t$ the matrix $A(L)$ be less than full rank. Computations are again burdensome. The forecasting accuracy of neither method has been assessed carefully.

A second approach is to use prior information in a probabilistic way. This information might reflect economic theory, or might be based on purely statistical considerations. In the former case Bayes estimates of parameters and posterior odds ratios for the hypothesis of unidirectional causality will result, whereas in the latter case final estimates are more naturally interpreted as Stein estimates. In both cases, computation is simple if priors on parameters in a finite order autoregression are linear and one is content to use Theil–Goldberger mixed estimates. Litterman (1980) has constructed mixed estimates of a seven-variable autoregression, using six lags and 87 observations: 301 parameters are estimated, beginning with 609 degrees of freedom. The mean of the mixed prior is 1.0 for the seven coefficients on own lags and zero for all others, and variances decline as lag increases. His out-of-sample forecasts are better than those of the system estimated by classical methods and those issued by the proprietors of two large econometric models. This second approach is attractive relative to the first because of its computational simplicity, and because Bayesian or Stein estimators are methodologically better suited to the investigator's predicament in the estimation of large vector autoregressions than are exact restrictions.

References

Akaike, H. (1974) "A New Look at the Statistical Model Identification", *IEEE Transactions on Automatic Control*, AC-19, 716–723.

Amameiya, T. (1980) "Selection of Regressors", *International Economic Review*, forthcoming.

Amemiya, T. (1973) "Generalized Least Squares with an Estimated Autocovariance Matrix", *Econometrica*, 41, 723–732.

Anderson, T. W. (1958) *An Introduction to Multivariate Statistical Analysis*. New York: Wiley.

Anderson, T. W. (1959) "On Asymptotic Distributions of Estimates of Parameters of Stochastic Difference Equations", *Annals of Mathematical Statistics*, 30, 676–687.

Bahadur, (1960) "Stochastic Comparisons of Tests", *Annals of Mathematical Statistics*, 38, 303–324.

Barth, J. R. and J. T. Bennett (1974) "The Role of Money in the Canadian Economy: An Empirical Test", *Canadian Journal of Economics*, 7, 306–311.

Basmann, R. L. (1965) "A Note on the Statistical Testability of 'Explicit Causal Chains' Against the Class of 'Interdependent Models'", *Journal of the American Statistical Association*, 60, 1080–1093.

Berndt, E. R. and N. E. Savin (1977) "Conflict Among Criteria for Testing Hypotheses in the Multivariate Linear Regression Model", *Econometrica*, 45, 1263–1278.

Blalock, H. M., Jr. (1961) *Causal Inferences in Non-experimental Research*. Chapel Hill: University of North Carolina Press.

Bunge, M. (1959) *Causality: The Place of the Causal Principle in Modern Science*. Cambridge: Harvard University Press.

Christ, C. F. (1966) *Econometric Models and Methods*. Wiley: New York.

Doob, J. (1953) *Stochastic Processes*. New York: Wiley.

Engle, R., D. Hendry, and J.-F. Richard (1983) "Exogeneity" *Econometrica*, 51, 277–304.

Feigel, H. (1953) "Notes on Causality", in Feigl and Brodbeck, eds., *Readings in the Philosophy of Science*.

Fuller, W. A. (1976) *Introduction to Statistical Time Series*. New York: Wiley.

Gel'fand, I. M. and A. M. Yaglom (1959) "Calculation of the Amount of Information about a Random Function Contained in Another Such Function", *American Mathematical Society Translations Series 2*, 12, 199–264.

Geweke, J. (1978) "Testing the Exogeneity Specification in the Complete Dynamic Simultaneous Equation Model", *Journal of Econometrics*, 7, 163–185.

Geweke, J. (1981) "The Approximate Slopes of Econometric Tests", *Econometrica*, 49, 1427–1442.

Geweke, J. (1982a) "Measurement of Linear Dependence and Feedback Between Time Series", *Journal of the American Statistical Association*, 77, 304–324.

Geweke, J. (1982b) "The Neutrality of Money in the United States, 1867–1981: An Interpretation of the Evidence", Carnegie-Mellon University Working Paper.

Geweke, J. and R. Meese (1981) "Estimating Regression Models of Finite But Unknown Order", *International Economic Review*, forthcoming.

Geweke, J., R. Meese and W. Dent (1983) "Comparing Alternative Tests of Causality in Temporal Systems: Analytic Results and Experimental Evidence", *Journal of Econometrics*, 21, 161–194.

Granger, C. W. J. (1963) "Economic Processes Involving Feedback", *Information and Control*, 6, 28–48; also Chapter 7 of Granger, C. W. J. and M. Hatanaka, *Spectral Analysis of Economic Time Series*. Princeton: University Press.

Granger, C. W. J. (1969) "Investigating Causal Relations by Econometric Models and Cross-Spectral Methods", *Econometrica*, 37, 424–438.

Grenander, U. and G. Szego (1958) *Toeplitz Forms and Their Applications*., Berkeley: University of California Press.

Guilkey, D. K. and M. K. Salemi (1982) "Small Sample Properties of Three Tests for Granger-Causal Ordering in a Bivariate Stochastic System", *Review of Economics and Statistics*, 64, 68–80.

Hannan, E. J. (1963) "Regression for Time Series", in: M. Rosenblatt, ed., *Proceedings of a Symposium on Time Series Analysis*. New York: Wiley.

Hannan, E. J. (1970) *Multiple Time Series*. New York: Wiley.

Haugh, L. D. (1976) "Checking the Independence of Two Covariance-Stationary Time Series: A Univariate Residual Cross Correlation Approach", *Journal of the American Statistical Association*, 71, 378–385.

Hosoya, Y. (1977) "On the Granger Condition for Non-causality", *Econometrica*, 45, 1735–1736.
Hsiao, C. (1979a) "Autoregressive Modeling of Canadian Money and Income Data", *Journal of the American Statistical Association*, 74, 553–560.
Hsiao, C. (1979b) "Causality Tests in Econometrics", *Journal of Economic Dynamics and Control*, 1, 321–346.
Hurwicz, L. (1962) "On the Structural Form of Interdependent Systems", in: Nagel, E., et al., *Logic, Methodology and the Philosophy of Science*. Palo Alto: Stanford University Press.
Koopmans, T. C. (1950) "When is an Equation System Complete for Statistical Purposes?" in: Koopmans, T. C., ed., *Statistical Inference in Dynamic Economic Models*. New York: Wiley.
Koopmans, T. C. and W. C. Hood (1953) "The Estimation of Simultaneous Economic Relationships", in: Hood, W. C. and T. C. Koopmans, eds., *Studies in Econometric Method*. New Haven: Yale University Press.
Litterman, R., "A Bayesian Procedure for Forecasting with Vector Autoregressions" (MIT manuscript).
Mallows, C. L. (1973) "Some Comments on C_p", *Technometrics*, 15, 661–675.
Neftci, S. N. (1978) "A Time-Series Analysis of the Real Wages–Employment Relationship", *Journal of Political Economy*, 86, 281–292.
Nelson, C. R. and G. W. Schwert (1982) "Tests for Predictive Relationships between Time Series Variables: A Monte Carlo Investigation", *Journal of the American Statistical Association*, 77, 11–18.
Nerlove, M., D. M. Grether, and J. K. Carvalho (1979) *Analysis of Economic Time Series: A Synthesis*. New York: Academic Press.
Oberhofer, W. and J. Kmenta (1974) "A General Procedure for Obtaining Maximum Likelihood Estimates in Generalized Regression Models", *Econometrica*, 42, 579–590.
Osborn, D. R. (1977) "Exact and Approximate Maximum Likelihood Estimators for Vector Moving Average Processes", *Journal of the Royal Statistical Society* , Series B, 39, 114–118.
Parzen, E. (1977) "Multiple Time Series: Determining the Order of Approximating Autoregressive Schemes", in: P. Krishnaiah, ed., *Multivariate Analysis* IV, Amsterdam: North-Holland.
Pierce, D. A. (1977) "Relationships—and the Lack Thereof—Between Economic Time Series, with Special Reference to Money and Interest Rates", *Journal of the American Statistical Association*, 72, 11–22.
Pierce, D. A. (1979) "R^2 Measures for Time Series", *Journal of the American Statistical Association*, 74, 901–910.
Pierce, D. A. and L. D. Haugh (1977) "Causality in Temporal Systems: Characterizations and a Survey", *Journal of Econometrics*, 5, 265–293.
Rao, M. M. (1961) "Consistency and Limit Distributions of Estimators of Parameters in Explosive Stochastic Difference Equations", *Annals of Mathematical Statistics*, 32, 195–218.
Robinson, P. (1978) "Distributed Lag Approximations to Linear Time-Invariant Systems," *Annals of Statistics* 6, 507–515.
Rozanov, Y. A. (1967) *Stationary Random Processes*. San Francisco: Holden-Day.
Sargent, T. J. and C. A. Sims (1977) "Business Cycle Modeling with Pretending to Have Too Much a priori Economic Theory", in: C. A. Sims, ed., *New Methods in Business Cycle Research: Proceedings from a Conference*. Minneapolis: Federal Reserve Bank of Minneapolis.
Schwarz, G. (1978) "Estimating the Dimension of a Model", *Annals of Statistics*, 6, 461–464.
Simon, H. A. (1952) "On the Definition of the Causal Relation", *Journal of Philosophy*, 49, 517–527; reprinted in H. A. Simon, ed., 1957, *Models of Man*. New York: Wiley.
Sims, C. A. (1972) "Money, Income and Causality", *American Economic Review*, 62, 540–552.
Sims, C. A. (1974) "Output and Labor Input in Manufacturing", *Brookings Papers on Economic Activity 1974*, 695–736.
Sims, C. A. (1977a) "Exogeneity and Causal Ordering in Macroeconomic Models", in: C. A. Sims, ed., *New Methods in Business Cycle Research: Proceedings from a Conference*. Minneapolis: Federal Reserve Bank of Minneapolis.
Sims, C. A. (1977b) "Comment", *Journal of the American Statistical Association*, 72, 23–24.
Strotz, R. H. and H. O. A. Wold (1960) "Recursive vs. Non-recursive Systems: An Attempt at Synthesis", *Econometrica*, 28, 417–427.
Theil, H. (1971) *Principles of Econometrics*. New York: John Wiley.
Wald, A. (1943) "Tests of Statistical Hypotheses Concerning Several Parameters When the Number of Observations is Large", *Transactions of the American Mathematical Society*, 54, 426–485.

Wallis, K. F. (1977) "Multiple Time Series Analysis and the Final Form of Econometric Models", *Econometrica*, 45, 1481–1498.
Whittle, P. (1983) *Prediction and Regulation by Linear Least-Square Methods*.
Wiener, N. (1956) "The Theory of Prediction", in: E. F. Beckenback, ed., *Modern Mathematics for Engineers*.
Wold, H. (1938) *The Analysis of Stationary Time Series* Uppsala: Almquist and Wicksell.
Zellner, A. (1979) "Causality and Econometrics".
Zemanian, A. H. (1972) *Realizability Theory for Continuous Linear Systems*. New York: Academic Press.

Chapter 20

CONTINUOUS TIME STOCHASTIC MODELS AND ISSUES OF AGGREGATION OVER TIME

A. R. BERGSTROM

University of Essex

Contents

Handbook of Econometrics, Volume II, Edited by Z. Griliches and M.D. Intriligator

1. Introduction

Since the publication of the influential articles of Haavelmo (1943) and Mann and Wald (1943) and the subsequent work of the Cowles Commission [see, especially, Koopmans (1950a)], most econometric models of complete economies have been formulated as systems of simultaneous stochastic difference equations and fitted to either quarterly or annual data. Models of this sort, which are discussed in Chapter 7 of this Handbook, can be written in either the structural form:

$$\Gamma y_t + B_0 x_t + \sum_{r=1}^{k} B_r y_{t-r} = u_t, \tag{1}$$

or the reduced form:

$$y_t = \Pi_0 x_t + \sum_{r=1}^{k} \Pi_r y_{t-r} + v_t, \tag{2}$$

where y_t is an $n \times 1$ vector of observable random variables (endogenous variables), x_t is an $m \times 1$ vector of observable non-random variables (exogenous variables), u_t is a vector of unobservable random variables (disturbances), Γ is an $n \times n$ matrix of parameters, B_0 is an $n \times m$ matrix of parameters, $B_1,\dots,B_k$ are $n \times n$ matrices of parameters, $\Pi_r = -\Gamma^{-1}B_r$, $r = 0,\dots,k$, and $v_t = \Gamma^{-1}u_t$. It is usually assumed that $\mathrm{E}(u_t) = 0$, $\mathrm{E}(u_s u_t') = 0$, $s \neq t$, and $\mathrm{E}(u_t u_t') = \Sigma$, implying that $\mathrm{E}(v_t) = 0$, $\mathrm{E}(v_s v_t') = 0$, $s \neq t$, and $\mathrm{E}(v_t v_t') = \Omega = \Gamma^{-1}\Sigma\Gamma'^{-1}$.

The variables $x_{1t},\dots,x_{mt}, y_{1t},\dots,y_{nt}$ will usually include aggregations over a quarter (or year) of flow variables such as output, consumption, exports, imports and investment as well as levels at the beginning or end of the quarter (or year) of stock variables representing inventories, fixed capital and financial assets. They will also include indices of prices, wages and interest rates. These may relate to particular points of time, as will usually be the case with an index of the yield on bonds, or to time intervals as will be the case with implicit price deflators of the components of the gross national product.

Although the reduced form (2) is all that is required for the purpose of prediction under unchanged structure, the structural form (1) is the means through which a priori information derived from economic theory is incorporated in the model. This information is introduced by placing certain a priori restrictions on the matrices $B_0,\dots,B_k$ and Γ. The number of these restrictions is, normally, such as to imply severe restrictions on the matrices $\Pi_0,\dots,\Pi_k$ of reduced form coefficients. Because of the smallness of the samples available to the econometrician, these restrictions play a very important role in reducing the variances of the estimates of these coefficients and the resulting variances of the predictions obtained from the reduced form. The most common form of restriction on the matrices $B_0,\ldots,B_k$ and Γ is that certain elements of these matrices are zero, representing the assumption that each endogenous variable is directly dependent (in a causal sense) on only a few of the other variables in the model. But Γ is not assumed to be a diagonal matrix.

The simultaneity in the unlagged endogenous variables implied by the fact that Γ is not a diagonal matrix is the distinguishing feature of this set of models as compared with models used in the natural sciences. It is necessary in order to avoid the unrealistic assumption that the minimum lag in any causal dependency is not less than the observation period. But there are obvious difficulties in interpreting the general simultaneous equations model as a system of unilateral causal relations in which each equation describes the response of one variable to the stimulus provided by other variables. For this reason Wold (1952, 1954, 1956) advocated the use of recursive models, these being models in which Γ is a triangular matrix and Σ is a diagonal matrix.

One way of interpreting the more general simultaneous equations model, which is not recursive, is to assume that the economy moves in discrete jumps between successive positions of temporary equilibrium at intervals whose length coincides with the observation period. We might imagine, for example, that on the first day of each quarter the values of both the exogenous variables and the disturbances affecting decisions relating to that quarter become known and that a new temporary equilibrium is established, instantaneously, for the duration of the quarter. But this is clearly a very unrealistic interpretation. For, if it were practicable to make accurate daily measurements of such variables as aggregate consumption, investment, exports, imports, inventories, the money supply and the implicit price deflators of the various components of the gross national product, these variables would undoubtedly be observed to change from day to day and be approximated more closely by a continuous function of time than by a quarterly step function.

A more realistic interpretation of the general simultaneous equations model is that it is derived from an underlying continuous time model. This is a more basic system of structural equations in which each of the variables is a function of a continuous time parameter t. The variables in this model will, therefore, be $y_1(t),\ldots,y_n(t)$, $x_1(t),\ldots,x_m(t)$ where t assumes all real values. The relation between each of these variables and the corresponding variable in the simultaneous equations model will depend on the type of variable. If the variable is a flow variable like consumption, in which case $y_i(t)$ is the instantaneous rate of consumption at time t, then the corresponding variable in the simultaneous equations model is the integral of $y_i(t)$ over an interval whose length equals the observation period, so that, if we identify the unit of time with the observation period, we can write $y_{it} = \int_{t-1}^{t} y_i(r)\,\mathrm{d}r$. If $y_i(t)$ is a stock variable like the money supply then the corresponding variable in the simultaneous equations model will be the value of $y_i(t)$ at an integer value of t so that we have $y_{it} = y_i(t)$, $t = 1, 2, \ldots, T$.

It is intuitively obvious that, if the simultaneous equations system (1) is derived from an underlying continuous time model it will, generally, be no more than an approximation, even when the continuous time model holds exactly. One of the main considerations of this chapter will be the consequences, for estimation, of this sort of approximation and what is the best sort of approximation to use. This involves a precise specification of the continuous time model and a rigorous study of the properties of the discrete vector process generated by such a model.

If the underlying continuous time model is a system of linear stochastic differential equations with constant coefficients, and the exogenous variables and disturbances satisfy certain conditions, then, as we shall see, the discrete observations will satisfy, exactly, a system of stochastic difference equations in which each equation includes the lagged values of all variables in the system, and not just those which occur in the corresponding equation of the continuous time model. The disturbance vector in this exact discrete model is generated by a vector moving average process with coefficient matrices depending on the structural parameters of the continuous time model. A system such as this will be satisfied by the discrete observations whether they are point observations, integral observations or a mixture of the two, as they will be if the continuous time model contains a mixture of stock and flow variables. If there are no exogenous variables, so that the continuous time model is a closed system of stochastic differential equations, then the exact discrete model can be written in the form

$$
\begin{aligned}
y_t &= \sum_{r=1}^{k} F_r(\theta)\, y_{t-r} + \eta_t, \\
\eta_t &= \sum_{r=0}^{l} C_r(\theta)\, \varepsilon_{t-r}, \\
&\mathrm{E}(\varepsilon_t) = 0, \qquad \mathrm{E}(\varepsilon_t \varepsilon_t') = K(\theta), \qquad \mathrm{E}(\varepsilon_s \varepsilon_t') = 0, \qquad s \neq t,
\end{aligned}
\tag{3}
$$

where the elements of the matrices $F_1(\theta), \ldots, F_k(\theta), C_1(\theta), \ldots, C_l(\theta)$ and $K(\theta)$ are functions of a vector θ of structural parameters of the continuous time model.

It is a remarkable fact that the discrete observations satisfy the system (3) even though neither the integral $\int_{t-1}^{t} y(r)\,\mathrm{d}r$ nor the pair of point observations $y(t)$ and $y(t-1)$ conveys any information about the way in which $y(t)$ varies over the interval $(t-1, t)$ and the pattern of variation of $y(t)$ over a unit interval varies both as between different realizations (corresponding to different elementary events in the space on which the probability measure is defined), for a given interval, and as between different intervals for a given realization. Moreover, the form of the system (3) does not depend on the observation period, but only on the form of the underlying continuous time model. That is to say the integers k and l do not depend on the observation period, and the matrices $F_1(\theta),\dots,F_k(\theta), C_1(\theta),\dots,C_l(\theta)$ and $K(\theta)$ depend on the observation period only to the extent that they will involve a parameter δ to represent this period, if it is not identified with the unit of time. The observation period is, therefore, of no importance except for the fact that the shorter the observation period the more observations there will be and the more efficient will be the estimates of the structural parameters.

The exact discrete model (3) plays a central role in the statistical treatment of continuous time stochastic models, for two reasons. First, a comparison of the exact discrete model with the reduced form of an approximating simultaneous model provides the basis for the study of the sampling properties of parameter estimates obtained by using the latter model and may suggest more appropriate approximate models. Secondly, the exact discrete model provides the means of obtaining consistent and asymptotically efficient estimates of the parameters of the continuous time model.

For the purpose of predicting future observations, when the structure of the continuous time model is unchanged, all that we require is the system (3). But, even for this purpose, the continuous time model plays a very important role. For it is the means through which we introduce the a priori restrictions derived from economic theory. Provided that the economy is adjusting continuously, there is no simple way of inferring the appropriate restrictions on (3) to represent even such a simple implication of our theory as the implication that certain variables have no direct causal influence on certain other variables. For, in spite of this causal independence, all of the elements in the matrices $F_1,\dots,F_k$ in the system (3) will generally be non-zero. In this respect forecasting based on a continuous time model derived from economic theory has an important advantage over the methods developed by Box and Jenkins (1970) while retaining the richer dynamic structure assumed by their methods as compared with that incorporated in most discrete time econometric models.

For a fuller discussion of some of the methodological issues introduced above and an outline of the historical antecedents [among which we should mention Koopmans (1950b), Phillips (1959) and Durbin (1961)] and development of the theory of continuous time stochastic models in relation to econometrics, the reader is referred to Bergstrom (1976, ch. 1). Here we remark that the study of these models has close links with several other branches of econometrics and statistics. First, as we have indicated, it provides a new way of interpreting simultaneous equations models and suggests a more careful specification of such models. Secondly, it provides a further contribution to the theory of causal chain models as developed by Wold and others. Finally, as we shall see, it provides a potentially important application of recent developments in the theory of vector time series models. To some extent these developments have been motivated by the needs of control engineering. But it seems likely that their most important application in econometrics will be to continuous time models.

In the following section we shall deal fairly throughly with closed first order systems of linear stochastic differential or integral equations, proving a number of basic theorems and discussing various methods of estimation. We shall deal with methods of estimation based on both approximate and exact discrete models and their application to both stock and flow data. The results and methods discussed in this section will be extended to higher order systems in Section 3. In Section 4 we shall discuss the treatment of exogenous variables and more general forms of continuous time stochastic models.

2. Closed first-order systems of differential and integral equations

2.1. Stochastic limit operations and stochastic differential equations

Before getting involved with the problems associated with stochastic limit operations, it will be useful to consider the non-stochastic differential equation:

$$\mathrm{D}y(t)=ay(t)+b+\phi(t), \tag{4}$$

where D is the ordinary differential operator $\mathrm{d}/\mathrm{d}t$, a and b are constants and $\phi(t)$ is a continuous function of t (time). It is easy to see that the solution to (4) subject to the condition that $y(0)$ is a given constant is:

$$y(t)=\int_0^t \mathrm{e}^{a(t-r)}\phi(r)\,\mathrm{d}r+\left(y(0)+\frac{b}{a}\right)\mathrm{e}^{at}-\frac{b}{a}. \tag{5}$$

For, by differentiating (5) we obtain:

$$\begin{aligned}\mathrm{D}y(t)&=\frac{\mathrm{d}}{\mathrm{d}t}\left[\mathrm{e}^{at}\int_0^t \mathrm{e}^{-ar}\phi(r)\,\mathrm{d}r+\left(y(0)+\frac{b}{a}\right)\mathrm{e}^{at}-\frac{b}{a}\right]\\&=a\mathrm{e}^{at}\int_0^t \mathrm{e}^{-ar}\phi(r)\,\mathrm{d}r+\phi(t)+a\left(y(0)+\frac{b}{a}\right)\mathrm{e}^{at}\\&=ay(t)+b+\phi(t).\end{aligned}$$

From (5) we obtain:

$$y(t)=\int_0^{t-1}e^{a(t-r)}\phi(r)\,dr+\left(y(0)+\frac{b}{a}\right)e^{at}-\frac{b}{a}+\int_{t-1}^{t}e^{a(t-r)}\phi(r)\,dr$$

$$=e^a\int_0^{t-1}e^{a(t-1-r)}\phi(r)\,dr+e^a\left(y(0)+\frac{b}{a}\right)e^{a(t-1)}-e^a\frac{b}{a}$$

$$+(e^a-1)\frac{b}{a}+\int_{t-1}^{t}e^{a(t-r)}\phi(r)\,dr$$

$$=e^a y(t-1)+(e^a-1)\frac{b}{a}+\int_{t-1}^{t}e^{a(t-r)}\phi(r)\,dr.$$

We have shown, therefore, that the solution to the differential equation (4) satisfies the difference equation:

$$y(t)=fy(t-1)+g+\psi_t, \tag{6}$$

where

$$f=e^a,\qquad g=(e^a-1)\frac{b}{a}$$

and

$$\psi_t=\int_{t-1}^{t}e^{a(t-r)}\phi(r)\,dr.$$

In order to apply the above argument to an equation in which $\phi(t)$ is replaced by a random disturbance function it is necessary to define stochastic differentiation and integration. We can do this by making use of the concept of convergence in mean square. The sequence ξ_n, $n=1,2,\dots$, of random variables is said to *converge in mean square* to a random variable η if:

$$\lim_{n\to\infty}E(\xi_n-\eta)^2=0.$$

In this case η is said to be the *limit in mean square* of ξ_n and we write:

$$\underset{n\to\infty}{\text{l.i.m.}}\,\xi_n=\eta.$$

Suppose now that $\{\xi(t)\}$ is a family of random variables, there being one member of the family for each value of t (time). We shall call $\{\xi(t)\}$ a *continuous time random process* if t takes on all real values and a *discrete time random process* if t

takes on all integer values, the words "continuous time" or "discrete time" being omitted when it is clear from the context which case is being considered. A random process $\{\xi(t)\}$ is said to be *mean square continuous* on an interval $[c, d]$ if:

$$E[\xi(t)-\xi(t-h)]^2 \to 0$$

uniformly in t, on $[c, d]$, as $h \to 0$. And it is said to be *mean square differentiable* if there is a random process $\{\eta(t)\}$ such that:

$$\lim_{h \to 0} E\left\{\frac{\xi(t+h)-\xi(t)}{h} - \eta(t)\right\}^2 = 0.$$

In the latter case we shall write:

$$D\xi(t) = \frac{d\xi(t)}{dt} = \eta(t),$$

and call $D = d/dt$ the *mean square differential operator*.

In order to define integration we can follow a procedure similar to that used in defining the Lebesgue integral of a measurable function [see Kolmogorov and Fomin (1961, ch. 7)]. We start by considering a simple random process which can be integrated by summing over a sequence of measurable sets. The random process $\{\xi(t)\}$ is said to be *simple* on an interval $[c, d]$ if there is a finite or countable disjoint family of measurable sets Δ_k, $k = 1, 2, \ldots$, whose union is the interval $[c, d]$ and corresponding random variables ξ_k, $k = 1, 2, \ldots$, such that:

$$\xi(t) = \xi_k, \qquad t \in \Delta_k, \qquad k = 1, 2, \ldots.$$

Let $|\Delta_k|$ be the Lebesgue measure of Δ_k. Then the simple random process $\{\xi(t)\}$ is said to be *integrable in the wide sense* on $[c, d]$ if the series $\sum_{k=1}^{\infty} \xi_k |\Delta_k|$ converges in mean square. The *integral* of $\{\xi(t)\}$ over $[c, d]$ is then defined by:

$$\int_c^d \xi(t)\,dt = \sum_{k=1}^{\infty} \xi_k |\Delta_k| = \underset{n \to \infty}{\text{l.i.m.}} \sum_{k=1}^{n} \xi_k |\Delta_k|.$$

We turn now to the integration of an arbitrary random process. We say that a random process $\{\xi(t)\}$ is *integrable in the wide sense* on the interval $[c, d]$ if there exists a sequence $\{\xi_n(t)\}$, $n = 1, 2, \ldots$, of simple integrable processes which converges uniformly to $\{\xi(t)\}$ on $[c, d]$, i.e.

$$E[\xi(t)-\xi_n(t)]^2 \to 0,$$

as $n \to \infty$, uniformly in t, on $[c, d]$. It can be shown [see Rozanov (1967, p. 11)] that, in this case, the sequence $\int_c^d \xi_n(t)\,\mathrm{d}t$, $n = 1,2,\ldots$, of random variables has a limit in mean square which we call the *integral* of $\{\xi(t)\}$ over $[c, d]$. We can then write:

$$\int_c^d \xi(t)\,\mathrm{d}t = \text{l.i.m.} \int_c^d \xi_n(t)\,\mathrm{d}t.$$

If the random process $\{\xi(t)\}$ is mean square continuous over a finite interval $[c, d]$, then it is integrable over that interval. For we can then define the simple random process $\{\xi_n(t)\}$ by dividing $[c, d]$ into n equal subintervals $\Delta_{n1},\ldots,\Delta_{nn}$, letting $\xi_{n1},\ldots,\xi_{nn}$ be the random variables defining $\xi(t)$ at, say, the midpoints of $\Delta_{n1},\ldots,\Delta_{nn}$, respectively, and putting:

$$\xi_n(t) = \xi_{nk}, \qquad t \in \Delta_{nk}, \qquad k = 1,\ldots,n.$$

The simple random process $\{\xi_n(t)\}$ is obviously integrable, its integral being $\sum_{k=1}^n \xi_{nk}|\Delta_{nk}|$. Moreover, it follows directly from the definition of mean square continuity and the fact that the length of the intervals Δ_{nk} tends to zero as $t \to \infty$ that the sequence $\{\xi_n(t)\}$, $n = 1,2,\ldots$, of simple random processes converges, uniformly on $[c, d]$, to $\{\xi(t)\}$. We have shown, by this argument, that if a random process is mean square continuous, then it is integrable over an interval, not only in the wide sense defined above, but also in a stricter sense corresponding to the Riemann integral. A much weaker sufficient condition for a random process to be integrable in the wide sense is given by Rozanov (1967, theorem 2.3).

It is easy to show that the integral defined above has all the usual properties. In particular, if $\{\xi_1(t)\}$ and $\{\xi_2(t)\}$ are two integrable random processes and a_1 and a_2 are constants, then:

$$\int_c^d [a_1\xi_1(t) + a_2\xi_2(t)]\,\mathrm{d}t = a_1\int_c^d \xi_1(t)\,\mathrm{d}t + a_2\int_c^d \xi_2(t)\,\mathrm{d}t.$$

And if $\{\xi(t)\}$ is mean square continuous in a neighbourhood of t, then:

$$\frac{\mathrm{d}}{\mathrm{d}t}\int_{t_1}^t \xi(r)\,\mathrm{d}r = \xi(t),$$

where $\mathrm{d}/\mathrm{d}t$ is the mean square differential operator.

In addition to the various limit operations defined above we shall use the assumption of stationarity. A random process $\{\xi(t)\}$ is said to be *stationary in the wide sense* if it has an expected value $\mathrm{E}[\xi(t)]$ not depending on t and a correlation function $B(t,s)$ defined by:

$$B(t,s)=\mathrm{E}[\xi(t)\xi(s)],$$

depending only on the difference $t-s$, so that we can write:

$$B(t,s)=\gamma(t-s).$$

A wide sense stationary process is said to be *ergodic* if the time average $(1/T)\int_0^T \xi(t)\mathrm{d}t$ converges in mean square to the expected value $\mathrm{E}[\xi(t)]$ as $T\to\infty$. A random process $\{\xi(t)\}$ is said to be *strictly stationary* if, for any numbers $t_1,\ldots,t_k$, the joint distribution of $\xi(t_1+r),\ldots,\xi(t_k+r)$ does not depend on r.

A necessary and sufficient condition for a wide sense stationary process to be mean square continuous is that its correlation function is a continuous function of $(t-s)$ at the origin (i.e. when $t=s$). For we have:

$$\begin{aligned}\mathrm{E}[\xi(t)-\xi(t-h)]^2 &= \mathrm{E}[\xi(t)]^2+\mathrm{E}[\xi(t-h)]^2-2\mathrm{E}[\xi(t)\xi(t-h)]\\ &= 2\mathrm{E}[\xi(t)]^2-2\mathrm{E}[\xi(t)\xi(t-h)]\\ &= 2[B(t,t)-B(t,t-h)]\\ &= 2[\gamma(0)-\gamma(h)].\end{aligned}$$

We shall now consider the *stochastic differential equation*:

$$\mathrm{D}y(t)=ay(t)-b+\xi(t), \tag{7}$$

where D is the mean square differential operator and $\{\xi(t)\}$ is a mean square continuous wide sense stationary process. Our ultimate aim is to consider the estimation of the parameters a and b from a sequence $y(1), y(2),\ldots,$ of observations when $\xi(t)$ is an unobservable disturbance. For this purpose it will be necessary to place further restrictions on $\{\xi(t)\}$. But we shall not concern ourselves with these at this stage. Our immediate aim is to find a solution to (7) and show that this solution satisfies a difference equation whose relation to (7) is similar to that of the non-stochastic difference equation (6) to the differential equation (4).

Theorem 1

If $\{\xi(t)\}$ is a mean square continuous wide sense stationary process, then, for any given $y(0)$, (7) has a solution:

$$y(t)=\int_0^t \mathrm{e}^{a(t-r)}\xi(r)\mathrm{d}r+\left[y(0)+\frac{b}{a}\right]\mathrm{e}^{at}-\frac{b}{a}, \tag{8}$$

and this solution satisfies the stochastic difference equation:

$$y(t) = fy(t-1) + g + \varepsilon_t, \tag{9}$$

where

$$f = e^a, \qquad g = (e^a - 1)\frac{b}{a}$$

and

$$\varepsilon_t = \int_{t-1}^{t} e^{a(t-r)}\xi(r)\,dr.$$

Proof

The random process $\{e^{at}\xi(t)\}$ is mean square continuous on any finite interval. For:

$$\begin{aligned}E\left[e^{at}\xi(t) - e^{a(t-h)}\xi(t-h)\right]^2 &= E\{e^{at}[\xi(t)-\xi(t-h)] \\ &\quad + e^{at}(1-e^{-ah})\xi(t-h)\}^2 \\ &= e^{2at}E[\xi(t)-\xi(t-h)]^2 + e^{2at}(1-e^{-ah})^2\gamma(0) \\ &\quad + 2e^{2at}(1-e^{-ah})[\gamma(h)-\gamma(0)].\end{aligned}$$

And, since e^{2at} is bounded on any finite interval while $E[\xi(t)-\xi(t-h)]^2 \to 0$, uniformly in t, as $h \to 0$, the right-hand side of the last equation converges to zero, uniformly in t, as $h \to 0$.

It follows that the integral $\int_0^t e^{-ar}\xi(r)\,dr$ exists. Moreover:

$$\int_0^t e^{a(t-r)}\xi(r)\,dr = e^{at}\int_0^t e^{-ar}\xi(r)\,dr$$

and

$$\frac{d}{dt}\left[\int_0^t e^{-ar}\xi(r)\,dr\right] = e^{-at}\xi(t).$$

All the operations that were performed in showing that (5) is a solution to (4) and that this solution satisfies (6) are valid in the mean square sense, therefore, when $\phi(t)$ is replaced by $\xi(t)$. It follows that (8) is a solution to (7) and that this solution satisfies (9). ■

We turn now to a preliminary consideration of the problem of estimation. It would be very convenient if, in addition to being wide sense stationary and mean square continuous the random process $\{\xi(t)\}$ had the property that its integrals over any pair of disjoint intervals were uncorrelated. For then the disturbances ε_t, $t=1,2,\ldots$, in (9) would be uncorrelated and, provided that they satisfied certain additional conditions (e.g. that they are distributed independently and identically), the least squares estimates f^* and g^* of the constants in this equation would be consistent. We could then obtain consistent estimates of a and b from

$$a^* = \log f^* \quad \text{and} \quad b^* = \frac{a^* g^*}{f^* - 1}.$$

But it is easy to see that, if $\{\xi(t)\}$ is mean square continuous, it cannot have the property that its integrals over any pair of disjoint intervals are uncorrelated. For the integrals $\int_{t-h}^{t}\xi(r)\mathrm{d}r$ and $\int_{t}^{t+h}\xi(r)\mathrm{d}r$ will obviously be correlated if h is sufficiently small. The position is worse than this, however. We shall now show that there is no wide sense stationary process (whether mean square continuous or otherwise) which is integrable in the wide sense and whose integrals over every pair of disjoint intervals are uncorrelated.

Suppose that the wide sense stationary process $\{\xi(t)\}$ is integrable in the wide sense and that its integrals over every pair of disjoint intervals are uncorrelated. We may assume, without loss of generality, that $\mathrm{E}[\xi(t)]=0$. Let $\mathrm{E}[\int_{t-1}^{t}\xi(r)\mathrm{d}r]^2 = c$ and let $h=1/n$, where n is an integer greater than 1. We shall consider the set of n integrals:

$$\int_{t-h}^{t}\xi(r)\mathrm{d}r, \int_{t-2h}^{t-h}\xi(r)\mathrm{d}r, \ldots, \int_{t-1}^{t-(n-1)h}\xi(r)\mathrm{d}r.$$

By hypothesis these integrals are uncorrelated, and by the assumption of stationarity their variances are equal. It follows that:

$$\mathrm{E}\left[\int_{t-h}^{t}\xi(r)\mathrm{d}r\right]^2 = ch,$$

and hence:

$$\mathrm{E}\left[\frac{1}{h}\int_{t-h}^{t}\xi(r)\mathrm{d}r\right]^2 = \frac{c}{h} \to \infty, \quad \text{as } n \to \infty,$$

i.e. the variance of the mean value, over an interval, of a realization of $\xi(t)$ tends to infinity as the length of the intervals tends to zero. But this is impossible since, for any random process which is integrable in the wide sense, the integrals must satisfy [see Rozanov (1967, p. 10)] the inequality:

$$\mathrm{E}\left[\int_{t-h}^{t} \xi(r)\,\mathrm{d}r\right]^2 \le \left\{\int_{t-h}^{t} \left[\mathrm{E}(\xi(r))^2\right]^{1/2} \mathrm{d}r\right\}^2. \tag{10}$$

And, if the process is stationary in the wide sense, the right-hand side of (10) equals $\gamma(0)h^2$. It follows that:

$$\mathrm{E}\left[\frac{1}{h}\int_{t-h}^{t} \xi(r)\,\mathrm{d}r\right]^2 \le \gamma(0).$$

This contradiction shows that the integrals over sufficiently small adjacent intervals must be correlated.

Although it is not possible for the integrals over every pair of disjoint intervals, of a wide sense stationary process, to be uncorrelated, their correlations can, for intervals of given length, be arbitrarily close to zero. They will be approximately zero if, for example, the correlation function is given by:

$$\gamma(t-s) = \sigma^2 \mathrm{e}^{-\beta|t-s|},$$

where σ^2 and β are positive numbers and β is large. A stationary process with this correlation function does (as we shall see) exist and, because the correlation function is continuous at the origin, it is mean square continuous and hence integrable over a finite interval. If β is sufficiently large the disturbances ε_t, $t = 1, 2, \dots$, in (9) can, for all practical purposes, be treated as uncorrelated, and we may expect the least squares estimates f^* and g^* to be approximately consistent.

Heuristically we may think of an improper random process $\{\zeta(t)\}$ called "white noise" which is obtained by letting $\beta \to \infty$ in a wide sense stationary process with the above correlation function. For practical purposes we may regard white noise as indistinguishable from a process in which β is finite but large. But this is not a very satisfactory basis for the rigorous development of the theory of estimation. For this purpose we shall need to define white noise more precisely.

2.2. *Random measures and systems with white noise disturbances*

A precise definition of white noise can be given by defining a random set function which has the properties that we should expect the integral of $\zeta(t)$ to have under our heuristic interpretation. That is to say we define a random set function ζ which associates with each semi-interval $\Delta = [s, t)$ (or $\Delta = (s, t]$) on the real line a random variable $\zeta(\Delta)$ and has the properties:

$$\zeta(\Delta_1 \cup \Delta_2) = \zeta(\Delta_1) + \zeta(\Delta_1),$$

when Δ_1 and Δ_2 are disjoint, i.e. it is additive,

$$\mathrm{E}[\zeta(\Delta)]^2 = \sigma^2[t-s],$$

where σ^2 is a positive constant,

$$\mathrm{E}[\zeta(\Delta_1)\zeta(\Delta_2)] = 0,$$

when Δ_1 and Δ_2 are disjoint.

A set function with these properties is a special case of a type of set function called a random measure. The concept of a random measure is of fundamental importance, not only in the treatment of white noise, but also (in the more general form) in the spectral representation of a stationary process, which will be used in Section 4. We shall now digress, briefly, to discuss the concept more generally and define integration with respect to a random measure and properties of the integral which we shall use in the sequel.

Let R be some semiring of subsets of the real line (e.g. the left closed semi-intervals, or the Borel sets, or the sets with finite Lebesgue measure [see Kolmogorov and Fomin (1961, ch. 5)]. And let Φ be a random set function which associates with any subset $\Delta \in R$ a random variable $\Phi(\Delta)$ (generally complex valued) and has the properties:

$$\Phi(\Delta_1 \cup \Delta_2) = \Phi(\Delta_1) + \Phi(\Delta_2),$$

if Δ_1 and Δ_2 are disjoint, i.e. it is additive:

$$\mathrm{E}|\Phi(\Delta)|^2 = F(\Delta) < \infty,$$

$$\mathrm{E}[\Phi(\Delta_1)\bar{\Phi}(\Delta_2)] = 0,$$

when Δ_1 and Δ_2 are disjoint. Then Φ is said to be a *random measure*. If, in addition,

$$\Phi(\Delta) = \sum_k \Phi(\Delta_k)$$

for every $\Delta \in R$ which is the union of disjoint subsets Δ_k and the series on the right-hand side converges in mean square, then the random measure Φ is said to be *σ-additive*.

It can be shown [Rozanov (1967, theorem 2.1)] that a σ-additive random measure defined on some semiring on the real line can be extended to the ring of all measurable sets in the σ-ring generated by that semiring. This implies that if we give the random measure ζ, defined above, the additional property of σ-additivity so that:

$$\zeta(\Delta)=\sum_k \zeta(\Delta_k),$$

whenever the semi-interval Δ is the union of disjoint semi-intervals Δ_k then it can be extended to the ring of all Borel sets on the real line with finite Lebesque measure. We shall define *white noise* by the random measure ζ extended in this way to the measurable sets on the real line representing the time domain.

We turn now to the definition of the integral of a (non-random but, generally, complex valued) measurable function $f(x)$ with respect to a random measure Φ which is defined on the Borel sets of some interval $[c,d]$ (where c and d may have the values $-\infty$ and ∞, respectively). We start by defining the integral of a simple function. The measurable function $f(x)$ is said to be *simple* on the interval $[c,d]$ if it assumes a finite or countable set of values f_k on disjoint sets Δ_k whose union is $[c,d]$. And a simple function is said to be *integrable* with respect to the random measure Φ on the interval $[c,d]$ if the series $\sum_k f_k\Phi(\Delta_k)$ converges in mean square. The *integral* of $f(x)$ with respect Φ over $[c,d]$ is then defined as the limit in mean square to which this sum converges and we write:

$$\int_c^d f(x)\Phi(\mathrm{d}x)=\sum_{k=1}^{\infty} f_k\Phi(\Delta_k).$$

An arbitrary measurable function $f(x)$ is said to be *integrable* with respect to Φ on the interval $[c,d]$ if there is a sequence $f_n(x)$, $n=1,2,\ldots,$ of simple integrable measurable functions which converges in mean square to $f(x)$ on $[c,d]$, i.e.

$$\int_a^b |f_n(x)-f(x)|^2F(\mathrm{d}x)\to 0,$$

as $n\to\infty$, where the integral is defined in the Lebesgue–Stieltjes sense [see Cramér (1951, ch. 7)]. It can be shown [Rozanov (1967, p. 7)], that, in this case, the sequence $\int_c^d f_n(x)\Phi(\mathrm{d}x)$, $n=1,2,\ldots,$ has a limit in mean square which we call the *integral* of $f(x)$ over $[c,d]$. We can then write:

$$\int_c^d f(x)\Phi(\mathrm{d}x)=\underset{n\to\infty}{\mathrm{l.i.m.}}\int_c^d f_n(x)\Phi(\mathrm{d}x).$$

If $\Phi(\Delta)$ is undefined on $[-\infty, \infty]$ we can define $\int_{-\infty}^{\infty} f(x)\Phi(\mathrm{d}x)$ by:

$$\int_{-\infty}^{\infty} f(x)\Phi(\mathrm{d}x) = \underset{\substack{d\to\infty \\ c\to-\infty}}{\mathrm{l.i.m.}} \int_c^d f(x)\Phi(\mathrm{d}x),$$

provided that the limit on the right-hand side of the equation exists.

A necessary and sufficient condition for the existence of the integral $\int_c^d f(x)\Phi(\mathrm{d}x)$, where $f(x)$ is an arbitrary measurable function (and c and d may assume the values $-\infty$ and ∞, respectively), is:

$$\int_c^d |f(x)|^2 F(\mathrm{d}x) < \infty. \tag{11}$$

If this condition is satisfied, then [Rozanov (1967, p. 7)]:

$$\mathrm{E}\left|\int_c^d f(x)\Phi(\mathrm{d}x)\right|^2 = \int_c^d |f(x)|^2 F(\mathrm{d}x). \tag{12}$$

And, if the measurable functions $f(x)$ and $g(x)$ each satisfy condition (11), then [Rozanov (1967, p. 7)]:

$$\mathrm{E}\left\{\int_c^d f(x)\Phi(\mathrm{d}x)\overline{\int_c^d g(x)\,\Phi(\mathrm{d}x)}\right\} = \int_c^d f(x)\overline{g(x)}\,F(\mathrm{d}x). \tag{13}$$

When Φ is the random measure, ζ, by which we have defined white noise, $F(\Delta)$ has the simple form $\sigma^2|\Delta|$, where

$$\sigma^2 = \mathrm{E}\left\{\int_{t-1}^{t} \zeta(\mathrm{d}r)\right\}^2,$$

and $|\Delta|$ is the Lebesgue measure of Δ. A necessary and sufficient condition for the existence of the integral $\int_c^d f(r)\zeta(\mathrm{d}r)$, where $f(r)$ is a measurable function (and c and d may assume the values $-\infty$ and ∞, respectively), is:

$$\int_c^d \{f(r)\}^2 \mathrm{d}r < \infty, \tag{14}$$

the integral in (14) being the ordinary Lebesgue integral [which will be equal to the Riemann integral if $f(r)$ is a continuous function]. If this condition is satisfied, then [as a special case of (12)]:

$$\mathrm{E}\left\{\int_c^d f(r)\zeta(\mathrm{d}r)\right\}^2 = \sigma^2 \int_c^d \{f(r)\}^2 \mathrm{d}r. \tag{15}$$

And, if the measurable functions $f(x)$ and $g(x)$ each satisfy condition (14), then [as a special case of (13)]:

$$\mathrm{E}\left\{\int_c^d f(r)\zeta(\mathrm{d}r)\int_c^d g(r)\zeta(\mathrm{d}r)\right\}=\sigma^2\int_c^d f(r)g(r)\mathrm{d}r. \tag{16}$$

We note incidentally that $\int_0^t f(r)\zeta(\mathrm{d}r)$ is a random process whose increments are uncorrelated, i.e. a random process with *orthogonal increments* [see Doob (1953, ch. 9) for a full discussion of such processes].

Before applying the above results in the treatment of stochastic differential equations with white noise disturbances, we shall illustrate their application by proving the existence of a wide sense stationary process with the correlation function $\sigma^2\mathrm{e}^{-\beta|t-s|}$, as assumed in the heuristic introduction to the concept of white noise given at the end of the last subsection. The function $f(r)$, defined by

$$f(r)=\sigma(2\beta)^{1/2}\mathrm{e}^{-\beta(t-r)},$$

is integrable, with respect to ζ, over the interval $[-\infty,t]$, since

$$\int_{-\infty}^t\left[\sigma(2\beta)^{1/2}\mathrm{e}^{-\beta(t-r)}\right]^2\mathrm{d}r=\sigma^2,$$

i.e. condition (14) is satisfied. Using (15):

$$\mathrm{E}\left[\int_{-\infty}^t\sigma(2\beta)^{1/2}\mathrm{e}^{-\beta(t-r)}\zeta(\mathrm{d}r)\right]^2=\sigma^2.$$

And, if $s<t$:

$$\begin{aligned}
&\mathrm{E}\left[\int_{-\infty}^s\sigma(2\beta)^{1/2}\mathrm{e}^{-\beta(s-r)}\zeta(\mathrm{d}r)\int_{-\infty}^t\sigma(2\beta)^{1/2}\mathrm{e}^{-\beta(t-r)}\zeta(\mathrm{d}r)\right]\\
&=2\beta\sigma^2\mathrm{E}\left[\int_{-\infty}^s\mathrm{e}^{-\beta(s-r)}\zeta(\mathrm{d}r)\int_{-\infty}^s\mathrm{e}^{-\beta(t-r)}\zeta(\mathrm{d}r)\right]\\
&\quad+2\beta\sigma^2\mathrm{E}\left[\int_{-\infty}^s\mathrm{e}^{-\beta(s-r)}\zeta(\mathrm{d}r)\int_s^t\mathrm{e}^{-\beta(t-r)}\zeta(\mathrm{d}r)\right]\\
&=2\beta\sigma^2\mathrm{e}^{-\beta(t-s)}\mathrm{E}\left[\int_{-\infty}^s\mathrm{e}^{-\beta(s-r)}\right]^2\\
&=\sigma^2\mathrm{e}^{-\beta(t-s)}.
\end{aligned}$$

It follows that $\{\xi(t)\}$, where

$$\xi(t)=\int_{-\infty}^t\sigma(2\beta)^{1/2}\mathrm{e}^{-\beta(t-r)}\zeta(\mathrm{d}r)$$

is a wide sense stationary process with the correlation function $\sigma^2\mathrm{e}^{-\beta|t-s|}$.

A stochastic differential equation with a white noise disturbance is sometimes written like (7) with $\mathrm{D}y(t)$ (or $\mathrm{d}y(t)/\mathrm{d}t$) on the left-hand side and $\zeta(t)$ in place of $\xi(t)$ on the right-hand side. It is then understood that $\zeta(t)$ is defined only by the properties of its integral and that $y(t)$ is not mean square differentiable. We shall not use that notation in this chapter since we wish to reserve the use of the operator D for random processes which are mean square differentiable, as is $y(t)$ in (7). A first-order, linear stochastic differential equation with constant coefficients and a white noise disturbance will be written:

$$\mathrm{d}y(t)=(ay(t)+b)\,\mathrm{d}t+\zeta(\mathrm{d}t), \tag{17}$$

which will be interpreted as meaning that the random process $y(t)$ satisfies the *stochastic integral equation*:

$$y(t)-y(0)=\int_0^t[ay(r)+b]\,\mathrm{d}r+\int_0^t\zeta(\mathrm{d}r), \tag{18}$$

for all t.

Equation (17) is a special case of the equation

$$\mathrm{d}y(t)=m(t,y(t))\,\mathrm{d}t+\sigma(t,y(t))\zeta(\mathrm{d}t), \tag{19}$$

in which the functions $m(t, y)$ and $\sigma(t, y)$ can be non-linear in t and y, and which is interpreted [see Doob (1953, ch. 6)] as meaning that $y(t)$ satisfies the stochastic integral equation:

$$y(t)-y(0)=\int_0^t m(r,y(r))\,\mathrm{d}r+\int_0^t\sigma(r,y(r))\zeta(\mathrm{d}r), \tag{20}$$

for all t on some interval $[0, T]$. It has been shown [see Doob (1953, ch. 6), which modifies the work of Ito (1946) and (1950)] that, under certain conditions, there exists a random process $y(t)$ satisfying (20), for all t on an interval $[0, T]$, and that, for any given $y(0)$, this solution is unique in the sense that, if $\hat{y}(t)$ is any other solution:

$$P[y(t)-\hat{y}(t)=0]=1,\qquad 0\leq t\leq T. \tag{21}$$

The conditions, assumed in Doob (1953), are that the process $\{\int_0^t\zeta(\mathrm{d}r)\}$ is Gaussian and that the functions m and σ satisfy a Lipschitz condition and certain other conditions on $[0, T]$. A random process $\{\xi(t)\}$ is said to be *Gaussian* if the random variables $\xi(t)$ are normally distributed. The assumption that $\{\int_0^t\zeta(\mathrm{d}r)\}$ is Gaussian implies, therefore, that $\zeta(\Delta_1)$ and $\zeta(\Delta_2)$ are independent if Δ_1 and Δ_2 are disjoint and identically distributed if $|\Delta_1|=|\Delta_2|$.

In discussing the solution to (18) we shall not need to assume that $\{\int_0^t\zeta(\mathrm{d}r)\}$ is Gaussian. We shall now show that this equation has a solution, which will be given explicitly, that this solution is unique, in the sense of (21), in the class of mean square continuous processes, and that it satisfies a difference equation with serially uncorrelated disturbances.

Theorem 2

If ζ is a random measure, defined on all subsets of the line $-\infty < t < \infty$ with finite Lebesgue measure, such that:

$$E[\zeta(\mathrm{d}t)] = 0, \qquad E[\zeta(\mathrm{d}t)]^2 = \sigma^2 \mathrm{d}t,$$

then

(a) for any given $y(0)$ (18) has a solution:

$$y(t) = \int_0^t e^{a(t-r)}\zeta(\mathrm{d}r) + \left[y(0) + \frac{b}{a}\right]e^{at} - \frac{b}{a}; \tag{22}$$

(b) the solution (22) is unique in the class of mean square continuous processes, i.e. if $\hat{y}(t)$ is any other mean square continuous process satisfying (18) and $\hat{y}(0) = y(0)$, then (21) holds for any interval $[0, T]$;

(c) the solution (22) satisfies the stochastic difference equation:

$$y(t) = fy(t-1) + g + \varepsilon_t, \tag{23}$$

where

$$f = e^a, \qquad g = (e^a - 1)\frac{b}{a}, \qquad E(\varepsilon_t) = 0,$$

$$E(\varepsilon_t^2) = \frac{\sigma^2}{2a}(e^{2a} - 1), \qquad E(\varepsilon_s \varepsilon_t) = 0, \qquad s \neq t.$$

Proof

(a) We first note that the integral $\int_0^t e^{a(t-r)}\zeta(\mathrm{d}r)$ exists, since condition (14) is satisfied. Now let $y(t)$ be defined by (22) and let h be any positive number. Then

$$\begin{aligned}
y(t) - y(t-h) &= \int_0^t e^{a(t-r)}\zeta(\mathrm{d}r) - \int_0^t e^{a(t-h-r)}\zeta(\mathrm{d}r) \\
&\quad + \left[y(0) + \frac{b}{a}\right](e^{at} - e^{a(t-h)}) + \int_{t-h}^t e^{a(t-h-r)}\zeta(\mathrm{d}r) \\
&= (1 - e^{-ah})y(t) + (1 - e^{-ah})\frac{b}{a} + e^{-ah}\int_{t-h}^t (e^{a(t-r)} - 1)\zeta(\mathrm{d}r) \\
&\quad + (e^{-ah} - 1)\int_{t-h}^t \zeta(\mathrm{d}r) + \int_{t-h}^t \zeta(\mathrm{d}r) \\
&= ahy(t) + bh + \int_{t-h}^t \zeta(\mathrm{d}r) + u(t,h),
\end{aligned} \tag{24}$$

where

$$u(t,h)=\left(-\frac{a^2h^2}{2}+\frac{a^3h^3}{6}-\cdots\right)\left[y(t)+\frac{b}{a}\right]$$
$$+\mathrm{e}^{-ah}\int_{t-h}^{t}(\mathrm{e}^{a(t-r)}-1)\zeta(\mathrm{d}r)+\left(-ah+\frac{a^2h^2}{2}-\cdots\right)\int_{t-h}^{t}\zeta(\mathrm{d}r).$$

But

$$\mathrm{E}\left[\int_{t-h}^{t}(\mathrm{e}^{a(t-r)}-1)\zeta(\mathrm{d}r)\right]^2=\int_{t-h}^{t}(\mathrm{e}^{a(t-r)}-1)^2\mathrm{d}r$$
$$=\frac{1}{2a}(\mathrm{e}^{2ah}-1)-\frac{2}{a}(\mathrm{e}^{ah}-1)+h$$
$$=O(h^3).$$

Therefore

$$\mathrm{E}[u(t,h)]^2=O(h^3).$$

Now let $h=t/n$, where n is a positive integer. Then

$$y(t)-y(0)=[y(t)-y(t-h)]+[y(t-h)-y(t-2h)]+\cdots$$
$$+[y(t-(n-1)h)-y(0)]$$
$$=a\left[\frac{t}{n}\sum_{r=1}^{n}y\left(\frac{rt}{n}\right)\right]+bt+\int_0^t\zeta(\mathrm{d}r)+\sum_{r=1}^{n}u\left(\frac{rt}{n},\frac{t}{n}\right).\tag{25}$$

Now it is clear from (24) that $y(t)$ is mean square continuous, and hence:

$$\operatorname*{l.i.m.}_{n\to\infty}\left[\frac{t}{n}\sum_{r=1}^{n}y\left(\frac{rt}{n}\right)\right]=\int_0^t y(r)\,\mathrm{d}r.$$

Moreover, since

$$\mathrm{E}\left[u\left(\frac{rt}{n},\frac{t}{n}\right)\right]^2=O\left(\frac{1}{n^3}\right),$$

$$\operatorname*{l.i.m.}_{n\to\infty}\sum_{r=1}^{n}u\left(\frac{rt}{n},\frac{t}{n}\right)=0.$$

Since (25) holds for all positive integers n, it must hold when each term is replaced by its limit in mean square as $n \to \infty$, i.e.

$$y(t)-y(0)=\int_0^t [ay(t)+b]\,\mathrm{d}r+\int_0^t \zeta(\mathrm{d}r).$$

(b) Let $\hat{y}(t)$ be any other mean square continuous process satisfying (18), on $[0, T]$, and $\hat{y}(0)=y(0)$. Define:

$$\xi(t)=\hat{y}(t)-y(t).$$

Then

$$\xi(t)=a\int_0^t \xi(r)\,\mathrm{d}r, \qquad 0\le t\le T \tag{26}$$

Since $y(t)$ and $\hat{y}(t)$ are mean square continuous, $\xi(t)$ is mean square continuous. Therefore $\mathrm{E}[\xi(t)]^2$ is a continuous function of t, since

$$\begin{aligned}
\left|\mathrm{E}[\xi(t)]^2-\mathrm{E}[\xi(t-h)]^2\right| &\le \mathrm{E}\left|[\xi(t)]^2-[\xi(t-h)]^2\right| \\
&= \mathrm{E}[\,|\xi(t)+\xi(t-h)|\,|\xi(t)-\xi(t-h)|\,] \\
&\le \left\{\mathrm{E}[\xi(t)+\xi(t-h)]^2\mathrm{E}[\xi(t)-\xi(t-h)]^2\right\}^{1/2} \\
&\to 0, \quad \text{as } h\to 0.
\end{aligned}$$

Let n be a positive integer such that $\tau=T/n<1/a$. Since $\mathrm{E}[\xi(t)]^2$ is continuous it has a maximum $\mathrm{E}[\xi(\tau_1)]^2$ on the closed interval $[0,\tau]$, and $0\le\tau_1\le\tau\le 1/a$. Therefore, using (10) and (26):

$$\begin{aligned}
\mathrm{E}[\xi(\tau_1)]^2 &= a^2\mathrm{E}\left[\int_0^{\tau_1}\xi(r)\,\mathrm{d}r\right]^2 \\
&\le a^2\left\{\int_0^{\tau_1}\left[\mathrm{E}(\xi(r))^2\right]^{1/2}\mathrm{d}r\right\}^2 \\
&\le a^2\tau_1^2\mathrm{E}[\xi(\tau_1)]^2.
\end{aligned}$$

but $a^2\tau_1^2<1$. Therefore:

$$\mathrm{E}[\xi(\tau_1)]^2=0.$$

Therefore:

$$P(\xi(t)=0)=1, \qquad 0\le t\le\tau.$$

Since a similar relation holds for each of the n intervals, of length τ, whose union is $[0, T]$ we have:

$$P(\xi(t)=0)=1, \qquad 0\le t\le T.$$

(c) Let $y(t)$ be the random process defined by (22). Then

$$\begin{aligned} y(t) &= \mathrm{e}^{a}\int_0^{t-1}\mathrm{e}^{a(t-1-r)}\zeta(\mathrm{d}r)+\mathrm{e}^{a}\left[y(0)+\frac{b}{a}\right]\mathrm{e}^{a(t-1)} \\ &\quad -\mathrm{e}^{a}\left(\frac{b}{a}\right)+(\mathrm{e}^{a}-1)\frac{b}{a}+\int_{t-1}^{t}\mathrm{e}^{a(t-r)}\zeta(\mathrm{d}r) \\ &= fy(t-1)+g+\varepsilon_t, \end{aligned}$$

where

$$\varepsilon_t=\int_{t-1}^{t}\mathrm{e}^{a(t-r)}\zeta(\mathrm{d}r). \tag{27}$$

It is clear from the definition of the integral that:

$$\mathrm{E}(\varepsilon_t)=\mathrm{E}\left[\int_{t-1}^{t}\mathrm{e}^{a(t-r)}\zeta(\mathrm{d}r)\right]=0.$$

And, using (15) and (16):

$$\begin{aligned} \mathrm{E}(\varepsilon_t)^2 &= \mathrm{E}\left[\int_{t-1}^{t}\mathrm{e}^{a(t-r)}\zeta(\mathrm{d}r)\right]^2=\sigma^2\int_{t-1}^{t}\mathrm{e}^{2a(t-r)}\mathrm{d}r \\ &= \sigma^2\int_0^1\mathrm{e}^{2ar}\mathrm{d}r=\frac{\sigma^2}{2a}(\mathrm{e}^{2a}-1), \end{aligned}$$

$$\mathrm{E}(\varepsilon_s\varepsilon_t)=\mathrm{E}\left[\int_{s-1}^{s}\mathrm{e}^{a(s-r)}\zeta(\mathrm{d}r)\int_{t-1}^{t}\mathrm{e}^{a(t-r)}\zeta(\mathrm{d}r)\right]=0, \qquad s\neq t,$$

where s and t are integers. ■

In order to prove, in the simplest form, certain results which will be used throughout this chapter, we have dealt very fully with a single stochastic differential equation with a white noise disturbance. But, from the point of view of econometrics, our main interest is with systems of equations. These introduce new problems. For the coefficients of a system of stochastic differential equations, representing a system of causal adjustment relations, will be subject to certain a priori restrictions derived from economic theory, and these will imply certain restrictions on the coefficients of the derived system of difference equations used for estimation purposes. Because of the complexity of the latter restrictions and the fact that they cannot be inferred directly from economic theory, the continuous time formulation of the model is important, even if our ultimate aim is only to predict the future discrete observations of the variables.

We shall consider the system:

$$\mathrm{d}y(t)=[A(\theta)y(t)+b(\theta)]\,\mathrm{d}t+\zeta(\mathrm{d}t), \tag{28}$$

where $y(t)=[y_1(t),\dots,y_n(t)]'$ is a vector whose elements are random processes, $A(\theta)$ is an $n\times n$ matrix whose elements are functions of a vector $\theta=[\theta_1,\dots,\theta_p]$ of structural parameters and $b(\theta)$ is a vector whose elements are functions of θ. We assume that $p<n^2$, so that the matrix A is restricted. In the simplest case, where the only a priori restrictions are that certain variables have no direct causal influence on certain other variables, the only restrictions on A are that certain specified elements of this matrix are zero, and θ is then the vector of unrestricted elements of A. With regard to the disturbance vector $\zeta(\mathrm{d}t)$ we introduce the following assumption.

Assumption 1

$$\zeta=[\zeta_1,\dots,\zeta_n]$$

is a vector of random measures defined on all subsets of the line $-\infty<t<\infty$ with finite Lebesgue measure, such that

$$\mathrm{E}[\zeta(\mathrm{d}t)]=0,\qquad \mathrm{E}[\zeta(\mathrm{d}t)\zeta'(\mathrm{d}t)]=(\mathrm{d}t)\Sigma,$$

where Σ is a positive definite matrix.

Equation (28) will be interpreted as meaning that the vector random process $y(t)$ satisfies the system:

$$y(t)-y(0)=\int_0^t[A(\theta)y(r)+b(\theta)]\,\mathrm{d}r+\int_0^t\zeta(\mathrm{d}r), \tag{29}$$

for all t. With respect to this system, we shall now prove a theorem which generalizes Theorem 2.

Theorem 3

If ζ satisfies Assumption 1, then:

(a) for any given $n\times 1$ vector $y(0)$, the system (29) has a solution:

$$y(t)=\int_0^t e^{(t-r)A(\theta)}\zeta(\mathrm{d}r)+e^{tA(\theta)}[y(0)+A^{-1}(\theta)b(\theta)]$$
$$-A^{-1}(\theta)b(\theta), \tag{30}$$

where, for any matrix A, e^A is defined by

$$e^A = I + \sum_{r=1}^{\infty} \frac{1}{r!} A^r;$$

(b) the solution (30) is unique in the class of mean square continuous vector processes, i.e. if $\hat{y}(t)$ is any other vector of mean square continuous processes satisfying (29) and $\hat{y}(0) = y(0)$, then (21) holds on any interval $[0, T]$;

(c) the solution (30) satisfies the system

$$y(t) = F(\theta) y(t-1) + g(\theta) + \varepsilon_t \tag{31}$$

of stochastic difference equations, where

$$F = e^A, \qquad g = (e^A - I)A^{-1}b, \qquad E(\varepsilon_t) = 0,$$
$$E(\varepsilon_t \varepsilon_t') = \int_0^1 e^{rA} \Sigma e^{rA'} dr = \Omega, \qquad E(\varepsilon_s \varepsilon_t') = 0, \qquad s \neq t.$$

Proof

(a) For the purpose of the proof we shall assume that A has distinct eigenvalues, $\lambda_1, \dots, \lambda_n$, although this is not essential for the validity of the theorem. We then have:

$$A = H^{-1} \Lambda H,$$

where

$$\Lambda = \begin{bmatrix} \lambda_1 & 0 \cdots & 0 \\ 0 & \lambda_2 \cdots & 0 \\ \cdot & \cdot \;\cdots & \cdot \\ 0 & 0 \cdots & \lambda_n \end{bmatrix}$$

and H is a matrix whose columns are eigenvectors of A. Now define:

$$z(t) = Hy(t).$$

Then, from (29):

$$z(t) - z(0) = \int_0^t [\Lambda z(r) + Hb] \, dr + \int_0^t H\zeta(dr). \tag{32}$$

Clearly, $H\zeta$ is a vector of random measures, such that:

$$\mathrm{E}\big[(H\zeta(\mathrm{d}t)(H\zeta(\mathrm{d}t))'\big] = (\mathrm{d}t)H\Sigma H'.$$

Each equation in the system (32) satisfies the conditions of Theorem 2, therefore, and, by a direct application of that theorem to each equation in the system, we obtain the solution:

$$z(t)=\int_0^t \mathrm{e}^{(t-r)\Lambda}H\zeta(\mathrm{d}r)+\mathrm{e}^{t\Lambda}\big[z(0)+\Lambda^{-1}Hb\big]-\Lambda^{-1}Hb. \tag{33}$$

Then, premultiplying (33) by H^{-1}, we obtain:

$$y(t)=\int_0^t H^{-1}\mathrm{e}^{(t-r)\Lambda}H\zeta(\mathrm{d}r)+H^{-1}\mathrm{e}^{t\Lambda}H\big[y(0)+H^{-1}\Lambda^{-1}Hb\big]$$
$$-H^{-1}\Lambda^{-1}Hb. \tag{34}$$

But:

$$H^{-1}\mathrm{e}^{t\Lambda}H=H^{-1}\left[I+t\Lambda+\frac{t^2}{2}\Lambda^2+\cdots\right]H$$
$$=I+tH\Lambda H+\frac{t^2}{2}H^{-1}\Lambda HH^{-1}\Lambda H+\cdots$$
$$=I+tA+\frac{t^2}{2}A^2+\cdots$$
$$=\mathrm{e}^{tA}.$$

Equation (34) can, therefore, be written as (30).

(b) It follows from Theorem (2) that (33) is a unique solution to (32), and, hence, (34) or (30) is a unique solution to (29) in the class of mean square continuous vector processes.

(c) Let $z(t)$ be the vector random process defined by (33). Then

$$z(t)=\mathrm{e}^{\Lambda}\int_0^{t-1}\mathrm{e}^{(t-1-r)\Lambda}H\zeta(\mathrm{d}r)+\mathrm{e}^{\Lambda}\mathrm{e}^{(t-1)\Lambda}\big[z(0)+\Lambda^{-1}Hb\big]$$
$$-\mathrm{e}^{\Lambda}\Lambda^{-1}Hb+(\mathrm{e}^{\Lambda}-I)\Lambda^{-1}Hb+\int_{t-1}^t\mathrm{e}^{(t-r)\Lambda}H\zeta(\mathrm{d}r)$$
$$=\mathrm{e}^{\Lambda}z(t-1)+(\mathrm{e}^{\Lambda}-I)\Lambda^{-1}b+\int_{t-1}^t\mathrm{e}^{(t-r)\Lambda}H\zeta(\mathrm{d}r). \tag{35}$$

Premultiplying (35) by H^{-1}, we obtain:

$$y(t)=Fy(t-1)+g+\varepsilon_t,$$

where

$$\varepsilon_t=\int_{t-1}^{t}e^{(t-r)A}\zeta(dr). \tag{36}$$

It is clear from the definition of the integral that:

$$E(\varepsilon_t)=E\left[\int_{t-1}^{t}e^{(t-r)A}\zeta(dr)\right]=0.$$

And, using the generalizations of (15) and (16) for a vector random measure, we obtain:

$$\begin{aligned}E(\varepsilon_t\varepsilon_t')&=E\left[\int_{t-1}^{t}e^{(t-r)A}\zeta(dr)\int_{t-1}^{t}e^{(t-r)A'}\zeta(dr)\right]\\&=\int_{t-1}^{t}\left[e^{(t-r)A}\Sigma e^{(t-r)A'}\right]dr\\&=\int_0^1 e^{rA}\Sigma e^{rA'}dr=\Omega,\\E(\varepsilon_s\varepsilon_t')&=E\left[\int_{s-1}^{s}e^{(s-r)A}\zeta(dr)\int_{t-1}^{t}e^{(t-r)A'}\zeta(dr)\right]\\&=0,\qquad s\neq t.\quad\blacksquare\end{aligned}$$

We shall refer to the system (31) as the *exact discrete model* corresponding to the continuous time model (29). It should be emphasized that, unlike the continuous time model from which it is derived, the exact discrete model is not a system of structural relations. It cannot be interpreted as a system of causal relations in which each equation describes the direct response of one variable to the stimulus provided by other variables in the system. For each coefficient in the matrix F will reflect the interaction between all variables in the system during the observation period. Even if the only a priori restrictions on the matrix A are that certain elements of this matrix are zero, in which case θ is a vector whose elements are the unrestricted elements of A, the elements of F will be complicated transcendental functions of the elements of θ and will, generally, be all non-zero. And, even if Σ is a diagonal matrix, the elements of Ω will, generally, be all non-zero.

The relation of the exact discrete model (31) to the continuous time model (29) is rather similar, therefore, to that of the reduced form of a simultaneous equations model to the structural form of the model. And, as we shall see, the relation between the exact discrete model of a continuous time model and the reduced form of a simultaneous equations model, used to approximate the continuous time model, plays an important role in the analysis of the properties of various estimators.

2.3. Estimation

It is easy to see that a necessary and sufficient condition for the identifiability of the parameter vector θ in the model (29) is that the correspondence between θ and $[F(\theta), g(\theta)]$ is one to one. But this condition is more restrictive than it might, at first sight, appear to be. It is more restrictive than the condition that the correspondence between θ and $[A(\theta), b(\theta)]$ is one to one. For the equation

$$e^{A} = F \tag{37}$$

will, generally, not have a unique solution unless A is restricted. This is because, if A is a matrix satisfying (37) and some of its eigenvalues are complex, then by adding to each pair of conjugate complex eigenvalues the imaginary numbers $2in\Pi$ and $-2in\Pi$, respectively, where n is an integer, we obtain another matrix satisfying (37). For identifiability the restrictions on A must be sufficient to exclude any other matrix obtained in this way.

The real problem here is that, unless our model incorporates sufficient a priori restrictions we cannot distinguish between structures generating oscillations whose frequencies differ by integer multiples of the observation period. This phenomenon is known as *aliasing*. The identification problem is more complicated for continuous time models, therefore, than it is for discrete time models. For a fuller discussion of the identification problem the reader is referred to Phillips (1973) who derives a rank condition for identifiability in the case in which each a priori restriction on A is a linear homogeneous relation between the elements of a row of A.[1] We shall assume throughout the rest of this section that θ is identifiable.

In the discussion of estimation methods we shall assume, initially, that the sample is of the form $y(1),\dots,y(T)$ as it would be if all variables were stock variables or prices at points of time. The complications arising when some or all of the variables are observable only as integrals will be discussed later.

The problem of estimating θ is equivalent to the problem of estimating $[F, g]$ subject to the restriction that this matrix can be written as $[F(\theta), g(\theta)]$ for some vector θ in p-dimensional space (or the subset of this space to which θ is required to belong). As we have seen this restriction is very complicated, even in the

[1] See also the recent contributions of Hansen and Sargent (1981, 1983).

simplest cases, and the computational problem of obtaining a consistent estimate of θ in a large model is such that it is worth considering methods based on an approximate discrete model. Such methods are likely to be useful in any research programme, at least for the preliminary screening of hypotheses.

An obvious approximation can be obtained from (29) by using $\frac{1}{2}[y(t)+y(t-1)]$ as an approximation for $\int_{t-1}^{t} y(r)\mathrm{d}r$. This gives the approximate simultaneous equations model:

$$y(t)-y(t-1)=\tfrac{1}{2}A(\theta)[y(t)+y(t-1)]+b(\theta)+u_t, \tag{38}$$
$$\mathrm{E}(u_t)=0, \qquad \mathrm{E}(u_t u_t')=\Sigma, \qquad \mathrm{E}(u_s u_t')=0, \qquad s\neq t, \qquad s,t=1,2,\ldots.$$

The model is approximate because, if u_t is defined in such a way that (38) holds exactly, then the condition $\mathrm{E}(u_s u_t)=0$, $s\neq t$, will be only approximately satisfied.

We can write the model (38) in the reduced form:

$$y(t)=\Pi_1(\theta)y(t-1)+\Pi_0(\theta)+v_t, \tag{39}$$

where

$$\Pi_1=[I-\tfrac{1}{2}A]^{-1}[I+\tfrac{1}{2}A],$$
$$\Pi_0=[I-\tfrac{1}{2}A]^{-1}b,$$
$$v_t=[I-\tfrac{1}{2}A]^{-1}u_t,$$

so that

$$\mathrm{E}(v_t)=0, \qquad \mathrm{E}(v_t v_t')=[I-\tfrac{1}{2}A]^{-1}\Sigma[I-\tfrac{1}{2}A']^{-1}$$
$$\mathrm{E}(v_s v_t')=0, \qquad s\neq t, \qquad s,t=1,2,\ldots.$$

The use of the approximate simultaneous equations model (38) is particularly convenient when the elements of $A(\theta)$ are linear functions of θ. For then we can estimate θ by applying a non-iterative procedure, such as two-stage least squares or three-stage least squares, to this model, as if it were the true model. But even the application of the full information quasi-maximum likelihood procedure to (38) is computationally simpler than the application of the same procedure to the exact discrete model (31). Estimates obtained by any of these methods will, of course, be asymptotically biased because of the error of specification in the model (38). It is important, therefore, that we should investigate the sampling properties of these estimators when the data have been generated by the continuous time model (29) or, equivalently, by the exact discrete model (31).

Such an investigation was undertaken by Bergstrom (1966). The central idea which was put forward in this article, and further discussed in Bergstrom (1967, ch. 9), is that the restrictions on the matrix $[\Pi_1, \Pi_0]$ of reduced form coefficients of the approximate simultaneous equations model can be regarded as convenient approximations to the restrictions on the matrix $[F, g]$ of coefficients of the exact discrete model. In particular, if the elements of $[A, b]$ are linear functions of θ, then the elements of $[\Pi_1, \Pi_0]$ will be rational functions of θ whereas the elements of $[F, g]$ will be complicated transcendental functions of θ. Some idea of the goodness of the approximation can be obtained by comparing the power series expansions:

$$\begin{aligned}\Pi_1 &= \left[I + \tfrac{1}{2}A + \tfrac{1}{4}A^2 + \tfrac{1}{8}A^3 + \dots\right]\left[I + \tfrac{1}{2}A\right] \\ &= I + A + \tfrac{1}{2}A^2 + \tfrac{1}{4}A^3 + \dots\end{aligned}$$

and

$$F = I + A + \tfrac{1}{2}A^2 + \tfrac{1}{6}A^3 + \dots.$$

It should be noted, however, that whereas the power series expansion of F is convergent for any matrix A that of Π_1 is convergent only if the eigenvalues of A lie within the unit circle [see Halmos (1958, ch. 4)].

We shall now introduce two more assumptions.

Assumption 2

The vector process $\int_0^t \zeta(\mathrm{d}r)$ is Gaussian.

Assumption 3

The eigenvalues of $A(\theta^0)$ (where θ^0 is the true parameter vector) have negative real parts.

Assumption 2 is introduced in order to ensure that the disturbance vectors ε_t, $t = 1, 2, \dots$, in the system (31) are independently and identically distributed. The fact that it implies that they are normally distributed is incidental. Once we have assumed that the orthogonal increments (corresponding to a sequence of intervals of equal length) in the process $\int_0^t \zeta(\mathrm{d}r)$ are independently and identically distributed we are committed to assuming that they are normal. This can be seen by dividing the interval $[0, t]$ into n equal subintervals and applying the Lindberg-Levy central limit theorem [see Cramér (1951, p. 215)] to the sum $\sum_{i=1}^n \int_{(i-1)t/n}^{it/n} \zeta(\mathrm{d}r)$, when $n \to \infty$.

Assumption 3 implies that the eigenvalues of $F(\theta^0)$ lie within the unit circle. It follows, by applying the results of Mann and Wald (1943) to the system (31), that, under Assumptions 2 and 3, the sample mean vector $(1/T)\sum_{t=1}^T y(t)$ and sample moment matrices $(1/T)\Sigma_{t=1}^{T}y(t)y'(t)$ and $(1/T)\Sigma_{t=1}^{T}y(t)y'(t-1)$ converge in probability, as $T\to\infty$, to limits which do not depend on $y(0)$ and that $(1/\sqrt{T})\Sigma_{t=1}^{T}y(t)\varepsilon_t'$ has a limiting normal distribution. In establishing these results Mann and Wald assumed that $\varepsilon_1, \varepsilon_2, \ldots$ have finite fourth moments. Although Assumption 2 ensures that this condition is satisfied it is now known to be unnecessary [see Anderson (1959) and Hannan (1970, ch. 6)].

Since the probability limits of the sample moments of $y(t)$ can be expressed as functions of F, g and Ω, and hence as functions of θ and Σ, we can, in principle, find a formula for the asymptotic bias of any estimator of θ which can be expressed as a vector of rational functions of the sample moments. This is the case with the estimator obtained by applying two-stage least squares or three-stage least squares to the approximate simultaneous equations model (38). The formula would express the asymptotic bias of such an estimator in terms of the parameters of the continuous time model, i.e. the elements of θ and Σ. It would, of course, be very cumbersome if written out explicitly. But it is implicit in the calculations of Bergstrom (1966) who derives the asymptotic bias and approximate sampling variances of the estimates obtained by applying three-stage least squares to the approximate simultaneous equations model when the data are generated by a three equation continuous time model of the form (29).

In this example it is assumed that $b=0$, $\Sigma=I$ and that the only restrictions on A are that three of its elements are zero so that θ is a vector of the unrestricted elements of A. The assumed matrix A and derived matrix F are:

$$A=\begin{bmatrix}-1.0 & 0.8 & 0.0\\ 0.0 & -0.5 & 0.2\\ 0.1 & 0.0 & -0.2\end{bmatrix},$$

$$F=\begin{bmatrix}0.369 & 0.382 & 0.046\\ 0.006 & 0.608 & 0.142\\ 0.056 & 0.023 & 0.820\end{bmatrix}.$$

The interpretation of A, assuming that the time unit is 3 months, is that y_1 is causally dependent on y_2, y_2 on y_3 and y_3 on y_1 with mean time lags of 3 months, 6 months and 15 months, respectively. The probability limits of the estimators $\hat{A}$ and $\hat{\Pi}_1$ obtained by applying three-stage least squares to the approximate simultaneous equations model of the form (38) are:

$$\text{plim}\,\hat{A}=\begin{bmatrix}-0.922 & 0.710 & 0.000\\ 0.000 & -0.488 & 0.193\\ 0.098 & 0.000 & -0.199\end{bmatrix},$$

$$\text{plim}\,\hat{\Pi}_1=\begin{bmatrix}0.370 & 0.391 & 0.034\\ 0.005 & 0.609 & 0.141\\ 0.061 & 0.017 & 0.821\end{bmatrix}.$$

It is interesting to note that the estimated reduced form matrix $\hat{\Pi}_1$ provides a remarkably good estimator of the matrix F of coefficients in the exact discrete model, whereas $\hat{A}$ is a somewhat less satisfactory estimator of A. A heuristic explanation of this is that, even if there were no a priori restrictions on A, $\hat{A}$ would be an astymptotically biased estimator of this matrix whereas $\hat{\Pi}_1$ would, in this case, be identical with the least squares estimator F^* and, therefore, a consistent estimator of F. [See Bergstrom (1966, 1967) for a further discussion of this point and a proposed two stage estimator of A based on $\hat{\Pi}_1$.]

Since it is the matrix F which is of interest for the purpose of predicting future discrete observations, it is important to consider the question of whether or not it would be better, for this purpose, to use the least squares estimator F^* when A is restricted. Since F^* is a consistent estimator of F while $\hat{\Pi}_1$ is not, it would always be better to use F^* rather than $\hat{\Pi}_1$ if the sample size were sufficiently large. But with smaller samples the bias in any element of $\hat{\Pi}_1$ (as an estimator of the corresponding element of F) will be more than outweighed by its lower variance as compared with the variance of the corresponding element of F^*. Calculations presented in Bergstrom (1966) show that for the above example with a sample of 100 observations the reduction in the variance obtained by using $\hat{\Pi}_1$ rather than F^* heavily outweighs the squared asymptotic bias in any element of $\hat{\Pi}_1$.

The results of the above study suggest that the simultaneous equations model (38) is likely to be a useful approximation for the purpose of estimating the parameters of the underlying continuous time model from quarterly observations, and that the predictions obtained from the reduced form of this model, when the structural parameters are estimated by three-stage least squares, are likely to be better than those obtained from the ordinary least squares estimates of the coefficient of the exact discrete model ignoring the a priori restrictions. But there is, clearly, a need for a more general study, comparing the sampling properties of various estimators, applied to various approximate discrete models. An important step in this direction was taken by Sargan (1974, 1976). He generalizes the model (29) by including exogenous variables and considers the asymptotic bias of estimators obtained by applying the methods of two-stage least squares, three-stage least squares and full information maximum likelihood to the approximate simultaneous equations model (38), extended to include exogenous variables. He shows, in particular, that the proportional asymptotic bias of all of these estimators is of the same order of smallness as the square of the observation period as this tends to zero.

The econometrician cannot, of course, obtain observations of macroeconomic variables at arbitrary small intervals of time. He must, generally, do the best that he can with quarterly observations of such variables as the gross national product and its components. But the results of the study by Bergstrom (1966), which assumes a realistic pattern of time lags and quarterly observations, suggest that Sargan's criterion may, nevertheless, be useful for the ranking of various estimators and various approximate discrete models. Since Sargan uses only one approximate discrete model, and the asymptotic bias of each of the three estimators considered by him is of the same order of smallness, the significance of his results could, easily, be overlooked. Before proving his basic result, therefore, we shall apply his method to an even simpler approximate discrete model, which has been more widely used than (38). This is the model:

$$y(t)-y(t-1)=A(\theta)y(t-1)+b(\theta)+u_t, \tag{40}$$

$$\mathrm{E}(u_t)=0, \quad \mathrm{E}(u_t u_t')=\Sigma, \quad \mathrm{E}(u_s u_t')=0, \quad s\neq t, \quad s,t=1,2,\ldots.$$

We shall show that estimates obtained from the model (40) will be inferior to those obtained from (38) if the observation period is sufficiently short and the data are generated by (29).

We assume, for this purpose, that $b=0$ and that the only other a priori restrictions are that certain elements of A are zero so that θ is the vector of unrestricted elements of A. The continuous time model (28) can then be written:

$$\mathrm{d}y_i(t)=\theta^{(i)}y^{(i)}(t)\,\mathrm{d}t+\zeta_i(\mathrm{d}t), \quad i=1,\ldots,n, \tag{41}$$

where $y_i(t)$ is the ith element of $y(t)$, $\theta^{(i)}$ is the vector of unrestricted elements of the ith row of A and $y^{(i)}(t)$ is a vector of the corresponding elements of $y(t)$. The system (29), by which we give a precise interpretation of (28), can be written:

$$y_i(t)-y_i(0)=\int_0^t\theta^{(i)}y^{(i)}(r)\,\mathrm{d}r+\int_0^t\zeta_i(\mathrm{d}r), \quad i=1,\ldots,n. \tag{42}$$

Following Sargan we shall keep the time unit constant and denote the observation period by δ so that we can consider the behaviour of our estimators as $\delta\to 0$ while keeping the elements of θ constant. Then, defining $y_r=y(r\delta)$, the exact discrete model is:

$$y_t=\mathrm{e}^{\delta A}y_{t-1}+\varepsilon_t, \tag{43}$$

$$\mathrm{E}(\varepsilon_t)=0, \quad \mathrm{E}(\varepsilon_t\varepsilon_t')=\int_0^\delta \mathrm{e}^{rA}\Sigma\mathrm{e}^{rA'}\,\mathrm{d}r,$$

$$\mathrm{E}(\varepsilon_s\varepsilon_t')=0, \quad s\neq t, \quad s,t=1,2,\ldots.$$

The approximate discrete model (40) can be written:

$$\frac{y_{it}-y_{i,t-1}}{\delta}=\theta^{(i)}y_{t-1}^{(i)}+u_{it}, \quad i=1,\ldots,n. \tag{44}$$

We can now show that the asymptotic bias of the estimator θ^* obtained by applying ordinary least squares to each equation of the system (44) is $O(\delta)$ as $\delta \to 0$.

Theorem 4

If θ^* is the estimator obtained from a sample $y_1, y_2, \dots, y_T$ [i.e. $y(\delta), y(2\delta), \dots, y(T\delta)$] of vectors generated by (42) by applying ordinary least squares to each equation of (44), then, under Assumptions 1, 2 and 3:

$$\operatorname*{plim}_{T \to \infty} \theta^* - \theta = O(\delta), \quad \text{as } \delta \to 0.$$

Proof

From (43) we obtain:

$$\begin{aligned} \frac{1}{\delta}(y_t - y_{t-1}) &= \left[A + \frac{\delta}{2}A^2 + \frac{\delta^2}{3!}A^3 + \cdots \right] y_{t-1} + \frac{1}{\delta}\varepsilon_t \\ &= Ay_{t-1} + Hy_{t-1} + \frac{1}{\delta}\varepsilon_t, \end{aligned} \tag{45}$$

where

$$H = O(\delta).$$

The system (45) can be written:

$$\frac{y_{it} - y_{i,t-1}}{\delta} = \theta^{(i)} y_{t-1}^{(i)} + h_i' y_{t-1} + \frac{\varepsilon_{it}}{\delta}, \qquad i = 1, \dots, n, \tag{46}$$

where h_i' is the ith row of H and ε_{it} is the ith element of ε_t. Then the estimator $\theta^{*(i)}$ obtained by applying ordinary least squares to the ith equation of (44) is:

$$\begin{aligned} \theta^{*(i)} &= \left[\frac{1}{T} \sum_{t=1}^{T} \left(\frac{y_{it} - y_{i,t-1}}{\delta} \right) y_{t-1}'^{(i)} \right] \left[\frac{1}{T} \sum_{t=1}^{T} y_{t-1}^{(i)} y_{t-1}'^{(i)} \right]^{-1} \\ &= \left[\frac{1}{T} \sum_{t=1}^{T} \left(\theta^{(i)} y_{t-1}^{(i)} + h_i' y_{t-1} + \frac{\varepsilon_{it}}{\delta} \right) y_{t-1}'^{(i)} \right] \left[\frac{1}{T} \sum_{t=1}^{T} y_{t-1}^{(i)} y_{t-1}'^{(i)} \right]^{-1} \\ &= \theta^{(i)} + \left[\frac{1}{T} \sum_{t=1}^{T} \left(h_i' y_{t-1} + \frac{\varepsilon_{it}}{\delta} \right) y_{t-1}'^{(i)} \right] \left[\frac{1}{T} \sum_{t=1}^{T} y_{t-1}^{(i)} y_{t-1}'^{(i)} \right]^{-1}. \end{aligned}$$

But, from the Mann and Wald results:

$$\operatorname*{plim}_{T \to \infty} \frac{1}{T} \sum_{t=1}^{T} y_{t-1}^{(i)} \varepsilon_{it} = 0$$

and

$$\underset{T\to\infty}{\text{plim}} \frac{1}{T} \sum_{t=1}^{T} y_{t-1} y'_{t-1} \text{ exists.}$$

Therefore:

$$\underset{T\to\infty}{\text{plim}}\, \theta^{*(i)} - \theta^{(i)} = O(h_i) = O(\delta), \quad \text{as } \delta \to 0. \quad \blacksquare$$

We shall consider, next, estimates obtained by using the approximate simultaneous equations model (38). When the continuous time model is (42), the system (38) can be written:

$$\frac{y_{it} - y_{i,t-1}}{\delta} = \theta^{(i)}\left[\tfrac{1}{2}\left(y_t^{(i)} + y_{t-1}^{(i)}\right)\right] + u_{it}, \qquad i = 1,\dots,n. \tag{47}$$

We shall prove a theorem which includes, as a special case, Sargan's basic theorem (when there are no exogenous variables).

Theorem 5

Let $\bar{\theta}^{(i)}$ be the instrumental variables estimator, defined by:

$$\bar{\theta}^{(i)} = \left[\frac{1}{T}\sum_{t=1}^{T}\left(\frac{y_{it} - y_{i,t-1}}{\delta}\right) z_t^{(i)}\right]\left[\frac{1}{T}\sum_{t=1}^{T}\tfrac{1}{2}\left(y_t^{(i)} + y_{t-1}^{(i)}\right) z_t^{(i)}\right]^{-1},$$

where $y_1,\dots,y_T$ [i.e. $y(\delta),\dots,y(T\delta)$] are vectors generated by (42) and $z_1^{(i)},\dots,z_T^{(i)}$ are random row vectors such that:

$$\underset{T\to\infty}{\text{plim}} \frac{1}{T} \sum_{t=1}^{T} \varepsilon_{it} z_t^{(i)} = 0,$$

while

$$\underset{T\to\infty}{\text{plim}} \frac{1}{T} \sum_{t=1}^{T} y_t^{(i)} z_t^{(i)} \quad \text{and} \quad \underset{T\to\infty}{\text{plim}} \frac{1}{T} \sum_{t=1}^{T} y_{t-1}^{(i)} z_t^{(i)} \text{ exist.}$$

Then, under Assumptions 1, 2 and 3:

$$\underset{T\to\infty}{\text{plim}}\, \bar{\theta}^{(i)} - \theta^{(i)} = O(\delta^2), \quad \text{as } \delta \to 0.$$

Proof

Using (43), we obtain:

$$\frac{1}{\delta}(y_t - y_t) = \left[A + \frac{\delta}{2}A^2 + \frac{\delta^2}{3!}A^3 + \cdots\right]y_{t-1} + \frac{1}{\delta}\varepsilon_t$$

$$= \tfrac{1}{2}Ay_{t-1} + \tfrac{1}{2}A\left[I + \delta A + \frac{\delta^2}{3}A^2 + \cdots\right]y_{t-1} + \frac{1}{\delta}\varepsilon_t$$

$$= \tfrac{1}{2}Ay_{t-1} + \tfrac{1}{2}Ae^{\delta A}y_{t-1} + Ly_{t-1} + \frac{1}{\delta}\varepsilon_t$$

$$= A\left[\tfrac{1}{2}(y_t + y_{t-1})\right] + Ly_{t-1} + \left(\frac{1}{\delta}I - \tfrac{1}{2}A\right)\varepsilon_t,$$

where

$$L = O(\delta^2).$$

Therefore

$$\frac{y_{it} - y_{i,t-1}}{\delta} = \tfrac{1}{2}\theta^{(i)}\left(y_t^{(i)} + y_{t-1}^{(i)}\right) + l_i'y_{t-1} + \frac{1}{\delta}\varepsilon_{it} - \tfrac{1}{2}\theta^{(i)}\varepsilon_t,$$

where l_i' is the ith row of L. And, hence:

$$\bar{\theta}^{(i)} = \left[\frac{1}{T}\sum_{t=1}^{T}\left\{\tfrac{1}{2}\theta^{(i)}\left(y_t^{(i)} + y_{t-1}^{(i)}\right) + l_i'y_{t-1} + \frac{1}{\delta}\varepsilon_{it} - \tfrac{1}{2}\theta^{(i)}\varepsilon_t\right\}z_t^{(i)}\right]$$

$$\times\left[\frac{1}{T}\sum_{t=1}^{T}\tfrac{1}{2}\left(y_t^{(i)} + y_{t-1}^{(i)}\right)z_t^{(i)}\right]^{-1}$$

$$= \theta^{(i)} + \left[\frac{1}{T}\sum_{t=1}^{T}\left(l_i'y_{t-1} + \frac{1}{\delta}\varepsilon_{it} - \tfrac{1}{2}\theta^{(i)}\varepsilon_t\right)z_t^{(i)}\right]\left[\frac{1}{T}\sum_{t=1}^{T}\tfrac{1}{2}\left(y_t^{(i)} + y_{t-1}^{(i)}\right)z_t^{(i)}\right]^{-1}.$$

Therefore

$$\operatorname*{plim}_{T\to\infty}\bar{\theta}^{(i)} - \theta^{(i)} = O(\delta^2), \quad \text{as } \delta \to 0. \quad \blacksquare$$

The two-stage least squares estimator is obtained as a special case of the estimator $\bar{\theta}^{(i)}$ by putting:

$$z_t^{(i)} = \tfrac{1}{2}\left(y_t^{*(i)} + y_{t-1}^{(i)}\right)',$$

where

$$y_t^{*(i)} = \left[\frac{1}{T}\sum_{t=1}^{T}y_t^{(i)}y_{t-1}'\right]\left[\frac{1}{T}\sum_{t=1}^{T}y_{t-1}y_{t-1}'\right]^{-1}y_{t-1}.$$

It is fairly obvious that the above argument can be extended to show that the asymptotic bias of the three-stage least squares estimator of θ is $O(\delta^2)$. Sargan (1976) shows that the asymptotic bias of the full information maximum likelihood estimator [applied to (47)] is also $O(\delta^2)$ and that the difference between the limits in probability of the three-stage least squares and full information maximum likelihood estimators is $O(\delta^5)$. He also finds sufficient conditions for these results to hold when the model contains exogenous variables. These conditions will be given in Section 4.

We turn now to the problem of finding consistent and asymptotically efficient estimators of θ from discrete data generated by (29). For this purpose the following additional assumptions are introduced.

Assumption 4

It is known that θ belongs to a compact subset Θ of p-dimensional space.

Assumption 5

Let Ψ be the subset of $n\times(n+1)$ matrices which can be written in the form $[F(\theta), g(\theta)]$ for some $\theta\in\Theta$, where

$$F(\theta)=\mathrm{e}^{A(\theta)},\qquad g(\theta)=[\mathrm{e}^{A(\theta)}-I]A^{-1}(\theta)b(\theta).$$

Then the mapping from Ψ to Θ defined by the inverse of $[F(\theta), g(\theta)]$ is one to one and continuous in a neighbourhood of the true parameter vector θ^0; i.e. every sequence θ^n, $n=1,2,\dots$, of vectors in Θ such that $[F(\theta^n), g(\theta^n)]\to[F(\theta^0), g(\theta^0)]$ converges to θ^0 as $T\to\infty$.

Assumption 6

The set Θ contains a neighbourhood of θ^0 in which the derivatives up to the third order of the elements of $[F(\theta), g(\theta)]$ are bounded. The vector θ^0 is not a singularity point of $[F(\theta), g(\theta)]$; i.e. there is no set of numbers $\lambda_1,\dots,\lambda_p$, not all zero, such that:

$$\sum_{k=1}^{p}\lambda_k\frac{\partial}{\partial\theta_k}[F(\theta^0), g(\theta^0)]=0.$$

We shall consider, first, the minimum distance estimator θ^{**} which is defined as the vector $\theta\in\Theta$ that minimises:

$$\frac{1}{T}\sum_{t=1}^{T}[y(t)-F(\theta)y(t-1)-g(\theta)]'M_{\varepsilon\varepsilon}^{*-1}[y(t)-F(\theta)y(t-1)-g(\theta)],$$

where

$$M_{\varepsilon\varepsilon}^{*}=\frac{1}{T}\sum_{t=1}^{T}[y(t)-F^{*}y(t-1)-g^{*}][y(t)-F^{*}y(t-1)-g^{*}]',$$

and F^* and g^* are the ordinary least squares estimators of F and g respectively. The properties of this estimator have been studied by Malinvaud (1970, ch. 9) when applied to the model:

$$y_t=A(\theta)x_t+\varepsilon_t, \tag{48}$$

where A is a matrix of non-linear functions of the parameter vector and x_t, $t=1,2,\dots$, is a sequence of non-random vectors. Since the model (48) contains no lagged dependent variables we cannot rely on Malinvaud's results for inferring the properties of the minimum distance estimator when applied to the model (31). But, by using the Mann and Wald results and modifying Malinvaud's proofs, we can prove Theorems 6 and 7 which, together, correspond to theorem 5 of Malinvaud (1970, ch. 9).

Theorem 6

Under Assumptions 1–5:

$$\operatorname*{plim}_{T\to\infty}\theta^{**}=\theta^{0}$$

Theorem 7

Under Assumptions 1–6, $\sqrt{T}(\theta^{**}-\theta^{0})$ has a limiting normal distribution whose covariance matrix is the limit is probability, as $T\to\infty$, of the inverse of the matrix whose (kl)th element is:

$$\operatorname{tr}\left\{\frac{\partial}{\partial\theta_k}[F(\theta),g(\theta)]'M_{\varepsilon\varepsilon}^{*-1}\frac{\partial}{\partial\theta_l}[F(\theta),g(\theta)]\right.$$
$$\left.\times\begin{bmatrix}\frac{1}{T}\sum_{t=1}^{T}y(t)y'(t) & \frac{1}{T}\sum_{t=1}^{T}y(t)\\ \frac{1}{T}\sum_{t=1}^{T}y'(t) & 1\end{bmatrix}\right\}. \tag{49}$$

Since the logarithm of the likelihood function is:

$$L(\theta,\Omega) = -\frac{T}{2}\log 2\Pi - \frac{1}{2}\log|\Omega| - \frac{1}{2}\sum_{t=1}^{T}[y(t)-F(\theta)y(t-1)-g(\theta)]'\Omega^{-1}$$
$$\times[y(t)-F(\theta)y(t-1)-g(\theta)], \qquad (50)$$

it follows from (49) that the covariance matrix of the limiting distribution of $\sqrt{T}(\theta^{**}-\theta^0)$ is:

$$-\left[\mathrm{E}\left(\frac{\partial^2 L}{\partial\theta_k\,\partial\theta_l}\right)\right]^{-1}.$$

The estimator θ^{**} is asymptotically efficient, therefore, in the sense defined by Cramér (1946, ch. 32).

For the purpose of predicting the future discrete observations of $y(t)$, we are interested in the estimator

$$[F^{**}, g^{**}] = [F(\theta^{**}), g(\theta^{**})]$$

of the matrix of coefficients of the exact discrete model (31). By using (49) and the argument of Malinvaud (1970, p. 357) we can show that the concentration ellipsoid E^{**} [in $n(n+1)$ dimensional space] of the limiting distribution of $[\sqrt{T}(F^{**}-F^0), \sqrt{T}(g^{**}-g^0)]$ is the set of $n\times(n+1)$ matrices $[F, g]$ that can be written in the form:

$$[F,g] = \sum_{k=1}^{p}\theta_k\frac{\partial}{\partial\theta_k}[F(\theta^0), g(\theta^0)], \qquad (51)$$

for some vector $[\theta_1,\ldots,\theta_p]$ and satisfy the inequality

$$\operatorname{tr}\left\{[F,g]'\Omega^{-1}[F,g]\begin{bmatrix}\operatorname{plim}\frac{1}{T}\sum_{t=1}^{T}y(t)y'(t) & \operatorname{plim}\frac{1}{T}\sum_{t=1}^{T}y(t)\\ \operatorname{plim}\frac{1}{T}\sum_{t=1}^{T}y'(t) & 1\end{bmatrix}\right\} \leq 1. \qquad (52)$$

Since the concentration ellipsoid E^* of the limiting distribution of $[\sqrt{T}(F^*-F^0), \sqrt{T}(g^*-g^0)]$ is the set of all matrices $[F, g]$ satisfying (52) we have:

$$E^{**} \subset E^*.$$

In geometrical terms, E^{**} is the intersection of E^* with the hyperplane of matrices defined by (51), i.e. it is the intersection of E^* with the hyperplane of matrices satisfying the restrictions implied by the continuous time model in the neighbourhood of $[F^0, g^0] = [F(\theta^0), g(\theta^0)]$.

This result implies that the asymptotic standard error of any element of $[F^{**}, g^{**}]$ (or any linear combination of such elements) is at least as small as the asymptotic standard error of the corresponding element of $[F^*, g^*]$ (or linear combination of such elements). For it follows from the invariance property of the concentration ellipsoid under a linear transformation [see Malinvaud (1970, ch. 5, Lemma 1)] that the asymptotic standard errors of identical linear combinations of elements of $[F^{**}, g^{**}]$ and $[F^*, g^*]$ can be compared by comparing the images of E^{**} and E^*, respectively, under a linear transformation which transforms $n(n+1)$ dimensional space into the appropriate one dimensional subspace. Provided, therefore, that the sample size is sufficiently large we can obtain better predictions of the future discrete observations by using the continuous time model than by using the unrestricted least squares estimates of the coefficients of the exact discrete model.

Finally we shall consider, very briefly, the maximum likelihood estimator. This is obtained by maximising $L(\theta, \Omega)$, as defined by (50), with respect to θ and Ω. We can do this in two stages. We first maximise $L(\theta, \Omega)$ with respect to Ω to obtain $\hat{\Omega}(\theta)$ and then substitute into $L(\theta, \Omega)$ to obtain the concentrated likelihood function:

$$L(\theta) = L(\theta, \hat{\Omega}(\theta)) = c - \log|M(\theta)|,$$

where c is a constant, $|M|$ is the determinant of M and

$$M(\theta) = \frac{1}{T}\sum_{t=1}^{T}[y(t) - F(\theta)y(t-1) - g(\theta)][y(t) - F(\theta)y(t-1) - g(\theta)]'.$$

Then the maximum likelihood estimator $\hat{\theta}$ is the $\theta \in \Theta$ that maximises $L(\theta)$. The estimation equations, obtained by equating to zero the partial derivatives of $L(\theta)$ with respect to $\theta_1, \dots, \theta_p$, are:

$$\operatorname{tr} M^{-1}(\theta)\frac{\partial}{\partial\theta_k}M(\theta) = 0, \qquad k = 1, \dots, p. \tag{53}$$

The estimation equations for the minimum distance estimator θ^{**} are:

$$\operatorname{tr} M_{\varepsilon\varepsilon}^{*-1}\frac{\partial}{\partial\theta_k}M(\theta) = 0, \qquad k = 1, \dots, p. \tag{54}$$

The system (54) is easier to solve than (53) since the matrix $M^{-1}(\theta)$ in (53) involves the unknown elements of θ, for which we are solving, whereas the matrix $M_{\varepsilon\varepsilon}^{*-1}$ in (54) can be computed as an initial step without iteration. Under Assumptions 1–6 $\sqrt{T}(\hat{\theta} - \theta^0)$ has the same limiting distribution as $\sqrt{T}(\theta^{**} - \theta^0)$.

This follows from the results of Dunsmuir and Hannan (1976) who consider a very general model which includes the exact discrete model (31) as a special case. We shall consider their results in more detail at a later stage.

It is of interest to compare the estimator θ^{**} (or $\hat{\theta}$) obtained by using the exact discrete model with an estimator obtained by applying either the three-stage least squares or full information maximum likelihood method to an approximate simultaneous equations model. We know that if the sample size is sufficiently large the estimator obtained by using the exact discrete model will be better, since it is consistent, whereas the estimator obtained by using the approximate simultaneous equations model has an asymptotic bias of the order δ^2, as we have seen. But, as the study of Bergstrom (1966) showed, for samples of the size available in practice the squared asymptotic bias in an estimator obtained by using the approximate simultaneous equations model can be small compared with the sampling variance.

A comparison of estimates obtained, from finite samples, by using the exact discrete model and an approximate simultaneous equations model was undertaken by Phillips (1972) who, for this purpose, wrote the first computer program for obtaining consistent and asymptotically efficient estimates of the parameters of a continuous time model, of the form (29), from discrete data. This program was for the computation of the minimum distance estimator θ^{**}, using the exact discrete model. The first program for the computation of the more complicated maximum likelihood estimator $\hat{\theta}$, using the exact discrete model, was written by Wymer (1974) and applied to the model of Bergstrom and Wymer (1976) which will be discussed later. The main difficulty in computing either θ^{**} or $\hat{\theta}$ as compared with estimators obtained from an approximate simultaneous equations model is that $F(\theta)$ must be expressed as a series of matrices and summed to a sufficient number of terms to give the desired degree of accuracy.

Phillips (1972) applied his program, in a Monte Carlo study, to a three equation trade cycle model based on the model of Phillips (1954) [see Bergstrom (1967, ch. 3)]. The model, in its deterministic form, is:

$$\mathrm{D}C(t)=\alpha[(1-s)Y(t)+a-C(t)], \tag{55}$$

$$\mathrm{D}Y(t)=\lambda[C(t)+\mathrm{D}K(t)-Y(t)], \tag{56}$$

$$\mathrm{D}K(t)=\gamma[vY(t)-K(t)]. \tag{57}$$

where $C=$ consumption, $Y=$ income and $K=$ capital. By adding white noise disturbances and substituting for DK from (57) into (56) the model can be written:

$$\mathrm{d}y(t)=A(\theta)y(t)\,\mathrm{d}t+b(\theta)+\zeta(\mathrm{d}t), \tag{58}$$

where

$$\theta=(\alpha,\gamma,\lambda,s,v,a).$$

A hundred synthetic samples each of 25 observations were generated by the exact discrete model derived from (58) and used in the estimation of θ, both by applying the minimum distance estimation procedure to the exact discrete model and applying three-stage least squares to an approximate simultaneous equations model. The results are shown in Table 2.1. As can be seen, the estimates obtained by using the exact discrete model are, generally, superior to those obtained by using the approximate simultaneous equations model. Moreover, considering the smallness of the sample, the number of times that the 5% confidence interval, computed from the estimated asymptotic standard errors, does not include the true value of the parameter is, for the estimates obtained by applying the minimum distance procedure to the exact discrete model, remarkably close to 5 (i.e. 7 for the parameters α, γ, λ and s, and 9 for v).

In the above example Phillips assumed the existence of point observations of all three variables. But the variables $C(t)$ and $Y(t)$ are flow variables and, in practice, could be observed only as the integrals $\int_{t-1}^{t} C(r)\mathrm{d}r$ and $\int_{t-1}^{t} Y(r)\mathrm{d}r$, $t = 1, 2, \ldots$. This does not, of course detract from the value of his study for the general purpose of comparing estimates derived from point observations using the exact discrete model and approximate simultaneous equations model. Moreover, at the time when the study was undertaken, the theoretical problems of obtaining consistent and asymptotically efficient estimates of the parameters of a continuous time stochastic model, of the form (29), from flow data had not been seriously studied. This is the problem to which we now turn. The essential difficulty is that, even when the continuous time model is a first-order system with white noise disturbances, the disturbances in the exact discrete model satisfied by the integral observations will be autocorrelated. The precise form of the autocorrelation is given in the following theorem.

Table 2.1

Parameter:	α	γ	λ	s	v
True value:	0.6	0.4	4.0	0.25	2.0
Minimum distance					
Mean of the estimates	0.5734	0.4016	4.0709	0.2537	2.0021
Standard deviation of the estimates	0.1410	0.0153	0.7077	0.0259	0.0149
Root mean square error	0.1435	0.0154	0.7112	0.0262	0.0150
Number of wrong intervals[a]	7	7	7	7	9
Three-stage least squares					
Mean of the estimates	0.6652	0.4182	2.7444	0.2767	1.995
Standard deviation of the estimates	0.1800	0.0241	0.8015	0.0937	0.0311
Root mean square error	0.1914	0.0302	1.4896	0.0974	0.0311
Number of wrong intervals[a]	10	3	62	17	3

[a]Intervals not containing the true parameter value.

Theorem 8

If $\zeta(dt)$ satisfies Assumption 1, $y(t)$ is the solution to (29) and y_t is defined by:

$$y_t = \int_{t-1}^{t} y(r)\,dr,$$

then y_t satisfies the system:

$$y_t = F(\theta)y_{t-1} + g(\theta) + \eta_t, \tag{59}$$

$$E(\eta_t) = 0, \qquad E(\eta_t\eta_t') = \Omega_0, \qquad E(\eta_t\eta_{t-1}') = \Omega_1,$$

$$E(\eta_s\eta_t') = 0, \qquad |s-t|>1, \qquad s,t = 1,2,\dots,$$

where

$$F = e^A, \qquad g = (e^A - I)A^{-1}b,$$

$$\Omega_0 = \int_0^1 A^{-1}(I - e^{rA})\Sigma(I - e^{rA'})A'^{-1}\,dr$$

$$+ \int_0^1 A^{-1}e^A(I - e^{(r-1)A})\Sigma(I - e^{(r-1)A'})e^{A'}A'^{-1}\,dr, \tag{60}$$

$$\Omega_1 = \int_0^1 A^{-1}e^A(I - e^{(r-1)A})\Sigma(e^{rA'} - I)A'^{-1}\,dr. \tag{61}$$

Proof

From (29) we obtain:

$$y(t) - y(t-1) = A\int_{t-1}^{t} y(r)\,dr + b + \int_{t-1}^{t}\zeta(dr), \tag{62}$$

and hence:

$$\int_{t-1}^{t} y(r)\,dr = A^{-1}[y(t) - y(t-1)] - A^{-1}b - A^{-1}\int_{t-1}^{t}\zeta(dr). \tag{63}$$

Then, substituting from (31) into (63) and using (36), we obtain:

$$\int_{t-1}^{t} y(r)\,dr = A^{-1}F[y(t-1) - y(t-2)]$$

$$+ A^{-1}\left[\int_{t-1}^{t} e^{(t-r)A}\zeta(dr) - \int_{t-2}^{t-1} e^{(t-1-r)A}\zeta(dr)\right]$$

$$- A^{-1}b - A^{-1}\int_{t-1}^{t}\zeta(dr), \tag{64}$$

and then, from (62) and (64):

$$\int_{t-1}^{t} y(r)\,\mathrm{d}r = A^{-1}FA\int_{t-2}^{t-1} y(r)\,\mathrm{d}r + A^{-1}Fb + A^{-1}F\int_{t-2}^{t-1}\zeta(\mathrm{d}r)$$
$$+ A^{-1}\left[\int_{t-1}^{t} \mathrm{e}^{(t-r)A}\zeta(\mathrm{d}r) - \int_{t-2}^{t-1}\mathrm{e}^{(t-1-r)A}\zeta(\mathrm{d}r)\right]$$
$$- A^{-1}b - A^{-1}\int_{t-1}^{t}\zeta(\mathrm{d}r),$$

which, since

$$A^{-1}FA = A^{-1}\left[I + A + \tfrac{1}{2}A^2 + \cdots\right]A = F$$

and

$$A^{-1}[F-I]b = [F-I]A^{-1}b = g,$$

reduces to (59) with η_t defined by:

$$\eta_t = \int_{t-1}^{t} A^{-1}(\mathrm{e}^{(t-r)A} - I)\zeta(\mathrm{d}r)$$
$$+ \int_{t-2}^{t-1} A^{-1}(\mathrm{e}^{A} - \mathrm{e}^{(t-1-r)A})\zeta(\mathrm{d}r). \tag{65}$$

Finally, using the generalizations of (15) and (16) for a vector random measure we obtain:

$$\mathrm{E}(\eta_t\eta_t') = \int_{t-1}^{t} A^{-1}(\mathrm{e}^{(t-r)A} - I)\Sigma(\mathrm{e}^{(t-r)A'} - I)A'^{-1}\,\mathrm{d}r$$
$$+ \int_{t-2}^{t-1} A^{-1}(\mathrm{e}^{A} - \mathrm{e}^{(t-1-r)A})\Sigma(\mathrm{e}^{A'} - \mathrm{e}^{(t-1-r)A'})A'^{-1}\,\mathrm{d}r$$
$$= \int_0^1 A^{-1}(I - \mathrm{e}^{rA})\Sigma(I - \mathrm{e}^{rA'})A'^{-1}\,\mathrm{d}r$$
$$+ \int_0^1 A^{-1}\mathrm{e}^{A}(I - \mathrm{e}^{(r-1)A})\Sigma(I - \mathrm{e}^{(r-1)A'})\mathrm{e}^{A'}A'^{-1}\,\mathrm{d}r,$$

$$\mathrm{E}(\eta_t\eta_{t-1}') = \int_{t-2}^{t-1} A^{-1}(\mathrm{e}^{A} - \mathrm{e}^{(t-1-r)A})\Sigma(\mathrm{e}^{(t-1-r)A'} - I)A'^{-1}\,\mathrm{d}r$$
$$= \int_0^1 A^{-1}\mathrm{e}^{A}(I - \mathrm{e}^{(r-1)A})\Sigma(\mathrm{e}^{rA'} - I)A'^{-1}\,\mathrm{d}r. \quad \blacksquare$$

It is clear from these results that $\{\eta_t\}$ is a vector moving average process of the form:

$$\eta_t = \varepsilon_t + C\varepsilon_{t-1}, \tag{66}$$

where

$$E(\varepsilon_t) = 0, \qquad E(\varepsilon_t\varepsilon_t') = K, \qquad E(\varepsilon_s\varepsilon_t') = 0, \qquad s \neq t, \qquad s,t = 1,2,\ldots,$$

and C and K satisfy the equations:

$$K + CKC' = \Omega_0 \tag{67}$$

and

$$CK = \Omega_1. \tag{68}$$

Equations (67) and (68) imply that the elements of C and K are functions of the elements of Ω_0 and Ω_1 and, hence, of the elements of A and Σ. The expressions (60) and (61) can be written as infinite series in ascending powers of A and A' by expressing the matrix exponential functions, in the integrals, in series form and integrating, term by term, with respect to r. Evaluating the terms up to the first power of A in A' we obtain:

$$\Omega_0 = \tfrac{2}{3}\Sigma + \tfrac{1}{3}(A\Sigma + \Sigma A') + \cdots, \tag{69}$$
$$\Omega_1 = \tfrac{1}{6}\Sigma + \tfrac{1}{8}A\Sigma + \tfrac{1}{24}\Sigma A' + \cdots. \tag{70}$$

Phillips (1978) shows that, if the observation period is δ, then:

$$\Omega_0 = \frac{2\delta^3}{3}\Sigma + \frac{\delta^4}{3}(A\Sigma + \Sigma A') + O(\delta^5), \tag{71}$$

$$\Omega_1 = \frac{\delta^3}{6}\Sigma + \frac{\delta^4}{8}A\Sigma + \frac{\delta^4}{24}\Sigma A' + O(\delta^5), \tag{72}$$

these equations being identical with (69) and (70) when $\delta = 1$. He also shows that:

$$C = \alpha I + \frac{\alpha\delta}{4}(A - \Sigma A'\Sigma^{-1}) + O(\delta^2), \tag{73}$$

$$K = \frac{\delta^3}{6\alpha}\left[\Sigma + \frac{\delta}{2}(A\Sigma + \Sigma A')\right] + O(\delta^5), \tag{74}$$

where α ($=0.268$) is a root of the equation:

$$z^2 - 4z + 1 = 0.$$

The first term on the right-hand side of (73) and (74) is easily obtained by substituting the first term on the right hand side of (69) and (70) into (67) and (68) respectively and solving for C and K.

It is convenient, at this stage, to write Σ as $\Sigma(\mu)$ where μ is a vector of parameters. If, as we have assumed so far, Σ is unrestricted then μ will have $n(n+1)/2$ elements. But we could, for example, require Σ to be a diagonal matrix in which case μ would have n elements. We can now obtain the exact discrete model corresponding to the continuous time model (29) for the case in which the observations are in integral form. Combining (59) and (66) we obtain:

$$y_t - F(\theta) y_{t-1} - g(\theta) = \varepsilon_t + C(\theta,\mu)\varepsilon_{t-1}, \tag{75}$$

$$\mathrm{E}(\varepsilon_t) = 0, \quad \mathrm{E}(\varepsilon_t \varepsilon_t') = K(\theta,\mu), \quad \mathrm{E}(\varepsilon_s \varepsilon_t') = 0, \quad s \neq t, \quad s,t = 1,2\ldots.$$

An important point to notice is that, even though the covariance matrix Σ of the disturbance vector in the continuous time model does not depend on θ, the covariance matrix K of the random vector ε_t in the exact discrete model depends on θ as well as μ, as is clear from (67) and (68).

If g is a zero vector (i.e. if b, in the continuous time model, is a zero vector), then (75) is a special case of the model:

$$y_t - \sum_{j=1}^{q} F_j(\theta,\mu) y_{t-j} = \varepsilon_t + \sum_{j=1}^{r} C_j(\theta,\mu)\varepsilon_{t-j}, \tag{76}$$

$$\mathrm{E}(\varepsilon_t) = 0, \quad \mathrm{E}(\varepsilon_t \varepsilon_t') = K(\theta,\mu), \quad \mathrm{E}(\varepsilon_s \varepsilon_t') = 0, \quad s \neq t, \quad s,t = 1,2,\ldots,$$

which was studied by Dunsmuir and Hannan (1976). They show, under certain assumptions, that the maximum likelihood estimator of θ, in the model (76), is strongly consistent (i.e. converges almost surely to θ). And, for the case in which the matrices $F_1,\ldots,F_q, C_1,\ldots,C_r$ do not depend on μ and K does not depend on θ, so that these matrices can be written $F_1(\theta),\ldots,F_q(\theta)$, $C_1(\theta),\ldots,C_r(\theta)$ and $K(\mu)$, they prove a central limit theorem. But, as we have shown, the matrices C and K in the model (75) depend on both θ and μ. The case in which the matrices $F_1,\ldots,F_q$, $C_1,\ldots,C_r$ and K, in the model (76), all depend on both θ and μ was further considered by Dunsmuir (1979). Here he proves a central limit theorem for an estimator obtained by maximising an approximate likelihood function expressed in terms of the discrete Fourier transform of the data. [This estimator was proved to be strongly consistent by Dunsmuir and Hannan (1976).] His results imply that when $\{\varepsilon_t\}$ is Gaussian this estimator is asymptotically efficient (in the Cramér sense). Kohn (1979) extends the model (76) by including exogenous variables (which cover the case in which g, in (75), is not a zero vector); but he assumes that K is unrestricted. Although the properties of the exact maximum likelihood estimator $(\hat{\theta}, \hat{\mu})$ of the parameter vector (θ, μ), in the model (75), cannot be definitely inferred from the results obtained in the above studies, these results leave little doubt that, under Assumptions 1–6 and the assumption that Σ is unrestricted, $(\hat{\theta}, \hat{\mu})$ is strongly consistent, asymptotically normal and asymptotically efficient.

In order to obtain the exact likelihood function for the parameters of (75) it is not necessary to find explicit formulae for $C(\theta, \mu)$ and $K(\theta, \mu)$. Instead we can work directly from the formulae for Ω_0 and Ω_1 which we have obtained from the continuous time model, thus avoiding the problem of solving (67) and (68). For this purpose we shall assume that $\{y(t)\}$ is stationary [i.e. that $y(0)$, in the model (29), is not a fixed vector but a random vector generated by the application of this model over $(-\infty, 0)$]. Then defining

$$m = \mathrm{E}y(t),$$

we obtain, from (59):

$$m = Fm + g, \tag{77}$$

and hence

$$m = m(\theta) = [I - F(\theta)]^{-1} g(\theta). \tag{78}$$

Then, from (59) and (77), we obtain:

$$y_t - m(\theta) = F(\theta)[y_{t-1} - m(\theta)] + \eta_t, \tag{79}$$

whose solution is:

$$y_t - m(\theta) = \sum_{j=0}^{\infty} [F(\theta)]^j \eta_{t-j}. \tag{80}$$

We note, incidentally, that (79) cannot be treated as a special case of the model (76), considered by Dunsmuir and Hannan, since $y_t - m(\theta)$ is not observable.

Since $E(\eta_s\eta_t') = 0$, $|s-t|>1$, we obtain, from (80):

$$\begin{aligned}E(y_s - m(\theta))(y_t - m(\theta))' = & \sum_{j=0}^{\infty} [F(\theta)]^j \Omega_0(\theta,\mu)[F'(\theta)]^{t-s+j} \\ & + \sum_{j=0}^{\infty} [F(\theta)]^j \Omega_1(\theta,\mu)[F'(\theta)]^{t-s+j+1} \\ & + \sum_{j=0}^{\infty} [F(\theta)]^{j+1} \Omega_1'(\theta,\mu)[F'(\theta)]^{t-s+j} \\ = & V_{st}(\theta,\mu), \qquad t \geq s. \end{aligned} \tag{81}$$

By using (60), (61) and the series expansion of $F(\theta) = e^{A(\theta)}$, the matrix $V_{st}(\theta,\mu)$ can be expressed as a power series in $A(\theta)$ and $A'(\theta)$ with each term involving $\Sigma(\mu)$. Then the maximum likelihood estimator $(\hat{\theta},\hat{\mu})$ is obtained by maximising:

$$\begin{aligned}L(\theta,\mu) = \log|V(\theta,\mu)| + & \left[(y_1 - m(\theta))',\ldots,(y_T - m(\theta))'\right][V(\theta,\mu)]^{-1} \\ & \times \begin{bmatrix} y_1 - m(\theta) \\ \vdots \\ y_T - m(\theta) \end{bmatrix}, \end{aligned} \tag{82}$$

where $V(\theta,\mu)$ is the $nT \times nT$ matrix whose (st)th $n \times n$ block is $V_{st}(\theta,\mu)$.

Because of the computational difficulty of maximising $L(\theta,\mu)$ it is useful to consider estimates based on approximate models, even if these are not consistent. A simple approximate model is obtained by replacing C by αI, which is the limit, as $\delta \to 0$, of the right-hand side of (73). In place of (75) we then have:

$$y_t - F(\theta)y_{t-1} - g(\theta) = \varepsilon_t + \alpha I \varepsilon_{t-1}. \tag{83}$$

Then we can define the vectors $y_1^{(1)},\ldots,y_T^{(1)}$ by the transformation:

$$\begin{bmatrix} y_1^{(1)} \\ y_2^{(1)} \\ \vdots \\ y_T^{(1)} \end{bmatrix} = \begin{bmatrix} I & 0 & \cdots & 0 \\ -\alpha I & I & \cdots & 0 \\ \vdots & \vdots & & \\ (-\alpha)^{T-1}I & (-\alpha)^{T-2}I & \cdots & I \end{bmatrix} \begin{bmatrix} y_1 \\ y_2 \\ \vdots \\ y_T \end{bmatrix} \tag{84}$$

and, assuming that $\varepsilon(0) = 0$, we obtain:

$$y_t^{(1)} = F(\theta)y_{t-1}^{(1)} + g(\theta) + \varepsilon_t, \tag{85}$$

which can be treated like (31). A procedure which is approximately equivalent to this was used by Bergstrom and Wymer (1976) who applied the transformation:

$$y_t^{(1)} = y_t - \alpha y_{t-1} + \alpha^2 y_{t-2} - \alpha^3 y_{t-3}. \tag{86}$$

We could also use the transformation (84) or (86) and then the approximate simultaneous equations model (38), which would be even simpler.

A method which can be expected to yield better estimates was studied by Phillips (1974b, 1978). His method is to obtain a preliminary estimate of A, ignoring the a priori restrictions on this matrix, from data transformed by (84) and then apply a second transformation:

$$\begin{bmatrix} y_1^{(2)} \\ y_2^{(2)} \\ \vdots \\ y_T^{(2)} \end{bmatrix} = \begin{bmatrix} I & 0 & \cdots & 0 \\ -C & I & \cdots & 0 \\ & & & \\ (-C)^{T-1} & (-C)^{T-2} & \cdots & I \end{bmatrix} \begin{bmatrix} y_1 \\ y_2 \\ \vdots \\ y_T \end{bmatrix}, \tag{87}$$

where C is computed from the first two terms on the right-hand side of (73), using the preliminary estimates of A. He shows that the proportional asymptotic bias in the estimates obtained in this way tends to zero as the observation period tends to zero.

Having dealt with the cases in which the data are all point observations or all integrals we can easily deal with the case in which the data are mixed, with some variables being observed at points of time and others as integrals. Suppose, for example, that the first m variables are stock variables and the remaining $n-m$ variables are flow variables, so that the observations are: $y_1(t),\dots,y_m(t)$, $\int_{t-1}^{t} y_{m+1}(r)\,\mathrm{d}r,\dots,\int_{t-1}^{t} y_n(r)\,\mathrm{d}r$, $t = 1,\dots,T$. Then we can solve the first m equations of the system (62) to express $\int_{t-1}^{t} y_1(r)\,\mathrm{d}r,\dots,$ $\int_{t-1}^{t} y_m(r)\,\mathrm{d}r$ in terms of $y_1(t) - y_1(t-1),\dots,y_m(t) - y_m(t-1)$, $\int_{t-1}^{t} y_{m+1}(r)\,\mathrm{d}r,\dots,\int_{t-1}^{t} y_n(r)\,\mathrm{d}r$, $\int_{t-1}^{t} \zeta_1(\mathrm{d}r),\dots,\int_{t-1}^{t} \zeta_n(\mathrm{d}r)$. Then substituting into (59), we obtain a system in which all variables are in the form in which they are observed and the disturbance vector is a vector moving average process whose autocorrelation properties can easily be obtained as in the proof of Theorem 8. The exact likelihood function can then be obtained in the same way as (82).

The feasibility of the methods discussed in this section has been demonstrated by Bergstrom and Wymer (1976) who applied them in the construction of a continuous time model of the United Kingdom. This model is a closed first order system of 13 non-linear differential equations with 35 parameters including three trend parameters to represent technical progress, the growth of the labour supply and growth in the demand for exports. For the purpose of estimation the model was represented by a system similar to (28), with the addition of trend terms, by taking a linear approximation about the sample means and adding white noise disturbances. The resulting matrix $[A(\theta), b(\theta)]$ implies quite complicated non-linear cross equation restrictions, derived from economic theory. The estimate of θ was obtained from quarterly data, for the years 1955–1966, by applying the method of full information maximum likelihood to a system similar to (85) (including derived trend terms) with the vector $y_t^{(1)}$ defined by (86).

An intensive mathematical study of the steady state and asymptotic stability properties of the original model (i.e. not the linear approximation used for estimation) shows that it generates plausible long-run behaviour for the estimated values of the parameters. Moreover, post sample predictions for the period 1969–1970 are remarkably accurate in view of the fact that the model contains no exogenous variables and the predictions are for a period up to eight quarters ahead of the latest data used in making them. But it should be possible to improve on this predictive performance by using a second or higher order system of differential equations. Such a system could represent more accurately the dynamics of the partial adjustment relations and allow a more satisfactory treatment of expectations.

3. Higher order systems[2]

We shall consider the system:

$$\begin{aligned} \mathrm{d}[\mathrm{D}^{k-1}y(t)] = {} & A_1(\theta)\mathrm{D}^{k-1}y(t)+\cdots+A_{k-1}(\theta)\mathrm{D}y(t) \\ & + A_k(\theta)y(t)+b(\theta)+\zeta(\mathrm{d}t), \end{aligned} \tag{88}$$

where $\{y(t)\}$ is a vector random process, $A_1(\theta),\dots,A_k(\theta)$ are $n\times n$ matrices whose elements are functions of the parameter vector θ, $b(\theta)$ is an $n\times 1$ vector whose elements are functions of θ and $\zeta(\mathrm{d}t)$ is a vector of white noise disturbances, i.e. a vector satisfying Assumption 1. The system (88) will be interpreted as meaning that $y(t)$ satisfies:

$$\begin{aligned} D^{k-1}y(t)-D^{k-1}y(0) = {} & \int_0^t \big[A_1(\theta)D^{k-1}y(r)+\cdots+A_{k-1}(\theta)Dy(r) \\ & + A_k(\theta)y(r)+b(\theta)\big]\,\mathrm{d}r + \int_0^t \zeta(\mathrm{d}r), \end{aligned} \tag{89}$$

for all t.

[2] For a more general and comprehensive treatment of higher order systems, see Bergstrom (1982).

The assumption that the disturbances are white noise is more easily justified in a higher order system than in a first order system. An econometric model comprising a system of stochastic differential equations is usually obtained by, heuristically, adding disturbance functions to a non-stochastic system of differential equations which may be derived from certain optimisation assumptions and would hold exactly under certain ideal conditions. These conditions might include, for example, the conditions that each agent's objective function remains constant and contains no variables other than those in the model, that his assumptions about the random processes generating these variables are constant and that the physical constraints on his behaviour are constant. The disturbances are added to take account of the fact that none of these things is really constant. Although it is difficult to justify the assumption that they are white noise disturbances, it is not unrealistic to assume that they are random processes generated by an unknown system of stochastic differential equations with white noise disturbances. Indeed the physical processes generating many of the non-economic variables that affect economic behaviour will be approximately of this form. We can, in this case, transform our original model into a higher order system of stochastic differential equations with white noise disturbances.

Suppose, for example, that the original model is a proper first order stochastic differential equation system:

$$\mathrm{D}y(t) = A(\theta)y(t) + \xi(t), \tag{90}$$

with a mean square continuous disturbance vector $\xi(t)$ generated by this system:

$$\mathrm{d}\xi(t) = Q\xi(t) + \zeta(\mathrm{d}t), \tag{91}$$

where Q is a $n \times n$ matrix of unknown constants and $\zeta(\mathrm{d}t)$ satisfies Assumption 1. The system (91) can be interpreted as meaning that

$$\xi(t) - \xi(0) - Q\int_0^t \xi(r)\,\mathrm{d}r = \int_0^t \zeta(\mathrm{d}r) \tag{92}$$

holds for all t. From (90) and (92) we obtain:

$$\mathrm{D}y(t) - \mathrm{D}y(0) - Q\int_0^t \mathrm{D}y(r)\,\mathrm{d}r = A(\theta)y(t) - A(\theta)y(0)$$
$$- QA(\theta)\int_0^t y(r)\,\mathrm{d}r + \int_0^t \zeta(\mathrm{d}r),$$

and hence

$$\mathrm{D}y(t)-\mathrm{D}y(0)=[Q+A(\theta)]\int_0^t \mathrm{D}y(r)\,\mathrm{d}r-QA(\theta)\int_0^t y(r)\,\mathrm{d}r + \int_0^t \zeta(\mathrm{d}r). \tag{93}$$

The system (93) is of the same form as (89) with $k=2$. It can be written as:

$$\mathrm{d}[\mathrm{D}y(t)]=[Q+A(\theta)]\,\mathrm{D}y(t)-QA(\theta)\,y(t)+\zeta(\mathrm{d}t), \tag{94}$$

which is of the same form as (88) and is interpreted as meaning that (93) holds for all t. Obviously Q and A will not be identifiable if A is unrestricted since, in this case, interchanging Q and A will not affect (93) and (94). But, in practice, A will be severely restricted by the requirement that its elements be known functions of θ.

Systems of stochastic differential equations of order $k>1$ are discussed by Wymer (1972) who, following Sargan (1974, 1976) (the main results of which were available in a preliminary mimeographed paper in 1970), considers the properties of an approximate simultaneous equations model when the observation period tends to zero. Here, for simplicity, we shall start by considering the second order system, which is likely to be of considerable practical importance. We shall consider estimates based on both the exact discrete model and the approximate simultaneous equations model and then indicate, briefly, how the results can be extended to systems of order greater than two.

The second-order system to be considered is:

$$\mathrm{d}[\mathrm{D}y(t)]=A_1(\theta)\,\mathrm{D}y(t)+A_2(\theta)\,y(t)+b(\theta)+\zeta(\mathrm{d}t), \tag{95}$$

which is interpreted as meaning that $y(t)$ satisfies:

$$\mathrm{D}y(t)-\mathrm{D}y(0)=\int_0^t [A_1(\theta)\,\mathrm{D}y(r)+A_2(\theta)\,y(r)+b(\theta)]\,\mathrm{d}r+\int_0^t \zeta(\mathrm{d}r), \tag{96}$$

for all t. We know, from Theorem 3, that the first-order system (29) has a solution which is unique (with probability 1) in the class of mean square continuous vector processes. It is natural, therefore, to seek a solution to (96) which is unique in the class of vector processes whose first derivatives $\{\mathrm{D}y(t)\}$ are mean square continuous. It is easy to see (from the definition of differentiation, integration and mean square continuity given in Section 2.1) that if $\{\mathrm{D}y(t)\}$ is mean square continuous, then

$$y(t)-y(0)=\int_0^t \mathrm{D}y(r)\,\mathrm{d}r. \tag{97}$$

Combining (96) and (97) we have:

$$\begin{bmatrix} \mathrm{D}y(t) \\ y(t) \end{bmatrix} - \begin{bmatrix} \mathrm{D}y(0) \\ y(0) \end{bmatrix} = \int_0^t \left\{ \begin{bmatrix} A_1 & A_2 \\ I & 0 \end{bmatrix} \begin{bmatrix} \mathrm{D}y(r) \\ y(r) \end{bmatrix} + \begin{bmatrix} b \\ 0 \end{bmatrix} \right\} \mathrm{d}r + \begin{bmatrix} \int_0^t \zeta(\mathrm{d}r) \\ 0 \end{bmatrix}, \tag{98}$$

which is a first-order system of the form (29) in the $2n \times 1$ vector:

$$\begin{bmatrix} \mathrm{D}y(t) \\ y(t) \end{bmatrix}.$$

By Theorem 3 the system (98) has a solution which, for any given pair of $n \times 1$ vectors $y(0)$ and $\mathrm{D}y(0)$, is unique in the class of random processes $\{y(t)\}$ such that $\{\mathrm{D}y(t)\}$ is mean square continuous (since if the process $\{\mathrm{D}y(t)\}$ exists the process $\{y(t)\}$ is, obviously, mean square continuous). And this solution satisfies the stochastic difference equation system:

$$\begin{bmatrix} \mathrm{D}y(t) \\ y(t) \end{bmatrix} = \begin{bmatrix} [\mathrm{e}^A]_{11} & [\mathrm{e}^A]_{12} \\ [\mathrm{e}^A]_{21} & [\mathrm{e}^A]_{22} \end{bmatrix} \begin{bmatrix} \mathrm{D}y(t-1) \\ y(t-1) \end{bmatrix} + \begin{bmatrix} g_1 \\ g_2 \end{bmatrix} + \int_{t-1}^t \begin{bmatrix} [\mathrm{e}^{(t-r)A}]_{11} & [\mathrm{e}^{(t-r)A}]_{12} \\ [\mathrm{e}^{(t-r)A}]_{21} & [\mathrm{e}^{(t-r)A}]_{22} \end{bmatrix} \begin{bmatrix} \zeta(\mathrm{d}r) \\ 0 \end{bmatrix}, \tag{99}$$

where

$$\begin{bmatrix} [\mathrm{e}^A]_{11} & [\mathrm{e}^A]_{12} \\ [\mathrm{e}^A]_{21} & [\mathrm{e}^A]_{22} \end{bmatrix} = \begin{bmatrix} I & 0 \\ 0 & I \end{bmatrix} + \begin{bmatrix} A_1 & A_2 \\ I & 0 \end{bmatrix} + \frac{1}{2}\begin{bmatrix} A_1 & A_2 \\ I & 0 \end{bmatrix}^2 + \frac{1}{3!}\begin{bmatrix} A_1 & A_2 \\ I & 0 \end{bmatrix}^3 + \cdots,$$

and

$$\begin{bmatrix} g_1 \\ g_2 \end{bmatrix} = \begin{bmatrix} [\mathrm{e}^A]_{11} - I & [\mathrm{e}^A]_{12} \\ [\mathrm{e}^A]_{21} & [\mathrm{e}^A]_{22} - I \end{bmatrix} \begin{bmatrix} A_1 & A_2 \\ I & 0 \end{bmatrix}^{-1} \begin{bmatrix} b \\ 0 \end{bmatrix}.$$

The exact discrete model (99) cannot be used as a basis for estimation since $\mathrm{D}y(t)$ is not observable, even at discrete intervals. This is the reason why the second order system (or any higher order system) cannot be treated as a trivial extension of the first order system as in the theory of ordinary linear differential equations. An exact discrete model which can be used for estimation purposes can be obtained by eliminating $Dy(t)$ from the system (99) to obtain a second order difference equation system in $y(t)$. The precise form and properties of this system are given in the following theorem.

Theorem 9

If $\zeta(dt)$ satisfies Assumption 1, then for any given pair of $n\times 1$ vectors $y(0)$ and $Dy(0)$ the system (96) has a solution which is unique in the class of random vector processes $\{y(t)\}$ such that the process $\{Dy(t)\}$ is mean square continuous, i.e. if $\hat{y}(t)$ is any other such solution then (21) holds for any interval $[0,T]$. This solution satisfies the stochastic difference equation:

$$\begin{aligned} &y(t)=F_1(\theta)y(t-1)+F_2(\theta)y(t-2)+g(\theta)+\eta_t, \\ &\mathrm{E}(\eta_t)=0, \qquad \mathrm{E}(\eta_t\eta_t')=\Omega_0, \qquad \mathrm{E}(\eta_t\eta_{t-1}')=\Omega_1, \\ &\mathrm{E}(\eta_s\eta_t')=0, \qquad |s-t|>1, \qquad s,t=1,2,\ldots, \end{aligned} \tag{100}$$

where

$$\begin{aligned} F_1&=[e^A]_{21}[e^A]_{11}[e^A]_{21}^{-1}+[e^A]_{22}, \\ F_2&=[e^A]_{21}[e^A]_{12}-[e^A]_{21}[e^A]_{11}[e^A]_{21}^{-1}[e^A]_{22}, \\ g&=[e^A]_{21}g_1+\{I-[e^A]_{21}[e^A]_{11}[e^A]_{21}^{-1}\}g_2, \\ \Omega_0&=\int_0^1\{[e^A]_{21}[e^{rA}]_{11}-[e^A]_{21}[e^A]_{11}[e^A]_{21}^{-1}[e^{rA}]_{21}\} \\ &\quad\times\Sigma\{[e^A]_{21}[e^{rA}]_{11}-[e^A]_{21}[e^A]_{11}[e^A]_{21}^{-1}[e^{rA}]_{21}\}'\,dr \\ &\quad+\int_0^1[e^{rA}]_{21}\Sigma[e^{rA}]_{21}'\,dr, \end{aligned} \tag{101}$$

$$\begin{aligned} \Omega_1&=\int_0^1\{[e^A]_{21}[e^{rA}]_{11}-[e^A]_{21}[e^A]_{11}[e^A]_{21}^{-1}[e^{rA}]_{21}\} \\ &\quad\times\Sigma[e^{rA}]_{21}'\,dr. \end{aligned} \tag{102}$$

Proof

The system (99) can be written:

$$Dy(t)=[e^A]_{11}Dy(t-1)+[e^A]_{12}y(t-1)+g_1+\int_{t-1}^t[e^{(t-r)A}]_{11}\zeta(dr), \tag{103}$$

$$y(t)=[e^A]_{21}Dy(t-1)+[e^A]_{22}y(t-1)+g_2+\int_{t-1}^t[e^{(t-r)A}]_{21}\zeta(dr). \tag{104}$$

From (103) we obtain:

$$\mathrm{D}y(t-1)=[\mathrm{e}^{A}]_{11}\mathrm{D}y(t-2)+[\mathrm{e}^{A}]_{12}y(t-2)+g_1 + \int_{t-2}^{t-1}[\mathrm{e}^{(t-1-r)A}]_{11}\zeta(\mathrm{d}r), \tag{105}$$

and from (104):

$$\mathrm{D}y(t-1)=[\mathrm{e}^{A}]_{21}^{-1}y(t)-[\mathrm{e}^{A}]_{21}^{-1}[\mathrm{e}^{A}]_{22}y(t-1)-[\mathrm{e}^{A}]_{21}^{-1}g_2 -[\mathrm{e}^{A}]_{21}^{-1}\int_{t-1}^{t}[\mathrm{e}^{(t-r)A}]_{21}\zeta(\mathrm{d}r), \tag{106}$$

$$\mathrm{D}y(t-2)=[\mathrm{e}^{A}]_{21}^{-1}y(t-1)-[\mathrm{e}^{A}]_{21}^{-1}[\mathrm{e}^{A}]_{22}y(t-2)-[\mathrm{e}^{A}]_{21}^{-1}g_2 -[\mathrm{e}^{A}]_{21}^{-1}\int_{t-2}^{t-1}[\mathrm{e}^{(t-1-r)A}]_{21}\zeta(\mathrm{d}r). \tag{107}$$

Substituting from (106) and (107), for $\mathrm{D}y(t-1)$ and $\mathrm{D}y(t-2)$, respectively, into (105) and premultiplying by $[\mathrm{e}^{A}]_{21}$, we obtain the system (100) with:

$$\eta_t=\int_{t-2}^{t-1}\{[\mathrm{e}^{A}]_{21}[\mathrm{e}^{(t-1-r)A}]_{11}-[\mathrm{e}^{A}]_{21}[\mathrm{e}^{A}]_{11}[\mathrm{e}^{A}]_{21}^{-1}[\mathrm{e}^{(t-1-r)A}]_{21}\}\zeta(\mathrm{d}r) +\int_{t-1}^{t}[\mathrm{e}^{(t-r)A}]_{21}\zeta(\mathrm{d}r). \tag{108}$$

Then, by using the generalizations of (15) and (16) for a vector random process, we obtain:

$$\begin{aligned}\mathrm{E}(\eta_t\eta_t')=&\int_{t-2}^{t-1}\{[\mathrm{e}^{A}]_{21}[\mathrm{e}^{(t-1-r)A}]_{11}-[\mathrm{e}^{A}]_{21}[\mathrm{e}^{A}]_{11}[\mathrm{e}^{A}]_{21}^{-1}[\mathrm{e}^{(t-1-r)A}]_{21}\}\\&\times\Sigma\{[\mathrm{e}^{A}]_{21}[\mathrm{e}^{(t-1-r)A}]_{11}-[\mathrm{e}^{A}]_{21}[\mathrm{e}^{A}]_{11}[\mathrm{e}^{A}]_{21}^{-1}[\mathrm{e}^{(t-1-r)A}]_{21}\}'\,\mathrm{d}r\\&+\int_{t-1}^{t}[\mathrm{e}^{(t-r)A}]_{21}\Sigma[\mathrm{e}^{(t-r)A}]_{21}'\,\mathrm{d}r\\=&\int_{0}^{1}\{[\mathrm{e}^{A}]_{21}[\mathrm{e}^{rA}]_{11}-[\mathrm{e}^{A}]_{21}[\mathrm{e}^{A}]_{11}[\mathrm{e}^{A}]_{21}^{-1}[\mathrm{e}^{rA}]_{21}\}\\&\times\Sigma\{[\mathrm{e}^{A}]_{21}[\mathrm{e}^{rA}]_{11}-[\mathrm{e}^{A}]_{21}[\mathrm{e}^{A}]_{11}[\mathrm{e}^{A}]_{21}^{-1}[\mathrm{e}^{rA}]_{21}\}'\,\mathrm{d}r\\&+\int_{0}^{1}[\mathrm{e}^{rA}]_{21}\Sigma[\mathrm{e}^{rA}]_{21}'\,\mathrm{d}r,\end{aligned}$$

$$\begin{aligned}
E(\eta_t\eta'_{t-1}) &= \int_{t-2}^{t-1}\{[e^A]_{21}[e^{(t-1-r)A}]_{11}-[e^A]_{21}[e^A]_{11}[e^A]_{21}^{-1}[e^{(t-1-r)A}]_{21}\}\\
&\quad\times\Sigma[e^{(t-1-r)A}]'\mathrm{d}r\\
&=\int_0^1\{[e^A]_{21}[e^{rA}]_{11}-[e^A]_{21}[e^A]_{11}[e^A]_{21}^{-1}[e^{rA}]_{21}\}\\
&\quad\times\Sigma[e^{rA}]'_{21}\mathrm{d}r,\\
E(\eta_s\eta'_t) &= 0,\qquad |s-t|>1,\qquad s,t=1,2,\dots. \quad\blacksquare
\end{aligned}$$

The expressions (101) and (102) can be written as infinite series by expressing the matrix exponential functions in the integrals in series form and integrating, with respect to r, term by term. Evaluating the terms not involving A_1 and A_2 we obtain:

$$\Omega_0=\tfrac{2}{3}\Sigma+\cdots, \tag{109}$$

$$\Omega_1=\tfrac{1}{6}\Sigma+\cdots. \tag{110}$$

It is interesting to note that the first terms on the right-hand sides of (109) and (110) are identical with the first terms on the right hand sides of (69) and (70), respectively, which were obtained for the first order system with flow data.

It is clear from Theorem 9 that, if $\Sigma=\Sigma(\mu)$, the exact discrete model corresponding to (96) is:

$$y(t)-F_1(\theta)y(t-1)-F_2(\theta)y(t-2)-g(\theta)=\varepsilon_t+C(\theta,\mu)\varepsilon_{t-1}, \tag{111}$$

$$E(\varepsilon_t)=0,\quad E(\varepsilon_t\varepsilon'_t)=K(\theta,\mu),\quad E(\varepsilon_s\varepsilon'_t)=0,\quad s\neq t,\quad s,t=1,2,\dots,$$

where $C(\theta,\mu)$ and $K(\theta,\mu)$ satisfy (67) and (68) with $\Omega_0(\theta,\mu)$ and $\Omega_1(\theta,\mu)$ given by (101) and (102), respectively. The exact likelihood function can be obtained in the same way as it was for the model (75). And, in view of the results of Dunsmuir and Hannan (1976) and Dunsmuir (1979), we can expect the maximum likelihood estimates to be strongly consistent, asymptotically normal and asymptotically efficient under fairly general assumptions.

The model (111) can, of course, be used only if we have point observations. An exact discrete model satisfied by the integral observations $y_t=\int_{t-1}^{t}y(r)\mathrm{d}r$, $t=1,2,\dots$, can be obtained by combining the arguments used in the proofs of Theorems 8 and 9. Since the derivation is straight forward, but somewhat tedious, we shall not set it out in detail. The first step is to derive the system which is related to (98) in the same way as (59) is related to (29) with η_t in (59) replaced by the expression on the right hand side of (65). This system is:

$$\int_{t-1}^{t}\begin{bmatrix} Dy(r) \\ y(r) \end{bmatrix}\mathrm{d}r = \begin{bmatrix} [e^A]_{11} & [e^A]_{12} \\ [e^A]_{21} & [e^A]_{22} \end{bmatrix}\int_{t-2}^{t-1}\begin{bmatrix} Dy(r) \\ y(r) \end{bmatrix}\mathrm{d}r + \begin{bmatrix} g_1 \\ g_2 \end{bmatrix}$$
$$+\int_{t-1}^{t}\begin{bmatrix} A_1 & A_2 \\ I & 0 \end{bmatrix}^{-1}\begin{bmatrix} [e^{(t-r)A}]_{11}-I & [e^{(t-r)A}]_{12} \\ [e^{(t-r)A}]_{21} & [e^{(t-r)A}]_{22}-I \end{bmatrix}$$
$$\times\begin{bmatrix} \zeta(\mathrm{d}r) \\ 0 \end{bmatrix}\mathrm{d}r$$
$$+\int_{t-2}^{t-1}\left\{\begin{bmatrix} A_1 & A_2 \\ I & 0 \end{bmatrix}^{-1}\begin{bmatrix} [e^A]_{11} & [e^A]_{12} \\ [e^A]_{21} & [e^A]_{22} \end{bmatrix}\right.$$
$$\left.-\begin{bmatrix} A_1 & A_2 \\ I & 0 \end{bmatrix}^{-1}\begin{bmatrix} [e^{(t-1-r)A}]_{11} & [e^{(t-1-r)A}]_{12} \\ [e^{(t-1-r)A}]_{21} & [e^{(t-1-r)A}]_{22} \end{bmatrix}\right\}$$
$$\times\begin{bmatrix} \zeta(\mathrm{d}r) \\ 0 \end{bmatrix}\mathrm{d}r. \qquad (112)$$

The exact discrete model, satisfied by the observations, is then obtained from (112) in the same way as (100) was obtained from (99).

Clearly, the disturbances in the exact discrete model, satisfied by the observations, will involve integrals with respect to $\zeta(\mathrm{d}r)$ over the intervals $(t-3, t-2)$, $(t-2, t-1)$ and $(t-1, t)$, so that, in place of (100), we have:

$$\begin{aligned} &y_t = F_1(\theta)y_{t-1} + F_2(\theta)y_{t-2} + g(\theta) + \eta_t, \\ &E(\eta_t) = 0, \quad E(\eta_t\eta_t') = \Omega_0, \quad E(\eta_t\eta_{t-1}') = \Omega_1, \quad E(\eta_t\eta_{t-2}') = \Omega_2, \\ &E(\eta_s\eta_t') = 0, \quad |s-t| > 2, \quad s, t = 1, 2, \ldots, \end{aligned} \qquad (113)$$

where $F_1(\theta)$, $F_2(\theta)$ and $g(\theta)$ are defined in the same way as for (100) and Ω_0, Ω_1 and Ω_2 are derived, from the rather complicated expression which we obtain for η_t, in the same way as Ω_0 and Ω_1 were derived for (100). In place of (111) we have an exact discrete model of the form:

$$y_t - F_1(\theta)y_{t-1} - F_2(\theta)y_{t-2} - g(\theta) = \varepsilon_t + C_1(\theta,\mu)\varepsilon_{t-1} + C_2(\theta,\mu)\varepsilon_{t-2}, \qquad (114)$$
$$E(\varepsilon_t) = 0, \quad E(\varepsilon_t\varepsilon_t') = K(\theta,\mu), \quad E(\varepsilon_s\varepsilon_t') = 0, \quad s \neq t, \quad s, t = 1, 2\ldots.$$

Again the exact likelihood function can be derived in the same way as it was for (75).

An approximate discrete model can be obtained by first approximating the system (98) by the system:

$$\begin{bmatrix} \Delta \mathrm{D}y(t) \\ \Delta y(t) \end{bmatrix} = \begin{bmatrix} A_1 & A_2 \\ I & 0 \end{bmatrix} \begin{bmatrix} M\mathrm{D}y(t) \\ My(t) \end{bmatrix} + \begin{bmatrix} b \\ 0 \end{bmatrix} + \begin{bmatrix} u_{1t} \\ 0 \end{bmatrix}, \tag{115}$$

where

$$\Delta y(t) = y(t) - y(t-1), \qquad My(t) = \tfrac{1}{2}[y(t) + y(t-1)].$$

This is of the same form as (38), but cannot be used as a basis for estimation since $\mathrm{D}y(t)$ is not observable. Eliminating $\mathrm{D}y(t)$ we obtain the approximate simultaneous equations model:

$$\Delta^2 y(t) = A_1 M\Delta y(t) + A_2 M^2 y(t) + b + u_t. \tag{116}$$

Wymer (1972) shows that the disturbance vector u_t is approximately a moving average process with coefficient matrix αI where α is a root of $z^2 - 4z + 1 = 0$. We could, therefore, obtain a simultaneous equations model with an approximately serially uncorrelated disturbance by applying the transformation (84).

All of the above results can be extended to systems of order greater than two. We start by considering the system:

$$\begin{bmatrix} \mathrm{D}^{k-1}y(t) \\ \mathrm{D}^{k-2}y(t) \\ \vdots \\ y(t) \end{bmatrix} - \begin{bmatrix} \mathrm{D}^{k-1}y(0) \\ \mathrm{D}^{k-2}y(0) \\ \vdots \\ y(0) \end{bmatrix}$$

$$= \int_0^t \left\{ \begin{bmatrix} A_1 & A_2 & \cdots & A_{k-1} & A_k \\ I & 0 & \cdots & 0 & 0 \\ \vdots & \vdots & & \vdots & \vdots \\ 0 & 0 & & I & 0 \end{bmatrix} \begin{bmatrix} \mathrm{D}^{k-1}y(r) \\ \mathrm{D}^{k-2}y(r) \\ \vdots \\ y(r) \end{bmatrix} + \begin{bmatrix} b \\ 0 \\ \vdots \\ 0 \end{bmatrix} \right\} \mathrm{d}r + \begin{bmatrix} \int_0^t \zeta(\mathrm{d}r) \\ 0 \\ \vdots \\ 0 \end{bmatrix}, \tag{117}$$

which is obtained from (89). From (117) we obtain, for point observations, the exact discrete model:

$$y(t) - F_1 y(t-1) - \cdots - F_k y(t-k) - g = \varepsilon_t + C_1\varepsilon_{t-1} + \cdots + C_{k-1}\varepsilon_{t-k+1}, \tag{118}$$

which is a generalization of (111), and, for integral observations, the exact discrete model:

$$y_t - F_1 y_{t-1} - \cdots - F_k y_{t-k} - g = \varepsilon_t + C_1\varepsilon_{t-1} + \cdots + C_k\varepsilon_{t-k}, \quad (119)$$

which is a generalization of (114). It is understood that the elements of the matrices $F_1,\ldots,F_k$ in (118) and (119) are functions of the parameter vector θ while the elements of $C_1,\ldots,C_{k-1}$ in (118) and $C_1,\ldots,C_k$ in (119) are functions of the extended parameter vector (θ,μ). The corresponding matrices F_i, $i=1,\ldots,k$, in (118) and (119) are identical, whereas the corresponding matrices C_i, $i=1,\ldots,k-1$, in (118) and (119) are different.

We can also obtain the approximate simultaneous equations model:

$$\Delta^k y(t) = A_1 M\Delta^{k-1}y(t) + A_2 M^2\Delta^{k-2}y(t) + \cdots + A_k M^k y(t) + b + u_t, \quad (120)$$

in which u_t is approximately a vector moving average process of order $k-1$ [see Wymer (1972)], and a similar model for integral observations, with a disturbance vector which is approximately a vector moving average process of order k.

4. The treatment of exogenous variables and more general models

We shall now extend the model (28) by including a vector $x(t)=[x_1(t),\ldots,x_m(t)]'$ of exogenous variables. The $x_i(t)$, $i=1,\ldots,m$, can be either integrable non-random functions or integrable random processes satisfying the condition that $\mathrm{E}[x_i(t)\zeta(\mathrm{d}\tau)]=0$ for all real t and all intervals $\mathrm{d}\tau$ on the real line. In place of (28) we have:

$$\mathrm{d}y(t) = [A(\theta)y(t) + B(\theta)x(t)]\,\mathrm{d}t + \zeta(\mathrm{d}t), \quad (121)$$

which is interpreted as meaning that:

$$y(t) - y(0) = \int_0^t [A(\theta)y(r) + B(\theta)x(r)]\,\mathrm{d}r + \int_0^t \zeta(\mathrm{d}r) \quad (122)$$

holds for all t. The elements of the $n\times m$ matrix $B(\theta)$ are functions of the basic parameter vector θ. There is no need to include a vector of constants since this can be allowed for by letting $x_m(t)=1$.

We are interested in the problem of estimating θ from a sample of discrete observations of the $n+m$ variables $y_1(t),\ldots,y_n(t), x_1(t),\ldots,x_m(t)$ when, in general, the observations of some of these variables are point observations while the remainder are integral observations. The simplest case, in principle, is where the vector $x(t)$ is itself generated by a stochastic differential equation system:

$$\mathrm{d}x(t) = Rx(t)\,\mathrm{d}t + \zeta_x(\mathrm{d}t), \tag{123}$$

where $\zeta_x(\mathrm{d}t)$ is a white noise disturbance vector. We can then treat the system:

$$\begin{bmatrix} y(t) \\ x(t) \end{bmatrix} - \begin{bmatrix} y(0) \\ x(0) \end{bmatrix} = \int_0^t \begin{bmatrix} A(\theta) & B(\theta) \\ 0 & R \end{bmatrix} \begin{bmatrix} y(r) \\ x(r) \end{bmatrix} \mathrm{d}r + \int_0^t \begin{bmatrix} \zeta(r) \\ \zeta_x(r) \end{bmatrix} \mathrm{d}r \tag{124}$$

as a case of (29) by replacing θ by an extended parameter vector comprising θ and the m^2 elements of R. The exact likelihood function for the parameters of this extended system can be obtained as in Section 2.3, the simplest case being where the observations of all the variables are point observations and the most complicated where the observations of some variables are point observations while the remainder are integral observations. The assumption that $x(t)$ is generated by a system such as (123) will often be a good approximation even if $x(t)$ is not, in fact, generated by such a system. An even better approximation might be obtained by assuming that $x(t)$ is generated by a higher order system, which can then be combined with (121) and treated by the methods of Section 3. But, clearly, this would involve heavy computational costs.

We turn now to less costly approximate methods. An obvious approximation is the simultaneous equations model:

$$y(t) - y(t-1) = \tfrac{1}{2}A(\theta)[y(t) + y(t-1)] + \tfrac{1}{2}B(\theta)[x(t) + x(t-1)] + u_t, \tag{125}$$

$$\mathrm{E}(u_t) = 0, \quad \mathrm{E}(u_t u_t') = \Sigma, \quad \mathrm{E}(u_s u_t') = 0, \quad s \neq t,$$

$$\mathrm{E}(x_s u_t') = 0, \quad s, t = 1, 2, \ldots,$$

which is a natural extension of (38). This model is approximate in the sense that, if u_t is defined in such a way that (125) holds exactly, then the conditions $\mathrm{E}(u_s u_t') = 0$, $s \neq t$, $\mathrm{E}(x_s u_t') = 0$, $s, t = 1, 2, \ldots$, will be only approximately satisfied. The use of the model (125) is particularly convenient when the only restrictions on A and B are that certain elements of these matrices are zero (or some other specified numbers) so that θ is a vector of the unknown elements of A and B. For this case, Sargan (1976) has made a thorough study of the behaviour of the two-stage least squares, three-stage least squares and full information maximum likelihood estimators as the observation period δ tends to zero [for which purpose the model can be reformulated like (47)]. He introduces three alternative assumptions about the exogenous variables. Assumption 7 is the simpler of his two alternative assumptions for the case of non-stochastic exogenous variables while Assumption 8 is his assumption for the case of stochastic exogenous variables.

Assumption 7

(i) $\mathrm{d}^2x/\mathrm{d}t^2$ exists and is bounded and continuous for all t, except a countable set of points S. There exists a time period p such that the number of points of S lying in the time interval $(s, s+p)$ is less than or equal to d for any s.

(ii) $\mathrm{d}x/\mathrm{d}t$ exists and is bounded for all t except points of S.

(iii) $x(t)$ is bounded for all t. The size of the discontinuity of $x(t)$ (which can occur only at points of S) is bounded for all points of S.

Assumption 8

(i) $x(t)$ is generated by a strictly stationary ergodic process with $\mathrm{E}[x(t)x'(t+r)] = \Omega_x(r)$, all t.

(ii) $\Omega_x(r)$ has one-sided derivatives at the origin up to the fourth order, so that a one-sided Taylor series expansion of $\Omega_x(r)$ at the origin, up to the fourth power of r, exists.

(iii) $\Omega_{xy}(r) = \mathrm{E}[x(t)y(t+r)']$ has positive one-sided first and second derivatives at the origin, so that a one-sided Taylor series expansion of $\Omega_{xy}(r)$ at the origin, up to the second power of r, exists.

Sargan's results imply that, under Assumptions 1,2,3 and either 7 or 8, the asymptotic bias of the two-stage least squares, three-stage least squares and full information maximum likelihood estimates are $O(\delta^2)$ as $\delta \to 0$. Moreover, under these assumptions, the difference between the limits in probability of the three-stage least squares and full information maximum likelihood estimates are $O(\delta^5)$ as $\delta \to 0$.

Assumption 8 will be satisfied if $x(t)$ is generated by the system (123). But it is easy to prove directly, by an extension of the argument used in the proof of Theorem 5, that, in this case, the asymptotic bias of the two-stage least squares estimator is $O(\delta^2)$ as $\delta \to 0$. We can show that if A and ε_t, as used in Theorem 5, are redefined for the extended model (124), then the elements of the first n rows of the matrix:

$$\operatorname*{plim}_{T\to\infty} \sum_{t=1}^{T} \left(\frac{1}{\delta} I - \frac{1}{2} A \right) \varepsilon_t x_t',$$

are $O(\delta^2)$ as $\delta \to 0$, where $x_t = x(t\delta)$. It is obvious from this that Theorem 5 holds for the n equations of the model (125) when $z_t^{(i)}$, $i = 1,\ldots,n$, are the vectors of instruments that yield two-stage least squares estimates for these equations. Moreover it would not be difficult to extend this argument to three-stage least squares estimators.

The discussion of the closed model, in Section 2, suggests that we should next consider estimators based on the exact discrete model:

$$y(t)=\mathrm{e}^{A(\theta)}y(t-1)+\int_{t-1}^{t}\mathrm{e}^{(t-r)A(\theta)}x(t-r)\,\mathrm{d}r+\int_{t-1}^{t}\mathrm{e}^{(t-r)A(\theta)}\zeta(\mathrm{d}r), \quad (126)$$

which can be obtained by a fairly obvious revision of Theorems 2 and 3 to include exogenous variables. Since we do not have a continuous record of $x(t)$, the model (126) cannot be used directly as a basis for estimation. But it was used by Phillips (1974a, 1976) in order to obtain a more complicated approximate discrete model than that studied by Sargan. This is obtained by replacing $x(t-r)$ in the first integral on the right hand side of (126) by:

$$\hat{x}(t-r)=x(t)+\frac{r}{2}[3x(t)-4x(t-1)+x(t-2)]$$
$$+\frac{r^2}{2}[x(t)-2x(t-1)+x(t-2)],$$

which is the quadratic function of r chosen so that $\hat{x}(t)=x(t)$, $\hat{x}(t-1)=x(t-1)$ and $\hat{x}(t-2)=x(t-2)$. Evaluating the integral we obtain the approximate discrete model:

$$y(t)=F(\theta)y(t-1)+G_1(\theta)x(t)+G_2(\theta)x(t-1)+G_3(\theta)x(t-2)+v_t, \quad (127)$$

$$\mathrm{E}(v_t)=0, \quad \mathrm{E}(v_tv_t')=\Omega, \quad \mathrm{E}(v_sv_t')=0, \quad s\neq t,$$
$$\mathrm{E}(x_sv_t')=0, \quad s,t=1,2,\dots,$$

where

$$F=\mathrm{e}^A,$$
$$G_1=\left[\left(\tfrac{1}{2}A^{-2}+A^{-3}\right)\mathrm{e}^A-A^{-1}-\tfrac{3}{2}A^{-2}-A^{-3}\right]B,$$
$$G_2=\left[\left(A^{-1}-2A^{-3}\right)\mathrm{e}^A+2A^{-2}+2A^{-3}\right]B,$$
$$G_3=\left[\left(-\tfrac{1}{2}A^{-2}+A^{-3}\right)\mathrm{e}^A-\tfrac{1}{2}A^{-2}-A^{-3}\right]B.$$

In the above derivation we have identified the time unit with the observation period as we would in practical applications. But, for the purpose of considering the behaviour of the estimators as the observation period tends to zero, Phillips follows Sargan's procedure of introducing a parameter δ to represent the observation period. If $\delta\neq 1$, we must replace A and B by δA and δB, respectively, in the expressions for F, G_1, G_2 and G_3.

Phillips considers, first, the estimator of θ obtained by applying full information maximum likelihood to the model (127) as if this were the true model, i.e. the vector θ that minimises:

$$\det\left[\frac{1}{T}\sum_{t=1}^{T}\{y(t)-F(\theta)y(t-1)-G_1(\theta)x(t) \\ -G_2(\theta)x(t-1)-G_3(\theta)x(t-2)\} \\ \{y(t)-F(\theta)y(t-1)-G_1(\theta)x(t) \\ -G_2(\theta)x(t-1)-G_3(\theta)x(t-2)\}'\right].$$

He shows, under certain assumptions, that this estimator has a limiting normal distribution as $T\to\infty$ and that the asymptotic bias is $O(\delta^2)$ as $\delta\to 0$. But the assumptions made for this purpose are stronger than Assumptions 7 or 8 and require the exogenous variables to follow a smoother time path, whether random or non-random, than the latter assumptions. In particular, they rule out the case of exogenous variables generated by the first order stochastic differential equation system (123), with white noise disturbances.

Phillips then considers the properties of an instrumental variables estimator in which the vector $[y'(t-2), x'(t), x'(t-2), x'(t-3)]$ is used as a vector of instruments. He shows that the asymptotic bias of this estimator is $O(\delta^3)$ under much weaker assumptions which do not exclude exogenous variables generated by the system (123). This method can be expected to give better estimates, therefore, than the use of the approximate simultaneous equations model (125).

We turn, finally, to a powerful method due to Robinson (1976a). This makes use of a discrete Fourier transformation of the data and is applicable to a very general model which includes, as special cases, systems of stochastic differential equations and mixed systems of stochastic differential and difference equations. Moreover, it does not assume that the disturbances are white noise. They are assumed to be strictly stationary ergodic processes with unknown correlation of functions. But the method is not applicable to a closed model.

The model considered by Robinson can be written:

$$y(t)=B\int_{-\infty}^{\infty}\Gamma(r,\theta)x(t-r)\mathrm{d}r+\xi(t), \tag{128}$$

where $y(t)$ is an $n\times 1$ vector of endogenous variables, $x(t)$ is an $m\times 1$ vector of exogenous variables, $\xi(t)$ is a disturbance vector, B is an $n\times l$ matrix of parameters which are subject to specified linear restrictions (e.g. certain elements of B could be specified as zero), $\Gamma(r,\theta)$ is an $l\times m$ matrix of generalized functions and θ is a $1\times p$ vector of parameters. An even more general model in which Γ is not required to belong to a finite dimensional space was considered by Sims (1971). (His investigation is confined to the single equation case, but the results could be extended to a system of equations.) In this case Γ is not identifiable from discrete data. Moreover, the results obtained by Sims show that it can be very misleading to approximate $\Gamma(r,\theta)$ by smoothing the lag distribution in the equivalent discrete model.

At this stage we shall give a few results, relating to the spectral representation of a stationary process, which are essential for an understanding of Robinson's method. It should be remarked, however, that Robinson (1976) is concerned with the use of Fourier methods in the estimation of the parameters of a particular model formulated in the time domain and not with the more general spectral analysis of time series. The latter is discussed in Chapter 17 of this Handbook.

A wide sense stationary random vector process $\{x(t)\}$ has [Rozanov (1967, Theorem 4.2] the Cramér representation:

$$x(t)=\int_{-\infty}^{\infty}\mathrm{e}^{\mathrm{i}\lambda t}\Phi_x(\mathrm{d}\lambda), \tag{129}$$

where Φ_x is a complex valued random measure of the general type discussed in Section 2.2 and integration with respect to a random measure is defined as in Section 2.2. The random measure Φ_x is called the *random spectral measure* of the process $\{x(t)\}$, and if $F_{xx}(\mathrm{d}\lambda)$ is defined by:

$$F_{xx}(\mathrm{d}\lambda)=\mathrm{E}\left[\Phi_x(\mathrm{d}\lambda)\Phi_x^*(\mathrm{d}\lambda)\right],$$

where $\Phi_x^*(\mathrm{d}\lambda)$ denotes the complex conjugate of the transpose of $\Phi_x(\mathrm{d}\lambda)$, we call $F_{xx}(\mathrm{d}\lambda)$ the *spectral measure* of $\{x(t)\}$. If $F_{xx}(\mathrm{d}\lambda)$ is a matrix of absolutely continuous set functions with the derivative matrix:

$$f_{xx}(\lambda)=\lim_{d\lambda\to 0}\frac{1}{\mathrm{d}\lambda}F_{xx}(\mathrm{d}\lambda),$$

then $f_{xx}(\lambda)$ is called the *spectral density* of $\{x(t)\}$. The random spectral measure $\Phi_x(\mathrm{d}\lambda)$ can be obtained from $\{x(t)\}$ by [Rozanov (1967, p. 27)] the inverse Fourier transformation:

$$\Phi_x(\Delta)=\lim_{T\to\infty}\frac{1}{2\pi}\int_{-T}^{T}\left\{\frac{\mathrm{e}^{-\mathrm{i}\lambda_2 t}-\mathrm{e}^{-\mathrm{i}\lambda_1 t}}{-\mathrm{i}t}\right\}x(t)\,\mathrm{d}t, \tag{130}$$

which holds for any interval $\Delta=(\lambda_1,\lambda_2)$ such that:

$$\Phi_x(\lambda_1)=\Phi_x(\lambda_2)=0.$$

Returning now to Robinson's model, let $\tilde{\Gamma}(\lambda,\theta)$ be defined by:

$$\tilde{\Gamma}(\lambda,\theta)=\int_{-\infty}^{\infty}\mathrm{e}^{\mathrm{i}\lambda t}\Gamma(t,\theta)\,\mathrm{d}t.$$

Then it can be shown [Rozanov (1967, p. 38)] that the random spectral measure of $\int_{-\infty}^{\infty}\Gamma(r,\theta)x(t-r)\mathrm{d}r$ is $\tilde{\Gamma}(-\lambda,\theta)\Phi_x(\mathrm{d}\lambda)$. By replacing each of the terms in (128) by its Cramér representation and applying the inverse Fourier transformation (130), we obtain, therefore:

$$\lim_{T\to\infty}\frac{1}{2\pi}\int_{-T}^{T}\left\{\frac{\mathrm{e}^{-\mathrm{i}\lambda_2 t}-\mathrm{e}^{-\mathrm{i}\lambda_1 t}}{-\mathrm{i}t}\right\}y(t)\,\mathrm{d}t$$
$$=B\tilde{\Gamma}(-\lambda,\theta)\lim_{T\to\infty}\frac{1}{2\pi}\int_{-T}^{T}\left\{\frac{\mathrm{e}^{-\mathrm{i}\lambda_2 t}-\mathrm{e}^{-\mathrm{i}\lambda_1 t}}{-\mathrm{i}t}\right\}x(t)\,\mathrm{d}t+\Phi_\xi(\Delta). \qquad (131)$$

The equation system (131) holds exactly. Moreover, if $\Delta_1,\dots,\Delta_n$ are disjoint intervals the disturbance terms $\Phi_\xi(\Delta_1),\dots,\Phi_\xi(\Delta_n)$ are uncorrelated, although they are not homoscedastic when the intervals $\Delta_1,\dots,\Delta_n$ are of equal length. But we cannot estimate (131) directly since we cannot observe the integrals.

In order to derive an approximate model we first note that:

$$\frac{\mathrm{e}^{-\mathrm{i}\lambda_2 t}-\mathrm{e}^{-\mathrm{i}\lambda_1 t}}{-\mathrm{i}t}=\mathrm{e}^{-\mathrm{i}\lambda_1 t}(\lambda_2-\lambda_1)+O(\lambda_2-\lambda_1)^2. \qquad (132)$$

If we now divide the interval $(-\pi,\pi)$ into N subintervals $\Delta_1,\dots,\Delta_n$, each of length $2\pi/N$, and use (132), we obtain from (131) the approximate system:

$$\lim_{T\to\infty}\frac{1}{N}\int_{-T}^{T}\mathrm{e}^{-\mathrm{i}\lambda_s t}y(t)\,\mathrm{d}t=B\tilde{\Gamma}(-\lambda,\theta)\lim_{T\to\infty}\frac{1}{N}\int_{-T}^{T}\mathrm{e}^{-\mathrm{i}\lambda_s t}x(t)\,\mathrm{d}t+\Phi_\xi(\Delta_s),$$
$$\lambda_s=2\pi s/N,\qquad -\tfrac{1}{2}N<s\le\tfrac{1}{2}N. \qquad (133)$$

If we normalize (133) by dividing by $(2\pi/N)^{1/2}$ we obtain a system with a disturbance vector $(N/2\pi)^{1/2}\Phi_\xi(\Delta_s)$ whose covariance matrix is approximately the spectral density of $\xi(t)$ at the frequency λ_s. If we then conjugate and replace the integrals by discrete Fourier transforms of the observations we obtain:

$$w_y(s)=B\tilde{\Gamma}(\lambda_s,\theta)w_x(s)+w_\xi(s), \qquad (134)$$

where

$$w_x(s)=(2\pi N)^{-1/2}\sum_{n=1}^{N}x(n)\mathrm{e}^{\mathrm{i}n\lambda_s}.$$

The model (134) is the approximation used by Robinson for estimation purposes. Estimation is carried out in two stages. We first minimise the sums of squares of the errors which are then used to compute estimates of the spectral density of $\xi(t)$ in the frequency bands corresponding to various values of s. These estimates are then used in the construction of a Hermitian form in the errors in (134) which is minimised with respect to B and θ, subject to the restrictions, in order to obtain estimates of the parameters. In another article Robinson (1976b) considers the application of this general method specifically to a system of stochastic differential equations. The differential equations model is also treated by an instrumental variables method in Robinson (1976c). These two articles contain interesting Monte Carlo studies of the results of the application of the two methods.

Robinson (1976a) shows, under certain assumptions, that the estimation procedure described above, for the model (128) using the approximate discrete model (134), yields estimates which are strongly consistent, asymptotically normal and asymptotically efficient. The most restrictive of his assumptions is that the spectral density of $x(t)$ is zero outside the frequency range $(-\pi, \pi)$. This assumption is necessary when estimating the parameters of such a general model from equispaced discrete observations because of aliasing, to which we have already referred in Section 2.3. The assumption would not be satisfied if, for example, $x(t)$ were generated by the stochastic differential equation system (123), with white noise disturbances. But in this case, we can always extend the system, as we have seen, so that it can be treated as a closed model. And, if necessary, we can transform the model into a higher order system so that the assumption that the disturbances are white noise is approximately satisfied.

5. Conclusion

In this chapter we have described statistical methods which are applicable to a class of continuous time stochastic models and discussed the theoretical foundations of these methods. An important feature of the class of models considered is that such models allow for the incorporation of a priori restrictions, such as those derived from economic theory, through the structural parameters of the continuous time system. They can be used, therefore, to represent a dynamic system of causal relations in which each variable is adjusting continuously in response to the stimulus provided by other variables and the adjustment relations involve the basic structural parameters in some optimization theory. These structural parameters can be estimated from a sample comprising a sequence of discrete observations of the variables which will, generally, be a mixture of stock variables (observable at points of time) and flow variables (observable as integrals). In this way it is possible to take advantage of the a priori restrictions derived from economic theory (which are very important in econometric work, because of the smallness of the samples) without making the unrealistic assumption that the economy moves in discrete jumps between successive positions of temporary equilibrium.

The feasibility of constructing a continuous adjustment model of an economy, using the methods described in this chapter, was demonstrated by Bergstrom and Wymer (1976) to whose work we have referred in Section 2.3. The methods are now being widely used, and the Bergstrom–Wymer model has been used as a prototype for a larger econometric model of the United Kingdom [see Knight and Wymer (1978)] as well as for models of various other countries [see, for example, Jonson, Moses and Wymer (1977)]. There have also been some applications of the models, not only for forecasting, but also for the investigation of the effects of various types of policy feed-back [see, for example, Bergstrom (1978, 1984)]. And, in addition to these macroeconomic applications there have been applications to commodity and financial markets [see, for example, Richard (1978) and Wymer (1973)]. The results of these various studies, which are concerned with models formulated, mainly, as first order systems of stochastic differential equations, are very encouraging. They suggest that further empirical work with higher order systems of differential equations or more general continuous time models is a promising field of econometric research.

On the theoretical side an important and relatively unexplored field of research is in the development of methods of estimation for systems of non-linear stochastic differential equations. So far these have been treated by replacing the original model by an approximate system of linear stochastic differential equations, which is treated as if it were the true model for the purpose of deriving the "exact discrete model" or, alternatively, making a direct approximation to the non-linear system of differential equations with a non-linear simultaneous equations model. In some cases it may be possible to derive the exact likelihood function in terms of the discrete observations generated by a system of non-linear stochastic differential equations. But, more generally, we shall have to rely on approximate methods, possibly involving the use of numerical solutions to the non-linear differential equations system.

References

Anderson, T. W. (1959) "On Asymptotic Distributions of Estimates of Parameters of Stochastic Difference Equations", *Annals of Mathematical Statistics*, 30, 676–687.

Bergstrom, A. R. (1966) "Non-recursive Models as Discrete Approximations to Systems of Stochastic Differential Equations", *Econometrica*, 34, 173–182.

Bergstrom, A. R. (1967) *The Construction and Use of Economic Models*. London: English Universities Press; also published as: *Selected Economic Models and Their Analysis*. New York: American Elsevier.

Bergstrom, A. R. (1976) ed., *Statistical Inference in Continuous Time Economic Models*. Amsterdam: North-Holland.

Bergstrom, A. R. (1978) "Monetary Policy in a Model of the United Kingdom" in: Bergstrom, A. R., A. J. L. Catt, M. H. Peston, and B. D. J. Silverstone, eds., *Stability and Inflation*. New York: Wiley, 89–102.
Bergstrom, A. R. (1983) "Gaussian Estimation of Structural Parameters in Higher Order Continuous Time Dynamic Models", *Econometrica*, 51, 117–152.
Bergstrom, A.R. (1984) "Monetary, Fiscal and Exchange Rate Policy in a Continuous Time Econometric Model of the United Kingdom", in: Malgrange, P. and P. Muet, eds., *Contemporary Macroeconomic Modelling*. Oxford: Blackwell.
Bergstrom, A. R. and C. R. Wymer, (1976) "A Model of Disequilibrium Neoclassical Growth and Its Application to the United Kingdom", in: A. R. Bergstrom, ed., *Statistical Inference in Continuous Time Economic Models*. Amsterdam: North-Holland, 267–327.
Box, G. E. P. and G. M. Jenkins, (1971) *Time Series Analysis: Forecasting and Control*. Oakland: Holden-Day.
Cramér, H. (1946) *Mathematical Methods of Statistics*. Princeton University Press.
Doob, J. L. (1953) *Stochastic Processes*. New York: Wiley.
Dunsmuir, W. (1979) "A Central Limit Theorem for Parameter Estimation in Stationary Vector Time Series and its Application to Models for a Signal Observed with Noise", *Annals of Statistics* 7, 490–506.
Dunsmuir, W. and E. J. Hannan (1976) "Vector Linear Time Series Models", *Advances in Applied Probability*, 339–364.
Durbin, J. (1961) Efficient Fitting of Linear Models for Continuous Stationary Time Series from Discrete Data, Bulletin of the International Statistical Institute 38, 273–282.
Haavelmo, T. (1943) "The Statistical Implications of a System of Simultaneous Equations", *Econometrica* 11, 1–12.
Halmos, Paul R. (1958) *Finite-Dimensional Vector Spaces*. Princeton: Van Nostrand.
Hannan, E. J. (1970) *Multiple Time Series*. New York: Wiley.
Hansen, L. P. and T. J. Sargent (1981) "Identification of Continuous Time Rational Expectations Models from Discrete Data", unpublished manuscript.
Hansen, L. P. and T. J. Sargent (1983) "The Dimensionality of the Aliasing Problem in Models with Rational Spectral Densities", *Econometrica* 51, 377–388.
Ito, Kiyosi (1946) "On a Stochastic Integral Equation", *Proceedings of the Japanese Academy* 1, 32–35.
Ito, Kiyosi (1951) "On Stochastic Differential Equations", *Memoir of the American Mathematical Society* 4, 51pp.
Jonson, P. D., E. R. Moses, and C. R. Wymer (1977) "The RBA76 Model of the Australian Economy", in: Conference in Applied Economic Research (Reserve Bank of Australia).
Knight, Malcolm D. and Clifford R. Wymer (1978) "A Macroeconomic Model of the United Kingdom", IMF Staff Papers 25, 742–778.
Kohn, R. (1979) "Asymptotic Estimation and Hypothesis Testing Results for Vector Linear Time Series Models", *Econometrica* 47, 1005–1030.
Kolmogorov, A. N. and S. V. Fomin (1961) *Elements of the Theory of Functions and Functional Analysis*. Albany, New York: Graylock Press.
Koopmans, T. C. (1950a) ed., *Statistical Inference in Dynamic Economic Models*. New York: Wiley.
Koopmans, T. C. (1950b) "Models Involving a Continuous Time Variable", in: Koopmans, T. C., ed., *Statistical Inference in Dynamic Economic Models*, New York: Wiley, 384–392.
Malinvaud, E. (1970) *Statistical Methods of Econometrics*. Amsterdam: North-Holland.
Mann, H. B. and A. Wald (1943) "On the Statistical Treatment of Linear Stochastic Difference Equations", *Econometrica* 11, 173–220.
Phillips, A. W. (1954) "Stabilization Policy in a Closed Economy", *Economic Journal* 64, 283–299.
Phillips, A. W. (1959) "The Estimation of Parameters in Systems of Stochastic Differential Equations", *Biometrika* 46, 67–76.
Phillips, P. C. B. (1972) "The Structural Estimation of a Stochastic Differential Equation System", *Econometrica*, 40, 1021–1041.
Phillips, P. C. B. (1973) "The Problem of Identification in Finite Parameter Continuous Time Models", *Journal of Econometrics* 1, 351–362.
Phillips, P. C. B. (1974a) "The Estimation of Some Continuous Time Models", *Econometrica* 42, 803–824.

Phillips, P. C. B. (1974b) "The Treatment of Flow Data in the Estimation of Continuous Time Systems", presented at Econometric Society Meeting Grenoble and available as Discussion Paper, University of Essex.

Phillips, P. C. B. (1976) "The Estimation of Linear Stochastic Differential Equations with Exogenous Variables", in: Bergstrom, A. R., ed., *Statistical Inference in Continuous Time Economic Models* (North-Holland, Amsterdam) 135–173.

Phillips, P. C. B. (1978) "The Treatment of Flow Data in the Estimation of Continuous Time Systems", in: Bergstrom, A. R., A. J. L., Catt, M. H. Peston, and B. D. J. Silverstone, eds., *Stability and Inflation*. New York: Wiley.

Richard, D. M. (1978) "A Dynamic Model of the World Copper Industry", IMF Staff Papers 25, 779–833.

Robinson, P. M. (1976a) "Fourier Estimation of Continuous Time Models", in: Bergstrom, A. R., ed., *Statistical Inference in Continuous Time Economic Models*. Amsterdam: North-Holland, 215–266.

Robinson, P. M. (1976b) "The Estimation of Linear Differential Equations with Constant Coefficients", *Econometrica* 44, 751–764.

Robinson, P. M. (1976c) "Instrumental Variables Estimation of Differential Equations", *Econometrica* 44, 765–776.

Rozanov, Y. A. (1967) *Stationary Random Processes*. San Francisco: Holden-Day.

Sargan, J. D. (1974) "Some Discrete Approximations to Continuous Time Stochastic Models", *Journal of the Royal Statistical Society*, Series B, 36, 74–90.

Sargan, J. D. (1976) "Some Discrete Approximations to Continuous Time Stochastic Models", in: Bergstrom, A. R., ed., *Statistical Inference in Continuous Time Economic Models*. Amsterdam: North-Holland, 27–80.

Sims, C. A. (1971) "Discrete Approximations to Continuous Time Distributed Lag Models in Econometrics", *Econometrica* 39, 545–563.

Wold, H. O. A. (1952) *Demand Analysis*. Stockholm: Almqvist and Wicksell.

Wold, H. O. A. (1954) "Causality and Econometrics", *Econometrica* 22, 162–177.

Wold, H. O. A. (1956) "Causal Inference from Observational Data. A Review of Ends and Means", *Journal of the Royal Statistical Society*, Series A, 119, 28–50.

Wymer, C. R. (1972) "Econometric Estimation of Stochastic Differential Equation Systems", *Econometrica* 40, 565–577.

Wymer, C. R. (1973) "A Continuous Disequilibrium Adjustment Model of the United Kingdom Financial Market", in: Powell, A. A. and R. A. Williams, eds., *Econometric Studies of Macro and Monetary Relations* Amsterdam: North-Holland, 301–334.

Wymer, C. R. (1974) "Computer Programs: Discon Manual, and: Continuous Systems Manual", mimeo.

Chapter 21

RANDOM AND CHANGING COEFFICIENT MODELS

GREGORY C. CHOW*

Princeton University

Contents

*I would like to thank Zvi Griliches, Andrew Harvey, Michael Intriligator and Adrian Pagan for providing helpful comments on an earlier draft and acknowledge financial support from the National Science Foundation in the preparation of this chapter.

Handbook of Econometrics, Volume II, Edited by Z. Griliches and M.D. Intriligator

1. Introduction

The standard linear regression model has been a very attractive model to use in econometrics. If econometricians can uncover stable economic relations which satisfy at least approximately the assumptions of this model, they deserve the credit and the convenience of using it. Sometimes, however, econometricians are not lucky or ingenious enough to specify a stable regression relationship, and the relationship being studied is gradually changing. Under such circumstances, an option is to specify a linear regression model with stochastically evolving coefficients. For the purpose of parameter estimation, this model takes into account the possibility that the coefficients may be time-dependent and provides estimates of these coefficients at different points of time. For the purpose of forecasting, this model may have an advantage over the standard regression model in utilizing the estimates of the most up-to-date coefficients. From the viewpoint of hypothesis testing, this model serves as a viable alternative to the standard regression model for the purpose of checking the constancy of the coefficients of the latter model.

The basic linear regression model with a changing coefficient vector β_t is represented by:

$$y_t = x_t\beta_t + \varepsilon_t \qquad (t=1,\dots,T) \tag{1.1}$$

and

$$\beta_t = M\beta_{t-1} + \eta_t \qquad (t=1,\dots,T), \tag{1.2}$$

where x_t is a row vector of k fixed explanatory variables, ε_t is normally and independently distributed with mean 0 and variance s^2, and η_t is k-variate normal and independent with mean zero and covariance matrix $s^2P \equiv V$. When $V=0$ and $M=I$, this model is reduced to the standard normal regression model. We will be concerned with the estimation and statistical testing of β_t $(t=1,\dots,T)$, s^2, V and M using observations on (y_t, x_t). We are restricting our discussion to the case of fixed x_t. If x_t were to include lagged dependent variables, the log-likelihood function given at the beginning of Section 4 would no longer be valid since the individual terms would no longer be normal and serially uncorrelated.

Assuming tentatively that s^2, V and M are known, one may consider the problem of estimating β_t using information I_s up to time s. Denote by $\mathrm{E}(\beta_t|I_s) \equiv \beta_{t|s}$ the conditional expectation of β_t given I_s. The evaluation of $\beta_{t|t}$ is known as filtering. The evaluation of $\beta_{t|s}(s>t)$ is called smoothing, and the evaluation of $\beta_{t|s}(s<t)$ is called prediction. In Section 2 we will derive the filtered and smoothed estimates of β_t recursively for $t=1,2,\dots,$ by the use of a regression of $\beta_1,\dots,\beta_t$ on $y_1,\dots,y_s$. The basic results are due to Kalman (1960). Section 3 contains an alternative derivation of the same results using the method of

Aitken's generalized least squares applied to a regression of $y_1,\dots,y_s$ on $x_1,\dots,x_s$ with β_t as the regression coefficient. This exposition is due to Sant (1977). We will then study the problem of estimating s^2, V and M by the method of maximum likelihood in Section 4.

In Section 5 we consider a system of linear regressions with changing coefficients. In Sections 6 and 7, respectively, we treat a system of linear and non-linear simultaneous stochastic equations with changing parameters. Finally, in Section 8, we modify (1.2) by introducing a mean vector $\bar{\beta}$, thus replacing β_t and β_{t-1} in (1.2) by $\beta_t-\bar{\beta}$ and $\beta_{t-1}-\bar{\beta}$ respectively and assuming the characteristic roots of M to be smaller than one in absolute value. When $M=0$, a random-coefficient regression model results. Section 9 states some conditions for the identification of the parameters.

Besides estimation, an important problem is hypothesis testing especially using the null hypothesis $V=0$. Testing this null hypothesis is equivalent to testing the stability of a set of regression coefficients through time, with the model (1.1)–(1.2) serving as the alternative hypothesis. This topic is treated in Section 10. Section 11 concludes this survey by suggesting some problems for further research.

2. Derivation of $\beta_{t|s}$ by recursive regression of β_t on $y_1,\dots,y_s$

Consider the regression of β_t on y_t, conditioned on $y_1,\dots,y_{t-1}$. Denote $(y_1,\dots,y_t)$ by Y_t. The regression of interest is by definition:

$$\mathrm{E}(\beta_t|y_t,Y_{t-1})=\mathrm{E}(\beta_t|Y_{t-1})+K_t[y_t-\mathrm{E}(y_t|Y_{t-1})]. \tag{2.1}$$

This regression is linear because β_t and Y_t are jointly normal as a consequence of the normality of ε_t and η_t in the model (1.1)–(1.2). Taking expectation of y_t from (1.1) conditioned on Y_{t-1}, we have $y_{t|t-1}\equiv\mathrm{E}(y_t|Y_{t-1})=x_t\beta_{t|t-1}$. Equation (2.1) can be written as:

$$\beta_{t|t}=\beta_{t|t-1}+K_t[y_t-x_t\beta_{t|t-1}]. \tag{2.2}$$

K_t is a column vector or regression coefficients, originally derived by Kalman (1960). If this vector is known, we can use (2.2) to update our estimate $\beta_{t|t-1}$ to form $\beta_{t|t}$.

To derive K_t we apply the well-known formula for a vector of regression coefficients:

$$K_t=\left[\mathrm{E}(\beta_t-\beta_{t|t-1})(y_t-y_{t|t-1})'\right][\mathrm{cov}(y_t|Y_{t-1})]^{-1}. \tag{2.3}$$

Denoting the covariance matrix $\mathrm{cov}(\beta_t|Y_{t-1})$ by $\Sigma_{t|t-1}$ and using

$$y_t-y_{t|t-1}=x_t(\beta_t-\beta_{t|t-1})+\varepsilon_t,$$

we can write (2.3) as:

$$K_t = \Sigma_{t|t-1}x_t'\left[x_t\Sigma_{t|t-1}x_t' + \sigma^2\right]^{-1}. \tag{2.4}$$

$\Sigma_{t|t-1}$ can be computed recursively as follows. First, by evaluating the covariance matrix of each side of (1.2) conditioned on Y_{t-1}, we obtain:

$$\Sigma_{t|t-1} = M\Sigma_{t-1|t-1}M' + V. \tag{2.5}$$

Second, using (2.2) and (1.1) we write:

$$\beta_t - \beta_{t|t} = \beta_t - \beta_{t|t-1} - K_t\left[x_t(\beta_t - \beta_{t|t-1}) + \varepsilon_t\right]. \tag{2.6}$$

Taking the expectation of the product of (2.6) and its transpose and using (2.4), we obtain:

$$\begin{aligned}\Sigma_{t|t} &= \Sigma_{t|t-1} - K_t\left[x_t\Sigma_{t|t-1}x_t' + \sigma^2\right]K_t' \\ &= \Sigma_{t|t-1} - \Sigma_{t|t-1}x_t'\left[x_t\Sigma_{t|t-1}x_t' + \sigma^2\right]^{-1}x_t\Sigma_{t|t-1}.\end{aligned} \tag{2.7}$$

Equations (2.5) and (2.7) can be used to compute $\Sigma_{t|t}$ $(t = 1,2,\dots)$ successively given $\Sigma_{0|0}$, without using the observations y_t $(t = 1,2,\dots)$. Having computed $\Sigma_{t|t-1}$, we can use (2.4) to compute K_t and (2.2) to compute $\beta_{t|t}$ where, on account of (1.2):

$$\beta_{t|t-1} = M\beta_{t-1|t-1}. \tag{2.8}$$

Thus, $\beta_{t|t}$ can be computed from $\beta_{t-1|t-1}$ using (2.8) and (2.2). The estimates $\beta_{t|t}$ so obtained are known as estimates by the Kalman filter.

Although we have employed classical regression theory in deriving the Kalman filter, one should note that it can be derived by Bayesian methods. Given the prior density of β_{t-1} to be normal with mean $\beta_{t-1|t-1}$ and covariance matrix $\Sigma_{t-1|t-1}$, a prior density of β_t is found using (1.2), which has mean $\beta_{t|t-1} = M\beta_{t-1|t-1}$ and covariance matrix $\Sigma_{t|t-1} = M\Sigma_{t-1|t-1}M' + V$. The posterior density of β_t given y_t is normal, with mean $\beta_{t|t}$ and covariance matrix $\Sigma_{t|t}$ as given by the Kalman filter. See Ho and Lee (1964).

In order to utilize future observations $y_{t+1}, y_{t+2},\dots,y_{t+n}$ for the estimation of β_t, we first consider the regression of β_t on y_{t+1}, conditioned on Y_t. Analogous to (2.2) and (2.3) are:

$$\beta_{t|t+1} = \beta_{t|t} + D_{t|t+1}(y_{t+1} - y_{t+1|t}) \tag{2.9}$$

and

$$D_{t|t+1} = \left[\mathrm{E}(\beta_t - \beta_{t|t})(y_{t+1} - y_{t+1|t})'\right]\left[\mathrm{cov}(y_{t+1}|Y_t)\right]^{-1}. \tag{2.10}$$

Using (1.1) and (1.2), we write:

$$y_{t+1}-y_{t+1|t}=x_{t+1}\beta_{t+1}+\varepsilon_{t+1}-x_{t+1}\beta_{t+1|t}$$
$$=x_{t+1}M\beta_t+x_{t+1}\eta_{t+1}+\varepsilon_{t+1}-x_{t+1}M\beta_{t|t},$$

which, in conjunction with (2.10) implies:

$$D_{t|t+1}=\Sigma_{t|t}M'x'_{t+1}\left[x_{t+1}\Sigma_{t+1|t}x'_{t+1}+\sigma^2\right]^{-1}$$
$$=\Sigma_{t|t}M'\Sigma^{-1}_{t+1|t}K_{t+1} \tag{2.11}$$

Equations (2.9) and (2.11) can be used to evaluate $\beta_{t|t+1}$. With the aid of (2.11) and (2.2), (2.9) can be rewritten as:

$$\beta_{t|t+1}=\beta_{t|t}+\Sigma_{t|t}M'\Sigma^{-1}_{t+1|t}\left(\beta_{t+1|t+1}-\beta_{t+1|t}\right). \tag{2.12}$$

The smoothing formula (2.12) will be generalized to

$$\beta_{t|t+n}=\beta_{t|t+n-1}+H_t\left(\beta_{t+1|t+n}-\beta_{t+1|t+n-1}\right), \tag{2.13}$$

where $H_t\equiv\Sigma_{t|t}M'\Sigma^{-1}_{t+1|t}$. We will prove (2.13) by induction. Equation (2.13) holds for $n=1$. We now assume (2.13) to hold for $n-1$, which implies:

$$\beta_{t|t+n-1}=\beta_{t|t+n-2}+H_t\left(\beta_{t+1|t+n-1}-\beta_{t+1|t+n-2}\right)$$
$$=\beta_{t|t+n-2}+H_tH_{t+1}\left(\beta_{t+2|t+n-1}-\beta_{t+2|t+n-2}\right)$$
$$=\beta_{t|t+n-2}+H_tH_{t+1}\dots H_{t+n-2}\left(\beta_{t+n-1|t+n-1}-\beta_{t+n-1|t+n-2}\right)$$
$$=\beta_{t|t+n-2}+H_tH_{t+1}\dots H_{t+n-2}K_{t+n-1}\left(y_{t+n-1}-y_{t+n-1|t+n-2}\right). \tag{2.14}$$

Consider the regression of β_t on y_{t+n-1}, conditioned on Y_{t+n-2}. Analogous to (2.9) an (2.10) are:

$$\beta_{t|t+n-1}=\beta_{t|t+n-2}+D_{t|t+n-1}\left(y_{t+n-1}-y_{t+n-1|t+n-2}\right) \tag{2.15}$$

and

$$D_{t|t+n-1}=\left[\mathrm{E}\left(\beta_t-\beta_{t|t+n-2}\right)\left(y_{t+n-1}-y_{t+n-1|t+n-2}\right)'\right]$$
$$\left[\operatorname{cov}\left(y_{t+n-1}\,|\,Y_{t+n-2}\right)\right]^{-1}=H_tH_{t+1}\dots H_{t+n-2}K_{t+n-1}, \tag{2.16}$$

where the last equality sign results from comparing (2.14) and (2.15). Equation

(2.16) implies:

$$\begin{aligned}
&\mathrm{E}(\beta_t-\beta_{t|t+n-2})(y_{t+n-1}-y_{t+n-1|t+n-2})'\\
&\quad=\mathrm{E}(\beta_t-\beta_{t|t+n-2})(\beta_{t+n-1}-\beta_{t+n-1|t+n-2})'x'_{t+n-1}\\
&\quad=H_tH_{t+1}\dots H_{t+n-2}K_{t+n-1}\left(x_{t+n-1}\Sigma_{t+n-1|t+n-2}x'_{t+1}+\sigma^2\right)\\
&\quad=H_tH_{t+1}\dots H_{t+n-2}\Sigma_{t+n-1|t+n-2}x'_{t+n-1},
\end{aligned}\tag{2.17}$$

where (2.4) has been used.

To prove (2.13), we need to find the regression of β_t on y_{t+n}, conditioned on Y_{t+n-1}:

$$\beta_{t|t+n}=\beta_{t|t+n-1}+D_{t|t+n}(y_{t+n}-y_{t+n|t+n-1}).\tag{2.18}$$

To evaluate the vector of regression coefficients $D_{t|t+n}$ we write, using (2.15):

$$\begin{aligned}
\beta_t-\beta_{t|t+n-1}&=\beta_t-\beta_{t|t+n-2}-D_{t|t+n-1}(y_{t+n-1}-y_{t+n-1|t+n-2})\\
&=\beta_t-\beta_{t|t+n-2}-D_{t|t+n-1}\left[x_{t+n-1}(\beta_{t+n-1}-\beta_{t+n-1|t+n-2})\right.\\
&\qquad\left.+\varepsilon_{t+n-1}\right],
\end{aligned}\tag{2.19}$$

and using (2.2):

$$\begin{aligned}
y_{t+n}-y_{t+n|t+n-1}&=x_{t+n}M(\beta_{t+n-1}-\beta_{t+n-1|t+n-1})+x_{t+n}\eta_{t+n}+\varepsilon_{t+n}\\
&=x_{t+n}M\left[\beta_{t+n-1}-\beta_{t+n-1|t+n-2}-K_{t+n-1}\right.\\
&\qquad\left.\times(y_{t+n-1}-y_{t+n-1|t+n-2})\right]+x_{t+n}\eta_{t+n}+\varepsilon_{t+n}\\
&=x_{t+n}M\left[(I-K_{t+n-1}x_{t+n-1})(\beta_{t+n-1}-\beta_{t+n-1|t+n-2})\right.\\
&\qquad\left.-K_{t+n-1}\varepsilon_{t+n-1}\right]+x_{t+n}\eta_{t+n}+\varepsilon_{t+n}.
\end{aligned}\tag{2.20}$$

Equations (2.19) and (2.20) imply:

$$\begin{aligned}
&\mathrm{E}(\beta_t-\beta_{t|t+n-1})(y_{t+n}-y_{t+n|t+n-1})'\\
&\quad=\mathrm{E}(\beta_t-\beta_{t|t+n-2})(\beta_{t+n-1}-\beta_{t+n-1|t+n-2})'(I-x'_{t+n-1}K'_{t+n-1})M'x'_{t+n}\\
&\quad-D_{t|t+n-1}\left[x_{t+n-1}\Sigma_{t+n-1|t+n-2}(I-x'_{t+n-1}K'_{t+n-1})-\sigma^2K'_{t+n-1}\right]M'x'_{t+1}\\
&\quad=H_tH_{t+1}\dots H_{t+n-2}\Sigma_{t+n-1|t+n-1}M'x'_{t+n}\\
&\quad=H_tH_{t+1}\dots H_{t+n-1}K_{t+n}\left(x_{t+n}\Sigma_{t+n|t+n-1}x'_{t+n}+\sigma^2\right),
\end{aligned}\tag{2.21}$$

where the second equality sign results from using (2.16), (2.17) and (2.7), and the third equality sign is due to (2.4). Hence, the regression coefficient is:

$$D_{t|t+n} = \left[\mathrm{E}(\beta_t - \beta_{t|t+n-1})(y_{t+n} - y_{t+n|t+n-1})'\right]\mathrm{cov}(y_{t+n}|Y_{t+n-1})]^{-1}$$
$$= H_t H_{t+1} \dots H_{t+n-1} K_{t+n}. \tag{2.22}$$

Equation (2.22) generalizes the coefficient given by (2.16). Substituting (2.22) into (2.18) yields:

$$\beta_{t|t+n} = \beta_{t|t+n-1} + H_t H_{t+1} \dots H_{t+n-1}(\beta_{t+n|t+n} - \beta_{t+n|t+n-1})$$
$$= \beta_{t|t+n-1} + H_t(\beta_{t+1|t+n} - \beta_{t+1|t+n-1}), \tag{2.23}$$

where the last step is due to the third equality sign of (2.14) with t replaced by $t+1$. Equation (2.23) completes the proof. Equations (2.23) and (2.18) provide three alternative formulas to evaluate $\beta_{t|t+n}$.

To derive the covariance matrix $\Sigma_{t|t+n}$, we use (2.18) and (2.21):

$$\Sigma_{t|t+n} = \mathrm{E}(\beta_t - \beta_{t|t+n})(\beta_t - \beta_{t|t+n})'$$
$$= \mathrm{E}\left[\beta_t - \beta_{t|t+n-1} - D_{t|t+n}(y_{t+n} - y_{t+n|t+n-1})\right]$$
$$\times\left[\beta_t - \beta_{t|t+n-1} - D_{t|t+n}(y_{t+n} - y_{t+n|t+n-1})\right]'$$
$$= \Sigma_{t|t+n-1} - D_{t|t+n}\left(x_{t+n}\Sigma_{t+n|t+n-1}x'_{t+n} + \sigma^2\right)D'_{t|t+n}. \tag{2.24}$$

By (2.22), (2.4) and (2.7), the formula (2.24) can be written alternatively:

$$\Sigma_{t|t+n} = \Sigma_{t|t+n-1} - H_t \dots H_{t+n-1}K_{t+n}\left(x_{t+n}\Sigma_{t+n|t+n-1}x'_{t+n} + \sigma^2\right)$$
$$\times K'_{t+n}H'_{t+n-1}\dots H'_t$$
$$= \Sigma_{t|t+n-1} - H_t \dots H_{t+n-1}\Sigma_{t+n|t+n-1}x'_{t+n}\left(x_{t+n}\Sigma_{t+n|t+n-1}x'_{t+n} + \sigma^2\right)^{-1}$$
$$\times x_{t+n}\,\Sigma_{t+n|t+n-1}H'_{t+n-1}\dots H'_t$$
$$= \Sigma_{t|t+n-1} + H_t \dots H_{t+n-1}\left(\Sigma_{t+n|t+n} - \Sigma_{t+n|t+n-1}\right)H'_{t+n-1}\dots H'_t. \tag{2.25}$$

Equations (2.24) and (2.25) provide the covariance matrix of the smoothed estimate $\beta_{t|t+n}$ of β_t given the data up to $t+n$. The estimates $\beta_{t|s}$ and $\Sigma_{t|s}$ of this section require knowledge not only of the parameters σ^2, V and M, but also of the initial values $\beta_{0|0}$ and $\Sigma_{0|0}$ at time 0. Actually, dividing (2.5) and (2.7) through by σ^2 and using $R_t \equiv \sigma^{-2}\Sigma_{t|t-1}$ in (2.4) for K_t, one observes that $\beta_{t|\sigma}$ can be computed by using $P = \sigma^{-2}V$ and M. See (4.4) and (4.5).

3. Derivations of $\beta_{t|s}$ by regression of $y_1,\ldots,y_s$ on $x_1,\ldots,x_s$

Econometricians might find it more appealing to view $\beta_{t|s}$ as a coefficient vector in the regression of $y_1,\ldots,y_s$ on $x_1,\ldots,x_s$. This interpretation was given by Sant (1977) as follows. Applying (1.2) repeatedly, we have:

$$\begin{aligned}\beta_t &= M\beta_{t-1} + \eta_t = M^2\beta_{t-2} + \eta_t + M\eta_{t-1}\\ &= M^{t-1}\beta_1 + M^{t-2}\eta_2 + \cdots + M\eta_{t-1} + \eta_t,\end{aligned}$$

which can be used to express $\beta_1,\ldots,\beta_{t-1}$ as functions of β_t, provided M^{-1} exists. The observations $y_1,\ldots,t_t$ from (1.1) thus become:

$$\begin{bmatrix} y_1 \\ y_2 \\ \vdots \\ y_{t-1} \\ y_t \end{bmatrix} = \begin{bmatrix} x_1M^{-t+1} \\ x_2M^{-t+2} \\ \vdots \\ x_{t-1}M^{-1} \\ x_t \end{bmatrix}\beta_t + \begin{bmatrix} \varepsilon_1 \\ \varepsilon_2 \\ \vdots \\ \varepsilon_{t-1} \\ \varepsilon_t \end{bmatrix}$$

$$-\begin{bmatrix} x_1M^{-1} & x_1M^{-2} & \cdots & x_1M^{-t+1} \\ 0 & x_2M^{-1} & \cdots & x_2M^{-t+2} \\ \vdots & \vdots & \ddots & \\ 0 & 0 & \cdots & x_{t-1}M^{-1} \\ 0 & 0 & \cdots & 0 \end{bmatrix}\begin{bmatrix} \eta_2 \\ \eta_3 \\ \vdots \\ \eta_{t-1} \\ \eta_t \end{bmatrix}. \tag{3.1}$$

The filtered estimate $\beta_{t|t}$ is equivalent to the estimate of β_t in the regression model (3.1) by Aitken's generalized least squares. The covariance matrix of the residuals in this regression is $\sigma^2[I_t + A_t(I_{t-1}\otimes P)A_t']$ where $\sigma^2P = \operatorname{cov}\eta_i$ and A_t is the coefficient of $(\eta_2\ \eta_3\ \ldots\ \eta_t)'$ in (3.1).

Similarly, $\beta_{t+s} = M^s\beta_t + M^{s-1}\eta_{t+1} + \cdots + M\eta_{t+s-1} + \eta_{t+s}$ is a function of β_t and observations on y_{t+s} $(s = 1,\ldots,n)$ can be used to form a regression model with β_t as the coefficient:

$$\begin{bmatrix} y_{t+1} \\ y_{t+2} \\ \vdots \\ y_{t+n} \end{bmatrix} = \begin{bmatrix} x_{t+1}M \\ x_{t+2}M^2 \\ \vdots \\ x_{t+n}M^n \end{bmatrix} \beta_t + \begin{bmatrix} \varepsilon_{t+1} \\ \varepsilon_{t+2} \\ \vdots \\ \varepsilon_{t+n} \end{bmatrix}$$
$$- \begin{bmatrix} x_{t+1} & 0 & \cdots & 0 \\ x_{t+2}M & x_{t+2} & \cdots & 0 \\ & & \ddots & \\ x_{t+n}M^{n-1} & x_{t+n}M^{n-2} & \cdots & x_{t+n} \end{bmatrix} \begin{bmatrix} \eta_{t+1} \\ \eta_{t+2} \\ \vdots \\ \eta_{t+n} \end{bmatrix}. \tag{3.2}$$

The smoothed estimate $\beta_{t|t+n}$ is equivalent to the estimate of β_t in a regression model combining (3.1) and (3.2) by Aitken's generalized least squares.

A by-product of this interpretation should be noted. Whereas the recursive method of Section 2 requires the initial values $\beta_{0|0}$ and $\Sigma_{0|0}$, the GLS method of this section provides estimates of $\beta_{k|k}$ and $\Sigma_{k|k}$ by using the first k observations, β_t being a vector of k elements. Applying GLS to (3.1) for $t = k$, one obtains:

$$\Sigma_{k|k} = \sigma^2 \left\{ \left[M'^{-k+1}x_1' \quad M'^{-k+2}x_2' \ldots x_k' \right] \right.$$
$$\left. \times \left[I_k + A_k(I_{k-1} \otimes P)A_k' \right]^{-1} \begin{bmatrix} x_1 M^{-k+1} \\ x_2 M^{-k+2} \\ \vdots \\ x_k \end{bmatrix} \right\}^{-1} \tag{3.3}$$

and

$$\beta_{k|k} = \sigma^{-2}\Sigma_{k|k} \left[M'^{-k+1}x_1' \quad M'^{-k+2}x_2' \ldots x_k' \right] \left[I_k + A_k(I_{k+1} \otimes P)A_k' \right]^{-1} \begin{bmatrix} y_1 \\ \vdots \\ y_k \end{bmatrix} \tag{3.4}$$

$\Sigma_{k|k}$ is a function of σ^2, $P = \sigma^{-2}V$, and M; $\beta_{k|k}$ is a function of P and M since $\sigma^{-2}\Sigma_{k|k} = R_k^*$ can be computed from (2.5) and (2.7) using these parameters.

4. Maximum likelihood estimation of σ^2, V and M

To form the likelihood function, we note that:

$$y_t - y_{t|t-1} = x_t(\beta_t - \beta_{t|t-1}) + \varepsilon_t = y_t - x_t\beta_{t|t-1}$$

is normal and serially uncorrelated. $\beta_t - \beta_{t|t-1}$ is the residual in the regression of β_t on $y_{t-1}, y_{t-2},\dots,$ and is therefore uncorrelated with $y_{t-1} - y_{t-1|t-2}$. Hence, $y_t - y_{t|t-1}$ is uncorrelated with $y_{t-1} - y_{t-1|t-2}$. The log-likelihood function based on observations $(y_1,\dots,y_T)$ is:

$$\log L = \text{const} - \frac{1}{2}\sum_{t=k+1}^{T}\log\left(x_t\Sigma_{t|t-1}x_t' + \sigma^2\right)$$

$$-\frac{1}{2}\sum_{t=k+1}^{T}\left(y_t - x_t\beta_{t|t-1}\right)^2/\left(x_t\Sigma_{t|t-1}x_t' + \sigma^2\right).$$

The first k observations are used to compute $\beta_{k|k}$ and $\Sigma_{k|k}$ by (3.3) and (3.4) as functions of σ^2, V and M. Hence, the data $\beta_{t|t-1}$ and $\Sigma_{t|t-1}$ $(t = k+1,\dots,T)$ required to evaluate logL are functions of σ^2, V and M, as given by the Kalman filtering equations (2.5), (2.7), (2.8), (2.2) and (2.4).

To maximize logL with respect to σ^2, we define $P \equiv \sigma^{-2}V$, $R_t \equiv \sigma^{-2}\Sigma_{t|t-1}$, and $u_t \equiv (y_t - x_t\beta_{t|t-1})/(x_tR_tx_t' + 1)^{1/2}$. The log-likelihood function can be rewritten as:

$$\log L = \text{const} - \tfrac{1}{2}(T-k)\log\sigma^2 - \tfrac{1}{2}\sum_{t=k+1}^{T}\log\left(x_tR_tx_t' + 1\right) - \frac{1}{2\sigma^2}\sum_{t=k+1}^{T}u_t^2. \tag{4.1}$$

On maximizing (4.1) with respect to σ^2, we obtain:

$$\hat{\sigma}^2 = \frac{1}{T-k}\sum_{t=k+1}^{T}u_t^2. \tag{4.2}$$

The concentrated likelihood function, after the elimination of σ^2, is:

$$\log L^* = \text{const} - \tfrac{1}{2}(T-k)\log\left[\sum_{t=k+1}^{T}\left(y_t - x_t\beta_{t|t-1}\right)^2/\left(x_tR_tx_t' + 1\right)\right]$$

$$-\tfrac{1}{2}\sum_{t=k+1}^{T}\log\left(x_tR_tx_t' + 1\right), \tag{4.3}$$

where, by (2.5) and (2.7):

$$R_t = MR_{t-1}\left[I - x'_{t-1}\left(x_{t-1}R_{t-1}x'_{t-1}+1\right)^{-1}x_{t-1}R_{t-1}\right]M' + P \qquad (t = k+2,\dots,T), \quad (4.4)$$

and, by (2.8), (2.2) and (2.4):

$$\beta_{t|t-1} = M\left[\beta_{t-1|t-2} + R_{t-1}x'_{t-1}\left(x_{t-1}R_{t-1}x'_{t-1}+1\right)^{-1}\left(y_{t-1} - x_{t-1}\beta_{t-1|t-2}\right)\right] \qquad (t = k+2,\dots,T) \quad (4.5)$$

The initial conditions are:

$$R_{k+1} = \sigma^{-2}M\Sigma_{k|k}M' + P \quad (4.6)$$

and

$$\beta_{k+1|k} = M\beta_{k|k}, \quad (4.7)$$

with $\Sigma_{k|k}$ and $\beta_{k|k}$ given by (3.3) and (3.4). One would have to rely on a numerical method to maximize (4.3) with respect to the unknown parameters in P and M, P being symmetric, positive semidefinite. Garbade (1977) gives an example.

An alternative expression of the likelihood function can be obtained by using the normal regression model (3.1) for $t = T$, i.e.

$$y = Z\beta + \varepsilon - A\eta, \quad (4.8)$$

where

$$y = \begin{bmatrix} y_1 \\ y_2 \\ \vdots \\ y_T \end{bmatrix}, \qquad Z = \begin{bmatrix} x_1M^{-T+1} \\ x_2M^{T+2} \\ \vdots \\ x_T \end{bmatrix}, \qquad \varepsilon = \begin{bmatrix} \varepsilon_1 \\ \varepsilon_2 \\ \vdots \\ \varepsilon_T \end{bmatrix}, \qquad \eta = \begin{bmatrix} \eta_2 \\ \eta_3 \\ \vdots \\ \eta_T \end{bmatrix},$$

$\beta = \beta_T$ and $A = A_T$ as defined by the last coefficient matrix of (3.1) for $t = T$. The log-likelihood function of this model is:

$$\text{logL} = \text{const} - \tfrac{1}{2}\log|\sigma^2 I_T| - \tfrac{1}{2}\log|Q| - \tfrac{1}{2}(y - Z\beta)'Q^{-1}(y - Z\beta)/\sigma^2, \quad (4.9)$$

where

$$Q = I_T + A(I_{T-1}\otimes P)A'. \quad (4.10)$$

Maximization of (4.9) with respect to σ^2 yields:

$$\hat{\sigma}^2 = \frac{1}{T}(y - Z\beta)'Q^{-1}(y - Z\beta). \tag{4.11}$$

Maximization of (4.9) with respect to β yields:

$$\hat{\beta} = (Z'Q^{-1}Z)^{-1}Z'Q^{-1}y. \tag{4.12}$$

Differentiating (4.9) with respect to the unknown elements $p_{ij} = p_{ji}$ of P, one obtains

$$\frac{\partial \log L}{\partial p_{ij}} = -\operatorname{tr}\left(Q^{-1}\frac{\partial Q}{\partial p_{ij}}\right) - \sigma^2(y - Z\beta)'\frac{\partial Q^{-1}}{\partial p_{ij}}(y - Z\beta). \tag{4.13}$$

To evaluate $\partial Q^{-1}/\partial p_{ij}$, we differentiate both sides of $QQ^{-1} = I$ with respect to p_{ij} to get:

$$\frac{\partial Q^{-1}}{\partial p_{ij}} = -Q^{-1}\frac{\partial Q}{\partial p_{ij}}Q^{-1}. \tag{4.14}$$

Using the definition (4.10) for Q, we have:

$$\frac{\partial Q}{\partial p_{ij}} = A(I_{T-1}\otimes E_{ij})A', \tag{4.15}$$

where E_{ij} is an elementary $k \times k$ matrix with all zero elements except the $i-j$ and $j-i$ elements which equal unity. Substituting (4.11), (4.12), (4.14) and (4.15) into (4.13) gives:

$$\frac{\partial \log L}{\partial p_{ij}} = -\operatorname{tr}\left[Q^{-1}A(I_{T-1}\otimes E_{ij})A'\right] + \left(\frac{1}{T}y'N'Q^{-1}y\right)^{-1} y'N'Q^{-1}A\left(I_{T-1}\otimes E_{ij}A'\right)Q^{-1}Ny, \tag{4.16}$$

where N denotes $I - Z(Z'Q^{-1}Z)^{-1}Z'Q^{-1}$. Equation (4.16) is useful for the maximization of (4.19) when a numerical method requiring analytical first derivatives is applied. Furthermore, in econometric applications M is frequently assumed to be an identity matrix and P to be diagonal. In this important special case, the only unknown parameters in (4.9) are $p_{11}\dots p_{kk}$. One can start with zero as the initial value for each p_{ii} and increase its value if $\partial \log L/\partial p_{ii}$ as evaluated by (4.16) is positive.

For the general case, numerical methods can be applied to maximize (4.9) after the elimination of σ^2 and β by (4.11) and (4.12), i.e. $-\frac{1}{2}\log|(y'N'Q^{-1}y)Q|$, with respect to the unknown parameters in P and M. For a discussion of conditions for the identifiability of these parameters and the asymptotic distribution of the maximum likelihood estimator, the reader is referred to Section 9 below.

5. System of linear regressions with changing coefficients

A generalization of the regression model (1.1)–(1.2) is a system of m linear regressions with changing coefficients:

$$\begin{bmatrix} y_{1t} \\ \vdots \\ y_{mt} \end{bmatrix} = \begin{bmatrix} x_{1t} & & 0 \\ & \ddots & \\ 0 & & x_{mt} \end{bmatrix} \begin{bmatrix} \beta_{1t} \\ \vdots \\ \beta_{mt} \end{bmatrix} + \begin{bmatrix} \varepsilon_{1t} \\ \vdots \\ \varepsilon_{mt} \end{bmatrix} \qquad (t=1,\dots,T), \tag{5.1}$$

$$\begin{bmatrix} \beta_{1t} \\ \vdots \\ \beta_{mt} \end{bmatrix} = \begin{bmatrix} M_1 & & 0 \\ & \ddots & \\ 0 & & M_m \end{bmatrix} \begin{bmatrix} \beta_{1,t-1} \\ \vdots \\ \beta_{m,t-1} \end{bmatrix} + \begin{bmatrix} \eta_{1t} \\ \vdots \\ \eta_{mt} \end{bmatrix} \qquad (t=1,\dots,T). \tag{5.2}$$

Here x_{jt} is a row vector of k_j explanatory variables. $(\varepsilon_{1t},\dots,\varepsilon_{mt})$ is m-variate normal and independent with mean zero and covariance matrix $S=(\sigma_{ij})$. η_{jt} is k_j-variate normal and independent with mean 0 and covariance matrix $V_j=\sigma_{jj}P_j$, being independent of η_{it} for $i\neq j$. If σ_{ij} were zero for $i\neq j$, the m regression models will be treated separately, each by the methods previously presented. When $\sigma_{ij}\neq 0$, efficiency may be gained in the estimation of β_{jt} by combining the m regressions into a system.

If we write (5.1)–(5.2) more compactly as:

$$y'_{\cdot t} = X_t\beta_{\cdot t} + \varepsilon_{\cdot t} \qquad (t=1,\dots,T), \tag{5.3}$$

$$\beta_{\cdot t} = M\beta_{\cdot t-1} + \eta_{\cdot t} \qquad (t=1,\dots,T), \tag{5.4}$$

the filtering and smoothing equations of Section 2 remain entirely valid for this model, with σ, $V_1,\dots,V_m$ and M treated as given. The derivations are the same as in Section 2, with the scalar y_t replaced by the column vector $y'_{\cdot t}$, the row vector x_t by the $m\times(\Sigma_j k_j)$ matrix X_t, and the variance σ^2 by the covariance matrix Σ of the vector $\varepsilon_{\cdot t}$. K_t in (2.4) becomes a $(\Sigma_j k_j)\times m$ matrix.

For the estimation of Σ, $V_1,\dots,V_m$ and M, we write the T observations on the jth regression model with coefficient $\beta_{jT}=\beta_j$ in the notation of (4.8) as:

$$y_j = Z_j\beta_j + \varepsilon_j - A_j\eta_j, \tag{5.5}$$

where the residual vector $\varepsilon_j - A_j\eta_j$ has covariance matrix $\sigma_{jj}Q_j$, with

$$Q_j = I_T + A_j\left(I_{T-1}\otimes P_j\right)A_j'. \tag{5.6}$$

Combining the m regression models, we have:

$$\begin{bmatrix} y_1 \\ \vdots \\ y_m \end{bmatrix} = \begin{bmatrix} Z_1 & & 0 \\ & \ddots & \\ 0 & & Z_m \end{bmatrix}\begin{bmatrix} \beta_1 \\ \vdots \\ \beta_m \end{bmatrix} + \begin{bmatrix} \varepsilon_1 \\ \vdots \\ \varepsilon_m \end{bmatrix} - \begin{bmatrix} A_1\eta_1 \\ \vdots \\ A_m\eta_m \end{bmatrix}, \tag{5.7}$$

where the residual vector has covariance matrix:

$$\Phi = \begin{bmatrix} \sigma_{11}Q_1 & \sigma_{12}I & \cdots & \sigma_{1m}I \\ \sigma_{12}I & \sigma_{22}Q_2 & \cdots & \sigma_{2m}I \\ & \cdots & & \\ \sigma_{1m}I & \sigma_{2m}I & \cdots & \sigma_{mm}Q_m \end{bmatrix}; \quad \Phi^{-1} = \begin{bmatrix} \sigma^{11}Q_1^{-1} & \sigma^{12}I & \cdots & \sigma^{1m}I \\ \sigma^{12}I & \sigma^{22}Q_2^{-1} & \cdots & \sigma^{2m}I \\ & \cdots & & \\ \sigma^{1m}I & \sigma^{2m}I & \cdots & \sigma^{mm}Q_m^{-1} \end{bmatrix}, \tag{5.8}$$

with σ^{ij} denoting the $i-j$ element of Σ^{-1}.

The log-likelihood function for the model (5.7) is:

$$\log L = \text{const} - \tfrac{1}{2}\log|\Phi| - \tfrac{1}{2}\sum_{i=1}^{m}(y_1 - Z_i\beta_i)'Q^{-1}(y_i - Z_i\beta_i)\sigma^{ii}$$

$$- \sum_{i<j}^{m}(y_i - Z_i\beta_i)'(y_j - Z_j\beta_j)\sigma^{ij}. \tag{5.9}$$

Observing that

$$\frac{\partial\log|\Phi|}{\partial\sigma^{ii}} = -\frac{\partial\log|\Phi^{-1}|}{\partial\sigma^{ii}} = -\text{tr}\left(\Phi\frac{\partial\Phi^{-1}}{\partial\sigma^{ii}}\right) = -T\sigma_{ii}$$

and

$$\frac{\partial\log|\Phi|}{\partial\sigma^{ij}} = -2\,\text{tr}\left(\Phi\frac{\partial\Phi^{-1}}{\partial\sigma^{ij}}\right) = -2T\sigma_{ij} \qquad (\sigma_{ij} = \sigma_{ji},\ i\neq j),$$

we differentiate (5.9) with respect to σ^{ii} and σ^{ij} to obtain:

$$\hat{\sigma}_{ii} = \frac{1}{T}(y_i - Z_i\beta_i)'Q_i^{-1}(y_i - Z_i\beta_i) \tag{5.10}$$

and

$$\hat{\sigma}_{ij} = \frac{1}{T}(y_i - Z_i\beta_i)'(y_j - Z_j\beta_j) \qquad (i \neq j) \tag{5.11}$$

Equation (5.10) is identical with (4.11) and shows that $\hat{\sigma}_{ii}$ can be obtained from the residuals $y_i - Z_i\beta_i$ of the ith regression only, if the parameters P_i and M_i are known. Equation (5.11) shows that σ_{ij} $(i \neq j)$ is the sample covariance of the residuals in the ith and jth regressions and is independent of P_i and P_j.

Differentiating (5.9) with respect to β_i gives:

$$\frac{\partial \log L}{\partial \beta_i} = \sigma^{ii} Z_i' Q_i^{-1}(y_i - Z_i\beta_i) + Z_i' \sum_{j \neq i}^{m} \sigma^{ij}(y_j - Z_j\beta_j) = 0.$$

Combining the above equations for $i = 1,\dots,m$, we have:

$$\begin{bmatrix} \sigma^{11} Z_1' Q_1^{-1} Z_1 & \sigma^{12} Z_1' Z_2 & \cdots & \sigma^{1m} Z_1' Z_m \\ \vdots & & & \vdots \\ \sigma^{m1} Z_m' Z_1 & \sigma^{m2} Z_m' Z_2 & \cdots & \sigma^{mm} Z_m' Q_m^{-1} Z_m \end{bmatrix} \begin{bmatrix} \hat{\beta}_1 \\ \vdots \\ \hat{\beta}_m \end{bmatrix}$$

$$= \begin{bmatrix} Z_1'\left(\sigma^{11} Q_1^{-1} y_1 + \sum_{j \neq 1} \sigma^{ij} y_j\right) \\ \vdots \\ Z_m'\left(\sigma^{mm} Q_m^{-1} y_m + \sum_{j \neq m} \sigma^{mj} y_j\right) \end{bmatrix}. \tag{5.12}$$

Differentiating (5.9) with respect to the $i-j$ element $p_{k,ij}$ of P_k yields:

$$\frac{\partial \log L}{\partial p_{k,ij}} = -\operatorname{tr}\left(\Phi^{-1}\frac{\partial \Phi}{\partial p_{k,ij}}\right) - (y_k - Z_k\beta_k)'\frac{\partial Q_k^{-1}}{\partial p_{k,ij}}(y_k - Z_k\beta_k)$$

$$= -\operatorname{tr}\left(\sigma^{kk}\sigma_{kk} Q_k^{-1}\frac{\partial Q_k}{\partial p_{k,ij}}\right) + (y_k - Z_k\beta_k)' Q_k^{-1}\frac{\partial Q_k}{\partial p_{k,ij}} Q_k^{-1}(y_k - Z_k\beta_k)$$

$$= -\sigma^{kk}\sigma_{kk}\operatorname{tr}\left[Q_k^{-1} A_k (I_{T-1} \otimes E_{ij}) A_k'\right]$$

$$+ (y_k - Z_k\beta_k)' Q_k^{-1} A_k (I_{T-1} \otimes E_{ij}) A_k' Q_k^{-1}(y_k - Z_k\beta_k). \tag{5.13}$$

The maximization of logL with respect to P_k $(k = 1,\dots,m)$ can proceed iteratively as follows. First consider the important case with $M_1,\dots,M_m$ given. Starting with $P_k = 0$ for all k, which implies $Q_k = I_T$, solve (5.10)–(5.12) for $\hat{\sigma}_{ij}$

and $\hat{\beta}_i$ (all i, j) as a standard problem of estimating a system of linear regressions. Use (5.13) to find the gradient of logL with respect to the unknown elements $p_{k,ij}$, and apply a gradient method to maximize logL with $\hat{\sigma}_{ij}$ and $\hat{\beta}_i$ (all i, j) treated as fixed. By (5.9) logL equals const $-\frac{1}{2}\log|\Phi|$, where σ_{ij} is replaced by $\hat{\sigma}_{ij}$. Having revised $p_{k,ij}$, solve (5.10)–(5.12) again for $\hat{\sigma}_{ij}$ and $\hat{\beta}_i$, and so forth. In practice, the computational burden is not as heavy as it might appear because only a small number of equations k will contain time-varying coefficients and for those equations P_k are often assumed to be diagonal with a few unknown $p_{k,ij}$.

In the general case with unknown elements in M_k, the computational problem becomes more difficult. We will still treat σ_{ij} and β_i (all i, j) separately from the unknown parameters in M_k and P_k $(k=1,\dots,m)$. Given the latter, (5.10)–(5.12) will be used to estimate $\hat{\sigma}_{ij}$ and $\hat{\beta}_i$. Given $\hat{\sigma}_{ij}$ we maximize logL = const $-\frac{1}{2}\log|\Phi|$ with respect to the unknown parameters in M_k and P_k by some numerical method.

For both the special and the general case, an approximate solution to the problem of maximum likelihood is to ignore the interdependence of the m regressions due to the covariances σ_{ij} and treat the estimation of P_k and M_k separately for each regression k by the method of Section 4. This approximation is appealing because, without pooling the m regressions, the GLS estimate β_k for each regression k is still consistent; so are the associated estimates of the unknown elements of P_k and M_k. They correspond to the estimates of the parameters of one structural equation in a system of simultaneous equations by a limited-information method such as two-stage least squares. Having so obtained the estimates of P_k and M_k, one can pool the regression equations to estimate σ_{ij} and β_i by (5.10)–(5.12). Or, as a further approximation, treat the estimates of σ_{ij} as given and use (5.12) alone to reestimate β_i $(i=1,\dots,m)$. This procedure corresponds to three-stage least squares where the estimates of the covariance matrix of the residuals are obtained by two-stage least squares.

The above discussion has concentrated on the estimation of $\beta_k \equiv \beta_{kT}$ for the last period and the associated covariance matrices Σ and P_k. Given estimates of Σ and P_k, and the initial values $\beta_{k;\tau|\tau}$ and $\Sigma_{k;\tau|\tau}$ as obtained by (5.12) using τ observations, we can apply the filtering and smoothing algorithms to (5.3) and (5.4) to estimate $\beta_{k;t|t}$ and $\beta_{k;t|T}$ for $t>\tau$.

6. System of linear simultaneous equations

Let the tth observation of a system of m linear simultaneous equations with time-varying coefficients be written as:

$$y_{\cdot t}\Gamma_t + x_{\cdot t}B_t = -\varepsilon'_{\cdot t} \qquad (t=1,\dots,T), \tag{6.1}$$

where $y_{\cdot t} = (y_{1t}\cdots y_{mt})$ is a row vector of m endogenous variables; $x_{\cdot t}$ is a row vector of K exogenous variables; and the diagonal elements $\gamma_{ii,t}$ of Γ_t are normalized to be -1. The reduced form of (6.1) is:

$$y_{\cdot t} = -x_{\cdot t} B_t \Gamma_t^{-1} - \varepsilon'_{\cdot t} \Gamma_t^{-1} \equiv x_{\cdot t} \Pi_t + u'_{\cdot t} \qquad (t=1,\dots,T), \tag{6.2}$$

where we have defined Π_t as $-B_t\Gamma_t^{-1}$ and $u'_{\cdot t}$ as $-\varepsilon'_{\cdot t}\Gamma_t^{-1}$. The jth structural equation can be written as:

$$y_{jt} = y_{jt}^{*}\gamma_{jt} + x_{jt}^{*}\beta_{jt} + \varepsilon_{jt} \equiv z_{jt}^{*}\delta_{jt} + \varepsilon_{jt}, \tag{6.3}$$

where y_{jt}^{*} is a row vector of endogenous variables appearing in equation j, other than y_{jt}; γ_{jt} is a column vector of unknown coefficients in equation j which is composed of selected elements from the jth column $\gamma_{\cdot jt}$ of Γ_{jt}, and similarly for x_{jt}^{*} and β_{jt}. We have also defined z_{jt}^{*} as $(y_{jt}^{*}x_{jt}^{*})$ and δ'_{jt} as $(\gamma'_{jt}\beta'_{jt})$.

Corresponding to (5.1) and (5.2) are:

$$\begin{bmatrix} y_{1t} \\ \vdots \\ y_{mt} \end{bmatrix} = \begin{bmatrix} z_{1t}^{*} & & 0 \\ & \ddots & \\ 0 & & z_{mt}^{*} \end{bmatrix} \begin{bmatrix} \delta_{1t} \\ \vdots \\ \delta_{mt} \end{bmatrix} + \begin{bmatrix} \varepsilon_{1t} \\ \vdots \\ \varepsilon_{mt} \end{bmatrix} \qquad (t=1,\dots,T), \tag{6.4}$$

$$\begin{bmatrix} \delta_{1t} \\ \vdots \\ \delta_{mt} \end{bmatrix} = \begin{bmatrix} M_1 & & 0 \\ & \ddots & \\ 0 & & M_m \end{bmatrix} \begin{bmatrix} \delta_{1,t-1} \\ \vdots \\ \delta_{m,t-1} \end{bmatrix} + \begin{bmatrix} \eta_{1t} \\ \vdots \\ \eta_{mt} \end{bmatrix} \qquad (t=1,\dots,T), \tag{6.5}$$

as we are making similar assumptions about the evolution of δ_{jt}. However, the techniques of Section 5 cannot be applied directly to (6.4) and (6.5) because (6.4), in contrast with (5.1) or (1.1), is a non-linear function of the random coefficient vector $\delta'_t = (\delta'_{1t}\dots\delta'_{mt})$. The y_{jt}^{*} component of z_{jt}^{*} is itself a non-linear function of $(\gamma'_{1t}\cdots\gamma'_{mt})$ as seen from the reduced-form (6.2). In order to apply the techniques of Section 5, it is proposed to approximate the right-hand side of (6.2) by a linear function of δ_t and $\varepsilon_{\cdot t}$. This approach amounts to treating a regression problem involving a non-linear model $y = f(x,\beta,\varepsilon)$ by approximating f by a linear function of β and ε. In the control engineering literature, the resulting estimation method is known as an extended Kalman filter.

We will linearize the right-hand side of (6.2) about $\delta_t^0 = \delta_{t|t-1}$. Hence we will linearize Π_t about δ_t^0 and define:

$$\Pi_t^0 = -(B_{t|t-1})(\Gamma_{t|t-1})^{-1} = (\pi_{1t}^0\dots\pi_{mt}^0). \tag{6.6}$$

The jth column of Π_t is $\pi_{jt} = -B_t\gamma_t^{\cdot j}$, with $\gamma_t^{\cdot j}$ denoting the jth column of Γ_t^{-1}.

The linear approximation of π_{jt} is:

$$\pi_{jt} \simeq \pi_{jt}^0 + \left(\frac{\partial \pi_{jt}}{\partial \delta_t'}\right)_0 (\delta_t - \delta_{t|t-1}), \tag{6.7}$$

where the subscript zero indicates that the matrix $\partial \pi_{jt}/\partial \delta_t'$ is evaluated at $\delta_t^0 = \delta_{t|t-1}$. To evaluate the matrix $\partial \pi_{jt}/\partial \delta_t'$, we note:

$$\frac{\partial \pi_{jt}}{\partial \gamma_{ik,t}} = -B_t \frac{\partial \gamma_t^{\cdot j}}{\partial \gamma_{ik,t}} = B_t \gamma_t^{\cdot i} \gamma_t^{kj};$$

or, with $\gamma_{\cdot k,t}$ denoting the kth column of Γ_t:

$$\frac{\partial \pi_{jt}}{\partial \gamma'_{\cdot k,t}} = B_t \Gamma_t^{-1} \gamma_t^{kj} = -\Pi_t \gamma_t^{kj} \tag{6.8}$$

and

$$\frac{\partial \pi_{jt}}{\partial \beta_{ik,t}} = -\frac{\partial B_t}{\partial \beta_{ik,t}} \gamma_t^{\cdot j} = -\begin{bmatrix} 0 \\ \vdots \\ \gamma^{kj} \\ \vdots \\ 0 \end{bmatrix} \qquad (\gamma_t^{kj} \text{ in } i\text{th row});$$

or, with $\beta_{\cdot k,t}$ denoting the kth column of B_t:

$$\frac{\partial \pi_{jt}}{\partial \beta'_{\cdot k,t}} = -I_K \gamma_t^{kj}. \tag{6.9}$$

Hence, the matrix $(\partial \pi_{jt}/\partial \delta_t')_0$ of (6.7) can be evaluated by (6.8) and (6.9) with Π_t replaced by Π_t^0 and Γ_t^{-1} replaced by $\Gamma_t^{0-1} = \Gamma_{t|t-1}^{-1}$. Similarly, approximating $-\varepsilon'_{\cdot t}\Gamma_t^{-1}$ by a linear function of $\varepsilon_{\cdot t}$ and $(\gamma_{1t} \dots \gamma_{mt})$ about $\varepsilon_{\cdot t} = 0$ and $(\gamma_{1t} \dots \gamma_{mt}) = (\gamma_{1t|t-1} \cdots \gamma_{mt|t-1})$ yields $-\varepsilon'_{\cdot t}\Gamma_{t|t-1}^{-1}$. Combining this result with (6.7) and denoting $x_{\cdot t}\pi_{jt}^0$ by y_{jt}^0, we can write the linearized version of (the transpose of) (6.2) as:

$$\begin{bmatrix} y_{1t} \\ \vdots \\ y_{mt} \end{bmatrix} = \begin{bmatrix} y_{1t}^0 \\ \vdots \\ y_{mt}^0 \end{bmatrix} + \begin{bmatrix} x_{\cdot t}(\partial \pi_{1t}/\partial \delta_t')_0 \\ \vdots \\ x_{\cdot t}(\partial \pi_{mt}/\partial \delta_t')_0 \end{bmatrix} (\delta_t - \delta_{t|t-1}) - \Gamma_{t|t-1}'^{-1} \varepsilon_{\cdot t}, \tag{6.10}$$

or more compactly as:

$$y'_{\cdot t} = y^{0\prime}_{\cdot t} + W_t^0(\delta_t - \delta_{t|t-1}) - \Gamma'^{-1}_{t|t-1}\varepsilon_{\cdot t}, \tag{6.11}$$

where the jth row of W_t^0 is, by (6.8) and (6.9):

$$\begin{aligned} x_{\cdot t}\left(\frac{\partial \pi_{jt}}{\partial \delta'_t}\right) &= -\left[\left(y^{*0}_{1t} \quad x^*_{1t}\right)\gamma^{1j}_{t|t-1}\cdots\left(y^{*0}_{mt} \quad x_{mt}\right)\gamma^{mj}_{t|t-1}\right] \\ &= -\left[z^{*0}_{1t}\gamma^{1j}_{t|t-1}\cdots z^{*0}_{mt}\gamma^{mj}_{t|t-1}\right], \end{aligned} \tag{6.12}$$

with y^{*0}_{it} denoting a row vector composed of those elements of $y^0_{\cdot t} = x_{\cdot t}\Pi^0_t$ which correspond to γ_{it}.

The linearized model (6.11) will replace (5.1) or (5.3) for the purpose of deriving filtering equations. For the model (6.11)–(6.5), the derivations are exactly the same as in Section 5 or Section 2, with $\Sigma = \mathrm{E}\varepsilon_{\cdot t}\varepsilon'_{\cdot t}$, $V_i = \mathrm{E}\eta_{it}\eta'_{it}$ $(i = 1,\dots,m)$, and M_i $(i = 1,\dots,m)$ treated as given. From the linear model (6.11), one finds the conditional expectation $y'_{\cdot t|t-1}$ to be $y^{0\prime}_{\cdot t}$. Repeating the derivations from (2.2) to (2.8) one finds:

$$\delta_{t|t} = \delta_{t|t-1} + K_t\left[y'_{\cdot t} - y'_{\cdot t|t-1}\right], \tag{6.13}$$

and, denoting $\mathrm{E}(\delta_t - \delta_{t|t-1})(\delta_t - \delta_{t|t-1})'$ by $\Sigma_{t|t-1}$, etc.:

$$K_t = \Sigma_{t|t-1}W_t^{0\prime}\left[W_t^0\Sigma_{t|t-1}W_t^{0\prime} + \Gamma'^{-1}_{t|t-1}S\Gamma^{-1}_{t|t-1}\right]^{-1} \tag{6.14}$$

Corresponding to (2.5), (2.7) and (2.8) are, respectively,

$$\Sigma_{t|t-1} = M\Sigma_{t-1|t-1}M' + V, \tag{6.15}$$

M being the coefficient matrix of (6.5) and V being the covariance matrix of its residual $\eta'_t = (\eta'_{1t}\cdots\eta'_{mt})$:

$$\Sigma_{t|t} = \Sigma_{t|t-1} - K_t\left[W_t^0\Sigma_{t|t-1}W_t^{0\prime} + \Gamma'^{-1}_{t|t-1}S\Gamma^{-1}_{t|t-1}\right]K'_t \tag{6.16}$$

and

$$\delta_{t|t-1} = M\delta_{t|t-1}. \tag{6.17}$$

An alternative way of estimating $\delta_{T|T}$, given Σ, V_i and M_i, is to form a regression model analogous to (3.1) using (6.11) for $y'_{\cdot t}$ $(t=1,\dots,T)$ and denoting $y'_{\cdot t} - y^{0\prime}_{\cdot t} + W_t^0\delta_{t|t-1}$ by $\tilde{y}'_{\cdot t}$:

$$\begin{bmatrix} \tilde{y}'_{\cdot 1} \\ \tilde{y}'_{\cdot 2} \\ \vdots \\ \tilde{y}_{\cdot T-1} \\ \tilde{y}'_{\cdot T} \end{bmatrix} = \begin{bmatrix} W_1^0 M^{-T+1} \\ W_2^0 M^{-T+2} \\ \vdots \\ W_{T-1}^0 M^{-1} \\ M_T^0 \end{bmatrix} \delta_T - \begin{bmatrix} \Gamma_1^{0\prime-1}\varepsilon_{\cdot 1} \\ \Gamma_2^{0\prime-1}\varepsilon_{\cdot 2} \\ \vdots \\ \Gamma_{T-1}^{0\prime-1}\varepsilon_{\cdot T-1} \\ \Gamma_T^{0\prime-1}\varepsilon_{\cdot T} \end{bmatrix}$$
$$- \begin{bmatrix} W_1^0(M^{-1} & M^{-2} & \cdots & M^{-T+1}) \\ 0 & W_2^0(M^{-1} & \cdots & M^{-T+2}) \\ \vdots & \vdots & \ddots & \vdots \\ 0 & 0 & \cdots & W_{T-1}^0 M^{-1} \\ 0 & 0 & \cdots & 0 \end{bmatrix} \begin{bmatrix} \eta_2 \\ \eta_3 \\ \vdots \\ \eta_{T-1} \\ \eta_T \end{bmatrix} \tag{6.18}$$

The covariance matrix of the residual vector of (6.18) has an $i-j$ block:

$$\delta_{ij}\Gamma_i^{0\prime-1}S\Gamma_i^{0-1} + W_i^0 \sum_{t=\max(i,j)}^{T-1} M^{-(t-i+1)}VM^{\prime-(t-j+1)}W_j^{0\prime}, \tag{6.19}$$

where δ_{ij} is the Kronecker delta. Aitken's generalized least squares can be applied to estimate δ_T once the coefficients W_t^0 and Γ_t^0 of the linearized model (6.11) are evaluated. One can choose an initial guess δ_T^0 for $\delta_{T|T}$, and the associated $\delta_t^0 = M^{-(T-t)}\delta_T^0$ $(t=1,\dots,T)$. These initial values permit the evaluation of W_t^0, Γ_t^0, $y^0_{\cdot t} = -x_{\cdot t}B_t^0\Gamma_t^{0-1}$ and $\delta_{t|t-1} = \delta_t^0$. Equation (6.18) will be treated as a linear regression model to estimate $\delta_{T|T}$. The resulting estimate will be used to form a new initial guess δ_T^0 and the process continues iteratively.

In order to estimate the unknown parameters in Σ, V_i $(i=1,\dots,m)$ and M_i $(i=1,\dots,m)$, (6.18) can be used to form a likelihood function. However, unlike the situation with truly constant coefficients W_t^0 and Γ_t^0, the evaluation of the likelihood function requires iterative solution of $\delta_{T|T}$ as described in the last paragraph. The computational problem involved in maximizing the likelihood function is hence more burdensome than in the case of a truly linear (in contrast with a linearized) model. This problem deserves further study.

7. System of non-linear simultaneous equations

Let the tth observation of a system of m non-linear simultaneous equations with time-varying parameters be written as:

$$y'_{\cdot t} = \Phi(y_{\cdot t}, x_{\cdot t}, \delta_t) + \varepsilon_{\cdot t} \qquad (t = 1,\ldots,T), \tag{7.1}$$

where Φ is a vector function of m components and

$$\delta_t = M\delta_{t-1} + \eta_t \qquad (t = 1,\ldots,T), \tag{7.2}$$

which is identical with (6.5). Like the reduced-form (6.2) for a system of linear structural equations, (7.1) is a non-linear function of the parameter vector δ_t. The approach to be adopted is similar to the one used in Section 6. It amounts to linearizing the non-linear observation equation (7.1) about some δ_t^0 and the associated $y_{\cdot t}^0$ defined by:

$$y_{\cdot t}^{0\prime} = \Phi(y_{\cdot t}^0, x_{\cdot t}, \delta_t^0).$$

Given δ_t^0 and $x_{\cdot t}$, $y_{\cdot t}^0$ can be computed by the Gauss–Seidel method, for example.

Linearizing Φ in (7.1) about δ_t^0 and $y_{\cdot t}^0$, we have:

$$y'_{\cdot t} = y_{\cdot t}^{0\prime} + \left(\frac{\partial \Phi}{\partial y_{\cdot t}}\right)_0 (y'_{\cdot t} - y_{\cdot t}^{0\prime}) + \left(\frac{\partial \Phi}{\partial \delta_t}\right)_0 (\delta_t - \delta_t^0) + \varepsilon_{\cdot t},$$

where, as in Section 6, the sub-script zero indicates that the matrix of partial derivatives of Φ is evaluated at $y_{\cdot t}^0$ and δ_t^0. Solving for $y'_{\cdot t}$, we get:

$$\begin{aligned} y'_{\cdot t} &= y_{\cdot t}^{0\prime} + \left[I - \left(\frac{\partial \Phi}{\partial y_{\cdot t}}\right)_0\right]^{-1} \left(\frac{\partial \Phi}{\partial \delta'_t}\right)_0 (\delta_t - \delta_t^0) + \left[I - \left(\frac{\partial \Phi}{\partial y_{\cdot t}}\right)_0\right]^{-1} \varepsilon_{\cdot t} \\ &\equiv y_{\cdot t}^0 + W_t^0(\delta_t - \delta_t^0) + R_t \varepsilon_{\cdot t}, \end{aligned} \tag{7.3}$$

which replaces the linearized observation equation (6.11) of Section 6. The treatment of the model (7.1)–(7.2) is the same as in Section 6. The computational problem is only slightly more difficult because the linearization to achieve (7.3) requires the evaluation of the partial derivatives $(\partial \Phi / \partial y_{\cdot t})_0$ and $(\partial \Phi / \partial \delta'_t)_0$ whereas the linearization to obtain (6.11) requires matrix inversion only. These partial derivatives can be evaluated numerically, and their evaluation is computationally much simpler than the maximization of the likelihood function for the linearized model with respect to the parameters S, V_i $(i = 1,\ldots,m)$ and M_i $(i = 1,\ldots,m)$ as discussed at the end of Section 6.

8. Model with stationary coefficients

An alternative specification to (1.1) and (1.2) is:

$$y_t = x_t\beta_t + \varepsilon_t, \tag{8.1}$$

$$\beta_t - \bar{\beta} = M(\beta_{t-1} - \bar{\beta}) + \eta_t, \tag{8.2}$$

where all characteristic roots of M are assumed to be smaller than one in absolute value. In stochastic equilibrium β_t will have mean $\bar{\beta}$ and a covariance matrix Γ satisfying:

$$\Gamma = M\Gamma M' + V,$$

where, as before, V is the covariance matrix of η_t. In the special case with $M = 0$, the model (8.1)–(8.2) becomes a linear regression model with random coefficients.

The model (8.1)–(8.2) differs from (1.1)–(1.2) mainly by the introduction of the parameter vector $\bar{\beta}$. However, it can be rewritten in the same form as (1.1)–(1.2), so that our results in Section 2 are applicable here as well. Defining $\beta_t^* = \beta_t - \bar{\beta}$ and $\bar{\beta}_t = \bar{\beta}$ for all t, we write (8.1)–(8.2) as:

$$y_t = (x_t \quad x_t)\begin{pmatrix} \bar{\beta}_t \\ \beta_t^* \end{pmatrix} + \varepsilon_t, \tag{8.3}$$

$$\begin{bmatrix} \bar{\beta}_t \\ \beta_t^* \end{bmatrix} = \begin{bmatrix} I & 0 \\ 0 & M \end{bmatrix}\begin{bmatrix} \bar{\beta}_{t-1} \\ \beta_{t-1}^* \end{bmatrix}\begin{bmatrix} 0 \\ \eta_t \end{bmatrix}, \tag{8.4}$$

which is a special case of (1.1)–(1.2). In most applications, not all components of β_t in (8.2) are random. If only a sub-vector $\tilde{\beta}_t$ of β_t consisting of k_1 elements, say, is random, (8.3) and (8.4) will become:

$$y_t = (x_t \quad \tilde{x}_t)\begin{pmatrix} \bar{\beta}_t \\ \tilde{\beta}_t^* \end{pmatrix} + \varepsilon_t, \tag{8.5}$$

$$\begin{bmatrix} \bar{\beta}_t \\ \tilde{\beta}_t^* \end{bmatrix} = \begin{bmatrix} I & 0 \\ 0 & M \end{bmatrix}\begin{bmatrix} \bar{\beta}_{t-1} \\ \tilde{\beta}_{t-1}^* \end{bmatrix} + \begin{bmatrix} 0 \\ \eta_t \end{bmatrix}. \tag{8.6}$$

Since the model (8.5)–(8.6) is a special case of the model (1.1)–(1.2), all the filtering and smoothing equations of Section 2 and the log-likelihood functions (4.1) and (4.3) are applicable to this model. However, the estimation problem for this model deserves a special treatment. Because the roots of M are smaller than one in absolute value and the process generating $\tilde{\beta}_t^*$ is covariance-stationary, one may choose to estimate this model by assuming that the $\tilde{\beta}_t^*$ process starts in a stochastic equilibrium, rather than assuming a fixed, but unknown, initial value $\tilde{\beta}_1^*$ in period one. The latter assumption was made in (3.1), where we used the relation:

$$\tilde{\beta}_t^* = M\tilde{\beta}_{t-1}^* + \eta_t = M^{t-1}\tilde{\beta}_1^* + \eta_t + M\eta_{t-1} + \dots + M^{t-2}\eta_2,$$

and treated $\tilde{\beta}_1^*$ as fixed. In estimating the model (8.5)–(8.6), one may treat $\tilde{\beta}_1^*$ as random, with mean zero and covariance matrix Γ_0 satisfying:

$$\Gamma_0 = M\Gamma_0 M' + V. \tag{8.7}$$

The autocovariance matrix for the $\tilde{\beta}_t^*$ process is:

$$\Gamma_s = \mathrm{E}\tilde{\beta}_t^*\tilde{\beta}_{t-s}^{*\prime} = M^s\Gamma_0 = \Gamma'_{-s} \qquad (s \geq 0; t \geq 1). \tag{8.8}$$

If $\tilde{\beta}_1^*$ is regarded as fixed, instead of (8.7) and (8.8), the covariance matrix of $\tilde{\beta}_t^*$ and $\tilde{\beta}_{t-s}^{*\prime}$ is:

$$\begin{aligned}
&\mathrm{E}\left(\tilde{\beta}_t^* - M^{t-1}\tilde{\beta}_1^*\right)\left(\tilde{\beta}_{t-s}^* - M^{t-s-1}\tilde{\beta}_1^*\right)' \\
&\quad = \mathrm{E}\left(\eta_t + M\eta_{t-1} + \cdots + M^{t-2}\eta_2\right)\left(\eta_{t-s} + M\eta_{t-s-1} + \cdots + M^{t-s-2}\eta_2\right) \\
&\quad = \sum_{i=0}^{t-s-2} M^{s+i}VM'^{i} \qquad (s \geq 0; t \geq 1).
\end{aligned} \tag{8.9}$$

The difference in the treatment of $\tilde{\beta}_1^*$ has implications for estimation. When $\tilde{\beta}_1^*$ is regarded as fixed, all inferences are conditional on this assumption. When $\hat{\beta}_1^*$ is regarded as a random drawing from a distribution with mean zero and covariance matrix Γ_0 as specified by (8.7), the inferences are no longer conditional. Furthermore, to provide the initial estimates $\beta_{k|k}$ and $\Sigma_{k|k}$ to start up Kalman filtering equations for the evaluation of the log-likelihood functions (4.1) and (4.3), the two assumptions lead to different procedures. In the case of fixed $\tilde{\beta}_1^*$, we regard (8.5) as a special case of (1.1). Therefore, the number of initial observations required to perform a generalized least squares regression equals the number of elements in $\bar{\beta}_t$ and $\tilde{\beta}_t^*$, or $k + k_1$, say. (3.3) and (3.4) are applied to these $k + k_1$ observations, and the analysis proceeds as before.

In the case of random $\tilde{\beta}_1^*$, (8.5) can be written as:

$$y_t = x_t\bar{\beta} + \left(\tilde{x}_t\tilde{\beta}_t^* + \varepsilon_t\right) = x_t\bar{\beta} + u_t. \tag{8.10}$$

The term in parentheses or u_t is treated as a serially correlated residual satisfying:

$$\mathrm{E}u_t u'_{t-s} = \tilde{x}_t \Gamma_s \tilde{x}'_{t-s} + \delta_{t,t-s}\sigma^2, \tag{8.11}$$

where Γ_s is defined by (8.8) and $\delta_{t,t-s}$ is the Kronecker delta. Therefore, given (8.11) only k initial observations are required to obtain a GLS estimate $\bar{\beta}_{k|k}$ of $\bar{\beta}$ and its covariance matrix. Writing the first k observations of (8.10) as:

$$y = X\bar{\beta} + u, \tag{8.12}$$

where X is assumed to be a non-singular k by k matrix and $Euu' = W$ is given by (8.11), we have:

$$\bar{\beta}_{k|k} = (X'W^{-1}X)^{-1}X'W^{-1}y = X^{-1}y, \tag{8.13}$$

$$\operatorname{cov}(\bar{\beta}_{k|k} - \bar{\beta}) = (X'W^{-1}X)^{-1} = X^{-1}WX'^{-1}. \tag{8.14}$$

For $\tilde{\beta}^*_k$ in equilibrium, we set its mean equal to zero and its covariance matrix to Γ_0, i.e.

$$\tilde{\beta}^*_{k|k} = 0; \qquad \operatorname{cov}(\tilde{\beta}^*_k - \tilde{\beta}^*_{k|k}) = \Gamma_0. \tag{8.15}$$

The covariance of $\bar{\beta}_{k|k} - \bar{\beta}$ and $\tilde{\beta}^*_k - \tilde{\beta}^*_{k|k}$ is:

$$\mathrm{E}(\bar{\beta}_{k|k} - \bar{\beta})\tilde{\beta}^{*\prime}_k = \mathrm{E}X^{-1}u\tilde{\beta}^{*\prime}_k = X^{-1}\begin{bmatrix} \tilde{x}_1\Gamma'_{k-1} \\ \vdots \\ \tilde{x}_k\Gamma'_0 \end{bmatrix}. \tag{8.16}$$

Equations (8.13)–(8.16) provide the components of $\beta_{k|k}$ and $\Sigma_{k|k}$ to be used for the evaluation of the log-likelihood functions (4.1) and (4.3). They are to be contrasted with (3.3) and (3.4) for fixed β^*_1 which would require $k + k_1$ initial observations.

Once the likelihood function (4.3) can be evaluated, a numerical method can be applied to maximize it with respect to the unknown parameters in $V = \sigma^2 P$ and M. The computations will be simplified when P and M are diagonal, being $\operatorname{diag}\{p_i\}$ and $\operatorname{diag}\{m_i\}$, respectively. Equations (8.7) and (8.8) would become:

$$\gamma_{ii,0} = \mathrm{E}\beta^{*2}_{it} = \frac{\sigma^2 p_i}{1 - m_i^2} \qquad (i = 1, \ldots, k_1), \tag{8.17}$$

$$\gamma_{ii,s} = \mathrm{E}\beta^*_{it}\beta^*_{i,t-s} = m_i^s \gamma_{ii,0} \qquad (i = 1, \ldots, k_1), \tag{8.18}$$

and $E\beta_{ii}^*\beta_{j,t-s}^* = 0$ for $i \neq j$ and for all s. Accordingly the matrix Γ_s used in (8.11) is a diagonal matrix with elements given by (8.17) and (8.18). As an alternative to using the likelihood function (4.3), one can form a likelihood function using the regression model (8.12) for all T observations, as it was done by using the model (4.8) in Section 4.

For further discussion of the stationary-coefficient regression model, the reader is referred to Rosenberg (1973), Cooley and Prescott (1976), Harvey and Phillips (1982), and Pagan (1980). The exposition of this section has drawn from Harvey and Phillips (1982). For a survey of the random-coefficient model, the reader is referred to Swamy (1971, 1974). Swamy and Tinsley (1980) generalize the model (8.1)–(8.2) by replacing $\bar{\beta}$ by $\bar{B}z_t$, z_t being a vector of fixed variables. Kelejian (1974) treats linear simultaneous-equation models with random parameters.

9. Identifiability of parameters

Recently, Pagan (1980) has studied identifiability conditions for the parameters of a regression model with stationary parameters. His model is:

$$y_t = x_t\bar{\beta} + x_t\beta_t^* + \varepsilon_t = x_t\bar{\beta} + u_t, \tag{9.1}$$

$$\beta_t^* = \beta_t - \bar{\beta} = A^{-1}(L)e_t, \tag{9.2}$$

where $A(L)$ is a ratio of polynomials of orders p and q in the lag operator L and e_t is normal, independent and identically distributed, so that β_t^* follows an ARMA(p, q) process. Since an ARMA process can be written as a first-order AR process, as, for example

$$\beta_t^* = A_1\beta_{t-1}^* + A_2\beta_{t-2}^* + e_t + A_3e_{t-1}$$

can be written as

$$\begin{bmatrix} \beta_t^* \\ \beta_{t-1}^* \\ e_t \end{bmatrix} = \begin{bmatrix} A_1 & A_2 & A_3 \\ I & 0 & 0 \\ 0 & 0 & 0 \end{bmatrix} \begin{bmatrix} \beta_{t-1}^* \\ \beta_{t-2}^* \\ e_{t-1} \end{bmatrix} + \begin{bmatrix} e_t \\ 0 \\ e_t \end{bmatrix}$$

or

$$\tilde{\beta}_t^* = M\tilde{\beta}_{t-1}^* + \eta_t,$$

the model (9.1)–(9.2) is formally identical with our model (8.5)–(8.6), with $\tilde{x}_t$ in (8.5) denoting $(x_t \quad 0 \quad 0)$ in the above example. The parameters of (9.1)–(9.2) consist of $\bar{\beta}$, $\sigma^2 = \text{var}\,\varepsilon_t$, and all the parameters in the ARMA(p, q) process for β_t^*. By identification, Pagan means asymptotic local parametric identification, i.e. the non-singularity of $\lim_{T\to\infty} T^{-1}I(\theta^0)$, where I is the information matrix, under certain regularity conditions. Pagan has provided two sets of sufficient conditions for the identifiability of $\bar{\beta}$ and of all parameters, respectively.

First, under the assumptions ($A1$) the ARMA(p, q) process generating $\beta_t^* = \beta_t - \bar{\beta}$ is stationary and obeys the identification conditions set out in Hannan (1968); ($A2$) x_t has an upper bound for all elements $\forall t$, and ($A3$) $\lim_{T\to\infty} T^{-1}X'X$ is positive definite, $\bar{\beta}$ is identifiable.

To state the second set of sufficient conditions, let Γ_j denote $E\beta_t^*\beta_{t-j}^{*\prime}$ as in Section 8, and observe that if all Γ_j are known the parameters of the ARMA(p, q) process (9.2) can be determined by the Yule–Walker equations. Let a subset of all Γ_j ($j \in \psi$) be sufficient to determine the parameters of (9.2) uniquely. Then, if 0 is not in the set ψ, i.e. if Γ_0 is not required to determine the parameters of (9.2), a set of sufficient conditions for the identification of all parameters in (9.1)–(9.2) consists of ($A1$), ($A2$), ($A3$) and ($A4$): the non-singularity of

$$R_k = \lim_{T\to\infty} T^{-1}\sum_t x_{t-k}'x_{t-k}\otimes x_t'x_t,$$

for all $k \in \psi$.

To motivate the condition ($A4$), recall that

$$\begin{aligned} Eu_tu_{t-k} &= x_t\Gamma_k x_{t-k}' + \delta_{t,t-k}\sigma^2 \\ &= (x_{t-k}\otimes x_t)\text{vec}(\Gamma_k) + \delta_{t,t-k}\sigma^2. \end{aligned} \tag{9.3}$$

Thus, vec(Γ_k) is a vector of coefficients in the regression of u_tu_{t-k} on $x_{t-k}\otimes x_t$. TR_k is the cross-product matrix of the explanatory variables in this regression. If it is non-singular, the elements of Γ_k can be consistently estimated, but the knowledge of Γ_k ($k \in \psi$) is sufficient to identify the parameters of the model (9.2). In the case that (9.2) is a first-order autoregressive process $\beta_t^* = M\beta_{t-1}^* + \eta_t$ with diagonal M and $V = E\eta_t\eta_t'$, $\Gamma_k = (\gamma_{ij,k})$ is diagonal and

$$Eu_tu_{t-k} = \sum_j x_{jt}x_{j,t-k}\gamma_{jj,k} \qquad (k > 0).$$

The assumption ($A4$) in this case states that the matrix with

$$\lim_{T\to\infty} T^{-1}\sum_t (x_{it}x_{jt}x_{i,t-k}x_{j,t-k})$$

as its $i-j$ element be non-singular. If Γ_0 is in the set of Γ_k required to determine the parameters of (9.2), ($A4$) should be modified to include:

$$\lim_{T\to\infty} T^{-1}\sum_t w_t w_t' > 0, \quad \text{where } w_t = [x_t \otimes x_t, 1].$$

Pagan (1980) also shows that if (a) the model is locally asymptotically identified; (b) (i) x_t is uniformly bounded from above and non-stochastic; (ii) the model in state-space form is uniformly completely observable and uniformly completely controllable; (iii) the characteristic roots of M are smaller than one in absolute value, and (c) the true parameter vector θ^0 is an interior point of the permissible parameter space which is a subset of R^s, then the maximum likelihood estimator of θ^0 is consistent and has a limiting distribution which is normal with a covariance matrix equal to the inverse of the information matrix. If the transition matrix M is given, the conclusion holds for the maximum likelihood estimator of the remaining parameters, with assumption (b) (iii) deleted.

For the case of regression with non-stationary coefficients, i.e. M having characteristic roots equal to unity, conditions for the identifiability of the parameters remain to be further investigated. Hatanaka and Tanaka (1980) have studied the identifiability conditions under the assumption that β_0 has a known normal prior distribution.

10. Testing constancy of regression coefficients

An important question in regression analysis is whether the coefficients for different observations are identical. A test frequently employed is to divide the observations into two groups and test the null hypothesis of the equality of the entire set or a subset of coefficients in the two regressions using an F statistic. A number of other tests have been suggested for the null hypothesis of constancy of regression coefficients, partly depending on the alternative hypotheses to be compared. A useful alternative hypothesis is that the vector $\tilde{\beta}_t$ of k_1 regression coefficients of interest is generated by the process:

$$\tilde{\beta}_t = \tilde{\beta}_{t-1} + \eta_t, \tag{10.1}$$

where η_t is normally and independently distributed with mean zero and a diagonal covariance matrix V. The null hypothesis states that $V = 0$. Several tests of the null hypothesis have been suggested.

First, by the asymptotic normality of the maximum likelihood estimator of the elements of V, one can use a quadratic form in these elements weighted by the inverse of their covariance matrix (obtained from the information matrix) and approximate its distribution by a χ^2 distribution, but this approximation is crude.

Second is the likelihood ratio test. Let λ be the ratio of the maximum likelihood functions under the null and alternative hypotheses. However, the distribution of $-2\log\lambda$ under the null hypothesis $V=0$ is not well approximated by a χ^2 distribution because, with the restriction that the maximum likelihood estimates of the diagonal elements of V be non-negative, there is a density mass at $\hat{V}=0$ and, accordingly, the distribution of $-2\log\lambda$ is more concentrated toward the origin than the χ^2 distribution. Therefore, applying the χ^2 distribution will lead to rejecting the null hypothesis less frequently than the stated level of significance. Garbade (1977) has applied the likelihood ratio test and studied the sampling distribution of $-2\log\lambda$ by Monte Carlo experiments. It would be worthwhile to obtain a better approximation to the distribution of the likelihood ratio test statistic for the present problem when some parameters are subject to inequality constraints along the lines of Chernoff (1954), Moran (1970) and Gourieroux, Holly and Montfort (1982).

Third is an application of the Lagrangian multiplier test of Silvey (1959) or, equivalently, the score test of Rao (1973, p. 417). With $\tilde{\beta}_0$ regarded as fixed, we have:

$$\tilde{\beta}_t = \tilde{\beta}_0 + \sum_{s=1}^{t} \eta_s,$$

and the regression model is:

$$y_t = x_t\beta_t + \varepsilon_t = x_t\beta + \varepsilon_t + \tilde{x}_t\left(\sum_1^t \eta_s\right) = x_t\beta + u_t \qquad (t=1,\dots,T), \tag{10.2}$$

where β has a sub-vector $\tilde{\beta}_0$ corresponding to the time-varying coefficients and

$$\mathrm{E}u_t^2 = \sigma^2 + t\tilde{x}_t V\tilde{x}_t'. \tag{10.3}$$

For $V=\operatorname{diag}\{v_{ii}\}$, (10.3) becomes:

$$\mathrm{E}u_t^2 = \sigma^2 + t\sum_{i=1}^{k_1} x_{it}^2 v_{ii} \qquad (t=1,\dots,T) \tag{10.4}$$

Let $\hat{u}_t$ be the residuals of (10.2) estimated by OLS, and $\hat{\sigma}^2 = T^{-1}\Sigma\hat{u}_t^2$. According to a theorem of Breusch and Pagan (1979, p. 1288), the Lagrangian multiplier statistic for testing H_0: $v_{ii}=0$ $(i=1,\dots,k_1)$ in (10.4) equals one-half the explained sum of squares in a regression of $T(\Sigma\hat{u}_t^2)^{-1}\cdot\hat{u}_t^2$ on tx_{it}^2 $(i=1,\dots,k_1)$, and is asymptotically distributed as $\chi^2(k_1)$ when the null hypothesis is true. Although this test is based on the Lagrangian multiplier, its null hypothesis is formulated solely on the relations (10.4), and ignores the relations:

$$\mathrm{E}u_t u_s = \min(t,s)\sum_{i=1}^{k_1} x_{it}x_{is}v_{ii}, \qquad t \neq s,$$

for the parameters v_{ii}. Thus, the null hypothesis, though implied by $v_{ii}=0$, is not equivalent to it.

Fourth is a test proposed by Pagan and Tanaka (1979) based on the score test statistic of Rao (1973), namely, $S=\hat{d}_1'\hat{I}^{11}\hat{d}_1$, where $\hat{d}_1$ is a vector of first derivatives of the log-likelihood with respect to the parameters v_{ii} $(i=1,\dots,k_1)$ and $\hat{I}^{11}$ is the sub-matrix of the inverse of the information matrix which corresponds to these parameters, both evaluated at the maximum likelihood estimates obtained under H_0: $v_{ii}=0$ $(i=1,\dots,k_1)$. Pagan showed how the exact distribution of this statistic can be evaluated under H_0, and concluded by Monte Carlo experiments that, unlike the likelihood ratio test statistic $-2\log\lambda$, the score test statistic can be well approximated by the χ^2 distribution.

Fifth is a test suggested by LaMotte and McWhorter (1978). They assume $V=\sigma_\eta^2 D$, where the diagonal matrix D is known; therefore, in our notations, $\beta_t=\bar{\beta}_t$ with the constant coefficients corresponding to the zero diagonal elements of D. Consider T observations of the regression model (10.2). The covariance matrix of the vector u of its T residuals is:

$$\mathrm{E}uu' = I_T\sigma^2 + \sigma_\eta^2\Phi,$$

with the $t-s$ element of Φ equal to $\min(t,s)x_t D x_s'$. Define:

$$HH' = I_T - X(X'X)^{-1}X' \quad \text{and} \quad H'H = I_{T-k}.$$

Let $\lambda_1,\dots,\lambda_q$ be the distinct characteristic roots of $H'\Phi H$ with multiplicities $r_1,\dots,r_q$, respectively. Let P_i be a $(T-k)\times r_i$ matrix whose columns are orthonormal characteristic vectors of $H'\Phi H$, and Q_i be $y'HP_iP_i'H'y$. It can be shown that $(\sigma^2+\lambda_i\sigma_\eta^2)^{-1}Q_i$ is distributed as $\chi^2(r_i)$. Assume that $\lambda_1>\lambda_2>\dots>\lambda_q$ and partition $(Q_1,\dots,Q_q)$ into two sets $(Q_1,\dots,Q_g)$ and $(Q_{g+1},\dots,Q_q)$ for some $0<g<q$. Under the null hypothesis $\sigma_\eta^2=0$, LaMotte and McWhorter showed that

$$\left[\sum_{i=1}^{g} Q_1 \Big/ \sum_{i=1}^{q} r_i\right] \Big/ \left[y'HH'y \Big/ \left(T-k-\sum_{i=1}^{g} r_i\right)\right]$$

will have an F distribution with $\sum_1^g r_i$ and $T-k-\sum_1^g r_i$ degrees of freedom; they also suggested a procedure for selecting the value g to increase the power of the test, although a uniformly most powerful test is not available.

The above brief review of several tests of the constancy of regression coefficients indicates that further research is required to obtain a uniformly most powerful test statistic which has a known distribution in small samples and is also computationally simple.

11. Problems for research

In concluding this survey of models with time-varying coefficients, I would like to list several problems for research.

First, computational problems require further attention for several of the methods discussed in this paper: in the numerical maximization of the likelihood functions of the basic regression model in Section 4, of the model of system of linear regressions in Section 5, and of the models of linear and non-linear simultaneous equations in Sections 6 and 7.

Second, the identification problem remains to be further investigated for both models with roots of unity in the transition matrix and models with stationary coefficients, but especially the former. It would be desirable to find some useful conditions for the identifiability of the parameters of the models of Sections 4, 5, 6 and 7.

Third, the finite-sample distributions of many of the statistics used in the regression model with changing coefficients can be further examined. If σ^2, V and M are known, the estimate of β_t is a GLS estimate and is normal, best linear unbiased. When maximum likelihood estimates of σ^2, V and possibly M are used, not only their own sampling distributions but the sampling distributions of the estimates of β_t based on them in finite samples are not sufficiently known. In particular, the estimates of the diagonal elements of V are subject to non-negativity constraints and special attention needs to be given to their sampling distributions.

Fourth, the problem of estimating simultaneous-equation models with changing parameters deserves to be further studied. How good is the linearization approach suggested in Sections 6 and 7? What are the sampling distributions of the estimates obtained in finite samples? Other approaches than the linearization approach should be considered, including, for example, the use of second-order terms in the expansion of the model equations, the application of the method of instrumental variables, and the search for limited-information methods of estimation, as compared with the full-information method suggested.

Fifth, as mentioned in Section 10, the problem of testing the constancy of regression coefficients with the time-varying coefficient model serving as the alternative is by no means completely resolved, although several useful solutions have been suggested. Surely, the problem of testing the constancy of parameters in simultaneous-equation models, with the corresponding models with time-varying parameters serving as alternatives, is open.

Finally, the applications of the models discussed in this paper to economic problems will most likely continue. Applications to specific applied problems will generate problems of their own. A number of applications have appeared, several having been cited in LaMotte and McWhorter (1978, p. 816), for example. An illustration of an applied problem having its special features which deserve special treatment is the estimation of seasonal components in economic time-series, as discussed in Pagan (1975), Engle (1978), and Chow (1978).

References

Arora, Swarnjit S. (1973) "Error Components Regression Model and Their Applications", *Annals of Economic and Social Measurement*, 2, 451–462.

Belsley, David A. (1973) "On the Determination of Systematic Parameter Variation in the Linear Regression Model", *Annals of Economic and Social Measurement*, 2, 487–494.

Belsley, David A. (1973) "A Test for Systematic Variation in Regression Coefficients", *Annals of Economic and Social Measurement*, 2, 495–500.

Belsley, David A. (1973) "The Applicability of the Kalman Filter in the Determination of Systematic Parameter Variation", *Annals of Economic and Social Measurement*, 2, 531–534.

Belsley, David A. and Kuh, Edwin (1973) "Time-varying Parameter Structures: An Overview", *Annals of Economic and Social Measurement*, 2, 375–380.

Breusch, T. S. and Pagan, A. R. (1979) "A Simple Test for Heteroscedasticity and Random Coefficient Variation", *Econometrica*, 47, 1287–1294.

Brown, R. L., Durbin, J. and Evans J. M. (1975) "Techniques for Testing the Constancy of Regression Relationships Over Time, with Comments", *Journal of the Royal Statistical Society* B-37, 149–192.

Chernoff, H. (1954) "On the Distribution of the Likelihood Ratio", *Annals of Mathematical Statistics*, 25, 573–578.

Chow, G. C. (1960) "Tests of Equality Between Sets of Coefficients in Two Linear Regressions", *Econometrica*, 28, 591–605.

Chow, G. C. (1975) *Analysis and Control of Dynamic Systems*. New York: John Wiley and Sons, Inc.

Chow, G. C. (1978) "Comments on 'A Time Series Analysis of Seasonality in Econometric Models' by Charles Plosser", in A. Zellner (ed.), *Seasonal Analysis of Economic Time Series* (U.S. Department of Commerce, Bureau of the Census, Economic Research Report ER-1, 398–401.

Chow, G. C. (1981) *Econometric Analysis by Control Methods*. New York: John Wiley and Sons, Inc.

Cooley, T. F. and Prescott, E. (1973) "An Adaptive Regression Model", *International Economic Review*, 14, 364–371.

Cooley, T. F. and Prescott, E. (1973) "Tests of an Adaptive Regression Model", *Review of Economics and Statistics*, 55, 248–256.

Cooley, T. F. and Prescott, E. (1973) "Varying Parameter Regression: A Theory and Some Applications", *Annals of Economic and Social Measurement*, 2, 463–474.

Cooley, T. F. and Prescott, E. (1976) "Estimation in the Presence of Stochastic Parameter Variation", *Econometrica*, 44, 167–183.

Cooley, T. F., Rosenberg, B. and Wall, K. (1977) "A Note on Optimal Smoothing for Time-varying Coefficient Problems", *Annals of Economic and Social Measurement*, 6, 453–456.

Cooper, J. P. (1973) "Time-varying Regression Coefficients: A Mixed Estimation Approach and Operational Limitations of the General Markov Structure", *Annals of Economic and Social Measurement*, 2, 525–530.

Duncan, D. B. and Horn, S. D. (1972) "Linear Dynamic Recursive Estimation From the Viewpoint of Regression Analysis", *Journal of the American Statistical Association*, 67, 815–821.

Engle, R. F. (1978) "Estimating Structural Models of Seasonality", in A. Zellner (ed.), *Seasonal Analysis of Economic Time Series* (U.S. Department of Commerce, Bureau of the Census, Economic Research Report ER-1), 281–297.

Garbade, K. (1977) "Two Methods for Examining the Stability of Regression Coefficients", *Journal of the American Statistical Association*, 72, 54–63.

Goldfeld, S. M. and Quandt, R. E. (1973) "The Estimation of Structural Shifts by Switching Regressions", *Annals of Economic and Social Measurement*, 2, 475–486.

Gourieroux, C., Holly, A. and Montfort, A. (1982) "Likelihood Ratio Test, Wald Test and Kühn-Tucker Test in Linear Models with Inequality Constraints on the Regression Parameters", *Econometrica*, 50, 63-80.

Hannan, E. J. (1968) "The Identification of Vector Mixed Autoregressive–Moving Average Systems", *Biometrika*, 56: 223–225.

Harvey, A. and Phillips, G. D. A. (1976) "The Maximum Likelihood Estimation of Autoregressive-Moving Average Models by Kalman Filtering". University of Kent at Canterbury, Working Paper No. 3, S.S.R.S. Supported Project on Testing for Specification Error in Econometric Models.

Harvey, A. and Collier, P. (1977) "Testing for Functional Misspecification in Regression Analysis", *Journal of Econometrics*, 6, 103–120.

Harvey, A. C. (1978) "The Estimation of Time-Varying Parameters from Panel Data", *Annales de l'INSEE*, 30–31, 203–226.

Harvey, A. C. and Phillips, G. D. A. (1979) "Maximum Likelihood Estimation of Regression Models with Autoregressive-Moving Average Disturbances", *Biometrika*, 66.

Harvey, A.C. and Phillips, G.D A. (1982) "The Estimation of Regression Models with Time-Varying Parameters", in: M. Deistler, E. Fürst and G. Schwödiauer, eds., *Games, Economic Dynamics and Time Series Analysis*. Cambridge, Mass.: Physica-Verlag.

Hatanaka, M. (1980) "A Note on the Applicability of the Kalman Filter to Regression Models with Some Parameters Time-Varying and Others Invarying", *Australian Journal of Statistics*, 22, 298–306.

Hatanaka, M. and Tanaka, K. (1980) "On the Estimability of the Covariance Matrix in the Random Walk Representing the Time-Changing Parameters of the Regression Models" Osaka University, Faculty of Economics, Discussion Paper No. 28.

Ho, Y. C. and Lee, R. C. K. (1964) "A Bayesian Approach to Problems in Stochastic Estimation and Control", *IEEE Transactions on Automatic Control*, AC-9, 333–339.

Kalman, R. E. (1960) "A New Approach to Linear Filtering and Prediction Problems", *Transactions of ASME* (American Society of Mechanical Engineers), Series D: *Journal of Basic Engineering*, 82, 35–45.

Kelejian, H. H. (1974) "Random Parameters in Simultaneous Equation Framework: Identification and Estimation", *Econometrica*, 42, 517–527.

LaMotte, L. R. and McWhorter, A., Jr. (1978) "An Exact Test for the Presence of Random Walk Coefficients in a Linear Regression Model", *Journal of the American Statistical Association*, 73, 816–820.

Ljung, L. (1979) "Asymptotic Behavior of the Extended Kalman Filter as a Parameter Estimator for Linear Systems", *IEEE Transactions on Automatic Control*, AC-24, 36–50.

Mehra, R. K. (1970) "On the Identification of Variances and Adaptive Kalman Filtering", *IEEE Transactions on Automatic Control*, AC15, 175–184.

Mehra, R. K. (1971) "Identification of Stochastic Linear Dynamic Systems Using Kalman Filter Representation", *AIAA Journal*, 9, 28–31.

Mehra, R. K. "Kalman Filters and Their Applications to Forecasting", in: *TIMS Studies in the Management Sciences*. Amsterdam: North-Holland, 75–94.

Moran, P. A. P. (1970) "On Asymptotically Optimal Tests of Composite Hypotheses", *Biometrika*, 57, 47–55.

Pagan, A. R. (1975) "A Note on the Extraction of Components From Time Series", *Econometrica*, 43, 163–168.

Pagan, A. R. (1978) "A Unified Approach to Estimation and Inference for Stochastically Varying Coefficient Regression Models", Center for Operations Research and Econometrics, Louvain-la-Neuve, CORE Discussion Paper 7814.

Pagan, A. R. (1980) "Some Identification and Estimation Results for Regression Models with Stochastically Varying Coefficients", *Journal of Econometrics*, 13, 341–363.

Pagan, A. R. and Tanaka, K. (1979) "A Further Test for Assessing the Stability of Regression Coefficients". Australian National University, Canberra, unpublished manuscript.

Phillips, G. D. A. and Harvey, A. C. (1974) "A Simple Test for Serial Correlation in Regression Analysis", *Journal of the American Statistical Association*, 69, 935–939.

Poirier, D. (1976) *The Econometrics of Structural Change*. Amsterdam: North-Holland Publishing Company.

Rao, C. R. (1973) *Linear Statistical Inference and its Applications*. 2nd Edition, New York: John Wiley and Sons, Inc.

Rosenberg, B. (1977) "Estimation Error Covariance in Regression with Sequentially Varying Parameters", *Annals of Economic and Social Measurement*, 6, 457–462.

Rosenberg, B. (1973) "A Survey of Stochastic Parameter Regression", *Annals of Economic and Social Measurement*, 2, 381–398.

Rosenberg, B. (1973) "The Analysis of a Cross Section of Time Series by Stochastically Convergent Parameter Regression", *Annals of Economic and Social Measurement*, 2, 399–428.

Rosenberg, B. (1972) "Estimation of Stationary Stochastic Regression Parameters Reexamined", *Journal of the American Statistical Association*, 67, 650–654.

Sage, A. P. (1968) *Optimum Systems Control*. Englewood Cliffs: Prentice-Hall, Inc., 265–275.

Sant, D. (1977) "Generalized Least Squares Applied to Time-Varying Parameter Models", *Annals of Economic and Social Measurement*, 6, 301–314.

Sarris, A. H. (1973) "A Bayesian Approach to Estimation of Time-Varying Regression Coefficients", *Annals of Economic and Social Measurement*, 2, 501–524.

Schweder, T. (1976) "Some Optimal Methods to Detect Structural Shifts or Outliers in Regression", *Journal of the American Statistical Association*, 71, 491–501.

Schweppe, F. C. (1965) "Evaluation of Likelihood Functions for Gaussian Signals", *IEEE Transactions on Information Theory*, 11, 61–70.

Silvey, S. D. (1959) "The Lagrangian Multiplier Test", *Annals of Mathematical Statistics*, 30, 389–407.

Swamy, P. A. V. B. (1971) *Statistical Inference in Random Coefficient Regression Models*. New York: Springer Verlag.

Swamy, P. A. V. B., (1973) "Criteria, Constraints and Multicollinearity in Random Coefficient Regression Models", *Annals of Economic and Social Measurement*, 2, 429–450.

Swamy, P. A. V. B. (1974) "Linear Models with Random Coefficients", in P. Zarembka (ed.), *Frontiers in Econometrics*. New York: Academic Press, 143–168.

Swamy, P. A. V. B. (1979) "Relative Efficiencies of Some Simple Bayes Estimators of Coefficients in a Dynamic Equation with Serially Correlated Errors-II", *Journal of Econometrics*, 7, 245–258.

Swamy, P. A. V. B. and Mehta, J. S. (1975) "Bayesian and Non-Bayesian Analysis of Switching Regressions and of Random Coefficient Regression Models", *Journal of the American Statistical Association*, 70, 593–602.

Swamy, P. A. V. B. and Tinsley, P. (1980) "Linear Prediction and Estimation Methods for Regression Models with Stationary Stochastic Coefficients", *Journal of Econometrics*, 12, 103–142.

Taylor, L. (1970) "The Existence of Optimal Distributed Lags", *Review of Economic Studies*, 37, 95–106.

Tesfatsion, L. (1978) "A New Approach to Filtering and Adaptive Control", *Journal of Optimization Theory and Applications*, 25, 247–261.

Tesfatsion, L. (1979) "Direct Updating of Intertemporal Criterion Functions for a Class of Adaptive Control Problems", *IEEE Transactions on Systems, Man, and Cybernetics*, SMC-9, 143–151.

Wall, K. (1976) "Time-Varying Models in Econometrics: Identifiability and Estimation". Unpublished manuscript.

Zellner, A. (1970) "Estimation of Regression Relationships Containing Unobservable Variables", *International Economic Review*, 11, 441–454.

Chapter 22

PANEL DATA

GARY CHAMBERLAIN*
University of Wisconsin-Madison and NBER

Contents

*I am grateful to a number of individuals for helpful discussions. Financial support was provided by the National Science Foundation (Grants No. SOC-7925959 and No. SES-8016383), by the University of Wisconsin Graduate School, and by funds granted to the Institute for Research on Poverty at the University of Wisconsin-Madison by the Department of Health, Education and Welfare pursuant to the provisions of the Economic Opportunity Act of 1964.

Handbook of Econometrics, Volume II, Edited by Z. Griliches and M.D. Intriligator

1. Introduction and summary

The chapter has four parts: the specification of linear models; the specification of nonlinear models; statistical inference; and empirical applications. The choice of topics is highly selective. We shall focus on a few problems and try to develop solutions in some detail.

The discussion of linear models begins with the following specification:

$$y_{it} = \beta x_{it} + c_i + u_{it}, \tag{1.1}$$

$$\mathrm{E}(u_{it} | x_{i1}, \dots, x_{iT}, c_i) = 0 \qquad (i = 1, \dots, N;\, t = 1, \dots, T). \tag{1.2}$$

For example, in a panel of farms observed over several years, suppose that y_{it} is a measure of the output of the ith farm in the tth season, x_{it} is a measured input that varies over time, c_i is an unmeasured, fixed input reflecting soil quality and other characteristics of the farm's location, and u_{it} reflects unmeasured inputs that vary over time such as rainfall.

Suppose that data is available on $(x_{i1}, \dots, x_{iT}, y_{i1}, \dots, y_{iT})$ for each of a large number of units, but c_i is not observed. A cross-section regression of y_{i1} on x_{i1} will give a biased estimate of β if c is correlated with x, as we would expect it to be in the production function example. Furthermore, with a single cross section, there may be no internal evidence of this bias. If $T > 1$, we can solve this problem given the assumption in (1.2). The change in y satisfies:

$$\mathrm{E}(y_{i2} - y_{i1} | x_{i2} - x_{i1}) = \beta(x_{i2} - x_{i1}),$$

and the least squares regression of $y_{i2} - y_{i1}$ on $x_{i2} - x_{i1}$ provides a consistent estimator of β (as $N \to \infty$) if the change in x has sufficient variation. A generalization of this estimator when $T > 2$ can be obtained from a least squares regression with individual specific intercepts, as in Mundlak (1961).

The restriction in (1.2) is necessary for this result. For example, consider the following autoregressive specification:

$$y_{it} = \beta y_{i,t-1} + c_i + u_{it},$$

$$\mathrm{E}(u_{it} | y_{i,t-1}, c_i) = 0.$$

It is clear that a regression of $y_{it} - y_{i,t-1}$ on $y_{i,t-1} - y_{i,t-2}$ will not provide a consistent estimator of β, since $u_{it} - u_{i,t-1}$ is correlated with $y_{i,t-1} - y_{i,t-2}$. Hence, it is not sufficient to assume that:

$$\mathrm{E}(u_{it} | x_{it}, c_i) = 0.$$

Much of our discussion will be directed at testing the stronger restriction in (1.2).

Consider the (minimum mean-square error) linear predictor of c_i conditional on $x_{i1},\dots,x_{iT}$:

$$\mathrm{E}^*(c_i|x_{i1},\dots,x_{iT}) = \eta + \lambda_1 x_{i1} + \cdots + \lambda_T x_{iT}. \tag{1.3}$$

Given the assumptions that variances are finite and that the distribution of $(x_{i1},\dots,x_{iT},c_i)$ does not depend upon i, there are no additional restrictions in (1.3); it is simply notation for the linear predictor. Now consider the linear predictor of y_{it} given $x_{i1},\dots,x_{iT}$:

$$\mathrm{E}^*(y_{it}|x_{i1},\dots,x_{iT}) = \zeta_t + \pi_{t1}x_{i1} + \cdots + \pi_{tT}x_{iT}.$$

Form the $T \times T$ matrix $\boldsymbol{\Pi}$ with π_{ts} as the (t,s) element. Then the restriction in (1.2) implies that $\boldsymbol{\Pi}$ has a distinctive structure:

$$\boldsymbol{\Pi} = \beta \boldsymbol{I} + \boldsymbol{l}\boldsymbol{\lambda}',$$

where $\boldsymbol{I}$ is the $T \times T$ identity matrix, $\boldsymbol{l}$ is a $T \times 1$ vector of ones, and $\boldsymbol{\lambda}' = (\lambda_1,\dots,\lambda_T)$. A test for this structure could usefully accompany estimators of β based on change regressions or on regressions with individual specific intercepts. Moreover, this formulation suggests an alternative estimator for β, which is developed in the inference section.

This test is an exogeneity test and it is useful to relate it to Granger (1969) and Sims (1972) causality. The novel feature is that we are testing for noncausality conditional on a latent variable. Suppose that $t=1$ is the first period of the individual's (economic) life. Within the linear predictor context, a Granger definition of "y does not cause x conditional on a latent variable c" is:

$$\mathrm{E}^*(x_{i,t+1}|x_{i1},\dots,x_{it},y_{i1},\dots,y_{it},c_i) = \mathrm{E}^*(x_{i,t+1}|x_{i1},\dots,x_{it},c_i) \qquad (t=1,2,\dots).$$

A Sims definition is:

$$\mathrm{E}^*(y_{it}|x_{i1},x_{i2},\dots,c_i) = \mathrm{E}^*(y_{it}|x_{i1},\dots,x_{it},c_i) \qquad (t=1,2,\dots).$$

In fact, these two definitions imply identical restrictions on the covariance matrix of $(x_{i1},\dots,x_{iT},y_{i1},\dots,y_{iT})$. The Sims form fits directly into the $\boldsymbol{\Pi}$ matrix framework and implies the following restrictions:

$$\boldsymbol{\Pi} = \boldsymbol{B} + \boldsymbol{\gamma}\boldsymbol{\lambda}', \tag{1.4}$$

where $\boldsymbol{B}$ is a lower triangular matrix and $\boldsymbol{\gamma}$ is a $T \times 1$ vector. We show how these nonlinear restrictions can be transformed into linear restrictions on a standard simultaneous equations model.

A Π matrix in the form (1.4) occurs in the autoregressive model of Balestra and Nerlove (1966). The $\gamma\lambda'$ term is generated by the projection of the initial condition onto the x's. We also consider autoregressive models in which a time-invariant omitted variable is correlated with the x's.

The methods we shall discuss rely on the measured x_{it} changing over time whereas the unmeasured c_i is time invariant. It seems plausible to me that panel data should be useful in separating the effects of x_{it} and c_i in this case. An important limitation, however, is that measured, time-invariant variables (z_i) can be absorbed into c_i. Their effects are not identified without further restrictions that distinguish them from c_i. Some solutions to this problem are discussed in Chamberlain (1978) and in Hausman and Taylor (1981).

In Section 3 we use a multivariate probit model to illustrate the new issues that arise in models that are nonlinear in the variables. Consider the following specification:

$$\tilde{y}_{it} = \beta x_{it} + c_i + u_{it},$$
$$y_{it} = 1, \quad \text{if } \tilde{y}_{it} \geq 0,$$
$$= 0, \quad \text{otherwise} \qquad (i=1,\dots,N;\ t=1,\dots,T),$$

where, conditional on $x_{i1},\dots,x_{iT},c_i$, the distribution of $(u_{i1},\dots,u_{iT})$ is multivariate normal ($N(\mathbf{0},\Sigma)$) with mean $\mathbf{0}$ and covariance matrix $\Sigma = (\sigma_{jk})$. We observe $(x_{i1},\dots,x_{iT},y_{i1},\dots,y_{iT})$ for a large number of individuals, but we do not observe c_i. For example, in the reduced form of a labor force participation model, y_{it} can indicate whether or not the ith individual worked during period t, x_{it} can be a measure of the presence of young children, and c_i can capture unmeasured characteristics of the individual that are stable at least over the sample period. In the certainty model of Heckman and MaCurdy (1980), c_i is generated by the single life-time budget constraint.

If we treat the c_i as parameters to be estimated, then there is a severe incidental parameter problem. The consistency of the maximum likelihood estimator requires that $T \to \infty$, but we want to do asymptotic inference with $N \to \infty$ for fixed T, which reflects the sample sizes in the panel data sets we are most interested in. So we consider a random effects estimator, which is based on the following specification for the distribution of c conditional on x:

$$c_i = \eta + \lambda_1 x_{i1} + \cdots + \lambda_T x_{iT} + v_i, \tag{1.5}$$

where the distribution of v_i conditional on $x_{i1},\dots,x_{iT}$ is $N(0,\sigma_v^2)$. This is similar to our specification in (1.3) for the linear model, but there is an important difference; (1.3) was just notation for the linear predictor, whereas (1.5) embodies substantive restrictions. We are assuming that the regression function of c on the

x's is linear and that the residual variation is homoskedastic and normal. Given these assumptions, our analysis runs parallel to the linear case. There is a matrix Π of multivariate probit coefficients which has the following structure:

$$\Pi = \text{diag}\{\alpha_1,\dots,\alpha_T\}[\beta I + l\lambda'],$$

where $\text{diag}\{\alpha_1,\dots,\alpha_T\}$ is a diagonal matrix of normalization factors with $\alpha_t = (\sigma_{tt} + \sigma_v^2)^{-1/2}$. We can impose these restrictions to obtain an estimator of $\alpha_t \beta$ which is consistent as $N \to \infty$ for fixed T. We can also test whether Π in fact has this structure.

A quite different treatment of the incidental parameter problem is possible with a logit functional form for $P(y_{it} = 1|x_{it}, c_i)$. The sum $\sum_{t=1}^T y_{it}$ provides a sufficient statistic for c_i. Hence we can use the distribution of $y_{i1},\dots,y_{iT}$ conditional on $x_{i1},\dots,x_{iT}, \sum_t y_{it}$ to obtain a conditional likelihood function that does not depend upon c_i. Maximizing it with respect to β provides an estimator that is consistent as $N \to \infty$ for fixed T, and the other standard properties for maximum likelihood hold as well. The power of the procedure is that it places no restrictions on the conditional distribution of c given x. It is perhaps the closest analog to the change regression in the linear model. A shortcoming is that the residual covariance matrix is constrained to be equicorrelated. Just as in the probit model, a key assumption is:

$$P(y_{it} = 1|x_{i1},\dots,x_{iT}, c_i) = P(y_{it} = 1|x_{it}, c_i), \tag{1.6}$$

and we discuss how it can be tested.

It is natural to ask whether (1.6) is testable without imposing the various functional form restrictions that underlie our tests in the probit and logit cases. First, some definitions. Suppose that $t = 1$ is the initial period of the individual's (economic) life; an extension of Sims' condition for x to be strictly exogenous is that y_t is independent of $x_{t+1}, x_{t+2},\dots$ conditional on $x_1,\dots,x_t$. An extension of Granger's condition for "y does not cause x" is that x_{t+1} is independent of $y_1,\dots,y_t$ conditional on $x_1,\dots,x_t$. Unlike the linear predictor case, now strict exogeneity is weaker than noncausality. Noncausality requires that y_t be independent of $x_{t+1}, x_{t+2},\dots$ conditional on $x_1,\dots,x_t$ *and* on $y_1,\dots,y_{t-1}$. If x is strictly exogenous and in addition y_t is independent of $x_1,\dots,x_{t-1}$ conditional on x_t, then we shall say that the relationship of x to y is static.

Then our question is whether it is restrictive to assert that there exists a latent variable c such that the relationship of x to y is static conditional on c. We know that this is restrictive in the linear predictor case, since the weaker condition that x be strictly exogenous conditional on c is restrictive. Unfortunately, there are no restrictions when we replace zero partial correlation by conditional independence. It follows that conditional strict exogeneity is restrictive only when combined with specific functional forms—a truly nonparametric test cannot exist.

Section 4 presents our framework for inference. Let $\boldsymbol{r}_i' = (1, x_{i1}, \dots, x_{iT}, y_{i1}, \dots, y_{iT})$ and assume that $\boldsymbol{r}_i$ is independent and identically distributed (i.i.d.) for $i = 1, 2, \dots$. Let $\boldsymbol{w}_i$ be the vector formed from the squares and cross-products of the elements in $\boldsymbol{r}_i$. Our framework is based on a simple observation: the matrix $\boldsymbol{\Pi}$ of linear predictor coefficients is a function of $\mathrm{E}(\boldsymbol{w}_i)$; if $\boldsymbol{r}_i$ is i.i.d. then so is $\boldsymbol{w}_i$; hence our problem is to make inferences about a function of a population mean under random sampling. This is straightforward and provides an asymptotic distribution theory for least squares that does not require a linear regression function or homoskedasticity.

Stack the columns of $\boldsymbol{\Pi}'$ into a vector $\boldsymbol{\pi}$ and let $\boldsymbol{\pi} = \boldsymbol{h}(\boldsymbol{\mu})$, where $\boldsymbol{\mu} = \mathrm{E}(\boldsymbol{w}_i)$. Then the limiting distribution for least squares is normal with covariance matrix:

$$\boldsymbol{\Omega} = \frac{\partial \boldsymbol{h}}{\partial \boldsymbol{\mu}'} V(\boldsymbol{w}_i) \frac{\partial \boldsymbol{h}'}{\partial \boldsymbol{\mu}}.$$

We impose restrictions on $\boldsymbol{\Pi}$ by using a minimum distance estimator. The restrictions can be expressed as $\boldsymbol{\mu} = \boldsymbol{g}(\boldsymbol{\theta})$, where $\boldsymbol{\theta}$ is free to vary within some set Υ. Given the sample mean $\bar{\boldsymbol{w}} = \sum_{i=1}^{N} \boldsymbol{w}_i / N$, we choose $\hat{\boldsymbol{\theta}}$ to minimize the distance between $\bar{\boldsymbol{w}}$ and $\boldsymbol{g}(\boldsymbol{\theta})$, using the following distance function:

$$\min_{\boldsymbol{\theta} \in \Upsilon} [\bar{\boldsymbol{w}} - \boldsymbol{g}(\boldsymbol{\theta})]' \hat{V}^{-1}(\boldsymbol{w}_i) [\bar{\boldsymbol{w}} - \boldsymbol{g}(\boldsymbol{\theta})],$$

where $\hat{V}(\boldsymbol{w}_i)$ is a consistent estimator of $V(\boldsymbol{w}_i)$. This is a generalized least squares estimator for a multivariate regression model with nonlinear restrictions on the parameters; the only explanatory variable is a constant term. The limiting distribution of $\hat{\boldsymbol{\theta}}$ is normal with covariance matrix:

$$\left[\frac{\partial \boldsymbol{g}'}{\partial \boldsymbol{\theta}} V^{-1}(\boldsymbol{w}_i) \frac{\partial \boldsymbol{g}}{\partial \boldsymbol{\theta}'} \right]^{-1}.$$

An asymptotic distribution theory is also available when we use some matrix other than $\hat{V}^{-1}(\boldsymbol{w}_i)$ in the distance function. This theory shows that $\hat{V}^{-1}(\boldsymbol{w}_i)$ is the optimal choice. However, by using suboptimal norms, we can place a number of commonly used estimators within this framework.

The results on efficient estimation have some surprising consequences. The simplest example is a univariate linear predictor: $\mathrm{E}^*(y_i | x_{i1}, x_{i2}) = \pi_0 + \pi_1 x_{i1} + \pi_2 x_{i2}$. Consider imposing the restriction that $\pi_2 = 0$; we do not want to maintain any other restrictions, such as linear regression, homoskedasticity, or normality. How shall we estimate π_1? Let $\hat{\boldsymbol{\pi}}' = (\hat{\pi}_1, \hat{\pi}_2)$ be the estimator obtained from the least squares regression of y on x_1, x_2. We want to find a vector of the form $(\theta, 0)$ as close as possible to $(\hat{\pi}_1, \hat{\pi}_2)$, using $\hat{V}^{-1}(\hat{\boldsymbol{\pi}})$ in the distance function. Since we are not using the conventional estimator of $V(\hat{\boldsymbol{\pi}})$, the answer to this minimization problem is not, in general, to set $\hat{\theta} = b_{yx_1}$, the estimator obtained from the least squares regression of y on x_1. We can do better by using $b_{yx_1} + \tau\hat{\pi}_2$; the asymptotic mean of $\hat{\pi}_2$ is zero if $\pi_2 = 0$, and if b_{yx_1} and $\hat{\pi}_2$ are correlated, then we can choose τ to reduce the asymptotic variance below that of b_{yx_1}.

This point has a direct counterpart in the estimation of simultaneous equations. The restrictions on the reduced form can be imposed using a minimum distance estimator. This is more efficient than conventional estimators since it is using the optimal norm. In addition, there are generalizations of two- and three-stage least squares that achieve this efficiency gain at lower computational cost.

A related application is to the estimation of restricted covariance matrices. Here the assumption to be relaxed is multivariate normality. We show that the conventional maximum likelihood estimator, which assumes normality, is asymptotically equivalent to a minimum distance estimator. But that minimum distance estimator is not, in general, using the optimal norm. Hence, there is a feasible minimum distance estimator that is at least as good as the maximum likelihood estimator; it is strictly better in general for non-normal distributions.

The minimum distance approach has an application to the multivariate probit model of Section 3. We begin by estimating T separate probit specifications in which all leads and lags of x are included in the specification for each y_{it}:

$$P(y_{it} = 1 | x_{i1}, \dots, x_{iT}) = F(\pi_{t0} + \pi_{t1}x_{i1} + \cdots + \pi_{tT}x_{iT}),$$

where F is the standard normal distribution function. Each of the T probit specifications is estimated using a maximum likelihood program for univariate probit analysis. There is some sacrifice of efficiency here, but it may be outweighed by the advantage of avoiding numerical integration. Given the estimator for Π, we derive its asymptotic covariance matrix and then impose and test restrictions by using the minimum distance estimator.

Section 5 presents two empirical applications, which implement the specifications discussed in Sections 2 and 3 using the inference procedures from Section 4. The linear example is based on the panel of Young Men in the National Longitudinal Survey (Parnes); y_t is the logarithm of the individual's hourly wage and x_t includes variables to indicate whether or not the individual's wage is set by collective bargaining; whether or not he lives in an SMSA; and whether or not he lives in the South. We present unrestricted least squares regressions of y_t on $x_1, \dots, x_T$, and we examine the form of the Π matrix. There are significant leads and lags, but there is evidence in favor of a static relationship conditional on a latent variable; the leads and lags could be interpreted as just due to c, with $E(y_t | x_1, \dots, x_T, c) = \beta x_t + c$. The estimates of β that control for c are smaller in absolute value than the cross-section estimates. The union coefficient declines by 40%, with somewhat larger declines for the SMSA and region coefficients.

The second application presents estimates of a model of labor force participation. It is based on a sample of married women in the Michigan Panel Study of

Income Dynamics. We focus on the relationship between participation and the presence of young children. The unrestricted Π matrix for the probit specification has significant leads and lags; but, unlike the wage example, there is evidence here that the leads and lags are not generated just by a latent variable. If we do impose this restriction, then the resulting estimator of β indicates that the cross-section estimates overstate the negative effect of young children on the woman's participation probability.

The estimates for the logit functional form present some interesting contrasts to the probit results. The cross-section estimates, as usual, are in close agreement with the probit estimates. But when we use the conditional maximum likelihood estimator to control for c, the effect of an additional young child on participation becomes substantially more negative than in the cross-section estimates; so the estimated sign of the bias is opposite to that of the probit results. Here the estimation method is having a first order effect on the results. There are a variety of possible explanations. It may be that the unrestricted distribution for c in the logit form is the key. Or, since there is evidence against the restriction that:

$$P(y_{it}|x_{i1},\dots,x_{iT},c_i) = P(y_{it}|x_{it},c_i),$$

perhaps we are finding that imposing this restriction simply leads to different biases in the probit and logit estimates.

2. Specification and identification: Linear models

2.1. A production function example

We shall begin with a production function example, due to Mundlak (1961).[1] Suppose that a farmer is producing a product with a Cobb–Douglas technology:

$$y_{it} = \beta x_{it} + c_i + u_{it} \qquad (0 < \beta < 1;\ i = 1,\dots,N;\ t = 1,\dots,T),$$

where y_{it} is the logarithm of output on the ith farm in season t, x_{it} is the logarithm of a variable input (labor), c_i represents an input that is fixed over time (soil quality), and u_{it} represents a stochastic input (rainfall), which is not under the farmer's control. We shall assume that the farmer knows the product price (P) and the input price (W), which do not depend on his decisions, and that he knows c_i. The factor input decision, however, is made before knowing u_{it}, and we shall assume that x_{it} is chosen to maximize expected profits. Then the factor demand equation is:

$$x_t = \{\ln\beta + \ln[\mathrm{E}(\mathrm{e}^{u_t}|\mathscr{I}_t)] + \ln(P_t/W_t) + c\}/(1-\beta), \tag{2.1}$$

[1] This example is also discussed in Mundlak (1963) and in Zellner, Kmenta, and Drèze (1966).

where $\mathscr{J}_t$ is the information set available to the farmer when he chooses x_t, and we have suppressed the i subscript.

Assume first that u_t is independent of $\mathscr{J}_t$, so that the farmer cannot do better than using the unconditional mean. In that case we have:

$$\mathrm{E}(y_t|x_1,\ldots,x_T,c)=\beta x_t+c.$$

So if c is observed, only one period of data is needed; the least squares regression of y_1 on x_1, c provides a consistent estimator of β as $N\to\infty$.

Now suppose that c is not observed by the econometrician, although it is known to the farmer. Consider the least squares regression of y_1 on x_1, using just a single cross-section of the data. The population counterpart is:

$$\mathrm{E}^*(y_1|x_1)=\pi_0+\pi x_1,$$

where E* is the minimum mean-square error linear predictor (the wide-sense regression function):

$$\pi=\mathrm{cov}(y_1,x_1)/V(x_1),\qquad \pi_0=\mathrm{E}(y_1)-\pi\mathrm{E}(x_1).$$

We see from (2.1) that c and x_1 are correlated; hence $\pi\neq\beta$ and the least squares estimator of β does not converge to β as $N\to\infty$. Furthermore, with a single cross section, there may be no internal evidence of this omitted-variable bias.

Now the panel can help to solve this problem. Mundlak's solution was to include farm specific indicator variables: a least squares regression of y_{it} on x_{it}, $\boldsymbol{d}_{it}$ $(i=1,\ldots,N;\ t=1,\ldots,T)$, where $\boldsymbol{d}_{it}$ is an $N\times 1$ vector of zeros except for a one in the ith position. So this solution treats the c_i as a set of parameters to be estimated. It is a "fixed effects" solution, which we shall contrast with "random effects". The distinction is that under a fixed effects approach, we condition on the c_i, so that their distribution plays no role. A random effects approach invokes a distribution for c. In a Bayesian framework, β and the c_i would be treated symmetrically, with a prior distribution for both. Since I am only going to use asymptotic results on inference, however, a "gentle" prior distribution for β will be dominated. That this need not be true for the c_i is one of the interesting aspects of our problem.

We shall do asymptotic inference as N tends to infinity for fixed T. Since the number of parameters (c_i) is increasing with sample size, there is a potential "incidental parameters" problem in the fixed effects approach. This does not, however, pose a deep problem in our example. The least squares regression with the indicator variables is algebraically equivalent to the least squares regression of $y_{it}-\bar{y}_i$ on $x_{it}-\bar{x}_i$ $(i=1,\ldots,N;\ t=1,\ldots,T)$, where $\bar{y}_i=\sum_{t=1}^{T}y_{it}/T$, $\bar{x}_i=$

$\sum_{t=1}^{T} x_{it}/T$. If $T = 2$, this reduces to a least squares regression of $y_{i2} - y_{i1}$ on $x_{i2} - x_{i1}$. Since

$$\mathrm{E}(y_{i2} - y_{i1} | x_{i2} - x_{i1}) = \beta(x_{i2} - x_{i1}),$$

the least squares regression will provide a consistent estimator of β if there is sufficient variation in $x_{i2} - x_{i1}$.[2]

2.2. *Fixed effects and incidental parameters*

The incidental parameters can create real difficulties. Suppose that u_{it} is independently and identically distributed (i.i.d.) across farms and periods with $V(u_{it}) = \sigma^2$. Then under a normality assumption, the maximum likelihood estimator of σ^2 converges (almost surely) to $\sigma^2(T-1)/T$ as $N \to \infty$ with T fixed.[3] The failure to correct for degrees of freedom leads to a serious inconsistency when T is small. For another example, consider the following autoregression:

$$y_{i1} = \beta y_{i0} + c_i + u_{i1},$$
$$y_{i2} = \beta y_{i1} + c_i + u_{i2}.$$

Assume that u_{i1} and u_{i2} are i.i.d. conditional on y_{i0} and c_i, and that they follow a normal distribution ($N(0, \sigma^2)$). Consider the likelihood function corresponding to the distribution of (y_{i1}, y_{i2}) conditional on y_{i0} and c_i. The log-likelihood function is quadratic in $\beta, c_1, \dots, c_N$ (given σ^2), and the maximum likelihood estimator of β is obtained from the least squares regression of $y_{i2} - y_{i1}$ on $y_{i1} - y_{i0}$ $(i = 1, \dots, N)$. Since u_{i1} is correlated with y_{i1}, and

$$y_{i2} - y_{i1} = \beta(y_{i1} - y_{i0}) + u_{i2} - u_{i1},$$

it is clear that

$$\mathrm{E}(y_{i2} - y_{i1} | y_{i1} - y_{i0}) \neq \beta(y_{i1} - y_{i0}),$$

and the maximum likelihood estimator of β is not consistent. If the distribution of y_{i0} conditional on c_i does not depend on β or c_i, then the likelihood function based on the distribution of (y_{i0}, y_{i1}, y_{i2}) conditional on c_i gives the same inconsistent maximum likelihood estimator of β. If the distribution of (y_{i0}, y_{i1}, y_{i2})

[2]We shall not discuss methods for eliminating omitted-variable bias when x does not vary over time ($x_{it} = x_i$). See Chamberlain (1978) and Hausman and Taylor (1981).
[3]This example is discussed in Neyman and Scott (1948).

is stationary, then the estimator obtained from the least squares regression of $y_{i2} - y_{i1}$ on $y_{i1} - y_{i0}$ converges, as $N \to \infty$, to $(\beta - 1)/2$.[4]

2.3. *Random effects and specification analysis*

We have seen that the success of the fixed effects estimator in the production function example must be viewed with some caution. The incidental parameter problem will be even more serious when we consider nonlinear models. So we shall consider next a random effects treatment of the production function example; this will also provide a convenient framework for specification analysis.[5]

Assume that there is some joint distribution for $(x_{i1},\dots,x_{iT},c_i)$, which does not depend upon i, and consider the regression function that does not condition on c:

$$\mathrm{E}(y_{it}|x_{i1},\dots,x_{iT}) = \beta x_{it} + \mathrm{E}(c_i|x_{i1},\dots,x_{iT}).$$

The regression function for c_i given $\boldsymbol{x}_i = (x_{i1},\dots,x_{iT})$ will generally be some nonlinear function. But we can specify a minimum mean-square error linear predictor:[6]

$$\mathrm{E}^*(c_i|x_{i1},\dots,x_{iT}) = \psi + \lambda_1 x_{i1} + \cdots + \lambda_T x_{iT} = \psi + \boldsymbol{\lambda}'\boldsymbol{x}_i, \tag{2.2}$$

where $\boldsymbol{\lambda} = V^{-1}(\boldsymbol{x}_i)\mathrm{cov}(\boldsymbol{x}_i, c_i)$. No restrictions are being imposed here—(2.2) is simply giving our notation for the linear predictor.

Now we have:

$$\mathrm{E}^*(y_{it}|\boldsymbol{x}_i) = \psi + \beta x_{it} + \boldsymbol{\lambda}'\boldsymbol{x}_i.$$

Combining these linear predictors for the T periods gives the following multivariate linear predictor:[7]

$$\begin{aligned} &\mathrm{E}^*(\boldsymbol{y}_i|\boldsymbol{x}_i) = \boldsymbol{\pi}_0 + \boldsymbol{\Pi}\boldsymbol{x}_i, \\ &\boldsymbol{\Pi} = \mathrm{cov}(\boldsymbol{y}_i, \boldsymbol{x}_i')V^{-1}(\boldsymbol{x}_i) = \beta \boldsymbol{I} + \boldsymbol{l}\boldsymbol{\lambda}', \end{aligned} \tag{2.3}$$

where $\boldsymbol{y}_i' = (y_{i1},\dots,y_{iT})$, $\boldsymbol{I}$ is the $T \times T$ identity matrix, and $\boldsymbol{l}$ is a $T \times 1$ vector of ones.

[4]See Chamberlain (1980) and Nickell (1981).

[5]In our notation, Kiefer and Wolfowitz (1956) invoke a distribution for c to pass from the distribution of (y, x) conditional on c to the marginal distribution of (y, x). Note that they did not assume a parametric form for the distribution of c.

[6]Mundlak (1978) uses a similar specification, but with $\lambda_1 = \cdots = \lambda_T$. The appropriateness of these equality constraints is discussed in Chamberlain (1980, 1982a).

[7]We shall not discuss the problems caused by attrition. See Griliches, Hall and Hausman (1978) and Hausman and Wise (1979).

The Π matrix is a useful tool for analyzing this model. Consider first the estimation of β; if $T=2$ we have:

$$\Pi = (\pi_{jk}) = \begin{pmatrix} \beta+\lambda_1 & \lambda_2 \\ \lambda_1 & \beta+\lambda_2 \end{pmatrix}.$$

Hence,

$$\beta = \pi_{11} - \pi_{21} = \pi_{22} - \pi_{12}.$$

So given a consistent estimator for Π, we can obtain a consistent estimator for β. The estimation of Π is almost a standard problem in multivariate regression; but, due to the nonlinearity in $\mathrm{E}(c_i|\boldsymbol{x}_i)$, we are estimating only a wide-sense regression function, and some care is needed. It turns out that there is a way of looking at the problem which allows a straightforward treatment, under very weak assumptions. We shall develop this in the section on inference.

We see in (2.3) that there are restrictions on the Π matrix. The off-diagonal elements within the same column of Π are all equal. The T^2 elements of Π are functions of the $T+1$ parameters $\beta, \lambda_1, \dots, \lambda_T$. This suggests an obvious specification test. Or, backing up a bit, we could begin with the specification that $\Pi = \beta \boldsymbol{I}$. Then passing to (2.3) would be a test for whether there is a time-invariant omitted variable that is correlated with the x's. The test of $\Pi = \beta\boldsymbol{I} + \boldsymbol{l}\lambda'$ against an unrestricted Π would be an omnibus test of a variety of misspecifications, some of which will be considered next.[8]

Suppose that there is serial correlation in u, with $u_t = \rho u_{t-1} + w_t$, where w_t is independent of $\mathscr{I}_t$ and we have suppressed the i subscripts. Now we have:

$$\mathrm{E}(\mathrm{e}^{u_t}|\mathscr{I}_t) = \mathrm{e}^{\rho u_{t-1}}\mathrm{E}(\mathrm{e}^{w_t}).$$

So the factor demand equation becomes:

$$x_t = \{\ln\beta + \ln[\mathrm{E}(\mathrm{e}^{w_t})] + \ln(P_t/W_t) + \rho u_{t-1} + c\}/(1-\beta).$$

Suppose that there is no variation in prices across the farms, so that the P_t/W_t term is captured in period specific intercepts, which we shall suppress. We can solve for u_t in terms of x_{t+1} and c, and substitute this solution into the y_t equation. Then we have:

$$\mathrm{E}^*(y_t|x_1,\dots,x_T) = \beta x_t + (1-\rho^{-1})(\lambda_1 x_1 + \cdots + \lambda_T x_T) + \varphi x_{t+1},$$

[8] This specification test was proposed in Chamberlain (1978a,1979). The restrictions are similar to those in the MIMIC model of Jöreskog and Goldberger (1975); also see Goldberger (1974a), Griliches (1974), Jöreskog and Sörbom (1977), Chamberlain (1977), and Jöreskog (1978). There are also connections with the work on sibling data, which is surveyed in Griliches (1979).

where $\varphi = \rho^{-1}(1-\beta)$. So the Π matrix would indicate a distributed lead, even after controlling for c. If instead there is a first order moving average, $u_t = w_t + \rho w_{t-1}$, then:

$$\mathrm{E}(\mathrm{e}^{u_t} | \mathscr{I}_t) = \mathrm{e}^{\rho w_{t-1}} \mathrm{E}(\mathrm{e}^{w_t}),$$

and a bit of algebra gives:

$$\mathrm{E}(y_t | x_1, \ldots, x_T) = x_t - \rho^{-1}(\lambda_1 x_1 + \cdots + \lambda_T x_T) + \varphi x_{t+1}.$$

Once again there is a distributed lead, but now β is not identified from the Π matrix.

2.4. *A consumer demand example*

2.4.1. *Certainty*

We shall follow Ghez and Becker (1975), Heckman and MaCurdy (1980), and MaCurdy (1981) in presenting a life-cycle model under certainty. Suppose that the consumer is maximizing

$$V = \sum_{t=1}^{\tau} \rho^{(t-1)} U_t(C_t)$$

subject to

$$\sum_{t=1}^{\tau} \gamma^{-(t-1)} P_t C_t \le B, \; C_t \ge 0 \qquad (t = 1, \ldots, \tau),$$

where $\rho^{-1} - 1$ is the rate of time preference, $\gamma - 1$ is the (nominal) interest rate, C_t is consumption in period t, P_t is the price of the consumption good in period t, and B is the present value in the initial period of lifetime income. In this certainty model, the consumer faces a single lifetime budget constraint.

If the optimal consumption is positive in every period, then

$$U_t'(C_t) = (\gamma\rho)^{-(t-1)} (P_t / P_1) U_1'(C_1).$$

A convenient functional form is $U_t(C) = A_t C^{\delta}/\delta$ ($A_t > 0$, $\delta < 1$); then we have:

$$y_t = \beta x_t + \varphi(t-1) + c + u_t, \tag{2.4}$$

where $y_t = \ln C_t$, $x_t = \ln P_t$, $c = (\delta - 1)^{-1}\ln[U_1'(C_1)/P_1]$, $u_t = (1-\delta)^{-1}\ln A_t$, $\beta = (\delta - 1)^{-1}$, and $\varphi = (1-\delta)^{-1}\ln(\gamma\rho)$. Note that c is determined by the marginal utility of initial wealth: $U_1'(C_1)/P_1 = \partial V/\partial B$.

We shall assume that A_t is not observed by the econometrician, and that it is independent of the P's. Then the model is similar to the production function example if there is price variation across consumers as well as over time. There will generally be correlation between c and $(x_1,\dots,x_T)$. As before we have the prediction that $\Pi = \beta I + l\lambda'$, which is testable. A consistent estimator of β can be obtained with only two periods of data since

$$y_t - y_{t-1} = \beta(x_t - x_{t-1}) + \varphi + u_t - u_{t-1}.$$

We shall see next how these results are affected when we allow for some uncertainty.

2.4.2. *Uncertainty*

We shall present a highly simplified model in order to obtain some explicit results in the uncertainty case. The consumer is maximizing

$$\mathrm{E}\left[\sum_{t=1}^{\tau} \rho^{t-1} U_t(C_t)\right]$$

subject to

$$P_1 C_1 + S_1 \le B,$$
$$P_t C_t + S_t \le \gamma S_{t-1}, \qquad C_t \ge 0, \qquad S_t \ge 0 \qquad (t = 1,\dots,\tau).$$

The only source of uncertainty is the future prices. The consumer is allowed to borrow against his future income, which has a present value of B in the initial period. The consumption plan must have C_t a function only of information available at date t.

It is convenient to set $\tau = \infty$ and to assume that P_{t+1}/P_t is i.i.d. $(t = 1,2,\dots)$. If $U_t(C) = A_t C^{\delta}/\delta$, then we have the following optimal plan:[9]

$$\begin{aligned} &C_1 = d_1 B/P_1,\ S_1 = (1-d_1)B, \\ &C_t = d_t \gamma S_{t-1}/P_t,\ S_t = (1-d_t)\gamma S_{t-1} \qquad (t = 2,3,\dots), \end{aligned} \tag{2.5}$$

[9]We require $\rho\kappa g < 1$, where $A_t \le Mg^t$ for some constant M. Phelps (1962) obtained explicit solutions for models of this type. The derivation of (2.5) can be obtained by following Levhari and Srinivasan (1969) or Dynkin and Yushkevich (1979, Ch. 6.9).

where

$$d_t = [1 + f_{t+1} + (f_{t+1} f_{t+2}) + \cdots]^{-1},$$
$$f_t = (\rho\kappa A_t / A_{t-1})^{[1/(1-\delta)]}, \qquad \kappa = \gamma^{\delta} \mathrm{E}\left[(P_{t-1}/P_t)^{\delta}\right].$$

It follows that:

$$y_t - y_{t-1} = (-1)(x_t - x_{t-1}) + \zeta + u_t - u_{t-1},$$

where y, x, u are defined as in (2.4) and $\zeta = (1-\delta)^{-1}\ln(\rho\kappa) + \ln\gamma$.

We see that, in this particular example, the appropriate interpretation of the change regression is very sensitive to the amount of information available to the consumer. In the uncertainty case, a regression of $(\ln C_t - \ln C_{t-1})$ on $(\ln P_t - \ln P_{t-1})$ does not provide a consistent estimator of $(\delta - 1)^{-1}$; in fact, the estimator converges to -1, with the implied estimator of δ converging to 0.

2.4.3. *Labor supply*

We shall consider a certainty model in which the consumer is maximizing

$$V = \sum_{t=1}^{\tau} \rho^{(t-1)} U_t(C_t, L_t) \tag{2.6}$$

subject to

$$\sum_{t=1}^{\tau} \gamma^{-(t-1)}(P_t C_t + W_t L_t) \le B + \sum_{t=1}^{\tau} \gamma^{-(t-1)} W_t \bar{L},$$
$$C_t \ge 0, \qquad 0 \le L_t \le \bar{L} \qquad (t = 1, \ldots, \tau),$$

where L_t is leisure, W_t is the wage rate, B is the present value in the initial period of nonlabor income, and $\bar{L}$ is the time endowment. We shall assume that the inequality constraints on L are not binding; the participation decision will be discussed in the section on nonlinear models. If U_t is additively separable:

$$U_t(C, L) = U_t^*(C) + \tilde{U}_t(L),$$

and if $\tilde{U}_t(L) = A_t L^{\delta}/\delta$, then we have:

$$y_t = \beta x_t + \varphi(t-1) + c + u_t, \tag{2.7}$$

where $y_t = \ln L_t$, $x_t = \ln W_t$, $c = (\delta - 1)^{-1}\ln[\tilde{U}_1'(L_1)/W_1]$, $u_t = (1-\delta)^{-1}\ln A_t$, $\beta =$

$(\delta - 1)^{-1}$, and $\varphi = (1-\delta)^{-1}\ln(\gamma\rho)$. Once again c is determined by the marginal utility of initial wealth: $\tilde{U}_1'(L_1)/W_1 = \partial V/\partial B$.

We shall assume that A_t is not observed by the econometrician. There will generally be a correlation between c and $(x_1,\ldots,x_T)$, since L_1 depends upon wages in all periods. If A_t is independent of the W's, then we have the prediction that $\boldsymbol{\Pi} = \beta \boldsymbol{I} + \boldsymbol{l}\boldsymbol{\lambda}'$. If, however, wages are partly determined by the quantity of previous work experience, then there will be lags and leads in addition to those generated by c, and $\boldsymbol{\Pi}$ will not have this simple structure.[10]

It would be useful at this point to extend the uncertainty model to incorporate uncertainty about future wages. Unfortunately, a comparably simple explicit solution is not available. But we may conjecture that the correct interpretation of a regression of $(\ln L_t - \ln L_{t-1})$ on $(\ln W_t - \ln W_{t-1})$ is also sensitive to the amount of information available to the consumer.

2.5. *Strict exogeneity conditional on a latent variable*

We shall relate the specification analysis of $\boldsymbol{\Pi}$ to the causality definitions of Granger (1969) and Sims (1972). Consider a sample in which $t=1$ is the first period of the individual's (economic) life.[11] A Sims definition of "x is strictly exogenous" is:

$$E^*(y_t|x_1, x_2,\ldots) = E^*(y_t|x_1,\ldots,x_t) \qquad (t=1,2,\ldots).$$

In this case $\boldsymbol{\Pi}$ is lower triangular: the elements above the main diagonal are all zero. This fails to hold in the models we have been considering, due to the omitted variable c. But, in some cases, we do have the following property:

$$E^*(y_t|x_1, x_2,\ldots,c) = E^*(y_t|x_1,\ldots,x_t,c) \qquad (t=1,2,\ldots). \tag{2.8}$$

It was stressed by Granger (1969) that the assessment of noncausality depends crucially on what other variables are being conditioned on. The novel feature of (2.8) is that we are asking whether there exists some latent variable (c) such that x is strictly exogenous conditional on c. The question is not vacuous since c is restricted to be time invariant.

[10] See Blinder and Weiss (1976) and Heckman (1976) for life-cycle labor supply models with human capital accumulation.

[11] We shall not discuss the problems that arise from truncating the lag distribution. See Griliches and Pakes (1980).

Let us examine what restrictions are implied by (2.8). Define the following linear predictors:[12]

$$\begin{aligned} &y_t = \beta_{t1}x_1 + \cdots + \beta_{tT}x_T + \gamma_t c + u_t, \\ &\mathrm{E}^*(u_t|x_1,\dots,x_T,c) = 0 \qquad (t=1,\dots,T). \end{aligned}$$

Then (2.8) is equivalent to $\beta_{ts} = 0$ for $s > t$. If $\gamma_1 \neq 0$, we can choose a scale normalization for c such that $\gamma_1 \equiv 1$. Then we can rewrite the system with $\beta_{ts} = 0$ $(s > t)$ as follows:

$$\begin{aligned} &y_t = \tilde{\beta}_{t1}x_1 + \beta_{t2}x_2 + \cdots + \beta_{tt}x_t + \gamma_t y_1 + \tilde{u}_t, \\ &\tilde{\beta}_{t1} = \beta_{t1} - \gamma_t\beta_{11},\ \tilde{u}_t = u_t - \gamma_t u_1, \\ &\mathrm{E}(x_s\tilde{u}_t) = 0 \qquad (s=1,\dots,T;\ t=2,\dots,T). \end{aligned} \tag{2.9}$$

Consider the "instrumental variable" orthogonality conditions implied by $\mathrm{E}(x_s\tilde{u}_t) = 0$. In the y_T equation, we have $T+1$ unknown coefficients: $\tilde{\beta}_{T1}, \beta_{T2},\dots,\beta_{TT},\gamma_T$, and T orthogonality conditions. So these coefficients are not identified. In the y_{T-1} equation, however, we have just enough orthogonality conditions; and in the y_{T-j} equation $(j \leq T-2)$, we have $j-1$ more than we need since there are $T-j+1$ unknown coefficients: $\tilde{\beta}_{T-j,1}, \beta_{T-j,2},\dots, \beta_{T-j,T-j}, \gamma_{T-j}$, and T orthogonality conditions: $\mathrm{E}(x_s\tilde{u}_{T-j}) = 0$ $(s=1,\dots,T)$. It follows that, subject to a rank condition, we can identify β_{ts}, γ_t, and $\tilde{\beta}_{t1}$ for $2 \leq s \leq t \leq T-1$. In addition, the hypothesis in (2.8) implies that if $T \geq 4$, there are $(T-3)(T-2)/2$ over identifying restrictions.

Consider next a Granger definition of "y does not cause x conditional on c":

$$\mathrm{E}^*(x_{t+1}|x_1,\dots,x_t,y_1,\dots,y_t,c) = \mathrm{E}^*(x_{t+1}|x_1,\dots,x_t,c) \qquad (t=1,\dots,T-1). \tag{2.10}$$

Define the following linear predictors:

$$\begin{aligned} &x_{t+1} = \psi_{t1}x_1 + \cdots + \psi_{tt}x_t + \varphi_{t1}y_1 + \cdots + \varphi_{tt}y_t + \zeta_{t+1}c + v_{t+1}, \\ &\mathrm{E}^*(v_{t+1}|x_1,\dots,x_t,y_1,\dots,y_t,c) = 0 \qquad (t=1,\dots,T-1). \end{aligned}$$

Then (2.10) is equivalent to $\varphi_{ts} = 0$. We can rewrite the system, imposing $\varphi_{ts} = 0$, as follows:

$$\begin{aligned} &x_{t+1} = \tilde{\psi}_{t1}x_1 + \cdots + \tilde{\psi}_{t,t-1}x_{t-1} + \tau_t x_t + \tilde{v}_{t+1}, \\ &\tilde{\psi}_{ts} = \psi_{ts} - (\zeta_{t+1}/\zeta_t)\psi_{t-1,s}, \qquad \tau_t = \psi_{tt} + (\zeta_{t+1}/\zeta_t), \\ &\tilde{v}_{t+1} = v_{t+1} - (\zeta_{t+1}/\zeta_t)v_t, \qquad \mathrm{E}(x_s\tilde{v}_{t+1}) = \mathrm{E}(y_s\tilde{v}_{t+1}) = 0 \\ &\quad (s \leq t-1;\ t=2,\dots,T-1). \end{aligned} \tag{2.11}$$

[12]We are suppressing the period specific intercepts.

In the equation for x_{t+1}, there are t unknown parameters, $\tilde{\psi}_{t1},\ldots,\tilde{\psi}_{t,t-1}$, τ_t, and $2(t-1)$ orthogonality conditions. Hence, there are $t-2$ restrictions $(3 \le t \le T-1)$.

It follows that the Granger condition for "y does not cause x conditional on c" implies $(T-3)(T-2)/2$ restrictions, which is the same number of restrictions implied by the Sims condition. In fact, it is a consequence of Sims' (1972) theorem, as extended by Hosoya (1977), that the two sets of restrictions are equivalent; this is not immediately obvious from a direct comparison of (2.9) and (2.11).

In terms of the Π matrix, conditional strict exogeneity implies that:

$$\Pi = B + \gamma\lambda',$$

$$B = \begin{bmatrix} \beta_{11} & 0 & \cdots & & 0 \\ \beta_{21} & \beta_{22} & 0 & \cdots & 0 \\ \vdots & & & & \\ \beta_{T1} & \beta_{T2} & \cdots & & \beta_{TT} \end{bmatrix}, \quad \gamma = \begin{pmatrix} \gamma_1 \\ \vdots \\ \gamma_T \end{pmatrix}, \lambda = \begin{pmatrix} \lambda_1 \\ \vdots \\ \lambda_T \end{pmatrix}.$$

These nonlinear restrictions can be imposed and tested using the minimum distance estimator to be developed in the inference section. Alternatively, we can use the transformations in (2.9) or in (2.11). These transformations give us "simultaneous equations" systems with linear restrictions; (2.9) can be estimated using three-stage least squares. A generalization of three-stage least squares, which does not require homoskedasticity assumptions, is developed in the inference section. It is asymptotically equivalent to imposing the nonlinear restrictions directly on Π, using the minimum distance estimator.

2.6. *Lagged dependent variables*

For a specific example, write the labor supply model in (2.7) as follows:

$$\begin{aligned} &y_t = \delta_1 x_t + \delta_2 x_{t-1} + \delta_3 y_{t-1} + v_t, \\ &\mathrm{E}^*(v_t | x_1,\ldots,x_T) = 0 \qquad (t=1,\ldots,T); \end{aligned} \tag{2.12}$$

this reduces to (2.7) if $\delta_2 = -\delta_1$ and $\delta_3 = 1$. If we assume that $v_t = w + e_t$, where w is uncorrelated with the x's and e_t is i.i.d. and uncorrelated with the x's and w, then we have the autoregressive, variance-components model of Balestra and Nerlove (1966).[13] In keeping with our general approach, we shall avoid placing

[13]Estimation in variance-components models is discussed in Nerlove (1967, 1971, 1971a), Wallace and Hussain (1969), Amemiya (1971), Madalla (1971), Madalla and Mount (1973), Harville (1977), Mundlak (1978), Mazodier and Trognon (1978), Trognon (1978), Lee (1979), and Taylor (1980).

restrictions on the serial correlation structure of v_t; our inference procedures will be based on the strict exogeneity condition that $E^*(v_t|x_1,\dots,x_T)=0$.

We can fit this model into the $\boldsymbol{\Pi}$ matrix framework by using recursive substitution to obtain the reduced form:

$$y_t = \beta_{t1}x_1 + \cdots + \beta_{tt}x_t + \gamma_t c + u_t,$$
$$E^*(u_t|x_1,\dots,x_T)=0,$$

where

$$\beta_{ts} = (\delta_2+\delta_3\delta_1)\delta_3^{t-s-1}, \qquad \beta_{tt}=\delta_1,\ \gamma_t=\delta_3^{t-1},$$
$$c=\delta_2 x_0+\delta_3 y_0, \qquad u_t = v_t+\delta_3 v_{t-1}+\cdots+\delta_3^{t-1}v_1$$
$$(1\le s\le t-1,\ t=1,\dots,T).$$

[We are assuming that (2.12) holds for $t\ge 1$, but data on (x_0, y_0) are not available.] Hence, this model satisfies the conditional strict exogeneity restrictions:

$$\boldsymbol{\Pi} = \boldsymbol{B} + \boldsymbol{\gamma}\boldsymbol{\lambda}',$$

where $\boldsymbol{B}$ is lower triangular. The $\boldsymbol{\gamma}\boldsymbol{\lambda}'$ term is generated by the projection of the initial condition $(\delta_2 x_0+\delta_3 y_0)$ on $x_1,\dots,x_T$.[14]

Estimation can proceed by using the minimum distance procedure to impose the nonlinear restrictions on $\boldsymbol{\Pi}$. Alternatively, we can complete the system in (2.12) with:

$$y_1 = \varphi_1 x_1 + \cdots + \varphi_T x_T + v_1;$$

this is just notation for the identity:

$$y_1 = E^*(y_1|x_1,\dots,x_T)+\left[y_1 - E^*(y_1|x_1,\dots,x_T)\right].$$

Then we can apply the generalized three-stage least squares estimator to be developed in the inference section. It achieves the same limiting distribution at lower computational cost, since the restrictions in this form are linear and can be imposed without requiring iterative optimization techniques.

Now consider a second-order autoregression:

$$y_t = \delta_1 x_t + \delta_2 x_{t-1} + \delta_3 y_{t-1} + \delta_4 y_{t-2} + v_t,$$
$$E^*(v_t|x_1,\dots,x_T)=0 \qquad (t=1,\dots,T).$$

[14]The treatment of initial conditions in linear models is also discussed in Anderson and Hsiao (1981, 1982) and MaCurdy (1982). The difficulties that arise in nonlinear models are discussed in Heckman (1981a).

Recursive substitution gives:

$$y_t = \beta_{t1}x_1 + \cdots + \beta_{tt}x_t + \gamma_{t1}c_1 + \gamma_{t2}c_2 + u_t,$$
$$\mathrm{E}^*(u_t|x_1,\ldots,x_T) = 0 \qquad (t=1,\ldots,T),$$

where

$$c_1 = \delta_2 x_0 + \delta_3 y_0 + \delta_4 y_{-1}, \qquad c_2 = y_0,$$

and there are nonlinear restrictions on the parameters. The Π matrix has the following form:

$$\Pi = \boldsymbol{B} + \gamma_1\lambda_1' + \gamma_2\lambda_2', \tag{2.13}$$

where $\boldsymbol{B}$ is lower triangular, $\gamma_j' = (\gamma_{1j},\ldots,\gamma_{Tj})$, and $\mathrm{E}^*(c_j|\boldsymbol{x}) = \lambda_j'\boldsymbol{x}$ $(j=1,2)$.

This specification suggests a natural extension of the conditional strict exogeneity idea, with the conditioning set indexed by the number of latent variables. We shall say that "x is strictly exogenous conditional on c_1, c_2" if:

$$\mathrm{E}^*(y_t|\ldots,x_{t-1},x_t,x_{t+1},\ldots,c_1,c_2) = \mathrm{E}^*(y_t|x_t,x_{t-1},\ldots,c_1,c_2).$$

We can also introduce a Granger version of this condition and generalize the analysis in Section 2.5.

Finally, consider an autoregressive model with a time-invariant omitted variable that is correlated with x:

$$y_t = \delta_1 x_t + \delta_2 y_{t-1} + c + v_t,$$

where $\mathrm{E}^*(v_t|x_1,\ldots,x_T) = 0$. Recursive substitution gives:

$$y_t = \beta_{t1}x_1 + \cdots + \beta_{tt}x_t + \gamma_{t1}c_1 + \gamma_{t2}c_2 + u_t,$$
$$\mathrm{E}^*(u_t|x_1,\ldots,x_T) = 0 \qquad (t=1,\ldots,T),$$

where $c_1 = y_0$, $c_2 = c$, and there are nonlinear restrictions on the parameters. So y is strictly exogenous conditional on c_1, c_2, and setting $\mathrm{E}^*(c_j|\boldsymbol{x}) = \psi_j + \lambda_j'\boldsymbol{x}$ $(j=1,2)$ gives a Π matrix in the (2.13) form.

We can impose the restrictions on Π directly, using a minimum distance estimator. There is, however, a transformation of the model that allows a computationally simpler instrumental variable estimator:

$$y_t - y_{t-1} = \delta_1(x_t - x_{t-1}) + \delta_2(y_{t-1} - y_{t-2}) + v_t - v_{t-1},$$
$$\mathrm{E}^*(v_t - v_{t-1}|\boldsymbol{x}) = 0 \qquad (t = 3,\ldots,T);$$
$$y_2 = \varphi_{21}x_1 + \cdots + \varphi_{2T}x_T + w_2,$$
$$y_1 = \varphi_{11}x_1 + \cdots + \varphi_{1T}x_T + w_1,$$
$$\mathrm{E}^*(w_j|\boldsymbol{x}) = 0 \qquad (j = 1,2),$$

where $\mathrm{E}^*(y_j|\boldsymbol{x}) = \boldsymbol{\varphi}_j'\boldsymbol{x}$ is unrestricted since $\mathrm{E}^*(c_j|\boldsymbol{x})$ is unrestricted ($j = 1,2$). Now we can apply the generalized three-stage least squares estimator. This is computationally simple since the parameter restrictions are linear. The estimator is asymptotically equivalent to applying the minimum distance procedure directly to Π. Since the linear predictor equations for y_1 and y_2 are unrestricted, the limiting distribution of $\hat{\delta}_1$ and $\hat{\delta}_2$ is not affected if we drop these equations when we form the generalized three-stage least squares estimator. (See the Appendix.)

2.7. *Serial correlation or partial adjustment?*

Griliches (1967) considered the problem of distinguishing between the following two models: a partial adjustment model,[15]

$$y_t = \beta x_t + \gamma y_{t-1} + v_t, \tag{2.14}$$

and a model with no structural lagged dependent variable but with a residual following a first-order Markov process:

$$\begin{aligned} y_t &= \beta x_t + u_t, \\ u_t &= \rho u_{t-1} + e_t, \qquad e_t \text{ i.i.d.}; \end{aligned} \tag{2.15}$$

in both cases x is strictly exogenous:

$$\mathrm{E}^*(v_t|x_1,\ldots,x_T) = \mathrm{E}^*(u_t|x_1,\ldots,x_T) = 0 \qquad (t = 1,\ldots,T).$$

In the serial correlation case, we have:

$$y_t = \beta x_t - \rho\beta x_{t-1} + \rho y_{t-1} + e_t;$$

[15]See Nerlove (1972) for distributed lag models based on optimizing behavior in the presence of uncertainty and costs of adjustment.

as Griliches observed, the least squares regression will have a distinctive pattern —the coefficient on lagged x equals (as $N \to \infty$) minus the product of the coefficients on current x and lagged y.

I want to point out that this prediction does not rest on the serial correlation structure of u. It is a direct implication of the assumption that u is uncorrelated with $x_1, \dots, x_T$:

$$\begin{aligned}
E^*(y_t|x_t, x_{t-1}, y_{t-1}) &= \beta x_t + E^*(u_t|x_t, x_{t-1}, y_{t-1}) \\
&= \beta x_t + E^*(u_t|u_{t-1}) \\
&= \beta x_t + \varphi_t u_{t-1} \\
&= \beta x_t - \varphi_t \beta x_{t-1} + \varphi_t y_{t-1}.
\end{aligned}$$

Here $\varphi_t u_{t-1}$ is simply notation for the linear predictor. In general u_t is not a first-order process ($E^*(u_t|u_{t-1}, u_{t-2}) \neq E^*(u_t|u_{t-1})$), but this does not affect our argument.

Within the $\boldsymbol{\Pi}$ matrix framework, the distinction between the two models is that (2.15) implies a diagonal $\boldsymbol{\Pi}$ matrix, with no distributed lag, whereas the partial adjustment specification in (2.14) implies that $\boldsymbol{\Pi} = \boldsymbol{B} + \gamma\lambda'$, with a distributed lag in the lower triangular $\boldsymbol{B}$ matrix and a rank one set of lags and leads in $\gamma\lambda'$.

We can generalize the serial correlation model to allow for an individual specific effect that may be correlated with x:

$$y_t = \beta x_t + c + u_t, \qquad E^*(u_t|x_1, \dots, x_T) = 0.$$

Now both the serial correlation and the partial adjustment models have a rank one set of lags and leads in $\boldsymbol{\Pi}$, but we can distinguish between them because only the partial adjustment model has a distributed lag in the $\boldsymbol{B}$ matrix. So the absence of structural lagged dependent variables is signalled by the following special case of conditional strict exogeneity:

$$E^*(y_t|x_1, \dots, x_T, c) = E^*(y_t|x_t, c).$$

In this case the relationship of x to y is "static" conditional on c. We shall pursue this distinction in nonlinear models in Section 3.3.

2.8. *Residual covariances: Heteroskedasticity and serial correlation*

2.8.1. *Heteroskedasticity*

If $E(c_i|\boldsymbol{x}_i) \neq E^*(c_i|\boldsymbol{x}_i)$, then there will be heteroskedasticity, since the residual will contain $E(c_i|\boldsymbol{x}_i) - E^*(c_i|\boldsymbol{x}_i)$. Another source of heteroskedasticity is random coefficients:

$$y_{it} = b_i x_{it} + c_i + u_{it},$$

$$b_i = \beta + w_i, \qquad \mathrm{E}(w_i) = 0,$$

$$y_{it} = \beta x_{it} + c_i + (w_i x_{it} + u_{it}).$$

If w is independent of x, then $\boldsymbol{\Pi} = \beta \boldsymbol{I} + \boldsymbol{l}\boldsymbol{\lambda}'$, and our previous discussion is relevant for the estimation of β. We shall handle the heteroskedasticity problem in the inference section by allowing $\mathrm{E}[(y_i - \boldsymbol{\Pi}\boldsymbol{x}_i)(y_i - \boldsymbol{\Pi}\boldsymbol{x}_i)'|\boldsymbol{x}_i]$ to be an arbitrary function of $\boldsymbol{x}_i$.[16]

2.8.2. *Serial correlation*

It may be of interest to impose restrictions on the residual covariances, such as a variance-components structure together with an autoregressive-moving average scheme.[17] Consider the homoskedastic case in which

$$\boldsymbol{\Omega} = \mathrm{E}[(y_i - \boldsymbol{\Pi}\boldsymbol{x}_i)(y_i - \boldsymbol{\Pi}\boldsymbol{x}_i)'|\boldsymbol{x}_i]$$

does not depend upon $\boldsymbol{x}_i$. Then the restrictions can be expressed as $\Omega_{jk} = g_{jk}(\boldsymbol{\theta})$, where the g's are known functions and $\boldsymbol{\theta}$ is an unrestricted parameter vector. We shall discuss a minimum distance procedure for imposing such restrictions in Section 4.

2.9. *Measurement error*

Suppose that

$$y_{it} = \beta x^*_{it} + u_{it},$$

$$x_{it} = x^*_{it} + v_{it} \qquad (i = 1,\dots,N;\ t = 1,\dots,T),$$

where x^*_{it} is not observed. We assume that the measurement error v_{it} satisfies $\mathrm{E}^*(v_{it}|\boldsymbol{x}_i) = 0$. If $\mathrm{E}^*(u_{it}|\boldsymbol{x}_i) = 0$, then $\mathrm{E}^*(y_i|\boldsymbol{x}_i) = \boldsymbol{\Pi}\boldsymbol{x}_i$, with

$$\boldsymbol{\Pi} = \beta V(\boldsymbol{x}^*_i) V^{-1}(\boldsymbol{x}_i). \tag{2.16}$$

[16] Anderson (1969,1970), Swamy (1970,1974), Hsiao (1975), and Mundlak (1978a) discuss estimators that incorporate the particular form of heteroskedasticity that is generated by random coefficients.

[17] Such models for the covariance structure of earnings have been considered by Hause (1977, 1980), Lillard and Willis (1978), Lillard and Weiss (1979), MaCurdy (1982), and others.

Since $V(\boldsymbol{x}_i)$ and $V(\boldsymbol{x}_i^*)$ will generally not be diagonal matrices, (2.16) provides an alternative interpretation of lags and leads in the Π matrix. The Π matrix in (2.16) generally does not have the form $\tau_1 \boldsymbol{I} + \tau_2 \boldsymbol{l}\boldsymbol{\lambda}'$; nevertheless, it may be difficult to distinguish between measurement error and a time-invariant omitted variable if T is small. For example, if the covariance matrices of $\boldsymbol{x}_i$ and $\boldsymbol{x}_i^*$ have the form $\varphi_1 \boldsymbol{I} + \varphi_2 \boldsymbol{l}\boldsymbol{l}'$ (equicorrelated), then Π has this form also and no distinction is possible. Although $\operatorname{cov}(x_{it}, x_{is})$ generally declines as $|t-s|$ increases, the equicorrelated approximation may be quite good for small T.

It has been noted in other contexts that the bias from measurement error can be aggravated by analysis of covariance techniques.[18] Consider the following example with $T = 2$:

$$y_{i2} - y_{i1} = \beta(x_{i2} - x_{i1}) + u_{i2} - u_{i1} - \beta(v_{i2} - v_{i1}),$$

so that $\mathrm{E}^*(y_{i2} - y_{i1} | x_{i2} - x_{i1}) = \tilde{\beta}(x_{i2} - x_{i1})$ with

$$\tilde{\beta} = \beta\left(1 - \frac{V(v_{i2} - v_{i1})}{V(x_{i2} - x_{i1})}\right).$$

If $V(v_{i1}) = V(v_{i2})$ and $V(x_{i1}) = V(x_{i2})$, then we can rewrite this as:

$$\tilde{\beta} = \beta\left(1 - \frac{V(v_{i1})\left(1 - r^2_{v_1 v_2}\right)}{V(x_{i1})\left(1 - r^2_{x_1 x_2}\right)}\right),$$

where $r_{v_1 v_2}$ denotes the correlation between v_{i1} and v_{i2}. If x_{i1} and x_{i2} are highly correlated but v_{i1} and v_{i2} are not, then a modest bias from measurement error in a cross-section regression can become large when we relate the change in y to the change in x. On the other hand, if $v_{i1} = v_{i2}$, then the change regression eliminates the bias from measurement error. Data from reinterview surveys should be useful in distinguishing between these two cases.

3. Specification and identification: Nonlinear models

3.1. A random effects probit model

Our treatment of individual effects carries over with some important qualifications to nonlinear models. We shall illustrate with a labor force participation example. If the upper bound on leisure is binding in (2.6), then

$$\rho^{(t-1)} \tilde{U}_t'(\bar{L}) > m\gamma^{-(t-1)} W_t,$$

[18] See Griliches (1979) for example.

where m is the Lagrange multiplier corresponding to the lifetime budget constraint (the marginal utility of initial wealth) and $\tilde{U}_t(L) = A_t L^\delta/\delta$. Let $y_{it} = 1$ if individual i works in period t, $y_{it} = 0$ otherwise. Let:

$$\ln W_{it} = \varphi_1 x_{it} + e_{1it},$$
$$\ln A_{it} = \varphi_2 x_{it} + e_{2it},$$

where x_{it} contains measured variables that predict wages and tastes for leisure. We shall simplify the notation by supposing that x_{it} consists of a single variable. Then, $y_{it} = 1$ if:

$$(\varphi_1 - \varphi_2)x_{it} - (t-1)\ln(\gamma\rho) + \ln m_i + (1-\delta)\ln \bar{L} + e_{1it} - e_{2it} \geq 0,$$

which we shall write as:

$$\beta x_{it} + \varphi(t-1) + c_i + u_{it} \geq 0. \tag{3.1}$$

Now we need a distributional assumption for the u's. We shall assume that $(u_1,\dots,u_T)$ is independent of c and the x's, with a multivariate normal distribution $(N(\mathbf{0},\boldsymbol{\Sigma}))$. So we have a probit model (suppressing the i subscripts and period-specific intercepts):

$$P(y_t = 1|x_1,\dots,x_T,c) = F\left[\sigma_{tt}^{-1/2}(\beta x_t + c)\right],$$

where $F(\cdot)$ is the standard normal distribution function and σ_{tt} is the tth diagonal element of $\boldsymbol{\Sigma}$.

Next we shall specify a distribution for c conditional on $\boldsymbol{x} = (x_1,\dots,x_T)$:

$$c = \psi + \lambda_1 x_1 + \cdots + \lambda_T x_T + v,$$

where v is independent of the x's and has a normal distribution $(N(0,\sigma_v^2))$. There is a very important difference in this step compared with the linear case. In the linear case it was not restrictive to decompose c into its linear projection on $\boldsymbol{x}$ and an orthogonal residual. Now, however, we are assuming that the regression function $\mathrm{E}(c|\boldsymbol{x})$ is actually linear, that v is independent of $\boldsymbol{x}$, and that v has a normal distribution. These are restrictive assumptions and there may be a payoff to relaxing them.

Given these assumptions, the distribution for y_t conditional on $x_1,\dots,x_T$ but marginal on c also has a probit form:

$$P(y_t = 1|x_1,\dots,x_T) = F\left[\alpha_t(\beta x_t + \lambda_1 x_1 + \cdots + \lambda_T x_T)\right],$$
$$\alpha_t = \left(\sigma_{tt} + \sigma_v^2\right)^{-1/2}.$$

Combining these T specifications gives the following matrix of coefficients:[19]

$$\Pi = \operatorname{diag}\{\alpha_1,\ldots,\alpha_T\}[\beta I_T + l\lambda']. \tag{3.2}$$

This differs from the linear case only in the diagonal matrix of normalization factors α_t. There are now nonlinear restrictions on Π, but the identification analysis is still straightforward. We have:

$$\alpha_t\beta = \frac{\alpha_t}{\alpha_1}\pi_{11} - \pi_{t1} = \pi_{tt} - \frac{\alpha_t}{\alpha_1}\pi_{1t},$$

$$\frac{\alpha_t}{\alpha_1} = \frac{(\pi_{tt}+\pi_{t1})}{(\pi_{11}+\pi_{1t})} \qquad (t=2,\ldots,T),$$

if $\beta+\lambda_1+\lambda_t \neq 0$. Then, as in the linear case, we can solve for $\alpha_1\beta$ and $\alpha_1\lambda$. Only ratios of coefficients are identified, and so we can use a scale normalization such as $\alpha_1 \equiv 1$.

As for inference, a computationally simple approach is to estimate T cross-sectional probit specifications by maximum likelihood, where $x_1,\ldots,x_T$ are included in each of the T specifications. This gives $\hat{\pi}_t$ $(t=1,\ldots,T)$ and we can use a Taylor expansion to derive the covariance matrix of the asymptotic normal distribution for $(\hat{\pi}_1,\ldots,\hat{\pi}_T)$. Then restrictions can be imposed on Π using a minimum distance estimator, just as in the linear case.

We shall conclude our discussion of this model by considering the interpretation of the coefficients. We began with the probit specification that

$$P(y_t=1|x_1,\ldots,x_T,c) = F\left[\sigma_{tt}^{-1/2}(\beta x_t + c)\right].$$

So one might argue that the correct measure of the effect of x_t is based on $\sigma_{tt}^{-1/2}\beta$, whereas we have obtained $(\sigma_{tt}+\sigma_v^2)^{-1/2}\beta$, which is then an underestimate. But there is something curious about this argument, since the "omitted variable" v is independent of $x_1,\ldots,x_T$. Suppose that we decompose u_t in (3.1) into $u_{1t}+u_{2t}$ and that measurements on u_{1t} become available. Then this argument implies that the correct measure of the effect of x_t is based on $[V(u_{2t})]^{-1/2}\beta$. As the data collection becomes increasingly successful, there is less and less variance left in the residual u_{2t}, and $[V(u_{2t})]^{-1/2}$ becomes arbitrarily large.

The resolution of this puzzle is that the effect of x_t depends upon the value of c, and the effect evaluated at the average value for c is not equal to the average of the effects, averaging over the distribution for c. Consider the effect on the probability that $y_t=1$ of increasing x_t from x' to x''; using the average value for c

[19] This approach to analysis of covariance in probit models was proposed in Chamberlain (1980). For other applications of multivariate probit models to panel data, see Heckman (1978, 1981).

gives:

$$F\left[\sigma_{tt}^{-1/2}(\beta x''+\mathrm{E}(c))\right]-F\left[\sigma_{tt}^{-1/2}(\beta x'+\mathrm{E}(c))\right].$$

The problem with this measure is that it may be relevant for only a small fraction of the population. I think that a more appropriate measure is the mean effect for a randomly drawn individual:

$$\int\left[P(y_t=1|x_t=x'',c)-P(y_t=1|x_t=x',c)\right]\mu(\mathrm{d}c),$$

where $\mu(\mathrm{d}c)$ gives the population probability measure for c.

We shall see how to recover this measure within our framework. Let $z=\lambda_1x_1+\cdots+\lambda_Tx_T$; let $\mu(\mathrm{d}z)$ and $\mu(\mathrm{d}v)$ give the population probability measures for the independent random variables z and v. Then:

$$\begin{aligned}P(y_t=1|x_t,c)&=P(y_t=1|x_1,\dots,x_T,c)\\&=P(y_t=1|x_t,z,v);\end{aligned}$$

$$\begin{aligned}\int P(y_t=1|x_t,z,v)\mu(\mathrm{d}z)\mu(\mathrm{d}v)&\\=\int P(y_t=1|x_t,z,v)\mu(\mathrm{d}v|x_t,z)\mu(\mathrm{d}z)&\\=\int P(y_t=1|x_t,z)\mu(\mathrm{d}z),&\end{aligned}$$

where $\mu(\mathrm{d}v|x_t,z)$ is the conditional probability measure, which equals the unconditional measure since v is independent of x_t and z. [It is important to note that the last integral does *not*, in general, equal $P(y_t=1|x_t)$. For if x_t and z are correlated, as they are in our case, then

$$\begin{aligned}P(y_t=1|x_t)&=\int P(y_t=1|x_t,z)\mu(\mathrm{d}z|x_t)\\&\neq\int P(y_t=1|x_t,z)\mu(\mathrm{d}z).]\end{aligned}$$

We have shown that:

$$\begin{aligned}&\int\left[P(y_t=1|x_t=x'',c)-P(y_t=1|x_t=x',c)\right]\mu(\mathrm{d}c)\\&\qquad=\int\left[P(y_t=1|x_t=x'',z)-P(y_t=1|x_t=x',z)\right]\mu(\mathrm{d}z).\end{aligned}\tag{3.3}$$

The integration with respect to the marginal distribution for z can be done using the empirical distribution function, which gives the following consistent (as

$N \to \infty$) estimator of (3.3):

$$\frac{1}{N}\sum_{i=1}^{N}\left\{F\left[\alpha_t\left(\beta x'' + \lambda_1 x_{i1} + \cdots + \lambda_T x_{iT}\right)\right]\right.$$
$$\left.- F\left[\alpha_t\left(\beta x' + \lambda_1 x_{i1} + \cdots + \lambda_T x_{iT}\right)\right]\right\}. \tag{3.4}$$

3.2. *A fixed effects logit model: Conditional likelihood*

A weakness in the probit model was the specification of a distribution for c conditional on $\boldsymbol{x}$. A convenient form was chosen, but it was only an approximation, perhaps a poor one. We shall discuss a technique that does not require us to specify a particular distribution for c conditional on $\boldsymbol{x}$; it will, however, have its own weaknesses.

Consider the following specification:

$$P(y_t = 1 | x_1,\dots,x_T, c) = G(\beta x_t + c), \qquad G(z) = e^z/(1+e^z), \tag{3.5}$$

where $y_1,\dots,y_T$ are independent conditional on $x_1,\dots,x_T, c$. Suppose that $T = 2$ and compute the probability that $y_2 = 1$ conditional on $y_1 + y_2 = 1$:

$$P(y_2 = 1 | x_1, x_2, c, y_1 + y_2 = 1) = G[\beta(x_2 - x_1)], \tag{3.6}$$

which does not depend upon c. Given a random sample of individuals, the conditional log-likelihood function is:

$$L = \sum_{i \in B}\{w_i \ln G[\beta(x_{i2} - x_{i1})] + (1 - w_i)\ln G[-\beta(x_{i2} - x_{i1})]\},$$

where

$$w_i = \begin{cases} 1, & \text{if } (y_{i1}, y_{i2}) = (0,1), \\ 0, & \text{if } (y_{i1}, y_{i2}) = (1,0), \end{cases}$$
$$B = \{i | y_{i1} + y_{i2} = 1\}.$$

This conditional likelihood function does not depend upon the incidental parameters. It is in the form of a binary logit likelihood function in which the two outcomes are (0,1) and (1,0) with explanatory variables $x_2 - x_1$. This is the analog of differencing in the two period linear model. The conditional maximum likelihood (ML) estimate of β can be obtained simply from a ML binary logit program. This conditional likelihood approach was used by Rasch (1960, 1961) in his model for intelligence tests.[20]

The conditional ML estimator of β is consistent provided that the conditional likelihood function satisfies regularity conditions, which impose mild restrictions on the c_i. These restrictions, which are satisfied if the c_i are a random sample from some distribution, are discussed in Andersen (1970). Furthermore, the inverse of the information matrix based on the conditional likelihood function provides a covariance matrix for the asymptotic ($N \to \infty$) normal distribution of the conditional ML estimator of β.

These results should be contrasted with the inconsistency of the standard fixed effects ML estimator, in which the likelihood function is based on the distribution of $y_1,\dots,y_T$ conditional on $x_1,\dots,x_T, c$. For example, suppose that $T = 2$, $x_{i1} = 0$, $x_{i2} = 1$ ($i = 1,\dots,N$). The following limits exist with probability one if the c_i are a random sample from some distribution:

$$\lim_{N\to\infty} \frac{1}{N} \sum_{i=1}^{N} \mathrm{E}[y_{i1}(1-y_{i2})|c_i] = \varphi_1,$$

$$\lim_{N\to\infty} \frac{1}{N} \sum_{i=1}^{N} \mathrm{E}[(1-y_{i1})y_{i2}|c_i] = \varphi_2,$$

where

$$\mathrm{E}[y_{i1}(1-y_{i2})|c_i] = G(c_i)G(-\beta - c_i),$$
$$\mathrm{E}[(1-y_{i1})y_{i2}|c_i] = G(-c_i)G(\beta + c_i).$$

Andersen (1973, p. 66) shows that the ML estimator of β converges with probability one to 2β as $N \to \infty$. A simple extension of his argument shows that if G is replaced by any distribution function ($\tilde{G}$) corresponding to a symmetric, continuous, nonzero probability density, then the ML estimator of β converges

[20] In Rasch's model, the probability that person i gives a correct answer to item number t is $\exp(\beta_t + c_i)/[1+\exp(\beta_t + c_i)]$; this is a special case in which x_{it} is a set of dummy indicator variables. An algorithm for maximum likelihood estimation in this case is described in Andersen (1972). The use of conditional likelihood in incidental parameter problems is discussed in Andersen (1970, 1973), Kalbfleisch and Sprott (1970), and Barndorff–Nielson (1978). The conditional likelihood approach in the logit case is closely related to Fisher's (1935) exact test for independence in a 2×2 table. This exact significance test has been extended by Cox (1970) and others to the case of several contingency tables. Additional references are in Cox (1970) and Bishop et al. (1975). Chamberlain (1979) develops a conditional likelihood estimator for a point process model based on duration data, and Griliches, Hall and Hausman (1981) apply conditional likelihood techniques to panel data on counts.

with probability one to:

$$2\tilde{G}^{-1}\left(\frac{\varphi_2}{\varphi_1+\varphi_2}\right).$$

The logit case is special in that $\varphi_2/\varphi_1 = e^{\beta}$ for any distribution for c. In general the limit depends on this distribution; but if all of the $c_i = 0$, then once again we obtain convergence to 2β as $N \to \infty$.

For general T, conditioning on $\sum_t y_{it}$ $(i=1,\dots,N)$ gives the following conditional log-likelihood function:

$$L = \sum_{i=1}^{N} \ln\left[\exp\left(\beta \sum_{t=1}^{T} x_{it} y_{it}\right) \Big/ \sum_{d \in B_i} \exp\left(\beta \sum_{t=1}^{T} x_{it} d_t\right)\right],$$

$$B_i = \left\{ \boldsymbol{d} = (d_1,\dots,d_T) \,|\, d_t = 0 \text{ or } 1 \text{ and } \sum_{t=1}^{T} d_t = \sum_{t=1}^{T} y_{it} \right\}.$$

L is in the conditional logit form considered by McFadden (1974), with the alternative set (B_i) varying across the observations. Hence, it can be maximized by standard programs. There are $T+1$ distinct alternative sets corresponding to $\sum_t y_{it} = 0,1,\dots,T$. Groups for which $\sum_t y_{it} = 0$ or T contribute zero to L, however, and so only $T-1$ alternative sets are relevant. The alternative set for the group with $\sum_t y_{it} = s$ has $\binom{T}{s}$ elements, corresponding to the distinct sequences of T trials with s successes. For example, with $T=3$ and $s=1$ there are three alternatives with the following conditional probabilities:

$$P\left(1,0,0|\boldsymbol{x}_i, c_i, \sum_t y_{it} = 1\right) = \exp[\beta(x_{i1} - x_{i3})]/D,$$

$$P\left(0,1,0|\boldsymbol{x}_i, c_i, \sum_t y_{it} = 1\right) = \exp[\beta(x_{i2} - x_{i3})]/D,$$

$$P\left(0,0,1|\boldsymbol{x}_i, c_i, \sum_t y_{it} = 1\right) = 1/D,$$

$$D = \exp[\beta(x_{i1} - x_{i3})] + \exp[\beta(x_{i2} - x_{i3})] + 1.$$

A weakness in this approach is that it relies on the assumption that the y_t are independent conditional on $\boldsymbol{x}$, c, with an identical form for the conditional probability each period: $P(y_t = 1|\boldsymbol{x}, c) = G(\beta x_t + c)$. In the probit framework, these assumptions translate into $\boldsymbol{\Sigma} = \sigma^2 \boldsymbol{I}$, so that $v + u_t$ generates an equicorrelated matrix: $\sigma_v^2 \boldsymbol{ll}' + \sigma^2 \boldsymbol{I}$. We have seen that it is straightforward to allow $\boldsymbol{\Sigma}$ to be unrestricted in the probit framework; that is not true here.

An additional weakness is that we are limited in the sorts of probability statements that can be made. We obtain a clean estimate of the effect of x_t on the log odds:

$$\ln\left[\frac{P(y_t=1|x_t=x'',c)}{P(y_t=0|x_t=x'',c)}\Big/\frac{P(y_t=1|x_t=x',c)}{P(y_t=0|x_t=x',c)}\right]=\beta(x''-x');$$

the special feature of the logistic functional form is that this function of the probabilities does not depend upon c; so the problem of integrating over the marginal distribution of c (instead of the conditional distribution of c given $\boldsymbol{x}$) does not arise. But this is not the only function of the probabilities that one might want to know. In the probit section we considered

$$P(y_t=1|x_t=x'',c)-P(y_t=1|x_t=x',c),$$

which depends upon c for probit or logit, and we averaged over the marginal distribution for c:

$$\int\left[P(y_t=1|x_t=x'',c)-P(y_t=1|x_t=x',c)\right]\mu(\mathrm{d}c). \tag{3.7}$$

This requires us to specify a marginal distribution for c, which is what the conditioning argument trys to avoid. We cannot estimate (3.7) if all we have is the conditional ML estimate of β.

Our specification in (3.5) asserts that y_t is independent of $x_1,\dots,x_{t-1},x_{t+1},\dots,x_T$ conditional on x_t, c. This can be relaxed somewhat, but the conditional likelihood argument certainly requires more than

$$P(y_t=1|x_t,c)=G(\beta x_t+c);$$

to see this, try to derive (3.6) with $x_2=y_1$. We can, however, implement the following specification (with $\boldsymbol{x}'=(x_1,\dots,x_T)$):

$$P(y_t=1|\boldsymbol{x},c)=G(\beta_{t0}+\beta_{t1}x_1+\cdots+\beta_{tt}x_t+c), \tag{3.8}$$

where $y_1,\dots,y_T$ are independent conditional on $\boldsymbol{x},c$. This corresponds to our specification of "x is strictly exogenous conditional on c" in Section 2.5, except that $\gamma_t=1$ in the term $\gamma_t c$—it is not straightforward to allow a time-varying coefficient on c in the conditional likelihood approach. The extension of (3.6) is:

$$P(y_t=1|\boldsymbol{x},c,y_1+y_t=1)=G(\tilde{\beta}_{t0}+\tilde{\beta}_{t1}x_1+\beta_{t2}x_2+\cdots+\beta_{tt}x_t)$$
$$(t=2,\dots,T), \tag{3.9}$$

where $\tilde{\beta}_{tj} = \beta_{tj} - \beta_{1j}$ $(j=0,1)$. So if x has sufficient variation, we can obtain consistent estimates of $\tilde{\beta}_{t0}$, $\tilde{\beta}_{t1}$, and β_{ts} $(s=2,\dots,t)$. Only these parameters are identified, since we can transform the model replacing c by $\tilde{c} = \beta_{10} + \beta_{11}x_1 + c$ without violating any restrictions.

The restrictions in (3.5) or in (3.8) can be tested against the following alternative:

$$P(y_t = 1 | x, c) = G(\pi_{t0} + \pi_{t1}x_1 + \cdots + \pi_{tT}x_T + c). \tag{3.10}$$

We can identify only $\pi_{tj} - \pi_{1j}$ and so we can normalize $\pi_{1j} = 0$ $(j=0,\dots,T;$ $t=2,\dots,T)$. The maximized values of the conditional log-likelihoods can be used to form χ^2 statistics.[21] There are $(T-2)(T-1)/2$ restrictions in passing from (3.10) to (3.8), and (3.5) imposes an additional $(T-1)(T+4)/2 - T$ restrictions.

3.3. *Serial correlation and lagged dependent variables*

Consider the following two models:

$$y_t = \begin{cases} 1, & \text{if } y_t^* = u_t \geq 0, \\ 0, & \text{otherwise; } u_t = \rho u_{t-1} + e_t; \end{cases} \tag{3.11b}$$

$$y_t = \begin{cases} 1, & \text{if } y_t^* = u_t \geq 0, \\ 0, & \text{otherwise; } u_t = \rho u_{t-1} + e_t; \end{cases} \tag{3.11b}$$

in both cases e_t is i.i.d. $N(0, \sigma^2)$. Heckman (1978) observed that we can distinguish between these two models.[22] In the first model,

$$P(y_t = 1 | y_{t-1}, y_{t-2}, \dots) = P(y_t = 1 | y_{t-1}) = F(\gamma y_{t-1}/\sigma),$$

where $F(\cdot)$ is the standard normal distribution function. In the second model, however, $P(y_t = 1 | y_{t-1}, y_{t-2}, \dots)$ depends upon the entire history of the process. If we observed u_{t-1}, then previous outcomes would be irrelevant. In fact, we observe only whether $u_{t-1} \geq 0$; hence conditioning in addition on whether $u_{t-2} \geq 0$ affects the distribution of u_{t-1} and y_t. So the lagged y implies a Markov chain whereas the Markov assumption for the probit residual does not imply a Markov chain for the binary sequence that it generates.

[21] Conditional likelihood ratio tests are discussed in Andersen (1971).
[22] Also see Heckman (1981, 1981b).

There is an analogy with the following linear models:

$$y_t = \gamma y_{t-1} + e_t, \tag{3.12a}$$

$$y_t = u_t, u_t = e_t + \rho e_{t-1}, \tag{3.12b}$$

where e_t is i.i.d. $N(0, \sigma^2)$. We know that if $u_t = \rho u_{t-1} + e_t$, then no distinction would be possible, without introducing more structure, since both models imply a linear Markov process. With the moving average residual, however, the serial correlation model implies that the entire past history is relevant for predicting y. So the distinction between the two models rests on the order of the dependence on previous realizations of y_t.

We can still distinguish between the two models in (3.11) even when $(u_1, \dots, u_T)$ has a general multivariate normal distribution ($N(\mu, \Sigma)$). Given normalizations such as $V(u_t) = 1$ $(t = 1, \dots, T)$, the serial correlation model has $T(T+1)/2$ free parameters. Hence, if $T \geq 3$, there are restrictions on the $2^T - 1$ parameters of the multinomial distribution for $(y_1, \dots, y_T)$. In particular, the most general multivariate probit model cannot generate a Markov chain. So we can add a lagged dependent variable and identify γ.

This result relies heavily on the restrictive nature of the multivariate probit functional form. A more robust distinction between the two models is possible when there is variation over time in x_t. We shall pursue this after first presenting a generalization of strict exogeneity and noncausality for nonlinear models.

Let $t = 1$ be the first period of the individual's (economic) life. An extension of Granger's definition of "y does not cause x" is that x_{t+1} is independent of $y_1, \dots, y_t$ conditional on $x_1, \dots, x_t$. An extension of Sims' strict exogeneity condition is that y_t is independent of $x_{t+1}, x_{t+2}, \dots$ conditional on $x_1, \dots, x_t$. In contrast to the linear predictor case, these two definitions are no longer equivalent.[23] For consider the following counterexample: let y_1, y_2 be independent Bernoulli random variables with $P(y_t = 1) = P(y_t = -1) = 1/2$ $(t = 1, 2)$. Let $x_3 = y_1 y_2$. Then y_1 is independent of x_3 and y_2 is independent of x_3. Let all of the other random variables be degenerate (equal to zero, say). Then x is strictly exogenous but x_3 is clearly not independent of y_1, y_2 conditional on x_1, x_2. The counterexample works for the following reason: if a random variable is uncorrelated with each of two other random variables, then it is uncorrelated with every linear combination of them; but if it is independent of each of the other random variables, it need not be independent of every function of them.

Consider the following modification of Sims' condition: y_t is independent of $x_{t+1}, x_{t+2}, \dots$ conditional on $x_1, \dots, x_t, y_1, \dots, y_{t-1}$ $(t = 1, 2, \dots)$. Chamberlain (1982) shows that, subject to a regularity condition, this is equivalent to our

[23] See Chamberlain (1982) and Florens and Mouchart (1982).

extended definition of Granger noncausality. The regularity condition is trivially satisfied whenever y_t has a degenerate distribution prior to some point. So it is satisfied in our case since $y_0, y_{-1},\dots$ have degenerate distributions.

It is straightforward to introduce a time-invariant latent variable into these definitions. We shall say that "*y does not cause x conditional on a latent variable c*" if either:

x_{t+1} *is independent of* $y_1,\dots,y_t$ *conditional on* $x_1,\dots,x_t, c$ $(t=1,2,\dots)$,

or

y_t *is independent of* $x_{t+1}, x_{t+2},\dots$ *conditional on* $x_1,\dots,x_t, y_1,\dots,y_{t-1}, c$ $(t=1,2,\dots)$;

they are equivalent. We shall say that "*x is strictly exogenous conditional on a latent variable c*" if:

y_t *is independent of* $x_{t+1}, x_{t+2},\dots$ *conditional on* $x_1,\dots,x_t, c$ $(t=1,2,\dots)$.

Now let us return to the problem of distinguishing between serial correlation and structural lagged dependent variables. Assume throughout the discussion that x_t and y_t are not independent. We shall say that the relationship of x to y is *static* if:

x is strictly exogenous and y_t *is independent of* $x_1,\dots,x_{t-1}$ *conditional on* x_t.

Then I propose the following distinctions:

There is residual serial correlation if y_t *is not independent of* $y_1,\dots,y_{t-1}$ *conditional on* $x_1,\dots,x_t$;

If the relationship of x to y is static, then there are no structural lagged dependent variables.

Suppose that y_t and x_t are binary and consider the probability that $y_2=1$ conditional on $(x_1,x_2)=(0,0)$ and conditional on $(x_1,x_2)=(1,0)$. Since y_t and x_t are assumed to be dependent, the distribution of y_1 is generally different in the two cases. If y_1 has a structural effect on y_2, then the conditional probability of $y_2=1$ should differ in the two cases, so that y_2 is not independent of x_1 conditional on x_2.

Note that this condition is one-sided: I am only offering a condition for there to be no structural effect of y_{t-1} on y_t. There can be distributed lag relationships in which we would not want to say that y_{t-1} has a structural effect on y_t. Consider the production function example with serial correlation in rainfall; assume for the moment that there is no variation in c. If the serial correlation in rainfall is not incorporated in the farmer's information set, then our definitions assert that there is residual serial correlation but no structural lagged dependent variables, since the relationship of x to y is static. Now suppose that the farmer does use previous rainfall to predict future rainfall. Then the relationship of x to y is not static since

x is not strictly exogenous. But we may not want to say that the relationship between y_{t-1} and y_t is structural, since the technology does not depend upon y_{t-1}.

How are these distinctions affected by latent variables? It should be clear that a time-invariant latent variable can produce residual serial correlation. A major theme of the paper has been that such a latent variable can also produce a failure of strict exogeneity. So consider conditional versions of these properties:

There is residual serial correlation conditional on a latent variable c if y_t is not independent of $y_1,\dots,y_{t-1}$ conditional on $x_1,\dots,x_t, c$;

The relationship of x to y is static conditional on a latent variable c if x is strictly exogenous conditional on c and if y_t is independent of $x_1,\dots,x_{t-1}$ conditional on x_t, c;

If the relationship of x to y is static conditional on a latent variable c, then there are no structural lagged dependent variables.

A surprising feature of the linear predictor definition of strict exogeneity is that it is restrictive to assert that there exists some time-invariant latent variable c such that x is strictly exogenous conditional on c. This is no longer true when we use conditional independence to define strict exogeneity. For a counterexample, suppose that x_t is a binary variable and consider the conditional strict exogeneity question: "Does there exist a time-invariant random variable c such that y_t is independent of $x_1,\dots,x_T$ conditional on $x_1,\dots,x_t, c$?" The answer is "yes" since we can order the 2^T possible outcomes of the binary sequence $(x_1,\dots,x_T)$ and set $c = j$ if the jth outcome occurs $(j = 1,\dots,2^T)$. Now y_t is independent of $x_1,\dots,x_T$ conditional on c!

For a nondegenerate counterexample, let y and x be binary random variables with:

$$P(y=\alpha_j, x=\alpha_k)=\tau_{jk}>0, \qquad \sum_{j,k=1}^{2}\tau_{jk}=1,$$

where $\alpha_1 = 1$, $\alpha_2 = 0$. Let $\gamma' = (\tau_{11}, \tau_{12}, \tau_{21}, \tau_{22})$. Then we can set:

$$\gamma=\sum_{m=1}^{4}\gamma_m e_m, \qquad \gamma_m>0, \qquad \sum_{m=1}^{4}\gamma_m=1,$$

where e_m is a vector of zeros except for a one in the mth component. Hence γ is in the interior of the convex hull of $\{e_m, m = 1,\dots,4\}$. Now consider the vector:

$$y(\delta,\lambda)=\begin{bmatrix}\delta\lambda\\ \delta(1-\lambda)\\ (1-\delta)\lambda\\ (1-\delta)(1-\lambda)\end{bmatrix}$$

The components of $y(\delta,\lambda)$ give the probabilities $P(y=\alpha_j,\ x=\alpha_k)$ when y and x are independent with $P(y=1)=\delta$, $P(x=1)=\lambda$. Set $e_m^*=y(\delta_m,\lambda_m)$ with $0<\delta_m<1, 0<\lambda_m<1$. Then γ will be in the interior of the convex hull of $\{e_m^*, m=1,\dots,4\}$ if we choose δ_m,λ_m so that e_m^* is sufficiently close to e_m. Hence:

$$\gamma=\sum_{m=1}^{4}\gamma_m^* e_m^*,\qquad \gamma_m^*>0,\qquad \sum_{m=1}^{4}\gamma_m^*=1.$$

Let the components of e_m^* be $(\tau_{11}^m,\tau_{12}^m,\tau_{21}^m,\tau_{22}^m)$. Let c be a random variable with $P(c=m)=\gamma_m^*$ $(m=1,\dots,4)$, and set

$$P(y=\alpha_j,\ x=\alpha_k\,|\,c=m)=\tau_{jk}^m.$$

Now y is independent of x conditional on c, and the conditional distributions are nondegenerate.

If $(x_1,\dots,x_T,y_1,\dots,y_T)$ has a general multinomial distribution, then a straightforward extension of this argument shows that there exists a random variable c such that $(y_1,\dots,y_T)$ is independent of $(x_1,\dots,x_T)$ conditional on c, and the conditional distributions are nondegenerate.

A similar point applies to factor analysis. Consider a linear one-factor model. The specification is that there exists a latent variable c such that the partial correlations between $y_1,\dots,y_T$ are zero given c. This is restrictive if $T>3$. But we now know that it is not restrictive to assert that there exists a latent variable c such that $y_1,\dots,y_T$ are independent conditional on c.

It follows that we cannot test for conditional strict exogeneity without imposing functional form restrictions; nor can we test for a conditionally static relationship without restricting the functional forms.

This point is intimately related to the fundamental difficulties created by incidental parameters in nonlinear models. The labor force participation example is assumed to be static conditional on c. We shall present some tests of this in Section 5, but we shall be jointly testing that proposition and the functional forms —a truly nonparametric test cannot exist. We stressed in the probit model that the specification for the distribution of c conditional on x is restrictive; we avoided such a restrictive specification in the logit model but only by imposing a restrictive functional form on the distribution of y conditional on x, c.

3.4. *Duration models*

In many problems the basic data is the amount of time spent in a state. For example, a complete description of an individual's labor force participation history is the duration of the first spell of participation and the date it began, the duration of the following spell of nonparticipation, and so on. This complete history will generate a binary sequence when it is cut up into fixed length periods, but these periods may have little to do with the underlying process.[24]

In particular, the measurement of serial correlation depends upon the period of observation. As the period becomes shorter, the probability that a person who worked last period will work this period approaches one. So finding significant serial correlation may say very little about the underlying process. Or consider a spell that begins near the end of a period; then it is likely to overlap into the next period, so that previous employment raises the probability of current employment.

Consider the underlying process of time spent in one state followed by time spent in the other state. If the individual's history does not help to predict his future given his current state, then this is a Markov process. Whereas serial independence in continuous time has the absurd implication that mean duration of a spell is zero, the Markov property does provide a fruitful starting point. It has two implications: the individual's history prior to the current spell should not affect the distribution of the length of the current spell; and the amount of time spent in the current state should not affect the distribution of remaining time in that state.

So the first requirement of the Markov property is that durations of the spells be independent of each other. Assuming stationarity, this implies an alternating renewal process. The second requirement is that the distribution of duration be exponential, so that we have an alternating Poisson process. We shall refer to departures from this model as duration dependence.

A test of this Markov property using binary sequences will depend upon what sampling scheme is being used. The simplest case is point sampling, where each period we determine the individual's state at a particular point in time, such as July 1 of each year. Then if an individual is following an alternating Poisson process, her history prior to that point is irrelevant in predicting her state at the next interview. So the binary sequence generated by point sampling should be a Markov chain.

It is possible to test this in a fixed effects model that allows each individual to have her own two exponential rate parameters (c_{i1}, c_{i2}) in the alternating Poisson process. The idea is related to the conditional likelihood approach in the fixed effects logit model. Let s_{ijk} be the number of times that individual i is observed making a transition from state j to state k ($j, k = 1,2$). Then the initial state and these four transition counts are sufficient statistics for the Markov chain. Sequences with the same initial state and the same transition counts should be equally likely. This is the Markov form of de Finetti's (1975) partial exchangeabil-

[24] This point is discussed in Singer and Spilerman (1974, 1976).

ity.[25] So we can test whether the Markov property holds conditional on c_{i1}, c_{i2} by testing whether there is significant variation in the sample frequencies of sequences with the same transition counts.

This analysis is relevant if, for example, each year the survey question is: "Did you have a job on July 1?" In the Michigan Panel Study of Income Dynamics, however, the most commonly used question for generating participation sequences is: "Did your wife do any work for money last year?" This interval sampling leads to a more complex analysis, since even if the individual is following an alternating Poisson process, the binary sequence generated by this sampling scheme is not a Markov chain. Suppose that $y_{t-1}=1$, so that we know that the individual worked at some point during the previous period. What is relevant, however, is the individual's state at the end of the period, and y_{t-2} will affect the probability that the spell of work occurred early in period $t-1$ instead of late in the period.

Nevertheless, it is possible to test whether the underlying process is alternating Poisson. The reason is that if $y_{t-1}=0$, we know that the individual never worked during period $t-1$, and so we know the state at the end of that period; hence $y_{t-2}, y_{t-3},\dots$ are irrelevant. So we have:

$$\begin{aligned}
&P(y_t=1|c_1,c_2,y_{t-1},y_{t-2},\dots)\\
&\quad=P(y_t=1|c_1,c_2,y_{t-1}=\cdots=y_{t-d}=1,y_{t-d-1}=0)\\
&\quad=P(y_t=1|c_1,c_2,d),
\end{aligned}$$

where d is the number of consecutive preceding periods that the individual was in state 1.

Let s_{01} be the number of times in the sequence that 1 is preceded by 0; let s_{011} be the number of times that 1 is preceded by 0,1; etc. Then sufficient statistics are $s_{01}, s_{011},\dots,$ as well as the number of consecutive ones at the beginning (n_1) and at the end (n_T) of a sequence.[26] For an example with $T=5$, let $n_1=0$, $n_5=0$, $s_{01}=1$, $s_{011}=1$, $s_{0111}=\cdots=0$; then we have

$$\begin{aligned}
&P(0,1,1,0,0|\boldsymbol{c})\\
&\quad=P(y_1=0|\boldsymbol{c})P(1|0,\boldsymbol{c})P(1|0,1,\boldsymbol{c})P(0|0,1,1,\boldsymbol{c})P(0|0,\boldsymbol{c});\\
&P(0,0,1,1,0|\boldsymbol{c})\\
&\quad=P(y_1=0|\boldsymbol{c})P(0|0,\boldsymbol{c})P(1|0,\boldsymbol{c})P(1|0,1,\boldsymbol{c})P(0|0,1,1,\boldsymbol{c}),
\end{aligned}$$

[25] We are using the fact that partial exchangeability is a necessary condition for the distribution to be a mixture of Markov chains. Diaconis and Freedman (1980) study the sufficiency of this condition. Heckman (1978) used exchangeability to test for serial independence in a fixed effects model.

[26] This test was presented in Chamberlain (1978a, 1979). It has been applied to unemployment sequences by Corcoran and Hill (1980). For related tests and extensions, see Lee (1980).

where $c=(c_1,c_2)$. Thus these two sequences are equally likely conditional on c, and letting μ be the probability measure for c gives:

$$P(0,1,1,0,0)=\int P(0,1,1,0,0|c)\mu(\mathrm{d}c)$$

$$=\int P(0,0,1,1,0|c)\mu(\mathrm{d}c)=P(0,0,1,1,0).$$

So the alternating Poisson process implies restrictions on the multinomial distribution for the binary sequence.

These tests are indirect. The duration dependence question is clearly easier to answer using surveys that measure durations of spells. Such duration data raises a number of new econometric problems, but we shall not pursue them here.[27] I would simply like to make one connection with the methods that we have been discussing.

Let us simplify to a one state process; for example, y_{it} can be the duration of the time interval between the starting date of the ith individual's tth job and his $(t+1)$th job. Suppose that we observe $T>1$ jobs for each of the N individuals, a not innocuous assumption. Impose the restriction that $y_{it}>0$ by using the following specification:

$$y_{it}=\exp(\beta x_{it}+c_i+u_{it}),$$
$$\mathrm{E}^*(u_{it}|x_i)=0 \qquad (t=1,\dots,T),$$

where $x_i'=(x_{i1},\dots,x_{iT})$. Then:

$$\mathrm{E}^*(\ln y_{it}|x_i)=\beta x_{it}+\lambda' x_i,$$

and our Section 2 analysis applies. The strict exogeneity assumption has a surprising implication in this context. Suppose that x_{it} is the individual's age at the beginning of the tth job. Then $x_{it}-x_{i,t-1}=y_{i,t-1}$—age is not strictly exogenous.[28]

4. Inference

Consider a sample $r_i'=(x_i',y_i')$, $i=1,\dots,N$, where $x_i'=(x_{i1},\dots,x_{iK})$, $y_i'=(y_{i1},\dots,y_{iM})$. We shall assume that r_i is independent and identically distributed (i.i.d.) according to some multivariate distribution with finite fourth moments

[27]See Tuma (1979, 1980), Lancaster (1979), Nickell (1979), Chamberlain (1979), Lancaster and Nickell (1980), Heckman and Borjas (1980), Kiefer and Neumann (1981), and Flinn and Heckman (1982, 1982a).

[28]This example is based on Chamberlain (1979).

and $E(x_i x_i')$ nonsingular. Consider the minimum mean-square error linear predictors,[29]

$$E^*(y_{im}|x_i) = \pi_m' x_i \qquad (m=1,\dots,M),$$

which we can write as:

$$E^*(y_i|x_i) = \Pi x_i, \qquad \Pi = E(y_i x_i')[E(x_i x_i')]^{-1}.$$

We want to estimate Π subject to restrictions and to test those restrictions. For example, we may want to test whether a submatrix of Π has the form $\beta I + l\lambda'$.

We shall not assume that the regression function $E(y_i|x_i)$ is linear. For although $E(y_i|x_i, c_i)$ may be linear (indeed, we hope that it is), there is generally no reason to insist that $E(c_i|x_i)$ is linear. So we shall present a theory of inference for linear predictors. Furthermore, even if the regression function is linear, there may be heteroskedasticity—due to random coefficients, for example. So we shall allow $E[(y_i - \Pi x_i)(y_i - \Pi x_i)'|x_i]$ to be an arbitrary function of x_i.

4.1. *The estimation of linear predictors*

Let w_i be the vector formed from the distinct elements of $r_i r_i'$ that have nonzero variance.[30] Since $r_i' = (x_i', y_i')$ is i.i.d., it follows that w_i is i.i.d. This simple observation is the key to our results. Since Π is a function of $E(w_i)$, our problem is to make inferences about a function of a population mean, under random sampling.

Let $\mu = E(w_i)$ and let π be the vector formed from the columns of Π' ($\pi = \text{vec}(\Pi')$). Then π is a function of μ: $\pi = h(\mu)$. Let $\bar{w} = \sum_{i=1}^N w_i/N$; then $\hat{\pi} = h(\bar{w})$ is the least squares estimator:

$$\hat{\pi} = \text{vec}\left[\left(\sum_{i=1}^N x_i x_i'\right)^{-1} \sum_{i=1}^N x_i y_i'\right].$$

By the strong law of large numbers, $\bar{w}$ converges almost surely to μ^0 as $N \to \infty$ $\left(\bar{w} \xrightarrow{\text{a.s.}} \mu^0\right)$, where μ^0 is the true value of μ. Let $\pi^0 = h(\mu^0)$. Since $h(\mu)$ is continuous at $\mu = \mu^0$, we have $\hat{\pi} \xrightarrow{\text{a.s.}} \pi^0$. The central limit theorem implies that:

$$\sqrt{N}(\bar{w} - \mu^0) \xrightarrow{D} N(0, V(w_i)).$$

[29] This agrees with the definition in Section 2 if x_i includes a constant.
[30] Sections 4.1–4.4 are taken from Chamberlain (1982a).

Since $h(\mu)$ is differentiable at $\mu = \mu^0$, the δ-method gives

$$\sqrt{N}(\hat{\pi} - \pi^0) \xrightarrow{D} N(0, \Omega),$$

where

$$\Omega = \frac{\partial h(\mu^0)}{\partial \mu'} V(w_i) \frac{\partial h'(\mu^0)}{\partial \mu}.^{31}$$

We have derived the limiting distribution of the least squares estimator. This approach was used by Cramèr (1946) to obtain limiting normal distributions for sample correlation and regression coefficients (p. 367); he presents an explicit formula for the variance of the limiting distribution of a sample correlation coefficient (p. 359). Kendall and Stuart (1961, p. 293) and Goldberger (1974) present the formula for the variance of the limiting distribution of a simple regression coefficient.

Evaluating the partial derivatives in the formula for Ω is tedious. That calculation can be simplified since $\hat{\pi}$ has a "ratio" form. In the case of simple regression with a zero intercept, we have $\pi = \mathrm{E}(y_i x_i)/\mathrm{E}(x_i^2)$ and

$$\sqrt{N}(\hat{\pi} - \pi^0) = \left(\sum_{i=1}^{N} y_i x_i - \pi^0 \sum_{i=1}^{N} x_i^2 \right) \Big/ \left[\sqrt{N} \left(\sum_{i=1}^{N} x_i^2 / N \right) \right].$$

Since $\sum_{i=1}^{N} x_i^2/N \xrightarrow{\text{a.s.}} \mathrm{E}(x_i^2)$, we obtain the same limiting distribution by working with

$$\sum_{i=1}^{N} \left[(y_i - \pi^0 x_i) x_i \right] / \left[\sqrt{N}\, \mathrm{E}(x_i^2) \right].$$

The definition of π^0 gives $\mathrm{E}[(y_i - \pi^0 x_i) x_i] = 0$, and so the central limit theorem implies that:

$$\sqrt{N}(\hat{\pi} - \pi^0) \xrightarrow{D} N\left\{0, \mathrm{E}\left[(y_i - \pi^0 x_i)^2 x_i^2 \right] / \left[\mathrm{E}(x_i^2) \right]^2 \right\}.$$

This approach was used by White (1980) to obtain the limiting distribution for univariate regression coefficients.[32] In the Appendix (Proposition 6) we follow

[31] See Billingsley (1979, example 29.1, p. 340) or Rao (1973, p. 388).
[32] Also see White (1980a,b).

White's approach to obtain:

$$\boldsymbol{\Omega} = \mathrm{E}\left[\left(\boldsymbol{y}_i - \boldsymbol{\Pi}^0\boldsymbol{x}_i\right)\left(\boldsymbol{y}_i - \boldsymbol{\Pi}^0\boldsymbol{x}_i\right)' \otimes \boldsymbol{\Phi}_x^{-1}\left(\boldsymbol{x}_i\boldsymbol{x}_i'\right)\boldsymbol{\Phi}_x^{-1}\right], \tag{4.1}$$

where $\boldsymbol{\Phi}_x = \mathrm{E}(\boldsymbol{x}_i\boldsymbol{x}_i')$. A consistent estimator of $\boldsymbol{\Omega}$ is readily available from the corresponding sample moments:

$$\hat{\boldsymbol{\Omega}} = \frac{1}{N}\sum_{i=1}^{N}\left[\left(\boldsymbol{y}_i - \hat{\boldsymbol{\Pi}}\boldsymbol{x}_i\right)\left(\boldsymbol{y}_i - \hat{\boldsymbol{\Pi}}\boldsymbol{x}_i\right)' \otimes \boldsymbol{S}_x^{-1}\left(\boldsymbol{x}_i\boldsymbol{x}_i'\right)\boldsymbol{S}_x^{-1}\right] \xrightarrow{\text{a.s.}} \boldsymbol{\Omega}, \tag{4.2}$$

where $\boldsymbol{S}_x = \sum_{i=1}^{N}\boldsymbol{x}_i\boldsymbol{x}_i'/N$.

If $\mathrm{E}(\boldsymbol{y}_i|\boldsymbol{x}_i) = \boldsymbol{\Pi}\boldsymbol{x}_i$, so that the regression function is linear, then:

$$\boldsymbol{\Omega} = \mathrm{E}\left[V(\boldsymbol{y}_i|\boldsymbol{x}_i) \otimes \boldsymbol{\Phi}_x^{-1}\left(\boldsymbol{x}_i\boldsymbol{x}_i'\right)\boldsymbol{\Phi}_x^{-1}\right].$$

If $V(\boldsymbol{y}_i|\boldsymbol{x}_i)$ is uncorrelated with $\boldsymbol{x}_i\boldsymbol{x}_i'$, then:

$$\boldsymbol{\Omega} = \mathrm{E}\left[V(\boldsymbol{y}_i|\boldsymbol{x}_i)\right] \otimes \boldsymbol{\Phi}_x^{-1}.$$

If the conditional variance is homoskedastic, so that $V(\boldsymbol{y}_i|\boldsymbol{x}_i) = \boldsymbol{\Sigma}$ does not depend on $\boldsymbol{x}_i$, then:

$$\boldsymbol{\Omega} = \boldsymbol{\Sigma} \otimes \boldsymbol{\Phi}_x^{-1}.$$

4.2. *Imposing restrictions: The minimum distance estimator*

Since $\boldsymbol{\Pi}$ is a function of $\mathrm{E}(\boldsymbol{w}_i)$, restrictions on $\boldsymbol{\Pi}$ imply restrictions on $\mathrm{E}(\boldsymbol{w}_i)$. Let the dimension of $\boldsymbol{\mu} = \mathrm{E}(\boldsymbol{w}_i)$ be q.[33] We shall specify the restrictions by the condition that $\boldsymbol{\mu}$ depends only on a $p \times 1$ vector $\boldsymbol{\theta}$ of unknown parameters: $\boldsymbol{\mu} = \boldsymbol{g}(\boldsymbol{\theta})$, where $\boldsymbol{g}$ is a known function and $p \le q$. The domain of $\boldsymbol{\theta}$ is Υ, a subset of p-dimensional Euclidean space (R^p) that contains the true value $\boldsymbol{\theta}^0$. So the restrictions imply that $\boldsymbol{\mu}^0 = \boldsymbol{g}(\boldsymbol{\theta}^0)$ is confined to a certain subset of R^q.

We can impose the restrictions by using a minimum distance estimator: choose $\hat{\boldsymbol{\theta}}$ to

$$\min_{\boldsymbol{\theta} \in \Upsilon} \sum_{i=1}^{N}\left[\boldsymbol{w}_i - \boldsymbol{g}(\boldsymbol{\theta})\right]' \boldsymbol{A}_N \left[\boldsymbol{w}_i - \boldsymbol{g}(\boldsymbol{\theta})\right],$$

[33] If there is one element in $\boldsymbol{r}_i\boldsymbol{r}_i'$ with zero variance, then $q = [(K+M)(K+M+1)/2] - 1$.

where $A_N \xrightarrow{\text{a.s.}} \Psi$ and Ψ is positive definite.[34] This minimization problem is equivalent to the following one: choose $\hat{\theta}$ to

$$\min_{\theta \in \Upsilon} [\bar{w} - g(\theta)]' A_N [\bar{w} - g(\theta)].$$

The properties of $\hat{\theta}$ are developed, for example, in Malinvaud (1970, ch. 9). Since g does not depend on any exogenous variables, the derivation of these properties can be simplified considerably, as in Chiang (1956) and Ferguson (1958).[35]

For completeness, we shall state a set of regularity conditions and the properties that they imply:

Assumption 1

$a_N \xrightarrow{\text{a.s.}} g(\theta^0)$; Υ is a compact subset of R^p that contains θ^0; g is continuous on Υ, and $g(\theta) = g(\theta^0)$ for $\theta \in \Upsilon$ implies that $\theta = \theta^0$; $A_N \xrightarrow{\text{a.s.}} \Psi$, where Ψ is positive definite.

Assumption 2

$\sqrt{N}[a_N - g(\theta^0)] \xrightarrow{\text{D}} N(0, \Delta)$; Υ contains a neighborhood Ξ_0 of θ^0 in which g has continuous second partial derivatives; rank $(G) = p$, where $G = \partial g(\theta^0)/\partial \theta'$.

Choose $\hat{\theta}$ to

$$\min_{\theta \in \Upsilon} [a_N - g(\theta)]' A_N [a_N - g(\theta)].$$

Proposition 1

If Assumption 1 is satisfied, then $\hat{\theta} \xrightarrow{\text{a.s.}} \theta^0$.

Proposition 2

If Assumptions 1 and 2 are satisfied, then $\sqrt{N}(\hat{\theta} - \theta^0) \xrightarrow{\text{D}} N(0, \Lambda)$, where

$$\Lambda = (G'\Psi G)^{-1} G'\Psi\Delta\Psi G (G'\Psi G)^{-1}.$$

If Δ is positive definite, then $\Lambda - (G'\Delta^{-1}G)^{-1}$ is positive semi-definite; hence an optimal choice for Ψ is Δ^{-1}.

[34] This application of nonlinear generalized least squares was proposed in Chamberlain (1980a).
[35] Some simple proofs are collected in Chamberlain (1982a).

Proposition 3

If Assumptions 1 and 2 are satisfied, if $\boldsymbol{\Delta}$ is a $q \times q$ positive-definite matrix, and if $\boldsymbol{A}_N \overset{\text{a.s.}}{\to} \boldsymbol{\Delta}^{-1}$, then:

$$N[\boldsymbol{a}_N - \boldsymbol{g}(\hat{\boldsymbol{\theta}})]'\boldsymbol{A}_N[\boldsymbol{a}_N - \boldsymbol{g}(\hat{\boldsymbol{\theta}})] \overset{\text{D}}{\to} \chi^2(q-p).$$

Now consider imposing additional restrictions, which are expressed by the condition that $\boldsymbol{\theta} = \boldsymbol{f}(\boldsymbol{\alpha})$, where $\boldsymbol{\alpha}$ is $s \times 1 (s \leq p)$. The domain of $\boldsymbol{\alpha}$ is Υ_1, a subset of R^s that contains the true value $\boldsymbol{\alpha}^0$. So $\boldsymbol{\theta}^0 = \boldsymbol{f}(\boldsymbol{\alpha}^0)$ is confined to a certain subset of R^p.

Assumption 2′

Υ_1 is a compact subset of R^s that contains $\boldsymbol{\alpha}^0$; $\boldsymbol{f}$ is a continuous mapping from Υ_1 into Υ; $\boldsymbol{f}(\boldsymbol{\alpha}) = \boldsymbol{\theta}^0$ for $\boldsymbol{\alpha} \in \Upsilon_1$ implies $\boldsymbol{\alpha} = \boldsymbol{\alpha}^0$; Υ_1 contains a neighborhood of $\boldsymbol{\alpha}^0$ in which $\boldsymbol{f}$ has continuous second partial derivatives; rank $(\boldsymbol{F}) = s$, where $\boldsymbol{F} = \partial \boldsymbol{f}(\boldsymbol{\alpha}^0)/\partial \boldsymbol{\alpha}'$.

Let $\boldsymbol{h}(\boldsymbol{\alpha}) = \boldsymbol{g}[\boldsymbol{f}(\boldsymbol{\alpha})]$. Choose $\hat{\boldsymbol{\alpha}}$ to

$$\min_{\boldsymbol{\alpha} \in \Upsilon_1} [\boldsymbol{a}_N - \boldsymbol{h}(\boldsymbol{\alpha})]'\boldsymbol{A}_N[\boldsymbol{a}_N - \boldsymbol{h}(\boldsymbol{\alpha})].$$

Proposition 3′

If Assumptions 1, 2, and 2′ are satisfied, if $\boldsymbol{\Delta}$ is positive definite, and if $\boldsymbol{A}_N \overset{\text{a.s.}}{\to} \boldsymbol{\Delta}^{-1}$, then $d_1 - d_2 \overset{\text{D}}{\to} \chi^2(p-s)$, where

$$d_1 = N[\boldsymbol{a}_N - \boldsymbol{h}(\hat{\boldsymbol{\alpha}})]'\boldsymbol{A}_N[\boldsymbol{a}_N - \boldsymbol{h}(\hat{\boldsymbol{\alpha}})],$$
$$d_2 = N[\boldsymbol{a}_N - \boldsymbol{g}(\hat{\boldsymbol{\theta}})]'\boldsymbol{A}_N[\boldsymbol{a}_N - \boldsymbol{g}(\hat{\boldsymbol{\theta}})].$$

Furthermore, $d_1 - d_2$ is independent of d_2 in their limiting joint distribution.

Suppose that the restrictions involve only $\boldsymbol{\Pi}$. We specify the restrictions by the condition that $\boldsymbol{\pi} = \boldsymbol{f}(\boldsymbol{\delta})$, where $\boldsymbol{\delta}$ is $s \times 1$ and the domain of $\boldsymbol{\delta}$ is Υ_1, a subset of R^s that includes the true value $\boldsymbol{\delta}^0$. Consider the following estimator of $\boldsymbol{\delta}^0$: choose $\hat{\boldsymbol{\delta}}$ to

$$\min_{\boldsymbol{\delta} \in \Upsilon_1} [\hat{\boldsymbol{\pi}} - \boldsymbol{f}(\boldsymbol{\delta})]'\hat{\boldsymbol{\Omega}}^{-1}[\hat{\boldsymbol{\pi}} - \boldsymbol{f}(\boldsymbol{\delta})],$$

where $\hat{\boldsymbol{\Omega}}$ is given in (4.2), and we assume that $\boldsymbol{\Omega}$ in (4.1) is positive definite. If Υ_1

and f satisfy Assumptions 1 and 2, then $\hat{\boldsymbol{\delta}} \overset{\text{a.s.}}{\to} \boldsymbol{\delta}^0$,

$$\sqrt{N}\,(\hat{\boldsymbol{\delta}} - \boldsymbol{\delta}^0) \overset{\text{D}}{\to} N\big(\boldsymbol{0},[\boldsymbol{F}'\boldsymbol{\Omega}^{-1}\boldsymbol{F}]^{-1}\big),$$

and

$$N[\hat{\boldsymbol{\pi}} - \boldsymbol{f}(\hat{\boldsymbol{\delta}})]'\hat{\boldsymbol{\Omega}}^{-1}[\hat{\boldsymbol{\pi}} - \boldsymbol{f}(\hat{\boldsymbol{\delta}})] \overset{\text{D}}{\to} \chi^2(MK - s),$$

where $\boldsymbol{F} = \partial \boldsymbol{f}(\boldsymbol{\delta}^0)/\partial \boldsymbol{\delta}'$.

We can also estimate $\boldsymbol{\delta}^0$ by applying the minimum distance procedure to $\bar{\boldsymbol{w}}$ instead of to $\hat{\boldsymbol{\pi}}$. Suppose that the components of $\boldsymbol{w}_i$ are arranged so that $\boldsymbol{w}_i' = (\boldsymbol{w}_{i1}', \boldsymbol{w}_{i2}')$, where $\boldsymbol{w}_{i1}$ contains the components of $\boldsymbol{x}_i\,\boldsymbol{y}_i'$. Partition $\boldsymbol{\mu} = \mathrm{E}(\boldsymbol{w}_i)$ conformably: $\boldsymbol{\mu}' = (\boldsymbol{\mu}_1', \boldsymbol{\mu}_2')$. Set $\boldsymbol{\theta}' = (\boldsymbol{\theta}_1', \boldsymbol{\theta}_2') = (\boldsymbol{\delta}', \boldsymbol{\mu}_2')$. Assume that $V(\boldsymbol{w}_i)$ is positive definite. Now choose $\hat{\boldsymbol{\theta}}$ to

$$\min_{\boldsymbol{\theta} \in \Upsilon} [\bar{\boldsymbol{w}} - \boldsymbol{g}(\boldsymbol{\theta})]'\boldsymbol{A}_N[\bar{\boldsymbol{w}} - \boldsymbol{g}(\boldsymbol{\theta})],$$

where $\boldsymbol{A}_N \overset{\text{a.s.}}{\to} V^{-1}(\boldsymbol{w}_i)$,

$$\boldsymbol{g}(\boldsymbol{\theta}) = \begin{bmatrix} \boldsymbol{g}_1[\boldsymbol{f}(\boldsymbol{\delta}), \boldsymbol{\mu}_2] \\ \boldsymbol{\mu}_2 \end{bmatrix},$$

and $\boldsymbol{g}_1(\boldsymbol{\pi}, \boldsymbol{\mu}_2) = \boldsymbol{\mu}_1$. Then $\hat{\boldsymbol{\theta}}_1$ gives an estimator of $\boldsymbol{\delta}^0$; it has the same limiting distribution as the estimator $\hat{\boldsymbol{\delta}}$ that we obtained by applying the minimum distance procedure to $\hat{\boldsymbol{\pi}}$.[36]

This framework leads to some surprising results on efficient estimation. For a simple example, we shall use a univariate linear predictor model,

$$\mathrm{E}^*(y_i | x_{i1}, x_{i2}) = \pi_0 + \pi_1 x_{i1} + \pi_2 x_{i2}.$$

Consider imposing the restriction $\pi_2 = 0$. Then the conventional estimator of π_1 is b_{yx_1}, the slope coefficient in the least squares regression of y on x_1. We shall show that this estimator is generally less efficient than the minimum distance estimator if the regression function is nonlinear or if there is heteroskedasticity.

Let $\hat{\pi}_1, \hat{\pi}_2$ be the slope coefficients in the least squares multiple regression of y on x_1, x_2. The minimum distance estimator of π_1 under the restriction $\pi_2 = 0$ can be obtained as $\hat{\delta} = \hat{\pi}_1 + \tau\hat{\pi}_2$, where τ is chosen to minimize the (estimated)

[36]See Chamberlain (1982a, proposition 9).

variance of the limiting distribution of $\hat{\delta}$; this gives:

$$\hat{\delta} = \hat{\pi}_1 - \frac{\hat{\omega}_{12}}{\hat{\omega}_{22}}\hat{\pi}_2,$$

where $\hat{\omega}_{jk}$ is the estimated covariance between $\hat{\pi}_j$ and $\hat{\pi}_k$ in their limiting distribution. Since $\hat{\pi}_1 = b_{yx_1} - \hat{\pi}_2 b_{x_2x_1}$, we have:

$$\hat{\delta} = b_{yx_1} - \left(b_{x_2x_1} + \frac{\hat{\omega}_{12}}{\hat{\omega}_{22}}\right)\hat{\pi}_2.$$

If $E(y_i|x_{i1}, x_{i2})$ is linear and if $V(y_i|x_{i1}, x_{i2}) = \sigma^2$, then $\omega_{12}/\omega_{22} = -\operatorname{cov}(x_{i1}, x_{i2})/V(x_{i1})$ and $\hat{\delta} = b_{yx_1}$. But in general $\hat{\delta} \neq b_{yx_1}$ and $\hat{\delta}$ is more efficient than b_{yx_1}. The source of the efficiency gain is that the limiting distribution of $\hat{\pi}_2$ has a zero mean (if $\pi_2 = 0$), and so we can reduce variance without introducing any bias if $\hat{\pi}_2$ is correlated with b_{yx_1}. Under the assumptions of linear regression and homoskedasticity, b_{yx_1} and $\hat{\pi}_2$ are uncorrelated; but this need not be true in the more general framework that we are using.

4.3. *Simultaneous equations: A generalization of three-stage least squares*

Given the discussion on imposing restrictions, it is not surprising that two-stage least squares is not, in general, an efficient procedure for combining instrumental variables. Also, three-stage least squares, viewed as a minimum distance estimator, is using the wrong norm in general.

Consider the standard simultaneous equations model:

$$y_i = \boldsymbol{\Pi} \boldsymbol{x}_i + \boldsymbol{u}_i, \qquad E(\boldsymbol{u}_i \boldsymbol{x}_i') = \boldsymbol{0},$$
$$\boldsymbol{\Gamma} y_i + \boldsymbol{B}\boldsymbol{x}_i = \boldsymbol{v}_i,$$

where $\boldsymbol{\Gamma\Pi} + \boldsymbol{B} = \boldsymbol{0}$ and $\boldsymbol{\Gamma u}_i = \boldsymbol{v}_i$. We are continuing to assume that y_i is $M \times 1$, $\boldsymbol{x}_i$ is $K \times 1$, $\boldsymbol{r}_i' = (\boldsymbol{x}_i', y_i')$ is i.i.d. according to a distribution with finite fourth moments ($i = 1,\dots,N$), and that $E(\boldsymbol{x}_i\boldsymbol{x}_i')$ is nonsingular. There are restrictions on $\boldsymbol{\Gamma}$ and $\boldsymbol{B}$: $\boldsymbol{m}(\boldsymbol{\Gamma}, \boldsymbol{B}) = \boldsymbol{0}$, where $\boldsymbol{m}$ is a known function. Assume that the implied restrictions on $\boldsymbol{\Pi}$ can be specified by the condition that $\boldsymbol{\pi} = \operatorname{vec}(\boldsymbol{\Pi}') = \boldsymbol{f}(\boldsymbol{\delta})$, where the domain of $\boldsymbol{\delta}$ is Υ_1, a subset of R^s that includes the true value $\boldsymbol{\delta}^0 (s \leq MK)$. Assume that Υ_1 and $\boldsymbol{f}$ satisfy assumptions 1 and 2; these properties could be derived from regularity conditions on $\boldsymbol{m}$, as in Malinvaud (1970, proposition 2, p. 670).

Choose $\hat{\boldsymbol{\delta}}$ to

$$\min_{\boldsymbol{\delta} \in \Upsilon_1} [\hat{\boldsymbol{\pi}} - \boldsymbol{f}(\boldsymbol{\delta})]'\hat{\boldsymbol{\Omega}}^{-1}[\hat{\boldsymbol{\pi}} - \boldsymbol{f}(\boldsymbol{\delta})],$$

where $\hat{\Omega}$ is given in (4.2) and we assume that Ω in (4.1) is positive definite. Let $F = \partial f(\delta^0)/\partial\delta'$. Then we have $\sqrt{N}(\hat{\delta} - \delta^0) \xrightarrow{D} N(0, \Lambda)$, where $\Lambda = (F'\Omega^{-1}F)^{-1}$. This generalizes Malinvaud's minimum distance estimator (p. 676); it reduces to his estimator if $u_i^0 u_i^{0\prime}$ is uncorrelated with $x_i x_i'$, so that $\Omega = \mathrm{E}(u_i^0 u_i^{0\prime}) \otimes [\mathrm{E}(x_i x_i')]^{-1}$ $(u_i^0 = y_i - \Pi^0 x_i)$.

Now suppose that the only restrictions on Γ and B are that certain coefficients are zero, together with the normalization restrictions that the coefficient of y_{im} in the mth structural equation is one. Then we can give an explicit formula for Λ. Write the mth structural equation as:

$$y_{im} = \delta_m' z_{im} + v_{im},$$

where the components of z_{im} are the variables in y_i and x_i that appear in the mth equation with unknown coefficients. Let there be M structural equations and assume that the true value Γ^0 is nonsingular. Let S_{zx} be the following block-diagonal matrix:

$$S_{zx} = \mathrm{diag}\left\{\frac{1}{N}\sum_{i=1}^{N} z_{i1}x_i', \dots, \frac{1}{N}\sum_{i=1}^{N} z_{iM}x_i'\right\},$$

and $s_{xy} = N^{-1}\sum_{i=1}^{N} y_i \otimes x_i$. Let $v_i^{0\prime} = (v_{i1}^0, \dots, v_{iM}^0)$, where $v_{im}^0 = y_{im} - \delta_m^{0\prime} z_{im}$ and δ_m^0 is the true value; let $\Phi_{zx} = \mathrm{E}(S_{zx})$ and $\Phi_x = \mathrm{E}(x_i x_i')$. Let $\delta' = (\delta_1', \dots, \delta_M')$. Then we have:

$$\Lambda = \left\{\Phi_{zx}\left[\mathrm{E}\left(v_i^0 v_i^{0\prime} \otimes x_i x_i'\right)\right]^{-1}\Phi_{zx}'\right\}^{-1}.\text{[37]} \tag{4.3}$$

If $u_i^0 u_i^{0\prime}$ is uncorrelated with $x_i x_i'$, then this reduces to:

$$\Lambda = \left\{\Phi_{zx}\left[\mathrm{E}^{-1}\left(v_i^0 v_i^{0\prime}\right) \otimes \Phi_x^{-1}\right]\Phi_{zx}'\right\}^{-1},$$

which is the conventional asymptotic covariance matrix for three-stage least squares [Zellner and Theil (1962)].

There is a generalization of three-stage least squares that has the same limiting distribution as the generalized minimum distance estimator. Let $\hat{\Psi} = N^{-1}\sum_{i=1}^{N}(\hat{v}_i\hat{v}_i' \otimes x_i x_i')$, where $\hat{v}_i = \hat{\Gamma} y_i + \hat{B} x_i$ and $\hat{\Gamma} \xrightarrow{\text{a.s.}} \Gamma^0$, $\hat{B} \xrightarrow{\text{a.s.}} B^0$. Define:

$$\hat{\delta}_{G3} = \left(S_{zx}\hat{\Psi}^{-1}S_{zx}'\right)^{-1}\left(S_{zx}\hat{\Psi}^{-1}s_{xy}\right).$$

[37] See Chamberlain (1982a).

The limiting distribution of this estimator is derived in the Appendix (Proposition 6). We record it as:

Proposition 4

$\sqrt{N}(\hat{\boldsymbol{\delta}}_{G3} - \boldsymbol{\delta}^0) \xrightarrow{D} N(\boldsymbol{0}, \boldsymbol{\Lambda})$, where $\boldsymbol{\Lambda}$ is given in (4.3). This generalized three-stage least squares estimator is asymptotically efficient within the class of minimum distance estimators.

Our derivation of the limiting distribution of $\hat{\boldsymbol{\delta}}_{G3}$ relies on linearity. For a generalized nonlinear three-stage least squares estimator, see Hansen (1982).[38]

4.4. *Asymptotic efficiency: A comparison with the quasi-maximum likelihood estimator*

Assume that $\boldsymbol{r}_i$ is i.i.d. ($i = 1, 2, \dots$) from a distribution with $\mathrm{E}(\boldsymbol{r}_i) = \boldsymbol{\tau}$, $V(\boldsymbol{r}_i) = \boldsymbol{\Sigma}$, where $\boldsymbol{\Sigma}$ is a $J \times J$ positive-definite matrix; the fourth moments are finite. Suppose that we wish to estimate functions of $\boldsymbol{\Sigma}$ subject to restrictions. Let $\boldsymbol{\sigma} = \mathrm{vec}(\boldsymbol{\Sigma})$ and express the restrictions by the condition that $\boldsymbol{\sigma} = \boldsymbol{g}(\boldsymbol{\theta})$, where $\boldsymbol{g}$ is a function from Υ into R^q with a domain $\Upsilon \subset R^p$ that contains the true value $\boldsymbol{\theta}^0$ ($q = J^2$; $p \le J(J+1)/2$). Let

$$\bar{\boldsymbol{S}} = \frac{1}{N} \sum_{i=1}^{N} (\boldsymbol{r}_i - \bar{\boldsymbol{r}})(\boldsymbol{r}_i - \bar{\boldsymbol{r}})',$$

and let $\bar{\boldsymbol{s}} = \mathrm{vec}(\bar{\boldsymbol{S}})$.

If the distribution of $\boldsymbol{r}_i$ is multivariate normal, then the log-likelihood function is:

$$L = \frac{N}{2} \ln|\boldsymbol{\Sigma}^{-1}| - \frac{N}{2} \mathrm{tr}\left\{\boldsymbol{\Sigma}^{-1}\left[\bar{\boldsymbol{S}} + (\bar{\boldsymbol{r}} - \boldsymbol{\tau})(\bar{\boldsymbol{r}} - \boldsymbol{\tau})'\right]\right\}.$$

If there are no restrictions on $\boldsymbol{\tau}$, then the maximum likelihood estimator of $\boldsymbol{\theta}^0$ is a solution to the following problem: Choose $\hat{\boldsymbol{\theta}}$ to solve:

$$\frac{\partial \boldsymbol{g}'(\boldsymbol{\theta})}{\partial \boldsymbol{\theta}} \left[\boldsymbol{\Sigma}^{-1}(\boldsymbol{\theta}) \otimes \boldsymbol{\Sigma}^{-1}(\boldsymbol{\theta})\right] (\bar{\boldsymbol{s}} - \boldsymbol{g}(\boldsymbol{\theta})) = \boldsymbol{0}.$$

We shall derive the properties of this estimator when the distribution of $\boldsymbol{r}_i$ is not necessarily normal; in that case we shall refer to the estimator as a quasi-maximum likelihood estimator ($\hat{\boldsymbol{\theta}}_{\mathrm{QML}}$).[39]

[38] There are generalizations of two-stage least squares in Chamberlain (1982a) and White (1982a).

[39] The quasi-maximum likelihood terminology was used by the Cowles Commission; see Malinvaud (1970, p. 678).

MaCurdy (1979) considered a version of this problem and showed that, under suitable regularity conditions, $\sqrt{N}(\hat{\boldsymbol{\theta}}_{\mathrm{QML}}-\boldsymbol{\theta}^0)$ has a limiting normal distribution; the covariance matrix, however, is not given by the standard information matrix formula. We would like to compare this distribution with the distribution of the minimum distance estimator.

This comparison can be readily made by using theorem 1 in Ferguson (1958). In our notation, Ferguson considers the following problem: Choose $\hat{\boldsymbol{\theta}}$ to solve

$$\boldsymbol{W}(\bar{\boldsymbol{s}},\boldsymbol{\theta})[\bar{\boldsymbol{s}}-\boldsymbol{g}(\boldsymbol{\theta})]=\boldsymbol{0}.$$

He derives the limiting distribution of $\sqrt{N}(\hat{\boldsymbol{\theta}}-\boldsymbol{\theta}^0)$ under regularity conditions on the functions $\boldsymbol{W}$ and $\boldsymbol{g}$. These regularity conditions are particularly simple in our problem since $\boldsymbol{W}$ does not depend on $\bar{\boldsymbol{s}}$. We can state them as follows:

Assumption 3

$\Xi_0\subset R^p$ is an open set containing $\boldsymbol{\theta}^0$; $\boldsymbol{g}$ is a continuous, one-to-one mapping of Ξ_0 into R^q with a continuous inverse; $\boldsymbol{g}$ has continuous second partial derivatives in Ξ_0; rank $[\partial\boldsymbol{g}(\boldsymbol{\theta})/\partial\boldsymbol{\theta}']=p$ for $\boldsymbol{\theta}\in\Xi_0$; $\boldsymbol{\Sigma}(\boldsymbol{\theta})$ is nonsingular for $\boldsymbol{\theta}\in\Xi_0$.

In addition, we shall need $\bar{\boldsymbol{s}}\overset{\text{a.s.}}{\to}\boldsymbol{g}(\boldsymbol{\theta}^0)$ and the central limit theorem result that $\sqrt{N}(\bar{\boldsymbol{s}}-\boldsymbol{g}(\boldsymbol{\theta}^0))\overset{\text{D}}{\to}N(\boldsymbol{0},\boldsymbol{\Delta})$, where $\boldsymbol{\Delta}=V[(\boldsymbol{r}_i-\boldsymbol{\tau}^0)\otimes(\boldsymbol{r}_i-\boldsymbol{\tau}^0)]$.

Then Ferguson's theorem implies that the likelihood equations almost surely have a unique solution within Ξ_0 for sufficiently large N, and $\sqrt{N}(\hat{\boldsymbol{\theta}}_{\mathrm{QML}}-\boldsymbol{\theta}^0)\overset{\text{D}}{\to}N(\boldsymbol{0},\boldsymbol{\Lambda})$, where

$$\boldsymbol{\Lambda}=(\boldsymbol{G}'\boldsymbol{\Psi}\boldsymbol{G})^{-1}\boldsymbol{G}'\boldsymbol{\Psi}\boldsymbol{\Delta}\boldsymbol{\Psi}\boldsymbol{G}(\boldsymbol{G}'\boldsymbol{\Psi}\boldsymbol{G})^{-1},$$

and $\boldsymbol{G}=\partial\boldsymbol{g}(\boldsymbol{\theta}^0)/\partial\boldsymbol{\theta}'$, $\boldsymbol{\Psi}=(\boldsymbol{\Sigma}^0\otimes\boldsymbol{\Sigma}^0)^{-1}$. It will be convenient to rewrite this, imposing the symmetry restrictions on $\boldsymbol{\Sigma}$. Let $\boldsymbol{\sigma}^*$ be the $J(J+1)/2\times 1$ vector formed by stacking the columns of the lower triangle of $\boldsymbol{\Sigma}$. We can define a $J^2\times[J(J+1)/2]$ matrix $\boldsymbol{T}$ such that $\boldsymbol{\sigma}=\boldsymbol{T}\boldsymbol{\sigma}^*$. The elements in each row of $\boldsymbol{T}$ are all zero except for a single element which is one; $\boldsymbol{T}$ has full column rank. Let $\bar{\boldsymbol{s}}=\boldsymbol{T}\bar{\boldsymbol{s}}^*$, $\boldsymbol{g}(\boldsymbol{\theta})=\boldsymbol{T}\boldsymbol{g}^*(\boldsymbol{\theta})$, $\boldsymbol{G}^*=\partial\boldsymbol{g}^*(\boldsymbol{\theta}^0)/\partial\boldsymbol{\theta}'$, $\boldsymbol{\Psi}^*=\boldsymbol{T}'\boldsymbol{\Psi}\boldsymbol{T}$; then $\sqrt{N}[\bar{\boldsymbol{s}}^*-\boldsymbol{g}^*(\boldsymbol{\theta}^0)]\overset{\text{D}}{\to}N(\boldsymbol{0},\boldsymbol{\Delta}^*)$, where $\boldsymbol{\Delta}^*$ is the covariance matrix of the vector formed from the columns of the lower triangle of $(\boldsymbol{r}_i-\boldsymbol{\tau}^0)(\boldsymbol{r}_i-\boldsymbol{\tau}^0)'$. Now we can set

$$\boldsymbol{\Lambda}=(\boldsymbol{G}'^*\boldsymbol{\Psi}^*\boldsymbol{G}^*)^{-1}(\boldsymbol{G}'^*\boldsymbol{\Psi}^*\boldsymbol{\Delta}^*\boldsymbol{\Psi}^*\boldsymbol{G}^*)(\boldsymbol{G}'^*\boldsymbol{\Psi}^*\boldsymbol{G}^*)^{-1}.$$

Consider the following minimum distance estimator. Choose $\hat{\boldsymbol{\theta}}_{\mathrm{MD}}$ to

$$\min_{\boldsymbol{\theta}\in\Upsilon}[\bar{\boldsymbol{s}}^*-\boldsymbol{g}^*(\boldsymbol{\theta})]'\boldsymbol{A}_N[\bar{\boldsymbol{s}}^*-\boldsymbol{g}^*(\boldsymbol{\theta})],$$

where Υ is a compact subset of Ξ_0 that contains a neighborhood of $\boldsymbol{\theta}^0$ and $\boldsymbol{A}_N \xrightarrow{\text{a.s.}} \boldsymbol{\Psi}^*$. Then the following result is implied by Proposition 2.

Proposition 5

If Assumption 3 is satisfied, then $\sqrt{N}(\hat{\boldsymbol{\theta}}_{\text{QML}} - \boldsymbol{\theta}^0)$ has the same limiting distribution as $\sqrt{N}(\hat{\boldsymbol{\theta}}_{\text{MD}} - \boldsymbol{\theta}^0)$.

If $\boldsymbol{\Delta}^*$ is nonsingular, an optimal minimum distance estimator has $\boldsymbol{A}_N \xrightarrow{\text{a.s.}} \zeta\boldsymbol{\Delta}^{*-1}$, where ζ is an arbitrary positive real number. If the distribution of $\boldsymbol{r}_i$ is normal, then $\boldsymbol{\Delta}^{*-1} = (1/2)\boldsymbol{\Psi}^*$; but in general $\boldsymbol{\Delta}^{*-1}$ is not proportional to $\boldsymbol{\Psi}^*$, since $\boldsymbol{\Delta}^*$ depends on fourth moments and $\boldsymbol{\Psi}^*$ is a function of second moments. So in general $\hat{\boldsymbol{\theta}}_{\text{QML}}$ is less efficient than the optimal minimum distance estimator that uses

$$\boldsymbol{A}_N = \left[\frac{1}{N}\sum_{i=1}^{N}(\boldsymbol{s}_i^* - \bar{\boldsymbol{s}}^*)(\boldsymbol{s}_i^* - \bar{\boldsymbol{s}}^*)'\right]^{-1}, \tag{4.4}$$

where $\boldsymbol{s}_i^*$ is the vector formed from the lower triangle of $(\boldsymbol{r}_i - \bar{\boldsymbol{r}})(\boldsymbol{r}_i - \bar{\boldsymbol{r}})'$.

More generally, we can consider the class of consistent estimators that are continuously differentiable functions of $\bar{\boldsymbol{s}}^*: \hat{\boldsymbol{\theta}} = \hat{\boldsymbol{\theta}}(\bar{\boldsymbol{s}}^*)$. Chiang (1956) shows that the minimum distance estimator based on $\boldsymbol{\Delta}^{*-1}$ has the minimal asymptotic covariance matrix within this class. The minimum distance estimator based on $\boldsymbol{A}_N$ in (4.4) attains this lower bound.

4.5. *Multivariate probit models*

Suppose that

$$\begin{aligned} y_{im} &= 1, \quad \text{if } \boldsymbol{\pi}_m'\boldsymbol{x}_i + u_{im} \geq 0, \\ &= 0, \quad \text{otherwise} \qquad (i = 1,\dots,N;\ m = 1,\dots,M), \end{aligned}$$

where the distribution of $\boldsymbol{u}_i' = (u_{i1},\dots,u_{iM})$ conditional on $\boldsymbol{x}_i$ is multivariate normal, $N(\boldsymbol{0}, \boldsymbol{\Sigma})$. There may be restrictions on $\boldsymbol{\pi}' = (\boldsymbol{\pi}_1',\dots,\boldsymbol{\pi}_M')$, but we want to allow $\boldsymbol{\Sigma}$ to be unrestricted, except for the scale normalization that the diagonal elements of $\boldsymbol{\Sigma}$ are equal to one. In that case, the maximum likelihood estimator has the computational disadvantage of requiring numerical integration over $M-1$ dimensions.

Our strategy is to avoid numerical integration. We estimate $\boldsymbol{\pi}_m$ by maximizing the marginal likelihood function that is based on the distribution of y_{im} conditional on $\boldsymbol{x}_i$:

$$P(y_{im} = 1 | \boldsymbol{x}_i) = F(\boldsymbol{\pi}_m'\boldsymbol{x}_i),$$

where F is the standard normal distribution function. Then under standard assumptions we have $\hat{\pi}_m \xrightarrow{\text{a.s.}} \pi_m^0$, the true value. If $\sqrt{N}(\hat{\pi}-\pi^0) \xrightarrow{\text{D}} N(\mathbf{0},\boldsymbol{\Omega})$, then we can impose the restriction that $\pi = f(\delta)$ by choosing $\hat{\delta}$ to minimize

$$[\hat{\pi}-f(\delta)]'\hat{\Omega}^{-1}[\hat{\pi}-f(\delta)].$$

We only need to derive a formula for $\boldsymbol{\Omega}$.[40]

Our estimator of π is solving the following equation:

$$s(\hat{\pi}) = \frac{\partial Q(\hat{\pi})}{\partial \pi} = \mathbf{0},$$

where

$$Q(\pi) = \sum_{i=1}^{N}\left\{\sum_{m=1}^{M} y_{im}\ln F(\pi_m' x_i) + (1-y_{im})\ln[1-F(\pi_m' x_i)]\right\}.$$

Hence, the asymptotic distribution of $\hat{\pi}$ can be obtained from the theory of "*M*-estimators". Huber (1967) provides general results, which do not impose differentiability restrictions on $s(\pi)$. His results cover, for example, regression estimators based on minimizing the residual sum of absolute deviations. We shall not need this generality here and shall sketch the derivation for the simpler, differentiable case. This case has been considered by Hansen (1982), MaCurdy (1981a), and White (1982).[41]

Let z_i be i.i.d. according to a distribution with support $Z \subset R^q$. Let $\boldsymbol{\Theta}$ be an open, convex subset of R^p and let $\psi(z,\theta)$ be a function from $Z\times\boldsymbol{\Theta}$ into R^p; its kth component is $\psi_k(z,\theta)$. For each $\theta\in\boldsymbol{\Theta}$, ψ is a measurable function of z, and there is a $\theta^0\in\boldsymbol{\Theta}$ with:

$$\mathrm{E}[\psi(z_1,\theta^0)] = \mathbf{0}, \qquad \mathrm{E}[\psi(z_1,\theta^0)\psi'(z_1,\theta^0)] = \Delta < \infty.$$

For each $z\in Z$, ψ is a twice continuously differentiable function of θ. In addition:

$$J = \mathrm{E}\left[\frac{\partial\psi(z_1,\theta^0)}{\partial\theta'}\right]$$

is nonsingular, and

$$\left|\frac{\partial\psi_k^2(z,\theta)}{\partial\theta_l\partial\theta_m}\right| \le h(z) \qquad (k,l,m=1,\dots,p)$$

for $\theta\in\boldsymbol{\Theta}$, where $\mathrm{E}[h(z_1)]<\infty$.

[40]For an alternative approach to multivariate probit models, see Avery, Hansen and Hotz (1981).
[41]Also see Rao (1973, problem 9, p. 378).

Suppose that we have a (measurable) estimator $\hat{\boldsymbol{\theta}}_N \in \boldsymbol{\Theta}$ such that $\hat{\boldsymbol{\theta}}_N \xrightarrow{\text{a.s}} \boldsymbol{\theta}^0$ and

$$\sum_{i=1}^{N} \boldsymbol{\psi}(z_i, \hat{\boldsymbol{\theta}}_N) = \boldsymbol{0}$$

for sufficiently large N a.s. By Taylor's theorem:

$$\frac{1}{\sqrt{N}} \sum_{i=1}^{N} \psi_k(z_i, \boldsymbol{\theta}^0) + \left[\boldsymbol{j}'_{Nk} + \tfrac{1}{2}(\hat{\boldsymbol{\theta}}_N - \boldsymbol{\theta}^0)' \boldsymbol{C}_{Nk} \right] \left[\sqrt{N}(\hat{\boldsymbol{\theta}}_N - \boldsymbol{\theta}^0) \right] = \boldsymbol{0},$$

where

$$\boldsymbol{j}_{Nk} = \frac{1}{N} \sum_{i=1}^{N} \frac{\partial \psi_k(z_i, \boldsymbol{\theta}^0)}{\partial \boldsymbol{\theta}}, \qquad \boldsymbol{C}_{Nk} = \frac{1}{N} \sum_{i=1}^{N} \frac{\partial^2 \psi_k(z_i, \boldsymbol{\theta}^*_{Nk})}{\partial \boldsymbol{\theta}\, \partial \boldsymbol{\theta}'},$$

and $\boldsymbol{\theta}^*_{Nk}$ is on the line segment joining $\hat{\boldsymbol{\theta}}_N$ and $\boldsymbol{\theta}^0$ $(k = 1, \dots, p)$. [The measurability of $\boldsymbol{\theta}^*_{Nk}$ follows from lemma 3 of Jennrich (1969).] By the strong law of large numbers, $\boldsymbol{j}'_{Nk}$ converges a.s. to the kth row of $\boldsymbol{J}$, and

$$\left| \frac{1}{N} \sum_{i=1}^{N} \frac{\partial^2 \psi_k(z_i, \boldsymbol{\theta}^*_{Nk})}{\partial \theta_l\, \partial \theta_m} \right| \le \frac{1}{N} \sum_{i=1}^{N} h(z_i) \xrightarrow{\text{a.s.}} E[h(z_1)]$$

$(k, l, m = 1, \dots, p)$. Hence $(\hat{\boldsymbol{\theta}}_N - \boldsymbol{\theta}^0)' \boldsymbol{C}_{Nk} \to \boldsymbol{0}$ a.s. and

$$\sqrt{N}(\hat{\boldsymbol{\theta}}_N - \boldsymbol{\theta}^0) = -\boldsymbol{D}_N^{-1} \left[\frac{1}{\sqrt{N}} \sum_{i=1}^{N} \boldsymbol{\psi}(z_i, \boldsymbol{\theta}^0) \right]$$

for N sufficiently large a.s. where $\boldsymbol{D}_N \xrightarrow{\text{a.s.}} \boldsymbol{J}$. By the central limit theorem,

$$\frac{1}{\sqrt{N}} \sum_{i=1}^{N} \boldsymbol{\psi}(z_i, \boldsymbol{\theta}^0) \xrightarrow{\text{D}} N(\boldsymbol{0}, \boldsymbol{\Delta}).$$

Hence:

$$\sqrt{N}(\hat{\boldsymbol{\theta}}_N - \boldsymbol{\theta}^0) \xrightarrow{\text{D}} N(\boldsymbol{0}, \boldsymbol{J}^{-1} \boldsymbol{\Delta} \boldsymbol{J}'^{-1}).$$

Applying this result to our multivariate probit estimator gives:

$$\sqrt{N}(\hat{\boldsymbol{\pi}} - \boldsymbol{\pi}^0) \xrightarrow{\text{D}} N(\boldsymbol{0}, \boldsymbol{J}^{-1} \boldsymbol{\Delta} \boldsymbol{J}^{-1}),$$

where $J = \operatorname{diag}\{J_1, \ldots, J_M\}$ is a block-diagonal matrix with:

$$J_m = \mathrm{E}\left[\left\{(F')^2/[F(1-F)]\right\} x_1 x_1'\right]$$

(F and its derivative F' are evaluated at $\pi_m^{0\prime} x_1$); and

$$\Delta = \mathrm{E}\left[H \otimes x_1 x_1'\right],$$

where the m, n element of the $M \times M$ matrix H is $h_{mn} = e_m e_n$ with

$$e_m = \frac{y_{1m} - F}{F(1-F)} F' \qquad (m = 1, \ldots, M)$$

(F and F' are evaluated at $\pi_m^{0\prime} x_1$). We obtain a consistent estimator ($\hat{\Omega}$) of $J^{-1} \Delta J^{-1}$ by replacing expectations by sample means and using $\hat{\pi}$ in place of π^0. Then we can apply the minimum distance theory of Section 4.2 to impose restrictions on π.

5. Empirical applications

5.1. Linear models: Union wage effects

We shall present an empirical example that illustrates some of the preceding results.[42] The data come from the panel of Young Men in the National Longitudinal Survey (Parnes). The sample consists of 1454 young men who were not enrolled in school in 1969, 1970, or 1971, and who had complete data on the variables listed in Table 5.1. Table 5.2(a) presents an unrestricted least squares regression of the logarithm of wage in 1969 on the union, *SMSA*, and region variables for all three years. The regression also includes a constant, schooling, experience, experience squared, and race. This regression is repeated using the 1970 wage and the 1971 wage.

In Section 2 we discussed the implications of a random intercept (c). If the leads and lags are due just to c, then the submatrices of Π corresponding to the union, *SMSA*, or region coefficients should have the form $\beta I + l\lambda'$. Consider, for example, the 3×3 submatrix of union coefficients—the off-diagonal elements in each column should be equal to each other. So we compare 0.048 to 0.046, 0.042 to 0.041, and -0.009 to 0.010; not bad.

[42] This application is taken from Chamberlain (1982a).

Table 5.1
Characteristics of National Longitudinal Survey Young Men, not enrolled in school in 1969, 1970, 1971: Means and standard deviations, $N = 1454$.

Variable	Mean	Standard deviation
LW1	5.64	0.423
LW2	5.74	0.426
LW3	5.82	0.437
U1	0.336	—
U2	0.362	—
U3	0.364	—
U1U2	0.270	—
U1U3	0.262	—
U2U3	0.303	—
U1U2U3	0.243	—
SMSA1	0.697	—
SMSA2	0.627	—
SMSA3	0.622	—
RNS1	0.409	—
RNS2	0.404	—
RNS3	0.410	—
S	11.7	2.64
EXP*69*	5.11	3.71
EXP69^2	39.8	46.6
RACE	0.264	—

Notes:
LW1, LW2, LW3—logarithm of hourly earnings (in cents) on the current or last job in 1969, 1970, 1971; *U1, U2, U3*—1 if wages on current or last job set by collective bargaining, 0 if not, in 1969, 1970, 1971; *SMSA1, SMSA2, SMSA3*—1 if respondent in *SMSA*, 0 if not, in 1969, 1970, 1971; *RNS1, RNS2, RNS3*—1 if respondent in South, 0 if not, in 1969, 1970, 1971; *S*—years of schooling completed; EXP*69*—(age in 1969$-S-6$); *RACE*—1 if respondent black, 0 if not.

In Table 5.2(b) we add a complete set of union interactions, so that, for the union variables at least, we have a general regression function. Now the submatrix of union coefficients is 3×7. If it equals $(\beta I_3, 0)+l\lambda'$, then in the first three columns, the off-diagonal elements within a column should be equal; in the last four columns, all elements within a column should be equal.

I first imposed the restrictions on the *SMSA* and region coefficients, using the minimum distance estimator. $\boldsymbol{\Omega}$ is estimated using the formula in (4.2), and $A_N = \hat{\boldsymbol{\Omega}}^{-1}$. The minimum distance statistic (Proposition 3) is 6.82, which is not a surprising value from a $\chi^2(10)$ distribution. If we impose the restrictions on the union coefficients as well, then the 21 coefficients in Table 5.2(b) are replaced by 8: one β and seven λ's. This gives an increase in the minimum distance statistic (Proposition 3′) of $19.36-6.82=12.54$, which is not a surprising value from a $\chi^2(13)$ distribution. So there is no evidence here against the hypothesis that all the lags and leads are generated by c. In the terminology of Section 3.3, the (linear predictor) relationship of x to y appears to be static conditional on c.

Consider a transformation of the model in which the dependent variables are *LW1*, *LW2-LW1*, and *LW3-LW2*. Start with a multivariate regression on all of the lags and leads (and union interactions); then impose the restriction that U, *SMSA*, and *RNS* appear in the *LW2-LW1* and *LW3-LW2* equations only as contemporaneous changes ($\mathrm{E}(y_t - y_{t-1} | x_1, x_2, x_3) = \beta(x_t - x_{t-1})$). This is equivalent to the restriction that c generates all of the lags and leads, and we have seen that it is supported by the data. I also considered imposing all of the restrictions with the single exception of allowing separate coefficients for entering and leaving

Table 5.2
Unrestricted least squares regressions.
(a)

Dependent variable	Coefficients (and standard errors) of: *U1*	*U2*	*U3*	*SMSA1*	*SMSA2*	*SMSA3*	*RNS1*	*RNS2*	*RNS3*
LW1	0.171	0.042	−0.009	0.135	−0.001	0.032	−0.016	−0.020	−0.108
	(0.025)	(0.026)	(0.025)	(0.028)	(0.055)	(0.054)	(0.081)	(0.081)	(0.070
LW2	0.048	0.150	0.010	0.086	0.053	0.020	0.065	−0.039	−0.155
	(0.023)	(0.028)	(0.026)	(0.027)	(0.065)	(0.061)	(0.099)	(0.109)	(0.092
LW3	0.046	0.041	0.132	0.083	0.003	0.088	0.074	0.056	−0.232
	(0.023)	(0.030)	(0.030)	(0.031)	(0.058)	(0.056)	(0.079)	(0.093)	(0.078

Notes:
All regressions include $(1, S, \mathrm{EXP69}, \mathrm{EXP69}^2, RACE)$. The standard errors are calculated using $\hat{\Omega}$ in (4.2).

(b)

Dependent variable	Coefficients (and standard errors) of: *U1*	*U2*	*U3*	*U1U2*	*U1U3*	*U2U3*	*U1U2U3*
LW1	0.127	−0.047	−0.072	0.128	0.092	0.156	−0.182
	(0.044)	(0.042)	(0.041)	(0.072)	(0.075)	(0.070)	(0.104)
LW2	−0.019	0.014	−0.085	0.181	0.118	0.227	−0.229
	(0.040)	(0.045)	(0.040)	(0.074)	(0.092)	(0.066)	(0.116)
LW3	−0.050	−0.072	−0.022	0.110	0.264	0.246	−0.256
	(0.037)	(0.053)	(0.052)	(0.079)	(0.081)	(0.079)	(0.113)

Notes:
All regressions include $(SMSA1, SMSA2, SMSA3, RNS1, RNS2, RNS3, 1, S, \mathrm{EXP69}, \mathrm{EXP69}^2, RACE)$. The standard errors are calculated using $\hat{\Omega}$ in (4.2).

union coverage in the wage change equations. The estimates (standard errors) are 0.097 (0.019) and −0.119 (0.022). The standard error on the sum of the coefficients is 0.024, so again there is no evidence against the simple model with $E(y_t|x_1, x_2, x_3, c) = \beta x_t + c$.[43]

Table 5.3(a) exhibits the estimates that result from imposing the restrictions using the optimal minimum distance estimator.[44] We also give the conventional generalized least squares estimates. They are minimum distance estimates in which the weighting matrix (A_N) is the inverse of

$$\hat{\Omega}_s = \frac{1}{N}\sum_{i=1}^{N}(y_i - \hat{\Pi}x_i)(y_i - \hat{\Pi}x_i)' \otimes \left(\frac{1}{N}\sum_{i=1}^{N} x_i x_i'\right)^{-1}. \tag{5.1}$$

We give the conventional standard errors based on $(F'\hat{\Omega}_s^{-1}F)^{-1}$ and the standard errors calculated according to Proposition 2, which do not require an assumption of homoskedastic linear regression. These standard errors are larger than the conventional ones, by about 30%. The estimated gain in efficiency from using the appropriate metric is not very large; the standard errors calculated according to Proposition 2 are about 10% larger when we use conventional GLS instead of the optimum minimum distance estimator.

Table 5.3(a) also presents the estimated λ's. Consider, for example, an individual who was covered by collective bargaining in 1969. The linear predictor of c increases by 0.089 if he is also covered in 1970, and it increases by an additional 0.036 if he is covered in all three years. The predicted c for someone who is always covered is higher by 0.102 than for someone who is never covered.

Table 5.3(b) presents estimates under the constraint that $\lambda = 0$. The increment in the distance statistic is $89.08 - 19.36 = 69.72$, which is a surprisingly large value to come from a $\chi^2(13)$ distribution. If we constrain only the union λ's to be zero, then the increment is $57.06 - 19.36 = 37.7$, which is surprisingly large coming from a $\chi^2(7)$ distribution. So there is strong evidence for heterogeneity bias.

The union coefficient declines from 0.157 to 0.107 when we relax the $\lambda = 0$ restriction. The least squares estimates for the separate cross sections, with no

[43] Using May–May CPS matches for 1977–1978, Mellow (1981) reports coefficients (standard errors) of 0.087 (0.018) and −0.069 (0.020) for entering and leaving union membership in a wage change regression. The sample consists of 6602 males employed as nonagricultural wage and salary workers in both years. He also reports results for 2177 males and females whose age was ≤ 25. Here the coefficients on entering and leaving union membership are quite different: 0.198 (0.031) and −0.035 (0.041); it would be useful to reconcile these numbers with our results for young men. Also see Stafford and Duncan (1980).

[44] We did not find much evidence for nonstationarity in the slope coefficients. If we allow the union β to vary over the three years, we get 0.105, 0.103, 0.114. The distance statistic declines to 18.51, giving $19.36 - 18.51 = 0.85$; this is not a surprising value from a $\chi^2(2)$ distribution. If we also free up β for *SMSA* and *RNS*, then the decline in the distance statistic is $18.51 - 13.44 = 5.07$, which is not a surprising value from a $\chi^2(4)$ distribution.

Table 5.3
Restricted estimates.
(a)

	Coefficients (and standard errors) of:		
	U	$SMSA$	RNS
$\hat{\boldsymbol{\beta}}$:	0.107	0.056	−0.082
	(0.016)	(0.020)	(0.045)
$\hat{\boldsymbol{\beta}}_{\text{GLS}}$:	0.121	0.050	−0.085
	(0.013)	(0.017)	(0.040)
	(0.018)	(0.021)	(0.052)

	$U1$	$U2$	$U3$	$U1U2$	$U1U3$	$U2U3$	$U1U2U3$
$\hat{\boldsymbol{\lambda}}$:	−0.023	−0.067	−0.082	0.156	0.152	0.195	−0.229
	(0.030)	(0.040)	(0.037)	(0.057)	(0.062)	(0.059)	(0.085)
	$SMSA1$	$SMSA2$		$SMSA3$	$RNS1$	$RNS2$	$RNS3$
	0.086	−0.008		0.032	0.100	−0.021	−0.128
	(0.025)	(0.046)		(0.046)	(0.072)	(0.077)	(0.068)

$\chi^2(23) = 19.36$

(b) Restrict $\boldsymbol{\lambda} = \mathbf{0}$

	Coefficients (and standard errors) of:		
	U	$SMSA$	RNS
$\hat{\boldsymbol{\beta}}$:	0.157	0.120	−0.150
	(0.012)	(0.013)	(0.016)

$\chi^2(36) = 89.08$

Notes:

$E^*(y|\boldsymbol{x}) = \boldsymbol{\Pi x} = \boldsymbol{\Pi}_1\boldsymbol{x}_1 + \boldsymbol{\Pi}_2\boldsymbol{x}_2$; $\boldsymbol{x}_1' = (U1, U2, U3, U1U2, U1U3, U2U3, U1U2U3, SMSA1, SMSA2, SMSA3, RNS1, RNS2, RNS3)$; $\boldsymbol{x}_2' = (1, S, \text{EXP69}, \text{EXP69}^2, RACE)$. $\boldsymbol{\Pi}_1 = (\beta_u \boldsymbol{I}_3, \boldsymbol{0}, \beta_{SMSA}\boldsymbol{I}_3, \beta_{RNS}\boldsymbol{I}_3) + \boldsymbol{l\lambda}'$; $\boldsymbol{\Pi}_2$ is unrestricted. The restrictions are expressed as $\boldsymbol{\pi} = \boldsymbol{F\delta}$, where $\boldsymbol{\delta}$ is unrestricted. $\hat{\boldsymbol{\beta}}$ and $\hat{\boldsymbol{\lambda}}$ are minimum distance estimates with $A_N^{-1} = \hat{\boldsymbol{\Omega}}$ in (4.2); $\hat{\boldsymbol{\beta}}_{\text{GLS}}$ and $\hat{\boldsymbol{\lambda}}_{\text{GLS}}$ are minimum distance estimates with $A_N^{-1} = \hat{\boldsymbol{\Omega}}_s$ in (5.1) ($\hat{\boldsymbol{\lambda}}_{\text{GLS}}$ is not shown in the table). The first standard error for $\hat{\boldsymbol{\beta}}_{\text{GLS}}$ is the conventional one based on $(\boldsymbol{F}'\hat{\boldsymbol{\Omega}}_s^{-1}\boldsymbol{F})^{-1}$; the second standard error for $\hat{\boldsymbol{\beta}}_{\text{GLS}}$ is based on $(\boldsymbol{F}'\hat{\boldsymbol{\Omega}}_s^{-1}\boldsymbol{F})^{-1}\boldsymbol{F}'\hat{\boldsymbol{\Omega}}_s^{-1}\hat{\boldsymbol{\Omega}}\hat{\boldsymbol{\Omega}}_s^{-1}\boldsymbol{F}(\boldsymbol{F}'\hat{\boldsymbol{\Omega}}_s^{-1}\boldsymbol{F})^{-1}$ (Proposition 2). The χ^2 statistics are computed from $N[\hat{\boldsymbol{\pi}} - \boldsymbol{F}\hat{\boldsymbol{\delta}}]'\hat{\boldsymbol{\Omega}}^{-1}[\hat{\boldsymbol{\pi}} - \boldsymbol{F}\hat{\boldsymbol{\delta}}]$ (Proposition 3).

leads or lags, give union coefficients of 0.195, 0.189, and 0.191 in 1969, 1970, and 1971.[45] So the decline in the union coefficient, when we allow for heterogeneity bias, is 32% or 44% depending on which biased estimate (0.16 or 0.19) one uses. The *SMSA* and region coefficients also decline in absolute value. The least squares estimates for the separate cross sections give an average *SMSA* coefficient of 0.147 and an average region coefficient of -0.131. So the decline in the *SMSA* coefficient is either 53% or 62%, and the decline in absolute value of the region coefficient is either 45% or 37%.

5.2. *Nonlinear models: Labor force participation*

We shall illustrate some of the results in Section 3. The sample consists of 924 married women in the Michigan Panel Study of Income Dynamics. The sample selection criteria and the means and standard deviations of the variables are in Table 5.4. Participation status is measured by the question: "Did _____ do any work for money last year?" We shall model participation in 1968, 1970, 1972, and 1974.

In terms of the model described in Section 3.1, the wage predictors are schooling, experience, and experience squared, where experience is measured as age minus schooling minus six; the tastes for nonmarket time are predicted by these variables and by children. The specification for children is a conventional one that uses the number of children of age less than six (YK) and the total number of children in the family unit (K).[46] Variables that affect only the lifetime budget constraint in this certainty model are captured by c. In particular, nonlabor income and the husband's wage are assumed to affect the wife's participation only through the lifetime budget constraint. The individual effect (c) will also capture unobserved permanent components in wages or in tastes for nonmarket time.

Table 5.5 presents maximum likelihood (ML) estimates of cross-section probit specifications for each of the four years. Table 5.6 presents unrestricted ML estimates for all lags and leads in YK and K. If the residuals (u_{it}) in the latent

[45] Using the NLS Young Men in 1969 ($N = 1362$), Griliches (1976) reports a union membership coefficient of 0.203. Using the NLS Young Men in a pooled regression for 1966–1971 and 1973 ($N = 470$), Brown (1980) reports a coefficient of 0.130 on a variable measuring the probability of union coverage. (The union coverage question was asked only in 1969, 1970, and 1971; so this variable is imputed for the other four years.) The coefficient declines to .081 when individual intercepts are included in the regression. His regressions also include a large number of occupation and industry specific job characteristics.

[46] Some of the work on participation and fertility is in Mincer (1963), Willis (1973), Gronau (1973, 1976, 1977), Hall (1973), Ben-Porath (1973), Becker and Lewis (1973), Mincer and Polachek (1974), Heckman (1974, 1980), Heckman and Willis (1977), Cain and Dooley (1976), Schultz (1980), Hanoch (1980), and Rosenzweig and Wolpin (1980).

variable model (3.1) have constant variance, then $\alpha_1 = \cdots = \alpha_4$ in (3.2), and the submatrices of Π corresponding to YK and K should have the form $\beta I + l\lambda'$. There may be some indication of this pattern in Table 5.6, but it is much weaker than in the wage regressions in Table 5.2.

We allow for unequal variances and provide formal tests by using the minimum distance estimator developed in Section 4.5. In Table 5.7(a) we impose the restrictions that

$$\Pi = \mathrm{diag}\{\alpha_1,\ldots,\alpha_4\}\left[\beta_{YK}\boldsymbol{I}_4 + \boldsymbol{l}\lambda'_{YK},\ \beta_K\boldsymbol{I}_4 + \boldsymbol{l}\lambda'_K\right].$$

The minimum distance statistic is 53.8, which is a very surprising value coming from a $\chi^2(19)$ distribution. So the latent variable c does not appear to provide an

Table 5.4
Characteristics of Michigan Panel Study of Income Dynamics married women: Means and standard deviations, $N = 924$.

Variable	Mean	Standard deviation
LFP1	0.499	—
LFP2	0.530	—
LFP3	0.529	—
LFP4	0.566	—
YK1	0.969	1.200
YK2	0.764	1.069
YK3	0.551	0.895
YK4	0.363	0.685
K1	2.38	1.69
K2	2.30	1.64
K3	2.11	1.61
K4	1.84	1.52
S	12.1	2.1
EXP*68*	17.2	8.5
EXP*68*^{2}	368.	301.

Notes:
LFP1,..., *LFP4*—1 if answered "yes" to "Did work for money last year?", 0 otherwise, referring to 1968, 1970, 1972, 1974; *YK1*,..., *YK4*—number of children of age less than six in 1968, 1970, 1972, 1974; *K1*,..., *K4*—number of children of age less than eighteen living in the family unit in 1968, 1970, 1972, 1974; *S*—years of schooling completed; EXP*68*—(age in 1968 − *S* − 6). The sample selection criteria required that the women be married to the same spouse from 1968 to 1976; not part of the low income subsample; between 20 and 50 years old in 1968; white; out of school from 1968 to 1976; not disabled. We required complete data on the variables in the table, and that there be no inconsistency between reported earnings and the answer to the participation question.

Table 5.5
ML probit cross-section estimates.

Dependent variable	Coefficients (and standard errors) of:							
	YK1	*YK2*	*YK3*	*YK4*	*K1*	*K2*	*K3*	*K4*
LFP1	−0.246 (0.046)	—	—	—	−0.063 (0.031)	—	—	—
LFP2	—	−0.293 (0.055)	—	—	—	−0.075 (0.031)	—	—
LFP3	—	—	−0.342 (0.067)	—	—	—	−0.077 (0.032)	—
LFP4	—	—	—	−0.366 (0.081)	—	—	—	−0.069 (0.034)

Notes:
Separate ML estimates each year. All specifications include $(1, S, \text{EXP}68, \text{EXP}68^2)$.

adequate interpretation of the unrestricted leads and lags. It may be that the distributed lag relationship between current participation and previous births is more general than the one implied by summing over the previous six years (YK) and over the previous eighteen years (K). It may be fruitful to explore this in more detail in future work. Perhaps strict exogeneity conditional on c will hold when we use a more general specification for lagged births. But we must keep in mind that this question is intrinsically tied to the functional form restrictions—we saw in Section 3.3 that there always exist specifications in which y_t is independent of $x_1,\dots,x_T$ conditional on c.

Table 5.6
Unrestricted ML probit estimates.

Dependent variable	Coefficients (and standard errors) of:							
	YK1	*YK2*	*YK3*	*YK4*	*K1*	*K2*	*K3*	*K4*
LFP1	−0.205 (0.081)	−0.017 (0.119)	−0.160 (0.141)	0.420 (0.144)	0.176 (0.076)	−0.142 (0.100)	−0.196 (0.110)	0.063 (0.090)
LFP2	−0.047 (0.079)	−0.238 (0.117)	−0.047 (0.140)	0.093 (0.142)	0.320 (0.077)	−0.278 (0.102)	−0.250 (0.110)	0.177 (0.090)
LFP3	−0.254 (0.080)	0.214 (0.116)	−0.190 (0.139)	−0.209 (0.141)	0.204 (0.077)	−0.210 (0.102)	−0.045 (0.112)	0.030 (0.090)
LFP4	−0.195 (0.079)	0.252 (0.118)	−0.211 (0.139)	−0.282 (0.138)	0.020 (0.075)	0.083 (0.100)	−0.181 (0.110)	0.058 (0.090)

Notes:
Separate ML estimates each year. All specifications include $(1, S, \text{EXP}68, \text{EXP}68^2)$.

Table 5.7
Restricted estimates.

(a)

	Coefficients (and standard errors) of:							
	YK	K						
$\alpha_4\hat{\beta}$	−0.121 (0.046)	−0.058 (0.029)						
	$YK1$	$YK2$	$YK3$	$YK4$	$K1$	$K2$	$K3$	$K4$
$\alpha_4\hat{\lambda}$	−0.042 (0.041)	0.038 (0.060)	−0.050 (0.070)	0.087 (0.077)	0.194 (0.056)	−0.118 (0.062)	−0.146 (0.073)	0.090 (0.056)
	$\hat{\alpha}_1$	$\hat{\alpha}_2$	$\hat{\alpha}_3$	$\hat{\alpha}_4$				
$\hat{\alpha}$	1.585 (0.392)	1.758 (0.375)	1.279 (0.231)	1.0 (−)				

$\chi^2(19) = 53.8$

(b) Restrict $\lambda = 0$

	YK	K		
$\alpha_4\hat{\beta}$	−0.273 (0.065)	−0.073 (0.023)		
	$\hat{\alpha}_1$	$\hat{\alpha}_2$	$\hat{\alpha}_3$	$\hat{\alpha}_4$
$\hat{\alpha}$	0.821 (0.198)	0.930 (0.205)	0.920 (0.191)	1.0 (−)

$\chi^2(27) = 78.4$

(c) Restrict $\alpha_t = 1$ $(t = 1,\dots,4)$

	Coefficients (and standard errors) of:							
	YK	K						
$\hat{\beta}$	−0.193 (0.043)	−0.070 (0.031)						
	$YK1$	$YK2$	$YK3$	$YK4$	$K1$	$K2$	$K3$	$K4$
$\hat{\lambda}$	−0.077 (0.062)	0.082 (0.082)	−0.098 (0.102)	0.102 (0.110)	0.203 (0.063)	−0.108 (0.083)	−0.157 (0.098)	0.072 (0.081)

$\chi^2(22) = 61.6$

(d) Restrict $\alpha_t = 1$; β_t unrestricted $(t = 1,\dots,4)$

	Coefficients (and standard errors) of:							
	$YK1$	$YK2$	$YK3$	$YK4$	$K1$	$K2$	$K3$	$K4$
$\hat{\beta}$	−0.107 (0.054)	−0.216 (0.059)	−0.198 (0.067)	−0.277 (0.086)	−0.107 (0.040)	−0.047 (0.035)	−0.046 (0.039)	−0.017 (0.043)
	$YK1$	$YK2$	$YK3$	$YK4$	$K1$	$K2$	$K3$	$K4$
$\hat{\lambda}$	−0.111 (0.063)	0.085 (0.083)	−0.102 (0.102)	0.126 (0.111)	0.213 (0.064)	−0.113 (0.083)	−0.155 (0.099)	0.052 (0.082)

$\chi^2(16) = 52.7$

Notes:
$\Pi_1 = \text{diag}\{\alpha_1,\dots,\alpha_4\}[\beta_{YK}I_4 + l\lambda'_{YK}, \beta_K I_4 + l\lambda'_K]$; Π_2 is unrestricted. In Table 5.7(d) $\beta_{YK}I_4$ and $\beta_K I_4$ are replaced by diagonal matrices with no restrictions on the diagonal elements. All restrictions are imposed by applying the minimum distance procedure to the unrestricted estimates of Π_1 in Table 5.6. The asymptotic covariance matrix of $\hat{\Pi}_1$ is obtained as in Section 4.5. α_4 is normalized to equal one.

If we do impose the restrictions in Table 5.7(a), then there is strong evidence that $\lambda \neq 0$. Constraining $\lambda = 0$ in Table 5.7(b) gives an increase in the distance statistic of $78.4 - 53.8 = 24.6$, which is surprisingly large to come from a $\chi^2(8)$ distribution.

In Table 5.7(c) we constrain all of the residual variances to be equal ($\alpha_t = 1$). An alternative interpretation of the time varying coefficients is provided in Table 5.7(d), where β_{YK} and β_K vary freely over time and $\alpha_t = 1$. In principle, we could also allow the α_t to vary freely, since they can be identified from changes over time in the coefficients of c. In fact that model gives very imprecise results and it is difficult to ensure numerical accuracy.

We shall interpret the coefficients on YK and K by following the procedure in (3.4). Table 5.8 presents estimates of the expected change in the participation probability when we assign an additional young child to a randomly chosen family, so that YK and K increase by one. We compute this measure for the models in Tables 5.7(a), 5.7(c) and 5.7(d). The average change in the participation probability is -0.096. We can get an indication of omitted variable bias by comparing these estimates with the ones based on Table 5.7(b), where λ is constrained to be zero. Now the average change in the participation probability is -0.122, so that the decline in absolute value when we control for c is 21%. An alternative comparison can be based on the cross-section estimates, with no leads or lags, in Table 5.5. Now the average change in the participation probability is -0.144, giving an omitted variable bias of 33%.

Next we shall consider estimates from the logit framework of Section 3.2. Table 5.9 presents (standard) maximum likelihood estimates of cross-section logit specifications for each of the four years. We can use the cross-section probit results in Table 5.5 to construct estimates of the expected change in the log odds of participation when we add a young child to a randomly chosen family. Doing this in each of the four years gives -0.502, -0.598, -0.683, and -0.703. With the logit estimates, we simply add together the coefficients on YK and K in Table 5.9; this gives -0.507, -0.612, -0.691, and -0.729. The average over the four years is -0.621 for probit and -0.635 for logit. So at this point there is little difference between the two functional forms.

Now allow for the latent variable (c). Table 5.10 presents the conditional maximum likelihood estimates for the fixed effects logit model. The striking result here is that, unlike the probit case, allowing for c leads to an *increase* in the absolute value of the children coefficients. If we constrain β_{YK} and β_K to be constant over time (Table 5.10(a)), the estimated change in the log odds of participation when we add an additional young child is -0.909. If we allow β_{YK} and β_K to vary freely over time (Table 5.10(b)), the average of the estimated changes is -0.879. So the absolute value of the estimates increases by about 40% when we control for c using the logit framework. The estimation method is having a first order effect on the results.

It is commonly found that probit and logit specifications, when properly interpreted, give very similar results; our cross-section estimates are an example of this. But our attempt to incorporate latent variables has turned up marked differences between the probit and logit specifications. There are a number of possible explanations for this. The probit specification restricts c to have a normal distribution conditional on $\boldsymbol{x}$ with a linear regression function and constant variance. The conditional likelihood approach in the logit model does not impose this possibly false restriction. On the other hand, the probit model has a more general specification for the residual covariance matrix.

Table 5.8
Estimated effects of an additional young child.

8.7(a)	Unrestricted $\boldsymbol{\lambda}$ and α_t $(t=1,\dots,4)$			
	$\mathrm{E}(P_{1t}-P_{0t})$			
	1968	1970	1972	1974
	−0.105	−0.116	−0.087	−0.069
8.7(b)	Restrict $\boldsymbol{\lambda}=\mathbf{0}$			
	−0.108	−0.123	−0.122	−0.134
8.7(c)	Restrict $\alpha_t=1$ $(t=1,\dots,4)$			
	−0.098	−0.099	−0.099	−0.101
8.7(d)	Restrict $\alpha_t=1$; $\boldsymbol{\beta}_t$ unrestricted $(t=1,\dots,4)$			
	−0.081	−0.099	−0.092	−0.112
8.5	Cross-section estimates			
	−0.116	−0.139	−0.157	−0.166

Notes:

$$\hat{P}_{0it}=F\left[\hat{\alpha}_t\left(\hat{\boldsymbol{\beta}}_t'\boldsymbol{x}_{1it}+\hat{\boldsymbol{\lambda}}'\boldsymbol{x}_{1i}\right)+\hat{\boldsymbol{\pi}}_{2t}'\boldsymbol{x}_{2i}\right],$$

$$\hat{P}_{1it}=F\left[\hat{\alpha}_t\left(\hat{\boldsymbol{\beta}}_t'\boldsymbol{x}_{1it}^*+\hat{\boldsymbol{\lambda}}'\boldsymbol{x}_{1i}\right)+\hat{\boldsymbol{\pi}}_{2t}'\boldsymbol{x}_{2i}\right],$$

where $F(\cdot)$ is the standard normal distribution function; $\boldsymbol{x}_{1it}'=(YKt,Kt)_i$;

$$\boldsymbol{x}_{1it}^{*\prime}=(YKt+1,Kt+1)_i\qquad(t=1,\dots,4);$$

$$\boldsymbol{x}_{1i}'=(YK1,YK2,YK3,YK4,K1,K2,K3,K4)_i;$$

$\boldsymbol{x}_{2i}'=(1,S,\mathrm{EXP}68,\mathrm{EXP}68^2)_i$. The estimate of $\mathrm{E}(P_{1t}-P_{0t})$ is:

$$\frac{1}{N}\sum_{i=1}^{N}(\hat{P}_{1it}-\hat{P}_{0it}).$$

The estimates of $\alpha_t,\boldsymbol{\beta}_t,\boldsymbol{\lambda},\boldsymbol{\pi}_{2t}$ used in Tables 8.7(a),...,8.7(d) are based on the specifications in Tables 5.7(a)–5.7(d); the estimates in Table 8.5 are based on the specification in Table 5.5.

Table 5.9
ML logit cross-section estimates.

Dependent variable	Coefficients (and standard errors) of: YK1	YK2	YK3	YK4	K1	K2	K3	K4
LFP1	−0.404 (0.077)	—	—	—	−0.103 (0.051)	—	—	—
LFP2	—	−0.494 (0.095)	—	—	—	−0.118 (0.035)	—	—
LFP3	—	—	−0.568 (0.114)	—	—	—	−0.123 (0.051)	—
LFP4	—	—	—	−0.617 (0.138)	—	—	—	−0.112 (0.055)

Notes:
Separate ML estimates each year. All specifications include (1, S, EXP*68*, EXP*68*^{2}).

We have seen that the restrictions on the probit Π matrix, which underlie our estimate of β, appear to be false. An analogous test in the logit framework is based on (3.10). We use conditional ML to estimate a model that includes $YK_s \cdot D_t$, $K_s \cdot D_t$ ($s = 1,\dots,4$; $t = 2,3,4$), where D_t is a dummy variable that is one in period t and zero otherwise. It is not restrictive to exclude $YK_s \cdot D_1$ and $K_s \cdot D_1$, since they can be absorbed in c. We include also D_t, $S \cdot D_t$, EXP*68*$\cdot D_t$, and EXP*68*$^2 \cdot D_t$ ($t = 2,3,4$). Then comparing the maximized conditional likelihoods

Table 5.10
Conditional ML estimates of the fixed effects logit model.

(a)

	Coefficients (and standard errors) of: YK	K
$\hat{\boldsymbol{\beta}}$	−0.573 (0.115)	−0.336 (0.120)

(b) $\boldsymbol{\beta}_t$ unrestricted ($t = 1,\dots,4$)

	Coefficients (and standard errors) of: YK1	YK2	YK3	YK4	K1	K2	K3	K4
$\hat{\boldsymbol{\beta}}$	−0.336 (0.144)	−0.679 (0.172)	−0.780 (0.205)	−0.967 (0.242)	−0.315 (0.135)	−0.178 (0.145)	−0.141 (0.155)	−0.120 (0.165)

Notes:
A conditional likelihood ratio test of $\boldsymbol{\beta}_1 = \cdots = \boldsymbol{\beta}_4$ gives $\chi^2(6) = 8.7$. The specifications in Tables 10(a) and 10(b) include dummy variables for 1970, 1972, 1974 (D_t, $t = 2,3,4$) and the interactions $S \cdot D_t$, EXP*68*$\cdot D_t$, EXP*68*$^2 \cdot D_t$ ($t = 2,3,4$). (Due to the presence of the fixed effect c_i, it is not restrictive to exclude D_1, S, EXP*68*, EXP*68*^{2}, $S \cdot D_1$, EXP*68*$\cdot D_1$, EXP*68*$^2 \cdot D_1$.)

for this specification and the specification in Table 5.10(b) gives a conditional likelihood ratio statistic of 53.9, which is a very surprising value to come from a $\chi^2(16)$ distribution. So the restrictions underlying our logit estimates of β also appear to be false. It may be that the false restrictions simply imply different biases in the probit and logit specifications.

6. Conclusion

Our discussion has focused on models that are static conditional on a latent variable. The panel aspect of the data has primarily been used to control for the latent variable. Much work needs to be done on models that incorporate uncertainty and interesting dynamics. Exploiting the martingale implications of time-additive utility seems fruitful here, as in Hall (1978) and Hansen and Singleton (1982). There is, however, a potentially important distinction between time averages and cross-section averages. A time average of forecast errors over T periods should converge to zero as $T \to \infty$. But an average of forecast errors across N individuals surely need not converge to zero as $N \to \infty$; there may be common components in those errors, due to economy-wide innovations. The same point applies when we consider covariances of forecast errors with variables that are in the agents' information sets. If those conditioning variables are discrete, we can think of averaging over subsets of the forecast errors; as $T \to \infty$, these averages should converge to zero, but not necessarily as $N \to \infty$.

As for controlling for latent variables, I think that future work will have to address the lack of identification that we have uncovered. It is not restrictive to assert that $(y_1,\dots,y_T)$ and $(x_1,\dots,x_T)$ are independent conditional on some latent variable c.

Appendix

Let $\boldsymbol{r}_i' = (\boldsymbol{x}_i', \boldsymbol{y}_i')$, $i=1,\dots,N$, where $\boldsymbol{x}_i' = (x_{i1},\dots,x_{iK})$ and $\boldsymbol{y}_i' = (y_{i1},\dots,y_{iM})$. Write the mth structural equation as:

$$y_{im} = \boldsymbol{\delta}_m' \boldsymbol{z}_{im} + v_{im} \qquad (m=1,\dots,M),$$

where the components of $\boldsymbol{z}_{im}$ are the variables in $\boldsymbol{y}_i$ and $\boldsymbol{x}_i$ that appear in the mth equation with unknown coefficients. Let $\boldsymbol{S}_{zx}$ be the following block-diagonal matrix:

$$\boldsymbol{S}_{zx} = \operatorname{diag}\left\{ \frac{1}{N}\sum_{i=1}^{N} \boldsymbol{z}_{i1}\boldsymbol{x}_i', \dots, \frac{1}{N}\sum_{i=1}^{N} \boldsymbol{z}_{iM}\boldsymbol{x}_i' \right\},$$

and

$$s_{xy} = \frac{1}{N}\sum_{i=1}^{N} y_i \otimes x_i.$$

Let $v_i'^0 = (v_{i1}^0,\dots,v_{iM}^0)$, where $v_{im}^0 = y_{im} - \delta_m'^0 z_{im}$ and δ_m^0 is the true value of δ_m; let $\Phi_{zx} = \mathrm{E}(S_{zx})$. Let $\delta' = (\delta_1',\dots,\delta_M')$ be $s\times 1$ and set

$$\hat{\delta} = \left(S_{zx} D^{-1} S_{zx}'\right)^{-1}\left(S_{zx} D^{-1} s_{xy}\right).$$

Proposition 6

Assume that (1) r_i is i.i.d. according to some distribution with finite fourth moments; (2) $\mathrm{E}[x_1(y_{1m} - \delta_m'^0 z_{1m})] = 0$ $(m = 1,\dots,M)$; (3) rank $(\Phi_{zx}) = s$; (4) $D \xrightarrow{\text{a.s.}} \Psi$ as $N\to\infty$, where Ψ is a positive-definite matrix. Then $\sqrt{N}(\hat{\delta} - \delta^0) \xrightarrow{D} N(0,\Lambda)$, where

$$\Lambda = \left(\Phi_{zx}\Psi^{-1}\Phi_{zx}'\right)^{-1}\Phi_{zx}\Psi^{-1}\left[\mathrm{E}\left(v_1^0 v_1'^0 \otimes x_1 x_1'\right)\right]\Psi^{-1}\Phi_{zx}'\left(\Phi_{zx}\Psi^{-1}\Phi_{zx}'\right)^{-1}.$$

Proof

$$\sqrt{N}(\hat{\delta} - \delta^0) = \left(S_{zx} D^{-1} S_{zx}'\right)^{-1} S_{zx} D^{-1} \sum_{i=1}^{N}\left(v_i^0 \otimes x_i\right)/\sqrt{N}.$$

By the strong law of large numbers, $S_{zx} \xrightarrow{\text{a.s.}} \Phi_{zx}$; $\Phi_{zx}\Psi^{-1}\Phi_{zx}'$ is an $s\times s$ positive-definite matrix since rank $(\Phi_{zx}) = s$. So we obtain the same limiting distribution by considering

$$\left(\Phi_{zx}\Psi^{-1}\Phi_{zx}'\right)^{-1}\Phi_{zx}\Psi^{-1}\sum_{i=1}^{N}\left(v_i^0 \otimes x_i\right)/\sqrt{N}.$$

Note that $v_i^0 \otimes x_i$ is i.i.d. with $\mathrm{E}(v_1^0 \otimes x_1) = 0$, $V(v_1^0 \otimes x_1) = \mathrm{E}(v_1^0 v_1'^0 \otimes x_1 x_1')$. Then applying the central limit theorem gives $\sqrt{N}(\hat{\delta} - \delta^0) \xrightarrow{D} N(0,\Lambda)$. Q.E.D.

This result includes as special cases a number of the commonly used estimators. If $z_{im} = x_i$ $(m = 1,\dots,M)$ and $D = I$, then $\hat{\delta}$ is the least squares estimator and Λ reduces to the formula for Ω given in (4.1). If $\Psi = \mathrm{E}(v_1^0 v_1'^0)\otimes \mathrm{E}(x_1 x_1')$, then Λ is the asymptotic covariance matrix for the three-stage least squares estimator. If

$\Psi = \mathrm{E}(v_1^0 v_1'^0 \otimes x_1 x_1')$, then Λ is the asymptotic covariance matrix for the generalized three-stage least squares estimator (4.3).

Consider applying the generalized three-stage least squares estimator to the first J equations ($J < M$). If $\mathrm{E}(z_{1j} x_1')$ is nonsingular for $j = J+1,\dots,M$, then this estimator for $(\delta_1',\dots,\delta_J')$ has the same asymptotic covariance matrix as the estimator obtained by applying the generalized three-stage least squares estimator to the full set of M equations. This follows from examining the partitioned inverse of (4.3).

References

Amemiya, T. (1971) "The Estimation of Variances in a Variance-Components Model", *International Economic Review*, 12, 1–13.

Andersen, E. B. (1970) "Asymptotic Properties of Conditional Maximum Likelihood Estimators", *Journal of the Royal Statistical Society*, Series B, 32, 283–301.

Andersen, E. B. (1971) "Asymptotic Properties of Conditional Likelihood Ratio Tests", *Journal of the American Statistical Association*, 66, 630–633.

Andersen, E. B. (1972) "The Numerical Solution of a Set of Conditional Estimation Equations", *Journal of the Royal Statistical Society*, Series B, 34, 42–54.

Andersen, E. B. (1973) *Conditional Inference and Models for Measuring*. Copenhagen: Mentalhygiejnisk Forlag.

Anderson, T. W. (1969) "Statistical Inference for Covariance Matrices with Linear Structure", in P. R. Krishnaiah, ed., *Proceedings of the Second International Symposium on Multivariate Analysis*, Academic Press, New York.

Anderson, T. W. (1970) "Estimation of Covariance Matrices Which are Linear Combinations or Whose Inverses are Linear Combinations of Given Matrices", *Essays in Probability and Statistics*, University of North Carolina Press, Chapel Hill.

Anderson, T. W. and C. Hsiao (1981) "Estimation of Dynamic Models with Error Components", *Journal of the American Statistical Association*, 76, 598–606.

Anderson, T. W. and C. Hsiao (1982) "Formulation and Estimation of Dynamic Models Using Panel Data", *Journal of Econometrics*, forthcoming.

Avery, R. B., L. P. Hansen, and V. J. Hotz (1981) "Multiperiod Probit Models and Orthogonality Condition Estimation", Carnegie-Mellon University, Graduate School of Industrial Administration Working Paper No. 62-80-81.

Balestra, P. and M. Nerlove (1966) "Pooling Cross Section and Time Series Data in the Estimation of a Dynamic Model: The Demand for Natural Gas", *Econometrica*, 34, 585–612.

Barndorff-Nielson, O. (1978) *Information and Exponential Families in Statistical Theory*. New York: Wiley.

Basmann, R. L. (1965) "On the Application of the Identifiability Test Statistic and Its Exact Finite Sample Distribution Function in Predictive Testing of Explanatory Economic Models", unpublished manuscript.

Becker, G. S. and H. G. Lewis (1973) "On the Interaction Between the Quantity and Quality of Children", *Journal of Political Economy*, 81, S279–S288.

Ben-Porath, Y. (1973) "Economic Analysis of Fertility in Israel: Point and Counter-Point", *Journal of Political Economy*, 81, S202–S233.

Billingsley, P. (1979) *Probability and Measure*, Wiley, New York.

Bishop, Y. M. M., S. E. Fienberg and P. W. Holland (1975) *Discrete Multivariate Analysis: Theory and Practice*, Cambridge, MA: MIT Press.

Blinder, A. S. and Y. Weiss (1976) "Human Capital and Labor Supply: A Synthesis", *Journal of Political Economy*, 84, 449–472.

Brown, C. (1980) "Equalizing Differences in the Labor Market", *Quarterly Journal of Economics*, 94, 113–134.

Cain, G. G. and M. D. Dooley (1976) "Estimation of a Model of Labor Supply, Fertility, and Wages of Married Women", *Journal of Political Economy*, 84, S179–S199.

Chamberlain, G. (1977) "An Instrumental Variable Interpretation of Identification in Variance-Components and MIMIC Models", in P. Taubman (ed.), *Kinometrics: The Determinants of Socioeconomic Success Within and Between Families*, North-Holland Publishing Company, Amsterdam.

Chamberlain, G. (1978) "Omitted Variable Bias in Panel Data: Estimating the Returns to Schooling", *Annales de l'INSEE*, 30/31, 49–82.

Chamberlain, G. (1978a) "On the Use of Panel Data", unpublished manuscript.

Chamberlain, G. (1979) "Heterogeneity, Omitted Variable Bias, and Duration Dependence", Harvard Institute for Economic Research Discussion Paper No. 691.

Chamberlain, G. (1980) "Analysis of Covariance with Qualitative Data", *Review of Economic Studies*, 47, 225–238.

Chamberlain, G. (1980a) "Studies of Teaching and Learning in Economics: Discussion", *American Economic Review*, Papers and Proceedings, 69, 47–49.

Chamberlain, G. (1982) "The General Equivalence of Granger and Sims Causality", *Econometrica*, 50, 569–581.

Chamberlain, G. (1982a) "Multivariate Regression Models for Panel Data", *Journal of Econometrics*, 18, 5–46.

Chiang, C. L. (1956) "On Regular Best Asymptotically Normal Estimates", *Annals of Mathematical Statistics*, 27, 336–351.

Corcoran, M. and M. S. Hill (1980) "Persistence in Unemployment Among Adult Men", in G. J. Duncan and J. N. Morgan (eds.), *Five Thousand American Families — Patterns of Economic Progress*, Institute for Social Research, University of Michigan, Ann Arbor.

Cox, D. R. (1970) *Analysis of Binary Data*, London, Methuen.

Cramèr, H. (1946) *Mathematical Methods of Statistics*, Princeton University Press, Princeton.

deFinetti, B. (1975) *Theory of Probability*, Vol. 2, Wiley, New York.

Diaconis, P. and D. Freedman (1980) "deFinetti's Theorem for Markov Chains", *The Annals of Probability*, 8, 115–130.

Dynkin, E. B. and A. A. Yushkevich (1979) *Controlled Markov Processes*, Springer-Verlag, New York.

Ferguson, T. S. (1958) "A Method of Generating Best Asymptotically Normal Estimates with Application to the Estimation of Bacterial Densities", *Annals of Mathematical Statistics*, 29, 1046–1062.

Fisher, R. A. (1935) "The Logic of Inductive Inference", *Journal of the Royal Statistical Society*, Series B, 98, 39–54.

Flinn, C. J. and J. J. Heckman (1982) "Models for the Analysis of Labor Force Dynamics", in G. Rhodes and R. L. Basmann (eds.), *Advances in Econometrics*, Vol. 1, *JAI Press*, Greenwich.

Flinn, C. J. and J. J. Heckman (1982a) "New Methods for Analyzing Structural Models of Labor Force Dynamics", *Journal of Econometrics*, 18, 115–168.

Florens, J. P. and M. Mouchart (1982) "A Note on Noncausality", *Econometrica*, 50, 583–591.

Ghez, G. R. and G. S. Becker (1975) *The Allocation of Time and Goods Over the Life Cycle*, Columbia University Press, New York.

Goldberger, A. S. (1974) "Asymptotics of the Sample Regression Slope", unpublished lecture notes, No. 12.

Goldberger, A. S. (1974a) "Unobservable Variables in Econometrics", in P. Zarembka (ed.), *Frontiers in Econometrics*, Academic Press, New York.

Granger, C. W. J. (1969) "Investigating Causal Relations by Econometric Models and Cross-Spectral Methods", *Econometrica*, 37, 424–438.

Griliches, Z. (1967) "Distributed Lags: A Survey", *Econometrica*, 35, 16–49.

Griliches, Z. (1974) "Errors in Variables and Other Unobservables", *Econometrica*, 42, 971–998.

Griliches, Z. (1976) "Wages of Very Young Men", *Journal of Political Economy*, 84, S69–S85.

Griliches, Z. (1979) "Sibling Models and Data in Economics: Beginnings of a Survey", *Journal of Political Economy*, 87, S37–S64.

Griliches, Z., B. H. Hall, and J. A. Hausman (1978) "Missing Data and Self-Selection in Large Panels", *Annales de l'INSEE*, 30/31, 137–176.

Griliches, Z., B. H. Hall, and J. A. Hausman (1981) "Econometric Models for Count Data with an Application to the Patents—R & D Relationship", National Bureau of Economic Research Technical Paper No. 17.

Griliches, Z. and A. Pakes (1980) "The Estimation of Distributed Lags in Short Panels", National Bureau of Economic Research Technical Paper No. 4.

Gronau, R. (1973) "The Effect of Children on the Housewife's Value of Time", *Journal of Political Economy*, 81, S168–S199.

Gronau, R. (1976) "The Allocation of Time of Israeli Women", *Journal of Political Economy*, 84, S201–S220.

Gronau, R. (1977) "Leisure, Home Production, and Work—the Theory of the Allocation of Time Revisited", *Journal of Political Economy*, 85, 1099–1123.

Hall, R. E. (1973) "Comment", *Journal of Political Economy*, 81, S200–S201.

Hall, R. E. (1978) "Stochastic Implications of the Life Cycle—Permanent Income Hypothesis: Theory and Evidence", *Journal of Political Economy*, 86, 971–988.

Hanoch, G. (1980) "A Multivariate Model of Labor Supply: Methodology and Estimation", in J. P. Smith (ed.), *Female Labor Supply: Theory and Estimation*, Princeton University Press, Princeton.

Hansen, L. P. (1982) "Large Sample Properties of Generalized Method of Moments Estimators", *Econometrica*, 50, 1029–1054.

Hansen, L. P. and K. J. Singleton (1982) "Generalized Instrumental Variable Estimation of Nonlinear Rational Expectations Models", *Econometrica*, 50, 1269–1286.

Harville, D. A. (1977) "Maximum Likelihood Approaches to Variance Component Estimation and to Related Problems", *Journal of the American Statistical Association*, 72, 320–338.

Hause, J. (1977) "The Covariance Structure of Earnings and the On-the-Job Training Hypothesis", *Annals of Economic and Social Measurement*, 6, 335–365.

Hause, J. (1980) "The Fine Structure of Earnings and the On-the-Job Training Hypothesis", *Econometrica*, 48, 1013–1029.

Hausman, J. A. and D. A. Wise (1979) "Attrition Bias in Experimental and Panel Data: The Gary Income Maintenance Experiment", *Econometrica*, 47, 455–473.

Hausman, J. A. and W. E. Taylor (1981) "Panel Data and Unobservable Individual Effects", *Econometrica*, 49, 1377–1398.

Heckman, J. J. (1974) "Shadow Prices, Market Wages, and Labor Supply", *Econometrica*, 42, 679–694.

Heckman, J. J. (1976) "A Life-Cycle Model of Earnings, Learning, and Consumption", *Journal of Political Economy*, 84, S11–S44.

Heckman, J. J. (1978) "Simple Statistical Models for Discrete Panel Data Developed and Applied to Test the Hypothesis of True State Dependence Against the Hypothesis of Spurious State Dependence", *Annales de l'INSEE*, 30/31, 227–269.

Heckman, J. J. (1980) "Sample Selection Bias as a Specification Error", in J. P. Smith (ed.), *Female Labor Supply: Theory and Estimation*, Princeton University Press, Princeton.

Heckman, J. J. (1981) "Statistical Models for Discrete Panel Data", in C. Manski and D. McFadden (eds.), *Structural Analysis of Discrete Data with Econometric Applications*, MIT Press, Cambridge.

Heckman, J. J. (1981a) "The Incidental Parameters Problem and the Problem of Initial Conditions in Estimating Discrete Time—Discrete Data Stochastic Processes and Some Monte Carlo Evidence", in C. Manski and D. McFadden (eds.), *Structural Analysis of Discrete Data with Econometric Applications*, MIT Press, Cambridge.

Heckman, J. J. (1981b) "Heterogeneity and State Dependence", in S. Rosen (ed.), *Conference on Labor Markets*, University of Chicago Press, Chicago.

Heckman, J. J. and G. Borjas (1980) "Does Unemployment Cause Future Unemployment? Definitions, Questions and Answers from a Continuous Time Model of Heterogeneity and State Dependence", *Econometrica*, 47, 247–283.

Heckman, J. J. and T. E. MaCurdy (1980) "A Life-Cycle Model of Female Labor Supply", *Review of Economic Studies*, 47, 47–74.

Heckman, J. J. and R. J. Willis (1977) "A Beta-Logistic Model for the Analysis of Sequential Labor Force Participation by Married Women", *Journal of Political Economy*, 85, 27–58.

Hosoya, Y. (1977) "On the Granger Condition for Non-Causality", *Econometrica*, 45, 1735–1736.

Hsiao, C. (1975) "Some Estimation Methods for a Random Coefficient Model", *Econometrica*, 43,

305–325.
Huber, P. J. (1967) "The Behavior of Maximum Likelihood Estimates Under Nonstandard Conditions", *Proceedings of the Fifth Berkeley Symposium on Mathematical Statistics and Probability*, Vol. 1, University of California Press, Berkeley and Los Angeles.
Jennrich, R. I. (1969) "Asymptotic Properties of Non-Linear Least Squares Estimators", *The Annals of Mathematical Statistics*, 40, 633–643.
Jöreskog, K. G. (1978) "An Econometric Model for Multivariate Panel Data", *Annales de l'INSEE*, 30/31, 355–366.
Jöreskog, K. G. and A. S. Goldberger (1975) "Estimation of a Model with Multiple Indicators and Multiple Causes of a Single Latent Variable", *Journal of the American Statistical Association*, 70, 631–639.
Jöreskog, K. G. and D. Sörbom (1977) "Statistical Models and Methods for Analysis of Longitudinal Data", in D. J. Aigner and A. S. Goldberger (eds.), *Latent Variables in Socio-economic Models*, North Holland, Amsterdam.
Kalbfleisch, J. D. and D. A. Sprott (1970) "Application of Likelihood Methods to Models Involving Large Numbers of Parameters", *Journal of the Royal Statistical Society*, Series B, 32, 175–208.
Kendall, M. G. and A. Stuart (1961) *The Advanced Theory of Statistics*, Vol. 2, Griffin, London.
Kiefer, J. and J. Wolfowitz (1956) "Consistency of the Maximum Likelihood Estimator in the Presence of Infinitely Many Incidental Parameters", *Annals of Mathematical Statistics*, 27, 887–906.
Kiefer, N. and G. Neumann (1981) "Individual Effects in a Nonlinear Model", *Econometrica*, 49, 965–980.
Lancaster, T. (1979) "Econometric Methods for the Duration of Unemployment", *Econometrica*, 47, 939–956.
Lancaster, T. and S. Nickell (1980) "The Analysis of Re-Employment Probabilities for the Unemployed", *Journal of the Royal Statistical Society*, Series A, 143, 141–165.
Lee, L. F. (1979) "Estimation of Error Components Model with ARMA (P, Q) Time Component—An Exact GLS Approach", unpublished manuscript.
Lee, L. F. (1980) "Analysis of Econometric Models for Discrete Panel Data in the Multivariate Log Linear Probability Models", unpublished manuscript.
Levhari, D. and T. N. Srinivasan (1969) "Optimal Savings Under Uncertainty", *Review of Economic Studies*, 36, 153–163.
Lillard, L. and Y. Weiss (1979) "Components of Variation in Panel Earnings Data: American Scientists 1960–1970", *Econometrica*, 47, 437–454.
Lillard, L. and R. J. Willis (1978) "Dynamic Aspects of Earnings Mobility", *Econometrica*, 46, 985–1012.
MaCurdy, T. E. (1979) "Multiple Time Series Models Applied to Panel Data: Specification of a Dynamic Model of Labor Supply", unpublished manuscript.
MaCurdy, T. E. (1981) "An Empirical Model of Labor Supply in a Life-Cycle Setting", *Journal of Political Economy*, 89, 1059–1085.
MaCurdy, T. E. (1981a) "Asymptotic Properties of Quasi-Maximum Likelihood Estimators and Test Statistics", National Bureau of Economic Research Technical Paper No. 14.
MaCurdy, T. E. (1982) "The Use of Time Series Processes to Model the Error Structure of Earnings in a Longitudinal Data Analysis", *Journal of Econometrics*, 18, 83–114.
Madalla, G. S. (1971) "The Use of Variance Components Models in Pooling Cross Section and Time Series Data", *Econometrica*, 39, 341–358.
Madalla, G. S. and T. D. Mount (1973) "A Comparative Study of Alternative Estimators for Variance-Components Models Used in Econometric Applications", *Journal of the American Statistical Association*, 68, 324–328.
Malinvaud, E. (1970) *Statistical Methods of Econometrics*, North-Holland, Amsterdam.
Mazodier, P. and A. Trognon (1978) "Heteroskedasticity and Stratification in Error Components Models", *Annales de l'INSEE*, 30/31, 451–482.
McFadden, D. (1974) "Conditional Logit Analysis of Qualitative Choice Behavior", in P. Zarembka, (ed.), *Frontiers in Econometrics*, New York: Academic Press.
Mellow, W. (1981) "Unionism and Wages: A Longitudinal Analysis", *Review of Economics and Statistics*, 63, 43–52.
Mincer, J. (1963) "Market Prices, Opportunity Costs and Income Effects", in C. F. Christ, et. al.,

Measurement in Economics, Stanford University Press, Stanford.
Mincer, J. and S. Polachek (1974) "Family Investments in Human Capital: Earnings of Women", *Journal of Political Economy*, 82, S76–S108.
Mundlak, Y. (1961) "Empirical Production Function Free of Management Bias", *Journal of Farm Economics*, 43, 44–56.
Mundlak, Y. (1963) "Estimation of Production and Behavioral Functions From a Combination of Time Series and Cross Section Data", in C. F. Christ, et al., *Measurement in Economics*, Stanford University Press, Stanford.
Mundlak, Y. (1978) "On the Pooling of Time Series and Cross Section Data", *Econometrica*, 46, 69–85.
Mundlak, Y. (1978a) "Models with Variable Coefficients: Integration and Extension", *Annales de l'INSEE*, 30/31, 483–509.
Nerlove, M. (1967) "Experimental Evidence on the Estimation of Dynamic Economic Relations from a Time Series of Cross-Sections", *Economic Studies Quarterly*, 18, 42–74.
Nerlove, M. (1971) "Further Evidence on the Estimation of Dynamic Relations from a Time Series of Cross-Sections", *Econometrica*, 39, 359–382.
Nerlove, M. (1971a) "A Note on Error-Components Models", *Econometrica*, 39, 383–396.
Nerlove, M. (1972) "Lags in Economic Behavior", *Econometrica*, 40, 221–251.
Neyman, J. and E. L. Scott (1948) "Consistent Estimates Based on Partially Consistent Observations", *Econometrica*, 16, 1–32.
Nickell, S. (1979) "Estimating the Probability of Leaving Unemployment", *Econometrica*, 47, 1249–1266.
Nickell, S. (1981) "Biases in Dynamic Models with Fixed Effects", *Econometrica*, 49, 1417–1426.
Phelps, E. S. (1962) "The Accumulation of Risky Capital: A Sequential Utility Analysis", *Econometrica*, 30, 729–743.
Rao, C. R. (1973) *Linear Statistical Inference and Its Applications*, Wiley, New York.
Rasch, G. (1960) *Probabilistic Models for Some Intelligence and Attainment Tests*, Denmarks Paedagogiske Institute, Copenhagen.
Rasch, G. (1961) "On General Laws and the Meaning of Measurement in Psychology", *Proceedings of the Fourth Berkeley Symposium on Mathematical Statistics and Probability*, Vol. 4, University of California Press, Berkeley and Los Angeles.
Rosenzweig, M. R. and K. I. Wolpin (1980) "Life-Cycle Labor Supply and Fertility", *Journal of Political Economy*, 88, 328–348.
Rothenberg, T. J. (1973) *Efficient Estimation with A Priori Information*, Yale University Press, New Haven.
Schultz, T. P. (1980) "Estimating Labor Supply Functions for Married Women", in J. P. Smith (ed.), *Female Labor Supply: Theory and Estimation*, Princeton University Press, Princeton.
Sims, C. A. (1972) "Money, Income, and Causality", *The American Economic Review*, 62, 540–552.
Singer, B. and S. Spilerman (1974) "Social Mobility Models for Heterogeneous Populations", in H. L. Costner (ed.), *Sociological Methodology 1973–1974*, Jossey–Bass, Inc., San Francisco.
Singer, B. and S. Spilerman (1976) "Some Methodological Issues in the Analysis of Longitudinal Surveys", *Annals of Economic and Social Measurement*, 5, 447–474.
Stafford, F. P. and G. J. Duncan (1980) "Do Union Members Receive Compensating Wage Differentials?", *American Economic Review*, 70, 355–371.
Swamy, P. A. V. B. (1970) "Efficient Inference in a Random Coefficient Regression Model", *Econometrica*, 38, 311–323.
Swamy, P. A. V. B. (1974) "Linear Models with Random Coefficients", in P. Zarembka (ed.), *Frontiers in Econometrics*, Academic Press, New York.
Taylor, W. E. (1980) "Small Sample Considerations in Estimation from Panel Data", *Journal of Econometrics*, 13, 203–223.
Trognon, A. (1978) "Miscellaneous Asymptotic Properties of Ordinary Least Squares and Maximum Likelihood Estimators in Dynamic Error Components Models", *Annales de l'INSEE*, 30/31, 631–657.
Tuma, N. B., M. T. Hannan, and L. P. Groeneveld (1979) "Dynamic Analysis of Event Histories", *American Journal of Sociology*, 84, 820–854.
Tuma, N. B. and P. K. Robins (1980) "A Dynamic Model of Employment Behavior: An Application

to the Seattle and Denver Income Maintenance Experiments", *Econometrica*, 48, 1031–1052.
Wallace, T. D. and A. Hussain (1969) "The Use of Error Components Models in Combining Time Series with Cross Section Data", *Econometrica*, 37, 55–72.
White, H. (1980) "Using Least Squares to Approximate Unknown Regression Functions", *International Economic Review*, 21, 149–170.
White, H. (1980a) "Nonlinear Regression on Cross Section Data", *Econometrica*, 48, 721–746.
White, H. (1980b) "A Heteroskedasticity—Consistent Covariance Matrix Estimator and a Direct Test for Heteroskedasticity", *Econometrica*, 48, 817–838.
White, H. (1982) "Maximum Likelihood Estimation of Misspecified Models", *Econometrica*, 50, 1–25.
White, H. (1982a) "Instrumental Variable Regression with Independent Observations", *Econometrica*, 50, 483–499.
Willis, R. J. (1973) "A New Approach to the Economic Theory of Fertility Behavior", *Journal of Political Economy*, 81, S14–S64.
Zellner, A. and H. Theil (1962) "Three-Stage Least Squares: Simultaneous Estimation of Simultaneous Equations", *Econometrica*, 30, 54–78.
Zellner, A., J. Kmenta and J. Drèze (1966) "Specification and Estimation of Cobb–Douglas Production Function Models", *Econometrica*, 34, 784–795.

PART 6

SPECIAL TOPICS IN ECONOMETRICS—1

Chapter 23

LATENT VARIABLE MODELS IN ECONOMETRICS

DENNIS J. AIGNER

University of Southern California

CHENG HSIAO

University of Toronto

ARIE KAPTEYN

Tilburg University

TOM WANSBEEK*

Netherlands Central Bureau of Statistics

Contents

*The authors would like to express their thanks to Zvi Griliches, Hans Schneeweiss, Edward Leamer, Peter Bentler, Jerry Hausman, Jim Heckman, Wouter Keller, Franz Palm, and Wynand van de Ven for helpful comments on an early draft of this chapter and to Denzil Fiebig for considerable editorial assistance in its preparation. Sharon Koga has our special thanks for typing the manuscript. C. Hsiao also wishes to thank the Social Sciences and Humanities Research Council of Canada and the National Science Foundation, and Tom Wansbeek the Netherlands Organization for the Advancement of Pure Research (Z.W.O.) for research support.

Handbook of Econometrics, Volume II, Edited by Z. Griliches and M.D. Intriligator

1. Introduction

1.1. Background

Although it may be intuitively clear what a "latent variable" is, it is appropriate at the very outset of this discussion to make sure we all agree on a definition. Indeed, judging by a recent paper by a noted psychometrician [Bentler (1982)], the definition may not be so obvious.

The essential characteristic of a latent variable, according to Bentler, is revealed by the fact that the system of linear structural equations in which it appears cannot be manipulated so as to express the variable as a function of measured variables only. This definition has no particular implication for the ultimate identifiability of the parameters of the structural model itself. However, it does imply that for a linear structural equation system to be called a "latent variable model" there must be at least one more *independent* variable than the number of measured variables. Usage of the term "independent" variable as contrasted with "exogenous" variable, the more common phrase in econometrics, includes measurement errors and the equation residuals themselves. Bentler's more general definition covers the case where the covariance matrices of the independent and measured variables are singular.

From this definition, while the residual in an otherwise classical single-equation linear regression model is not a measured variable it is also not a latent variable because it can be expressed (in the population) as a linear combination of measured variables. There are, therefore, three sorts of variables extant: measured, unmeasured and latent. The distinction between an unmeasured variable and a latent one seems not to be very important except in the case of the so-called *functional* errors-in-variables model. For otherwise, in the *structural* model, the equation disturbance, observation errors, and truly exogenous but unmeasured variables share a similar interpretation and treatment in the identification and estimation of such models. In the functional model, the "true" values of exogenous variables are fixed variates and therefore are best thought of as nuisance parameters that may have to be estimated en route to getting consistent estimates of the primary structural parameters of interest.

Since 1970 there has been a resurgence of interest in econometrics in the topic of errors-in-variables models or, as we shall hereinafter refer to them, models involving latent variables. That interest in such models had to be restimulated at all may seem surprising, since there can be no doubt that economic quantities frequently are measured with error and, moreover, that many applications depend on the use of observable proxies for otherwise unobservable conceptual variables.

Yet even a cursory reading of recent econometrics texts will show that the historical emphasis in our discipline is placed on models without measurement error in the variables and instead with stochastic "shocks" in the equations. To the extent that the topic is treated, one normally will find a sentence alluding to the result that for a classical single-equation regression model, measurement error in the dependent variable, y, causes no particular problem because it can be subsumed within the equation's disturbance term.[1] And, when it comes to the matter of measurement errors in independent variables, the reader will usually be convinced of the futility of consistent parameter estimation in such instances unless repeated observations on y are available at each data point or strong a priori information can be employed. And the presentation usually ends just about there. We are left with the impression that the errors-in-variables "problem" is bad enough in the classical regression model; surely it must be worse in more complicated models.

But in fact this is not the case. For example, in a simultaneous equations setting one may employ overidentifying restrictions that appear in the system in order to identify observation error variances and hence to obtain consistent parameter estimates. (Not always, to be sure, but at least *sometimes*.) This was recognized as long ago as 1947 in an unpublished paper by Anderson and Hurwicz, referenced (with an example) by Chernoff and Rubin (1953) in one of the early Cowles Commission volumes. Moreover, dynamics in an equation can also be helpful in parameter identification, ceteris paribus. Finally, restrictions on a model's covariance structure, which are commonplace in sociometric and psychometric modelling, may also serve to aid identification. [See, for example, Bentler and Weeks (1980).] These are the three main themes of research with which we will be concerned throughout this essay. After brief expositions in this Introduction, each topic is treated in depth in a subsequent section.

1.2. *Our single-equation heritage (Sections 2 and 3)*

There is no reason to spend time and space at this point recreating the discussion of econometrics texts on the subject of errors of measurement in the independent variables of an otherwise conventional single-equation regression model. But the setting does provide a useful jumping-off-place for much of what follows.

Let each observation (y_i, x_i) in a random sample be generated by the stochastic relationships:

$$y_i = \eta_i + u_i, \tag{1.1}$$

$$x_i = \xi_i + v_i, \tag{1.2}$$

$$\eta_i = \alpha + \beta\xi_i + \varepsilon_i, \qquad i = 1,\dots n. \tag{1.3}$$

[1]That is to say, the presence of measurement error in y does not alter the properties of least squares estimates of regression coefficients. But the variance of the measurement error remains hopelessly entangled with that of the disturbance term.

Equation (1.3) is the heart of the model, and we shall assume $E(\eta_i|\xi_i) = \alpha + \beta\xi_i$, so that $E(\varepsilon_i) = 0$ and $E(\xi_i\varepsilon_i) = 0$. Also, we denote $E(\varepsilon_i^2) = \sigma_{\varepsilon\varepsilon}$. Equations (1.1) and (1.2) involve the measurement errors, and their properties are taken to be $E(u_i) = E(v_i) = 0$, $E(u_i^2) = \sigma_{uu}$, $E(v_i^2) = \sigma_{vv}$ and $E(u_iv_i) = 0$. Furthermore, we will assume that the measurement errors are each uncorrelated with ε_i and with the latent variables η_i and ξ_i. Inserting the expressions $\xi_i = x_i - v_i$ and $\eta_i = y_i - u_i$ into (1.3), we get:

$$y_i = \alpha + \beta x_i + w_i, \tag{1.4}$$

where $w_i = \varepsilon_i + u_i - \beta v_i$. Now since $E(v_i|x_i) \neq 0$, we readily conclude that least squares methods will yield biased estimates of α and β.

By assuming all random variables are normally distributed we eliminate any concern over estimation of the ξ_i's as "nuisance" parameters. This is the so-called *structural* latent variables model, as contrasted to the *functional* model, wherein the ξ_i's are assumed to be fixed variates (Section 2). Even so, under the normality assumption no consistent estimators of the primary parameters of interest exist. This can easily be seen by writing out the so-called "covariance" equations that relate consistently estimable variances and covariances of the observables (y_i and x_i) to the underlying parameters of the model. Under the assumption of joint normality, these equations exhaust the available information and so provide necessary and sufficient conditions for identification. They are obtained by "covarying" (1.4) with y_i and x_i, respectively. Doing so, we obtain:

$$\begin{aligned}\sigma_{yy} &= \beta\sigma_{yx} + \sigma_{\varepsilon\varepsilon} + \sigma_{uu},\\ \sigma_{yx} &= \beta\sigma_{xx} - \beta\sigma_{vv},\\ \sigma_{xx} &= \sigma_{\xi\xi} + \sigma_{vv}.\end{aligned} \tag{1.5}$$

Obviously, there are but three equations (involving three consistently estimable quantities, σ_{yy}, σ_{xx} and σ_{yx}) and five parameters to be estimated. Even if we agree to give up any hope of disentangling the influences of ε_i and u_i (by defining, say, $\sigma^2 = \sigma_{\varepsilon\varepsilon} + \sigma_{uu}$) and recognize that the equation $\sigma_{xx} = \sigma_{\xi\xi} + \sigma_{vv}$ will always be used to identify $\sigma_{\xi\xi}$ alone, we are still left with two equations in three unknowns (β, σ^2, and σ_{vv}).

The initial theme in the literature develops from this point. One suggestion to achieve identification in (1.5) is to assume we know something about σ_{vv} *relative* to σ^2 or σ_{vv} *relative* to σ_{xx}. Suppose this a priori information is in the form $\lambda = \sigma_{vv}/\sigma^2$. Then we have $\sigma_{vv} = \lambda\sigma^2$ and

$$\begin{aligned}\sigma_{yy} &= \beta\sigma_{yx} + \sigma^2,\\ \sigma_{yx} &= \beta\sigma_{xx} - \beta\lambda\sigma^2,\\ \sigma_{xx} &= \sigma_{\xi\xi} + \sigma_{vv}.\end{aligned} \tag{1.5a}$$

From this it follows that β is a solution to:

$$\beta^2\lambda\sigma_{yx} - \beta(\lambda\sigma_{yy} - \sigma_{xx}) - \sigma_{yx} = 0, \tag{1.6}$$

and that

$$\sigma^2 = \sigma_{yy} - \beta\sigma_{yx}. \tag{1.7}$$

Clearly this is but one of several possible forms that the prior information may take. In Section 3.2 we discuss various alternatives. A Bayesian treatment suggests itself as well (Section 3.11).

In the absence of such information, a very practical question arises. It is whether, in the context of a classical regression model where one of the independent variables is measured with error, that variable should be discarded or not, a case of choosing between two second-best states of the world, where inconsistent parameter estimates are forthcoming either from the errors-in-variables problem or through specification bias. As is well known, in the absence of an errors-of-observation problem in any of the independent variables, discarding one or more of them from the model may, in the face of severe multicollinearity, be an appropriate strategy under a mean-square-error (MSE) criterion. False restrictions imposed cause bias but reduce the variances on estimated coefficients (Section 3.6).

1.3. Multiple equations (Section 4)

Suppose that instead of having the type of information described previously to help identify the parameters of the simple model given by (1.1)–(1.3), there exists a z_i, observable, with the properties that z_i is correlated with x_i but uncorrelated with w_i. This is tantamount to saying there exists another equation relating z_i to x_i, for example,

$$x_i = \gamma z_i + \delta_i, \tag{1.8}$$

with $E(z_i\delta_i) = 0$, $E(\delta_i) = 0$ and $E(\delta_i^2) = \sigma_{\delta\delta}$. Treating (1.4) and (1.8) as our structure (multinormality is again assumed) and forming the covariance equations, we get, in addition to (1.5):

$$\begin{aligned} \sigma_{yz} &= \beta\sigma_{xz}, \\ \sigma_{xx} &= \gamma\sigma_{zx} + \sigma_{\delta\delta}, \\ \sigma_{zx} &= \gamma\sigma_{zz}. \end{aligned} \tag{1.9}$$

It is apparent that the parameters of (1.8) are identified through the last two of these equations. If, as before, we treat $\sigma_{\varepsilon\varepsilon} + \sigma_{uu}$ as a single parameter, σ^2, then (1.5) and the first equation of (1.9) will suffice to identify β, σ^2, σ_{vv}, and $\sigma_{\xi\xi}$.

This simple example serves to illustrate how additional equations containing the same latent variable may serve to achieve identification. This "multiple equations" approach, explored by Zellner (1970) and Goldberger (1972b), spawned the revival of latent variable models in the seventies.

1.4. Simultaneous equations (Section 5)

From our consideration of (1.4) and (1.8) together, we saw how the existence of an instrumental variable (equation) for an independent variable subject to measurement error could resolve the identification problem posed. This is equivalent to suggesting that an overidentifying restriction exists somewhere in the system of equations from which (1.4) is extracted that can be utilized to provide an instrument for a variable like x_i. But it is not the case that overidentifying restrictions can be traded-off against measurement error variances without qualification. Indeed, the *locations* of exogenous variables measured with error and overidentifying restrictions appearing elsewhere in the equation system are crucial. To elaborate, consider the following equation system, which is dealt with in detail in Section 5.2:

$$\begin{aligned} y_1 + \beta_{12} y_2 &= \gamma_{11}\xi_1 + \varepsilon_1, \\ \beta_{21} y_1 + y_2 &= \gamma_{22}\xi_2 + \gamma_{23}\xi_3 + \varepsilon_2, \end{aligned} \tag{1.10}$$

where ξ_j $(j=1,2,3)$ denote the latent exogenous variables in the system. Were the latent exogenous variables regarded as *observable*, the first equation is—conditioned on this supposition—overidentified (one overidentifying restriction) while the second equation is conditionally just-identified. Therefore, at most one measurement error variance can be identified.

Consider first the specifications $x_1 = \xi_1 + v_1$, $x_2 = \xi_2$, $x_3 = \xi_3$, and let σ_{11} denote the variance of v_1. The corresponding system of covariance equations turns out to be:

$$\begin{bmatrix} (\sigma_{y_1x_1} + \beta_{12}\sigma_{y_2x_1}) & (\sigma_{y_1x_2} + \beta_{12}\sigma_{y_2x_2}) & (\sigma_{y_1x_3} + \beta_{12}\sigma_{y_2x_3}) \\ (\beta_{21}\sigma_{y_1x_1} + \sigma_{y_2x_1}) & (\beta_{21}\sigma_{y_1x_2} + \sigma_{y_2x_2}) & (\beta_{21}\sigma_{y_1x_3} + \sigma_{y_2x_3}) \end{bmatrix}$$
$$- \begin{bmatrix} \gamma_{11}(\sigma_{x_1x_1} - \sigma_{11}) & \gamma_{11}\sigma_{x_1x_2} & \gamma_{11}\sigma_{x_1x_3} \\ (\gamma_{22}\sigma_{x_2x_1} + \gamma_{23}\sigma_{x_3x_1}) & (\gamma_{22}\sigma_{x_2x_2} + \gamma_{23}\sigma_{x_3x_2}) & (\gamma_{22}\sigma_{x_2x_3} + \gamma_{23}\sigma_{x_3x_3}) \end{bmatrix}$$
$$= \begin{bmatrix} \textcircled{1} & \textcircled{2} & \textcircled{3} \\ 0 & 0 & 0 \\ \textcircled{4} & \textcircled{5} & \textcircled{6} \\ 0 & 0 & 0 \end{bmatrix} \tag{1.11}$$

which, under the assumption of multinormality we have been using throughout the development, is sufficient to examine the state of identification of all parameters. In this instance, there are six equations available to determine the six unknowns, $\beta_{12}, \beta_{21}, \gamma_{11}, \gamma_{22}, \gamma_{23}$, and σ_{11}. It is clear that equations ② and ③ in (1.11) can be used to solve for β_{12} and γ_{11}, leaving ① to solve for σ_{11}. The remaining three equations can be solved for $\beta_{21}, \gamma_{22}, \gamma_{23}$, so in this case all parameters are identified. Were the observation error instead to have been associated with ξ_2, we would find a different conclusion. Under that specification, β_{12} and γ_{11} are overdetermined, whereas there are only three covariance equations available to solve for $\beta_{21}, \gamma_{22}, \gamma_{23}$, and σ_{22}. Hence, these latter four parameters [all of them associated with the second equation in (1.10)] are not identified.

1.5. *The power of a dynamic specification (Section 6)*

Up to this point in our introduction we have said nothing about the existence of dynamics in any of the equations or equation systems of interest. Indeed, the results presented and discussed so far apply only to models depicting *contemporaneous* behavior.

When dynamics are introduced into either the dependent or the independent variables in a linear model with measurement error, the results are usually beneficial. To illustrate, we will once again revert to a single-equation setting, one that parallels the development of (1.4). In particular, suppose that the sample at hand is a set of time-series observations and that (1.4) is instead:

$$\begin{aligned} \eta_t &= \beta\eta_{t-1} + \varepsilon_t, \\ y_t &= \eta_t + u_t, \qquad t = 1,\dots,T, \end{aligned} \tag{1.12}$$

with all the appropriate previous assumptions imposed, except that now we will also use $|\beta| < 1$, $\mathrm{E}(u_t) = \mathrm{E}(u_{t-1}) = 0$, $\mathrm{E}(u_t^2) = \mathrm{E}(u_{t-1}^2) = \sigma_{uu}$, and $\mathrm{E}(u_t u_{t-1}) = 0$. Then, analogous to (1.5) we have:

$$\begin{aligned} \sigma_{yy} &= \beta\sigma_{y,y-1} + \sigma_{\varepsilon\varepsilon} + \sigma_{uu}, \\ \sigma_{y,y-1} &= \beta(\sigma_{yy} - \sigma_{uu}), \end{aligned} \tag{1.13}$$

where $\sigma_{y,y-1}$ is our notation for the covariance between y_t and y_{t-1} and we have equated the variances of y_t and y_{t-1} by assumption. It is apparent that this variance identity has eliminated one parameter from consideration (σ_{vv}), and we are now faced with a system of two equations in only three unknowns. Unfortunately, we are not helped further by an agreement to let the effects of the equation disturbance term (ε_t) and the measurement error in the dependent variable (u_t) remain joined.

Fortunately, however, there is some additional information that can be utilized to resolve things: it lies in the covariances between current y_t and lags beyond one period (y_{t-s} for $s \geq 2$). These covariances are of the form:

$$\sigma_{y,y-s} = \beta\sigma_{y,y-s+1}, \qquad s \geq 2, \tag{1.14}$$

so that any one of them taken in conjunction with (1.13) will suffice to solve for β, $\sigma_{\varepsilon\varepsilon}$, and σ_{uu}.[2]

1.6. Prologue

Our orientation in this Chapter is primarily theoretical, and while that will be satisfactory for many readers, it may detract others from the realization that structural modelling with latent variables is not only appropriate from a conceptual viewpoint in many applications, it also provides a means to enhance marginal model specifications by taking advantage of information that otherwise might be misused or totally ignored.

Due to space restrictions, we have not attempted to discuss even the most notable applications of latent variable modelling in econometrics. And indeed there have been several quite interesting empirical studies since the early 1970's. In chronological order of appearance, some of these are: Griliches and Mason (1972), Aigner (1974a), Chamberlain and Griliches (1975, 1977), Griliches (1974, 1977), Chamberlain (1977a, 1977b, 1978), Attfield (1977), Kadane et al. (1977), Robinson and Ferrara (1977), Avery (1979), and Singleton (1980). Numerous others in psychology and sociology are not referenced here.

In the following discussion we have attempted to highlight interesting areas for further research as well as to pay homage to the historical origins of the important lines of thought that have gotten us this far. Unfortunately, at several points in the development we have had to cut short the discussion because of space constraints. In these instances the reader is given direction and references in order to facilitate his/her own completion of the topic at hand. In particular we abbreviate our discussions of parameter identification in deference to Hsiao's chapter on that subject in Volume I of this Handbook.

2. Contrasts and similarities between structural and functional models

In this section we analyze the relation between functional and structural models and compare the identification and estimation properties of them. For expository reasons we do not aim at the greatest generality possible. The comparison takes place within the context of the multiple linear regression model. Generalizations are considered in later sections.

[2] The existence of a set of solvable covariance equations should not be surprising. For, combining (1.12) to get the reduced form expression, $y_t = \alpha + \beta y_{t-1} + (\varepsilon_t + u_t) - \beta u_{t-1}$, which is in the form of an autoregressive/moving-average (ARMA) model.

2.1. ML estimation in structural and functional models

Consider the following multiple linear regression model with errors in variables:

$$y_i = \boldsymbol{\xi}_i'\boldsymbol{\beta} + \varepsilon_i, \tag{2.1}$$

$$\boldsymbol{x}_i = \boldsymbol{\xi}_i + \boldsymbol{v}_i, \qquad i = 1,\ldots,n, \tag{2.2}$$

where $\boldsymbol{\xi}_i$, $\boldsymbol{x}_i$, $\boldsymbol{v}_i$, and $\boldsymbol{\beta}$ are k-vectors, and y_i and ε_i are scalars. The $\boldsymbol{\xi}_i$'s are unobservable variables; instead $\boldsymbol{x}_i$ is observed. $\boldsymbol{v}_i$ is unobservable and we assume $\boldsymbol{v}_i \sim N(\mathbf{0}, \Omega)$ for all i. ε_i is assumed to follow a $N(0, \sigma^2)$ distribution. $\boldsymbol{v}_i$ and ε_i are mutually independent and independent of $\boldsymbol{\xi}_i$.

In the *functional* model the above statements represent all the assumptions one has to make, except for the possible specification of prior knowledge with respect to the parameters $\boldsymbol{\beta}$, σ^2 and Ω. The elements of $\boldsymbol{\xi}_i$ are considered to be unknown constants. For expository simplicity we assume that Ω is non-singular. The likelihood of the observable random variables y_i and $\boldsymbol{x}_i$ is then:

$$L_1 \propto \exp\left\{ -\tfrac{1}{2}\left[\operatorname{tr}(X - \Xi)\Omega^{-1}(X - \Xi)' + \sigma^{-2}(\boldsymbol{y} - \Xi\boldsymbol{\beta})'(\boldsymbol{y} - \Xi\boldsymbol{\beta})\right]\right\}, \tag{2.3}$$

where X and Ξ are $n \times k$-matrices with ith rows $\boldsymbol{x}_i'$ and $\boldsymbol{\xi}_i'$, respectively, and $\boldsymbol{y} \equiv (y_1, y_2, \ldots, y_n)'$. The unknown parameters in (2.3) are $\boldsymbol{\beta}$, Ω, σ^2 and the elements of Ξ. Since the order of Ξ is $n \times k$, the number of unknown parameters increases with the number of observations. The parameters $\boldsymbol{\beta}$, σ^2, and Ω are usually referred to as *structural parameters*, whereas the elements of Ξ are called *incidental parameters* [Neyman and Scott (1948)]. The occurrence of incidental parameters poses some nasty problems, as we shall soon see.

In the *structural* model one has to make an explicit assumption about the distribution of the vector of latent variables, $\boldsymbol{\xi}_i$. A common assumption is that $\boldsymbol{\xi}_i$ is normally distributed: $\boldsymbol{\xi}_i \sim N(\mathbf{0}, K)$, say. Consequently $\boldsymbol{x}_i \sim N(\mathbf{0}, A)$, where $A \equiv K + \Omega$. We assume K, hence A, to be positive definite. Under these assumptions we can write down the simultaneous likelihood of the random variables in y_i, $\boldsymbol{\xi}_i$ and $\boldsymbol{x}_i$. This appears as:

$$L_2 \propto \exp\left\{ -\tfrac{1}{2}\operatorname{tr}(\boldsymbol{y}, X, \Xi)S^{-1}\begin{pmatrix}\boldsymbol{y}'\\ X'\\ \Xi'\end{pmatrix}\right\}, \tag{2.4}$$

where S is the variance–covariance matrix:

$$S \equiv \mathrm{E}\begin{pmatrix} y_i \\ \boldsymbol{x}_i \\ \boldsymbol{\xi}_i \end{pmatrix}(y_i, \boldsymbol{x}_i', \boldsymbol{\xi}_i') = \begin{pmatrix} \boldsymbol{\beta}'K\boldsymbol{\beta} + \sigma^2 & \boldsymbol{\beta}'K & \boldsymbol{\beta}'K \\ K\boldsymbol{\beta} & A & K \\ K\boldsymbol{\beta} & K & K \end{pmatrix}. \tag{2.5}$$

In order to show the relationship between the functional and the structural models it is instructive to elaborate upon (2.4). It can be verified by direct multiplication that:

$$S^{-1}=\begin{bmatrix} \sigma^{-2} & 0 & -\sigma^{-2}\beta' \\ 0 & \Omega^{-1} & -\Omega^{-1} \\ -\sigma^{-2}\beta & -\Omega^{-1} & \Omega^{-1}+K^{-1}+\sigma^{-2}\beta\beta' \end{bmatrix} \tag{2.6}$$

Inserting this result in (2.4) we obtain:

$$L_2 \propto \exp\left\{-\tfrac{1}{2}\left[\operatorname{tr}(X-\Xi)\Omega^{-1}(X-\Xi)'+\sigma^{-2}(y-\Xi\beta)'(y-\Xi\beta)\right.\right.$$
$$\left.\left.+\operatorname{tr}\Xi K^{-1}\Xi'\right]\right\}, \tag{2.7}$$

which is proportional to $L_1 \cdot L_3$, where L_1 has been defined by (2.3) and L_3 is proportional to $\exp\{-\frac{1}{2}\operatorname{tr}\Xi K^{-1}\Xi'\}$. Obviously, L_3 is the marginal likelihood of ξ_i. Thus the simultaneous likelihood L_2 is the product of the likelihood of the functional model and the marginal likelihood of the latent variables. This implies that the likelihood of the functional model, L_1, is the likelihood of y_i and x_i *conditional upon* the latent variables ξ_i.

In the structural model estimation takes place by integrating out the latent variables. That is, one maximizes the *marginal likelihood* of y_i and x_i. This marginal likelihood, L_4, is:

$$L_4 \propto \exp\left\{-\tfrac{1}{2}\operatorname{tr}(y, X)\Sigma^{-1}\begin{pmatrix} y' \\ X' \end{pmatrix}\right\}, \tag{2.8}$$

Σ being the $(k+1)\times(k+1)$ variance–covariance matrix of y_i and x_i.

Using the fact that:

$$\Sigma^{-1}=\begin{pmatrix} \gamma^{-1} & -\gamma^{-1}\alpha' \\ -\gamma^{-1}\alpha & A^{-1}+\gamma^{-1}\alpha\alpha' \end{pmatrix}, \tag{2.9}$$

where $\alpha \equiv A^{-1}K\beta$ and $\gamma \equiv \sigma^2+\alpha'\Omega\beta$, (2.8) can be written as:

$$L_4 \propto \exp\left\{-\tfrac{1}{2}\left[\gamma^{-1}(y-X\alpha)'(y-X\alpha)+\operatorname{tr}XA^{-1}X'\right]\right\}. \tag{2.10}$$

So, the likelihood of the observable variables in the functional model is a conditional likelihood—conditional upon the incidental parameters, whereas the likelihood in the structural model is the marginal likelihood obtained by integrating out the incidental parameters. Indeed, Leamer (1978b, p. 229) suggests that the functional and structural models be called the "conditional" and "marginal" models, respectively. Although our demonstration of this relationship between the likelihood functions pertains to the linear regression model with measurement errors, its validity is not restricted to that case, neither is it dependent on the various normality assumptions made, since parameters (in this case the incidental parameters) can always be interpreted as random variables on which the model in which they appear has been conditioned. These conclusions remain essentially the same if we allow for the possibility that some variables are measured without error. If there are no measurement errors, the distinction between the functional and structural interpretations boils down to the familiar distinction between fixed regressors ("conditional upon X") and stochastic regressors [cf. Sampson (1974)].

To compare the functional and structural models a bit further it is of interest to look at the properties of ML estimators for both models, but for reasons of space we will not do that here. Suffice it to say that the structural model is underidentified. A formal analysis follows in Sections 2.2 and 2.3. As for the functional model, Solari (1969) was the first author to point out that the complete log-likelihood has no proper maximum.[3] She also showed that the stationary point obtained from the first order conditions corresponds to a saddle point of the likelihood surface. Consequently, the conditions of Wald's (1949) consistency proof are not fulfilled. The solution to the first order conditions is known to produce inconsistent estimators and the fact that the ML method breaks down in this case has been ascribed to the presence of the incidental parameters [e.g. Malinvaud (1970, p. 387), Neyman and Scott (1948)]. In a sense that explanation is correct. For example, Cramér's proof of the consistency of ML [Cramér (1946, pp. 500 ff.)] does not explicitly use the fact that the first order conditions actually generate a maximum of the likelihood function. He does assume, however, that the number of unknown parameters remains fixed as the number of observations increases.

Maximization of the likelihood in the presence of incidental parameters is not always impossible. If certain identifying restrictions are available, ML estimators can be obtained, but the resulting estimators still need not be consistent, as will be discussed further in Section 3.4. ML is not the only estimation method that breaks down in the functional model. In the next subsection we shall see that without additional identifying restrictions there does not exist a consistent estimator of the parameters in the functional model.

2.2. *Identification*

Since ML in the structural model appears to be perfectly straightforward, at least under the assumption of normality, identification does not involve any new conceptual difficulties. As before, if the observable random variables follow a

[3] See also Sprent (1970) for some further comments on Solari. A result similar to Solari's had been obtained 13 years before by Anderson and Rubin (1956), who showed that the likelihood function of a factor analysis model with fixed factors does not have a maximum.

multivariate normal distribution all information about the unknown parameters is contained in the first and second moments of this distribution.

Although the assumption of normality of the latent variables may simplify the analysis of identification by focusing on the moment equations, it is at the same time a very unfortunate assumption. Under normality the first and second-order moments equations exhaust all sample information. Under different distributional assumptions one may hope to extract additional information from higher order sample moments. Indeed, for the simple regression model ($k=1$, ξ_i a scalar) Reiersøl (1950) has shown that under normality of the measurement error v_i and the equation error ε_i, normality of ξ_i is the *only* assumption under which β is not identified. Although this result is available in many textbooks [e.g. Malinvaud (1970), Madansky (1976), Schmidt (1976)], a generalization to the multiple linear regression model with errors in variables was given only recently by Kapteyn and Wansbeek (1983).[4] They show that the parameter vector $\boldsymbol{\beta}$ in the structural model (2.1)–(2.2) is identified if and only if there exists no linear combination of the elements of $\boldsymbol{\xi}_i$ which is normally distributed.

That non-identifiability of $\boldsymbol{\beta}$ implies the existence of a normally distributed linear combination of $\boldsymbol{\xi}_i$ has been proven independently by Aufm Kampe (1979). He also considers different concepts of non-normality of $\boldsymbol{\xi}_i$. Rao (1966, p. 256) has proven a theorem implying that an element of $\boldsymbol{\beta}$ is unidentified if the corresponding latent variable is normally distributed. This is obviously a specialization of the proposition. Finally, Willassen (1979) proves that if the elements of $\boldsymbol{\xi}_i$ are independently distributed, a necessary condition for $\boldsymbol{\beta}$ to be identified is that none of them is normally distributed. This is a special case of the proposition as well.

The proposition rests on the assumed normality of ε_i and $\boldsymbol{v}_i$. If $\boldsymbol{\varepsilon}_i$ and $\boldsymbol{v}_i$ follow a different distribution, a normally distributed $\boldsymbol{\xi}_i$ need not spoil identifiability. For the simple regression model, Reiersøl (1950) showed that if $\boldsymbol{\xi}_i$ is normally distributed, $\boldsymbol{\beta}$ is still identified if neither the distribution of $\boldsymbol{v}_i$ nor the distribution of ε_i is divisible by a normal distribution.[5] Since non-normal errors play a modest role in practice we shall not devote space to the generalization of his result to the multiple regression errors-in-variables model. Unless otherwise stated, we assume normality of the errors throughout.

Obviously, the proposition implies that if the latent variables follow a k-variate normal distribution, $\boldsymbol{\beta}$ is not identified. Nevertheless, non-normality is rarely assumed in practice, although a few instances will be dealt with in Section 3. In quite a few cases normality will be an attractive assumption (if only for reasons of

[4] Part of the result was stated by Wolfowitz (1952).

[5] If three random variables, u, w, and z, have characteristic functions $\varphi_u(t)$, $\varphi_w(t)$, and $\varphi_z(t)$ satisfying $\varphi_u(t)=\varphi_w(t)\cdot\varphi_z(t)$, we say that the distribution of u is divisible by the distribution of w and divisible by the distribution of z.

tradition) and even if in certain instances normality is implausible, alternative assumptions may lead to mathematically intractable models. Certainly for applications the argument for tractability is most persuasive.

Due to a result obtained by Deistler and Seifert (1978), identifiability of a parameter in the structural model is equivalent to the existence of a consistent estimator of that parameter (see in particular their Remark 7, p. 978). In the functional model there is no such equivalence. It appears that the functional model is identified, but there do not exist consistent estimators of the parameters $\boldsymbol{\beta}^2$, σ^2, or Ω. Let us first look at the identification result.

According to results obtained by Rothenberg (1971) and Bowden (1973), a vector of parameters is identified if the information matrix is non-singular. So, in order to check identification we only have to compute the information matrix defined as:

$$\Psi \equiv -\mathrm{E}\left[\frac{\partial^2 \log L_1}{\partial \boldsymbol{\theta}\, \partial \boldsymbol{\theta}'}\right], \tag{2.11}$$

where $\log L_1$ is given by:

$$\begin{aligned}\log L_1 = & -\frac{n}{2}\log\sigma^2 - \frac{n}{2}\log|\Omega| - \tfrac{1}{2}\operatorname{tr}(X-\Xi)\Omega^{-1}(X-\Xi)' \\ & -\tfrac{1}{2}\sigma^{-2}(y-\Xi\beta)'(y-\Xi\beta) - \tfrac{1}{2}nk\log 2\pi,\end{aligned} \tag{2.12}$$

and $\boldsymbol{\theta}$ is the $[k^2+(n+1)k+1]$-vector of structural and incidental parameters given by $\boldsymbol{\theta} \equiv (\boldsymbol{\beta}', \sigma^2, \boldsymbol{\varphi}', \xi')'$, where $\boldsymbol{\varphi} \equiv \operatorname{vec}\Omega$, $\boldsymbol{\xi} \equiv \operatorname{vec}\Xi$. After some manipulation we find that:

$$\boldsymbol{\Psi} = \begin{bmatrix} \sigma^{-2}(\Xi'\Xi) & 0 & 0 & \sigma^{-2}(\boldsymbol{\beta}'\otimes\Xi') \\ 0 & \frac{n}{2}\sigma^{-4} & 0 & 0 \\ 0 & 0 & \frac{n}{2}(\Omega^{-1}\otimes\Omega^{-1}) & 0 \\ \sigma^{-2}(\boldsymbol{\beta}\otimes\Xi) & 0 & 0 & (\Omega^{-1}+\sigma^{-2}\boldsymbol{\beta}\boldsymbol{\beta}')\otimes I_n \end{bmatrix}. \tag{2.13}$$

In general the matrix Ψ is positive definite and hence both the structural and the incidental parameters are identified. But this result does not help us obtain reasonable estimates of the parameters since no consistent estimators exist.

To see why this is true we use a result obtained by Wald (1948). In terms of the functional model his result is that the likelihood (2.3) admits a consistent estimate of a parameter of the model (i.e. σ^2 or an element of $\boldsymbol{\beta}$ or Ω) if and only if the marginal likelihood of y_i and $\boldsymbol{x}_i$ admits a consistent estimate of this parameter *for any arbitrary choice of the distribution of* $\boldsymbol{\xi}_i$. To make sure that $\boldsymbol{\beta}$ can be consistently estimated, we have therefore to make sure that it is identified under normality of the incidental parameters (if no linear combination of the latent variables were normally distributed it would be identified according to the proposition). The same idea is exploited by Nussbaum (1977) to prove that in the functional model without additional restrictions no consistent estimator of the parameters exists.

This result is of obvious practical importance since it implies that, under the assumption of normally distributed $\boldsymbol{v}_i$ and ε_i, investigation of (consistent) estimability of parameters can be restricted to the structural model with normally distributed incidental parameters. If $\boldsymbol{v}_i$ and ε_i are assumed to be distributed other than normally, the proposition does not apply and investigation of the existence of consistent estimators has to be done on a case-by-case basis.

Some authors [e.g. Malinvaud (1970, p. 401n)] have suggested that in the functional model the relevant definition of identifiability of a parameter should be that there exists a consistent estimator of the parameter. We shall follow that suggestion from now on, observing that in the structural model the definition is equivalent to the usual definition (as employed in Reiersøl's proof). This convention permits us to say that, under normality of $\boldsymbol{v}_i$ and ε_i, identification of $\boldsymbol{\beta}$ in the structural model with normally distributed latent variables is equivalent to identification of $\boldsymbol{\beta}$ in the functional model.

The establishment of the identifiability of the parameters in the functional model via the rank of the information matrix is a bit lengthy, although we shall use the information matrix again below, in Section 2.3. The identifiability of parameters in the functional model can be seen more directly by taking expectations in (2.2) and (2.1). $\boldsymbol{\xi}_i$ is identifiable via $\boldsymbol{\xi}_i = \mathrm{E}\boldsymbol{x}_i$ and $\boldsymbol{\beta}$ via $\mathrm{E}y_i = \boldsymbol{\xi}_i'\boldsymbol{\beta}$, as long as the columns of Ξ are linearly independent. Furthermore, σ^2 and Ω are identified by $\sigma^2 = \mathrm{E}(y - \boldsymbol{\xi}_i'\boldsymbol{\beta})'(y - \boldsymbol{\xi}_i'\boldsymbol{\beta})$ and $\Omega = \mathrm{E}(\boldsymbol{x}_i - \boldsymbol{\xi}_i)(\boldsymbol{x}_i - \boldsymbol{\xi}_i')'$. Although these moment equations establish identifiability, it is clear that the estimators suggested by the moment equations will be inconsistent. (For example, Ω will always be estimated as a zero-matrix.)

2.3. *Efficiency*

The investigation of efficiency properties of estimators in the structural model does not pose new problems beyond the ones encountered in econometric models where all variables of interest are observable. In particular ML estimators are, under the usual regularity conditions, consistent and asymptotically efficient [see, for example, Schmidt (1976, pp. 255–256)].

With respect to the functional model, Wolfowitz (1954) appears to have shown that in general there exists no estimator of the structural parameters which is efficient for each possible distribution of the incidental parameters.[6] Thus, no unbiased estimator will attain the Cramér–Rao lower bound and no consistent estimator will attain the lower bound asymptotically. Nevertheless, it may be worthwhile to compute the asymptotic Cramér-Rao lower bound and check if an estimator comes close to it, asymptotically. For model (2.1)–(2.2) we already know that the information matrix is given by (2.13). The Cramér–Rao lower bound is given by the inverse of this matrix. The problem with Ψ^{-1} as a lower bound to the asymptotic variance–covariance matrix of an estimator is that its dimension grows with the number of observations. To obtain an asymptotic lower bound for the variance–covariance matrix of the estimators of the structural parameters we invert Ψ and only consider the part of Ψ^{-1} pertaining to $\delta \equiv (\beta, \sigma^2, \varphi')'$. This part is easily seen to be:

$$R_n \equiv \begin{bmatrix} (\sigma^2 + \beta'\Omega\beta)(\Xi'\Xi)^{-1} & 0 & 0 \\ 0 & \dfrac{2\sigma^4}{n} & 0 \\ 0 & 0 & \dfrac{2}{n}(\Omega \otimes \Omega) \end{bmatrix}. \tag{2.14}$$

R_n is a lower bound to the variance of any unbiased estimator of δ. A lower bound to the asymptotic variance of any consistent estimator of δ is obtained as $R \equiv \lim_{n\to\infty} nR_n$. Since no consistent estimator of the structural parameters exists without further identifying restrictions, R has to be adjusted in any practical application depending on the precise specification of the identifying restrictions. See Section 3.4 for further details.

2.4. *The ultrastructural relations*

As an integration of the simple functional and structural relations, Dolby (1976b) proposes the following model:

$$\eta_{ij} = \alpha + \beta\xi_{ij} + \varepsilon_{ij}, \tag{2.15}$$

$$x_{ij} = \xi_{ij} + \delta_{ij}, \qquad i = 1,\dots,n; \quad j = 1,\dots,r, \tag{2.16}$$

where $\delta_{ij} \sim N(0, \theta)$, $\varepsilon_{ij} \sim N(0, \tau)$, and $\xi_{ij} \sim N(\mu_i, \varphi)$. Dolby derives the likelihood

[6] The result quoted here is stated briefly in Wolfowitz (1954), but no conditions or proof are given. We are not aware of a subsequent publication containing a full proof.

equations for this model as well as the information matrix. Since the case $r=1$ yields a model which is closely related to the functional model, the analysis in the previous section would suggest that in this case the inverse of the information matrix does not yield a consistent estimate of the asymptotic variance–covariance matrix, even if sufficient identifying assumptions are made. This is also pointed out by Patefield (1978).

3. Single-equation models

For this section the basic model is given by (2.1) and (2.2), although the basic assumptions will vary over the course of the discussion. We first discuss the structural model with non-normally distributed latent variables when no extraneous information is available. Next we consider an example of a non-normal model with extraneous information. Since normal structural models and functional models have the same identification properties they are treated in one section, assuming that sufficient identifying restrictions are available. A variety of other topics comprise the remaining sub-sections, including non-linear models, prediction and aggregation, repeated observations, and Bayesian methods.

3.1. *Non-normality and identification: An example*

Let us specialize (2.1) and (2.2) to the following simple case:

$$y_i = \beta\xi_i + \varepsilon_i, \tag{3.1}$$

$$x_i = \xi_i + v_i, \qquad i=1,\dots,n, \tag{3.2}$$

where y_i, ξ_i, ε_i, x_i, and v_i are scalar random variables with zero means; also, v_i, ε_i, and ξ_i are mutually independent. Denote moments by subscripts, e.g. $\sigma_{xxxx} \equiv \mathrm{E}(x_i^4)$. Assuming that ξ_i is not normally distributed, not all information about its distribution is contained in its second moment. Thus, we can employ higher order moments, if such moments exist. Suppose ξ_i is symmetrically distributed around zero and that its second and fourth moments exist. Instead of three moment equations in four unknowns, we now have eight equations in five unknowns (i.e. four plus the kurtosis of ξ_i). Ignoring the overidentification, one possible solution for β can easily be shown to be:

$$\beta = \frac{\sigma_{yxxx} - 3\sigma_{xx}\sigma_{yx}}{\sigma_{xxxx} - 3\sigma_{xx}^2}. \tag{3.3}$$

One observes that the closer the distribution of ξ_i comes to a normal distribution, the closer $\sigma_{xxxx} - 3\sigma_{xx}^2$ (the kurtosis of the distribution of x_i) is to zero. In that case the variance of the estimator defined by (3.3) may become so large as to make it useless.

As an illustration of the results obtained in Section 2.2, the example shows how identification is achieved by non-normality. Two comments can be made. First, as already observed in Section 2.2, underidentification comes from the fact that *both* ξ_i and v_i are normally distributed. The denominator in (3.3) does not vanish if ξ_i is normally distributed but v_i is not. Secondly, let us extend the example by adding a latent variable ζ_i so that (3.1) becomes:

$$y_i = \beta\xi_i + \gamma\zeta_i + \varepsilon_i. \tag{3.4}$$

The measured value of ζ_i is z_i, generated by $z_i = \zeta_i + w_i$, where w_i is normally distributed and independent of $v_i, \varepsilon_i, \xi_i, \zeta_i$; ζ_i is assumed to be normally distributed, with mean zero, independent of $\xi_i, v_i, \varepsilon_i$. Applying the proposition of Kapteyn and Wansbeek (1983) (cf. Section 2.2) we realize that there *is* a linear combination of ξ_i and ζ_i, namely ζ_i itself, which is normally distributed. Thus, overidentification due to the non-normal distribution of ξ_i does not help in identifying γ, as one can easily check by writing down the moment equations.

3.2. *Estimation in non-normal structural models*

If the identification condition quoted in Section 2.2 is satisfied, various estimation methods can be used. The most obvious method is maximum likelihood (ML). If one is willing to assume a certain parametric form for the distribution of the latent variables, ML is straightforward in principle, although perhaps complicated in practice.

If one wants to avoid explicit assumptions about the distribution of the latent variables, the method of moments provides an obvious estimation method as has been illustrated above. In general the model will be overidentified so that the moment equations will yield different estimators depending on the choice of equations used to solve for the unknown parameters. In fact that number of equations may become infinite. One may therefore decide to incorporate only moments of lowest possible order and, if more than one possible estimator emerges as a solution of the moment equations, as in the example, to choose some kind of minimum variance combination of these estimators. It seems that both the derivation of such an estimator and the establishment of its properties can become quite complicated.[7]

[7]Scott (1950) gives a consistent estimator of β in (3.1) by using the third central moment of the distribution of ξ_i. Rather than seeking a minimum variance combination, Pal (1980) considers various moment-estimators and compares their asymptotic variances.

A distribution-free estimation principle related to the method of moments is the use of product cumulants, as suggested by Geary (1942, 1943). A good discussion of the method is given in Kendall and Stuart (1979, pp. 419–422). Just as the method of moments replaces population moments by sample moments, the method of product cumulants replaces population product cumulants by sample product cumulants. Also here there is no obvious solution for overidentification. For the case where one has to choose between two possible estimators, Madansky (1959) gives a minimum variance linear combination. A generalization to a minimum variance linear combination of more than two possible estimators appears to be feasible but presumably will be quite tedious.

A third simple estimator with considerable intuitive appeal is the method of grouping due to Wald (1940); see, for example, Theil (1971) for a discussion. In a regression context, this is nothing more than an instrumental variables technique with classification dummy variables as instruments.

The idea is to divide the observations into two groups, where the rule for allocating observations to the groups should be independent of ε_i and v_i. For both groups, mean values of y_i and x_i are computed, say $\bar{y}_1$, $\bar{x}_1$, $\bar{y}_2$, and $\bar{x}_2$. The parameter β in (3.1) is then estimated by:

$$\hat{\beta} \equiv \frac{\bar{y}_2 - \bar{y}_1}{\bar{x}_2 - \bar{x}_1}. \tag{3.5}$$

One sees that as an additional condition, $\text{plim}(\bar{x}_2 - \bar{x}_1)$ should be non-zero for $\hat{\beta}$ to exist asymptotically. If this condition and the condition for the allocation rule is satisfied, $\hat{\beta}$ is a consistent estimator of β. Wald also gives confidence intervals. The restrictive aspect of the grouping method is the required independence of the allocation rule from the errors ε_i and v_i.[8] If no such rule can be devised, grouping has no advantages over OLS. Pakes (1982) shows that under normality of the ξ_i and a grouping rule based on the observed values of the x_i, the grouping estimator has the same asymptotic bias as the OLS estimator. Indeed, as he points out, this should be expected since the asymptotic biases of the two estimators depend on unknown parameters. If the biases were different, this could be used to identify the unknown parameters.

If the conditions for the use of the grouping estimator are satisfied, several variations are possible, like groups of unequal size and more than two groups. [See, for example, Bartlett (1949), Dorff and Gurland (1961a), Ware (1972) and Kendall and Stuart (1979, p. 424 ff.). Small sample properties are investigated by Dorff and Gurland (1961b).]

[8]These are sufficient conditions for consistency; Neyman and Scott (1951) give slightly weaker conditions that are necessary and sufficient.

The three estimators discussed so far can also be used in the functional model under a somewhat different interpretation. The assumptions on cumulants or moments are now not considered as pertaining to the distribution of ξ_i but as assumptions on the behavior of sequences of the fixed variables. An example of the application of the method of moments to a functional model can be found in Drion (1951). Richardson and Wu (1970) give the exact distribution of grouping estimators for the case that the groups contain an equal number of observations.

In conclusion, we mention that Kiefer and Wolfowitz (1956) have suggested a maximum likelihood estimator for the non-normal structural model with one regressor. A somewhat related approach for the same model appears in Wolfowitz (1952). Until recently, it was not clear how these estimators could be computed, so they have not been used in practice.[9] Neyman (1951) provides a consistent estimator for the non-normal structural model with one regressor for which explicit formulas are given, but these are complicated and lack an obvious interpretation.

It appears that there exist quite a few consistent estimation methods for non-normal structural models, that is, structural models satisfying the proposition of Section 2.2. Unfortunately, most of these methods lack practical value, whereas a practical method like the method of product cumulants turns out to have a very large estimator variance in cases where it has been applied [Madansky (1959)]. These observations suggest that non-normality is not such a blessing as it appears at first sight. To make progress in practical problems, the use of additional identifying information seems almost indispensable.

3.3. *A non-normal model with extraneous information*

Consider the following model:

$$y_i = \beta_0 + \beta_1 \xi_i + \varepsilon_i, \tag{3.6}$$

where ε_i is normal i.i.d., with variance σ_ε^2.

The variable ξ_i follows a binomial distribution; it is equal to unity with probability p and to zero with probability q, where $p + q = 1$. But ξ_i is unobservable. Instead, x_i is observed. That is, $x_i = \xi_i + v_i$, where v_i is either zero (x_i measures ξ_i correctly) or minus one if ξ_i equals one, or one if ξ_i equals zero (x_i measures ξ_i incorrectly). There is, in other words, a certain probability of misclassification. Since the possible values of v_i depend on ξ_i, the measurement error is correlated with the latent variable. The pattern of correlation can be conveniently depicted in a joint frequency table of v_i and x_i, as has been done by Aigner (1973).

[9] For a recent operationalization, see, for example, Heckman and Singer (1982).

To check identification we can again write down moments (around zero):

$$\mathrm{E}(y_i^k) = p(\beta_0 + \beta_1)^k + q\beta_0^k + \mathrm{E}(\varepsilon_i^k). \tag{3.7}$$

Since the moments of ε_i are all a function of σ_ε^2, one can easily generate equations like (3.7) to identify the unknown parameters $p, \beta_0, \beta_1, \sigma_\varepsilon^2$. The model is thus identified even without using the observed variable x_i! The extraneous information used here is that we know the distribution function from which the latent variable has been drawn, although we do not known its unknown parameter p.

The identification result remains true if we extend model (3.6) by adding observable exogenous variables to the right-hand side. Such a relation may for example occur in practice if y represents an individual's wage income, ξ_i indicates whether or not he has a disease, which is not always correctly diagnosed, and the other explanatory variables are years of schooling, age, work experience, etc. In such an application we may even have more information available, like the share of the population suffering from the disease, which gives us the parameter p. This situation has been considered by Aigner (1973), who uses this knowledge to establish the size of the inconsistency of the OLS estimator (with x_i instead of the unobservable ξ_i) and then to correct for the inconsistency to arrive at a consistent estimator of the parameters in the model.

Mouchart (1977) has provided a Bayesian analysis for Aigner's model. A fairly extensive discussion of errors of misclassification outside regression contexts has been given by Cochran (1968).

3.4. *Identifying restrictions in normal structural and functional models*

Rewrite the model (2.1)–(2.2) in matrix form:

$$y = \Xi\beta + \varepsilon, \tag{3.8}$$

$$X = \Xi + V; \tag{3.9}$$

$\varepsilon \equiv (\varepsilon_1 \dots \varepsilon_n)'$ and V is the $(n \times k)$-matrix with v_i' as its ith row. In this section we assume the rows of Ξ either to be fixed or normally distributed. To remedy the resulting underidentification, $m \geq k^2$ identifying restrictions are supposed to be available:

$$F(\beta, \sigma^2, \Omega) = 0, \tag{3.10}$$

F being an m-vector of functions. If appropriate, we take these functions to be continuously differentiable.

Under the structural interpretation with normally distributed ξ_i, estimation of the model can take place by means of maximum likelihood where the restrictions (3.10) are incorporated in the likelihood function (2.10). The estimator will asymptotically attain the Cramér–Rao bound. The inverse of the information matrix hence serves as a consistent estimator of the variance–covariance matrix of the estimator of $\boldsymbol{\beta}$, σ^2 and Ω. Some special cases have been dealt with in the literature, like the simple regression model with errors-in-variables, where the variances of both the measurement error and the error in the equation are known [Birch (1964), Barnett (1967), Dolby (1976a)], or where one of the two variances is known [Birch (1964), Kendall and Stuart (1979, p. 405)].

Although the identifying restrictions (3.10) also make it possible to construct a consistent estimator of the parameters in the functional model, it is a little less obvious *how* to construct such an estimator. In Section 2.1 we saw that without identifying restrictions ML is not possible. In light of the findings of Section 2.2 this is not surprising, because without identifying restrictions a consistent estimator does not exist. It is of interest to see if unboundedness of the likelihood function persists in the presence of identifying restrictions.

Recall (2.12). In order to study the behavior of $\log L_1$, we first observe that a choice of Ξ such that $(X-\Xi)'(X-\Xi)$ and $(y-\Xi\boldsymbol{\beta})'(y-\Xi\boldsymbol{\beta})$ are both zero is only possible if y and X in the sample satisfy $y = X\boldsymbol{\beta}$. This event has zero probability so we assume that either $(X-\Xi)'(X-\Xi)$ or $(y-\Xi\boldsymbol{\beta})'(y-\Xi\boldsymbol{\beta})$ is non-zero. Next assume that $\boldsymbol{F}(\boldsymbol{\beta},\sigma^2,\Omega)$ is such that $\sigma^2 \to 0$ if and only if $|\Omega| \to 0$ and both converge to zero at the same rate. Obviously, for positive finite values of σ^2 and $|\Omega|$, $\log L_1$ is finite-valued. If σ^2 or $|\Omega|$ go to infinity, $\log L_1$ approaches minus infinity. Finally, consider the case where both σ^2 and $|\Omega|$ go to zero. Without loss of generality we assume that Ξ is chosen such that $X-\Xi$ is zero. The terms $-(n/2)\log\sigma^2$ and $-(n/2)\log|\Omega|$ go to infinity, but these terms are dominated by $-\frac{1}{2}\sigma^{-2}(y-\Xi\boldsymbol{\beta})'(y-\Xi\boldsymbol{\beta})$, which goes to minus infinity. Thus, under the assumption with respect to $\boldsymbol{F}(\boldsymbol{\beta};\sigma^2,\Omega)$, the log-likelihood is continuous and bounded from above, so that a proper maximum of the likelihood function exists.

A well-known example is the case where $\sigma^{-2}\Omega$ is known. While that case has received considerable attention in the literature, we have chosen to exclude a detailed treatment here because there seems to be little or no practical relevance to it. Some references are Moberg and Sundberg (1978), Copas (1972), Van Uven (1930), Sprent (1966), Dolby (1972), Höschel (1978), Casson (1974), Kapteyn and Wansbeek (1983), Robertson (1974), Schneeweiss (1976), Kapteyn and Wansbeek (1981), Fuller and Hidiroglou (1978), DeGracie and Fuller (1972), and Fuller (1980).

No definitive analyses exist of overidentified functional models. A promising approach appears to be to compute the ML estimator as if the model were structural with normally distributed latent variables and to study its properties under functional assumptions. Kapteyn and Wansbeek (1981) show that the ML estimator is asymptotically normally distributed with a variance–covariance matrix identical to the one obtained under structural assumptions. Also, the distributions of certain test statistics appear to be the same under functional and structural assumptions. They also show that a different estimator developed by Robinson (1977) has the same asymptotic distribution under functional and structural assumptions.

Let us next consider the (asymptotic) efficiency of estimators in the functional model with identifying restrictions. It has been observed in Section 2.3 that no estimator will attain the Cramér–Rao lower bound, but still the lower bound can be used as a standard of comparison, As before, $\boldsymbol{\varphi} \equiv \operatorname{vec} \Omega$ and $\boldsymbol{\delta} \equiv (\boldsymbol{\beta}', \sigma^2, \boldsymbol{\varphi}')'$. Furthermore, define the matrix of partial derivatives:

$$F_{\boldsymbol{\delta}} \equiv \left(\frac{\partial \boldsymbol{F}}{\partial \boldsymbol{\beta}'} \frac{\partial \boldsymbol{F}}{\partial \sigma^2} \frac{\partial \boldsymbol{F}}{\partial \boldsymbol{\varphi}'} \right), \tag{3.11}$$

where $\boldsymbol{F}$ has been defined in (3.10). Using the formula for the Cramér–Rao lower bound for a constrained estimator [Rothenberg (1973b, p. 21)] we obtain as an asymptotic lower bound for the variance of any estimator of $\boldsymbol{\delta}$:

$$P \equiv R - RF_{\boldsymbol{\delta}}'\left(F_{\boldsymbol{\delta}}RF_{\boldsymbol{\delta}}'\right)^{-1}F_{\boldsymbol{\delta}}R, \tag{3.12}$$

where $R \equiv \lim_{n \to \infty} nR_n$, R_n being given by (2.14).

The estimators discussed so far have been described by the large sample properties of consistency, asymptotic efficiency and asymptotic distribution. For some simple cases there do exist exact finite sample results that are worth mentioning.

One would suspect that the construction of exact distributions is simplest in the structural model since in that case the observable variables follow a multivariate normal distribution and the distributions of various statistics that are transforms of normal variates are known. This knowledge is used by Brown (1957) to derive simultaneous confidence intervals for the simple structural relation:

$$y_i = \beta_0 + \beta_1 \xi_i + \varepsilon_i, \qquad i = 1, \ldots, n, \tag{3.13}$$

with ξ_i and ε_i independently normally distributed variables, and where their variances are assumed to be known. The confidence intervals are based on a χ^2-distribution. For the same model with the ratio of the variances known, Creasy (1956) gives confidence intervals based on a t-distribution.[10] Furthermore, she shows that a confidence interval obtained in the structural model can be used as a conservative estimate of the corresponding confidence interval in the functional

[10] See Schneeweiss (1982) for an improved proof.

model, in the sense that the confidence interval in the functional model will actually be smaller.

The cases considered by Creasy and by Brown are rather special and simple. One would like to know, therefore, how good the asymptotic approximations in more general cases will be. Some optimism in this respect can be gleaned from results obtained by Richardson and Wu (1970), who present the exact distribution of the least squares estimator in model (3.13) under both functional and structural assumptions. It is found that the asymptotic approximations for the variance of the OLS estimator of β_1 in the functional model are very good. No asymptotic approximation is needed for the structural case as the exact expression is already quite simple. In light of the results obtained in Section 2.1, this is what one would expect.

3.5. *Non-linear models*

The amount of work done on non-linear models comprising latent variables is modest, not surprising in view of the particular difficulties posed by these models [Griliches and Ringstad (1970)]. In line with the sparse literature on the subject we only pay attention to one-equation models:

$$y_i = f(\boldsymbol{\xi}_i, \boldsymbol{\beta}) + \varepsilon_i, \qquad i = 1, \dots, n, \tag{3.14}$$

$$\boldsymbol{x}_i = \boldsymbol{\xi}_i + \boldsymbol{v}_i, \tag{3.15}$$

where $\boldsymbol{\xi}_i$, $\boldsymbol{x}_i$, $\boldsymbol{v}_i$, and $\boldsymbol{\beta}$ are k-vectors, y_i and ε_i are scalars; $\boldsymbol{v}_i \sim N(\mathbf{0}, \Omega)$, with Ω non-singular. There is statistical independence across observations. The function f is assumed to be twice continuously differentiable. Furthermore, $\mathrm{E}\boldsymbol{v}_i\varepsilon_i = 0$.

Let us consider the functional model.[11] The likelihood of the observable random variables y_i and $\boldsymbol{x}_i$ is given by:

$$L_1 \propto \exp\left\{-\tfrac{1}{2}\left[\operatorname{tr}(X - \Xi)\Omega^{-1}(X - \Xi)' + \sigma^{-2}(y - \boldsymbol{F}(\Xi, \boldsymbol{\beta}))'(y - \boldsymbol{F}(\Xi, \boldsymbol{\beta}))\right]\right\}. \tag{3.16}$$

The n-vector $\boldsymbol{F}(\Xi, \boldsymbol{\beta})$ has $f(\boldsymbol{\xi}_i, \boldsymbol{\beta})$ as its ith element. As in Section 2.2 identifiability of the functional model can be checked by writing down the information matrix corresponding to this likelihood. Again, identifiability does not guarantee the existence of consistent estimators of $\boldsymbol{\beta}$, Ω, and σ^2. No investigations have been carried out regarding conditions under which such consistent estimators exist. Dolby (1972) maximizes L_1 with respect to Ξ and $\boldsymbol{\beta}$, assuming σ^2 and Ω to

[11]We are unaware of any studies that deal with a non-linear structural model.

be known. He does not prove consistency of the resulting estimator. He claims that the inverse of the information matrix is the asymptotic variance–covariance matrix of the maximum likelihood estimator. This claim is obviously incorrect, a conclusion which follows from the result by Wolfowitz (1954). Dolby and Lipton (1972) apply maximum likelihood to (3.14)–(3.15), without assuming σ^2 and Ω to be known. Instead, they assume replicated observations to be available. A similar analysis is carried out by Dolby and Freeman (1975) for the more general case that the errors in (3.14)–(3.15) may be correlated across different values of the index i.

A troublesome aspect of the maximum likelihood approach in practice is that in general no closed form solutions for Ξ and $\boldsymbol{\beta}$ can be found so that one has to iterate over all $k(n+1)$ unknown parameters. For sample sizes large enough to admit conclusions on the basis of asymptotic results, that may be expected to be an impossible task. Also, Egerton and Laycock (1979) find that the method of scoring often does not yield the global maximum of the likelihood.

If more specific knowledge is available about the shape of the function f, the numerical problems may simplify considerably. O'Neill, Sinclair and Smith (1969) describe an iterative method to fit a polynomial for which computation time increases only linearly with the number of observations. They also assume the variance–covariance matrix of the errors to be known. The results by O'Neill, Sinclair and Smith suggest that it may be a good strategy in practice to approximate $f(\boldsymbol{\xi}_i, \boldsymbol{\beta})$ by a polynomial of required accuracy and then to apply their algorithm. Obviously a lot more work has to be done, particularly on the statistical properties of ML, before any definitive judgment can be made on the feasibility of estimating non-linear functional models.

3.6. *Should we include poor proxies?*

Rewrite (2.1) as:

$$y_i = \boldsymbol{\xi}_{i1}'\boldsymbol{\beta}_1 + \xi_{ik}\beta_k + \varepsilon_i, \tag{3.17}$$

with $\boldsymbol{\xi}_{i1}$ and $\boldsymbol{\beta}_1$ being $(k-1)$-vectors containing the first $(k-1)$ elements of $\boldsymbol{\xi}_i$ and $\boldsymbol{\beta}$; the scalars ξ_{ik} and β_k are the kth elements of $\boldsymbol{\xi}_i$ and $\boldsymbol{\beta}$. The vector $\boldsymbol{\xi}_{i1}$ is measured without error. For ξ_{ik} we have a proxy, x_{ik}, with observational error independent of $\boldsymbol{\xi}_i$ and ε_i. Suppose we are mainly interested in estimating $\boldsymbol{\beta}_1$. Wickens (1972) and McCallum (1972) compare two possible estimation methods: OLS with ξ_{ik} in (3.17) replaced by x_{ik}, or OLS after omitting ξ_{ik} from (3.17). They show that if ξ_{ik} correlates with $\boldsymbol{\xi}_{i1}$ the first method always gives an asymptotic bias which is smaller than that of the second method. If ξ_{ik} does not correlate with $\boldsymbol{\xi}_{i1}$, both estimation methods are, of course, unbiased. Thus one should always include a proxy, however poor it may be.

No such clear-cut conclusion can be obtained if also one or more elements of $\boldsymbol{\xi}_{i1}$ are measured with error [Barnow (1976) and Garber and Klepper (1980)], or if the measurement error in ξ_{ik} is allowed to correlate with $\boldsymbol{\xi}_{i1}$ [Frost (1979)]. Aigner (1974b) considers mean square error rather than asymptotic bias as a criterion to compare estimators in McCallum's and Wickens' model. He gives conditions under which the mean square error of OLS with omission is smaller than OLS with the proxy included. Giles (1980) turns the analyses of McCallum, Wickens and Aigner upside down by considering the question whether it is advisable to omit correctly measured variables if our interest is in the coefficient of the mismeasured variable.

McCallum's and Wickens' result holds true for both the functional and structural model. Aigner's conditions refer only to the structural model with normally distributed latent variables. It would be of interest to see how his conditions modify for a functional model.

3.7. *Prediction and aggregation*

It is a rather remarkable fact that in the structural model the inconsistent OLS estimator can be used to construct consistent predictors, as shown by Johnston (1972, pp. 290, 291). The easiest way to show this is by considering (2.10): y_i and $\boldsymbol{x}_i$ are simultaneously normally distributed with variance–covariance matrix Σ as defined in (2.5). Using a well-known property of the normal distribution we obtain for the conditional distribution of y_i given $\boldsymbol{x}_i$:

$$f(y_i|\boldsymbol{x}_i)=\frac{1}{\gamma\sqrt{2\pi}}\exp\left\{-\tfrac{1}{2}\gamma^{-1}\left(y_i-\boldsymbol{x}_i'\boldsymbol{\alpha}\right)^2\right\}, \tag{3.18}$$

with γ and $\boldsymbol{\alpha}$ defined with respect to (2.9). Therefore, $\mathrm{E}(\boldsymbol{y}|X)=X\boldsymbol{\alpha}$. This implies that $\hat{\boldsymbol{\alpha}}$, the OLS estimator of $\boldsymbol{\alpha}$ is unbiased given X, and $\mathrm{E}(X\hat{\boldsymbol{\alpha}}|X)=X\boldsymbol{\alpha}=\mathrm{E}(\boldsymbol{y}|X)$. We can predict $\boldsymbol{y}$ unbiasedly (and consistently) by the usual OLS predictor, ignoring the measurement errors. As with the preceding omitted variable problem, we should realize that the conclusion only pertains to prediction *bias*, not to precision.

The conclusion of unbiased prediction by OLS does not carry over to the functional model. There we have:

$$f(y_i|\boldsymbol{x}_i,\xi_i)=\frac{1}{\sigma\sqrt{2\pi}}\exp\left\{-\tfrac{1}{2}\sigma^{-2}\left(y_i-\boldsymbol{\xi}_i'\boldsymbol{\beta}\right)^2\right\}, \tag{3.19}$$

so that $E(y|X, \Xi) = \Xi\boldsymbol{\beta}$, which involves both the incidental parameters and the unidentified parameter vector $\boldsymbol{\beta}$. OLS predictions are biased in this case, cf. Hodges and Moore (1972).

A somewhat different approach to prediction (and estimation) was taken by Aigner and Goldfeld (1974). They consider the case where exogenous variables in micro equations are measured with error but not so the corresponding aggregated quantities in macro equations. That situation may occur if the aggregated quantities have to satisfy certain exact accounting relationships which do not have to hold on the micro level. The authors find that under certain conditions the aggregate equations may yield consistent predictions whereas the micro equations do not. Similar results are obtained with respect to the estimation of parameters.

In a sense this result can be said to be due to the identifying restrictions that are available on the macro level. The usual situation is rather the reverse, i.e. a model which is underidentified at the aggregate level may be overidentified if disaggregated data are available. An example is given by Hester (1976).

Finally, an empirical case study of the effects of measurement error in the data on the quality of forecasts is given by Denton and Kuiper (1965).

3.8. *Bounds on parameters in underidentified models*

The maximum-likelihood equations that correspond to the full log-likelihood L_4 [recall (2.10)] are:

$$\hat{\boldsymbol{\alpha}} = (X'X)^{-1}X'y \tag{3.20}$$

$$\hat{\gamma} = n^{-1}(y - X\hat{\boldsymbol{\alpha}})'(y - X\hat{\boldsymbol{\alpha}}), \tag{3.21}$$

$$\hat{A} = n^{-1}X'X. \tag{3.22}$$

Without further restrictions we cannot say very much about the parameters of main interest, $\boldsymbol{\beta}$. An easy-to-accept restriction would be that the estimates of σ^2 and the diagonal elements of K and Ω should be non-negative. If in addition we assume that Ω is diagonal we obtain the following results.

Denote by $\boldsymbol{\omega}$ the k-vector of the diagonal elements of Ω and by $\boldsymbol{k}$ the k-vector of diagonal elements of K; B is the $k \times k$ diagonal matrix with the elements of $\boldsymbol{\beta}$ on its main diagonal. From (3.20)–(3.22) we derive as estimators for σ^2, $\boldsymbol{\omega}$ and $\boldsymbol{k}$ (given $\boldsymbol{\beta}$):

$$\hat{\boldsymbol{\omega}} = B^{-1}\hat{A}\boldsymbol{\beta} - B^{-1}\hat{A}\hat{\boldsymbol{\alpha}}, \tag{3.23}$$

$$\hat{\boldsymbol{k}} = \operatorname{diag}\hat{A} - \hat{\boldsymbol{\omega}}, \tag{3.24}$$

$$\hat{\sigma}^2 = \frac{1}{n}(y'y - \boldsymbol{\beta}'X'y), \tag{3.25}$$

where $\operatorname{diag}\hat{A}$ is the k-vector of diagonal elements of $\hat{A}$.

In order to actually compute these estimators we have to choose a value for $\boldsymbol{\beta}$. The restrictions $\hat{\boldsymbol{\omega}} > 0$, $\hat{\boldsymbol{k}} > 0$, $\hat{\sigma}^2 > 0$ imply that this value, $\hat{\boldsymbol{\beta}}$, has to satisfy:

$$\operatorname{diag} X'X \geq \hat{B}^{-1}X'X\hat{\boldsymbol{\beta}} - \hat{B}^{-1}X'\boldsymbol{y} \geq 0, \tag{3.26}$$

$$\boldsymbol{y}'\boldsymbol{y} \geq \hat{\boldsymbol{\beta}}'X'\boldsymbol{y}. \tag{3.27}$$

Let us first look at the case where $k = 1$. Then (3.26) reads:

$$X'X \geq X'X - X'\boldsymbol{y}\hat{\beta}^{-1} \geq 0. \tag{3.28}$$

So $|\hat{\beta}| \geq |(X'X)^{-1}X'\boldsymbol{y}| = |\hat{\alpha}|$ and $\hat{\beta}$ must have the same sign as $\hat{\alpha}$. Inequality (3.27) implies for this case $|\hat{\beta}| \leq [(\boldsymbol{y}'\boldsymbol{y})^{-1}X'\boldsymbol{y}]^{-1}$. Thus, a consistent estimator for β must have the same sign as the OLS estimator and its absolute value has to be between the OLS estimator and the reciprocal of the OLS regression coefficient of the regression of X on $\boldsymbol{y}$.

For $k > 1$, such simple characterizations are no longer possible, since they depend in particular on the structure of $X'X$ and the signs of the elements of $\boldsymbol{\beta}$. The only result that seems to be known is that if one computes the $k+1$ regressions of each of the variables $y_i, x_{i1}, \dots, x_{ik}$ on the other k variables and all these regressions are in the same orthant, then $\hat{\boldsymbol{\beta}}$ has to lie in the convex hull of these regressions. [Frisch (1934), Koopmans (1937), Klepper and Leamer (1984); see Patefield (1981) for an elegant proof using the Frobenius theorem]. Klepper and Leamer (1984) show that if the $k+1$ regressions are not all in the same orthant, if $\boldsymbol{\lambda}$ is a k-vector not equal to $(1/n)X'\boldsymbol{y}$ or the zero vector, and if $(X'X)^{-1}$ has no zero elements, then the set $\{\boldsymbol{\lambda}'\boldsymbol{\beta} | \boldsymbol{\beta}$ satisfying (3.26) and (3.27)$\}$ is the set of real numbers. Obviously, if one is willing to specify further prior knowledge, bounds can also be derived for $k > 1$. For example, Levi (1973, 1977) considers the case where only one of the exogenous variables is measured with error and obtains bounds for the coefficient of the mismeasured variable. Different prior knowledge is considered by Klepper and Leamer (1984).

A related problem is whether the conventional t-statistics are biased towards zero. Cooper and Newhouse (1972) find that for $k = 1$ the t-statistic of the OLS regression coefficient is asymptotically biased toward zero. For $k > 1$ no direction of bias can be determined.

Although inequalities (3.26) and (3.27) were derived from the maximum likelihood equations of the structural model, the same inequalities are derived in the functional model, because $\hat{\boldsymbol{\alpha}}$ is simply the OLS estimator and $\hat{\gamma}$ the residual variance estimator resulting from OLS. In fact, Levi only considers the OLS estimator $\hat{\boldsymbol{\alpha}}$ and derives bounds for a consistent estimator by considering the inconsistency of the OLS estimator.

Notice that the bounds obtained are not confidence intervals but merely bounds on the numerical values of estimates. These bounds can be transformed into confidence intervals by taking into account the (asymptotic) distribution of the OLS estimator [cf. Rothenberg (1973a), Davies and Hutton (1975), Kapteyn and Wansbeek (1983)]. One can also use the asymptotic distribution of the OLS estimator and a prior guess of the order of magnitude of measurement error to derive the approximate bias of the OLS estimator and to judge whether it is sizable relative to its standard error. This gives an idea of the possible seriousness of the errors-in-variables bias. This procedure has been suggested by Blomqvist (1972) and Davies and Hutton (1975).

3.9. *Tests for measurement error*

Due to the underidentification of errors-in-variables models, testing for the presence of measurement error can only take place if additional information is available. Hitherto the literature has invariably assumed that this additional information comes in the form of instrumental variables. Furthermore, all tests proposed deal with the functional model; testing in a structural model (i.e. a structural multiple indicator model, cf. Section 4) would seem to be particularly simple since, for example, ML estimation generates obvious likelihood ratio tests.

For the single-equation functional model, various tests have been proposed, by Liviatan (1961, 1963), Wu (1973), and Hausman (1978), all resting upon a comparison of the OLS estimator and the IV estimator. Under the null-hypothesis, H_0, that none of the variables is measured with error, the OLS estimator is more efficient than the IV estimator, and both are unbiased and consistent. If H_0 is not true the IV estimator remains consistent whereas OLS becomes inconsistent. Thus, functions of the difference between both estimators are obvious choices as test-statistics.

To convey the basic idea, we sketch the development of Wu's second test statistic for the model (3.8)–(3.9). The stochastic assumptions are the same as in Sections 2.1 and 3.4. Let there be available an $(n \times k)$-matrix W of instrumental variables that do not correlate with ε or V. In so far as certain columns of Ξ are supposed to be measured without error, corresponding columns of Ξ and W may be identical.

A possible statistic to test the null-hypothesis that none of the columns of Ξ has been measured with error is:

$$T \equiv \frac{Q^*/k}{\bar{Q}/(n-2k)}, \tag{3.29}$$

where

$$Q^* \equiv (\boldsymbol{b}-\hat{\boldsymbol{\beta}}_{\mathrm{IV}})'\left[(W'X)^{-1}W'W(X'W)^{-1}-(X'X)^{-1}\right]^{-1}(\boldsymbol{b}-\hat{\boldsymbol{\beta}}_{\mathrm{IV}}), \tag{3.30}$$

$$\boldsymbol{b} \equiv (X'X)^{-1}X'\boldsymbol{y}, \tag{3.31}$$

$$\hat{\boldsymbol{\beta}}_{\mathrm{IV}} \equiv (W'X)^{-1}W'\boldsymbol{y}, \tag{3.32}$$

$$\overline{Q} \equiv Q' - Q^*, \tag{3.33}$$

and

$$Q' \equiv (\boldsymbol{y}-X\boldsymbol{b})'(\boldsymbol{y}-X\boldsymbol{b}). \tag{3.34}$$

Note that $\boldsymbol{b}$ is the OLS estimator of $\boldsymbol{\beta}$ and $\hat{\boldsymbol{\beta}}_{\mathrm{IV}}$ is the IV estimator of $\boldsymbol{\beta}$.

Wu shows that Q^* and $\overline{Q}$ are mutually independent χ^2 distributed random variables with degrees of freedom equal to k and $n-2k$, respectively. Consequently, T follows a central F-distribution with k and $n-2k$ degrees of freedom. This knowledge can be used to test H_0.

Conceivably T is not the only possible statistic to test H_0. Wu (1973) gives one other statistic based on the small sample distribution of $\boldsymbol{b}$ and $\hat{\boldsymbol{\beta}}_{\mathrm{IV}}$ and two statistics based on asymptotic distributions. Two different statistics are proposed by Hausman (1978).

3.10. *Repeated observations*

Hitherto we have only discussed models with single indexed variables. As soon as one has more than one observation for each value of the latent variable the identification situation improves substantially. We shall illustrate this fact by a few examples. We do not pay attention to matters of efficiency of estimation, because estimation of these models is discussed extensively in the variance components literature. [See for example, Amemiya (1971).] Consider the following model:

$$y_{ij} = z_{ij}\beta + \xi_i\lambda + \varepsilon_{ij}, \qquad i=1,\dots,n; \quad j=1,\dots,m. \tag{3.35}$$

The variables z_{ij} and ξ_i are for simplicity taken to be scalars; z_{ij} is observable, ξ_i is not. A model like (3.35) may occur in panel studies, where n is the number of individuals in the panel and m is the number of periods in which observations on the individuals are obtained. Alternatively, the model may describe a controlled experiment in which the index i denotes a particular treatment with m observations per treatment.

As to the information regarding ξ_i we can distinguish among three different situations. The first situation is that where there are no observations on ξ_i. In a single-indexed model, that fact is fatal for the possibility of obtaining a consistent estimator for β unless z_{ij} and ξ_i are uncorrelated. In the double-indexed model, however, we can run the regression:

$$y_{ij} = z_{ij}\beta + \alpha_i + \varepsilon_{ij}, \tag{3.36}$$

where the $\{\alpha_i\}$ are binary indicators. The resulting estimate of β is unbiased and consistent. Although it is not possible to estimate λ, the estimates of α_i are unbiased estimates of $\xi_i\lambda$ so that the treatment effects are identified. A classical example of this situation is the correction for management bias [Mundlak (1961)]: if (3.36) represents a production function and ξ_i is the unobservable quality of management in the ith firm, omission of ξ_i would bias β, whereas formulation (3.36) remedies the bias.

A second situation which may occur is that for each latent variable there is one fallible measurement: $x_i = \xi_i + v_i$, $i = 1,\dots,n$. One measurement per ξ_i allows for identification of all unknown parameters but does not affect the estimator of β, as can be seen readily by writing out the required covariance equations.

The third situation we want to consider is where there are m measurements of ξ_i:

$$x_{ij} = \xi_i + v_{ij}, \qquad i = 1,\dots,n; \quad j = 1,\dots,m. \tag{3.37}$$

Now there is overidentification, and allowing for correlation between v_{ij} and v_{il}, $l \neq j$, does not alter that conclusion. Under the structural interpretation, ML is the obvious estimation method for this overidentified case. In fact, (3.35) and (3.37) provide an example of the multiple equation model discussed in the next section, where ML estimation will also be considered.

ML estimation for the functional model with replicated observations has been considered by Villegas (1961), Barnett (1970), Dolby and Freeman (1975), and Cox (1976). Barnett restricts his attention to the case with only one independent variable. Cox analyzes the same model, but takes explicitly into account the required non-negativity of estimates of variances. Villegas finds that apart from a scalar factor the variance–covariance matrix of the errors is obtained as the usual analysis-of-variance estimator applied to the multivariate counterpart of (3.37). The structural parameters are next obtained from the usual functional ML equations with known error matrix. Healy (1980) considers ML estimation in a multivariate extension of Villegas' model (actually a more general model of which the multivariate linear functional relationship is a special case). Dolby and Freeman (1975) generalize Villegas' analysis by allowing the errors to be correlated across different values of i. They show that, given the appropriate estimator for the variance–covariance matrix of the errors, the ML estimator of the structural parameters is identical to a generalized least squares estimator. Both Barnett (1970) and Dolby and Freeman (1975) derive the information matrix and use the elements of the partitioned inverse of the information matrix corresponding to the structural parameters as asymptotic approximations to the variance of the estimator. In light of the result obtained by Wolfowitz (1954) (cf. Section 2.3) these approximations would seem to underestimate the true asymptotic variance of the estimator. Regarding Barnett's paper, this is shown explicitly by Patefield (1977).

Villegas (1964) provides confidence regions for parameters in the linear functional relation if there are replicated measurements for each variable. His analysis has been generalized to a model with r linear relations among p latent variables ($p > r$) by Basu (1969). For $r > 2$ the confidence regions are not exact.

3.11. *Bayesian analysis*

As various latent variables models suffer from underidentification, and hence require additional prior information, a Bayesian analysis would seem to be particularly relevant to this type of model. Still, the volume of the Bayesian literature on latent variables models has remained modest hitherto. We mention Lindley and El-Sayyad (1968), Zellner (1971, ch. V), Florens, Mouchart and Richard (1974), Mouchart (1977), and Leamer (1978b, ch. 7) as the main contributions in this area. As far as identification is concerned, a Bayesian approach is only one of many possible ways to employ extraneous information. The use of auxiliary relations (Section 4) provides an alternative way to tackle the same problem. The choice of any of these approaches to identification in practical situations will depend on the researcher's preferences and the kind of extraneous information available.

As noted by Zellner (1971, p. 145) and Florens et al. (1974), the distinction between functional and structural models becomes a little more subtle in a Bayesian context. To illustrate, reconsider model (2.1), (2.2). Under the functional interpretation, ξ_i, β, σ^2, and Ω are constants. A Bayesian analysis requires prior densities for each of these parameters. The prior density for ξ_i makes the model look like a structural relationship. Florens, Mouchart and Richard (1974, p. 429) suggest that the difference is mainly a matter of interpretation, i.e. one can interpret Ξ as random because it is subject to sampling fluctuations or because it is not perfectly known. In the structural model, in a Bayesian context one has to specify in addition a prior distribution for the parameters that governs the distribution of the incidental parameters. Of course, also in the functional model where one has specified a prior distribution for the incidental parameters, one may next specify a second stage prior for the parameters of the prior distribution of the incidental parameters. The parameters of the second stage distributions are sometimes called hyperparameters.

The Bayesian analysis of latent variables models has mainly been restricted to the simple linear regression model with errors-in-variables [i.e. (2.1) is simplified to $y_i = \beta_0 + \beta_1 \xi_i + \varepsilon_i$, with β_0, β_1, ξ_i scalars], although Florens et al. (1974) make some remarks on possible generalizations of their analysis to the multiple regression model with errors-in-variables.

The extent to which Bayesian analysis remedies identification problems depends on the strength of the prior beliefs expressed in the prior densities. This is illustrated by Lindley and El-Sayyad's analysis. In the simple linear regression model with errors in the variables they specify a normal prior distribution for the latent variables, i.e. the $\boldsymbol{\xi}_i$ are i.i.d. normal with mean zero and variance τ, and next a general prior for the hyperparameter τ and the structural parameters. Upon deriving the posterior distribution they find that some parts of it depend on the sample size n, whereas other parts do not. Specifically, the marginal posterior distribution of the structural parameters and the hyperparameter does not depend on n. Consequently, this distribution does not become more concentrated when n goes to infinity.

This result is a direct consequence of the underidentification of the model. When repeating the analysis conditional on a given value of the ratio of the error variances with a diffuse prior for the variance of the measurement error, the posterior distribution of the structural parameters does depend on n and becomes more and more concentrated if n increases. The marginal posterior distribution of β_1 concentrates around the functional ML value. This is obviously due to the identification achieved by fixing the ratio of the error variances at a given value.

The analyses by Zellner (1971, ch. V) and Florens et al. (1974) provide numerous variations and extensions of the results sketched above: if one imposes exact identifying restrictions on the parameters, the posterior densities become more and more concentrated around the true values of the parameters when the number of observations increases. If prior distributions are specified for an otherwise unidentified model, the posterior distributions will not degenerate for increasing n and the prior distributions exert a non-vanishing influence on the posterior distributions for any number of observations.

4. Multiple equations

To introduce the ideas to be developed in this section, let us momentarily return to the simple bivariate regression model (3.1)–(3.2) in vector notation:

$$y = \boldsymbol{\xi}\beta + \boldsymbol{\varepsilon}, \tag{4.1}$$

$$x = \boldsymbol{\xi} + \boldsymbol{v}, \tag{4.2}$$

with $\boldsymbol{y}$, $\boldsymbol{x}$, $\boldsymbol{\xi}$, $\boldsymbol{\varepsilon}$ and $\boldsymbol{v}$ being $(n\times 1)$-vectors and β a scalar. As before, $\boldsymbol{y}$ and $\boldsymbol{x}$ are observable, and $\boldsymbol{\xi}$, $\boldsymbol{\varepsilon}$ and $\boldsymbol{v}$ are not. For most of this section, we consider the *structural* model, i.e. $\boldsymbol{\xi}$ is random. The elements of $\boldsymbol{\xi}$, $\boldsymbol{\varepsilon}$, and $\boldsymbol{v}$ are assumed to be normally i.i.d. distributed with zero means and variances $\sigma_{\xi\xi}$, σ^2 and σ_{vv}, respectively.

As we have seen, there is no way of obtaining consistent estimators for this model without additional information. In this section it is assumed that the available additional information takes on either of two forms:

$$z = \xi\gamma + \delta, \tag{4.3}$$

with z an observable $(n\times 1)$ vector, γ a scalar parameter, and $\boldsymbol{\delta}$ an $(n\times 1)$ vector of independent disturbances following an $N(\mathbf{0}, \sigma_{\delta\delta}I_n)$ distribution, independent of $\boldsymbol{\varepsilon}$, $\boldsymbol{v}$ and $\boldsymbol{\xi}$; or:

$$\boldsymbol{\xi} = W\boldsymbol{\alpha} + \boldsymbol{u}, \tag{4.4}$$

with W an $(n\times m)$ matrix of observable variables, $\boldsymbol{\alpha}$ an $(m\times 1)$ vector of coefficients, and $\boldsymbol{u}$ an $(n\times 1)$ vector of independent disturbances following an $N(\mathbf{0}, \sigma_{uu}I_n)$ distribution, independent of $\boldsymbol{\varepsilon}$ and $\boldsymbol{v}$. Also, models will be considered that incorporate both types of additional equations at the same time.

An interpretation of (4.3) is that z is an *indicator* of $\boldsymbol{\xi}$; just like $\boldsymbol{y}$ and $\boldsymbol{x}$, z is proportional to the unobservable $\boldsymbol{\xi}$, apart from a random error term, and therefore contains information on $\boldsymbol{\xi}$. Relation (4.4) may be interpreted such that the variables in W are considered to be the *causes* of $\boldsymbol{\xi}$, again apart from a random error term. In any case, the model is extended by the introduction of one or more equations, hence the description "multiple equations" for this type of approach to the measurement error problem. Note that no simultaneity is involved.

Additional information in the form of an extra indicator being available for an unobservable variable is the most frequently considered cure for the errors-in-variables identification problem, popularized in particular by the work of Goldberger (1971, 1974) and Goldberger and Duncan (1973). It is in fact, nothing but the instrumental variables (IV) approach to the problem [Reiersøl (1945)]. Section 4.1 deals with the IV method, whereas Section 4.2 discusses factor analysis in its relation to IV. Section 4.3 discusses models with additional causes, and models both with additional causes and indicators.

4.1. Instrumental variables

Due to the assumption of joint normality for $\boldsymbol{\varepsilon}$, $\boldsymbol{v}$ and $\boldsymbol{\xi}$, all sample information relating to the parameters in the model (4.1), (4.2) and (4.3) is contained in the six covariance equations [(recall (1.5) and (1.9)]:

$$
\begin{aligned}
\sigma_{yy} &= \sigma^2 + \sigma_{\xi\xi}\beta^2,\\
\sigma_{yx} &= \sigma_{\xi\xi}\beta,\\
\sigma_{yz} &= \sigma_{\xi\xi}\beta\gamma,\\
\sigma_{xx} &= \sigma_{\xi\xi} + \sigma_{vv},\\
\sigma_{xz} &= \sigma_{\xi\xi}\gamma,\\
\sigma_{zz} &= \sigma_{\xi\xi}\gamma^2 + \sigma_{\delta\delta}.
\end{aligned}
\tag{4.5}
$$

This system of six equations in six unknowns can easily be solved to yield consistent estimators of $\sigma_{\xi\xi}$, β, γ, σ^2, σ_{vv} and $\sigma_{\delta\delta}$. So, the introduction of the indicator variable (or instrumental variable) z renders the model identified.

Since the number of equations in (4.5) is equal to the number of parameters, the moment estimators are in principle also the ML estimators. This statement is subject to a minor qualification when ML is applied and the restriction of non-negativity of the error variances is explicitly imposed. Leamer (1978a) has shown that the ML estimator of β is the median of S_{yz}/S_{xz}, S_{yx}/S_{xx} and S_{yy}/S_{yx} where S indicates the sample counterpart of σ, if these three quantities have the same sign.

In the multivariate errors-in-variables [cf. (3.8), (3.9)] model we need at least $l \geq k$ indicator variables (or instrumental variables) in order to identify the parameter vector $\boldsymbol{\beta}$. The following relation is then assumed to hold:

$$
Z = \Xi\Gamma' + \Delta, \tag{4.6}
$$

with Z the $(n \times l)$ matrix of indicator variables, Γ an $(l \times k)$ matrix of coefficients and Δ an $(n \times l)$ matrix of disturbances, each row of which is l-dimensional normally distributed, independent of $\boldsymbol{\varepsilon}$, V and Ξ, with zero expectation and variance–covariance matrix Θ. No restrictions are imposed on Θ. This means that the instrumental variables are allowed to show an arbitrary correlation pattern, correlate with Ξ (and hence X), but are independent of the disturbance $\boldsymbol{\varepsilon} - V\boldsymbol{\beta}$ in the regression of $\boldsymbol{y}$ on X. Note that in particular this makes it possible to use the columns of Ξ that are measured without error as instrumental variables.

Let Ω be the $(k \times k)$ variance–covariance matrix of a row of V, and let $K \equiv \mathrm{E}n^{-1}\Xi'\Xi$. Then, in an obvious notation, the covariance equations (4.5) now read:

$$\sigma_{yy} = \sigma^2 + \boldsymbol{\beta}' K \boldsymbol{\beta}, \tag{4.7}$$

$$\Sigma_{Xy} = K\boldsymbol{\beta}, \tag{4.8}$$

$$\Sigma_{Zy} = \Gamma K \boldsymbol{\beta}, \tag{4.9}$$

$$\Sigma_{XX} = K + \Omega, \tag{4.10}$$

$$\Sigma_{ZX} = \Gamma K, \tag{4.11}$$

$$\Sigma_{ZZ} = \Gamma K \Gamma' + \Theta. \tag{4.12}$$

The identification of this system can be assessed somewhat heuristically as follows. Equations (4.7), (4.10) and (4.12) serve to identify the error variances σ^2, Ω and Θ for given Γ, K and $\boldsymbol{\beta}$. Substitution of (4.11) into (4.9) yields:

$$\Sigma_{yZ} = \Sigma_{XZ}\boldsymbol{\beta}, \tag{4.13}$$

which shows that $l \geq k$ is a necessary condition for the identification of $\boldsymbol{\beta}$. When $l > k$, $\boldsymbol{\beta}$ is generally overidentified. For the identification of the other parameters, K and Γ, only (4.8) and (4.11) remain; these contain in general insufficient information, whether $k = l$ or $l > k$, so these parameters are not identified. This is basically due to the fact that Γ occurs only in conjunction with K. The only exception is when only one column of Ξ is unobservable. In that case Γ and K each contain k unknown elements that can be obtained from (4.8) and (4.11). More discussion of this point will be given in Section 4.2 below.

In the case $l > k$, i.e. there are more instrumental variables than regressors in the original model, (4.13) does not produce an estimator for $\boldsymbol{\beta}$ unambiguously. A way to reconcile the conflicting information in (4.13) is to reduce it to a system of k equations by premultiplication with some $(k \times l)$-matrix, G say. A possible choice for G is:

$$G = \Sigma'_{XZ}\Sigma_{ZZ}^{-1}. \tag{4.14}$$

Replacing the Σ's by their sample counterparts, indicated by a corresponding S, the estimator for $\boldsymbol{\beta}$ then is:

$$\begin{aligned}\hat{\boldsymbol{\beta}} &= \left(S'_{XZ}S_{ZZ}^{-1}S_{XZ}\right)^{-1}S'_{XZ}S_{ZZ}^{-1}S_{yX} \\ &= \left(X'Z(Z'Z)^{-1}Z'X\right)^{-1}X'Z(Z'Z)^{-1}Z'\boldsymbol{y}.\end{aligned} \tag{4.15}$$

For $l = k$, this reduces to the well-known formula:

$$\hat{\boldsymbol{\beta}} = (Z'X)^{-1}Z'\boldsymbol{y}. \tag{4.16}$$

Sargan (1958) has shown that the weighting matrix G is optimal in the sense that it has minimal asymptotic variance in the class of all linear combinations of estimators which can be derived from (4.13). [See also Malinvaud (1970, section 20.5).] The asymptotic variance–covariance matrix of $\hat{\beta}$ is, both for $l=k$ and $l>k$:

$$\mathscr{V}(\hat{\beta}) = (\sigma^2 + \beta'\Omega\beta)\left(\Sigma'_{XZ}\Sigma_{ZZ}^{-1}\Sigma_{XZ}\right)^{-1}. \tag{4.17}$$

When the researcher is in the happy situation that he has more instruments than error-ridden variables (i.e. $l>k$), he may also consider applying ML to the full model after imposing a sufficient number of identifying restrictions on (at least) Γ and K. The LISREL program (see Section 5.3) is well-suited for this purpose.

The major problem involved with IV in the non-dynamic single equation context, however, is to find instrumental variables. Columns of X without measurement errors can be used as instruments, but it is often difficult to find variables that are correlated with a variable in X and are not already explanatory variables in the model under consideration. The method of grouping, discussed in Section 3.2, can be considered as a special case of IV, where the instrument consists of a vector of $+1$'s and -1's, allocating observations to the two groups. The instrument should be uncorrelated with the measurement error in order to have a consistent estimator of the slope parameters. This is the case, for instance, when the size of the measurement error is bounded from above and the population consists of two subsets separated by an interval at least as great as twice this maximum. This situation is unlikely to occur in practice.

4.2. Factor analysis

Factor analysis (FA), a method for dealing with latent variables with a venerable history in psychometrics, is closely related to instrumental variables. In this section we will discuss some aspects of FA as far as it is relevant in the present context without the pretension of coming anywhere near a complete survey. For a more comprehensive coverage see, for example, Gorsuch (1974), Lawley and Maxwell (1971), Harman (1967) or Bentler and Weeks (1980); econometricians will find the book by Mulaik (1972) highly readable because of its notation.

The connection between the FA and IV models is as follows. Let, in (3.19), the measurement error between the columns of Ξ be uncorrelated, i.e. the matrix Ω of measurement error variances and covariances is diagonal, and let the coefficient matrix of Ξ, so far implicitly taken to be the unit matrix, be arbitrary. This means that (i) the correlation between different columns of X is attributable to Ξ only, and not to the measurement error, and (ii) X is no longer considered to be a direct but erroneous measurement of Ξ. Likewise, let the variance–covariance matrix Θ of the rows of Δ in (4.6) be diagonal also; so, the correlation between different columns of Z is attributable to Ξ only, and not to the disturbances Δ.

Under the new interpretation, the three equations (3.8), (3.9), (4.6) constituting the IV model have become formally isomorphous, and there is no reason for distinguishing between them anymore. We may thus dispense with (3.8) and (3.9) without loss of generality and take (4.6) as the FA model, under the following interpretation: Z is the $(n \times l)$-matrix of indicator variables of the k latent variables (or *common factors*) grouped in Ξ of order $(n \times k)$; Δ is an $(n \times l)$-matrix of disturbances (or *unique factors*.) The common factors account for the correlation between the indicators, and the unique factors take account of the remaining variance. Γ is again an $(l \times k)$-matrix of regression coefficients, or *factor loadings* or *factor pattern*, in the FA patois.

Although the formal analogy between the IV and FA models is apparent, there are interpretative differences between the two. In FA, the latent variables are fully conceptual variables and are not, as in the econometrics literature on measurement error, supposed to be observable in a direct (i.e. outside the model) albeit erroneous way; indeed, the number of factors need not be a given magnitude and becomes a parameter itself [e.g. Gorsuch (1974, ch. 8)].

Given the stochastic assumptions, the covariance equation corresponding to (4.6) is:

$$\Sigma_{ZZ} = \Gamma K \Gamma' + \Theta, \tag{4.18}$$

and the estimation problem is to derive estimators for K, Γ and Θ from the observed covariance matrix S_{ZZ}. Without further information, the model is clearly underidentified since postmultiplication of Γ by any non-singular $(k \times k)$-matrix T and replacing K by $T^{-1}K(T')^{-1}$ leads to the same value of Σ. There are several ways to cope with this indeterminacy, each of which identifies a main branch of factor analysis distinguished in the literature. [See, for example, Elffers et al. (1978).]

An extreme case arises if k is taken equal to l. Then Γ and K are of the same order as Σ_{ZZ}. This obviates the error term Δ, so Θ is put equal to 0. Next, the indeterminacy may be solved by taking Γ to be the matrix of eigenvectors of Σ_{ZZ}, and K is the diagonal matrix containing the k eigenvalues of Σ_{ZZ} on its main diagonal. The matrix $Z\Gamma$ is called the matrix of *principal components* of Z. [See, for example, Anderson (1958, ch. 12) and Kendall and Stuart (1979, ch. 43).] This relation between principal components and FA is a matter of mathematics only; conceptually, there is the essential difference that principal components is not based on a statistical model; it is a data reduction technique.[12]

[12]Principal components is sometimes used in econometrics when the number of observations is deficient and one wants to reduce the number of regressors. Kloek and Mennes (1960) and Amemiya (1966) explore this idea for simultaneous equations and propose using principal components of predetermined variables.

Apart from the principal components case, the number k of underlying factors is usually set at a (much) lower value than l. There are two different approaches to the normalization problem. In *confirmatory factor analysis*, the researcher has a number of a priori restrictions on Γ, K or Θ at his disposal that derive from say, the interpretation of the factors [like the implicit unit coefficient restriction in equation (3.9), where the factors correspond to phenomena that are in principle observable] or an extension of the model whereby the latent variables are, in turn, regressed on other, observable variables (an example of which is to be discussed below). These restrictions may serve to remove all indeterminacy in the parameters. In *exploratory factor analysis*, however, the researcher is unsure about the meaning of the factors and would like to treat them in a symmetric way. The usual approach then is to choose T such that $T^{-1}K(T')^{-1}$ is the unit matrix, i.e. the factors are uncorrelated. For $\tilde{\Gamma} \equiv \Gamma T$:

$$\Sigma_{ZZ} = \tilde{\Gamma}\tilde{\Gamma}' + \Theta. \tag{4.19}$$

There is still some indeterminacy left, since the columns of $\tilde{\Gamma}$ may be reweighted with any orthonormal matrix without affecting Σ_{ZZ}. This freedom may be used to make $\tilde{\Gamma}'\Theta^{-1}\tilde{\Gamma}$ a diagonal matrix, which is convenient in the course of ML estimation of the parameters [Jöreskog (1967)], or can be used at will to obtain some desired pattern in Γ. Such a reweighting is called a *rotation* by factor analysts, and a huge literature has evolved around the pros and cons of all possible types of rotations. Shapiro (1982) has investigated the identification of the exploratory FA model. He shows that it is identified (apart from the indeterminacies in Γ) if and only if $(l-k)^2 \geq l+k$.

Again, it should be stressed that the above treatment of FA is meant only to show its relation to the measurement error problem and to show that "factor analysis is just a generalization of the classical errors-in-the-variables model" [Goldberger (1972a, p. 992)].

4.3. *The MIMIC model and extensions*

In this section we will consider models where identifying information is of the form given in (4.4); the unobservable variable depends on other exogenous variables.

Recall (4.1), (4.2) and (4.4). By eliminating $\boldsymbol{\xi}$, we obtain the reduced form:

$$(\boldsymbol{y}, \boldsymbol{x}) = W(\boldsymbol{\alpha}\beta, \boldsymbol{\alpha}) + \boldsymbol{\psi}, \tag{4.20}$$

with

$$\boldsymbol{\psi} = (\boldsymbol{\varepsilon} + \beta\boldsymbol{u}, \boldsymbol{v} + \boldsymbol{u}). \tag{4.21}$$

The variance–covariance matrix $\Sigma_{\psi\psi}$ of ψ is:

$$\Sigma_{\psi\psi} = \begin{pmatrix} \sigma^2 + \beta^2\sigma_{uu} & \beta\sigma_{uu} \\ \beta\sigma_{uu} & \sigma_{vv} + \sigma_{uu} \end{pmatrix}. \tag{4.22}$$

The reduced form coefficients are restricted in that there is a proportionality restriction on the regression coefficients whenever l, the dimension of $\boldsymbol{\alpha}$, exceeds unity. For all $l \geq 1$, $\boldsymbol{\alpha}$ and β are identified and hence also the parameters in $\Sigma_{\psi\psi}$. So, an additional relation which "explains" the latent variable renders the model identifiable.

As a somewhat more general case, consider the case where an unspecified (r, say) number of indicators of the latent variable $\boldsymbol{\xi}$ is available, i.e. (4.1) and (4.2) are replaced by:

$$Z = \boldsymbol{\xi}\boldsymbol{\gamma}' + \Delta, \tag{4.23}$$

with Z and Δ being $(n \times r)$ matrices as in (4.6), and $\boldsymbol{\gamma}$ is an $(r \times 1)$ vector of regression coefficients. As with FA, Θ, the variance–covariance matrix of the rows of Δ, is assumed to be diagonal.

This model [i.e. (4.4) and (4.23)] is known as the Multiple Indicator–Multiple Cause (MIMIC) model relating a single unobservable to a number of indicators and a number of exogenous variables, and was introduced in the econometrics literature by Goldberger (1972a). In reduced form, it reads:

$$Z = W\boldsymbol{\alpha}\boldsymbol{\gamma}' + \Delta + \boldsymbol{u}\boldsymbol{\gamma}'. \tag{4.24}$$

The model has two kinds of restrictions on its parameters. First, the coefficient matrix has rank unity, and the disturbances have a variance–covariance matrix:

$$\Sigma \equiv \Theta + \sigma_{uu}\boldsymbol{\gamma}\boldsymbol{\gamma}', \tag{4.25}$$

which is the sum of a diagonal matrix and a matrix of unit rank.

There is an indeterminacy in the coefficients $\boldsymbol{\alpha}$ and $\boldsymbol{\gamma}$: the product $\boldsymbol{\alpha}\boldsymbol{\gamma}'$ remains the same when $\boldsymbol{\alpha}$ is multiplied by an arbitrary constant and $\boldsymbol{\gamma}$ is divided by the same constant. This indeterminacy can be removed, for example, by the normalization $\sigma_{uu} = 1$. Jöreskog and Goldberger (1975) discuss ML estimation of the MIMIC model, and Chen (1981) discusses iterative estimation via the EM algorithm.

The MIMIC model comprises several models as special cases. When no "cause" relation is present, we have the one-factor FA model. If in (4.4) $\boldsymbol{u} = \boldsymbol{0}$, i.e. the latent variable is an exact linear function of a set of explanatory variables, we are back to a model introduced by Zellner (1970). This model was inspired by the well-known problem of dealing with permanent income as an explanatory variable. In this model, y denotes consumption, x observed income and ξ permanent income. By expressing permanent income as a function of "such variables as house value, educational attainment, age, etc." [Zellner (1970, p. 442)], permanent income can be removed from the relation altogether; but simultaneous estimation of the complete reduced form of the model increases the precision of the estimates.

Zellner's paper also contains a discussion of limited-information estimation methods. Since full-information ML is now generally available (see Section 5.3), there seems to be little use left for limited-information methods and we will not attempt to present a summary of these methods. [See also Goldberger (1972b).]

A restriction of the MIMIC model is the diagonality of Θ, the variance–covariance matrix of the rows of Δ. This means that the indicators satisfy the factor analysis assumption that they are correlated only via the latent variable. This assumption may be unduly strong, and we may consider an unrestricted Θ as an alternative, as was the case in the original instrumental variables model. As is apparent from (4.25), this introduces an indeterminacy since:

$$\Theta + \sigma_{uu}\gamma\gamma' = \Theta + \phi\gamma\gamma' + (\sigma_{uu} - \phi)\gamma\gamma' \tag{4.26}$$

for any scalar ϕ. This indeterminacy may be solved by fixing σ_{uu} at some non-negative value, e.g. $\sigma_{uu} = 0$. This means that, in the case of Θ unrestricted, the model is operationally equivalent to a model without an error in the cause equation.

The MIMIC model relates a single latent variable to a number of indicators and a number of causes. The extension to a more general multiple equations model is obvious. A very general formulation is the following one, proposed by Robinson (1974):

$$Z = \Xi\Gamma' + W_1B_1 + W_2B_2 + \Delta, \tag{4.27}$$

$$\Xi = W_0A_0 + W_1A_1 + U, \tag{4.28}$$

with Z, Ξ and Γ defined as before; Δ and U are $(n \times l)$ and $(n \times k)$ matrices of disturbances, each row of which is taken to be normally, independently distributed with variance Θ and Ψ, respectively. No a priori restrictions are imposed on these matrices. W_0, W_1 and W_2 are $(n \times m_i)$, $i = 0,1,2$, matrices of observable exogenous variables, and A_0, A_1, B_1 and B_2 are conformable matrices of regression coefficients.

This model allows for structuring elaborate causal chains between variables. The indicators Z are determined not only by the latent variables Ξ but also by a set of exogenous variables. The latent variables in turn are determined by a set of exogenous variables, some of which (W_1) may also occur in the indicator equation. Note that there is no simultaneity in the model: the causal chain is in one direction, the W's determining Z directly and, after a detour, via Ξ. For this model, Robinson (1974) discusses identification and presents a (limited information) estimation method. The problems involved are apparent from the reduced form of (4.27) and (4.28):

$$Z = W_0 A_0 \Gamma' + W_1\left(A_1 \Gamma' + B_1\right) + W_2 B_2 + \Delta + U\Gamma, \tag{4.29}$$

where each row has variance–covariance matrix $\Theta + \Gamma\psi\Gamma'$. The model has, just like the MIMIC model, patterned coefficient matrices and a patterned variance–covariance matrix. Some of the coefficients are clearly underidentified. After imposing appropriate restrictions, overidentification may result. Instead of Robinson's method, one might estimate the (appropriately restricted) model by FIML, using (for instance) the LISREL computer program (see Section 5.3).

What should be clear from the development in this section (especially this subsection) is that an important convergence in methodology between psychometrics, sociometrics and econometrics has taken place over the last decade. The input into econometrics from the other two social sciences induced a breakthrough in the measurement error problem; in return, econometrics can contribute rigor in the fields of identification, estimation and hypothesis testing, areas where psychological and sociological researchers tend to be somewhat more casual than econometricians.

5. Simultaneous equations

Stripped to its bare essentials, the linear simultaneous equations model with latent variables is the following. Let Z be an $(n \times L)$-matrix of observations on an $(L \times 1)$-vector with n data points. Let Z be generated by an unobservable, "true" part $\tilde{Z}$ of order $(n \times L)$ and an $(n \times L)$-matrix U of measurement errors, each row of which is independently $N(\mathbf{0}, \Omega)$ distributed, with Ω an $(L \times L)$-matrix:

$$Z = \tilde{Z} + U. \tag{5.1}$$

The latent matrix $\tilde{Z}$ is subject to R linear constraints, $R \leq L$:

$$\tilde{Z}\Gamma = 0, \tag{5.2}$$

with Γ an $(L \times R)$-matrix of coefficients which has to be estimated. (The zero restriction is for convenience only and can be relaxed at some notational cost.)

When $R = 1$, (5.1) and (5.2) constitute the single-equation errors-in-variables model, where all variables are treated in a symmetric way. If some row of U

happens to be uncorrelated with the other rows of U, it may be interpreted as an error-in-equation, and the usual one-equation errors-in-variables model arises. For $R > 1$, (5.1) and (5.2) constitute a simultaneous errors-in-variables model.

In the early days of econometrics, attention focused on the case where Ω is known (or known up to a scalar factor) and there are no restrictions on the coefficients apart from normalization. In Section 5.1 we will briefly dwell on this case, mainly because of its historical interest. Then, in Section 5.2, we will pick up the "mainstream" approach to dealing with simultaneity. Section 5.3 discusses the LISREL computer program, which is well-suited to estimate linear equations systems with latent variables.

5.1. *The case of Ω known*

When Ω is known, Γ can be estimated by maximizing the likelihood of U subject to $\tilde{Z}\Gamma = 0$. [See (5.1) and (5.2).] The main results are due to Tintner (1945), extending results for the case $R = 1$ due to Van Uven (1930), which became well known to econometricians mainly through the work of Koopmans (1937, ch. 5). Also for $R = 1$, Malinvaud (1970, section 10.5) derives the variance–covariance matrix of the asymptotic distribution of $\hat{\Gamma}$.

An important special case arises when only the first G columns of $\tilde{Z}$ are unobservable, the last $K \equiv L - G$ being observable. Konijn (1962) discusses identification and estimation of this model.

Konijn's work may be viewed as the culmination point of a research direction that at present is dormant. Since the early 1950s, the emphasis in econometrics has focused on identification based on a priori restrictions on Γ rather than on Ω being known, as the empirical value of the latter case seems to be limited. Still, it might be a fruitful research project to make the communalities and differences between the two approaches explicit, e.g. by translating restrictions on Γ into restrictions on Ω. An attempt to use the errors-in-variables approach for the simultaneous equations model was made by Keller (1975), for instance, who demonstrates that several well-known limited information estimators correspond to the errors-in-variables estimator for particular choices of Ω.

5.2. *Identification and estimation*

In the non-simultaneous multiple-equations model, identification is achieved by employing auxiliary relations linking the latent variables to observable variables. The outstanding feature of the simultaneous equations model containing latent variables is that identification may be achieved without such additional information, because sometimes overidentifying information already present in the model can be used to remedy the underidentifiability caused by measurement error. In this section we will discuss some equivalent ways of assessing the identifiability of a simultaneous equations model containing latent variables. Complications arise when there are latent variables which enter into more than one equation or when the measurement error of latent variables in different equations is correlated. Then, identification cannot be settled on an equation-by-equation basis anymore and the structure of the total model has to be taken into consideration.

When an exogenous variable is measured with error, its observed value is no longer independent of the equation's disturbance and may be considered as an additional endogenous variable. Accordingly, we may expand the model by an additional relation. This approach is due to Chernoff and Rubin (1953) and is also used by Hausman (1977). As an example, consider the two-equation model of Section 1.4 (in vector notation):

$$\begin{aligned} y_1 + \beta_{12} y_2 &= \gamma_{11}\boldsymbol{\xi}_1 \qquad\qquad\qquad + \boldsymbol{\varepsilon}_1, \\ \beta_{21} y_1 + y_2 &= \qquad\quad \gamma_{22}\boldsymbol{\xi}_2 + \gamma_{23}\boldsymbol{\xi}_3 + \boldsymbol{\varepsilon}_2, \end{aligned} \tag{5.3}$$

where y_1 and y_2 are $(n\times 1)$-vectors of observations on the endogenous variables, and $\boldsymbol{\xi}_1$, $\boldsymbol{\xi}_2$ and $\boldsymbol{\xi}_3$ are $(n\times 1)$-vectors of observations on the exogenous variables; $\boldsymbol{\varepsilon}_1$ and $\boldsymbol{\varepsilon}_2$ are $(n\times 1)$-vectors of disturbances, independent for different observations, with zero expectations, variance $\sigma_{11}I_n$ and $\sigma_{22}I_n$, respectively, and covariance $\sigma_{12}I_n$. Let $\boldsymbol{\xi}_1$ be unobservable, and let $\boldsymbol{x}_1$ be a proxy:

$$\boldsymbol{x}_1 = \boldsymbol{\xi}_1 + \boldsymbol{v}_1, \tag{5.4}$$

with the measurement error $\boldsymbol{v}_1$ assumed to be distributed $N(\mathbf{0}, \sigma_{vv}I_n)$, independent of $\boldsymbol{\varepsilon}_1$ and $\boldsymbol{\varepsilon}_2$.

The translation of an unobservable into an additional endogenous variable can be made as follows. Let the elements of $\boldsymbol{\xi}_1$, $\boldsymbol{\xi}_2$ and $\boldsymbol{\xi}_3$ be jointly normally distributed. Then the regression of $\boldsymbol{\xi}_1$ on $\boldsymbol{\xi}_2$ and $\boldsymbol{\xi}_3$ can be written as:

$$\boldsymbol{\xi}_1 = \alpha_2\boldsymbol{\xi}_2 + \alpha_3\boldsymbol{\xi}_3 + \boldsymbol{u}_1, \tag{5.5}$$

with $\boldsymbol{u}_1$ distributed as $N(\mathbf{0}, \sigma_{uu}I_n)$, independent of $\boldsymbol{\xi}_2$, $\boldsymbol{\xi}_3$, $\boldsymbol{\varepsilon}_1$, $\boldsymbol{\varepsilon}_2$ and $\boldsymbol{v}_1$. Substitution of (5.5) into (5.4) and (5.4) into (5.3) yields the following three-equation system:[13]

$$\begin{aligned} y_1 + \beta_{12} y_2 - \gamma_{11}\boldsymbol{x}_1 &= \boldsymbol{\varepsilon}_1 - \gamma_{11}\boldsymbol{v}_1, \\ \beta_{21} y_1 + y_2 &= \gamma_{22}\boldsymbol{\xi}_2 + \gamma_{23}\boldsymbol{\xi}_3 + \boldsymbol{\varepsilon}_2, \\ \boldsymbol{x}_1 &= \alpha_2\boldsymbol{\xi}_2 + \alpha_3\boldsymbol{\xi}_3 + \boldsymbol{v}_1 + \boldsymbol{u}_1. \end{aligned} \tag{5.6}$$

[13]Not only is it possible to transform a model with errors in variables into one without mismeasured variables, one can also reformulate standard simultaneous equation models as functional models. For reasons of space we do not give the relationship between both models, but refer to Anderson (1976) instead. Among the results of exploring the link between functional and simultaneous models are asymptotic approximations to the distributions of various estimators. See Anderson (1976, 1980) and Patefield (1976) for details.

This reformulation of the system may be used to assess the state of identification. Still, this is no standard problem, since the variance–covariance matrix of the disturbances ($\tilde{\Sigma}$, say) of the extended *structural* model (5.6) is restricted:

$$\tilde{\Sigma}=\begin{pmatrix}\sigma_{11}+\gamma_{11}\sigma_{vv} & \sigma_{12} & \gamma_{11}\sigma_{vv}\\ \sigma_{21} & \sigma_{22} & 0\\ \gamma_{11}\sigma_{vv} & 0 & \sigma_{vv}+\sigma_{uu}\end{pmatrix}. \tag{5.7}$$

So, two elements of $\tilde{\Sigma}$ are restricted to be zero. Identification for this type of restricted model was studied by Wegge (1965) and Hausman and Taylor (1983), who present rank and order conditions for identification. Below, we will discuss identification of the simultaneous model with latent variables using a somewhat different approach.

Two features of this extension of the model should be noted. First, in order for (5.5) to make sense, the unobservable should be correlated with at least one other exogenous variable, i.e. α_2 or α_3 should be non-zero. Second, (5.5) fits in the Zellner–Goldberger approach of relating an unobservable to other, observable "causes". In the simultaneous equations context, such an additional relation comes off quite naturally from the model.

A direct approach to the assessment of the identification of the simultaneous equations model with latent variables is the establishment of a rank condition that generalizes the rank condition for the usual model without latent variables. Let the model be:

$$YB'=\Xi\Gamma'+U, \tag{5.8}$$

$$X=\Xi+V, \tag{5.9}$$

with Y and X being $(n\times G)$ and $(n\times K)$ matrices of observations, Ξ the $(n\times K)$-matrix of true values of X and V the $(n\times K)$-matrix of measurement errors, B and Γ $(G\times G)$ and $(G\times K)$ coefficient matrices and U an $(n\times G)$ disturbance matrix; U and V are mutually independent and their rows are independently normally distributed with variance–covariance matrices Σ and Ω, respectively.

The covariance equations corresponding to (5.8) and (5.9) are, in obvious notation:

$$\Sigma_{YY}=B^{-1}\Gamma\Sigma_{\Xi\Xi}\Gamma'(B')^{-1}+B^{-1}\Sigma(B')^{-1}, \tag{5.10}$$

$$\Sigma_{YX}=B^{-1}\Gamma\Sigma_{\Xi\Xi}, \tag{5.11}$$

$$\Sigma_{XX}=\Sigma_{\Xi\Xi}+\Omega. \tag{5.12}$$

(When we have a *structural* model, $\Sigma_{\Xi\Xi}$ denotes the variance–covariance matrix of a row of Ξ; when we have a *functional* model, it denotes $\lim(1/n)\Xi'\Xi$.) Rewrite (5.11) using (5.12):

$$\Sigma_{YX} = B^{-1}\Gamma(\Sigma_{XX} - \Omega). \tag{5.13}$$

When B, Γ and Ω are known, (5.10) and (5.12) serve to identify Σ and $\Sigma_{\Xi\Xi}$; so a priori information from (5.13) and identification of the full model is equivalent to the identification of B, Γ and Ω (e.g. normalizations, exclusions, and symmetry restrictions on Ω).

A necessary and sufficient rank condition for identification can now be developed as follows. Define $\boldsymbol{\alpha}_0 \equiv \operatorname{vec}(B, \Gamma)'$, $\boldsymbol{\omega} \equiv \operatorname{vec}\Omega$, and let $\boldsymbol{\alpha} \equiv (\boldsymbol{\alpha}_0', \boldsymbol{\omega}')'$ be the vector of all parameters. Then the a priori information can be written as:

$$R\boldsymbol{\alpha} = \boldsymbol{r}, \tag{5.14}$$

with R being an $(m \times (G(G+K)+K^2))$-matrix and r an $(m \times 1)$-vector of known constants. Now, $\boldsymbol{\alpha}$ is (locally) identifiable if and only if the Jacobian:

$$J \equiv \partial \begin{pmatrix} \operatorname{vec}\{B^{-1}\Gamma(\Sigma_{XX} - \Omega)\}' \\ R\boldsymbol{\alpha} \end{pmatrix} \Big/ \partial \boldsymbol{\alpha}' \tag{5.15}$$

has rank $G^2 + GK + K^2$, i.e. J has full column rank [and if $\boldsymbol{\alpha}$ is locally isolated—see, for example, Fisher (1966)]. It remains to evaluate J. Using standard matrix derivation methods, one readily obtains:

$$J = \begin{pmatrix} -B^{-1}\otimes(-\Sigma_{XY}, \Sigma_{\Xi\Xi}) & -B^{-1}\Gamma\otimes I_K \\ & R \end{pmatrix}. \tag{5.16}$$

This matrix has, of course, the same rank as:

$$\tilde{J} \equiv \begin{pmatrix} I_G\otimes(-\Sigma_{XY}, \Sigma_{\Xi\Xi}) & -\Gamma\otimes I_K \\ & R \end{pmatrix}. \tag{5.17}$$

As an example, consider the simple model (5.3). The a priori restrictions are $\beta_{11} = \beta_{22} = 1$, $\gamma_{12} = \gamma_{13} = \gamma_{21} = 0$, and, when uncorrelated measurement error is assumed, $\Omega = 0$ apart from Ω_{11}. So, there are $G(G+K)+K^2 = 19$ parameters on the one hand and $GK + m = 6+2+3+8 = 19$ restrictions on them. Denoting non-zero elements by a "+" for the sake of transparency, then $\tilde{J}$ is:

$$\begin{array}{ccccccccccccccccccc}
+ & + & + & + & + & 0 & 0 & 0 & 0 & 0 & + & 0 & 0 & 0 & 0 & 0 & 0 & 0 & 0 \\
+ & + & + & + & + & 0 & 0 & 0 & 0 & 0 & 0 & + & 0 & 0 & 0 & 0 & 0 & 0 & 0 \\
+ & + & + & + & + & 0 & 0 & 0 & 0 & 0 & 0 & 0 & + & 0 & 0 & 0 & 0 & 0 & 0 \\
 & & & & & & & & & & & & & & & & & & \\
0 & 0 & 0 & 0 & 0 & + & + & + & + & + & 0 & 0 & 0 & + & 0 & 0 & + & 0 & 0 \\
0 & 0 & 0 & 0 & 0 & + & + & + & + & + & 0 & 0 & 0 & 0 & + & 0 & 0 & + & 0 \\
0 & 0 & 0 & 0 & 0 & + & + & + & + & + & 0 & 0 & 0 & 0 & 0 & + & 0 & 0 & + \\
 & & & & & & & & & & & & & & & & & & \\
+ & 0 & 0 & 0 & 0 & 0 & 0 & 0 & 0 & 0 & 0 & 0 & 0 & 0 & 0 & 0 & 0 & 0 & 0 \\
0 & 0 & 0 & + & 0 & 0 & 0 & 0 & 0 & 0 & 0 & 0 & 0 & 0 & 0 & 0 & 0 & 0 & 0 \\
0 & 0 & 0 & 0 & + & 0 & 0 & 0 & 0 & 0 & 0 & 0 & 0 & 0 & 0 & 0 & 0 & 0 & 0 \\
0 & 0 & 0 & 0 & 0 & 0 & + & 0 & 0 & 0 & 0 & 0 & 0 & 0 & 0 & 0 & 0 & 0 & 0 \\
0 & 0 & 0 & 0 & 0 & 0 & 0 & + & 0 & 0 & 0 & 0 & 0 & 0 & 0 & 0 & 0 & 0 & 0 \\
 & & & & & & & & & & & & & & & & & & \\
0 & 0 & 0 & 0 & 0 & 0 & 0 & 0 & 0 & 0 & 0 & + & 0 & 0 & 0 & 0 & 0 & 0 & 0 \\
0 & 0 & 0 & 0 & 0 & 0 & 0 & 0 & 0 & 0 & 0 & 0 & + & 0 & 0 & 0 & 0 & 0 & 0 \\
0 & 0 & 0 & 0 & 0 & 0 & 0 & 0 & 0 & 0 & 0 & 0 & 0 & + & 0 & 0 & 0 & 0 & 0 \\
0 & 0 & 0 & 0 & 0 & 0 & 0 & 0 & 0 & 0 & 0 & 0 & 0 & 0 & + & 0 & 0 & 0 & 0 \\
0 & 0 & 0 & 0 & 0 & 0 & 0 & 0 & 0 & 0 & 0 & 0 & 0 & 0 & 0 & + & 0 & 0 & 0 \\
0 & 0 & 0 & 0 & 0 & 0 & 0 & 0 & 0 & 0 & 0 & 0 & 0 & 0 & 0 & 0 & + & 0 & 0 \\
0 & 0 & 0 & 0 & 0 & 0 & 0 & 0 & 0 & 0 & 0 & 0 & 0 & 0 & 0 & 0 & 0 & + & 0 \\
0 & 0 & 0 & 0 & 0 & 0 & 0 & 0 & 0 & 0 & 0 & 0 & 0 & 0 & 0 & 0 & 0 & 0 & +
\end{array} \qquad (5.18)$$

The rank of this matrix is easily assessed, as follows. The last 13 rows correspond to normalizations and exclusions (i.e. it shows the incidence of zero and non-zero elements in R); the columns in which non-zero elements occur are clearly linearly independent. So, the rank of $\tilde{J}$ equals 13 plus the rank of the matrix that remains after deleting the rows and columns in which these non-zero elements occur:

$$\begin{array}{cccccc}
+ & + & 0 & 0 & 0 & + \\
+ & + & 0 & 0 & 0 & 0 \\
+ & + & 0 & 0 & 0 & 0 \\
 & & & & & \\
0 & 0 & + & + & + & 0 \\
0 & 0 & + & + & + & 0 \\
0 & 0 & + & + & + & 0
\end{array} \qquad (5.19)$$

This matrix generally has rank 6, so the rank of $\tilde{J}$ equals 19. The model is hence identified.

Now suppose that ξ_2 instead of ξ_1 is unobservable. In terms of the scheme, this means that, in the last column of (5.19), the "+" moves from the first to the fifth position, introducing a linear dependence between the last four columns. Under this new specification, the model is underidentified.

This example serves to illustrate a few points. First, the identifiability of the model does not only depend on the number of unobservable variables, but also on their location. A measurement error in the first equation does not impair identifiability, since this equation is overidentified when all exogenous variables are measured accurately. This overidentification allows for identification of the measurement error variance of ξ_1. The second equation is just-identified and hence becomes underidentified when one of its exogenous variables cannot be observed.

Second, each exogenous variable occurs in exactly one equation. This means that the last column in the reduced "incidence" matrix in (5.19) contains just a single non-zero element. In such a situation, identification can still be assessed equation by equation. The situation becomes more complicated when a particular unobservable occurs in more than one equation. Then the identifiability of the equations sharing that unobservable becomes intertwined.

Third, the identifiability of the model depends basically on the pattern of zero and non-zero elements in J only. Further information as to their exact value is not needed. (It is assumed that Σ_{XY} and $\Sigma_{\Xi\Xi}$ have full rank and that the a priori information is in the form of exclusions and normalizations.) Note that the pattern of correlations between the ξ's does matter; if say ξ_1 is uncorrelated with ξ_2 and ξ_3, (5.19) becomes:

$$\begin{matrix} + & + & 0 & 0 & 0 & + \\ + & 0 & 0 & 0 & 0 & 0 \\ + & 0 & 0 & 0 & 0 & 0 \\ 0 & 0 & + & 0 & 0 & 0 \\ 0 & 0 & + & + & + & 0 \\ 0 & 0 & + & + & + & 0 \end{matrix} \tag{5.20}$$

where the second and sixth columns are proportional. So, the rank of $\tilde{J}$ is reduced by one. This problem has been noted already when discussing (5.5).

On the basis of the Jacobian, rank and order conditions for identification, both necessary and sufficient, can be derived, and a number of these results have been reported in the literature. They pertain to identification of the complete system as well as to identification of a single equation. Contrary to the situation with simultaneous equations without measurement error, this distinction is not trivial: a certain latent variable may enter into more than one equation, thereby tying together the identification of these equations. This problem occurs even when each latent variable enters into a single equation only, as soon as the measurement errors have a non-zero correlation.

In the first published paper on the problem, Hsiao (1976) presents a number of sufficient conditions for identification of a single equation of the model when the measurement errors are uncorrelated. For the correlated case, he derives on the basis of the Jacobian, a necessary and sufficient rank condition for a single equation, plus a derived necessary order condition. Geraci (1976) uses the Jacobian to derive, for the uncorrelated measurement error case, an "assignment condition" for identification of the complete model. This is a necessary condition, which can be verified solely on the basis of knowledge about the location of the latent variables and the number of overidentifying restrictions on each equation in the case of no measurement error. These "conditional" overidentifying restrictions can be used to identify variances of measurement error of exogenous variables in the equations where the restrictions apply. If it is possible to assign each error variance to a particular equation, the assignment condition is verified. In a recent paper, Geraci (1983) presents rank conditions for individual structural relations, both for a general model, where U and V may be correlated and Ω is non-diagonal, and for the restricted model with Ω diagonal.

Estimation of the simultaneous equations model with latent variables can be done by means of a program for the analysis of covariance structures, like LISREL (see Section 5.3). Under normality, LISREL delivers FIML estimates of the model parameters. (The newer versions of LISREL also have a least-squares option available.)

With the development of LISREL, the scope for alternative estimation methods seems to be limited. There are a few papers that propose other estimators. Geraci (1977) proposes three estimators that are all asymptotically equivalent to FIML but are likely to be simpler to compute. These estimators are based on the GLS approach due to Browne (1974), which leads to a simpler optimization criterion.[14] Hsiao (1976) presents, for the case of uncorrelated measurement error, a FIML estimator based on a transformation of the model, and a single-equation estimator.

5.3. *The analysis of covariance structures*

The breakthrough of latent variable modelling which has taken place in econometrics over the last decade has been accompanied by the availability of successive versions of the computer program LISREL. LISREL is particularly well-suited to deal with systems of linear structural multiple and simultaneous equations ("structural" in the sense of modelling the causal process, *not* as the opposite of functional!). This section describes the model handled by LISREL and discusses the importance for latent variable modelling in econometrics. For a full account, see Jöreskog and Sörbom (1977, 1981). LISREL (Linear Structural Relations—a registered trademark, but we will use the name to denote both the program and the model) is not the only program available, nor is it the most general linear

[14] See Jöreskog and Goldberger (1972) for a clear exposition of GLS vis-à-vis ML in the context of factor analysis.

model; yet its general availability and user-friendliness has made it perhaps the most important tool for handling latent variables at present.

The idea behind LISREL and similar programs is to compare a sample covariance matrix with the parametric structure imposed on it by the hypothesized model. Therefore, this type of analysis is frequently called the 'analysis of covariance structures' [e.g. Jöreskog (1970); see Bentler (1983) for an excellent overview].

The general format of the model to be analyzed by LISREL is as follows, using the notation of the LISREL manual. Let $\boldsymbol{\eta}$ and $\boldsymbol{\xi}$ be $(m\times 1)$ and $(n\times 1)$ vectors of latent dependent and independent variables, respectively, satisfying a system of linear structural relations:

$$B\boldsymbol{\eta}=\Gamma\boldsymbol{\xi}+\boldsymbol{\zeta}, \tag{5.21}$$

with B and Γ $(m\times m)$ and $(m\times n)$ coefficient matrices, B being non-singular, and $\boldsymbol{\zeta}$ an $(m\times 1)$-vector of disturbances. It is assumed that $\boldsymbol{\eta}$, $\boldsymbol{\xi}$ and $\boldsymbol{\zeta}$ have zero expectations, and that $\boldsymbol{\xi}$ and $\boldsymbol{\zeta}$ are uncorrelated. Instead of $\boldsymbol{\eta}$ and $\boldsymbol{\xi}$, $(p\times 1)$ and $(q\times 1)$-vectors $\boldsymbol{y}$ and $\boldsymbol{x}$ are observed such that:

$$\boldsymbol{y}=\Lambda_y\boldsymbol{\eta}+\boldsymbol{\varepsilon} \tag{5.22}$$

and

$$\boldsymbol{x}=\Lambda_x\boldsymbol{\xi}+\boldsymbol{\delta}, \tag{5.23}$$

with Λ_y and Λ_x $(p\times m)$ and $(q\times n)$ coefficient matrices, and $\boldsymbol{\varepsilon}$ and $\boldsymbol{\delta}$ $(p\times 1)$ and $(q\times 1)$ vectors of measurement errors, uncorrelated with $\boldsymbol{\eta}$, $\boldsymbol{\xi}$, $\boldsymbol{\zeta}$ and each other, but possibly correlated among themselves. The vectors $\boldsymbol{y}$ and $\boldsymbol{x}$ are measured as deviations from their means.

Let Φ and Ψ be the covariance matrices of $\boldsymbol{\xi}$ and $\boldsymbol{\zeta}$, respectively, and let Θ_ε and Θ_δ be true variance–covariance matrices of $\boldsymbol{\varepsilon}$ and $\boldsymbol{\delta}$, respectively. Then it follows from the above assumptions that the $(p+q)\times(p+q)$ variance–covariance matrix Σ of $(\boldsymbol{y}',\boldsymbol{x}')'$ is:

$$\Sigma=\begin{pmatrix}\Lambda_y\{B^{-1}(\Gamma\Phi\Gamma'+\Psi)(B')^{-1}\}\Lambda_y'+\Theta_\varepsilon & \Lambda_yB^{-1}\Gamma\Phi\Lambda_x' \\ \Lambda_x\Phi\Gamma'(B')^{-1}\Lambda_y' & \Lambda_x\Phi\Lambda_x'+\Theta_\delta\end{pmatrix} \tag{5.24}$$

The parameters occurring in Σ $(\Lambda_y, \Lambda_x, B, \Gamma, \Phi, \Psi, \Theta_\varepsilon, \Theta_\delta)$ are estimated on the basis of the $((p+q)\times(p+q))$-matrix S of second sample moments of $\boldsymbol{x}$ and $\boldsymbol{y}$. In order to render the model identified, restrictions on the parameters have to be imposed. LISREL can handle two kinds of restrictions: first, parameters may be set equal to other parameters. Given these restrictions and the structure that (5.24) imposes on the data, LISREL computes estimates of the parameters. These estimates are the FIML estimates when (y', x') is normally distributed, i.e. the criterion:

$$\ln|\Sigma| + \operatorname{tr}(S\Sigma^{-1}), \tag{5.25}$$

is minimized. (As mentioned above, newer versions have a least-squares option.)

The identification of the parameters is checked numerically by LISREL. It evaluates the information matrix on the basis of the starting values for the parameters in the iterations; when it is not positive definite, this is an indication of underidentification.

By imposing appropriate restrictions, the LISREL model reduces to any one of a number of well-known models. For instance, it is easy to see how (5.24) reduces to the FA model—one simply has to impose sufficient restrictions to retain only the part $\Gamma\Phi\Gamma' + \Psi$ in the NW-corner. From (5.21), the reduction to simultaneous equations is apparent. For a reduction to an econometric model, it is desirable to take $\boldsymbol{x}$ fixed, i.e. the analysis takes place conditional on $\boldsymbol{x}$. This is imposed by specifying $\Lambda_x = I$, $\Theta_\delta = 0$, and $\Phi = S_{xx}$, with S_{xx} the sample covariance matrix of $\boldsymbol{x}$. Measurement error in $\boldsymbol{x}$ is introduced by relaxing $\Theta_\delta = 0$.

Some *limitations* apply to the use of LISREL. It is limited to linear structures and it assumes independence of observations, rendering it unfit for the analysis of dynamic models, except some simple ones [Jöreskog (1978)]. The LISREL model is restricted in several ways and many extensions can be thought of; see, for example, Bentler and Weeks (1980) for a multi-level extension, Lee (1980) for a model with inequality constraints, and Muthén (1979) for an extension to probit analysis. A minor *caveat* applies to the numerical assessment of the identifiability of a particular model; an unfortunate choice of starting values may accidentally reduce the rank of the information matrix, as computed on the basis of these values (nothing can beat analytic insight, but the easy use of LISREL does not stimulate this). When the program indicates underidentification, it may still be difficult to indicate the troubled part of the model. Moreover, care must be taken in interpreting the goodness-of-fit of the model [Fornell (1983)]. Finally, the assumed multinormality may be troublesome, although the results by Kapteyn and Wansbeek (1981) suggest that the normality assumption regarding the latent variables can be replaced by functional assumptions without changing the asymptotic distribution of the estimators.

The B. Hall MOMENTS program (1979) generalizes LISREL and is no doubt more easily understood by economists, though it requires a more detailed specification by the user. Several recent applications attest to its usefulness.

6. Dynamic models

As discussed in previous sections, when variables are measured with error, an otherwise identified contemporaneous model may become unidentified in the sense that no consistent estimator of the parameter exists (see Section 2). But if a model contains a dynamic structure, whether in the form of a description of dynamic behavioral relations, or in the form of serially correlated exogenous variables, measurement error need not affect the identifiability of a model.

In this section we shall briefly illustrate how different dynamic assumptions affect the identification and estimation of a model.

We assume that all the variables are weakly stationary in the sense that the covariance sequence $\mathrm{E}\, y_i y_{i-s} = \sigma_{yy}(i-s)$ depends only upon $i-s$ and not upon i.[15] Since most estimation methods use second order quantities, we shall consider the problem of identification in terms of the covariances only.[16] We assume that the second-order moments of observables are estimable, thus, we shall assume that they are known precisely and ask what additional restrictions are required in order that the parameters of a model should be uniquely determined by these covariances. More detailed analysis of different dynamic models is contained in Engle (1974, 1980), Hannan (1963), Hsiao (1977, 1979), Hsiao and Robinson (1978), Maravall (1979), Maravall and Aigner (1977), and Nicholls, Pagan and Terrell (1975).

6.1. *Identification of single-equation models*

We first consider the problem of identification of a univariate process. We use a simple model to illustrate the effects of each one of the different dynamic assumptions, and then state the theorem for the general case. For details of the proof, see Maravall (1979) and Maravall and Aigner (1977).

Consider the following dynamic model:

$$\eta_i + \beta_i \eta_{i-1} + \cdots + \beta_p \eta_{i-p} + \gamma_0 \xi_i + \gamma_1 \xi_{i-1} + \cdots + \gamma_q \xi_{i-q} = \varepsilon_i, \tag{6.1}$$

[15] For a possible generalization of the results in this section to non-stationary cases, see, for example, Hannan (1971) and Maravall (1979).

[16] As discussed before, when variables are normally distributed all the information is contained in the first and second moments. When variables are not normally distributed, additional information may be contained in higher moments, which can be used to identify and estimate unknown parameters.

where ξ_i and ε_i are independent, and the roots of

$$|1+\beta_1 L+\cdots+\beta_p L^p|=0,$$
$$|\gamma_0+\gamma_1 L+\cdots+\gamma_q L^q|=0, \tag{6.2}$$

are greater than one in absolute value and the two sets have no roots in common.[17] The endogenous and exogenous variables, η_i and ξ_i, are assumed to be measured with error, according to (2.2) and

$$y_i=\eta_i+u_i, \tag{6.3}$$

where u_i is white noise with mean zero and constant variance σ_{uu}.

For simplicity, we shall for the moment assume that ε_i and ξ_i are white noise. Then:

$$y_i+\beta_1 y_{i-1}+\cdots+\beta_p y_{i-p}+\gamma_0 x_i+\cdots+\gamma_q x_{i-q}=w_i, \tag{6.4}$$

where

$$w_i=\varepsilon_i+u_i+\beta_1 u_{i-1}+\cdots+\beta_p u_{i-p}+\gamma_0 v_i+\cdots+\gamma_q v_{i-q}$$

will have the property:

$$\begin{aligned}
\sigma_{ww}(0)&\equiv \operatorname{var}(w_i)=\sigma_{\varepsilon\varepsilon}+\sigma_{uu}\left(1+\beta_1^2+\cdots+\beta_p^2\right)+\sigma_{vv}\left(\gamma_0^2+\gamma_1^2+\cdots+\gamma_q^2\right),\\
\sigma_{ww}(1)&\equiv \operatorname{cov}(w_i w_{i-1})=\sigma_{uu}\left(\beta_1+\beta_1\beta_2+\cdots+\beta_{p-1}\beta_p\right)\\
&\qquad+\sigma_{vv}\left(\gamma_0\gamma_1+\cdots+\gamma_{q-1}\gamma_q\right),\\
\sigma_{ww}(2)&\equiv \operatorname{cov}(w_i w_{i-2})=\sigma_{uu}\left(\beta_2+\beta_1\beta_3+\cdots+\beta_{p-2}\beta_p\right)\\
&\qquad+\sigma_{vv}\left(\gamma_0\gamma_2+\cdots+\gamma_{q-2}\gamma_q\right),\\
&\ \vdots\\
\sigma_{ww}(s)&\equiv \operatorname{cov}(w_i w_{i-s})=0,\quad \text{for } |s|>\max(p,q)\equiv\tau.
\end{aligned} \tag{6.5}$$

As this is the covariance function of a τth order moving average process, all information about the unknown parameters is contained in the variance and first

[17] For the generalization of results contained in this section to the non-stationary case, see Maravall (1979).

τ autocovariances of w_i. Thus, by (6.4) we know that the distribution of y is determined by the $0,1,\dots,\tau+p$, autocovariances of y and the $0,1,\dots,q$, cross-covariances between y and x.

From (6.1) we know that these cross- and autocovariances satisfy:

$$\begin{aligned}
&\sigma_{yx}(0)+\gamma_0\left[\sigma_{xx}(0)-\sigma_{vv}\right]=0,\\
&\sigma_{yx}(1)+\beta_1\sigma_{yx}(0)+\gamma_1\left[\sigma_{xx}(0)-\sigma_{vv}\right]=0,\\
&\vdots\\
&\sigma_{yx}(q)+\cdots+\beta_p\sigma_{yx}(q-p)+\gamma_q\left[\sigma_{xx}(0)-\sigma_{vv}\right]=0;
\end{aligned}\tag{6.6}$$

$$\begin{aligned}
&\left(\sigma_{yy}(0)-\sigma_{uu}\right)+\beta_1\sigma_{yy}(-1)+\cdots+\beta_p\sigma_{yy}(-p)+\gamma_0\sigma_{xy}(0)+\gamma_1\sigma_{xy}(-1)\\
&\qquad+\cdots+\gamma_q\sigma_{xy}(-q)=\sigma_{\varepsilon\varepsilon},\\
&\sigma_{yy}(1)+\beta_1\left[\sigma_{yy}(0)-\sigma_{uu}\right]+\cdots+\beta_p\sigma_{yy}(-p+1)+\gamma_0\sigma_{xy}(1)+\gamma_1\sigma_{xy}(0)\\
&\qquad+\cdots+\gamma_q\sigma_{xy}(-q+1)=0,\\
&\vdots\\
&\sigma_{yy}(\tau)+\beta_1\sigma_{yy}(\tau-1)+\cdots+\gamma_q\sigma_{xy}(-q+\tau)=0;
\end{aligned}\tag{6.7}$$

and

$$\begin{aligned}
&\sigma_{yy}(\tau+1)+\beta_1\sigma_{yy}(\tau)+\cdots+\beta_p\sigma_{yy}(\tau-p+1)=0,\\
&\sigma_{yy}(\tau+2)+\beta_1\sigma_{yy}(\tau+1)+\cdots+\beta_p\sigma_{yy}(\tau-p+2)=0,\\
&\vdots\\
&\sigma_{yy}(\tau+p)+\beta_1\sigma_{yy}(\tau+p-1)+\cdots+\beta_p\sigma_{yy}(\tau)=0;
\end{aligned}\tag{6.8}$$

where $\sigma_{yx}(j)=\sigma_{xy}(-j)=0$ for $j<0$.

The Jacobian of (6.6), (6.7), and (6.8) is of the form:

$$J=\begin{pmatrix} J_1 \\ \cdots\cdots \\ J_2 \\ \cdots\cdots \\ J_3 \,\vdots\, 0 \end{pmatrix},\tag{6.9}$$

where J_1, J_2, and J_3 are the partial derivatives of (6.6), (6.7), and (6.8) with respect to β's, γ's and σ_{vv}, σ_{uu}, $\sigma_{\varepsilon\varepsilon}$ and J_1 and J_2 have $p+q+4$ columns while J_3 has just p.

Defining a $(p\times(p+q+4))$ matrix $(J',\Xi')'$, it is easy to see that rank$(J)=$ rank $\begin{pmatrix} J \\ \Xi \end{pmatrix}$. It follows that the autoregressive parameters $\beta_1,\beta_2,\dots,\beta_p$ are identified. (For details of the identification conditions relying on the rank of a Jacobian matrix, see Chapter 4 in this Handbook by Hsiao.)

The Jacobian J_1 is of the form:

$$J_1=\begin{pmatrix} 0 & \cdots & 0 & \sigma_{xx}-\sigma_{vv} & 0 & \cdots & 0 & -\gamma_0 & 0 & 0 \\ \sigma_{yx}(0) & \cdots & 0 & 0 & \sigma_{xx}-\sigma_{vv} & \cdots & 0 & -\gamma_1 & 0 & 0 \\ & \cdots & \cdot & \cdot & \cdot & \cdots & \cdot & \cdot & \cdot & \cdot \\ \sigma_{yx}(q-1) & \cdots & \sigma_{yx}(q-p) & \cdot & \cdot & \cdots & \sigma_{xx}-\sigma_{vv} & -\gamma_q & 0 & 0 \end{pmatrix}. \tag{6.10}$$

The Jacobian J_2 is of the form:

$$J_2=\begin{pmatrix} \sigma_{yy}(-1) & \cdot & \cdots & \sigma_{yy}(-p) & \sigma_{xy}(0) & \sigma_{xy}(-1) & \cdot & \cdot & \sigma_{xy}(-q) & 0 & -1 & -1 \\ \sigma_{yy}(0)-\sigma_{vv} & \cdot & \cdots & \sigma_{yy}(-p+1) & 0 & \sigma_{xy}(0) & \cdot & \cdot & \sigma_{xy}(1-q) & 0 & -\beta_1 & 0 \\ \sigma_{yy}(1) & \sigma_{yy}(0)-\sigma_{uu} & \cdots & \sigma_{yy}(-p+2) & 0 & 0 & \sigma_{xy}(0) & \cdot & \sigma_{xy}(2-q) & 0 & -\beta_2 & 0 \\ \cdot & \cdot & \cdots & \cdot & \cdot & \cdot & \cdot & \cdot & \cdot & \cdot & \cdot & \cdot \\ \sigma_{yy}(p-1) & \cdot & \cdots & \sigma_{yy}(0)-\sigma_{uu} & 0 & \cdot & \cdot & \cdot & \sigma_{xy}(p-q) & 0 & -\beta_p & 0 \\ \sigma_{yy}(p) & \cdot & \cdots & \sigma_{yy}(1) & 0 & 0 & 0 & \cdot & \sigma_{xy}(p+1-q) & 0 & 0 & 0 \\ \sigma_{yy}(p+1) & \cdot & \cdots & \cdot & 0 & \cdot & \cdot & \cdot & \sigma_{xy}(p+2-q) & 0 & 0 & 0 \\ \cdot & \cdot & \cdots & \cdot & \cdot & 0 & 0 & \cdot & \cdot & 0 & 0 & 0 \end{pmatrix} \tag{6.11}$$

If $p=1$ and $q\le 1$ by elementary row and column operations we have rank $(J)=q+4$. The total number of parameters are $q+5$, therefore the complete model is not identifiable. On the other hand, if $q\ge 2$ or $p\ge 2$ and $q\ge 1$, rank $(J)=p+q+4$. Model (6.1) is locally identified if and only if either (i) $p\ge 2$ and $q\ge 1$ or (ii) $p\ge 1$ and $q\ge 2$.

Generalizing this result to the model:

$$\eta_i+\beta_1\eta_{i-1}+\cdots+\beta_p\eta_{i-p} + \sum_{k=1}^{K}\left(\gamma_{k0}\xi_{ki}+\gamma_{k1}\xi_{k,i-1}+\cdots+\gamma_{kq_k}\xi_{k,i-q_k}\right)=\varepsilon_k, \tag{6.12}$$

where $x_{ki}=\xi_{ki}+v_{ki}$, and ξ_{ki}, v_{ki}, are mutually independent white noises, Maravall (1979), Maravall and Aigner (1977) obtain the following result:

If the $(K+1)$ integers, $p,q_1,\dots,q_K$, are arranged in increasing order (ties are immaterial), and q_j^* denotes the one occupying the jth place in this new sequence,

(6.12) is locally identified if and only if $q_j^* \geq j$, for $j = 1, 2, \dots, K+1$.

When the shocks ε_i are serially correlated, the above results on identification will have to be modified. We first consider the case where ε_i is a sth order moving average process, $\varepsilon_i = a_i + \theta_1 a_{i-1} + \cdots + \theta_s a_{i-s}$, where $\theta_s \neq 0$ and a_i is white noise with mean zero and variance σ_{aa}.

The j-lag autocovariances of ε_i will be equal to zero for $j > s$. In other words, the $s+1$ unknown parameters $\theta_1, \dots, \theta_s$, and σ_{aa} only appear in the variance and first s-lag autocovariances of y. If other parameters of the model are identified, the autocovariance functions of y can be rewritten in terms of θ's and σ_{aa} as in the case of standard sth order moving average process. Thus a unique solution for them exists [for details, see Maravall (1979)]. However, if the variance and first s autocovariance functions of y are used to identify this set of parameters, it means that we have $(s+1)$ less equations to identify other parameters.[18] Assuming ε_i to be a sth order moving average process (6.12) is locally identified if and only if $q_j^* \geqq j + s, j = 1, 2, \dots, K+1$.

Alternatively, suppose we assume that the shocks, ε_i, follow a stationary rth order autoregressive process, $\varepsilon_i = \rho_1 \varepsilon_{i-1} + \cdots + \rho_r \varepsilon_{i-r} + a_i$. As we can see from (6.1), under this assumption the autocovariance functions of y alone can no longer be used to identify β. However, β can still be identified by the cross-covariance functions:

$$\sigma_{yx}(j) + \beta_1 \sigma_{yx}(j-1) + \cdots + \beta_p \sigma_{yx}(j-p) = 0, \tag{6.13}$$

for $j > \max(p, q)$ [or see (6.11)]. We also note that for $j > \max(p, q) + r$, the autocovariance function of y is:

$$\sigma_{yy}(j) + \beta_1 \sigma_{yy}(j-1) + \cdots + \beta_p \sigma_{yy}(j-p) = \sigma_{\varepsilon y}(j), \tag{6.14}$$

where

$$\sigma_{\varepsilon y}(j) = \rho_1 \sigma_{\varepsilon y}(j-1) + \cdots + \rho_r \sigma_{\varepsilon y}(j-r). \tag{6.15}$$

Once the β's are identified by (6.13), $\sigma_{\varepsilon y}(j)$ is identified also by (6.14). Therefore, the ρ's are identifiable by (6.15); hence σ_{aa}. Thus, contrary to the case of white noise shocks when $\sigma_{\varepsilon\varepsilon}$ only appears in the 0-lag autocovariance equations, now $\sigma_{\varepsilon\varepsilon}$ can be identified through the j-lag autocovariance equations of y when $j > \max(p, q)$. In a way, the autoregressive shocks help to identify a model by reducing the number of unknowns by one. Assuming ε_i to be a stationary rth

[18]Note now that this parameter set no longer needs to include $\sigma_{\varepsilon\varepsilon}$, which can be identified from θ's and σ_{aa}.

order autoregressive process, model (6.12) is locally identified if and only if $q_j^* \geq j-1$, $j=1,\dots,K+1$.

Combining these two results we have the general result with regard to autocorrelated shocks. If the ε_i follow a stationary autoregressive moving average process of order r and s, $\varepsilon_i = \rho_1\varepsilon_{i-1} + \cdots + \rho_r\varepsilon_{i-r} + a_i + \theta_1 a_{i-1} + \cdots + \theta_s a_{i-s}$, we have that model (6.12) is locally identified if and only if (a) when $r > s$, $q_j^* \geq j-1$; (b) when $r \leq s$, $q_j^* \geq j+s-r$ for $j=1,\dots,K+1$.

These results are based on the assumption that the exogenous variables are serially and mutually uncorrelated. If they are correlated, additional information will be available in the cross- and autocovariance functions of the y's and x's, and hence conditions for identification may be relaxed.

The main reason that a dynamic structure helps in identifying a model is because of our strong assumption that measurement errors are uncorrelated. This assumption means that cross- and autocovariances of the observed variables equal the corresponding ones of the unobserved variables. When measurement errors are autocorrelated, the problem becomes very complicated. For some examples, see Maravall (1979) and Nowak (1977).

6.2. *Identification of dynamic simultaneous equation models*

We have seen how a dynamic structure may affect the identification of a single equation model. The basic idea carries through to the dynamic simultaneous equation model. However, the problem is complicated by the interrelationships among variables, which means in general that stronger conditions are required than in the single equation model to ensure the proper rank of the Jacobian. We illustrate the problem by considering the following simple model:

$$B_0\boldsymbol{\eta}_i + B_1\boldsymbol{\eta}_{i-1} + \Gamma\boldsymbol{\xi}_i = \boldsymbol{\varepsilon}_i, \tag{6.16}$$

where $\boldsymbol{\eta}$ and $\boldsymbol{\xi}$ are $(G\times 1)$ and $(K\times 1)$ vectors of jointly dependent variables and exogenous variables, respectively; $\boldsymbol{\varepsilon}$ is a $(G\times 1)$ vector of disturbance terms with covariance matrix Σ. We assume that B_0 is non-singular and that the roots of $B_0 + B_1 L = 0$ lie outside the unit circle. We again assume that the exogenous variables $\boldsymbol{\xi}$ are stationary and disturbance $\boldsymbol{\varepsilon}$ is white noise. The $\boldsymbol{\eta}$ and $\boldsymbol{\xi}$ are unobservable. They are related to observable $\boldsymbol{y}$ and $\boldsymbol{x}$ by:

$$\boldsymbol{y}_i = \boldsymbol{\eta}_i + \boldsymbol{u}_i, \qquad \text{with } \mathrm{E}\boldsymbol{u}_i\boldsymbol{u}_i' = \Lambda, \quad \mathrm{E}\boldsymbol{u}_i\boldsymbol{\varepsilon}_j' = 0, \tag{6.17}$$

and (2.2).

Since the measurement errors are assumed to be serially uncorrelated, we know that $C_{yy}(\tau)$ and $C_{xx}(\tau)$ satisfy:

$$C_{yy}(0)=\mathrm{E}\boldsymbol{y}_i\boldsymbol{y}_i'=\mathrm{E}(\boldsymbol{\eta}_i+\boldsymbol{u}_i)(\boldsymbol{\eta}_i+\boldsymbol{u}_i)'=C_{\eta\eta}(0)+\Lambda,$$
$$C_{xx}(0)=\mathrm{E}\boldsymbol{x}_i\boldsymbol{x}_i'=\mathrm{E}(\boldsymbol{\xi}_i+\boldsymbol{\varepsilon}_i)(\boldsymbol{\xi}_i+\boldsymbol{\varepsilon}_i)'=C_{\xi\xi}(0)+\Omega; \tag{6.18}$$

and

$$C_{yy}(\tau)=\mathrm{E}\boldsymbol{y}_i\boldsymbol{y}_{i-\tau}'=\mathrm{E}\boldsymbol{\eta}_i\boldsymbol{\eta}_{i-\tau}=C_{\eta\eta}(\tau),$$
$$C_{xx}(\tau)=\mathrm{E}\boldsymbol{x}_i\boldsymbol{x}_{i-\tau}'=\mathrm{E}\boldsymbol{\xi}_i\boldsymbol{\xi}_{i-\tau}'=C_{\xi\xi}(\tau),\quad \text{for } \tau\neq 0.$$

Thus, the second-order moments satisfy:

$$B_0C_{yy}(1)+B_1\big(C_{yy}(0)-\Lambda\big)+\Gamma C_{xy}(1)=0, \tag{6.19}$$
$$B_0C_{yx}(0)+B_1C_{yx}(-1)+\Gamma\big(C_{xx}(0)-\Omega\big)=0; \tag{6.20}$$

and

$$B_0C_{yy}(\tau)+B_1C_{yy}(\tau-1)+\Gamma C_{xy}(\tau)=0,\qquad \tau=2,3,\dots, \tag{6.21}$$
$$B_0C_{yx}(\tau)+B_1C_{yx}(\tau-1)+\Gamma C_{xx}(\tau)=0,\qquad \tau=1,2,\dots. \tag{6.22}$$

We stack (B_0, B_1, Γ) into a $(1\times(2G^2+GK))$-vector $\boldsymbol{\lambda}'$ and assume that they satisfy R linear restrictions:

$$\Phi\boldsymbol{\lambda}=\mathbf{0}, \tag{6.23}$$

where Φ is an $(R\times(2G^2+GK))$-matrix with known elements. Let $\boldsymbol{\chi}'$ and $\boldsymbol{\omega}'$ denote the $(1\times n)$ and $(1\times l)$ vectors consisting of unknown elements of Λ and Ω. Letting $\boldsymbol{\alpha}'=(\boldsymbol{\lambda}',\boldsymbol{\chi}',\boldsymbol{\omega}')$, then $\boldsymbol{\alpha}'$ has to satisfy (6.19)–(6.23). Now we know that the $1\times(2G^2+GK+n+l)$ parameter vector $\boldsymbol{\alpha}'$ is locally identified if and only if the Jacobian

$$J=\left[\begin{array}{ccc:c:c}
 & \Phi & & 0 & 0\\
\hdashline
I_G\otimes C_{yy}(1) & I_G\otimes\big(C_{yy}(0)-\Lambda\big) & I_G\otimes C_{xy}(1) & H & 0\\
I_G\otimes C_{yy}(2) & I_G\otimes C_{yy}(1) & I_G\otimes C_{xy}(2) & 0 & 0\\
I_G\otimes C_{yy}(3) & I_G\otimes C_{yy}(2) & I_G\otimes C_{xy}(3) & 0 & 0\\
\vdots & \vdots & \vdots & \vdots & \vdots\\
\hdashline
I_G\otimes C_{yx}'(0) & I_G\otimes C_{yx}'(-1) & I_G\otimes\big(C_{xx}(0)-\Omega\big) & 0 & U\\
I_G\otimes C_{yx}'(1) & I_G\otimes C_{yx}'(0) & I_G\otimes C_{xx}(1) & 0 & 0\\
\vdots & \vdots & \vdots & \vdots & \vdots
\end{array}\right] \tag{6.24}$$

has rank $(2G^2 + GK + n + l)$ around its true value, where H is a $(G^2 \times n)$ matrix whose elements are either zero or elements of B_1, and U is a $(GK \times l)$ matrix whose elements are either zero or elements of Γ.

Unfortunately, this condition is usually difficult to check in practice. If we know that the matrix

$$\begin{bmatrix} C_{yy}(1) & C_{xy}(2) \\ C_{yy}(2) & C_{xy}(3) \\ \vdots & \vdots \\ \hdashline C'_{yx}(0) & C_{xx}(1) \\ C'_{yx}(1) & C_{xx}(2) \\ \vdots & \vdots \end{bmatrix} \tag{6.25}$$

has rank $(G+K)$, the G independent columns of $(B_0, B_1, \Gamma)'$ will form a basis of the column kernel of the transposes of (6.21) and (6.22). Then by an argument similar to Fisher's (1966), we can show that the usual order and rank conditions are necessary and sufficient to identify α'. However, because of the interrelation among G different variables, we need a stronger condition than the univariate case (Section 6.1) to ensure the rank of (6.25). Using the result of Hannan (1975, 1976) we know that one such condition is[19] to assume that B_1 is non-singular and that $C_{xi}(1)$ is non-singular, $C_{xi}(q) = 0$ for some $q \geq 2$.

Under these assumptions, the matrix

$$\begin{bmatrix} C_y(q) & \vdots & 0 \\ \hdashline C'_{yx}(0) & \vdots & C_x(1) \end{bmatrix} \tag{6.26}$$

has rank $(G+K)$, where matrix (6.26) is a submatrix of (6.25). Therefore, we have a necessary and sufficient condition to locally identify the coefficients of (6.16) is that rank $(M\Phi') = G^2$, where $M = (I_G \otimes B_0, I_G \otimes B_1, I_G \otimes \Gamma)$.

If instead of assuming ε_i to be serially uncorrelated, we assume that ε_i is stationary, then $C_y(\tau)$ for $\tau \geq 1$ no longer satisfies (6.19) and (6.21) and Λ is not identifiable. Now the parameter (λ', ω') is locally identified if and only if the rank

$$\begin{bmatrix} & \Phi & & \vdots & 0 \\ \hdashline I_G \otimes C'_{yx}(0) & I_G \otimes C'_{yx}(-1) & I_G \otimes (C_{xx}(0) - \Omega) & \vdots & U \\ I_G \otimes C'_{yx}(1) & I_G \otimes C'_{yx}(0) & I_G \otimes C_{xx}(1) & \vdots & 0 \\ \vdots & \vdots & & \vdots & \vdots \end{bmatrix} \tag{6.27}$$

is $2G^2 + GK + L$ around its true value.

[19]For other conditions, see Hsiao (1977).

Again, the rank of (6.27) is not easy to check. However, under certain conditions [Hsiao (1979)] the matrix

$$\begin{bmatrix} C'_{yx}(0) & C_x(1) \\ C'_{yx}(1) & C_x(2) \\ \vdots & \vdots \end{bmatrix} \tag{6.28}$$

has rank $(G+K)$, and hence the usual order and rank condition is necessary and sufficient to identify (6.16).

6.3. *Estimation of dynamic error–shock models*

Most literature on dynamic error–shock models deals with the identification problem only. Of course, if a model is identified, the unknown coefficients can be consistently estimated by solving the cross- and autocovariance equations. However, such a method is not efficient. In fact, it appears that an efficient, yet computationally simple estimation principle for a general error–shock model remains to be worked out. We shall in this section sketch some approaches to obtaining efficient estimates as background information for the development of future numerical studies.

We first consider the case where only dependent variables are observed with error (i.e. $\sigma_{vv}=0$, and $\boldsymbol{x}=\boldsymbol{\xi}$). As shown in Section 6.2 a dynamic model (6.1) under certain assumptions can be rewritten as a dynamic model with a moving average disturbance term (ARMAX). Many people have suggested methods for estimating ARMAX models [e.g. see Box and Jenkins (1970) and Phillips (1966) for time domain approaches and Hannan and Nicholls (1972) for a frequency domain approach; also see Nicholls, Pagan and Terrell (1975) for a survey]. However, such methods, although they remain consistent, are no longer efficient because they ignore the restrictions in the composite disturbance term. An efficient estimation method would have to take into account all prior restrictions. Unfortunately, the prior restrictions in this case are very complex and difficult to incorporate.

We illustrate this point by considering a simple case of (6.1) where $p=1$ and $q=0$. Now we have:

$$\eta_i+\beta\eta_{i-1}+\gamma\xi_i=\varepsilon_i, \tag{6.29}$$

$$y_i=\eta_i+u_i \tag{6.30}$$

and

$$x_i=\xi_i. \tag{6.31}$$

Rewriting (6.29) in terms of observables, we have:

$$y_i + \beta y_{i-1} + \gamma \xi_i = \varepsilon_i + u_i + \beta u_{i-1} = w_i. \tag{6.32}$$

Assuming ε_i to be white noise, the composite disturbance term w_i has variance and autocovariances:

$$\begin{aligned} \sigma_{ww}(0) &= \sigma_{\varepsilon\varepsilon} + (1+\beta^2)\sigma_{uu}, \\ \sigma_{ww}(1) &= \beta\sigma_{uu}, \\ \sigma_{ww}(j) &= 0, \quad \text{for } j \geq 2. \end{aligned} \tag{6.33}$$

Clearly, this has the property of a first order MA process. Establishing the equivalences:

$$w_i = \phi_i + \psi\phi_{i-1} = \varepsilon_i + u_i + \beta u_{i-1}, \tag{6.34}$$

we can solve for values of ψ, which are

$$\psi = (2\beta)^{-1}\left\{-(1+\mu+\beta^2) + \left[(1+\mu+\beta^2)^2 - 4\beta^2\right]^{1/2}\right\}, \tag{6.35}$$

where $\mu = \sigma_{\varepsilon\varepsilon}/\sigma_{uu}$. We choose the root which is greater than unity as the solution.

It is clear from this example that the restrictions are highly non-linear, and arise as the solution of the roots of a polynomial. It is not an easy matter to impose the requisite restrictions. Generally, it is impossible to derive an analytical solution for models with composite disturbance terms. Pagan (1973) has, therefore, resorted to numerical alternatives in order to obtain efficient estimates.[20]

Let $\boldsymbol{\alpha}$ denote the $m \times 1$ unknown parameters. To obtain an estimated $\boldsymbol{\alpha}$, Pagan (1973) adopts the Phillips/Box–Jenkins methodology by minimizing $\sum_{i=1}^{n}\phi_i^2$ with respect to $\boldsymbol{\alpha}$ with the aid of the Gauss–Newton algorithm, leading to the following iterative formula:

$$\hat{\boldsymbol{\alpha}}^{(j)} - \hat{\boldsymbol{\alpha}}^{(j-1)} = -\left(\frac{\partial\hat{\boldsymbol{\phi}}'}{\partial\boldsymbol{\alpha}}\frac{\partial\hat{\boldsymbol{\phi}}}{\partial\boldsymbol{\alpha}'}\right)^{-1}_{|\boldsymbol{\alpha}=\hat{\boldsymbol{\alpha}}^{(j-1)}} \frac{\partial\hat{\boldsymbol{\phi}}'}{\partial\boldsymbol{\alpha}}_{|\boldsymbol{\alpha}=\hat{\alpha}^{(j-1)}} \hat{\boldsymbol{\phi}}_{|\alpha=\hat{\alpha}^{(j-1)}}, \tag{6.36}$$

[20] Bar-Shalom (1972) has suggested a computationally simpler iterative scheme which involves solving the likelihood function as a system of non-linear equations with the parameters and unobservables η. The system of non-linear equations is then separated into two interconnected linear problems, one for the η, the other for the parameters. Besides the problem of the non-existence of the MLE in his approach, it is dubious that his method will have good convergence properties, although he did report so in his numerical examples.

where $\hat{\boldsymbol{\phi}}'$ denotes the disturbance vector $(\hat{\phi}_1,\dots,\hat{\phi}_n)$. Thus, the problem is shifted to one of computing derivatives.

Of course, to complete the algorithm we need to specify the process for determining $\boldsymbol{\psi}$ given $\boldsymbol{\alpha}$. One possibility would be to solve for the roots of the covariance generating function. However, Pagan (1973) reports that this approach revealed computational difficulties if the order of the moving average process was high. Hence, Wilson's (1969) method for factoring a covariance function into its moving average form was adopted.

The global minimum solution of the Pagan's (1973) method is asymptotically equivalent to that of the maximum likelihood method, and hence is consistent and asymptotically normally distributed. However, there is no guarantee that the convergent solution is a global minimum. Therefore it is advisable to start the iteration from a consistent estimate and perform a number of experiments with other starting values.

When exogenous variables are also measured with error (i.e. $x_i = \xi_i + v_i$ and $v_i \neq 0$), Pagan's (1973) method cannot be applied and neither can the iterative schemes suggested by Aoki and Yue (1970), Cox (1964), Levin (1964), Ljung (1977), etc. The main problem appears to be the correlation between the measured exogenous variables and the composite disturbance terms. If there is prior knowledge that measurement errors appear only at some frequencies [e.g. higher frequencies, Engle and Foley (1975)], or in other words that only a portion of the spectrum satisfies the model, Engle (1974, 1980) and Hannan (1963) have suggested a band spectrum approach. We illustrate their approach by considering model (6.29).

The spectrum approach to estimating $\boldsymbol{\alpha}$ involves first transforming the model by the $(n\times n)$ unitary matrix A with the j,lth element equal to:

$$A_{jl} = \frac{1}{\sqrt{2\pi n}}\exp\left[\zeta\left(\frac{2\pi j}{n}\right)l\right] = \frac{1}{\sqrt{2\pi n}}\exp(\zeta t_j l), \tag{6.37}$$

where $\zeta = \sqrt{-1}$. Ignoring the end-effects which are of order $1/\sqrt{n}$, we can write the log-likelihood function of (6.29) as:

$$L = -\frac{n}{2}\log 2\pi - \tfrac{1}{2}\sum_j \log f_w(t_j) - \tfrac{1}{2}\sum_j f_w(t_j)^{-1} I_w(t_j) + \sum_j \log(1+\beta e^{\zeta t_j}), \tag{6.38}$$

where $f_w(t_j)$ denotes the spectral density of w at frequency t_j:

$$\begin{aligned} 2\pi n I_w(t_j) &= \left[(1+\beta e^{\zeta t_j})y(t_j)+\gamma x(t_j)\right]\left[(1+\beta e^{\zeta t_j})y(t_j)+\gamma x(t_j)\right]^{\dagger}, \\ y(t_j) &= \frac{1}{\sqrt{2\pi n}}\sum_{i=1}^{n} y_i e^{\zeta t_j i}, \qquad x(t_j) = \frac{1}{\sqrt{2\pi n}}\sum_{i=1}^{n} x_i e^{\zeta t_j i}, \end{aligned} \tag{6.39}$$

and † denotes the complex conjugate of the transpose. Maximizing (6.38) with respect to unknowns we obtain the (full) spectrum estimates.

If only a subset of the full spectrum, say S, is assumed to satisfy the model, we can maximize (6.38) with respect to this set of frequencies, which leads to the estimator:

$$\begin{bmatrix}\hat{\beta}\\ \hat{\gamma}\end{bmatrix} = -\left\{\sum_{j\in s} f_w(t_j)^{-1}\begin{bmatrix} f_y(t_j)-\dfrac{f_w(t_j)}{1+2\beta\cos t_j+\beta^2} & f_{yx}(t_j)\mathrm{e}^{\zeta t_j}\\ f_{xy}(t_j)\mathrm{e}^{-\zeta t_j} & f_x(t_j)\end{bmatrix}\right\}^{-1}$$

$$\times\left\{\sum_{j\in s} f_w(t_j)^{-1}\begin{bmatrix} f_y(t_j)-\dfrac{f_w(t_j)}{1+2\beta\cos t_j+\beta^2}\mathrm{e}^{\zeta t_j}\\ f_{xy}(t_j)\end{bmatrix}\right\}, \tag{6.40}$$

where $f_{yx}(t_j)$ denotes the cross-spectral density between y and x. Under the assumption of smoothness of the spectral density, it can be shown that the band spectrum estimate (6.40) is consistent if a consistent estimate of f_w is available. One way to obtain a consistent estimate of f_w is by substituting a consistent estimate of β into:

$$f_w(t_j) = |1+\beta \mathrm{e}^{\zeta t_j}|^2\left(f_y(t_j) - f_{yx}(t_j) f_x(t_j)^{-1} f_{xy}(t_j)\right). \tag{6.41}$$

The band spectrum approach has the advantages that no explicit assumptions about the autocovariance structure of the measurement error are needed, and that it is somewhat easier computationally. However, the portion of the frequency band with a small signal-to-noise ratio may be rather large, and so if all these frequencies are omitted the resulting estimate may have a rather large variance. In particular, we have been assuming that the measurement error has a uniform spectrum (white noise) which may imply that there is no frequency for which the signal-to-noise ratio is really large. Also, there may be a problem in knowing S. A full spectrum method thus may be more desirable. Hannan (1963) has suggested such an approach for the case where no measurement error appears in y (i.e. $y_i = \eta_i$ and $u_i = 0$). His basic idea is to first estimate σ_{vv} by substituting consistent estimates of β and γ into the spectrum and cross spectrum of f_y, f_x and f_{yx} to obtain an estimated spectrum of ξ, then use an optimally weighting method to estimate β and γ. A generalization of Hannan's (1963) method to the case when both dependent and exogenous variables are observed with error in a single equation model seems highly desirable.

On the other hand, a full spectrum method can be applied to a simultaneous equation model without much problem. If a simultaneous equation model is identified, this is equivalent to the existence of a sufficient set of instruments. Hsiao and Robinson (1978) have made use of this idea to suggest a (full spectrum) instrumental variable method for estimating the unknown parameters. Their method may be summarized as follows.

The Fourier transform of (6.16) is:

$$B(t_j)y(t_j)+\Gamma x(t_j)=w(t_j), \tag{6.42}$$

where

$$B(t_j)=B_0+B_1\mathrm{e}^{\zeta t_j},$$
$$w_i=\varepsilon_i+B_0u_i+B_1u_{i-1}+\Gamma v_i,$$
$$w(t_j)=\frac{1}{\sqrt{2\pi n}}\sum_{i=1}^{n}w_i\mathrm{e}^{\zeta t_j i}.$$

Since it is known that under fairly general conditions:

$$\lim_{n\to\infty}\mathrm{E}w(t_j)x(t_j)^{\dagger}=f_{wx}(t_j)=\Gamma\Omega,\qquad t_j\neq 0, \tag{6.43}$$

we may rewrite (6.42) as:

$$\begin{aligned} y(t_j)&=\left[I_G-B(t_j)\right]y(t_j)-\left(I_K-\Omega f_x(t_j)\right)^{-1}x(t_j)+\tilde{w}(t_j),\\ \tilde{w}(t_j)&=w(t_j)-\Gamma\theta f_x(t_j)^{-1}x(t_j),\end{aligned} \tag{6.44}$$

where the coefficients of the gth equation are normalized to be unity. The transformed model (6.44) possesses (asymptotically) the classical property of orthogonality between the "exogenous variables" $x(t_j)$ and the "residual" $\tilde{w}(t_j)$.

We now stack (6.44) as:

$$y(t_j)=Z(t_j)L'\alpha+\tilde{w}(t_j), \tag{6.45}$$

where

$$Z'(t_j)=\left\{-\left[(1,\mathrm{e}^{\zeta t_j})\otimes y'(t_j)\otimes I_G\right],-x'(t_j)\otimes I_K,x'(t_j)f_x(t_j)^{-1}\otimes\Gamma\right\},$$

$$L=\begin{bmatrix}L_1&0&0\\0&L_2&0\\0&0&L_3\end{bmatrix},$$

$$\alpha'=(\beta',\gamma',\omega').$$

The matrices L_1, L_2, and L_3 and vectors $\boldsymbol{\beta}$, $\boldsymbol{\gamma}$, and $\boldsymbol{\omega}$, are obtained as follows. Suppose there are G_1 zero constraints on $B=[B_0-I_G, B_1]$. Then the unconstrained parameters may be rewritten as $\boldsymbol{\beta}=L_1\text{vec}(B)$, where L_1 is obtained from I_{2G^2} by eliminating the rows corresponding to zero elements. Likewise, if there are G_2 zero constraints on Γ we write the unconstrained parameters as $\boldsymbol{\gamma}=L_2\text{vec}(\Gamma)$, where L_2 is obtained from I_{GK} by eliminating the rows corresponding to zero elements. Also, we write $\boldsymbol{\omega}=L_3\text{vec}(\Omega)$, where L_3 is the $((K-F)\times K^2)$-matrix obtained from I_{K^2} by eliminating rows corresponding to the off-diagonal elements and the F $(0\le F\le K)$ a priori zero diagonal elements of Ω.

An instrumental variable method for (6.45) will be possible after we find an appropriate instrument for $y(t_j)$, and a consistent estimate of $f_{\tilde{w}}(t_j)=\lim_{n\to\infty}\mathrm{E}\tilde{\boldsymbol{w}}(t_j)\tilde{\boldsymbol{w}}^{\dagger}(t_j)$. A possible instrument for $\boldsymbol{y}(t_t)$ would be $\Delta(t_j)\boldsymbol{x}(t_j)$, where $\Delta(t_j)=\hat{f}_{yx}(t_j)\hat{f}_x(t_j)^{-1}$. A consistent estimate of $f_{\tilde{w}}(t_j)$ may be obtained from:

$$\hat{f}_{\tilde{w}}(t_j)=\hat{B}(t_j)\left[\hat{f}_y(t_j)-\hat{f}_{yx}(t_j)\hat{f}_x^{-1}(t_j)f_{xy}(t_j)\right]\hat{B}^{\dagger}(t_j), \tag{6.46}$$

where $\hat{B}(t_j)$ is some consistent estimate of $B(t_j)$, which may be obtained by solving the covariance equations.

We may now define our estimates as:

$$\hat{\boldsymbol{\alpha}}=(LDL')^{-1}L\boldsymbol{d}, \tag{6.47}$$

where

$$D=\frac{1}{T}\sum_j W^{\dagger}(t_j)Z(t_j),$$

$$\boldsymbol{d}=\frac{1}{T}\sum_j W^{\dagger}(t_j)\boldsymbol{y}(t_j),$$

$$W'(t_j)=\begin{bmatrix}(1,e^{\zeta t_j})\otimes\Delta(t_j)\boldsymbol{x}(t_j)\otimes\hat{f}_w^{-1}(t_j)\\ \left[I_K-\hat{\Omega}f_x^{-1}(t_j)\right]\boldsymbol{x}(t_j)\otimes\hat{f}_w^{-1}(t_j)\\ \hat{f}_x(t_j)\boldsymbol{x}(t_j)\otimes\hat{\Gamma}'\hat{f}_w(t_j)\end{bmatrix}.$$

If the spectrum is smooth, we can prove that (6.47) is consistent and asymptotically normally distributed. To obtain an efficient estimate it may be desirable to iterate (6.47). If $\boldsymbol{\varepsilon}$ is stationary then (6.47) is efficient in the sense that the limiting covariance matrix is the same as that of maximum likelihood estimates based on Gaussian $\hat{\boldsymbol{w}}(t_j)$ [Hsiao (1979)], and iteration produces no improvement in efficiency. If $\boldsymbol{\varepsilon}$ is a finite-order autoregressive moving average process, (6.47) is still consistent but will not be fully efficient [e.g. see Espasa (1979) and Hannan and Nicholls (1972)], and then iteration is probably desirable.

As one can see from the above description, the computation of the estimates for the dynamic error–shock model seems a formidable task, particularly if there is iteration. Yet on many occasions we would like to estimate behavioural relationships that are dynamic in character. It does seem desirable to devise some simple, yet reasonably efficient computational algorithms.

References

Aigner, D. J. (1974a) "An Appropriate Econometric Framework for Estimating a Labor-Supply Function from the SEO File", *International Economic Review*, 15, 59–68. Reprinted as Chapter 8 in D. J. Aigner and A. S. Goldberger, eds., *Latent Variables in Socio-Economic Models*. Amsterdam: North-Holland Publishing Company.

Aigner, D. J. (1974b) "MSE Dominance of Least Squares with Errors of Observation", *Journal of Econometrics*, 2, 365–72. Reprinted as Chapter 3 in D. J. Aigner and A. S. Goldberger, eds., *Latent Variables in Socio-Economic Models*. Amsterdam: North-Holland Publishing Company.

Aigner, D. J. (1973) "Regression with a Binary Independent Variable Subject to Errors of Observation", *Journal of Econometrics*, 1, 49–59.

Aigner, D. J. and S. M. Goldfeld (January 1974) "Estimation and Prediction from Aggregate Data when Aggregates are Measured More Accurately than Their Components", *Econometrica*, 42, 113–34.

Amemiya, T. (1971) "The Estimation of the Variances in a Variance-Components Model", *International Economic Review*, 12, 1–13.

Amemiya, T. (1966) "On The Use of Principal Components of Independent Variables in Two-Stage Least-Squares Estimation", *International Economic Review*, 7, 282–303.

Anderson, T. W. (1976) "Estimation of Linear Functional Relationships: Approximate Distributions and Connections with Simultaneous Equations in Econometrics", *Journal of the Royal Statistical Society*, Series B, 38, 1–20.

Anderson, T. W. (1980) "Recent Results in the Estimation of a Linear Functional Relationship", in P. R. Krishnaiah, ed., *Multivariate Statistics*, V. Amsterdam: North-Holland Publishing Company.

Anderson, T. W. and H. Rubin (1956) "Statistical Inference in Factor Analysis", in J. Neyman, ed., *Proceedings of the Third Berkeley Symposium on Mathematical Statistics and Probability*, 5. Berkeley: University of California Press.

Anderson, T. W. (1958) *Multivariate Statistical Analysis*. New York: Wiley.

Aoki, M. and P. C. Yue (1970) "On Certain Convergence Questions in System Identification", *SIAM Journal of Control*, 8, 239–256.

Attfield, C. L. F. (July 1977) "Estimation of a Model Containing Unobservable Variables Using Grouped Observations: An Application to the Permanent Income Hypothesis", *Journal of Econometrics*, 6, 51–63.

Aufm Kampe, H. (1979) *Identifizierbarkeit in Multivariaten Fehler-in-den-Variabelen-Modellen* (Identification in Multivariate Errors-in-Variables Models), Unpublished Masters thesis, University of Bonn.

Avery, Robert B. (1979) "Modelling Monetary Policy as an Unobserved Variable", *Journal of Econometrics*, 10, 291–311.

Barnett, V. D. (1967) "A Note on Linear Structural Relationships When Both Residual Variances are Known", *Biometrika*, 54, 670–672.

Barnett, V. D. (1970) "Fitting Straight Lines. The Linear Functional Relationship with Replicated Observations", *Applied Statistics*, 19, 135–144.

Barnow, Burt S. (August 1976) "The Use of Proxy Variables When One or Two Independent Variables are Measured with Error", *The American Statistician*, 30, 119–121.

Bar-Shalom, Y. (1972) "Optimal Simultaneous State Estimation and Parameter Identification in Linear Discrete-Time Systems", *IEEE Transactions on Automatic Control*, AC-17, 308–319.

Bartlett, M. S. (1949) "Fitting a Straight Line When Both Variables Are Subject to Error", *Biometrics*, 5, 207–212.

Basu, A. P. (1969) "On Some Tests for Several Linear Relations", *Journal of the Royal Statistical Society*, Series B, 31, 65–71.

Bentler, P. M. (1982) "Linear Systems with Multiple Levels and Types of Latent Variables", Chapter 5 in K. G. Jöreskog and H. Wold, eds., *Systems Under Indirect Observations Causality, Structure, Prediction*, Part I. Amsterdam: North-Holland Publishing Company, 101–130.

Bentler, P. M. (1983) "Simultaneous Equation Systems as Moment Structure Models, With an Introduction to Latent Variable Models", *Journal of Econometrics*, 22, 13–42.

Bentler, P. M. and D. G. Weeks (September 1980) "Linear Structural Equations with Latent Variables", *Psychometrika*, 45, 289–308.

Birch, M. W. (1964) "A Note on the Maximum Likelihood Estimation of a Linear Structural Relationship", *Journal of the American Statistical Association*, 59, 1175–1178.

Blomqvist, A. G. (1972) "Approximating the Least-Squares Bias in Multiple Regression with Errors in Variables", *The Review of Economics and Statistics*, 54, 202–204.

Bowden, R. (1973) "The Theory of Parametric Identification", *Econometrica*, 41, 1069–1074.

Box, G. E. P. and G. M. Jenkins (1970) *Time Series Analysis: Forecasting and Control*. San Francisco: Holden–Day.

Brown, R. L. (1957) "Bivariate Structural Relation", *Biometrika*, 44, 84–96.

Browne, M. W. (1974) "Generalized Least-Squares Estimators in the Analysis of Covariance Structures", *South African Statistical Journal*, 8, 1–24. Reprinted as Chapter 13 in D. J. Aigner and A. S. Goldberger, eds., *Latent Variables in Socio-Economic Models*. Amsterdam: North-Holland Publishing Company, 143–161.

Casson, M. C. (1974) "Generalized Errors in Variables Regression", *Review of Economic Studies*, 41, 347–352.

Chamberlain, Gary (1978) "Omitted Variables Bias in Panel Data: Estimating the Returns to Schooling", *Annales de l'INSEE 30–31*, 49–82

Chamberlain, Gary (1977a) "Education, Income and Ability Revisited", Chapter 10 in D. J. Aigner and A. S. Goldberger, eds., *Latent Variables in Socio-Economic Models*. Amsterdam: North-Holland Publishing Company, 143–161.

Chamberlain, Gary (1977b) "An Instrumental Variable Interpretation of Identification in Variance-Components and MIMIC Models", Chapter 7 in P. Taubman, ed., *Kinometrics: The Determinanets of Socio-Economic Success Within and Between Families*. Amsterdam: North-Holland Publishing Company.

Chamberlain, Gary and Zvi Griliches (1975) "Unobservables with a Variance-Components Structure: Ability, Schooling and the Economic Success of Brothers", *International Economic Review*, 16, 422–449. Reprinted as Chapter 15 in D. J. Aigner and A. S. Goldberger, eds., *Latent Variables in Socio-Economic Models*. Amsterdam: North-Holland Publishing Company.

Chamberlain, Gary and Zvi Griliches (1977) "More on Brothers", Chapter 4 in P. Taubman, ed., *Kinometrics: The Determinants of Socio-Economic Success Within and Between Families*. Amsterdam: North-Holland Publishing Company.

Chen, C.-F. (September 1981) "The EM Approach to the Multiple Indicators and Multiple Causes Model Via the Estimation of the Latent Variable", *Journal of the American Statistical Association*, 76, 704–708.

Chernoff, Herman and Herman Rubin (1953) "Asymptotic Properties of Limited-Information Estimates under Generalized Conditions", Chapter VII in W. C. Hood and T. C. Koopmans, eds., *Studies in Econometric Method*. New York: John Wiley and Sons.

Cochran, W. G. (1968) "Errors of Measurement in Statistics", *Technometrics*, 10, 637–666.

Cooper, R. V. and J. P. Newhouse (1972) "Further Results on the Errors in the Variables Problem", mimeo, The Rand Corporation, Santa Monica, Ca.

Copas, J. B. (1972) "The Likelihood Surface in the Linear Functional Relationship Problem", *Journal of the Royal Statistical Society*, Series B, 34, 274–278.

Cox, H. (1964) "On the Estimation of State Variables and Parameters for Noisy Dynamic Systems", *IEEE Transactions of Automatic Control*, AC-10, 5–12.

Cox, N. R. (1976) "The Linear Structural Relation for Several Groups of Data", *Biometrika*, 63, 231–237.

Cramér, H. (1946) *Mathematical Methods of Statistics*. Princeton: Princeton University Press.

Creasy, M. (1956) "Confidence Limits for the Gradient in the Linear Functional Relationship", *Journal of the Royal Statistical Society*, Series B, 18, 65–69.
Davies, R. B. and B. Hutton (1975) "The Effect of Errors in the Independent Variables in Linear Regression", *Biometrika*, 62, 383–391.
DeGracie, J. S. and W. A. Fuller (1972) "Estimation of the Slope and Analysis of Covariance When the Concomitant Variable is Measured with Error", *Journal of the American Statistical Association*, 67, 930–937.
Deistler, M. and H.-G. Seifert (1978) "Identifiability and Consistent Estimability in Econometric Models", *Econometrica*, 46, 969–980.
Denton, F. T. and J. Kuiper (1965) "The Effect of Measurement Errors on Parameter Estimates and Forecasts: A Case Study Based on the Canadian Preliminary National Accounts", *The Review of Economics and Statistics*, 47, 198–206.
Dolby, G. R. (1976a) "A Note on the Linear Structural Relation When Both Residual Variances are Known", *Journal of the American Statistical Association*, 71, 352–353.
Dolby, G. R. (1976b) "The Ultrastructural Relation: A Synthesis of the Functional and Structural Relations", *Biometrika*, 63, 39–50.
Dolby, G. R. (1972) "Generalized Least Squares and Maximum Likelihood Estimation of Nonlinear Functional Relationships", *Journal of the Royal Statistical Society*, Series B, 34, 393–400.
Dolby, G. R. and T. G. Freeman (1975) "Functional Relationships Having Many Independent Variables and Errors with Multivariate Normal Distribution", *Journal of Multivariate Analysis*, 5, 466–479.
Dolby, G. R. and S. Lipton (1972) "Maximum Likelihood Estimation of the General Nonlinear Relationship with Replicated Observations and Correlated Errors", *Biometrika*, 59, 121–129.
Dorff, M. and J. Gurland (1961a) "Estimation of the Parameters of a Linear Functional Relation", *Journal of the Royal Statistical Society*, Series B, 23, 160–170.
Dorff, M. and J. Gurland (1961b) "Small Sample Behavior of Slope Estimators in a Linear Functional Relation", *Biometrics*, 17, 283–298.
Drion, E. F. (1951) "Estimation of the Parameters of a Straight Line and of the Variances of the Variables, if They Are Both Subject to Error", *Indagationes Mathematicae*, 13, 256–260.
Egerton, M. F. and P. J. Laycock (1979) "Maximum Likelihood Estimation of Multivariate Non-Linear Functional Relationships", *Mathematische Operationsforschung und Statistik*, 10, 273–280.
Engle, R. F. (1974) "Band Spectrum Regression", *International Economic Review*, 15, 1–11.
Engle, R. F. (1980) "Exact Maximum Likelihood Methods for Dynamic Regressions and Band Spectrum Regressions", *International Economic Review*, 21, 391–408.
Engle R. F. and D. K. Foley (1975) "An Asset Price Model of Aggregate Investment", *International Economic Review*, 16, 625–47.
Elffers, H., J. G. Bethlehem and R. Gill (1978) "Indeterminacy Problems and the Interpretation of Factor Analysis Results", *Statistica Neerlandica*, 32, 181–199.
Espasa, A. (1979) *The Spectral Maximum Likelihood Estimation of Econometric Models with Stationary Errors*. Göttingen: Vandenhoeck und Ruprecht.
Fisher, F. M. (1966) *The Identification Problem in Econometrics*. New York: McGraw-Hill.
Florens, J.-P., M. Mouchart and J.-F. Richard (1974) "Bayesian Inference in Error-in-Variables Models", *Journal of Multivariate Analysis*, 4, 419–52.
Fornell, C. (1983) "Issues in the Application of Covariance Structure Analysis: A Comment", *Journal of Consumer Research*, 9, 443–448.
Frisch, R. (1934) *Statistical Confluence Analysis by Means of Complete Regression Systems*. Oslo: University Institute of Economics.
Frost, P. A. (1979) "Proxy Variables and Specification Bias", *The Review of Economics and Statistics*, 61, 323–325.
Fuller, W. A. (1980) "Properties of Some Estimators for the Errors-in-Variables Model", *The Annals of Statistics*, 8, 407–422.
Fuller, W. A. and M. A. Hidiroglou (1978) "Regression Estimation After Correcting for Attenuation", *Journal of the American Statistical Association*, 73, 99–104.
Garber, S. and S. Klepper (Sept. 1980) "Extending the Classical Normal Errors-in-Variables Model", *Econometrica*, 48, 1541–1546.
Geary, R. C. (1943) "Relations Between Statistics: The General and the Sampling Problem When the Samples are Large", *Proceedings of the Royal Irish Academy*, A, 49, 177–196.

Geary, R. C. (1942) "Inherent Relations Between Random Variables", *Proceedings of the Royal Irish Academy*, A, 47, 63–67.

Geraci, Vincent, J. (1976) "Identification of Simultaneous Equation Models with Measurement Error", *Journal of Econometrics*, 4, 263–283. Reprinted as Chapter 11 in D. J. Aigner and A. S. Goldberger, eds., *Latent Variables in Socio-Economic Models*. Amsterdam: North-Holland Publishing Company.

Geraci, Vincent, J. (1977) "Estimation of Simultaneous Equation Models with Measurement Error", *Econometrica*, 45, 1243–1255.

Geraci, Vincent, J. (1983) Errors in Variables and the Individual Structural Equation, *International Economic Review*, 24, 217–236.

Giles, R. L. (1980) "Error of Omission and Measurement: Estimating the Parameter of the Variable Subject to Error", Polytechnic of the South Bank, London.

Goldberger, A. S. (1974) "Unobservable Variables in Econometrics", Chapter 7 in P. Zarembka, ed., *Frontiers in Econometrics*. New York: Academic Press.

Goldberger, A. S. (November 1972a) "Structural Equation Methods in the Social Sciences", *Econometrica*, 40, 979–1001.

Goldberger, A. S. (1972b) "Maximum-Likelihood Estimation of Regressions Containing Unobservable Independent Variables", *International Economic Review*, 13, 1–15. Reprinted as Chapter 6 in D. J. Aigner and A. S. Goldberger, eds., *Latent Variables in Socio-Economic Models*. North-Holland Publishing Company.

Goldberger, A. S. (June 1971) "Econometrics and Psychometrics: A Survey of Communalities", *Psychometrika*, 36, 83–107.

Goldberger, A. S. and O. D. Duncan, eds. (1973) *Structural Equation Models in the Social Sciences*. New York: Seminar Press.

Gorsuch, S. A. (1974) *Factor Analysis*. Philadelphia: W. B. Saunders Company.

Griliches, Z. (January 1977) "Estimating the Returns to Schooling: Some Econometric Problems", *Econometrica*, 45, 1–22.

Griliches, Z. (1974) "Errors in Variables and Other Unobservables", *Econometrica*, 42, 971–998. Reprinted as Chapter 1 in D. J. Aigner and A. S. Goldberger, eds., *Latent Variables in Socio-Economic Models*. Amsterdam: North-Holland Publishing Company.

Griliches, Z. and Vidar Ringstad (March 1970) "Error-in-the-Variables Bias in Nonlinear Contexts", *Econometrica*, 38, 368–370.

Griliches, Z. and W. M. Mason (May 1972) "Education, Income and Ability", *Journal of Political Economy*, 80, 74–103.

Hall, Bronwyn (1979) *User's Guide to MOMENTS*. 204 Junipero Serra Blvd., Stanford, CA 94305.

Hannan, E. J. (1963) "Regression for Time Series with Errors of Measurement", *Biometrika*, 50, 293–302.

Hannan, E. J. (1971) "The Identification Problem for Multiple Equation Systems with Moving Average Errors", *Econometrica*, 39, 751–765.

Hannan, E. J. (1975) "The Estimation of ARMA Models", *The Annals of Statistics*, 3, 975–981.

Hannan, E. J. (1976) "The Identification and Parameterization of ARMAX and State Space Forms", *Econometrica*, 44, 713–723.

Hannan, E. J. and D. F. Nicholls (1972) "The Estimation of Mixed Regression, Autoregression, Moving Average, and Distributed Lag Models", *Econometrica*, 40, 529–547.

Harman, H. H. (1967) *Modern Factor Analysis*. Chicago: The University of Chicago Press.

Hausman, J. A. (1978) "Specification Tests in Econometrics", *Econometrica*, 46, 1251–1272.

Hausman, J. A. (May 1977) "Errors in Variables in Simultaneous Equation Models", *Journal of Econometrics*, 5, 389–401.

Hausman, J. A. and W. E. Taylor (1983) "Identification in Linear Simultaneous Equations Models with Covariance Restrictions: An Instrumental Variables Interpretation", *Econometrica*, 51, 1527–1549.

Healy, J. D. (1980) "Maximum Likelihood Estimation of a Multivariate Linear Functional Relationship", *Journal of Multivariate Analysis*, 10, 243–251.

Heckman, J. J. and B. Singer (1982) "The Identification Problem in Econometric Models for Duration Data", Chapter 2 in W. Hildenbrand, ed., *Advances in Econometrics*, Part II. Cambridge: Cambridge University Press.

Heise, D. R. (1975) *Causal Analysis*. New York: Wiley.

Hester, Donald D. (July 1976) "A Note on Identification and Information Loss Through Aggregation", *Econometrica*, 44, 815–818.

Hodges, S. D. and P. G. Moore (1972) "Data Uncertainties and Least Squares Regression", *Applied Statistics*, 21, 185–195.

Höschel, H.-P. (1978) "Generalized Least Squares Estimators of Linear Functional Relations with Known Error-Covariance", *Mathematische Operationsforschung und Statistik*, 9, 9–26.

Hsiao, C. (March 1979) "Measurement Error in a Dynamic Simultaneous Equation Model with Stationary Disturbances", *Econometrica*, 47, 475–494.

Hsiao, C. (February 1977) "Identification for a Linear Dynamic Simultaneous Error-Shock Model", *International Economic Review*, 18, 181–194.

Hsiao, C. (June 1976) "Identification and Estimation of Simultaneous Equation Models with Measurement Error", *International Economic Review*, 17, 319–339.

Hsiao, C. and P. M. Robinson (June 1978) "Efficient Estimation of a Dynamic Error-Shock Model", *International Economic Review*, 19, 467–480.

Johnston, J. (1972) *Econometric Methods*. New York: McGraw-Hill.

Jöreskog, K. G. (1978) "An Econometric Model for Multivariate Panel Data", *Annales de l'INSEE 30–31*, 355–366.

Jöreskog, K. G. and D. Sörbom (1981) *LISREL V User's Guide*. Chicago: National Educational Resources.

Jöreskog, K. G. (1970) "A General Method for Analysis of Covariance Structures", *Biometrika*, 57, 239–251. Reprinted as Chapter 12 in D. J. Aigner and A. S. Goldberger, eds., *Latent Variables in Socio-Economic Models*. Amsterdam: North-Holland Publishing Company.

Jöreskog, K. G. and A. S. Goldberger (1975) "Estimation of a Model with Multiple Indicators and Multiple Causes of a Single Latent Variable", *Journal of the American Statistical Association*, 70, 631–639.

Jöreskog, K. G. and Dag Sörbom (1977) "Statistical Models and Methods for Analysis of Longitudinal Data", Chapter 16 in D. J. Aigner and A. S. Goldberger, eds., *Latent Variables in Socio-Economic Models*. Amsterdam: North-Holland Publishing Company, 285–325.

Jöreskog, K. G. (1967) "Some Contributions to Maximum Likelihood Factor Analysis", *Psychometrika*, 32, 443–482.

Jöreskog, K. G. and A. S. Goldberger (1972) "Factor Analysis by Generalized Least Squares", *Psychometrika*, 37, 243–260.

Kadane, Joseph B., T. W. McGuire, P. R. Sanday and R. Staelin (1977) "Estimation of Environmental Effects on the Pattern of IQ Scores Over Time", Chapter 17 in D. J. Aigner and A. S. Goldberger, eds., *Latent Variables in Socio-Economic Models*. Amsterdam: North-Holland Publishing Company, 327–348.

Kapteyn, A. and T. J. Wansbeek (1983) "Identification in the Linear Errors in Variables Model", *Econometrica*, 51, 1847–1849.

Kapteyn, A. and T. J. Wansbeek (1983) "Errors in Variables: Consistent Adjusted Least Squares (CALS) Estimation", Netherlands Central Bureau of Statistics.

Kapteyn, A. and T. J. Wansbeek (1981) "Structural Methods in Functional Models", Modelling Research Group, University of Southern California.

Keller, W. J. (1975) "A New Class of Limited-Information Estimators for Simultaneous Equation Systems", *Journal of Econometrics*, 3, 71–92.

Kendall, M. G. and A. Stuart (1979) *The Advanced Theory of Statistics*, Fourth Edition. New York: Macmillan.

Kiefer, J. and J. Wolfowitz (1956) "Consistency of the Maximum Likelihood Estimator in the Presence of Infinitely Many Incidental Parameters", *Annals of Mathematical Statistics*, 27, 887–906.

Klepper, S. and E. E. Leamer (1984) "Consistent Sets of Estimates for Regressions with Errors in All Variables", *Econometrica*, 52, 163–183..

Kloek, T. and L. B. M. Mennes (1960) "Simultaneous Equations Estimation Based On Principal Components of Predetermined Variables", *Econometrica*, 28, 45–61.

Konijn, H. S. (1962) "Identification and Estimation in a Simultaneous Equations Model with Errors in the Variables", *Econometrica*, 30, 79–87.

Koopmans, T. C. (1937) *Linear Regression Analysis of Economic Time Series*. Haarlem: Netherlands Economic Institute, De Erven F. Bohn N.V.

Lawley, D. N. and A. E. Maxwell (1971) *Factor Analysis as a Statistical Method*. London: Butterworths.

Leamer, E. E. (1978a) "Least-Squares Versus Instrumental Variables Estimation in a Simple Errors in Variables Model", *Econometrica*, 46, 961–968.

Leamer, E. E. (1978b) *Specification Searches, Ad Hoc Inference with Nonexperimental Data*. New York: Wiley.

Lee, S.-Y. (September 1980) "Estimation of Covariance Structure Models with Parameters Subject to Functional Constraints", *Psychometrica*, 45, 309–324.

Levi, M. D. (1977) "Measurement Errors and Bounded OLS Estimates", *Journal of Econometrics*, 6, 165–171.

Levi, M. D. (1973) "Errors in the Variables Bias in the Presence of Correctly Measured Variables", *Econometrica*, 41, 985–986.

Levin, M. J. (1964) "Estimation of a System Pulse Transfer Function in the Presence of Noise", *IEEE Transactions on Automatic Control*, AC-9, 229–235.

Lindley, D. V. and G. M. El-Sayyad (1968) "The Bayesian Estimation of a Linear Functional Relationship", *Journal of the Royal Statistical Society*, Series B, 30, 198–202.

Liviatan, N. (1963) "Tests of the Permanent-Income Hypothesis Based on a Reinterview Savings Survey", in: C. F. Christ, ed., *Measurement in Economics*. Stanford: Stanford University Press.

Liviatan, N. (July 1961) "Errors in Variables and Engel Curve Analysis", *Econometrica*, 29, 336–362.

Ljung, L. (1977) "Positive Real Transfer Functions and the Convergence of Some Recursive Schemes", *IEEE Transactions of Automatic Control*, AC-22, 539–551.

Madansky, A. (1976) *Foundations of Econometrics*. Amsterdam: North-Holland Publishing Company.

Madansky, A. (1959) "The Fitting of Straight Lines When Both Variables are Subject to Error", *Journal of the American Statistical Association*, 54, 173–205.

Malinvaud, E. (1970) *Statistical Methods of Econometrics*, Second Revised Edition. Amsterdam: North-Holland Publishing Company.

Maravall, A. (1979) *Identification in Dynamic Shock-Error Models*. Berlin: Springer-Verlag.

Maravall, A. and D. J. Aigner (1977) "Identification of the Dynamic Shock–Error Model: The Case of Dynamic Regression", Chapter 18 in D. J. Aigner and A. S. Goldberger, eds., *Latent Variables in Socio-Economic Models*. Amsterdam: North-Holland Publishing Company, 349–363.

McCallum, B. T. (1977) "Relative Asymptotic Bias from Errors of Omission and Measurement", *Econometrica*, 40, 757–758. Reprinted as Chapter 2 in D. J. Aigner and A. S. Goldberger, eds., *Latent Variables in Socio-Economic Models*. Amsterdam: North-Holland Publishing Company.

Moberg, L. and R. Sundberg (1978) "Maximum Likelihood Estimation of a Linear Functional Relationship When One of the Departure Variances is Known", *Scandinavian Journal of Statistics*, 5, 61–64.

Mouchart, M. (1977) "A Regression Model with an Explanatory Variable Which Is Both Binary and Subject to Errors", Chapter 4 in D. J. Aigner and A. S. Goldberger, eds., *Latent Variables in Socio-Economic Models*. Amsterdam: North-Holland Publishing Company, 48–66.

Mulaik, S. D. (1972) *The Foundations of Factor Analysis*. New York: McGraw-Hill.

Mundlak, Y. (1961) "Empirical Production Function Free of Management Bias", *Journal of Farm Economics*, 43, 44–56.

Muthén, B. (December 1979) "A Structural Probit Model With Latent Variables", *Journal of the American Statistical Association*, 74, 807–811.

Neyman, J. (1951) "Existence of Consistent Estimates of the Directional Parameter in a Linear Structural Relation Between Two Variables", *Annals of Mathematical Statistics*, 22, 496–512.

Neyman, J. and Elizabeth L. Scott (1951) "On Certain Methods of Estimating the Linear Structural Relation", *Annals of Mathematical Statistics*, 22, 352–361.

Neyman, J. and Elizabeth L. Scott (1948) "Consistent Estimates Based on Partially Consistent Observations", *Econometrica*, 16, 1–32.

Nicholls, D. F., A. R. Pagan and R. D. Terrell (1975) "The Estimation and Use of Models with Moving Average Disturbance Terms: A Survey", *International Economic Review*, 16, 113–134.

Nowak, E. (1977) "An Identification Method for Stochastic Models of Time Series Analysis with Errors in the Variables", paper presented to the European Meeting of the Econometric Society, Vienna.

Nussbaum, M. (1977) "Asymptotic Optimality of Estimators of a Linear Functional Relation if the

Ratio of the Error Variances is Known", *Mathematische Operationsforschung und Statistik*, 8, 173–198.
O'Neill, I., G. Sinclair and F. J. Smith (1969) "Polynomial Curve Fitting When Abscissas and Ordinates are Both Subject to Error", *Computer Journal*, 12, 52–56.
Pagan, A. (1973) "Efficient Estimation of Models with Composite Disturbance Terms", *Journal of Econometrics*, 1, 329–340.
Pakes, A. (1982) "On the Asymptotic Bias of the Wald-Type Estimators of a Straight Line when Both Variables are Subject to Error", *International Economic Review*, 23, 491–497.
Pal, M. (1980) "Consistent Moment Estimators of Regression Coefficients in the Presence of Errors in Variables", *Journal of Econometrics*, 14, 349–364.
Patefield, W. M. (1981) "Multivariate Linear Relationships: Maximum Likelihood Estimation and Regression Bounds", *Journal of the Royal Statistical Society*, Series B, 43, 342–352.
Patefield, W. M. (1978) "The Unreplicated Ultrastructural Relation: Large Sample Properties", *Biometrika*, 65, 535–540.
Patefield, W. M. (1977) "On the Information Matrix in the Linear Functional Relationship Problem", *Applied Statistics*, 26, 69–70.
Patefield, W. M. (1976) "On the Validity of Approximate Distributions Arising in Fitting a Linear Functional Relationship", *Journal of Statistics, Computation and Simulation*, 5, 43–60.
Phillips, A. W. (1966) "The Estimation of Systems of Difference Equations with Moving Average Disturbances", paper presented at the Econometric Society meeting, San Francisco.
Rao, C. R. (1966) "Characterization of the Distribution of Random Variables in Linear Structural Relations", *Sankhyā*, 28, 251–260.
Reiersøl, Olav (1950) "Identifiability of a Linear Relation Between Variables Which are Subject to Error", *Econometrica*, 18, 375–389.
Reiersøl, Olav (1945) "Confluence Analysis by Means of Instrumental Sets of Variables", *Arkiv för Mathematik, Astronomi och Fysik*, 32A, 1–119.
Richardson, D. H. and D.-M. Wu (1970) "Least Squares and Grouping Method Estimators in the Errors-in-Variables Model", *Journal of the American Statistical Association*, 65, 724–748.
Robertson, C. A. (1974) "Large Sample Theory for the Linear Structural Relation", *Biometrika*, 61, 353–359.
Robinson, P. M. (1977) "The Estimation of a Multivariate Linear Relation", *Journal of Multivariate Analysis*, 7, 409–423.
Robinson, P. M. (1974) "Identification, Estimation and Large-Sample Theory for Regressions Containing Unobservable Variables", *International Economic Review*, 15, 680–692. Reprinted as Chapter 7 in D. J. Aigner and A. S. Goldberger, eds., *Latent Variables in Socio-Economic Models*. Amsterdam: North-Holland Publishing Company.
Robinson, P. M. and M. C. Ferrara (1977) "The Estimation of a Model for an Unobservable Variable with Endogenous Causes", Chapter 9 in D. J. Aigner and A. S. Goldberger, eds., *Latent Variables in Socio-Economic Models*. Amsterdam: North-Holland Publishing Company, 131–142.
Rothenberg, T. J. (1973a) "The Asymptotic Distribution of the Least-Squares Estimator in the Errors-in-Variables Model", mimeo, University of California, Berkeley.
Rothenberg, T. J. (1973b) *Efficient Estimation with A Priori Information*. New Haven: Yale University Press.
Rothenberg, T. J. (1971) "Identification in Parametric Models", *Econometrica*, 39, 577–592.
Sampson, A. R. (1974) "A Tale of Two Regressions", *Journal of the American Statistical Association*, 69, 682–689.
Sargan, J. D. (July 1958) "The Estimation of Economic Relationships Using Instrumental Variables", *Econometrica*, 26, 393–415.
Schmidt, P. (1976) *Econometrics*. New York: Marcel Dekker.
Schneeweiss, H. (1982) "Note on Creasy's Confidence Limits for the Gradient in the Linear Functional Relationship", *Journal of Multivariate Analysis*, 12, 155–158.
Schneeweiss, H. (1976) "Consistent Estimation of a Regression with Errors in the Variables", *Metrika*, 23, 101–115.
Scott, E. L. (1950) "Note on Consistent Estimates of the Linear Structural Relation Between Two Variables", *Annals of Mathematical Statistics*, 21, 284–288.
Shapiro, A. (1982) "Rank-Reducibility of a Symmetric Matrix and Sampling Theory of Minimum Trace Factor Analysis", *Psychometrika*, 47, 187–199.

Singleton, K. J. (October 1980) "A Latent Time Series Model of the Cyclical Behavior of Interest Rates", *International Economic Review*, 21, 559–576.

Solari, M. E. (1969) "The 'Maximum Likelihood Solution' of the Problem of Estimating a Linear Functional Relationship", *Journal of the Royal Statistical Society*, Series B, 31, 372–375.

Sprent, P. (1966) "A Generalized Least-Squares Approach to Linear Functional Relationships", *Journal of the Royal Statistical Society*, Series B, 28, 278–297.

Sprent, P. (1970) "The Saddlepoint of the Likelihood Surface for a Linear Functional Relationship", *Journal of the Royal Statistical Society*, Series B, 32, 432–434.

Theil, H. (1971) *Principles of Econometrics*. New York: Wiley.

Tintner, G. (1945) "A Note on Rank, Multicollinearity and Multiple Regression", *Annals of Mathematical Statistics*, 16, 304–308.

Van Uven, M. J. (1930) "Adjustment of N Points (in n-Dimensional Space) to the Best Linear $(n-1)$-Dimensional Space", *Koninklijke Akademie van Wetenschappen te Amsterdam*, *Proceedings of the Section of Sciences*, 33, 143–157, 307–326.

Villegas, C. (1961) "Maximum Likelihood Estimation of a Linear Functional Relationship", *Annals of Mathematical Statistics*, 32, 1040–1062.

Villegas, C. (1964) "Confidence Region for a Linear Relation", *Annals of Mathematical Statistics*, 35, 780–788.

Wald, A. (1949) "Note on the Consistency of the Maximum Likelihood Estimate", *Annals of Mathematical Statistics*, 20, 595–601.

Wald, A. (1948) "Estimation of a Parameter When the Number of Unknown Parameters Increases Indefinitely with Number of Observations", *Annals of Mathematical Statistics*, 19, 220–227.

Wald, A. (1940) "The Fitting of Straight Lines if Both Variables Are Subject to Error", *Annals of Mathematical Statistics*, 11, 284–300.

Ware, J. H. (1972) "The Fitting of Straight Lines When Both Variables are Subject to Error and the Ranks of the Means Are Known", *Journal of the American Statistical Association*, 67, 891–897.

Wegge, L. L. (1965) "Identifiability Criteria for a System of Equations as a Whole", *The Australian Journal of Statistics*, 7, 67–77.

Wickens, M. R. (1972) "A Note on the Use of Proxy Variables", *Econometrica*, 40, 759–761.

Wilson, G. J. (1969) "Factorization of the Generating Function of a Pure Moving Average Process", *SIAM Journal of Numerical Analysis*, 6, 1–7.

Willassen, Y. (1979) "Extension of Some Results by Reiersøl to Multivariate Models", *Scandinavian Journal of Statistics*, 6, 89–91.

Wolfowitz, J. (1954) "Estimation of Structural Parameters When The Number of Incidental Parameters is Unbounded" (abstract), *Annals of Mathematical Statistics*, 25, 811.

Wolfowitz, J. (1952) "Consistent Estimators of the Parameters of a Linear Structural Relation", *Skandinavisk Aktuarietidskrift*, 35, 132–151.

Wu, D.-M. (1973) "Alternative Tests of Independence Between Stochastic Regressors and Disturbances", *Econometrica*, 41, 733–750.

Zellner, Arnold (1970) "Estimation of Regression Relationships Containing Unobservable Independent Variables", *International Economic Review*, 11, 441–454. Reprinted as Chapter 5 in D. J. Aigner and A. S. Goldberger, eds., *Latent Variables in Socio-Economic Models*. Amsterdam: North-Holland Publishing Company.

Zellner, Arnold (1971) *An Introduction to Bayesian Inference in Econometrics*. New York: Wiley.

Chapter 24

ECONOMETRIC ANALYSIS OF QUALITATIVE RESPONSE MODELS

DANIEL L. McFADDEN

Massachusetts Institute of Technology

Contents

Handbook of Econometrics, Volume II, Edited by Z. Griliches and M.D. Intriligator

1. The problem

The subject of this chapter is the econometric analysis of qualitative endogenous variables. Such variables may result from economic behavior which is intrinsically categorical, such as choice of occupation, marriage partner, or entry into a product market. Alternatively, they may come from classification during observation, such as the coding of housing choices into "sub-standard" or "standard". Qualitative variables may be binomial (yes/no) or multinomial, and multinomial responses may be naturally ordered (number of telephone calls) or unordered (freight shipment mode). There may also be multivariate combinations of discrete and continuous variables (appliance portfolio and energy consumption). Binomial and multinomial models are the primary subject of this chapter. A final section considers extensions.

The most suitable statistical model for qualitative response data will depend on the nature of the economic behavior governing the response and on the objectives of the analysis. Consistency with utility or profit maximization may be imposed for consumer or firm decisions such as educational level or plant location. Other data such as resolution of labor disputes (strike/settlement) may involve behavior of multiple agents. In some cases such as industrial accident data, the response model may not have a behavioral, or even a causal, interpretation. The discrete response will be of primary interest in many problems, but may also indicate self-selection into a target population requiring correction of sampling bias. An example is potential bias in analysis of housing expenditure in a self-selected population of renters.

2. Binomial response models

2.1. *Latent variable specification*

The starting point for econometric analysis of a continuous response variable y is often a linear regression model:

$$y_t = x_t\beta - \varepsilon_t, \tag{2.1}$$

where x is a vector of exogenous variables, ε is an unobserved disturbance, and $t = 1,\dots,T$ indexes sample observations. The disturbances are usually assumed to have a convenient cumulative distribution function $F(\varepsilon|x)$ such as multivariate normal. The model is then characterized by the conditional distribution

$F(y - x\beta|x)$, up to the unknown parameters β and parameters of the distribution F. In economic applications, $x\beta$ may have a structure derived exactly or approximately from theory. For example, competitive firms may have $x\beta$ determined by Shephard's identity from a profit function.

The linear regression model is extended to binomial response by introducing an intermediate unobserved (latent) variable y^* with:

$$y_t^* = x_t\beta - \varepsilon_t, \tag{2.2}$$

and an indicator function:

$$y_t = z(y_t^*) = \begin{cases} 0, & \text{if } y_t^* < 0, \\ 1, & \text{if } y_t^* \geq 0. \end{cases} \tag{2.3}$$

If $F(\varepsilon|x)$ is the cumulative distribution function of the disturbances, then just as in the continuous case the model is characterized by the conditional distribution of y given x:

$$\begin{aligned} P_1 &= P(z(y^*) = 1|x) \\ &= P(y^* = x\beta - \varepsilon \geq 0) \\ &= F(x\beta|x), \end{aligned} \tag{2.4}$$

also termed the *response probability*.

2.2. *Functional forms*

The most common binomial models, which assume ε independent of x, are *logit* with

$$F(x\beta) = 1/(1 + e^{-x\beta}), \tag{2.5}$$

probit with

$$F(x\beta) = \Phi(x\beta), \tag{2.6}$$

where Φ is the standard cumulative normal, the *linear probability model* with

$$F(x\beta) = x\beta \qquad (0 \leq x\beta \leq 1), \tag{2.7}$$

and the *log linear model* with

$$F(x\beta) = e^{x\beta} \qquad (x\beta \leq 0). \tag{2.8}$$

The last two models require restrictions on the domain of the latent variable which may be difficult to enforce in estimation or forecasting.

The preceding models are derived from distribution functions with thin tails. Alternatives in which the response probabilities approach zero or one less rapidly can be constructed from the Student-t or Cauchy distributions; the latter yields the *arctan model*:

$$F(x\beta)=\tfrac{1}{2}+\frac{1}{\pi}\tan^{-1}(x\beta). \tag{2.9}$$

For a given latent variable model $y^* = x\beta + \varepsilon$, specification of the distribution function F for ε may change substantially the model's ability to fit data, particularly if restrictions are imposed on the domain of $x\beta$.[1] However respecification of the latent variable model can circumvent this problem. Suppose $F(\varepsilon)$ is any continuous cumulative distribution function, and $\tilde{x}\tilde{\beta} \doteq \ln(F(\beta x)/(1-F(\beta x)))$ is a linear (in parameters $\tilde{\beta}$) global approximation on a compact set[2] of βx satisfying $0 < F(\beta x) < 1$. Then to any desired level of accuracy, the response probability is logistic in the transformed latent variable model $\tilde{y}^* = \tilde{x}\tilde{\beta} + \varepsilon$:

$$F(x\beta)=1/(1+e^{-\tilde{x}\tilde{\beta}}). \tag{2.10}$$

Thus, the question of the appropriate F is recast as the question of the appropriate specification of arithmetic transformations $\tilde{x}$ of the data x in a logit model.[3]

2.3. *Estimation*

Consider a sample (y_t, x_t) with observations indexed $t=1,\ldots,T$, and a binomial model $P_{1t} = F(x_t\beta)$. Assume the sample is random[4] with independent observations. Then the log-likelihood normalized by sample size is:

$$L=\frac{1}{T}\sum_{t=1}^{T}\left[y_t \ln P_{1t}+(1-y_t)\ln P_{0t}\right], \tag{2.11}$$

[1] The logit and probit models however are rarely distinguishable empirically.

[2] The existence of such an approximation is guaranteed by the Weierstrauss approximation theorem. A constructive approximation theorem with explicit error bounds is given in McFadden (1981).

[3] Obviously, the logit base *cdf* could be replaced by any other continuous invertible *cdf* $G(\varepsilon)$, with $\tilde{x}\tilde{\beta} \doteq G^{-1}(F(x\beta))$.

[4] Specifically, the probability of being sampled is assumed independent of response; stratification with respect to x_t is permitted.

with $P_{1t} = F(x_t\beta)$ and $P_{0t} = 1 - P_{1t}$. The gradient of this function is:

$$L_\beta = \frac{1}{T}\sum_{t=1}^{T} w_t x_t (y_t - F(x_t\beta)), \tag{2.12}$$

with $w_t = F'(x_t\beta)/P_{0t}P_{1t}$, and the hessian is:

$$L_{\beta\beta} = -J_T + \frac{1}{T}\sum_{t=1}^{T} u_t x_t x_t' (y_t - F(x_t\beta)), \tag{2.13}$$

where

$$J_T = \frac{1}{T}\sum_{t=1}^{T} w_t^2 P_{0t} P_{1t} x_t x_t' = -EL_{\beta\beta} \tag{2.14}$$

is the information matrix, and

$$u_t = \left(F''(x_t\beta) + (P_{1t} - P_{0t})(F'(x_t\beta))^2\right)/P_{0t}P_{1t}. \tag{2.15}$$

Under mild regularity conditions, detailed in Section 3.2 below, the maximum likelihood estimator $\hat{\beta}$ of β is consistent, and $\sqrt{T}(\hat{\beta} - \beta)$ is asymptotically normal with mean zero and covariance matrix $J^{-1} = \lim_{T\to\infty} J_T^{-1}$. Solution of the normal equation (2.12) usually requires an iterative procedure. Optimizers such as Newton–Raphson, quadratic hill-climbing, or BHHH[5] work well if three cautions are observed:

(1) Accurate numerical approximations for $\ln F(x_t\beta)$ and $\ln(1 - F(x_t\beta))$ are needed in the tails of the distribution.

(2) There is a small (and vanishing) probability, in models where the domain of F is unbounded, that the maximum likelihood estimator will fail to exist and response is perfectly correlated with the sign of an index $x\bar{\beta}$. Adding a test for this condition during iteration permits detection of this case and estimation of the relative weights $\bar{\beta}$. For sample sizes of a few hundred, this outcome is extremely improbable unless the analyst has entered misspecified x variables which depend on y.

[5]See Berndt–Hausman–Hall–Hall (1974) and Goldfeld and Quandt (1972) for discussions of these algorithms. The largest component of computation cost in maximum likelihood estimation is usually evaluation of the response probabilities. Consequently, for maximum efficiency, the number of function evaluations and passes through the data should be minimized. This is usually achieved by using analytic derivatives calculated jointly with the likelihood for each observation. For initial search, it may be advantageous to calculate the hessian matrix required for the Newton–Raphson search direction rather than use the BHHH approximation. Methods such as Davidon–Fletcher–Powell which use numerical updates of the hessian matrix are not usually efficient for these problems. A careful interpolation along the direction of search (e.g. Davidon's linear search method which uses cubic interpolation) usually speeds convergence.

(3) The log-likelihood L need not be concave in the general case, and there may be local maxima. However, the logit, probit, and linear probability models for binomial response have strictly concave log-likelihood functions, provided the explanatory variables are linearly independent. A check of the condition number of the information matrix J_T during iteration should detect linear dependencies.

A family of consistent estimators of β can be derived by replacing w_t in (2.12) with other weight functions, which may depend on x_t and β but not the response y_t; for example $w_t = F'(x_t\beta)$ corresponds to non-linear least squares. These alternatives are usually inferior to maximum likelihood estimators in both computation and asymptotic statistical properties.

2.4. *Contingency table analysis*

In some economic applications, the number of configurations of explanatory variables is finite, and the data can be displayed in a contingency table with counts of responses in each cell. A variety of statistical methods are available for contingency table analysis; Goodman (1971) and Fienberg (1977) are general introductions. A common approach is to adopt a log-linear model of the *joint* distribution of (y, x) without imposing any structure of cause and response. The conditional probability of y given x will then have a logit form.

Log-linear models of contingency tables can be estimated by simple analysis-of-variance, and are often the most convenient method of obtaining a logit response probability when the dimension of x is not too large. It is difficult within this framework to impose prior restrictions from economic theory on the form of the response probability, a feature that most econometricians would consider a disadvantage.

2.5. *Minimum chi-square method*

Suppose the configurations of x in a contingency table are indexed $n = 1,\dots,N$, and let m_{in} denote the count in the cell with $y = i$ and configuration x_n. The log-likelihood function (2.1) in this notation becomes:

$$L = \frac{1}{T}\sum_{n=1}^{N}\left[m_{1n}\ln F(x_n\beta) + m_{0n}\ln(1 - F(x_n\beta)) + \ln C(m_{\cdot n}, m_{1n})\right], \quad (2.16)$$

with $m_{\cdot n} = m_{0n} + m_{1n}$, $C(m, r) = m!/r!(m-r)!$, and $T = \sum_{n=1}^{N} m_{\cdot n}$. Consistency of maximum likelihood estimates will follow whenever $T \to \infty$, provided a rank condition on the hessian is met. This can be accomplished by letting $N \to \infty$, all $m_{\cdot n} \to \infty$, or both, as long as N is at least the dimension of β.

When the $m_{\cdot n} \to \infty$, the cell frequencies $m_{1n}/m_{\cdot n}$ converge in probability to $P_{1n} = F(x_n\beta)$, suggesting an alternative method of estimating β. Let G denote the inverse of F, so that $G(P_{1n}) = x_n\beta$. A Taylor's expansion of $G(m_{1n}/m_{\cdot n})$ about P_{1n} yields:

$$G(m_{1n}/m_{\cdot n}) = x_n\beta + \varepsilon_n, \tag{2.17}$$

with

$$\varepsilon_n = G'(\tilde{P}_{1n})(m_{1n}/m_{\cdot n} - P_{1n}) \tag{2.18}$$

and $\tilde{P}_{1n}$ an intermediate point. If G' is continuous, then $m_{\cdot n} \to \infty$ implies:

$$\sqrt{m_{\cdot n}}\,\varepsilon_n \xrightarrow{d} N\left(0, (G'(P_{1n}))^2 P_{0n}P_{1n}\right). \tag{2.19}$$

Then β can be estimated consistently by applying least squares to (2.17). Large sample efficiency is improved by correcting for heteroscedasticity; note that $\sigma_n^2 = (G'(P_{1n}))^2 P_{0n}P_{1n}$ is estimated consistently by $(G'(m_{1n}/m_{\cdot n}))^2 m_{0n}m_{1n}/m_{\cdot n}^2$. This estimator was proposed for the logit model by Berkson (1955); further discussion of the approach is in Cox (1970), Goldberger (1964), Theil (1969), and McFadden (1973, 1976). Amemiya (1980) compares the second order asymptotic properties of the Berkson and maximum likelihood estimators.

Economic surveys seldom yield the large cell counts necessary for Berkson's method to have good statistical properties unless data are grouped. However, grouping introduces biases which in many cases are of unacceptable magnitude; see Domencich and McFadden (1975).

2.6. *Discriminant analysis*

When the response probability is deduced from an economic theory implying a causal relationship from x to y, then it is natural to parameterize this form directly as a function $P_1 = F(x\beta)$ consistent with the theory. In other cases it may be more natural to parameterize the joint distribution $H(y, x)$ of (y, x) or the conditional distribution $Q(x|y)$ of x given y. Letting $p(x)$ denote the marginal distribution of x and q_1 the marginal distribution of y, Bayes rule implies that the conditional distribution of y given x is:

$$P_1 = H(1,x)/(H(0,x) + H(1,x)) = Q(x|1)q_1/(Q(x|0)q_0 + Q(x|1)q_1). \tag{2.20}$$

This probability may in some cases have a parametric form commonly assumed for response models, and it may be tempting to give it a causal interpretation. However, a key property of a true causal response $P_1 = F(x\beta)$ is invariance with respect to the marginal distribution $p(x)$ of the explanatory variables. This invariance condition will be satisfied by (2.20) only if the parameterization of $H(y,x)$ or $Q(x|y)$ is "saturated" in x.[6]

Discriminant models parameterize the conditional distributions $Q(x|y)$, and may be motivated by an assumption of causality from y (subpopulation) to x (attributes of subpopulation members). For example, y may index subpopulations of sterile and fecund insects; then $Q(x|y)$ characterizes the distribution of observable attributes of these subpopulations and P_1 in (2.20) gives the probability that an insect with attributes x belongs to population 1. The commonly used normal linear discriminant model assumes the $Q(x|y)$ are normal with means μ_y and common covariance matrix Ω. This requires the x variables to be continuous and range over the real line. The conditional probability of y given x, from (2.20), then has a logit form:

$$P_1 = 1/(1+e^{-\alpha - x\beta}), \tag{2.21}$$

with $\beta = \Omega^{-1}(\mu_1 - \mu_0)$ and $\alpha = \frac{1}{2}(\mu_0'\Omega^{-1}\mu_0 - \mu_1'\Omega^{-1}\mu_1) + \ln(q_1/q_0)$. The parameters μ_y and Ω can be estimated using sub-sample means and pooled sample covariance, $\hat{\mu}_y$ and $\hat{\Omega}$. Alternatively, ordinary least squares applied to the "linear probability model",

$$y = a + xb + \nu, \tag{2.22}$$

yields an estimator $b = \lambda\hat{\Omega}^{-1}(\hat{\mu}_1 - \hat{\mu}_0) = \lambda\hat{\beta}$, where $\lambda = r_0 r_1/(1+(\hat{\mu}_1 - \hat{\mu}_0)'\hat{\Omega}^{-1}(\hat{\mu}_1 - \hat{\mu}_0))$ and r_i is the proportion of sub-population i in the pooled sample. This relation between logit and linear model parameters under the normality assumptions of discriminant analysis was noted by Fisher (1939); other references are Ladd (1966), Anderson (1958) and Chung and Goldberger (1982). It should be emphasized that the relations (2.21) and (2.22) obtained from the discriminant model do not imply a causal response structure despite the familiarity of the forms. Also, if there is in truth a logistic causal response model, it will be coincidental if the distribution of x is the precise mixture of normals consistent with the normal conditional distributions $Q(x|y)$ assumed in discriminant analysis. Otherwise, use of the discriminant sample moments will not yield consistent estimates of the logit model parameters. There is some evidence, however, that the

[6]A model is "saturated" in x if it has enough parameters to completely characterize the marginal distribution $p(x)$ without prior restrictions on $p(x)$. A full log-linear model for $H(y,x)$ has this property.

discriminant estimates of (2.21) are relatively robust with respect to some departures from the normality assumptions; see Domencich and McFadden (1975) and Amemiya (1981).

3. Multinomial response models

3.1. Foundations

The latent variable model (2.2) which generated a binomial response probability generalizes readily to systems in which a vector of latent variables is determined by the explanatory variables, and a generalized indicator function maps the latent variable vector into a vector indexing observed response. Suppose there are m mutually exclusive and exhaustive possible responses, indexed $i=1,\ldots,m$. Let y be an m-vector with $y_i=1$ if response i is observed and $y_i=0$ otherwise, and S_m the set of such vectors. Let

$$y^* = x\beta - \varepsilon \tag{3.1}$$

be a multivariate latent variable model with domain R^h for y^* and a cumulative distribution function $F(\varepsilon|x)$. The generalized indicator function is defined by a partition of R^h into subsets $A_1,\ldots,A_m$, and $z: R^h \to S_m$ with $y = z(y^*)$ satisfying $y_i=1$ if and only if $y^* \in A_i$. Then the response probability is:

$$P_i = P(y_i=1|x,\beta) = P(x\beta - \varepsilon \in A_i) = F(\{x\beta\} - A_i|x), \tag{3.2}$$

where $\{x\beta\} - A_i = \{\varepsilon | x\beta - \varepsilon \in A_i\}$. The latent variable model (3.1) may be a reduced form from a simultaneous equation system, and x may include lagged values of y and y^*, permitting a rich dynamic structure. The relationship between the dimensions h and m, the structure of the sets A_i, and the form of F can all be varied to fit the application. We give three examples.

(a) *Multinomial choice*. Suppose $h = m$, and y_i^* is interpreted as a measure of the utility of alternative i. This may be an index of desirability for consumers, or profit for firms. Suppose A_i is defined to be the set of y^* which have $y_i^* \geq y_j^*$ for $j=1,\ldots,m$, with some rules for breaking ties so that $A_1,\ldots,A_m$ define a partition. Then the response probabilities can be interpreted as the proportions of agents maximizing utility at each alternative when faced with a decision characterized by x. For example, i may index the brand of automobile purchased and y_i^* the maximum utility given brand i, achieved by optimizing on all remaining dimensions of choice.

(b) *Ordered choice*. Suppose $h=1$, and $y^* = x\beta + \varepsilon$ is the "ideal" demand for a commodity if it were available in continuous quantities. Suppose the commodity is actually available in discrete quantities λ_i, and $U(y^* - \lambda_i)$ is the utility of λ_i when the ideal is y^*. Define a_i so that $U(a_i - \lambda_i) = U(a_i - \lambda_{i-1})$ and $A_i = [a_i, a_{i+1})$. Then the response probability:

$$P_i = P(x\beta + \varepsilon \in A_i) = F(a_{i+1} - x\beta|x) - F(a_i - x\beta|x), \tag{3.3}$$

gives the proportion of agents for which quantity λ_i is optimal. This model might be appropriate for describing the choice of number of children or frequency of shopping trips.

(c) *Multivariate binomial choice*. Suppose a vector of h binomial choices $y = (y^1,\dots,y^h)$ is observed, with $y^j = 1$ if $y_j^* \geq 0$ and $y^j = 0$ otherwise. There are $m = 2^h$ possible observable vectors. In the general terminology, A_y is a cartesian product of half-lines, with term j equal to $(-\infty, 0)$ if $y^j = 0$, $[0, +\infty]$ otherwise, and $P_y = P(x\beta - \varepsilon \in A_y)$. If $\sum_{j=1}^{h} y_j^*$ is interpreted as an additively separable utility, with y_j^* the relative desirability of $y^j = 1$ over $y^j = 0$, then P_y gives the proportion of agents for which y is optimal. Dependence in the joint distribution $F(\varepsilon|x)$ generates dependence among the binomial choices. This model might be appropriate for describing holdings in a portfolio of household appliances, or for describing a sequence of binomial decisions over time such as participation in the labor force.

These examples should make clear that there is a rich variety of qualitative response models, drawing upon alternative latent variable structures and generalized indicator functions, which can be tailored for appropriateness and convenience in various applications. Multinomial, ordered, and multivariate responses can appear in any combination. In the third example above, multivariate binomial responses are rewritten as a single multinomial response. Conversely, a multinomial response can always be represented as a sequence of binomial responses. When observations extend over time, the system can be enriched further by treating ε as a stochastic process and permitting lagged responses ("state dependence") among the explanatory variables. With these elaborations, the full panoply of econometric techniques for linear models and time series problems can be brought to bear on qualitative response data. This development of the latent variable formulation of qualitative response models is due to Goldberger (1971), Heckman (1976), Amemiya (1976), and Lee (1981). The last paper also generalizes these systems to combinations of discrete, continuous, censored, and truncated variables. The examples above have been phrased in terms of optimizing behavior by economic agents. We shall develop this connection further to establish the link between stochastic factors surrounding agent decision-making and the structure of response probabilities. However, it should be noted that there are applications of qualitative response models where this framework is inappropriate, or where the analyst may not wish to impose it a priori. This will in general relax prior restrictions on the structure of $x\beta$ or the distribution $F(\varepsilon|x)$ in the latent variable model, but otherwise leave unchanged the latent variable system determining qualitative response. For example, the ordered response model (b) with the latent variable y^* interpreted as susceptibility and the a_i as thresholds for onset of a disease at varying degrees of severity is the Bradley–Terry model widely used in toxicology. Another example is the multivariate binomial model (c) applied to a sequence of outcomes of a collective bargaining process, with y_h^* interpreted as a measure of the relative strength of the opposing agents in period h.

Returning to the problem of qualitative response generated by optimization on the part of economic agents, consider the multinomial choice example (a). For concreteness, suppose the agent is a profit-maximizing firm deciding what product markets to enter or where to locate plants. Given a qualitative alternative i, the firm faces a technology T^i describing its feasible production plans. Maximization of profit subject to T^i yields a restricted profit function Π^i. The technology will depend on attributes t of the firm; the restricted profit function will consequently depend on t and on characteristics w of the firm's market environment, $\Pi^i(t,w)$. The firm will choose the alternative i which maximizes $\Pi^i(t,w)$.

The form of the restricted profit function Π^i will depend on prior assumptions on the technology and on the nature of the markets the firm faces. If, for example, the firm faces competitive markets and w is the vector of prices, then Π^i is a closed, convex, conical[7] function of w; see McFadden (1978a). In non-competitive markets, w summarizes the information available to the firm on strategies of other agents, and the form of Π^i is determined by a theory of non-competitive market behavior.

In empirical application, (t,w) will contain both observed and unobserved components, and the unobserved components will have some distribution over the population of firms. Let z denote the observed components of (t,w), and ν the unobserved components, and let $G(\nu|z)$ denote the distribution of the unobserved components, given z, in the population. Let $\overline{\Pi}^i(z)$ be the expectation of $\Pi^i(z,\nu)$ with respect to $G(\nu|z)$, or some other measure of location for the random function $\Pi^i(z,\cdot)$. Finally, let $x_i\beta$ be a linear-in-parameters global approximation to $\overline{\Pi}^i(z)$, where x is a vector of arithmetic functions of z, and define $\varepsilon_i = x_i\beta - \Pi^i(z,\nu)$. Then ε has a distribution $F(\varepsilon|x)$ induced by ν, and $y_i^* = x_i\beta - \varepsilon_i$ equals the maximum profit obtainable given discrete alternative i, written in the latent variable model notation. If all prices are observed and the function $\Pi^i(t,w)$ is closed, convex, and conical in prices, then the expectation $\overline{\Pi}^i(z)$ will have these properties. The approximation $x_i\beta$ to $\overline{\Pi}^i$ must then approximate these properties, although it need not have them exactly unless the family of functions $x(z)$ used in the approximation is selected to achieve this result. For example, a

[7]A function is conical if it is homogeneous of degree one; closed if the epigraph of the function is a closed set.

convex function $\overline{\Pi}^i$ can be approximated globally by a nonnegative linear combination of convex functions, or alternatively by a polynomial which may fail to be convex over some range; see McFadden (1978a). If it is important to the analysis to impose on the response model all the prior restrictions implied by the theory, as would be the case, for example, if the objective of the study were to test these restrictions, then an approximation should be chosen which inherits the prior restrictions and which does not in itself restrict the ability of the model to fit the data. Given the approximation $x_i\beta$, note that as a consequence of the definition of ε_i, the distribution $F(\varepsilon|x)$ will inherit some properties from the theory. For example, if Π^i and x_i are conical in prices, then $F(\varepsilon|x)$ must have a scale which is conical in prices.

The preceding paragraphs have described a path from the economic theory of behavior of a firm to properties of the latent variable model and associated response probability it generates. In applications it is often useful to reverse this path, writing down a convenient response probability model and then establishing that it meets sufficient conditions for derivation from the theory of the profit-maximizing firm. For the competitive case, a quite general sufficient condition is that $x_i\beta$ be closed, convex, and conical in prices and that ε be linear in prices; see Duncan (1980a) and McFadden (1979a).

Problems involving utility-maximizing consumers can be analyzed by methods paralleling the treatment of the firm, with Π^i replaced by the indirect utility function achieved for given i by optimizing in all remaining dimensions. However, this case is more complex since the expectation with respect to unobservables of the indirect utility function given i does not in general inherit all the properties of an indirect utility function. Consequently, known sufficient conditions for a specified response probability model to be derivable from a population of utility maximizers are quite restrictive, bearing a close relation to the sufficient conditions for individual preferences to aggregate to a social utility consistent with market demands; see McFadden (1981). Whether there is a practical general characterization of the response probability models consistent with a population of utility maximizers, analogous to the integrability theory for individual demand functions, remains an open question.

3.2. *Statistical analysis*

Consider a general multinomial response model with m alternatives, indexed $i=1,\dots,m$,

$$P_i = f^i(x,\theta), \tag{3.4}$$

generated by some latent variable model and generalized indicator function as in

(3.1) and (3.2). The x are observed explanatory variables, and θ is a vector of parameters. Consider an independent random sample with observations (y_t, x_t) for $t=1,\dots,T$. As indicated for the binomial case, maximum likelihood estimation is the most generally applicable and usually the most satisfactory approach to estimation of θ. Let

$$l(y_t, x_t, \theta) = \sum_{i=1}^{m} y_{it} \ln f^i(x_t, \theta) \tag{3.5}$$

denote the log-likelihood of observation t, and

$$L_T(\theta) = \frac{1}{T} \sum_{t=1}^{T} l(y_t, x_t, \theta) \tag{3.6}$$

the sample log-likelihood normalized by sample size. The following regularity conditions will be shown to imply that the maximum likelihood estimator is consistent and asymptotically normal.

(1) The domain of the explanatory variables is a measurable set X with a probability $p(x)$.

(2) The parameter space Θ is a subset of R^k, and the true parameter vector θ^* is in the interior of Θ.

(3) The response model $P_i = f^i(x, \theta)$ is measurable in x for each θ, and for x in a set X_1 with $p(X_1)=1$, $f^i(x,\theta)$ is continuous in θ.

(4) The model satisfies a global identification condition: given $\varepsilon > 0$, there exists $\delta > 0$ such that $|\theta - \theta^*| \geq \varepsilon$ implies:

$$\psi(\theta) \equiv \int \mathrm{d}p(x) \sum_{i=1}^{m} f^i(x,\theta^*) \ln[f^i(x,\theta^*)/f^i(x,\theta)] \geq \delta. \tag{3.7}$$

(5) For $x \in X_1$ with $p(X_1)=1$, and some neighborhood Θ_0 of θ^*, the derivative $\partial f^i(x,\theta)/\partial\theta$ exists and is measurable in x.

(6) For some neighborhood Θ_0 of θ^* and measurable functions $\alpha^i(x)$, $\beta^i(x)$, $\gamma^i(x)$, the following bounds hold:

(i) $f^i(x,\theta) \leq \alpha^i(x)$,

(ii) $|\partial \ln f^i(x,\theta)/\partial\theta| \leq \beta^i(x)$,

(iii) $|\partial \ln f^i(x,\theta)/\partial\theta - \partial \ln f^i(x,\theta')/\partial\theta| \leq \gamma^i(x)|\theta - \theta'|$,

(iv) $\int \mathrm{d}p(x)\alpha^i(x)\gamma^i(x)^2 < \infty$,

(v) $\int \mathrm{d}p(x)\alpha^i(x)\beta^i(x)\gamma^i(x) < \infty$,

(vi) $\int \mathrm{d}p(x)\alpha^i(x)\beta^i(x)^3 < \infty$.

(7) The information matrix $J(\theta^*)$, given by

$$\int \mathrm{d}p(x) \sum_{i=1}^{m} f^i(x,\theta^*)[\partial \ln f^i(x,\theta^*)/\partial\theta][\partial \ln f^i(x,\theta^*)/\partial\theta]', \tag{3.8}$$

is non-singular.

The main results are given by the following theorems.

Theorem 1

If conditions (1)–(4) hold, and $\bar{\theta}_T$ is any sequence of measurable estimators which satisfy

$$L_T(\bar{\theta}_T) \geq \sup_{\Theta} L_T(\theta) - 1/T \tag{3.9}$$

with probability one, then $\bar{\theta}_T$ converges almost surely to θ^*.

Theorem 2

If conditions (1)–(5) hold, then almost surely a unique maximum likelihood estimator $\hat{\theta}_T$ eventually exists and satisfies $\partial L_T(\hat{\theta}_T)/\partial\theta = 0$ and $\hat{\theta}_T \to \theta^*$.

Theorem 3

If conditions (1)–(7) hold, then $\sqrt{T}(\hat{\theta}_T - \theta^*)$ converges in distribution to a normal random vector with mean zero and covariance matrix $J(\theta^*)^{-1}$.

The following paragraphs discuss the regularity conditions and theorems; proof outlines are deferred to the Appendix. Note first that the theorems assume the explanatory variables are independently identically distributed for each observation. This is appropriate for sample survey data, but not necessarily for time-series data. Analogous theorems hold for the case of non-stochastic or jointly distributed explanatory variables, but require stronger bounds and a more complicated definition of the information matrix.

Conditions (1)–(3) are very mild and easily verified in most models. Note that the parameter space Θ is not required to be compact, nor is $\ln f^i(x,\theta)$ required to be bounded. Condition (4) is a substantive identification requirement which states that no parameter vector other than the true one can achieve as high a limiting value of the log-likelihood. Theorem 1 specializes a general consistency theorem of Huber (1965, theorem 1). It is possible to weaken conditions (1)–(4) further, with some loss of simplicity, and still utilize Huber's argument. Note that $L_T(\theta) \leq 0$ and, since $y_{it} = 1$ implies $f^i(x_t,\theta^*) > 0$ almost surely, $L_T(\theta^*) > -\infty$ almost surely. Hence, a sequence of estimators $\bar{\theta}_T$ satisfying (3.9) almost surely exists.

Condition (5), requiring differentiability of $L_T(\theta)$ in a neighborhood of θ^*, will be satisfied by most models. With this condition, Theorem 2 implies that a unique maximum likelihood estimator almost surely eventually exists and satisfies the first-order condition for an interior maximum. This result does *not* imply that every solution of the first-order conditions is consistent. Note that any strongly consistent estimator of θ^* almost surely eventually stays in any specified compact neighborhood of θ^*.

Condition (6) imposes uniform (in θ) bounds on the response probabilities and their first derivatives in a neighborhood of θ^*. Condition (6) (iii) requires that $\partial \ln f^i(x,\theta)/\partial\theta$ be Lipschitzian in a neighborhood of θ^*.

Condition (4) combined with (5) and (6) implies $J(\theta)$ is non-singular at some point in the intersection of each neighborhood of θ^* and line segment extending from θ^*. Hence, condition (7) excludes only pathological irregularities.

Theorem 3 establishes asymptotic normality for maximum likelihood estimates of discrete response models under substantially weaker conditions than are usually imposed. In particular, no assumptions are made regarding second or third derivatives. Theorem 3 extends an asymptotic normality argument of Rao (1972, 5e2) for the case of a multinomial model without explanatory variables.

To illustrate the use of these theorems, consider the multinomial logit model:

$$P_i = \mathrm{e}^{x_i\theta} \Big/ \sum_{j=1}^{m} \mathrm{e}^{x_j\theta}, \tag{3.10}$$

with $x = (x_1,\dots,x_m) \in R^{mk}$ and $\theta \in R^k$. This model is continuous in x and θ, and twice continuously differentiable in θ for each x. Hence, conditions (1)–(3) and (5) are immediately satisfied. Since

$$\partial \ln f^i(x,\theta)/\partial\theta = x_i - \sum_j x_j f^j(x,\theta) \equiv x_i - \bar{x}(\theta), \tag{3.11}$$

$\mathrm{E}|x|^3 < \infty$ is sufficient for condition (6). The information matrix is:

$$J(\theta^*) = \int \mathrm{d}p(x) \sum_{i=1}^{m} f^i(x,\theta)(x_i - \bar{x}(\theta^*))(x_i - \bar{x}(\theta^*))'; \tag{3.12}$$

its non-singularity in (7) is equivalent to a linear independence condition on $(x_1 - \bar{x}(\theta^*)),\dots,x_m - \bar{x}(\theta^*))$. The function $\ln f^i(x,\theta)$ is strictly concave in θ if condition (7) holds, implying that condition (4) is satisfied. Then Theorems 1–3 establish for this model that the maximum likelihood estimator $\hat{\theta}_T$ almost surely eventually exists and converges to θ^*, and $\sqrt{T}(\hat{\theta}_T - \theta^*)$ is asymptotically normal with covariance matrix $J(\theta^*)^{-1}$.

Since maximum likelihood estimators of qualitative response models fit within the general large sample theory for non-linear models, statistical inference is completely conventional, and Wald, Lagrange multiplier, or likelihood ratio statistics can be used for large sample tests. It is also possible to define summary measures of goodness of fit which are related to the likelihood ratio. Let g_t^i and f_t^i be two sequences of response probabilities for the sample points $t = 1,\ldots,T$, and define

$$I_T(g,f) = \frac{1}{T}\sum_{t=1}^{T}\sum_{i=1}^{m} g_t^i \ln\left(\frac{g_t^i}{f_t^i}\right) \tag{3.13}$$

to be the "average information in g beyond that in f". If g is the empirical distribution of the observed response and f is a parametric response model, then $I(g,f)$ is monotone in the likelihood function, and maximum likelihood estimation minimizes the average unexplained information. The better the model fits, the smaller $I_T(g,f)$. Note that for two models f_0 and f_1, the difference in average information $I_T(g,f_0) - I_T(g,f_1)$ is proportional to a likelihood ratio statistic. Goodness-of-fit measures related to (3.13) have been developed by Theil (1970); see also Judge et al. (1981). Related goodness of fit measures are discussed in Amemiya (1982). It is also possible to assess qualitative response models in terms of predictive accuracy; McFadden (1979b) defines prediction success tables and summary measures of predictive accuracy.

3.3. *Functional form*

The primary issues in choice of a functional form for a response probability model are computational practicality and flexibility in representing patterns of similarity across alternatives. Practical experience suggests that functional forms which allow similar patterns of inter-alternative substitution will give comparable fits to existing economic data sets. Of course, laboratory experimentation or more comprehensive economic observations may make it possible to differentiate the fit of function forms with respect to characteristics other than flexibility.

Currently three major families of concrete functional forms for response probabilities have been developed in the literature. These are multinomial logit models, based on the work of Luce (1959), multinomial probit models, based on the work of Thurstone (1927), and elimination models, based on the work of Tversky (1972). Figure 3.1 outlines these families; the members are defined in the following sections. We argue in the following sections that the multinomial logit model scores well on simplicity and computation, but poorly on flexibility. The multinomial probit model is simple and flexible, but scores poorly on computation. Variants of these models, the nested multinomial logit model and the factorial multinomial probit model, attempt to achieve both flexibility and computational practicality.

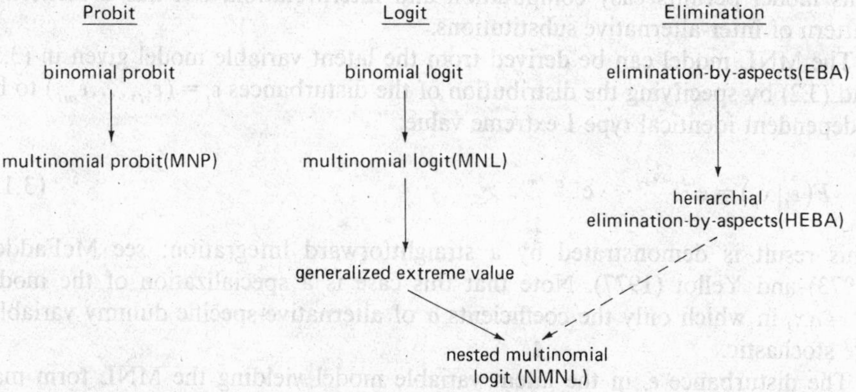

Figure 3.1. Functional forms for multinomial response probabilities.

In considering probit, logit, and related models, it is useful to quantify the hypothesis of an optimizing economic agent in the following terms. Consider a choice set $B = \{1,\dots,m\}$. Alternative i has a column vector of observed attributes x_i, and an associated utility $y_i^* = \alpha' x_i$, where α is a vector of taste weights. Assume α to have a parametric probability distribution with parameter vector θ, and let $\beta = \beta(\theta)$ and $\Omega = \Omega(\theta)$ denote the mean and covariance matrix of α. Let $x_B = (x_1,\dots,x_m)$ denote the array of observed attributes of the available alternatives. Then the vector of utilities $y_B^* = (y_1^*,\dots,y_m^*)$ has a multivariate probability distribution with mean $\beta' x_B$ and covariance matrix $x_B'\Omega x_B$. The response probability $f^i(x_B,\theta)$ for alternative i then equals the probability of drawing a vector y_B^* from this distribution such that $y_i^* \geq y_j^*$ for $j \in B$. For calculation, it is convenient to note that $y_{B-i}^* = (y_1^* - y_i^*,\dots,y_{i-1}^* - y_i^*, y_{i+1}^* - y_i^*,\dots,y_m^* - y_i^*)$ has a multivariate distribution with mean $\beta' x_{B-i}$ and covariance matrix $x_{B-i}'\Omega x_{B-i}$, where $x_{B-i} = (x_1 - x_i,\dots,x_{i-1} - x_i, x_{i+1} - x_i,\dots,x_m - x_i)$, and that $f^i(z_B,\theta)$ equals the non-positive orthant probability for this $(m-1)$-dimensional distribution.

The following sections review a series of concrete probabilistic choice models which can be derived from the structure above.

3.4. *The multinomial logit model*

The most widely used model of multinomial response is the multinomial logit (MNL) form:

$$f^i(x_B,\theta) = \mathrm{e}^{x_i\theta} / \sum_{j\in B} \mathrm{e}^{x_j\theta}. \tag{3.14}$$

This model permits easy computation and interpretation, but has a restrictive pattern of inter-alternative substitutions.

The MNL model can be derived from the latent variable model given in (3.1) and (3.2) by specifying the distribution of the disturbances $\varepsilon_t = (\varepsilon_{1t},\dots,\varepsilon_{mt})$ to be independent identical type I extreme value:

$$F(\varepsilon_t|x_t) = \mathrm{e}^{-\mathrm{e}^{-\varepsilon_{1t}}} \cdots \mathrm{e}^{-\mathrm{e}^{-\varepsilon_{mt}}}. \tag{3.15}$$

This result is demonstrated by a straightforward integration; see McFadden (1973) and Yellot (1977). Note that this case is a specialization of the model $y_i^* = \alpha x_i$ in which only the coefficients α of alternative-specific dummy variables are stochastic.

The disturbance ε_t in the latent variable model yielding the MNL form may have the conventional econometric interpretation of the impact of factors known to the decision-maker but not to the observer. However, it is also possible that a disturbance exists in the decision protocol of the economic agent, yielding stochastic choice behavior. These alternatives cannot ordinarily be distinguished unless the decision protocol is observable or individuals can be confronted experimentally with a variety of decisions.

Interpreted as a stochastic choice model, the MNL form is used in psychometrics and is termed the Luce strict utility model. In this literature, $v_{it} = x_{it}\beta$ is interpreted as a scale value associated with alternative i. References are Luce (1959, 1977) and Marschak (1960).

The vector of explanatory variables x_{it} in the MNL model can be interpreted as attributes of alternative i. Note that components of x_{it} which do not vary with i cancel out of the MNL formula (3.13), and the corresponding component of the parameter vector θ cannot be identified from observation on discrete response.

Some components of x_{it} may be alternative-specific, resulting from the interaction of a variable with a dummy variable for alternative i. This is meaningful if the alternatives are naturally indexed. For example, in a study of durable ownership the alternative of not holding the durable is naturally distinguished from all the alternatives where the durable is held. On the other hand, if there is no link between the true attributes of an alternative and its index i, as might be the case for the set of available dwellings in a study of housing purchase behavior, alternative dummies are meaningless.

Attributes of the respondent may enter the MNL model in interaction with attributes of alternatives or with alternative specific dummies. For example, income may enter a MNL model of the housing purchase decision in interaction with a dwelling attribute such as price, or with a dummy variable for the non-ownership alternative.

A case of the MNL model frequently encountered in sociometrics is that in which the variables in x_{it} are all interactions of respondent attributes and alternative-specific dummies. Let z_t be a $1 \times s$ vector of respondent attributes and δ_{im} be a dummy variable which is one when $i = m$, zero otherwise. Define the $1 \times sM$ vector of interactions,

$$x_{it} = (\delta_{i1} z_t, \ldots, \delta_{iM} z_t),$$

and let $\theta' = (\theta_1', \ldots, \theta_m')$ be a commensurate vector of parameters. Then

$$f^i(x_t, \theta) = \frac{e^{x_{it}\theta}}{e^{x_{1t}\theta} + \cdots + e^{x_{mt}\theta}}$$
$$= \frac{e^{z_t\theta_i}}{e^{z_t\theta_1} + \cdots + e^{z_t\theta_m}}. \tag{3.16}$$

An identifying normalization, say $\theta_1 = 0$, is required. This model is analyzed further by Goodman (1972) and Nerlove and Press (1976).

A convenient feature of the MNL model is that the hessian of the log-likelihood is everywhere negative definite (barring linear dependence of explanatory variables), so that any stationary value is a global maximum.

3.5. *Independence from irrelevant alternatives*

Suppose in the MNL model (3.13) that the vector x_{it} of explanatory variables associated with alternative i depends solely on the attributes of i, possibly interacted with attributes of the respondent. That is, x_{it} does not depend on the attributes of alternatives other than i. Then the MNL model has the Independence from Irrelevant Alternatives (IIA) property, which states that the odds of i being chosen over j is independent of the availability or attributes of alternatives other than i and j. In symbols, this property can be written:

$$\ln \frac{f^i(x_t, \theta)}{f^j(x_t, \theta)} = (x_{it} - x_{jt})\theta, \tag{3.17}$$

independent of x_{mt} for $m \neq i, j$. Equivalently, for $i \in A = \{1, \ldots, J\} \subseteq C = \{1, \ldots, M\}$:

$$f^i(x_{1t}, \ldots, x_{Mt}, \theta) = f^i(x_{1t}, \ldots, x_{Jt}, \theta) \cdot f^A(x_{1t}, \ldots, x_{Mt}, \theta),$$

where

$$f^A(x_{1t}, \ldots, x_{Mt}, \theta) \equiv \sum_{j \in A} f^j(x_{1t}, \ldots, x_{Mt}, \theta).$$

An implication of the IIA property is that the cross-elasticity of the probability of response i with respect to a component of x_{jt} is the same for all i with $i \neq j$. This property is theoretically implausible in many applications. Nevertheless, empirical experience is that the MNL model is relatively robust, as measured by goodness of fit or prediction accuracy, in many cases where the IIA property is theoretically implausible.

When the IIA property is valid, it provides a powerful and useful restriction on model structure. One of its implications is that response probabilities for choice in restricted or expanded choice sets are obtained from the basic MNL form (3.14) simply by deleting or adding terms in the denominator. Thus, for example, one can use the model estimated on existing alternatives to forecast the probability of a new alternative so long as no parameters unique to the new alternative are added.

One useful application of the IIA property is to data where preference rankings of alternatives are observed, or can be inferred from observed purchase order. If the probabilities for the most preferred alternatives in each choice set satisfy the IIA property, then they must be of the MNL form [see McFadden (1973)], and the probability of an observed ranking $1 > 2 > \cdots > m$ of the alternatives is the product of conditional probabilities of choice from successively restricted subsets:

$$P(1 > 2 > \cdots > m) = \frac{e^{x_1\beta}}{\sum_{i=1}^{m} e^{x_i\beta}} \cdot \frac{e^{x_2\beta}}{\sum_{i=2}^{m} e^{x_i\beta}} \cdot \ldots \cdot \frac{e^{x_{m-1}\beta}}{e^{x_{m-1}\beta} + e^{x_m\beta}}.$$

Thus, each selection of a next-ranked alternative from the subset of alternatives not previously ranked can be treated as an independent observation of choice from a MNL model. This formulation of ranking probabilities is due to Marschak (1960). An econometric application has been made by Beggs, Cardell and Hausman (1981); these authors use the method to estimate individual taste parameters and investigate the heterogeneity of these parameters across the population.

The restrictive IIA feature of the MNL model is present only when the vector x_{it} for alternative i is independent of the attributes of alternatives other than i. When this restriction is dropped, the MNL form is sufficiently flexible to approximate any continuous positive response probability model on a compact set of the explanatory variables. Specifically, if $f^i(x_t, \theta)$ is continuous, then it can be approximated globally to any desired degree of accuracy by a MNL model of the form:

$$\tilde{f}^i(x_t, \theta) = \frac{e^{z_{it}\theta}}{e^{z_{1t}\theta} + \cdots + e^{z_{mt}\theta}}, \tag{3.18}$$

where $z_{it} = z_{it}(x_t)$ is an arithmetic function of the attributes of *all* available alternatives, not just the attributes of alternative i. This approximation has been termed the universal logit model. The result follows easily from a global approximation of the vector of logs of choice probabilities by a multivariate Bernstein polynomial; details are given in McFadden (1981).

The universal logit model can describe any pattern of cross-elasticities. Thus, it is not the MNL form per se, but rather the restriction of x_{it} to depend only on attributes of i, which implies IIA restrictions. In practice, the global approximations yielding the universal logit model may be computationally infeasible or inefficient. In addition, the approximation makes it difficult to impose or verify consistency with economic theory. The idea underlying the universal logit model does suggest some useful specification tests; see McFadden, Tye and Train (1976).

3.6. *Limiting the number of alternatives*

When the number of alternatives is large, response probability models may impose heavy burdens of data collection and computation. The special structure of the MNL model permits a reduction in problem scale by either aggregating alternatives or by analyzing a sample of the full alternative set. Consider first the aggregation of relatively homogeneous alternatives into a smaller number of primary types.

Suppose elemental alternatives are doubly indexed ij, with i denoting primary type and j denoting alternatives within a type. Let M_i denote the number of alternatives which are of type i. Suppose choice among all alternatives is described by the MNL model. Then choice among primary types is described by MNL probabilities of the form:

$$f^i(x_t, \theta) = \frac{\exp(x_{it}\theta + \ln M_i + w_{it})}{\sum_k \exp(x_{kt}\theta + \ln M_k + w_{kt})}, \tag{3.19}$$

where x_{it} is the mean within type i of the vectors x_{ijt} of explanatory variables for the alternative ij, and w_{it} is a correction factor for heterogeneity within type i which satisfies:

$$w_{it} = \ln \frac{1}{M_i} \sum_{j=1}^{M_i} \exp\left[(x_{ijt} - x_{it})\theta\right]. \tag{3.20}$$

If the alternatives within a type are homogeneous, then $w_i = 0$.

A useful approximation to w_i can be obtained if the deviations $x_{ijt} - x_{it}$ within type i can be treated as independent random drawings from a multivariate distribution which has a cumulant generating function $W_{it}(\cdot)$. If the number of alternatives M_i is large, then the law of large numbers implies that w_i converges almost surely to $w_i = W_{it}(\theta)$. For example, if $x_{ijt} - x_{it}$ is multivariate normal with covariance matrix Ω_{it}, then $w_i \approx W_{it}(\theta) \equiv \theta'\Omega_{it}\theta/2$.

A practical method for estimation is to either assume within-type homogeneity, or to use the normal approximation to w_i, with Ω_{it} either fitted from data or treated as parameters with some identifying restrictions over i and t. Then θ can be estimated by maximum likelihood estimation of (3.19). The procedure can be iterated using intermediate estimates of θ in the exact formula for w_i. Data collection and processing can be reduced by sampling elemental alternatives to estimate w_i. However, it is then necessary to adjust the asymptotic standard errors of coefficients to include the effect of sampling errors on the measurement of w_i. Further discussion of aggregation of alternatives in a MNL model can be found in McFadden (1978b).

A second method of reducing the scale of data collection and computation in the MNL model when it has the IIA property is to sample a sub-set of the full set of alternatives. The IIA property implies that the *conditional* probabilities of choosing from a restricted subset of the full choice set equal the choice probabilities when the choice set equals the restricted set. Then the MNL model can be estimated from data on alternatives sampled from the full choice set. In particular, the MNL model can be estimated from data on binary conditional choices. Furthermore, subject to one weak restriction, biased sampling of alternatives can be compensated for within the MNL estimation.

Let $C = \{1,\dots,M\}$ denote the full choice set, and $D \subseteq C$ a restricted subset. The protocol for sampling alternatives is defined by a probability $\pi(D|i_t, x_t)$ that D will be sampled, given observed explanatory variables x_t and choice i_t. For example, the sampling protocol of selecting the chosen alternative plus one non-chosen alternative drawn at random satisfies

$$\pi(D|i_t, x_t) = \begin{cases} 1/(M-1), & \text{if } D = \{i_t, j\} \subseteq C, i_t \neq j, \\ 0, & \text{otherwise.} \end{cases} \tag{3.21}$$

Let D_t denote the subset for case t. The weak regularity condition is:

Positive conditioning property

If an alternative $j \in D_t$ were the observed choice, there would be a positive probability that the sampling protocol would select D_t; i.e. if $j \in D_t$, then $\pi(D_t|j, x_t) > 0$.

If the positive conditioning property and a standard identification condition hold, then maximization of the modified MNL log-likelihood function:

$$\frac{1}{T}\sum_{t=1}^{T}\ln\frac{\exp\left[x_{i_t}\theta+\ln\pi(D_t|i_t,x_t)\right]}{\sum_{j\in D_t}\exp\left[x_j\theta+\ln\pi(D_t|j,x_t)\right]} \tag{3.22}$$

yields consistent estimates of θ. This result is proved by showing that (3.22) converges in probability uniformly in θ to an expression which has a unique maximum at the true parameter vector; details are given in McFadden (1978). When π is the same for all $j\in D_t$, the terms involving π cancel out of the above expression. This is termed the *uniform* conditioning property; the example (3.21) satisfies this property.

Note that the modified MNL log-likelihood function (3.22) is simply the conditional log-likelihood of the i_t, given the D_t. The inverse of the information matrix for this conditional likelihood is a consistent estimator of the covariance matrix of the estimated coefficients, as usual.

3.7. *Specification tests for the MNL model*

The MNL model in which the explanatory variables for alternative i are functions solely of the attributes of that alternative satisfies the restrictive IIA property. An implication of this property is that the model structure and parameters are unchanged when choice is analyzed conditional on a restricted subset of the full choice set. This is a special case of uniform conditioning from the section above on sampling alternatives.

The IIA property can be used to form a specification test for the MNL model. Let C denote the full choice set, and D a proper subset of C. Let β_C and V_C denote parameter estimates obtained by maximum likelihood on the full choice set, and the associated estimate of the covariance matrix of the estimators. Let β_D and V_D be the corresponding expressions for maximum likelihood applied to the restricted choice set D. (If some components of the full parameter vector cannot be identified from choice within D, let β_C, β_D, V_C, and V_D denote estimates corresponding to the identifiable sub-vector.) Under the null hypothesis that the IIA property holds, implying the MNL specification, $\beta_D-\beta_C$ is a consistent estimator of zero. Under alternative model specifications where IIA fails, $\beta_D-\beta_C$ will almost certainly not be a consistent estimator of zero. Under the null hypothesis, $\beta_D-\beta_C$ has an estimated covariance matrix V_D-V_C. Hence, the statistic

$$S = (\beta_D - \beta_C)'(V_D - V_C)^{-1}(\beta_D - \beta_C) \tag{3.23}$$

is asymptotically chi-square with degrees of freedom equal to the rank of $V_D - V_C$.

This test is analyzed further in Hausman and McFadden (1984). Note that this is an omnibus test which may fail because of misspecifications other than IIA. Empirical experience and limited numerical experiments suggest that the test is not very powerful unless deviations from MNL structure are substantial.

3.8. *Multinomial probit*

Consider the latent variable model for discrete response, $y_t^* = x_t\theta + \varepsilon_t$ and $y_{mt} = 1$ if $y_{mt}^* \geq y_{nt}^*$ for $n = 1, \ldots, M$, from (3.1) and (3.2). If ε_t is assumed to be multivariate normal, the resulting discrete response model is termed the multinomial probit (MNP) model. The binary case has been used extensively in biometrics; see Finney (1971). The multivariate model has been investigated by Bock and Jones (1968), McFadden (1976), Hausman and Wise (1978), Daganzo (1980), Manski and Lerman (1981), and McFadden (1981).

A form of the MNP model with a plausible economic interpretation is $y_t^* = x_t\alpha_t$, where α_t is multivariate normal with mean β and covariance matrix Ω, and represents taste weights which vary randomly in the population. Note that this form implies $E\varepsilon_t = 0$ and $\mathrm{cov}(\varepsilon_t) = x_t\Omega x_t'$ in the latent variable model formulation. If x_t includes alternative dummies, then the corresponding components of α_t are additive random contributions to the latent values of the alternatives. Some normalizations are required in this model for identification.

When correlation is permitted between alternatives, so $\mathrm{cov}(\varepsilon_t)$ is not diagonal, the MNP model does not have the IIA or related restrictive properties, and permits very general patterns of cross-elasticities. This is true in particular for the random taste weight version of the MNP model when there are random components of α_t corresponding to attributes which vary across alternatives.

Evaluation of MNP probabilities for M alternatives generally requires evaluation of $(M-1)$-dimensional orthant probabilities. In the notation of subsection 3.3, $f^1(x_B; \beta, \Omega)$ is the probability that the $(M-1)$-dimensional normal random vector y_{B-1}^* with mean βx_{B-1} and covariance matrix $x_{B-1}\Omega X_{B-1}'$ is non-positive. For $M \leq 3$, the computation of these probabilities is comparable to that for the MNL model. However, for $M \geq 5$ and Ω unrestricted, numerical integration to obtain these orthant probabilities is usually too costly for practical application in iterative likelihood maximization for large data sets. An additional complication is that the hessian of the MNP model is not known to be negative definite; hence a search may be required to avoid secondary maxima.

For a multivariate normal vector $(y_1^*,\ldots,y_m^*)$, one can calculate the mean and covariance matrix of $(y_1^*,\ldots,y_{m-2}^*,\max(y_{m-1}^*,y_m^*))$; these moments involve only binary probits and can be computed rapidly. A quick, but crude, approximation to MNP probabilities can then be obtained by writing:

$$f^1(x,\beta,\Omega)=P(y_1^*>\max(y_2^*,\max(y_3^*,\ldots)\ldots)) \tag{3.24}$$

and approximating the maximum of two normal variates by a normal variate; see Clark (1961) and Daganzo (1980). This approximation is good for non-negatively correlated variates of comparable variance, but is poor for negative correlations or unequal variances. The method tends to overestimate small probabilities. For assessments of this method, see Horowitz, Sparmann and Daganzo (1981) and McFadden (1981).

A potentially rapid method of fitting MNP probabilities is to draw realizations of α_t repeatedly and use the latent variable model to calculate relative frequencies, starting from some approximation such as the Clark procedure. This requires a large number of simple computer tasks, and can be programmed quite efficiently on an array processor. However, it is difficult to compute small probabilities accurately by this method; see Lerman and Manski (1980).

One way to reduce the complexity of the MNP calculation is to restrict the structure of the covariance matrix Ω by adopting a "factor-analytic" specification of the latent variable model $y_i^*=\beta x_i+\varepsilon_i$. Take

$$\varepsilon_i=\eta_i+\sum_{j=1}^{J}\gamma_{ij}\nu_j, \tag{3.25}$$

with η_i and ν_j independent normal variates with zero means and variances σ_i^2 and 1 respectively. The "factor loading" γ_{ij} is in the most general case a parametric function of the observed attributes of alternatives, and can be interpreted as the level in alternative i of an unobserved characteristic j. With this structure, the response probability can be written:

$$f^1(x,\beta,\gamma,\sigma)=\int_{\eta_1=-\infty}^{+\infty}\int_{\nu_1=-\infty}^{+\infty}\cdots\int_{\nu_J=-\infty}^{+\infty}\frac{1}{\sigma_1}\phi\left(\frac{\eta_1}{\sigma_1}\right)\prod_{j=1}^{J}\phi(\nu_j)$$
$$\times\prod_{i=2}^{M}\Phi\left(\frac{\beta(x_1-x_i)+\sum_{j=1}^{J}(\gamma_{1j}-\gamma_{ij})\nu_j+\eta_1}{\sigma_i}\right)\mathrm{d}\eta_1,\mathrm{d}\nu_1\cdots\mathrm{d}\nu_J. \tag{3.26}$$

Numerical integration of this formula is easy for $J \leq 1$, but costly for $J \geq 3$. Thus, this approach is generally practical only for one or two factor models. The independent MNP model ($J = 0$) has essentially the same restrictions on cross-alternative substitutions as the MNL model; there appears to be little reason to prefer one of these models over the other. However, the one and two factor models permit moderately rich patterns of cross-elasticities, and are attractive practical alternatives in cases where the MNL functional form is too restrictive.

Computation is the primary impediment to widespread use of the MNP model, which otherwise has the elements of flexibility and ease of interpretation desirable in a general purpose qualitative response model. Implementation of a fast and accurate approximation to the MNP probabilities remains an important research problem.

3.9. *Elimination models*

An elimination model views choice as a process in which alternatives are screened from the choice set, using various criteria, until a single element remains. It can be defined by the probability of transition from a set of alternatives to any subset, $Q(D|C)$. If each transition probability is stationary throughout the elimination process, then the choice probabilities satisfy the recursion formula:

$$f^i(C) = \sum_D Q(D|C) f^i(D), \tag{3.27}$$

where $f^i(C)$ is the probability of choosing i from set C.

Elimination models were introduced by Tversky (1972) as a generalization of the Luce model to allow dependence between alternatives. An adaptation of Tversky's elimination by aspects (EBA) model suitable for econometric work takes transition probabilities to have a MNL form:

$$Q(D|C) = \mathrm{e}^{x_D \beta_D} \Big/ \sum_{\substack{A \subseteq C \\ A \neq C}} \mathrm{e}^{x_A \beta_A}, \tag{3.28}$$

where x_D is a vector of attributes common to and unique to the set of alternatives in D. When x_B is a null vector and by definition $\mathrm{e}^{x_B \beta_B} = 0$ for sets B of more than one element, this model reduces to the MNL model. Otherwise, it does not have restrictive IIA-like properties.

The elimination model is not known to have a latent variable characterization. However, it can be characterized as the result of maximization of random lexicographic preferences. The model defined by (3.27) and (3.28) has not been applied in economics. However, if the common unique attributes x_D can be defined in an application, this should be a straightforward and flexible functional form.

One elimination model which can be expressed in latent variable form is the *generalized extreme value* (GEV) model introduced by McFadden (1978, 1981). Let $H(w_1,\dots,w_m)$ be a non-negative, linear homogeneous function of non-negative $w_1,\dots,w_m$ which satisfies

$$\lim_{w_i \to \infty} H(w_1,\dots,w_m) = +\infty, \tag{3.29}$$

and has mixed partial derivatives of all orders, with non-positive even and non-negative odd mixed derivatives. Then,

$$F(\varepsilon_1,\dots,\varepsilon_m) = \exp\{-H(\mathrm{e}^{-\varepsilon_1},\dots,\mathrm{e}^{-\varepsilon_m})\} \tag{3.30}$$

is a multivariate extreme value cumulative distribution function. The latent variable model $y_i^* = x_i\beta + \varepsilon_i$ for $i \in B = \{1,\dots,m\}$ with $(\varepsilon_1,\dots,\varepsilon_m)$ distributed as (3.30) has response probabilities:

$$f^i(x,\beta) = \partial \ln H(\mathrm{e}^{x_1\beta},\dots,\mathrm{e}^{x_m\beta})/\partial(x_i\beta). \tag{3.31}$$

The GEV model reduces to the MNL model when

$$H(w_1,\dots,w_m) = \left(\sum_{i=1}^{M} w_i^{1/\lambda}\right)^{\lambda}, \tag{3.32}$$

with $0 < \lambda \le 1$. An example of a more general GEV function is:

$$H(w_1,\dots,w_m) = \sum_{C \subseteq B} a(C)\left\{\sum_{D \subseteq C} b(D,C)\left[\sum_{i \in D} w_i^{1/\lambda_{DC}\lambda_C}\right]^{\lambda_{DC}}\right\}^{\lambda_C}, \tag{3.33}$$

where $0 < \lambda_{DC}, \lambda_C \le 1$ and a and b are non-negative functions such that each i is contained in a D and C with $a(C), b(D,C) > 0$. The response probability for (3.33) can be written:

$$f^i(x,\theta) = \sum_{C \subseteq B}\sum_{D \subseteq C} Q(i|D,C)Q(D|C)Q(C|B), \tag{3.34}$$

where $i \in D \subseteq C \subseteq B$,

$$Q(i|D,C) = \exp(x_i\beta/\lambda_{DC}\lambda_C) \Big/ \sum_{j \in D} \exp(x_j\beta/\lambda_{DC}\lambda_C), \tag{3.35}$$

$$J(D,C) = \ln \sum_{j \in D} \exp(x_j\beta/\lambda_{DC}\lambda_C), \tag{3.36}$$

$$Q(D|C) = b(D,C)\exp[J(D,C)\lambda_{DC}] \Big/ \sum_{D' \subseteq C} b(D',C)\exp[J(D',C)\lambda_{D'C}], \tag{3.37}$$

$$I(C) = \ln \sum_{D' \subseteq C} b(D',C)\exp[J(D',C)\lambda_C], \tag{3.38}$$

$$Q(C) = a(C)\exp[I(C)\lambda_C] \Big/ \sum_{C' \subseteq B} a(C')\exp[I(C')\lambda_{C'}]. \tag{3.39}$$

This can be interpreted as an elimination model in which $a(C)$ and $b(D,C)$ determine the probability of various chains of sets of non-eliminated alternatives, and λ_{DC} and λ_C measure the degree of independence of the ε_i within the set D obtained from C, and within the set C, respectively. The expressions in (3.36) and (3.38) are termed *inclusive values* of the associated sets of alternatives.

When all the λ's are one, this model reduces to a simple MNL model. Alternatively, when λ_{DC} is near zero, the elimination model treats D essentially as if it contained a single alternative with a scale value equal to the maximum of the scale values of the elements in D.

Inspection of the two elimination models described above suggests that they are comparable in terms of flexibility and complexity. Other things equal, the GEV model will tend to imply sharper discrimination among similar alternatives than the EBA model. Limited numerical experiments suggest that the two models will be difficult to distinguish empirically.

3.10. *Hierarchical response models*

When asked to describe the decision process leading to qualitative choice, individuals often depict a hierarchical structure in which alternatives are grouped into clusters which are "similar". The decision protocol is then to eliminate clusters, proceeding until a single alternative remains. An example of a decision tree is given in Figure 3.2. Alternatives $a-e$ are in one primary cluster, f and g in a second, and $a-c$ are in a secondary cluster. Either of the elimination models described in the preceding section can be specialized to describe hierarchical response by permitting transitions from a node only to one of the nodes

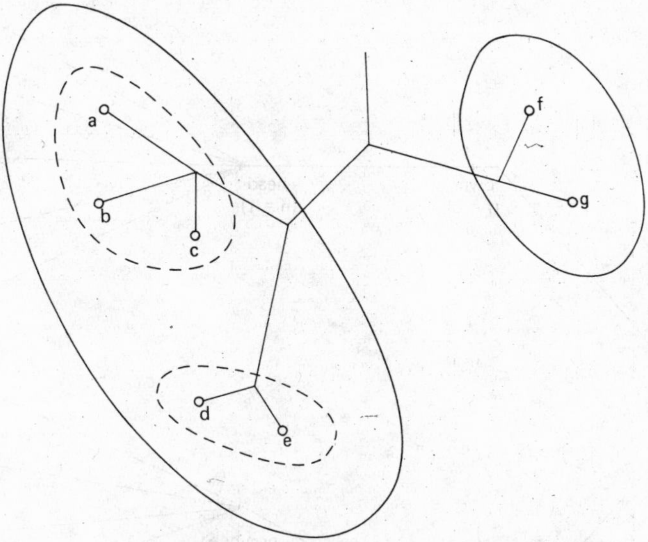

Figure 3.2. A hierarchical decision tree.

immediately below it in the tree. Hierarchical decision models are discussed further in Tversky and Sattath (1979), and McFadden (1981).

A hierarchical elimination model based on the generalized extreme value structure described earlier generalizes the MNL model to a nested multinomial logit (NMNL) structure. Each transition in the tree is described by a MNL model with one of the variables being an "inclusive value" which summarizes the attributes of alternatives below a node. An "independence" parameter at each node in the tree discounts the contribution to value of highly similar alternatives.

We shall discuss the structure of the NMNL model using an example of consumer choice of housing. As illustrated in Figure 3.3, the decision can be described in hierarchical form: first whether to own or rent, second if renting whether to be the head of household or to sublet from someone else (non-head), and finally what dwelling unit to occupy within the chosen cluster. Let $C = \{1,\dots,12\}$ index the final alternatives, $r = 0,1$ index the primary cluster for own and rent, and $h = 0,1$ index the secondary clusters for head and non-head. Define A_{rh} to be the set of final alternatives contained in the subcluster rh, and A_r to be the set of subclusters contained in the cluster r. For example, A_0 contains the (trivial) subcluster $h=1$; A_1 contains two subclusters $h=0$ and $h=1$; and $A_{11} = \{10,11,12\}$.

The response probability for the NMNL model can be written as a product of transition probabilities. For $i \in A_{rh}$ and $h \in A_r$:

$$f^i(x,\theta) = Q(i|A_{rh})Q(A_{rh}|A_r)Q(A_r). \tag{3.40}$$

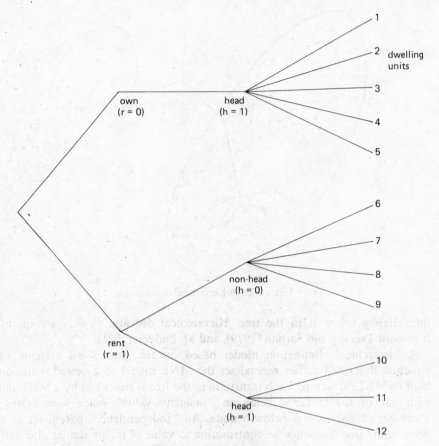

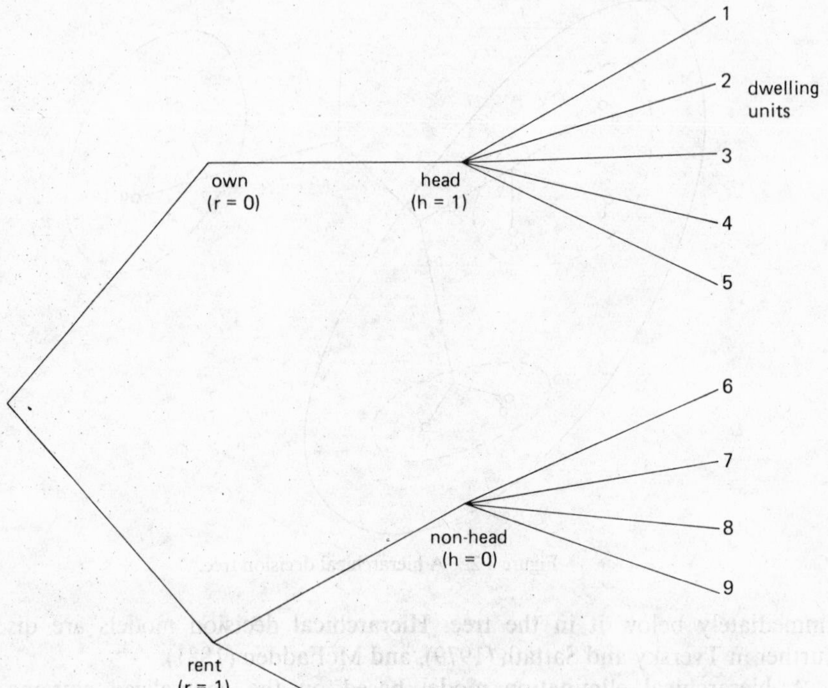

Figure 3.3. Housing choice.

Each transition probability has a NMNL form:

$$Q(i|A_{rh}) = e^{z_i\alpha} \Big/ \sum_{j \in A_{rh}} e^{z_j\alpha}, \tag{3.41}$$

$$Q(A_{rh}|A_r) = \exp(w_{rh}\beta + J_{rh}\kappa_{rh}) \Big/ \sum_{c \in A_r} \exp(w_{rc}\beta + J_{rc}\kappa_{rc}), \tag{3.42}$$

$$Q(A_r) = \exp(u_r\gamma + I_r\lambda_r) \Big/ \sum_{s=0}^{1} \exp(u_s\gamma + I_s\lambda_s). \tag{3.43}$$

Here, $x_i = (z_i, w_{rh}, u_r)$ is the vector of attributes associated with alternative $i \in A_{rh}$ and $h \in A_r$, with w_{rh} and u_r denoting components which are common within the clusters A_{rh} or A_r, respectively; $(\alpha, \beta, \gamma, \kappa_{rs}, \lambda_r) \equiv \theta$ are parameters; and J_{rh} and I_h

are inclusive values satisfying:

$$J_{rh} = \ln \sum_{i \in A_{rh}} e^{z_i \alpha}, \tag{3.44}$$

$$I_r = \ln \sum_{h \in A_r} \exp(w_{rh}\beta + J_{rh}\kappa_{rh}). \tag{3.45}$$

Note that J_{rh} and I_r are logs of the denominators in (3.41) and (3.42), respectively. For this example, note that $Q(A_{01}|A_0) = 1$ and $I_0 = w_{01}\beta + J_{01}\kappa_{01}$.

Consider the function:

$$H(e^{v_1}, \dots, e^{v_{12}}; \theta) = \sum_{r=0}^{1} \left[\sum_{h \in A_r} \left(\sum_{i \in A_{rh}} \exp(v_i / \kappa_{rh}\lambda_r) \right)^{\kappa_{rh}} \right]^{\lambda_r}, \tag{3.46}$$

with

$$v_i = z_i \alpha \kappa_{rh} \lambda_r + w_{rh} \beta \lambda_r + u_r \delta, \tag{3.47}$$

for $i \in A_{rh}$ and $h \in A_r$. This is a generating function for the response probabilities, satisfying $\partial \ln H / \partial v_i = f^i(x, \theta)$, and can be interpreted as a measure of social utility; see McFadden (1981). The parameters κ_{rh} and λ_r are measures of the "independence" of alternatives within subclusters and clusters respectively.

If $\kappa_{rh} = \lambda_r = 1$, then the NMNL model reduces to a simple MNL model. When $0 < \kappa_{rh}, \lambda_r \leq 1$, the NMNL model is consistent with a latent variable model with generalized extreme value distributed disturbances: $y_i^* = v_i + \varepsilon_i$ and

$$F(\varepsilon_1, \dots, \varepsilon_{12}) = \exp[-H(e^{-\varepsilon_1}, \dots, e^{-\varepsilon_{12}}; \theta)], \tag{3.48}$$

and is therefore consistent with an assumption of optimizing economic agents. It should be obvious that this structure generalizes to any number of alternatives and levels of clustering.

To interpret the impact of the independence parameters κ_{rh} and λ_r on cross-alternative substitutability, consider the cross-elasticity of the response probability for $i \in A_{rh}$ and $h \in A_r$ with respect to component k of the vector z_j of attributes of alternative $j \in A_{r'h'}$ and $h' \in A_{r'}$:

$$\begin{aligned} \partial \ln f^i(x,\theta) / \partial \ln z_{jk} &= \kappa_{r'h'}\lambda_{r'}\alpha_k z_{jk} f_i(x,\theta)^{-1} \partial^2 H / \partial v_i \partial v_j \\ &= \alpha_k z_{jk} \{ \delta_{ij} - \kappa_{r'h'}\lambda_{r'} f^j(x,\theta) \\ &\quad + \kappa_{rh'}(\lambda_r - 1)\delta_{rr'} Q(j|A_{rh'}) Q(A_{rh'}|A_r) \\ &\quad + (\kappa_{rh} - 1)\delta_{rr'}\delta_{hh'} Q(j|A_{rh}) \}. \end{aligned} \tag{3.49}$$

For $0 < \kappa_{rh}, \lambda_r < 1$, one obtains the plausible property that cross-elasticities are largest in magnitude for alternatives in the same r and h cluster, and smallest in magnitude when both r and h clusters differ. Note that values of κ_{rh} or λ_r outside the unit interval imply that one of the expected magnitude rankings is violated. Therefore, estimates of κ_{rh} or λ_r outside the unit interval may indicate a misspecified hierarchical structure, and the fitted cross-elasticity magnitude may identify a more appropriate structure.

It is of interest to compare the complexity and flexibility of the NMNL model, say in the form (3.40) corresponding to Figure 3.3, to a MNP model with a factorial structure which has the same pattern of similarities. This is achieved in the MNP model by introducing one factor for each node in the decision tree between the stem and the final "twigs". Thus, the clustering in Figure 3.3 requires four factors—own, rent, rent/head, and rent/non-head. An MNP model with this structure can be specified with a number of parameters comparable to the NMNL model by making the factor loadings uniform within each cluster, or can be made more flexible by allowing intra-cluster heterogeneity in loadings. However, as noted in the discussion of (3.26) computation of a four-factor model will be too costly in most applications. We conclude that the NMNL and factorial MNP are comparable in complexity, with some advantage to the latter in terms of flexibility and ease of interpretation. However, computational barriers currently limit use of the factorial MNP model to simple trees with one to three nodes.

The NMNL model can be estimated by direct maximum likelihood methods. The likelihood is not concave in all parameters, and is highly non-linear in the inclusive value coefficients. A simpler procedure which is consistent, but often fairly inefficient, is to estimate the transition probabilities (3.31)–(3.33) sequentially, using the data on transitions implied by the observed choices and inclusive values calculated from preceding stages. Each stage involves a concave MNL maximum likelihood problem. Beyond the first stage, standard errors are affected by the use of estimated coefficients in the calculation of inclusive values. Amemiya (1978d) and McFadden (1981) provide formulae for correcting the standard errors.

It is possible in principle to obtain asymptotically efficient estimates by carrying out one Newton–Raphson iteration on the full likelihood function, starting from the consistent sequential estimates. In practice, the strong non-linearity of the likelihood in the inclusive value coefficients and sensitivity of the estimates to model specification sometimes lead to full maximum likelihood estimates which are rather far from the initially consistent estimates, and to erratic results from the one-step procedure. Consequently, multiple iterations may be required to approximate the maximum of the full likelihood function. These problems seem to be particularly common when by other indications the decision tree is misspecified.

Under the null hypothesis that the NMNL model is correctly specified, the sequential estimator $\tilde{\theta}_T$ and full MLE $\hat{\theta}_T$ satisfy $\sqrt{T}(\tilde{\theta}_T - \theta) \xrightarrow{\text{L}} N(0, \tilde{\Omega})$ and $\sqrt{T}(\hat{\theta}_T - \theta) \xrightarrow{\text{L}} N(0, \Omega)$, and asymptotic efficiency implies $\sqrt{T}(\tilde{\theta}_T - \hat{\theta}_T) \xrightarrow{\text{L}}$

$N(0, \tilde{\Omega} - \Omega)$. Thus the statistic $T(\tilde{\theta}_T - \hat{\theta}_T)'(\tilde{\Omega} - \Omega)^{+}(\tilde{\theta}_T - \hat{\theta}_T)$, where $(\tilde{\Omega} - \Omega)^{+}$ is a generalized inverse of rank p, is asymptotically χ^2_p under the null. This statistic can then be used as an omnibus test of the NMNL specification.

Since the NMNL model reduces to a MNL model when the inclusive value coefficients are one, it can provide a basis for a classical Lagrange multiplier test of the IIA property of the MNL model. Consider testing $\lambda = 1$ in the model (3.30)–(3.33), with $\kappa = 1$ as a maintained hypothesis. Suppose A_1 is the set of rental alternatives. Let $\sigma = (\alpha\lambda, \beta\lambda, \gamma)$ and $\theta = (\sigma, \lambda)$, and let

$$L = \frac{1}{T}\sum_{t=1}^{T}\sum_{i\in C} y_{it}\ln f^i(x_t, \theta) \tag{3.50}$$

be the normalized log-likelihood function. The Lagrange multiplier statistic has the general form:

$$\mathrm{LM} = L'_{\lambda}\left[EL_{\lambda}L'_{\lambda} - (EL_{\lambda}L'_{\sigma})(EL_{\sigma}L'_{\sigma})^{-1}EL_{\sigma}L'_{\lambda}\right]^{-1}L_{\lambda}, \tag{3.51}$$

where the derivatives are evaluated at $\lambda = 1$ at which the model reduces to a MNL model. For this problem, letting $f_t^i = f^i(x_t, \theta)$:

$$L_{\sigma} = \frac{1}{T}\sum_{t=1}^{T}\sum_{i\in C}\left(y_{it} - f_t^i\right)x_{it}, \tag{3.52}$$

$$L_{\lambda} = \frac{1}{T}\sum_{t=1}^{T}\sum_{i\in A_1}\left(y_{it} - f_t^i\right)(I_t\lambda - x_{it}\sigma), \tag{3.53}$$

$$V_{\lambda\lambda} \equiv TEL_{\lambda}L'_{\lambda} = \frac{1}{T}\sum_{t=1}^{T}\left[\sum_{i\in A_1} f_t^i(I_t\lambda - x_{it}\sigma)^2 - \left(\sum_{i\in A_1} f_t^i(I_t\lambda - x_{it}\sigma)\right)^2\right], \tag{3.54}$$

$$V_{\sigma\sigma} \equiv TEL_{\sigma}L'_{\sigma} = \frac{1}{T}\sum_{t=1}^{T}\sum_{i\in C} f_t^i(x_{it} - x_{Ct})'(x_{it} - x_{Ct}), \tag{3.55}$$

$$V_{\lambda\sigma} = TEL_{\lambda}L'_{\sigma} = \frac{1}{T}\sum_{t=1}^{T}\sum_{i\in A_1} f_t^i(I_t\lambda - x_{it}\sigma)(x_{it} - x_{Ct}), \tag{3.56}$$

$$x_{Ct} = \sum_{i\in C} f_t^i x_{it}, \tag{3.57}$$

$$I_t = \ln\sum_{i\in A_1} \mathrm{e}^{x_{it}\sigma}. \tag{3.58}$$

Then the final form of the test statistic is:

$$\mathrm{LM} = \left[\sum_{t=1}^{T} \sum_{i \in A_1} (y_{it} - f_t^i)(I_t - x_{it}\sigma) \right]^2 \Big/ T\left(V_{\lambda\lambda} - V_{\lambda\sigma} V_{\sigma\sigma}^{-1} V_{\sigma\lambda}\right). \tag{3.59}$$

Under the null hypothesis, this statistic is asymptotically chi-square with one degree of freedom. Further discussion of specification tests for the MNL model and examples are given in Hausman and McFadden (1984).

3.11. *An empirical example*

To illustrate application of the MNL and NMNL models, and associated tests, we apply the housing decision tree given in Figure 3.3 and the associated NMNL model in (3.40)–(3.43) to data on the housing status of single elderly men from the 1977 U.S. Annual Housing Survey for the Albany–Schenectady–Troy, N.Y. SMSA. These results were prepared by Axel Boersch-Supan, and are a simplified version of some models estimated by Boersch-Supan and Pitkin (1982); this reference contains a detailed description of variables and analysis of other socioeconomic groups.

The sample contains 159 single elderly men, of whom 45.9% are owners, 30.2% are renter-heads, and 23.9% are renter non-heads. Selection of dwelling unit is not modeled, and units within a cluster are treated as homogeneous and sufficiently similar to be adequately characterized as a single typical unit.

Two price variables are considered for each person in the sample, out-of-pocket costs (*OPCOST*) and expected net return on equity in owned units (*RETURN*). Out-of-pocket costs are the operating costs of the housing unit. For rental units, this is gross rent including utilities as reported in the survey. For owner-occupied units, *OPCOST* consists of mortgage and real-estate tax payments, utility costs, insurance payments, and maintenance. These direct costs are reduced by estimated savings on federal income taxes resulting from the deductability of mortgage interest and property taxes. Consequently *OPCOST* is influenced in a non-linear way by income. Costs in dwellings with more than one nuclear family unit are assumed to be apportioned according to the total number of adults, children counting as half an adult.

The *RETURN* variable for owner-occupied dwellings is defined as expected appreciation less equity cost, and is taken to be a proportion of dwelling value determined by average annual appreciation in the area since 1970, equity as a fraction of value estimated from date of purchase, and a discount factor reflecting opportunity cost of equity.

The construction of *OPCOST* and *RETURN* for chosen alternatives is based on individual reported costs, while for non-chosen alternatives these variables are based on the average experience of recent movers. Consequently, the estimated models should be interpreted as "reduced form" state models which reflect the relationship between status and costs, taking into account the inertia and non-transferable discounts associated with tenure. These models may accurately forecast future status-cost patterns provided there is no structural change in turnover rates or tenure distributions. They should not be interpreted as transition probabilities from one dwelling state to another—the latter probabilities are likely to be strongly state dependent and display less sensitivity to costs.

In addition to *OPCOST* and *RETURN*, income enters as an explanatory variable in interaction with a dummy variable for owner (*YOWN*) and a dummy variable for non-head (*YNH*). Table 3.1 illustrates the structure of the explanatory variables.

First consider a MNL model fitted to these data. Table 3.2 gives the estimates, asymptotic standard errors and t-statistics, Table 3.4 gives elasticities for each response probability calculated at sample means (by alternative) of the explanatory variables. This model excludes alternative-specific dummy variables. Consequently, the coefficients of the income variables reflect both a correlation of response with income, and unobserved features specific to the associated alternative. The model suggests a strong positive association between ownership and return, and strong negative association between choice and out-of-pocket cost and between non-headship and income. The value of the log-likelihood at the maxi-

Table 3.1
Structure of the explanatory variables.

Person	Alternative[a]	*OPCOST*[b]	*RETURN*[c]	*YOWN*[d]	*YNH*[d]	*CHOICE*
1	1	1.24	2.79	6.1	0	1
1	2	1.13	0	0	6.1	0
1	3	2.33	0	0	0	0
2	1	5.48	5.07	4.3	0	0
2	2	1.13	0	0	4.3	0
2	3	0.93	0	0	0	1
⋮	⋮	⋮	⋮	⋮	⋮	⋮
Av.[e]	1	3.44	4.41	6.4	0	0.46
	2	0.97	0	0	6.4	0.24
	3	1.96	0	0	0	0.30

[a]Alternative 1 = own (73 cases); alternative 2 = rent/non-head (38 cases); alternative 3 = rent/head (48 cases).
[b]In thousand 1977 dollars.
[c]In thousand 1977 dollars.
[d]In thousand 1977 dollars.
[e]Sample average by alternative.

Table 3.2
Multinomial logit model of housing status

Variable	Parameter estimate	Standard error	t-Statistic
OPCOST	-4.544	1.011	-4.50
RETURN	2.506	0.747	3.35
YOWN	-0.055	0.074	-0.73
YNH	-0.838	0.202	-4.16

Auxiliary statistics
Sample size $=159$
Log-likelihood $=-15.91$
Estimation method: maximum likelihood, model (3.14).

mum likelihood estimates is -15.9, compared with a value of -174.7 when all coefficients are zero. Using the criterion of maximum probability, the model predicts correctly 97.5% of the observed states.

The MNL model specification can be tested using the procedure described in (3.28). Let S_{own} and S_{head} denote the test statistics obtained by deleting owner and renter-head alternatives, respectively, and estimating the reduced MNL model. Under the null hypothesis that the MNL specification is correct, S_{own} and S_{head} are asymptotically chi-square with three degrees of freedom. For this sample, $S_{\text{own}}=22.4$ and $S_{\text{head}}=1.7$. Then the first statistic rejects the MNL specification at the 0.001 significance level, the second does not reject. We conclude, with a significance level at most 0.002, that the MNL specification should be rejected.[8]

[8] The statistics S_{own} and S_{head} are not independent. However, the inequalities $\max(P(S_{\text{own}}>c), P(S_{\text{head}}>c)) \leq P(\max(S_{\text{own}}, S_{\text{head}})>c) \leq P(S_{\text{own}}>c)+P(S_{\text{head}}>c)$ can be used to bound the significance level and power curve of a criterion which rejects MNL if either of the statistics exceeds c. Alternatively, one can extend the analysis of Hausman and McFadden to establish an exact asymptotic joint test. Using the notation of subsection 3.7, let A and D be restricted subsets of the choice set C, β_D and β_C the restricted and full maximum likelihood estimates of the parameters identifiable from D, and α_A and α_C the restricted and full estimates of the parameters identifiable from A. There may be overlap between β_C and α_C. Let V_D and V_A denote the estimated covariance matrices of β_D and α_A respectively, and let $V_{C\alpha\alpha}$, $V_{C\alpha\beta}$, etc. denote submatrices of the estimated covariance matrix of the full maximum likelihood estimator. Define:

$$H=\frac{1}{T}\sum_{t=1}^{T}\sum_{i\in A\cap D}(x_{it}-x_{At})(z_{it}-z_{Dt})'P_{it},$$

where x and z are the variables corresponding to α and β; x_A and z_D are the probability-weighted averages of these variables within A and D; and P_{it} is the estimated response probability from the full choice set. Then $\sqrt{T}((\alpha_A-\alpha_C)',(\beta_D-\beta_C)')$ is, under the null hypothesis, asymptotically normal with mean zero and covariance matrix:

$$T\begin{bmatrix} V_A-V_{C\alpha\alpha} & V_AHV_D-V_{C\alpha\beta} \\ V_DH'V_A-V_{C\beta\alpha} & V_B-V_{C\beta\beta}\end{bmatrix}.$$

Then a quadratic form in these expressions is asymptotically chi-square under the null with degrees of freedom equal to the rank of the matrix.

Table 3.3
Nested multinomial logit model of housing status.

Variable	Parameter estimate	Standard error	t-Statistic
OPCOST	−2.334	0.622	−3.75
RETURN	0.922	0.479	1.93
YOWN	0.034	0.064	0.53
YNH	−0.438	0.121	−3.62
λ	0.179	0.109	1.64

Auxiliary statistics:
Sample size = 159
Log-likelihood = −10.79
Estimation method:
full maximum likelihood estimation, model (3.40) with lowest level of tree (unit choice) deleted

The MNL specification can be tested against a NMNL decision tree using the Lagrange multiplier statistic given in (3.59). We have calculated this test for the tree in Figure 3.3, with the rental alternatives in a cluster, and obtain a value LM = 15.73. This statistic is asymptotically chi-square with one degree of freedom under the null, leading to a rejection of the MNL specification at the .001 significance level.

We next estimate a nested MNL model of the form (3.40), but continue the simplifying assumption that dwelling units within a cluster are homogeneous. Then there are three alternatives: own, rent/non-head, and rent/head, and one inclusive value coefficient λ. Table 3.3 reports full maximum likelihood estimates of this model. Asymptotic standard errors and t-statistics are given. Table 3.4 gives elasticities calculated at the sample mean using (3.49). The estimated inclusive value coefficient $\lambda = 0.179$ is significantly less than one: the statistic $W = (1.0 - \lambda)^2/SE_\lambda^2 = 56.77$ is just the Wald statistic for the null hypothesis that the MNL model in Table 3.2 is correct, which is chi-square with one degree of freedom under the null. Hence, we again reject the MNL model at the 0.001 significance level. It is also possible to use a likelihood ratio test for the hypothesis. In the example, the likelihood ratio test statistic is LR = 10.24, leading to rejection at the 0.005 but not the 0.001 level. We note that for this example the Lagrange multiplier (LM), likelihood ratio (LR), and Wald (W) statistics differ substantially in value, with LR < LM < W. This suggests that for the sample size in the example the accuracy of the first-order asymptotic approximation to the tails of the exact distributions of these statistics may be low.

The impact of clustering of alternatives can be seen clearly by comparing elasticities between the MNL model and the NMNL model in Table 3.4. For example, the MNL model has equal elasticities of owner and renter-head response probabilities with respect to renter non-head *OPCOST*, as forced by the IIA

Table 3.4
Elasticities in the MNL and NMNL models.[a]

Variable	NML Model Alt.1 Own	NML Model Alt.2 Rent non-head	NML Model Alt.3 Rent head	MNML model Alt.1 Own	MNML model Alt.2 Rent non-head	MNML model Alt.3 Rent head
Owner *OPCOST*	−8.44	+7.19	+7.19	−4.33	+3.69	+3.69
Rental non-head *OPCOST*	+1.06	−3.35	+1.06	+0.54	−7.48	+5.15
Rent head *OPCOST*	+2.67	+2.67	−6.06	+1.37	+13.10	−12.41
RETURN	+5.97	−5.08	−5.08	+2.20	−1.87	−1.87
INCOME	+1.10	−3.92	+1.45	+0.79	−9.37	+6.29

[a]Elasticities are calculated at sample means of the explanatory variables; see formula (3.49).

property, whereas in the NMNL model the first of these elasticities is substantially decreased and the second is substantially increased.

A final comment on the NMNL model concerns the efficiency of sequential estimates and their suitability as starting values for iteration to full maximum likelihood. Sequential estimation of the NMNL model starts by fitting a MNL model to headship status among renters. Only *OPCOST* and *YNH* vary among renters, so that only the coefficients of these variables are identified. Next an inclusive value for renters is calculated. Finally, a MNL model is fitted to own-rent status, with *RETURN*, *YOWN*, and the calculated inclusive value as explanatory variables. The covariance matrix associated with sequential estimates is complicated by the use of calculated inclusive values; Amemiya (1978d) and McFadden (1981) give computational formulae. One Newton–Raphson iteration from the sequential estimates yields estimates which are asymptotically equivalent to full information maximum likelihood estimates.[9] In general sequential estimation may be quite inefficient, resulting in one-step estimators which are far from the full maximum likelihood estimates. However, in this example, the sequential estimates are quite good, agreeing with the full information estimates to the third decimal place. The log-likelihood for the sequential estimates is −10.7868, compared with −10.7862 at the full maximum.

The clustering of alternatives described by the preceding NMNL model could also be captured by a one-factor MNP model. With three alternatives, it is also feasible to estimate a MNP model with varying taste coefficients on all variables. We have not estimated these models. On the basis of previous studies [Hausman

[9]One must be careful to maintain consistent parameter definitions between the sequential procedure and the NMNL model likelihood specification; namely the parameter α from the first sequential step (3.41) is scaled to $\alpha\kappa_{rh}\lambda_r$ in the NMNL generating function (3.47).

and Wise (1978), Fischer–Nagin (1981)], we would expect the one-factor MNP model to give fits comparable to the NMNL model, and the MNP model with full taste variation to capture heterogeneities which the first two models miss.

4. Further topics

4.1. Extensions

Econometric analysis of qualitative response has developed in a number of directions from the basic problem of multinomial choice. This section reviews briefly developments in the areas of dynamic models, systems involving both discrete and continuous variables, self-selection and sampling problems, and statistical methods to improve the robustness of estimators or asymptotic approximations to finite-sample distributions.

4.2. Dynamic models

Many important economic applications of qualitative response models involve observations through time, often in a combined cross-section/time-series framework. The underlying latent variable model may then have a components of variance structure, with individual effects (population heterogeneity), autocorrelation, and dependence on lagged values of latent or indicator variables (state dependence). This topic has been developed in great depth by Heckman (1974, 1978b, 1981b, 1981c), Heckman and McCurdy (1982), Heckman and Willis (1975), Flinn and Heckman (1980), and Lee (1980b).

Dynamic discrete response models are special cases of systems of non-linear simultaneous equations, and their econometric analysis can utilize the methods developed for such systems, including generalized method of moments estimators; see Hausman (1982), Hansen (1982), and Newey (1982).

Most dynamic applications have used multivariate normal disturbances so that the latent variable model has a linear structure. This leads to MNP choice probabilities. As a consequence, computation limits problem size. Sometimes the dimension of the required integrals can be reduced by use of moments estimators [Hansen (1982) and Ruud (1982)]. Alternatively, one or two factor MNP models offer computational convenience when the implied covariance structure is appropriate. For example, consider the model:

$$y_{nt}^* = x_{nt}\beta + \varepsilon_{nt} - \nu_n \tag{4.1}$$

and $y_{nt}=1$ if $y^*_{nt}\geq 0$, $y_{nt}=-1$ otherwise, where $t=1,\ldots,T$; $n=1,\ldots,N$; ν_n is an individual random effect which persists over time; and ε_{nt} are disturbances independent of each other and of ν. If ε_{nt} and ν_n are normal, then the probability of a response sequence has the tractable form:

$$P(y_{n1},\ldots,y_{nT})=\int_{-\infty}^{+\infty}\frac{1}{\sigma_\nu}\phi\left(\frac{\nu}{\sigma_\nu}\right)\prod_{t=1}^{T}\Phi\left(\frac{y_{nt}(x_{nt}\beta+\nu)}{\sigma_\varepsilon}\right)\mathrm{d}\nu. \tag{4.2}$$

It is also possible to develop tractable dynamic models starting with extreme-value disturbances. If $(\nu_n,\varepsilon_{n1},\ldots,\varepsilon_{nT})$ has the generalized extreme value distribution

$$F(\nu_n,\varepsilon_{n1},\ldots,\varepsilon_{nT})=\exp\left\{-e^{-\nu_n}-\left[\sum_{t=1}^{T}\mathrm{e}^{-\varepsilon_{nt}/\lambda}\right]^{\lambda}\right\}, \tag{4.3}$$

with $0<\lambda\leq 1$, then the probability that $y_{nt}=1$ for all t in any subset A of $\{1,\ldots,T\}$ is:

$$P_A=1\bigg/\left(1+\left(\sum_{t\in A}\mathrm{e}^{-x_t\beta/\lambda}\right)^{\lambda}\right). \tag{4.4}$$

If A_n is the set of times with $y_{nt}=1$, and $k(B)$ is the cardinality of a set B, then

$$P(y_{n1},\ldots,y_{nT})=\sum_{B\subseteq A_n^c}(-1)^{k(B)}P_{A_n\cup B}. \tag{4.5}$$

For a more general discussion of models and functional forms for discrete dynamic models, see Heckman (1981b).

4.3. *Discrete–continuous systems*

In some applications, discrete response is one aspect of a larger system which also contains continuous variables. An example is a consumer decision on what model of automobile to purchase and how many kilometers to drive (VKT). It is important to account correctly for the joint determination of discrete and continuous choices in such problems. For example, regression of VKT on socioeconomic characteristics for owners of American cars is likely to be biased by self-selection into this sub-population.

A typical discrete–continuous model is generated by the latent variable model:

$$y_{it}^* = x_{it}\beta_{it} - \varepsilon_{it} \qquad (i = 1,2,3), \tag{4.6}$$

and generalized indicator function:

$$y_{1t} = \begin{cases} 1, & \text{if } y_{1t}^* \geq 0, \\ 0, & \text{otherwise,} \end{cases} \tag{4.7}$$

$$y_{2t} = y_{1t}y_{2t}^* + (1 - y_{1t})y_{3t}^*, \tag{4.8}$$

where $(\varepsilon_1, \varepsilon_2, \varepsilon_3)$ are multivariate normal with a mean zero and covariance matrix:

$$\begin{bmatrix} 1 & \sigma_{12} & \sigma_{13} \\ \sigma_{21} & \sigma_{22} & \sigma_{23} \\ \sigma_{31} & \sigma_{32} & \sigma_{33} \end{bmatrix}.$$

This model is sometimes termed a switching regression with observed regime.

Discrete–continuous models such as (4.6)–(4.7) can be estimated by maximum likelihood, or a computationally easier two-step procedure. The latter method first estimates the reduced form (marginal) equation for the discrete choice. Then the fitted probabilities are either used to construct instruments for the endogenous discrete choice variables in (4.8), or else used to augment (4.8) with a "hazard rate" whose coefficient absorbs the covariance of the disturbance and explanatory variables. Developments of these methods can be found in Quandt (1972), Heckman (1974), Amemiya (1974b), Lee (1980a), Maddala and Trost (1980), Hay (1979), Duncan (1980a), Dubin and McFadden (1980), and Poirier and Ruud (1980). A comprehensive treatment of models of this type can be found in Lee (1981). Empirical experience is that estimates obtained using augmentation by the hazard rate are quite sensitive to distributional assumptions; Lee (1981a) provides results on this question and develops a specification test for the usual normality assumption.

If discrete–continuous response is the result of economic optimization, then cross-equation restrictions are implied between the discrete and continuous choice equations. These conditions may be imposed to increase the efficiency of estimation, or may be used to test the hypothesis of optimization. These conditions are developed for the firm by Duncan (1980a) and for the consumer by Dubin and McFadden (1980).

4.4. *Self-selection and biased samples*

In the preceding section, we observed that joint discrete–continuous response could introduce bias in the continuous response equation due to self-selection into a target sub-population. This is one example of a general problem where ancillary responses lead to self-selection or biased sampling.

In general, it is possible to represent self-selection phenomena in a joint latent variable model which also determines the primary response. Then models with a mathematical structure similar to the discrete-continuous response models can be used to correct self-selection biases [Heckman (1976b), Hausman and Wise (1977), Maddala (1977a), Lee (1980a)].

Self-selection is a special case of biased or stratified sampling. In general, stratified sampling can be turned to the advantage of the econometrician by using estimators that correct bias and extract maximum information from samples [Manski and Lerman (1977), Manski and McFadden (1980), Cosslett (1980a), McFadden (1979c)]. To illustrate these approaches, consider the problem of analyzing multinomial response using self-selected or biased samples. Let $f^i(x,\beta^*)$ denote the true response probability and $p(x)$ the density of the explanatory variables in the population of interest. Self-selection or stratification can be interpreted as identifying an "exposed" sub-population from which the observations are drawn; let $\pi(i,x)$ denote the conditional probability that an individual with characteristics (i,x) is selected into the exposed sub-population. For example, π may be the probability that an individual agrees to be interviewed or is able to provide complete data on x, or the probability that the individual meets the screening procedures established by the sampling protocol (e.g. "terminate the interview on rental housing costs if respondent is an owner"). The selection probability may be known, particularly in the case of deliberately biased samples (e.g. housing surveys which over-sample rural households). Alternately, the selection process may be modeled as a function of a vector of unknown parameters γ^*. An example of a latent variable model yielding this structure is the recursive system $y_i^* = x_i\beta - \varepsilon_i$ and $y_i = 1$ if $y_i^* \geq y_j^*$, $y_i = 0$, otherwise, for $i = 1,\dots,m$; $y_0^* = x\gamma_0 + \sum_{i=1}^m y_i\gamma_i - \varepsilon_0$ and $y_0 = 1$ if $y_0^* \geq 0$ and $y_0 = 0$ otherwise; where $x = (x_1,\dots,x_m)$, $\gamma = (\gamma_0,\dots,\gamma_m)$, and y_0 is an indicator for selection into the exposed sub-population.

By Bayes' law, the likelihood of an observation (i,x) in the exposed population is:

$$h(i,x) = \pi(i,x,\gamma^*)f^i(x,\beta^*)p(x)/q(\beta^*,\gamma^*), \tag{4.9}$$

where

$$q(\beta^*,\gamma^*) = \sum_x \sum_i \pi(i,x,\gamma^*)f^i(x,\beta^*)p(x) \tag{4.10}$$

is the fraction of the target population which is exposed. Note that when γ^* is unknown, it may be impossible to identify all components of (γ^*, β^*), or identification may be predicated on arbitrary restrictions on functional form. Then auxiliary information on the selection process (e.g. follow-up surveys of non-respondents, or comparison of the distributions of selected variables between the exposed population and censuses of the target population) is necessary to provide satisfactory identification.

Stratified sampling often identifies several exposed sub-populations, and draws a sub-sample from each. The sub-populations need not be mutually exclusive; e.g. a survey of housing status may draw one stratum from lists of property-owners, a second stratum by drawing random addresses in specified census tracts. If the originating strata of observations are known, then the likelihood (4.9) specific to each stratum applies, with the stratum identifier s an additional argument of π and q. However, if the observations from different strata are pooled without identification, then (4.9) applies uniformly to all observations with π a mixture of the stratum-specific selection probabilities: if μ_s is the share of the sample drawn from stratum s, then except for an inessential constant:

$$\pi(i,x)=\sum_s \pi(i,x,s,\gamma^*)\mu_s/q_s. \tag{4.11}$$

The likelihood (4.9) depends on the density of the explanatory variables $p(x)$ which is generally unknown and of high dimensionality, making direct maximum likelihood estimation impractical. When the selection probability functions are known, one alternative is to form a likelihood function for the observations as if they were drawn from a random sample, and then weight the observations to obtain a consistent maximum "pseudo-likelihood" estimator. Specifically, if the selection probability $\pi(i,x)$ is positive for all i, then, under standard regularity conditions, consistent estimates of β are obtained when observation (i,x) is assigned the log pseudo-likelihood $(1/\pi(i,x))\ln f^i(x,\beta)$. This procedure can be applied stratum by stratum when there are multiple strata, provided the positivity of π is met. However, in general it is more efficient to pool strata and use the pooled selection rate (4.11).[10] Pooling is possible if the exposure shares of q_s are known or can be estimated from an auxiliary sample which grows in size at least as rapidly as the main sample. Further discussion of this method and derivation of the appropriate covariance matrices can be found in Manski and Lerman (1977) and Manski and McFadden (1981).

[10]Additional weighting of observations with weights depending on x, but not on (i,s), will in general not affect consistency, and may be used to improve efficiency. With appropriate redefinition of $\pi(i,x)$, it is also possible to reweight strata.

A second approach to estimation is to form the pooled sample conditional likelihood of response i and stratum s, given x,

$$l(i,s|x,\beta,\gamma)=\frac{f^i(x,\beta)\pi(i,x,s,\gamma)\mu_s/q_s}{\sum_{j=1}^{m}f^i(x,\beta)\sum_{t}\pi(j,x,t,\gamma)\mu_t/q_t}. \tag{4.12}$$

When the stratum s is not identified or there is a single stratum, this reduces to

$$l(i|x,\beta,\gamma)=\frac{f^i(x,\beta)\pi(i,x,\gamma)}{\sum_{j=1}^{m}f^j(x,\beta)\pi(j,x,\gamma)}. \tag{4.13}$$

With appropriate regularity conditions, plus the requirement that $\pi(i,x,\gamma)$ be positive for all i, maximum conditional likelihood estimates of (β,γ) are consistent.[11] Further discussion of this method and derivation of the appropriate covariance matrix is given in McFadden (1979) and Manski and McFadden (1981).

When the response model has the multinomial logit functional form, conditional maximum likelihood has a simple structure. For

$$f^i(x,\beta)=e^{x_i\beta}\Big/\sum_{j=1}^{m}e^{x_j\beta}, \tag{4.14}$$

(4.12) becomes:

$$l(i,s|x,\beta,\gamma)=\frac{\exp[x_i\beta+\ln(\pi(i,x,s,\gamma)\mu_s/q_s)]}{\sum_{(j,t)\in A}\exp[x_j\beta+\ln(\pi(j,x,t,\gamma)\mu_t/q_t)]}\quad(i,s)\in A, \tag{4.15}$$

where A is the set of pairs (i,s) with $\pi(i,x,s,\gamma)>0$. When $\pi(i,x,s,\gamma)\mu_s/q_s$ is known or can be estimated consistently from auxiliary data, or $\ln\pi(i,x,s,\gamma)$ is linear in unknown parameters, then the response and selection effects combine in a single MNL form, permitting simple estimation of the identifiable parameters.

[11] The exposure shares q_s may be known, or estimated consistently from an auxiliary sample. Alternatively, they can be estimated jointly with (β,γ). Identification in the absence of auxiliary information will usually depend on non-linearities of the functional form.

It is possible to obtain fully efficient variants of the conditional maximum likelihood method by incorporating side constraints (when q_s is known) and auxiliary information (when sample data on q_s is available); see Cosslett (1981a) and McFadden (1979).

4.5. Statistical methods

Econometric methods for qualitative response models have been characterized by heavy reliance on tractable but restrictive functional forms and error specifications, and on first-order asymptotic approximations. There are four areas of statistical investigation which have begun to relax these limits: development of a variety of functional forms for specialized applications, creation of batteries of specification tests, development of robust methods and identification of classes of problems where they are useful, and development of higher-order asymptotic approximations and selected finite-sample validations.

This chapter has surveyed the major lines of development of functional forms for general purpose multinomial response models. In a variety of applications with special structure such as longitudinal discrete response or serially ordered alternatives, it may be possible to develop specialized forms which are more appropriate. Consider, for example, the problem of modeling serially ordered data. One approach is to modify tractable multinomial response models to capture the pattern of dependence of errors expected for serial alternatives. This is the method adopted by Small (1982), who develops a generalized extreme value model with proximate dependence. A second approach is to generalize standard discrete densities to permit dependence of parameters on explanatory variables. For example, the Poisson density

$$P_k = \mathrm{e}^{-\lambda}\lambda^k/k! \qquad (k=0,\dots), \tag{4.16}$$

with $\lambda = \mathrm{e}^{x\beta}$, or the negative binomial density

$$P_k = \frac{\Gamma(r+k)}{\Gamma(r)k!}p^r(1-p)^k, \tag{4.17}$$

with $r>0$ and $p = 1/(1+\mathrm{e}^{-x\beta})$, provide relatively flexible forms. A good example of this approach and analysis of the relationships between functional forms is Griliches, Hall and Hausman (1982).

Specification tests for discrete response models have been developed primarily for multinomial response problems, using classical large-sample tests for nested hypotheses. Lagrange Multiplier and Wu–Hausman tests of the sort discussed in this chapter for testing the MNL specification clearly have much wider applicability. An example is Lee's (1981a) use of a Lagrange Multiplier test for a binomial probit model against the Pearson family. It is also possible to develop rather straightforward tests of non-nested models by applying Lagrange Multiplier tests to their probability mixtures. There has been relatively little development of non-parametric methods. McFadden (1973) and McFadden, Tye, and Train (1976) propose some tests based on standardized residuals; to date these have not proved useful. There is no finite sample theory, except for scattered Monte Carlo results, for specification tests.

The primary objective of the search for robust estimators for discrete response models is to preserve consistency when the shape of the error distribution is misspecified. This is a different, and more difficult, problem than is encountered in most discussions of linear model robust procedures where consistency is readily attained and efficiency is the issue. Consequently, results are sparse. Manski (1975) and Cosslett (1980) have developed consistent procedures for binomial response models of the form $P_1 = f^1(x\beta)$, where f^1 is known only to be monotone (with standardized location and scale); more general techniques developed by Manski (1981) may make it possible to extend these results.

To evaluate this approach, it is useful to consider the common sources of model misspecification: (1) incorrect assumptions on the error distribution, (2) reporting or coding errors in the discrete response, (3) omitted variables, and (4) measurement errors in explanatory variables. For concreteness, consider a simple binomial probit model with a latent variable representation $y^* = x^*\beta - \nu$, ν standard normal, $y = 1$ if $y^* \geq 0$ and $y = 0$ otherwise, so that in the absence of misspecification problems the response probability is $P_1 = \Phi(x^*\beta)$. Now suppose all the sources of misspecification are present: $x^* = (x_1^*, x_2^*)$ has x_2^* omitted and x_1^* measured with error, $x_1 = x_1^* + \eta_1$. Then $y^* = x_1^*\beta_1 - (\nu - x_2^*\beta_2 + \xi)$ with response error ξ. Suppose for analysis that $(x_1^*, x_2^*, \eta_1, \nu, \xi)$ is multivariate normal with mean $(\mu_1, \mu_2, 0, 0, 0)$ and covariance matrix:

$$\begin{bmatrix} \Sigma_{11} & \Sigma_{12} & 0 & 0 & 0 \\ \Sigma_{21} & \Sigma_{22} & 0 & 0 & 0 \\ 0 & 0 & \Omega & 0 & 0 \\ 0 & 0 & 0 & 1 & 0 \\ 0 & 0 & 0 & 0 & \sigma^2 \end{bmatrix}. \tag{4.18}$$

Then observations conform to the conditional probability of $y = 1$, given x_1:

$$P_1 = \Phi\left(\frac{x_1\beta_1 + \alpha + (x_1 - \mu_1)\gamma}{\lambda} \right), \tag{4.19}$$

where

$$\alpha = \mu_2 \beta_2,$$
$$\gamma = (\Omega_{11} + \Sigma_{11})^{-1}(\Sigma_{12}\beta_2 - \Omega_{11}\beta_1),$$
$$\lambda^2 = 1 + \sigma^2 + \beta_1'\Sigma_{12}\beta_2 + \beta_2'\left[\Sigma_{22} - \Sigma_{21}(\Omega_{11} + \Sigma_{11})^{-1}\Sigma_{12}\right]\beta_2 + \beta_1'\Omega_{11}(\Omega_{11} + \Sigma_{11})^{-1}[\Sigma_{11}\beta_1 + \Sigma_{12}\beta_2].$$

Estimating the probit model $P_1 = \Phi(x_1\tilde{\beta}_1)$ without allowance for errors of misspecification will lead to asymptotically biased estimates of relative coefficients in β_1 if x_1 is measured with error ($\Omega_{11} \neq 0$), or omitted variables are correlated with $x_1(\Sigma_{12}\beta_2 \neq 0)$ or make a non-zero contribution to the intercept ($\mu_2\beta_2 \neq 0$). These sources of error also change the scale λ. Reporting and coding errors in the response ($\sigma^2 \neq 0$) affect the scale λ, but do not affect the asymptotic bias of relative coefficients. Misspecification of the error distribution can always be reinterpreted, in light of the discussion in Section 2 on approximating response functions, as omission of the variables necessary to make the response a probit.

Consider an estimator of the Cosslett or Manski type which effectively estimates a model $P_1 = F(x_1\tilde{\beta}_1)$ with F a monotone function which is free to conform to the data. This approach can yield consistent estimates of relative coefficients in β_1 in the presence of response coding errors or an unknown error distribution, provided there are no omitted variables or measurement errors in x. However, the Cosslett–Manski procedures are ineffective against the last two error sources.[12] Furthermore, the non-linearity of (4.19) renders inoperative the instrumental variable methods which are effective for treatment of measurement error in linear models. How to handle measurement error in qualitative response models is an important unsolved problem. This topic is discussed further by Yatchew (1980).

Most applications of qualitative response problems to date have used statistical procedures based on first-order asymptotic approximations. Scattered Monte Carlo studies and second-order asymptotic approximations suggest that in many qualitative response models with sample sizes of a few hundred or more, first-order

[12] Manski and I have considered estimators obtained by maximizing a "pseudo-log-likelihood" in which observation (y_i, x) makes the contribution:

$$(y_i + a)(f^i(x,\beta) + a)^b/b - (f^i(x,\beta) + a)^{b+1}/(b+1),$$

with $a \geq 0$ and $b > -1$. In the absence of measurement errors, this method yields consistent estimators. For positive a and b, this procedure bounds the influence of extreme observations, and should reduce the impact of coding errors in y. [Note that this class defines a family of M-estimators in the terminology of Huber (1965); the cases $a = b = 0$ and $a = 0$, $b = 1$ yield maximum likelihood and non-linear least squares estimates respectively.] We find, however, that this approach does not substantially reduce asymptotic bias due to coding errors.

approximations are moderately accurate. Nevertheless, it is often worthwhile to make second-order corrections for bias of estimators and the size of tests for samples in this range. As a rule of thumb, sample sizes which yield less than thirty responses per alternative produce estimators which cannot be analyzed reliably by asymptotic methods. These issues are discussed further in Domencich and McFadden (1975), Amemiya (1980, 1981), Cavanaugh (1982), Hausman and McFadden (1982), Rothenberg (1982), and Smith, Savin, and Robertson (1982).

5. Conclusion

This chapter has surveyed the current state of econometric models and methods for the analysis of qualitative dependent variables. Several features of this discussion merit restatement. First, the models of economic optimization which are presumed to govern conventional continuous decisions are equally appropriate for the analysis of discrete response. While the intensive marginal conditions associated with many continuous decisions are not applicable, the characterization of economic agents as optimizers implies conditions at the extensive margin and substantive restrictions on functional form. Unless the tenets of the behavioral theory are themselves under test, it is good econometric practice to impose these restrictions as maintained hypotheses in the construction of discrete response models.

Second, as a formulation in terms of latent variable models makes clear, qualitative response models share many of the features of conventional econometric systems. Thus the problems and methods arising in the main stream of econometric analysis mostly transfer directly to discrete response. Divergences from the properties of the standard linear model arise from non-linearity rather than from discreteness of the dependent variable. Thus, most developments in the analysis of non-linear econometric systems apply to qualitative response models. In summary, methods for the analysis of qualitative dependent variables are part of the continuing development of econometric technique to match the real characteristics of economic behavior and data.

Appendix: Proof outlines for Theorems 1–3

Theorem 1

This result specializes a general consistency theorem of Huber (1965, theorem 1) which states that any sequence of estimators which almost surely approaches the suprema of the likelihood functions as $T \to \infty$ must almost surely converge to the true parameter vector θ^*. Assumptions (1)–(3) imply Huber's conditions A-1 and

A-2. The inequality $-e^{-1} \leq z \ln z \leq 0$ for $0 \leq z = f^i(x,\theta) \leq 1$ implies Huber's A-3, and assumption (4) and this inequality imply Huber's A-4 and A-5. It is possible to weaken assumptions (1)–(4) further and still utilize Huber's argument; the formulation of Theorem 1 is chosen for simplicity and ease in verification.

Theorem 2

Note first that $L_T(\theta) \leq 0$ and

$$\mathrm{E}L_T(\theta^*) = \int \mathrm{d}p(x) \sum_{i-1}^{m} f^i(x,\theta^*) \ln f^i(x,\theta^*) \geq -m/e$$

from the bound above on $z \ln z$ for $0 \leq z \leq 1$. Hence, $L_T(\theta^*) > -\infty$ almost surely, and a sequence of estimators satisfying (3.9) exists almost surely. Let Θ_2 be a compact subset of Θ_0 [assumption (5)] which contains a neighborhood Θ_1 of θ^*, and let $\hat{\theta}_T$ be a maximand of $L_T(\theta)$ on Θ_2. Choose $\tilde{\theta}_T \in \Theta \backslash \Theta_1$ such that $L_T(\tilde{\theta}_T) + 1/T \geq \sup_{\Theta \backslash \Theta_1} L_T(\theta)$. Define $\bar{\theta}_T = \tilde{\theta}_T$ if $L_T(\tilde{\theta}_T) \geq L_T(\hat{\theta}_T)$, and $\bar{\theta}_T = \hat{\theta}_T$ otherwise. Then $\bar{\theta}_T$ satisfies (3.9) and by Theorem 1 converges almost surely to θ^*, and therefore almost surely eventually stays in Θ_1. Hence, almost surely eventually $\bar{\theta}_T = \hat{\theta}_T \in \Theta_1$, implying $L_T(\theta_T) \geq L_T(\theta)$ on an open neighborhood of Θ_1, and therefore $\partial L_T(\hat{\theta}_T)/\partial\theta = 0$.

Theorem 3

This result extends a theorem of Rao (1973, 5e2) which establishes for a multinomial distribution without explanatory variables that maximum likelihood estimates are asymptotically normal. Assumptions (6) and (7) correspond to assumptions made by Rao, with the addition of bounds which are integrable with respect to the distribution of the explanatory variables. This proof avoids the assumption of continuous second derivatives usually made in general theorems on asymptotic normality [cf. Rao (1973, 5f.2(iii)), Huber (1965, theorem 3 corollary)].

Let Θ_1 be a neighborhood of θ^* with compact closure on which assumptions (5) and (6) hold. By Theorem 2, almost surely eventually $\hat{\theta}_T \in \Theta_1$ and

$$0 = \sum_{t=1}^{T} \sum_{i=1}^{m} y_{it} \frac{\partial \ln f^i(x_t, \hat{\theta}_T)}{\partial \theta}. \tag{A.1}$$

Noting that

$$\sum_{i=1}^{m} f^i(x_t,\theta) \frac{\partial \ln f^i(x_t,\theta)}{\partial \theta} \equiv 0, \tag{A.2}$$

one can rewrite (A.1) as $0 = A_T + B_T + C_T - D_T$ with:

$$A_T = \sum_{t=1}^{T} \sum_{i=1}^{m} \left(y_{it} - f^i(x_t, \theta^*) \right) \frac{\partial \ln f^i(x_t, \theta^*)}{\partial \theta},$$

$$B_T = \sum_{t=1}^{T} \sum_{i=1}^{m} \left(y_{it} - f^i(x_t, \theta^*) \right) \left[\frac{\partial \ln f^i(x_t, \hat{\theta}_T)}{\partial \theta} - \frac{\partial \ln f^i(x_t, \theta^*)}{\partial \theta} \right],$$

$$C_T = \sum_{t=1}^{T} \left\{ \sum_{i=1}^{m} \left(f^i(x_t, \theta^*) - f^i(x_t, \hat{\theta}_T) \right) \frac{\partial \ln f^i(x_t, \hat{\theta}_T)}{\partial \theta} + J(\theta^*)(\hat{\theta}_T - \theta^*) \right\},$$

$$D_T = TJ(\theta^*)(\hat{\theta}_T - \theta^*). \tag{A.3}$$

The steps in the proof are (1) show $B_T/\sqrt{T}(1+\sqrt{T}\,|\hat{\theta}_T - \theta^*|) \to 0$ in probability as $T \to \infty$, (2) show $C_T/\sqrt{T}(1+\sqrt{T}\,|\hat{\theta}_T - \theta^*|) \to 0$ in probability, (3) show $J(\theta^*)^{-1}(A_T - D_T)/\sqrt{T} \to 0$ in probability, and (4) show $J(\theta^*)^{-1}A_T/\sqrt{T}$ converges in distribution to a normal random vector with mean zero and covariance matrix $J(\theta^*)^{-1}$. With the result of Step 3, this proves Theorem 3.

Step 1. We use a fundamental lemma of Huber (1965, lemma 3). Define

$$\psi(y, x, \theta) = \sum_{i=1}^{m} \left(y_i - f^i(x, \theta^*) \right) \frac{\partial \ln f^i(x, \theta)}{\partial \theta} + \theta - \theta^*,$$

$$\lambda(\theta) = \mathrm{E}\psi(y, x, \theta) = \theta - \theta^* \tag{A.4}$$

$$u(y, x, \theta, d) = \sup_{|\theta' - \theta| \leq d} |\psi(y, x, \theta') - \psi(y, x, \theta)|.$$

Then assumption (6) (iii) implies:

$$u(y, x, \theta, d) \leq d\left(1 + \sum_{i=1}^{m} |y_i - f^i(x, \theta^*)| \cdot \gamma^i(x)\right), \tag{A.5}$$

and hence using the bounds (6) (iv):

$$\mathrm{E}u(y, x, \theta, d) \leq d\left(1 + \sum_{i=1}^{m} \int \mathrm{d}p(x)\alpha^i(x)\gamma^i(x)\right) \equiv A_1 d,$$

$$\mathrm{E}u(y, x, \theta, d)^2 \leq d^2\left(2A_1 + m\sum_{i=1}^{m} \mathrm{E}\left(y_i - f^i(x, \theta^*)\right)^2 \gamma^i(x)^2\right) \tag{A.6}$$

$$\leq d^2\left(2A_1 + m\sum_{i=1}^{m} \int \mathrm{d}p(x)\alpha^i(x)\gamma^i(x)^2\right) \equiv A_2 d^2.$$

These conditions imply Huber's assumptions ($N-1$) to ($N-3$). Define:

$$Z_T(\theta)=\frac{\left|\sum_{t=1}^{T}\left(\psi(y_t,x_t,\theta)-\psi(y_t,x_t,\theta^*)-\lambda(\theta)\right)\right|}{\sqrt{T}\left(1+\sqrt{T}\,|\theta-\theta^*|\right)}$$
$$=\frac{\left|\sum_{t=1}^{T}\sum_{i=1}^{m}\left(y_{it}-f^i(x_t,\theta^*)\right)\left[\frac{\partial\ln f^i(x,\theta)}{\partial\theta}-\frac{\partial\ln(f^i(x,\theta^*)}{\partial\theta}\right]\right|}{\sqrt{T}\left(1+\sqrt{T}\,|\theta-\theta^*|\right)}. \tag{A.7}$$

Then Huber's Lemma 3 states that:

$$\sup_{\Theta_0} Z_T(\theta)\to 0 \tag{A.8}$$

in probability as $T\to\infty$. But

$$Z_T(\hat{\theta}_T)=B_T/\sqrt{T}\left(1+\sqrt{T}\,|\hat{\theta}_T-\theta^*|\right), \tag{A.9}$$

and Step 1 is complete.

Step 2. Since f^i is differentiable on Θ_1, the mean value theorem implies:

$$f^i(x_t,\theta^*)-f^i(x_t,\hat{\theta}_T)=-\left[\frac{\partial f^i(x_t,\tilde{\theta}_t)}{\partial\theta}\right]'(\hat{\theta}_T-\theta^*), \tag{A.10}$$

where $\tilde{\theta}_t$ is some interior point on the line segment connecting θ^* and $\hat{\theta}_T$. Substituting this expression in C_T yields $C_T=(F_T+G_T)\cdot(\hat{\theta}_T-\theta^*)$, with

$$F_T=-\sum_{t=1}^{T}\sum_{i=1}^{m}f^i(x_t,\theta^*)\left[\frac{\partial\ln f^i(x_t,\theta^*)}{\partial\theta}\right]\left[\frac{\partial\ln f^i(x_t,\theta^*)}{\partial\theta}\right]'+TJ(\theta^*) \tag{A.11}$$

and

$$G_T=\sum_{t=1}^{T}\sum_{i=1}^{m}\left\{\frac{\partial\ln f^i(x_t,\theta^*)}{\partial\theta}\,\frac{\partial f^i(x_t,\theta^*)}{\partial\theta'}-\frac{\partial\ln f^i(x_t,\hat{\theta}_T)}{\partial\theta}\,\frac{\partial f^i(x_t,\tilde{\theta}_t)}{\partial\theta'}\right\}. \tag{A.12}$$

Then $F_T/T \to 0$ in probability by the law of large numbers and

$$|G_t/T| \leq |\hat{\theta}_T - \theta^*| \sum_{i=1}^{m} \int \mathrm{d}p(x)\alpha^i(x)\beta^i(x)\left[2\gamma^i(x)^2\right] \to 0 \tag{A.13}$$

in probability by Theorem 2. Since

$$\frac{C_T}{\sqrt{T}\left(1+\sqrt{T}\,|\hat{\theta}_T - \theta^*|\right)} = \left(\frac{F_T}{T} + \frac{G_T}{T}\right)\frac{\sqrt{T}\left(\hat{\theta}_T - \theta^*\right)}{1+\sqrt{T}\,|\hat{\theta}_T - \theta^*|}, \tag{A.14}$$

with the second term in the product stochastically bounded, this establishes Step 2.

Step 3. The first two steps establish that $(A_T - D_T)/\sqrt{T}(1+\sqrt{T}\,|\hat{\theta}_T - \theta^*|) \to 0$ in probability and hence $J(\theta^*)^{-1}(A_T - D_T)/\sqrt{T}(1+\sqrt{T}\,|\hat{\theta} - \theta^*|) \to 0$ in probability. Therefore given $\varepsilon > 0$, there exists T_ε such that for $T > T_\varepsilon$, the inequality

$$\left|J(\theta^*)^{-1}A_T/\sqrt{T} - \sqrt{T}\left(\hat{\theta}_T - \theta^*\right)\right| < \varepsilon\left(1+\sqrt{T}\,|\hat{\theta}_T - \theta^*|\right) \tag{A.15}$$

holds with probability at least $1-\varepsilon/2$. Chebyshev's inequality applied to $J(\theta^*)^{-1}A_T/\sqrt{T}$ implies [using assumptions (7) and (6) (iv)] that for some large constant K:

$$\left|J(\theta^*)^{-1}A_T/\sqrt{T}\right| < K \tag{A.16}$$

holds with probability at least $1-\varepsilon/2$. Then (A.15) and (A.16) imply

$$\sqrt{T}\,|\hat{\theta}_T - \theta^*| < (K+\varepsilon)/(1-\varepsilon), \tag{A.17}$$

and hence

$$\left|J(\theta^*)^{-1}A_T/\sqrt{T} - \sqrt{T}\left(\hat{\theta}_T - \theta^*\right)\right| \leq (K+1)\varepsilon/(1-\varepsilon) \tag{A.18}$$

with probability at least $1-\varepsilon$. Since ε can be made small, this establishes Step 3.

Step 4. The expression $J(\theta^*)^{-1}A_T/\sqrt{T}$ has mean zero and covariance matrix $J(\theta^*)^{-1}$, and satisfies the conditions of the Lindeberg–Levy central limit theorem. Therefore it converges in distribution to an asymptotically normal vector.

References

Adler, T. and M. Ben-Akiva (1975) "A Joint Frequency, Destination and Mode Choice Model for Shopping Trips", *Transportation Research Record*, 569, 136–150.

Aitchinson, J. and J. Bennett (1970) "Polychotomous Quantal Response by Maximum Indicant", *Biometrika*, 57, 253–262.

Aitchinson, J. and S. Silvey (1957) "The Generalization of Probit Analysis to the Case of Multiple Responses", *Biometrika*, 44, 131–140.
Amemiya, T. (1973) "Regression Analysis When the Dependent Variable is Truncated Normal", *Econometrica*, 41, 997–1016.
Amemiya, T. (1974a) "Bivariate Probit Analysis: Minimum Chi-Square Methods", *Journal of the American Statistical Association*, 69, 940–944.
Amemiya, T. (1974b) "Multivariate Regression of Simultaneous Equation Models When the Dependent Variables Are Truncated Normal", *Econometrica*, 42, 999–1012.
Amemiya, T. (1974c) "A Note on the Fair and Jaffee Model", *Econometrica*, 42, 759–762.
Amemiya, T. (1975) "Qualitative Response Models", *Annals of Economic and Social Measurement*, 4, 363–372.
Amemiya, T. (1976) "The Maximum Likelihood, the Minimum Chi-Square, and the Non-Linear Weighted Least Squares Estimator in the General Qualitative Response Model", *Journal of the American Statistical Association*, 71, 347–351.
Amemiya, T. (1978a) "The Estimation of a Simultaneous Equation Generalized Probit Model", *Econometrica*, 46, 1193–1205.
Amemiya, T. (1978b) "A Note on the Estimation of a Time Dependent Markov Chain Model", Department of Economics, Stanford University.
Amemiya, T. (1978c) "On a Two-Step Estimation of a Multivariate Logit Model", *Journal of Econometrics*, 8, 13–21.
Amemiya, T. (1979) "The Estimation of a Simultaneous Equation Tobit Model", *International Economic Review*, 20, 169–181.
Amemiya, T. (1980) "The n^2-Order Mean Squared Errors of the Maximum Likelihood and the Minimum Logit Chi-Square Estimates", *The Annals of Statistics*, 8, 488–505.
Amemiya, T. (1981) "Qualitative Response Models: A Survey", *Journal of Economic Literature*, 19, 1483–1536.
Amemiya, T. and F. Nold (1975) "A Modified Logit Model", *Review of Economics and Statistics*, 57, 255–257.
Amemiya, T. and J. Powell (1980) "A Comparison of the Logit Model and Normal Discriminant Analysis When the Independent Variables are Binary", Tech. Report 320, IMSSS, Stanford Univ.
Anas, A. (1981) "Discrete Choice Theory, Information Theory, and the Multinomial Logit and Gravity Models", manuscript, Northwestern Univ.
Anderson, T. (1958) *Introduction to Multivariate Statistical Analysis*. New York: Wiley.
Anscombe, E. J. (1956) "On Estimating Binomial Response Relations", *Biometrika*, 43, 461–464.
Antle, C., L. Klimko, and W. Harkness (1970) "Confidence Intervals for the Parameters of the Logistic Distribution", *Biometrika*, 57, 397–402.
Arabmazar, A. and P. Schmidt (1982) "An Investigation of the Robustness of the Tobit Estimator to Non-Normality", *Econometrica*, 50, 1055–1064.
Ashford, J. R. and R. R. Sowden (1970) "Multivariate Probit Analysis", *Biometrics*, 26, 535–546.
Ashton, W. (1972) *The Logit Transformation*. New York: Hafner.
Atkinson, A. (1972) "A Test of the Linear Logistic and Bradley-Terry Models", *Biometrika*, 59, 37–42.
Barlett, M. S. (1935) "Contingency Table Interactions", *Journal of the Royal Statistical Society* (Supplement) 2, 248–252.
Beggs, S., S. Cardell and J. Hausman (1981) "Assessing the Potential Demand for Electric Cars", *Journal of Econometrics*, 16, 1–19.
Ben-Akiva, M. (1973) "Structure of Passenger Travel Demand Models", *Transportation Research Board Record*, No. 526, Washington, D.C.
Ben-Akiva, M. (1974) "Multi-Dimensional Choice Models: Alternative Structures of Travel Demand Models", *Transportation Research Board Special Report 149*, Washington, D.C.
Ben-Akiva, M. (1977) "Choice Models with Simple Choice Set Generating Processes", Working Paper, MIT.
Ben-Akiva, M. and S. Lerman (1974) "Some Estimation Results of a Simultaneous Model of Auto Ownership and Mode Choice to Work", *Transportation*, 3, 357–376.
Ben-Akiva, M. and S. Lerman (1979) "Disaggregate Travel and Mobility Choice Modes and Measures of Accessibility". In *Behavior Travel Modelling* edited by D. Hensher and P. Stopher, pp. 654–679. London: Coom Helm.

Berkovec, J. and J. Rust (1982) "A Nested Logit Model of Automobile Holdings for One-Vehicle Households", Working Paper, Department of Economics, MIT.

Berkson, J. (1949) "Application of the Logistic Function to Bioassay", *Journal of the American Statistical Association*, 39, 357–365.

Berkson, J. (1951) "Why I Prefer Logits to Probits", *Biometrika*, 7, 327–339.

Berkson, J. (1953) "A Statistically Precise and Relatively Simple Method of Estimating the Bio-Assay with Quantal Response, Based on the Logistic Function", *Journal of the American Statistical Association*, 48, 565–599.

Berkson, J. (1955a) "Estimation of the Integrated Normal Curve by Minimum Normit Chi-Square with Particular Reference to Bio-Assay", *Journal of the American Statistical Association*, 50, 529–549.

Berkson, J. (1955b) "Maximum Likelihood and Minimum Chi-Square Estimations of the Logistic Function", *Journal of the American Statistical Association*, 50, 130–161.

Berndt, E., B. Hall, R. Hall, and J. Hausman (1974) "Estimation and Inference in Non-Linear Structural Models", *Annals of Economic and Social Measurement*, 3, 653–666.

Bishop, T., S. Fienberg, and P. Holland (1975) *Discrete Multivariate Analysis*. Cambridge: MIT Press.

Block, H. and J. Marschak (1960) "Random Orderings and Stochastic Theories of Response". In *Contributions to Probability and Statistics*, edited by I. Olkin, 97–132. Stanford University Press.

Bock, R. D. and L. Jones (1969) *The Measurement and Prediction of Judgement and Choice*. San Francisco: Holden-Day.

Boersch-Supan, A. and J. Pitkin (1982) "Multinomial Logit Models of Housing Choices". Working Paper, No. 79, Joint Center for Urban Studies, MIT and Harvard.

Boskin, M. (1974) "A Conditional Logit Model of Occupational Choice", *Journal of Political Economy*, 82, 389–398.

Boskin, M. (1975) "A Markov Model of Turnover in Aid to Families with Dependent Children", *Journal of Human Resources*, 10, 467–481.

Brownstone, D. (1978) "An Econometric Model of Consumer Durable Choice and Utilization Rate". Paper IP-258, Center for Research in Management Science, University of California, Berkeley.

Cardell, S. (1977) "Multinomial Logit with Correlated Stochastic Terms". Working Paper, Charles River Associates.

Carlton, D. (1979) "The Location and Employment Choices of New Firms: An Econometric Model with Discrete and Continuous Endogenous Variables". Working Paper, Univ. of Chicago.

Cavanagh, C. (1982) *Hypothesis Testing in Models with Discrete Dependent Variables*. Ph.D. Thesis, Dept. of Economics, Univ. of California, Berkeley.

Chambers, E. A. and D. R. Cox (1967) "Discrimination Between Alternative Binary Response Models", *Biometrika*, 54, 573–578.

Chung, C. and A. Goldberger (1982) "Proportional Projections in Limited Dependent Variable Models". Working Paper, Univ. of Wisconsin.

Clark, C. (1961) "The Greatest of a Finite Set of Random Variables", *Operations Research*, 9, 145–162.

Cook, P. (1975) "The Correctional Carrot: Better for Parolees", *Policy Analysis*, 1, 11–54.

Cosslett, S. (1978) "Efficient Estimation of Discrete-Choice Models from Choice-Based Samples". Ph.D. dissertation, Department of Economics, University of California, Berkeley.

Cosslett, S. (1980) "Estimation of Random Utility Models with Unknown Probability Distribution". Working Paper, Univ. of Florida.

Cosslett, S. (1981) "Efficient Estimators of Discrete Choice Models". In *Structural Analysis of Discrete Data*, edited by C. Manski and D. McFadden, 51–111. Cambridge: MIT Press.

Cosslett, S. (1981a) "Maximum Likelihood Estimator for Choice-Based Samples", *Econometrica*, 9, 1289–1316.

Cox, D. R. (1958) "The Regression Analysis of Binary Sequences", *Journal of the Royal Statistical Society* (Series B), 820, 215–242.

Cox, D. R. (1966) "Some Procedures Connected with the Logistic Response Curve". In *Research Papers in Statistics*, edited by F. David, 55–71. New York: Wiley.

Cox, D. R. (1970) *Analysis of Binary Data*. London: Methuen.

Cox, D. R. (1972) "The Analysis of Multivariate Binary Data", *Applied Statistics*, 21, 113–120.

Cox, D. R. and E. Snell (1968) "A General Definition of Residuals", *Journal of the Royal Statistical Society* (Series B), 30, 248–265.

Cox, D. R. and E. Snell (1971) "On Test Statistics Calculated from Residuals", *Biometrika*, 58, 589–594.

Cragg, J. G. (1971) "Some Statistical Models for Limited Dependent Variables with Application to the Demand for Durable Goods", *Econometrica*, 39, 829–844.

Cragg, J. G. and R. Uhler (1970) "The Demand for Automobiles", *Canadian Journal of Economics*, 3, 386–406.

Crawford, D. and R. Pollak (1982) "Order and Inference in Qualitative Response Models". Discussion Paper 82-4, Rutgers Univ.

Cripps, T. F. and R. J. Tarling (1974) "An Analysis of the Duration of Male Unemployment in Great Britain 1932–1973", *Economic Journal*, 84, 289–316.

Daganzo, C. (1980) *Multinomial Probit*. New York: Academic.

Daganzo, C., F. Bouthelier, and Y. Sheffi (1977) "Multinomial Probit and Qualitative Choice: A Computationally Efficient Algorithm", *Transportation Science*, 11, 338–358.

Dagenais, M. G. (1975) "Application of a Threshold Regression Model to Household Purchases of Automobiles", *Review of Economics and Statistics*, 57, 275–285.

Daly, A. and S. Zachary (1979) "Improved Multiple Choice Models". In *Identifying and Measuring the Determinants of Model Choice*, edited by D. Hensher and O. Dalvi, pp. 187–201.

Debreu, G. (1960) "Review of R. D. Luce Individual Choice Behavior", *American Economic Review*, 50, 186–188.

Diewert, E. (1978) "Duality Approaches to Microeconomic Theory". Technical Report 281, Institute of Mathematical Studies in the Social Sciences, Stanford University.

Domencich, T. and D. McFadden (1975) *Urban Travel Demand: A Behavioral Analysis*. Amsterdam: North-Holland.

Dubin, J. (1982) *Economic Theory and Estimation of the Demand for Consumer Durables and their Utilization*. Ph.D. Dissertation, MIT.

Dubin, J. and D. McFadden (1980) "An Econometric Analysis of Residential Electrical Appliance Holdings and Usage". Working Paper, Department of Economics, Massachusetts Institute of Technology.

Duncan, G. (1980a) "Formulation and Statistical Analysis of the Mixed Continuous/Discrete Model in Classical Production Theory", *Econometrica*, 48, 839–852.

Duncan, G. (1980b) "Mixed Continuous Discrete Choice Models in the Presence of Hedonic or Exogenous Price Functions". Working Paper, Department of Economics, Washington State University.

Durling, F. (1969) "Bivariate Probit, Logit, and Burrit Analysis". Tech. Report 41, Southern Methodist Univ.

Efron, B. (1975) "The Efficiency of Logistic Regression Compared to Normal Discrimination Analysis", *Journal of the American Statistical Association*, 70, 892–898.

Fair, R. C. and D. M. Jaffee (1972) "Methods of Estimation for Markets in Disequilibrium", *Econometrica*, 40, 497–514.

Farber, H. (1978) "Bargaining Theory, Wage Outcomes, and the Occurrence of Strikes", *American Economic Review*, 68, 262–271.

Farber, H. and D. Saks, (1981) "Why Workers Want Unions", *Journal of Public Economics*, to appear.

Fienberg, S. (1977) *The Analysis of Cross-Classified Data*. Cambridge: MIT Press.

Finney, D. (1964) *Statistical Method in Bio-Assay*. London: Griffin.

Finney, D. (1971) *Probit Analysis*. Cambridge University Press.

Fischer, G. and D. Nagin (1981) "Random versus Fixed Coefficient Quantal Choice Models" in C. Manski and D. McFadden, ed., *Structural Analysis of Discrete Data*. Cambridge: MIT.

Flinn, C. and J. Heckman (1982) "Models for the Analysis of Labor Force Dynamics", *Advances in Econometrics*, 1, 35–95.

Flinn, C. and J. Heckman (1982) "New Methods for Analyzing Structural Models of Labor Force Dynamics", *Journal of Econometrics*, 18, 115–168.

Friedman, P. (1973) "Suggestions for the Analysis of Qualitative Dependent Variables", *Public Finance Quarterly*, 1, 345–355.

Fuller, W., Manski, C. and D. Wise (1980) "New Evidence on the Economic Determinants of Post-Secondary Schooling Choices". J.F.K. School of Government Discussion Paper 9UD.

Gart, J. and J. Zweifel (1967) "On the Bias of Various Estimators of the Logit and its Variance", *Biometrika*, 54, 181–187.
Gilbert, C. (1979) "Econometric Models for Discrete Economic Processes", Working Paper, Oxford Univ.
Gillen, D. W. (1977) "Estimation and Specification of the Effects of Parking Costs on Urban Transport Model Choice", *Journal of Urban Economics*, 4, 186–199.
Goldberger, A. S. (1964) *Econometric Theory*. New York: Wiley.
Goldberger, A. S. (1971) "Econometrics and Psychometrics: A Survey of Communalities", *Psychometrika*, 36, 83–107.
Goldberger, A. S. (1973) "Correlations Between Binary Outcomes and Probabilistic Predictions", *Journal of the American Statistical Association*, 68, 84.
Goldfeld, S. M. and R. E. Quandt (1972) *Nonlinear Methods in Econometrics*. Amsterdam: North-Holland.
Goldfeld, S. and R. E. Quandt (1973) "The Estimation of Structural Shifts by Switching Regressions", *Annals of Economic and Social Measurement*, 2, 475–486.
Goldfeld, S. and R. E. Quandt (1976) "Techniques for Estimating Switching Regressions". In *Studies in Non-Linear Estimation*, edited by S. Goldfeld and R. E. Quandt, 3–37. Cambridge: Ballinger.
Goodman, L. A. (1970) "The Multivariate Analysis of Qualitative Data: Interactions Among Multiple Classifications", *Journal of the American Statistical Association*, 65, 226–256.
Goodman, L. A. (1971) "The Analysis of Multidimensional Contingency Tables: Stepwise Procedures and Direct Estimation Methods for Building Models for Multiple Classifications", *Technometrics*, 13, 33–61.
Goodman, L. A. (1972a) "A Modified Multiple Regression Approach to the Analysis of Dichotomous Variables", *American Sociological Review*, 37, 28–46.
Goodman, L. A. (1972b) "A General Model for the Analysis of Surveys", *American Journal of Sociology*, 77, 1035–1086.
Goodman, L. A. (1973) "Causal Analysis of Panel Study Data and Other Kinds of Survey Data", *American Journal of Sociology*, 78, 1135–1191.
Goodman, L. A. and W. H. Kruskal (1954) "Measures of Association for Cross Classifications", *Journal of the American Statistical Association*, 49, 732–764.
Goodman, L. A. and W. H. Kruskal (1954a) "Measures of Association for Cross Classification II, Further Discussion and References", *Journal of the American Statistical Association*, 54, 123–163.
Gourieroux, C. and A. Monfort (1981) "Asymptotic Properties of the Maximum Likelihood Estimator in Dichotomous Logit Models", *Journal of Econometrics*, 17, 83–97.
Gourieroux, C., J. Laffont, and A. Monfort (1980) "Coherency Conditions in Simultaneous Linear Equation Models with Endogenous Switching Regimes", *Econometrica*, 48, 675–695.
Gourieroux, C., A. Holly, and A. Monfort (1980) "Kuhn–Tucker, Likelihood Ratio, and Wald Tests for Non-Linear Models with Inequality Constraints on Parameters". Discussion Paper 770, Harvard Univ.
Greene, W. (1981) "Estimation of Some Limited Dependent Variable Models Using Least Squares and The Method of Moments". Working Paper 245, Cornell Univ.
Griliches, Z., B. Hall, and J. Hausman (1978) "Missing Data and Self-Selection in Large Panels", *Annals de l'Insee* 30–31, 137–176.
Griliches, Z., B. Hall, and J. Hausman (1981) "Econometric Models for Count Data with an Application to the Patents R&D Relationship", *Econometrica*, forthcoming.
Grizzle, J. (1962) "Asymptotic Power of Tests of Linear Hypotheses Using the Probit and Logit Transformations", *Journal of the American Statistical Association*, 57, 877–894.
Grizzle, J. (1971) "Multivariate Logit Analysis", *Biometrics*, 27, 1057–1062.
Gronau, R. (1973) "The Effect of Children on the Housewife's Value of Time", *Journal of Political Economy*, 81, 168–199.
Gronau, R. (1974) "Wage Comparisons—A Selectivity Bias", *Journal of Political Economy*, 82, 1119–1143.
Gurland, J., I. Lee, and P. Dahm (1960) "Polychotomous Quantal Response in Biological Assay", *Biometrics*, 16, 382–398.
Haberman, S. (1974) *The Analysis of Frequency Data*. University of Chicago Press.
Haldane, J. (1955) "The Estimation and Significance of the Logarithm of a Ratio of Frequencies",

Annals of Human Genetics, 20, 309–311.
Hall, R. (1970) "Turnover in the Labor Force", *Brookings Papers on Economic Activity*, 3, 709–756.
Hansen, L. (1982) "Large Sample Properties of Generalized Method of Moments Estimators", *Econometrica*, 50, 1029–1054.
Harter, J. and A. Moore (1967) "Maximum Likelihood Estimation, from Censored Samples, of the Parameters of a Logistic Distribution", *Journal of the American Statistical Association*, 62, 675–683.
Hausman, J. A. (1978) "Specification Tests in Econometrics", *Econometrica*, 46, 1251–1271.
Hausman, J. A. (1979) "Individual Discount Rates and the Purchase and Utilization of Energy Using Durables", *Bell Journal of Economics*, 10, 33–54.
Hausman, J. A. (1983) "Specification and Estimation of Simultaneous Equations Models", *Handbook of Econometrics*.
Hausman, J. A. and D. McFadden (1984) "A Specification Test for the Multinomial Logit Model", *Econometrica*, forthcoming.
Hausman, J. A. and D. A. Wise (1976) "The Evaluation of Results from Truncated Samples: The New Jersey Negative Income Tax Experiment", *Annals of Economic and Social Measurement*, 5, 421–445.
Hausman, J. A. and D. A. Wise (1977) "Social Experimentation, Truncated Distribution and Efficient Estimation", *Econometrica*, 45, 319–339.
Hausman, J. A. and D. A. Wise (1978) "A Conditional Probit Model for Qualitative Choice: Discrete Decisions Recognizing Interdependence and Heterogeneous Preferences", *Econometrica*, 46, 403–426.
Hausman, J. A. and D. Wise (1981) "Stratification on Endogenous Variables and Estimation: The Gary Experiment". In *Structural Analysis of Discrete Data*, edited by C. Manski and D. McFadden, 365–391. Cambridge: MIT Press.
Hay, J. (1979) "An Analysis of Occupational Choice and Income". Ph.D. dissertation, Yale University.
Heckman, J. (1974) "Shadow Prices, Market Wages, and Labor Supply", *Econometrica*, 42, 679–694.
Heckman, J. (1976a) "Simultaneous Equations Model with Continuous and Discrete Endogenous Variables and Structural Shifts". In *Studies in Non-Linear Estimation*, edited by S. Goldfeld and R. Quandt, 235–272. Cambridge: Ballinger.
Heckman, J. (1978a) "Dummy Exogenous Variables in a Simultaneous Equation System", *Econometrica*, 46, 931–959.
Heckman, J. (1978b) "Simple Statistical Models for Discrete Panel Data Developed and Applied to Test the Hypothesis of the True State Dependence Against the Hypothesis of Spurious State Dependence", *Annals de l'Insee*, 30–31, 227–269.
Heckman, J. (1979) "Sample Selection Bias as a Specification Error", *Econometrica*, 47, 153–161.
Heckman, J. (1981a) "Heterogeneity and State Dependence in Dynamic Models of Labor Supply". In *Conference on Low Income Labor Markets*, edited by S. Rosen. University of Chicago Press.
Heckman, J. (1981b) "Statistical Models for the Analysis of Discrete Panel Data". In *Structural Analysis of Discrete Data*, edited by C. Manski and D. McFadden, pp. 114–178. Cambridge: MIT Press.
Heckman, J. (1981c) "The Incidental Parameters Problem and the Problem of Initial Conditions in Estimating a Discrete Stochastic Process and Some Monte Carlo Evidence on Their Practical Importance". In *Structural Analysis of Discrete Data*, edited by C. Manski and D. McFadden, pp. 179–185. Cambridge: MIT Press.
Heckman, J. and R. Willis (1975) "Estimation of a Stochastic Model of Reproduction: An Econometric Approach". In *Household Production and Consumption*, edited by N. Terleckyj, pp. 99–138. New York: National Bureau of Economic Research.
Heckman, J. and R. Willis (1977) "A Beta Logistic Model for the Analysis of Sequential Labor Force Participation of Married Women", *Journal of Political Economy*, 85, 27–58.
Heckman, J. and T. McCurdy (1980) "A Dynamic Model of Female Labor Supply", *Review of Economic Studies*, 47, 47–74.
Heckman, J. and B. Singer (1980) *Longitudinal Labor Market Studies: Theory, Methods and Empirical Results*. Social Science Research Council Monograph. New York: Academic.
Heckman, J. and T. McCurdy (1982) "New Methods for Estimating Labor Supply Functions: A Survey", *Research in Labor Economics*, 4, 65–102.
Heckman, J. and B. Singer (1982) "The Identification Problem in Econometric Models for Duration

Data". Discussion Paper 82-6, N.O.R.C.

Henderson, J. and Y. Ioannides (1982) "Tenure Choice and the Demand for Housing". Working Paper, Dept. of Economics, Boston Univ.

Hockerman, I., Prashker, J. and M. Ben-Akiva (1982) "Estimation and Use of Dynamic Transaction Models of Automobile Ownership". Working Paper, Dept. of Civil Engineering, MIT.

Horowitz, J. (1979a) "Identification and Diagnosis of Specification Errors in the Multinomial Logit Model". Mimeograph, Environmental Protection Agency.

Horowitz, J. (1979b) "A Note on the Accuracy of the Clark Approximation for the Multinomial Probit Model". Mimeograph, Department of Transportation, Massachusetts Institute of Technology.

Horowitz, J. (1980) "The Accuracy of the Multinomial Logit Model as an Approximation to the Multinomial Probit Model of Travel Demand", *Transportation Research* (Part B).

Horowitz, J. (1981) "Sampling, Specification and Data Errors in Probabilistic Discrete Choice Models". In *Applied Discrete Choice Modeling*, edited by D. Hensher and L. Johnson, 417–435. London: Croom Helm.

Horowitz, J. (1981a) "Statistical Comparison of Non-Nested, Discrete-Choice, Random-Utility Models". Working Paper, Environmental Protection Agency.

Horowitz, J. (1981b) "Testing the Multinomial Logit Model Against the Multinomial Probit Model Without Estimating the Probit Parameters", *Transportation Science*, 15, 153–163.

Horowitz, J., J. Sparmonn, and C. Daganzo (1981) "An Investigation of the Accuracy of the Clark Approximation for the Multinomial Probit Model". Working Paper, EPA.

Huber, P. (1965) "The Behavior of Maximum Likelihood Estimates Under Non-Standard Conditions". *Fifth Berkeley Symposium on Statistics and Probability*, 1, 221–233.

Hulett, J. R. (1973) "On the Use of Regression Analysis with a Qualitative Dependent Variable". *Public Finance Quarterly*, 1, 339–344.

Ito, T. (1980) "Methods of Estimation for Multi-Market Disequilibrium Models", *Econometrica*, 48, 97–126.

Johnson, L. and D. Hensker (1981) "Application of Multinomial Probit to a Two-Period Panel Data Set". Working Paper, Macquarie Univ.

Joreskog, K. and A. Goldberger (1975) "Estimation of a Model with Multiple Indicators and Multiple Causes of a Single Latent Variable Model", *Journal of the American Statistical Association*, 70, 631–639.

Judge, G., W. Griffiths, R. Hill, and T. Lee (1980) *The Theory and Practice of Econometrics*. New York: Wiley, p. 601ff.

Kiefer, N. (1978) "Discrete Parametre Variation: Efficient Estimation of a Switching Regression Model", *Econometrica*, 46, 427–434.

Kiefer, N. (1980) "A Note on Switching Regressions and Discrimination", *Econometrica*, 48, 1065–1070.

Kiefer, N. M. (1979) "On the Value of Sample Separation Information", *Econometrica*, 47, 997–1003.

Kiefer, N. and G. Neumann (1979) "An Empirical Job Search Model with a Test of the Constant Reservation Wage Hypothesis", *Journal of Political Economy*, 87, 89–107.

King, M. (1980) "An Econometric Model of Tenure Choice and Demand for Housing as a Joint Decision". Working Paper, Department of Economics, University of Birmingham.

Klecka, W. (1980) *Discriminant Analysis*. Beverly Hills: Sage.

Knoke, D. and P. Burke (1980) *Log-Linear Models*. Beverly Hills: Sage.

Kohn, M., C. Manski, and D. Mundel (1976) "An Empirical Investigation of Factors Influencing College Going Behavior", *Annals of Economic and Social Measurement*, 5, 391–419.

Ladd, G. (1966) "Linear Probability Functions and Discriminant Functions", *Econometrica*, 34, 873–885.

Lave, C. (1970) "The Demand for Urban Mass Transit", *Review of Economics and Statistics*, 52, 320–323.

Lee, L. F. (1978) "Unionism and Wage Rates: A Simultaneous Equation Model with Qualitative and Limited Dependent Variables", *International Economic Review*, 19, 415–433.

Lee, L. F. (1979) "Identification and Estimation in Binary Choice Models with Limited (Censored) Dependent Variables", *Econometrica*, 47, 977–996.

Lee, L. F. (1980) "Fully Recursive Probability Models and Multivariate Log-Linear Probability Models for the Analysis of Qualitative Data", Discussion Paper 83-32, Univ. of Mass.

Lee, L. F. (1980a) "Specification Error in Multinomial Logit Models: Analysis of the Omitted Variable Bias". Working Paper, Department of Economics, University of Minnesota.

Lee, L. F. (1980b) "Statistical Analysis of Econometric Models of Discrete Panel Data". Working Paper, Department of Economics, University of Minnesota.

Lee, L. F. (1981a) "Estimation of Some Non-Normal Limited Dependent Variable Models". Discussion Paper, No. 43, Center for Econometrics and Decision Sciences, Univ. of Florida.

Lee, L. F. (1981b) "Simultaneous Equations Models with Discrete and Censored Variables". In *Structural Analysis of Discrete Data*, edited C. Manski and D. McFadden, 346–364. Cambridge MIT Press.

Lee, L. F. (1981c) "A Specification Test for Normality Assumption for the Truncated and Censored Tobit Models". Discussion Paper No. 44, Center for Econometrics and Decision Sciences", Univ. of Florida.

Lee, L. F., G. S. Maddala, and R. P. Trost (1979) "Testing for Structural Change by *D*-Methods in Switching Simultaneous Equation Models". *Proceedings of the American Statistical Association*, forthcoming.

Lee, L. F. and R. P. Trost (1978) "Estimation of Some Limited Dependent Variable Models with Applications to Housing Demand", *Journal of Econometrics*, 8, 357–382.

Lee, L. F., G. Maddala, and R. Trost (1980) "Asymptotic Covariance Matrices of Two-Stage Probit and Two-Stage Tobit Methods for Simultaneous Equations Models with Selectivity", *Econometrica*, 48, 491–504.

Lerman, S. R. (1977) "Location, Housing, Automobile Ownership and Model to Work: A Joint Choice Model". *Transportation Research Board Record*, No. 610, Washington, D.C.

Lerman, S. and C. Manski (1980) "Information Diffusion in Discrete Choice Contexts". Presented to CEME Conference on Discrete Econometrics.

Lerman, S. and C. Manski (1981) "On the Use of Simulated Frequencies to Approximate Choice Probabilities". In *Structural Analysis of Discrete Data*, edited by C. Manski and D. McFadden, 305–319. Cambridge: MIT Press.

Li, M. (1977) "A Logit Model of Home Ownership", *Econometrica*, 45, 1081–1097.

Little, R. E. (1968) "A Note on Estimation for Quantal Response Data", *Biometrika* 55, 578–579.

Loikkanen, H. (1982) "A Logit Model of Intra-Urban Mobility". Discussion Paper 22, Labour Institute for Economic Research, Helsinki.

Luce, R. D. (1959) *Individual Choice Behavior: A Theoretical Analysis*. New York: Wiley.

Luce, R. D. (1977) "The Choice Axiom After Twenty Years", *Journal of Mathematical Psychology*, 15, 215–233.

Luce, R. D. and P. Suppes (1965) "Preference, Utility, and Subjective Probability". In *Handbook of Psychology III*, edited by R. Luce, R. Bush, and E. Galanter, 249–410. New York: Wiley.

Maddala, G. S. (1977a) "Self-Selectivity Problem in Econometric Models". In *Applications of Statistics*, edited by K. Krishniah. Amsterdam: North-Holland.

Maddala, G. S. (1977b) "Identification and Estimation Problems in Limited Dependent Variable Models". In *Natural Resources, Uncertainty and General Equilibrium Systems: Essays in Memory of Rafael Lusky*, edited by A. S. Blinder and P. Friedman, 423–450. New York: Academic.

Maddala, G. S. (1978) "Selectivity Problems in Longitudinal Data", *Annals de l'Insee*, 30-1, 423–450.

Maddala, G. S. (1982) *Limited Dependent and Qualitative Variables in Econometrics*. New York: Cambridge Univ. Press.

Maddala, G. S. and L. F. Lee (1976) "Recursive Models with Qualitative Endogenous Variables", *Annals of Economic and Social Measurement*, 5, 525–545.

Maddala, G. and F. Nelson (1974) "Maximum Likelihood Methods for Markets in Disequilibrium", *Econometrica*, 42, 1013–1030.

Maddala, G. S. and F. D. Nelson (1975) "Switching Regression Models with Exogenous and Endogenous Switching". *Proceedings of the American Statistical Association* (Business and Economics Section), 423–426.

Maddala, G. S. and R. Trost (1978) "Estimation of Some Limited Dependent Variable Models with Application to Housing Demand", *Journal of Econometrics*, 8, 357–382.

Maddala, G. S. and R. Trost (1980) "Asymptotic Covariance Matrices of Two-Stage Probit and Two-Stage Tobit Methods for Simultaneous Equations Models with Selectivity", *Econometrica*, 48, 491–503.

Malhotra, N. (1982) "A Comparison of the Predictive Validity of Procedures for Analyzing Binary

Data". Working Paper, Georgia Institute of Technology.
Manski, C. (1975) "Maximum Score Estimation of the Stochastic Utility Model of Choice", *Journal of Econometrics*, 3, 205–228.
Manski, C. (1977) "The Structure of Random Utility Models", *Theory and Decision*, 8, 229–254.
Manski, C. (1981) "Closest Empirical Distribution Estimation". Working Paper, Hebrew Univ.
Manski, C. (1981a) "Structural Models for Discrete Data". *Sociological Methodology*, pp. 58–109.
Manski, C. (1982) "Analysis of Equilibrium Automobile Holdings in Israel with Aggregate Discrete Choice Models". Discussion Paper 82.06, Falk Institute.
Manski, C. and S. Lerman (1977) "The Estimation of Choice Probabilities from Choice-Based Samples", *Econometrica*, 45, 1977–1988.
Manski, C. and D. McFadden (1981) "Alterntative Estimates and Sample Designs for Discrete Choice Analysis". In *Structural Analysis of Discrete Data*, edited by C. Manski and D. McFadden, pp. 2–50. Cambridge: MIT Press.
Mardia, V. (1970) *Families of Bivariate Distributions*. Connecticut: Hafner.
Marschak, J. (1960) "Binary-Choice Constraints and Random Utility Indicators". In *Mathematical Methods in the Social Sciences*, edited by K. Arrow, S. Karlin, and P. Suppes, pp. 312–329. Stanford University Press.
McFadden, D. (1973) "Conditional Logit Analysis of Qualitative Choice Behavior". In *Frontiers in Econometrics*, edited by P. Zarembka, pp. 105–142. New York: Academic.
McFadden, D. (1974) "The Measurement of Urban Travel Demand", *Journal of Public Economics*, 3, 303–328.
McFadden, D. (1976a) "A Comment on Discriminant Analysis 'Versus' Logit Analysis", *Annals of Economics and Social Measurement*, 5, 511–523.
McFadden, D. (1976b) "Quantal Choice Analysis: A Survey", *Annals of Economic and Social Measurement*, 5, 363–390.
McFadden, D. (1976c) "The Revealed Preferences of a Public Bureaucracy", *Bell Journal*, 7, 55–72.
McFadden, D. (1978a) "Cost, Revenue, and Profit Functions, and the Generalized Linear Profit Function". In *Production of Economics*, edited by M. Fuss and D. McFadden, 3–110. Amsterdam: North-Holland.
McFadden, D. (1978b) "Modelling the Choice of Residential Location". In *Spatial Interaction Theory and Residential Location*, edited by A. Karlquist et al., 75–96. Amsterdam: North-Holland.
McFadden, D. (1979a) "Econometric Net Supply Systems for Firms with Continuous and Discrete Commodities". Working Paper, Department of Economics, Massachusetts Institute of Technology.
McFadden, D. (1979b) "Quantitative Methods for Analysing Travel Behavior of Individuals". *Behavioral Travel Modeling*, edited by D. Hensher and P. Stopher, 279–318. London: Croom Helm.
McFadden, D. (1979c) "Econometric Analysis of Discrete Data". Fisher–Schultz Lecture, Econometric Society, Athens.
McFadden, (1981) "Econometric Models of Probabilistic Choice". In *Structural Analysis of Discrete Data*, edited by C. Manski and D. McFadden, pp. 198–272. Cambridge: MIT Press.
McFadden, D., C. Puig, and D. Kirschner (1977) "Determinants of the Long-Run Demand for Electricity". *Proceedings of the American Statistical Association*, Business and Economics Section, Vol. 1.
McFadden, D. and F. Reid (1975) "Aggregate Travel Demand Forecasting from Disaggregated Behavioral Models". *Transportation Research Board Record*, 534, 24–37, Washington, D.C.
McFadden, D., A. Talvitie et al. (1977) "Demand Model Estimation and Validation". Final Report Series, Vol. 5. Urban Travel Demand Forecasting Project, Institute of Transportation Studies, University of California, Berkeley.
McFadden, C., W. Tye, and K. Train (1976) "An Application of Diagnostic Tests for the Independence from Irrelevant Alternatives Property of the Multinomial Logit Model". *Transportation Research Board Record* No. 637, pp. 39–45, Washington, D.C.
McKelvey, R. and W. Zovoina (1975) "A Statistical Model for the Analysis of Ordinal Level Dependent Variables", *Journal of Mathematical Sociology*, 4, 103–120.
Miller, L. and R. Radner (1970) "Demand and Supply in U.S. Higher Education", *American Economic Review*, 60, 326–334.
Mitchell, O. and G. Fields (1982) "The Effects of Changing the Budget Constraint: Parametric and Non-Parametric Approaches to Retirement Decisions". Working Paper, Cornell Univ.

Moore, D. H. (1973) "Evaluation of Five Discrimination Procedures for Binary Variables", *Journal of the American Statistical Association*, 68, 399–404.
Morimune, K. (1970) "Comparisons of Normal and Logistic Models in the Bivariate Dichotomous Analysis", *Econometrica*, 47, 957–975.
Moses, L., R. Beals, and M. Levy (1967) "Rationality and Migration in Ghana", *Review of Economics and Statistics*, 49, 480–486.
Mosteller, F. (1968) "Association and Estimation in Contingency Tables", *Journal of the American Statistical Association*, 63, 1–28.
Nelson, F. (1977) "Censored Regression Models with Unobserved Stochastic Censoring Thresholds", *Econometrica*, 6, 309–327.
Nelson, F. and L. Olsen (1978) "Specification and Estimation of a Simultaneous Equation Model with Limited Variables", *International Economic Review*, 19, 685–710.
Nerlove, M. (1978) "Econometric Analysis of Longitudinal Data: Approaches, Problems and Prospects, The Econometrics of Panel Data", *Annals de l'Insee*, 30:1, 7–22.
Nerlove, M. and J. Press (1973) "Univariable and Multivariable Log-Linear and Logistic Models". RAND report No. R-1306-EDA/NIH.
Nerlove, M. and J. Press (1976) "Multivariate and Log Linear Probability Models for the Analysis of Qualitative Data". Discussion Paper, Department of Economics, Northwestern University.
Newey, W. (1982) "Generalized Method of Moments Specification Testing". Working Paper, Dept. of Economics, MIT.
Nickell, S. (1979) "Estimating the Probability of Leaving Employment", *Econometrica*, 47, 1249–1266.
Oliveira, J. T., de (1958) "External Distributions". *Revista de Faculdada du Ciencia*, *Lisboa* (Serie A) 7, 215–227.
Olsen, R. (1978a) "Comment on 'The Effect of Unions of Earnings and Earnings on Unions: A Mixed Logit Approach'", *International Economic Review*, 19, 259–261.
Olsen, R. (1978b) "Tests for the Presence of Selectivity Bias and Their Relation to Specifications of Functional Form and Error Distribution". Working Paper No. 812, revised, Yale University.
Olsen, R. (1980a) "A Least Squares Correction for Selectivity Bias", *Econometrica*, 48, 1815–1820.
Olsen, R. (1980b) "Distributional Tests for Selectivity Bias and a More Robust Likelihood Estimator". Working Paper, Institute for Social and Policy Studies, Yale University.
Olsen, R. (1980c) "Estimating the Effect of Child Mortality on the Number of Births". Economic Growth Center, Yale University.
Palepu, (1982) *A Stochastic Model of Acquisitions*. Ph.D. thesis, Sloan School of Management, MIT.
Plackett, R. L. (1974) *The Analysis of Categorical Data*. London: Charles Griffin.
Poirier, D. J. (1977) "The Determinants of Home Buying" in *The New Jersey Income-Maintenance Experiment*". In Vol. II, *Expenditures, Health and Social Behavior, and the Quality of the Evidence*, edited by H. W. Watts and A. Rees, pp. 73–91. New York: Academic.
Poirier, D. J. (1978) "A Switching Simultaneous Equation Model of Physician Behavior in Ontario". In *Structural Analysis of Discrete Data: with Econometric Applications*, edited by D. McFadden and W. Manski, pp. 392–421. Cambridge: MIT Press.
Poirier, D. J. and P. Ruud (1980) "On the Appropriateness of Endogenous Switching", *J. Econometrics*, forthcoming.
Poirier, D. J. and P. Ruud (1981) "Conditional Minimum Distance Estimation and Autocorrelation in Limited Dependent Variable Models". Working paper, Univ. of Toronto.
Pollakowski, H. (1974) "The Effects of Local Public Service on Residential Location Decision: An Empirical Study of the San Francisco Bay Area". Ph.D. Dissertation, Department of Economics, University of California, Berkeley.
Pollakowski, H. (1980) *Residential Location and Urban Housing Markets*. Lexington, Mass.: D.C. Heath.
Poterba, J. and L. Summers (1982) "Unemployment Benefits, Labor Market Transitions, and Spurious Flows: A Multinomial Logit Model with Errors in Classification", Working Paper, MIT.
Powell, J. (1982) "Least Absolute Deviations Estimation for Censored and Truncated Regression Models". Ph.D. thesis, Department of Economics, Stanford Univ.
Pratt, J. (1977) "Concavity of the Log Likelihood of a Model of Ordinal Categorical Regression". Working Paper, Harvard Univ.
Press, J. and S. Wilson (1978) "Choosing Between Logistic Regression and Discriminant Analysis", *JASA*, 73, 699–705.

Quandt, R. E. (1956) "Probabilistic Theory of Consumer Behavior", *Quarterly Journal of Economics*, 70, 507–536.

Quandt, R. E. (1968) "Estimation of Model Splits", *Transportation Research*, 2, 41–50.

Quandt, R. E. (1970) *The Demand for Travel*. London: D.C. Heath.

Quandt, R. E. (1972) "A New Approach to Estimating Switching Regressions", *Journal of the American Statistical Association*, 67, 306–310.

Quandt, R. E. (1978) "Test of the Equilibrium vs Disequilibrium Hypothesis", *International Economic Review*, 19, 435–452.

Quandt, R. E. (1981) "Econometric Disequilibrium Models". Working Paper, Princeton Univ.

Quandt, R. E. (1982) "A Bibliography of Quantity Rationing and Disequilibrium Models", Working Paper, Princeton Univ.

Quandt, R. E. and W. Baumol (1966) "The Demand for Abstract Travel Modes: Theory and Measurement", *Journal of Regional Science*, 6, 13–26.

Quandt, R. E. and J. Ramsey (1978) "Estimating Mixtures of Normal Distributions and Switching Regressions", *Journal of the American Statistical Association*, 71, 730–752.

Quigley, J. (1976) "Housing Demand in the Short-Run: An Analysis of Polytomous Choice", *Explorations in Economic Research*, 3, 76–102.

Radner, R. and L. Miller (1975) *Demand and Supply in U.S. Higher Education*. New York: McGraw-Hill.

Rao, C. R. (1973) *Linear Statistical Inference and Its Applications*. New York: Wiley.

Richards, M. G. and M. Ben-Akiva (1974) "A Simultaneous Destination and Mode Choice Model for Shopping Trips", *Transportation*, 3, 343–356.

Robinson, P. (1982) "On the Asymptotic Properties of Estimators of Models Containing Limited Dependent Variables", *Econometrica*, 50, 27–41.

Rosen, H. and K. Small (1979) "Applied Welfare Economics with Discrete Choice Models". Working Paper 319, National Bureau of Economic Research.

Rothenberg, T. (1982) "Approximating the Distributions of Econometric Estimators and Test Statistics", manuscript, Univ. of California, Berkeley.

Ruud, P. (1982) "A Score Test of Consistency". Working Paper, Dept. of Economics, Univ. of California, Berkeley.

Sattath, S. and A. Tversky (1977) "Additive Similarity Trees", *Psychometrika*, 42, 319–345.

Schultz, T. P. (1975) "The Determinants of Internal Migration in Venezuela: An Application of the Polytomous Logistic Model". Presented at the Third World Congress of the Econometric Society, Toronto, Canada.

Sickles, R. C. and P. Schmidt (1978) "Simultaneous Equation Models with Truncated Dependent Variables: A Simultaneous Tobit Model", *Journal of Economics and Business*, 31, 11–21.

Sjoberg, L. (1977) "Choice Frequency and Similarity", *Scandinavian Journal of Psychology*, 18, 103–115.

Small, K. (1981) "Ordered Logit: A Discrete Choice Model with Proximate Covariance Among Alternatives". Working Paper, Department of Economics, Princeton Univ.

Smith, K., Savin, N. and J. Robertson (1982) "A Monte Carlo Study of Maximum Likelihood and Minimum Chi-Square Estimators in Logit Analysis". Working Paper, USDA, Forest Service, Pacific Southwest Forest and Range Experiment Station.

Spilerman, S. (1972) "Extensions of the Mover-Stayer Model", *American Journal of Sociology*, 78, 599–626.

Stewman, S. (1976) "Markov Models of Occupational Mobility: Theoretical Development and Empirical Support", *Journal of Mathematical Sociology*, 4, 201–245.

Swartz, C. (1976) "Screening in the Labor Market: A Case Study". Ph.D. dissertation, University of Wisconsin, Madison.

Tachibanaki, T. (1981) Education, Occupation, Hierarchy, and Earnings: A Recursive Logit Approach for Japan". Research paper, Stanford Univ.

Talvities, A. (1972) "Comparison of Probabilistic Modal-Choice Models; Estimation Methods and System Inputs". Highway Research Board Record 392, pp. 111–120.

Theil, H. (1969) "A Multinomial Extension of the Linear Logit Model", *International Economic Review*, 10, 251–259.

Theil, H. (1970) "On the Estimation of Relationships Involving Qualitative Variables", *American Journal of Sociology*, 76, 103–154.
Thurstone, L. (1927) "A Law of Comparative Judgement", *Psychological Review*, 34, 273–286.
Tobin, J. (1958) "Estimation of Relationships for Limited Dependent Variables", *Econometrica*, 26, 24–36.
Train, K. (1978) "A Validation Test of a Disaggregate Mode Choice Model", *Transportation Research*, 12, 167–174.
Train, K., and D. McFadden (1978) "The Goods/Leisure Tradeoff and Disaggregate Work Trip Mode Choice Models", *Transportation Research*, 12, 349–353.
Train, K. and M. Lohrer (1982) "Vehicle Ownership and Usage: An Integrated System of Disaggregate Demand Models". Final Report, Cambridge Systematics.
Trost, R. and L. Lee (1982) "Technical Training and Earnings: A Polychotomous Choice Model with Selectivity". Working Paper, Dept. of Economics, Univ. of Florida.
Tversky, A. (1972a) "Choice by Elimination", *Journal of Mathematical Psychology*, 9, 341–367.
Tversky, A. (1972b) "Elimination by Aspects: A Theory of Choice", *Psychological Review*, 79, 281–299.
Tversky, A. and S. Sattath (1979) "Preference Trees", *Psychology Review*, 86, 542–573.
Vuong, Q. (1982) "Probability Feedback in a Recursive System of Logit Models: Estimation". Working Paper, California Institute of Technology.
Walker, S. and D. Duncan (1967) "Estimation of the Probability of an Event as a Function of Several Independent Variables", *Biometrika*, 54, 167–179.
Warner, S. (1962) *Stochastic Choice of Mode in Urban Travel*. Evanston: North-Western University Press.
Warner, S. L. (1963) "Multivariate Regression of Dummy Variates under Normality Assumptions", *Journal of the American Statistical Association*, 58, 1054–1063.
Westin, R. B. (1974) "Predictions from Binary Choice Models", *Journal of Econometrics*, 2, 1–16.
Westin, R. B. and D. W. Gillen (1978) "Parking Location and Transit Demand: A Case Study of Endogenous Attributes in Disaggregate Mode Choice Functions", *Journal of Econometrics*, 8, 75–101.
Williams, H. (1977) "On the Formation of Travel Demand Models and Economic Evaluation Measures of User Benefit", *Environment Planning*, A9, 285–344.
Willis, R. and S. Rosen (1979) "Education and Self-Selection", *Journal of Political Economy*, 87, 507–536.
Winston, C. (1981) "A Disaggregate Model of the Demand for Intercity Freight Transport", *Econometrica*, 49, 981–1006.
Yatchew, A. (1980) *Design of Econometric Choice Models*. Ph.D. Thesis, Harvard Univ.
Yellot, J. (1977) "The Relationship Between Luce's Choice Axiom, Thurstone's Theory of Comparative Judgement, and the Double Exponential Distribution", *Journal of Mathematical Psychology*, 15, 109–144.
Zellner, A. and T. Lee (1965) "Joint Estimation of Relationships Involving Discrete Random Variables", *Econometrica*, 33, 382–394.

INDEX